T0214962

Lecture Notes in Computer Science 9563

Commenced Publication in 1973
Founding and Former Series Editors:
Gerhard Goos, Juris Hartmanis, and Jan van Leeuwen

Eyal Kushilevitz · Tal Malkin (Eds.)

Theory of Cryptography

13th International Conference, TCC 2016-A,
Tel Aviv, Israel, January 10–13, 2016
Proceedings, Part II

Springer

Editors
Eyal Kushilevitz
Department of Computer Science
Technion
Haifa
Israel

Tal Malkin
Department of Computer Science
Columbia University
New York, NY
USA

ISSN 0302-9743   ISSN 1611-3349   (electronic)
Lecture Notes in Computer Science
ISBN 978-3-662-49098-3   ISBN 978-3-662-49099-0   (eBook)
DOI 10.1007/978-3-662-49099-0

Library of Congress Control Number: 2015957796

LNCS Sublibrary: SL4 – Security and Cryptology

Printed on acid-free paper

This Springer imprint is published by SpringerNature
The registered company is Springer-Verlag GmbH Berlin Heidelberg

Preface

The 13th Theory of Cryptography Conference (TCC 2016-A) was held during January 10–13, 2016, at the Suzanne Dellal Center in Tel Aviv, Israel. It was sponsored by the International Association for Cryptographic Research (IACR). The general chairs of the conference were Ran Canetti and Iftach Haitner. We would like to thank them for their hard work in organizing the conference.

The conference received 112 submissions, of which the Program Committee (PC) selected 45 for presentation (with three pairs of papers sharing a single presentation slot per pair). Each submission was reviewed by at least three PC members, often more. The 24 PC members, all top researchers in our field, were helped by 112 external reviewers, who were consulted when appropriate. These proceedings consist of the revised version of the 45 accepted papers. The revisions were not reviewed, and the authors bear full responsibility for the content of their papers.

As in previous years, we used Shai Halevi's excellent web-review software, and are extremely grateful to him for writing it, and for providing fast and reliable technical support whenever we had any questions. Based on the experience from last year, we again made use of the interaction feature supported by the review software, where PC members may directly and anonymously interact with authors. This was used to ask specific technical questions that arise, such as suspected bugs. We felt this was efficient and successful, and are thankful to last year's chairs, Yevgeniy Dodis and Jesper Buus Nielsen, for suggesting this feature, and to Shai Halevi for implementing it.

This was the second year where TCC presented the Test of Time Award to an outstanding paper that was published at TCC at least eight years ago, making a significant contribution to the theory of cryptography, preferably with influence also in other areas of cryptography, theory, and beyond. This year the Test of Time Award Committee selected the following paper, published ten years ago at TCC 2006:

"Calibrating Noise to Sensitivity in Private Data Analysis," by Cynthia Dwork, Frank McSherry, Kobbi Nissim, and Adam Smith.

This paper was selected *for introducing the definition of differential privacy, providing a solid mathematical foundation for a vast body of subsequent work on private data analysis.* The authors were also invited to deliver a talk at TCC 2016-A. The conference also featured two other invited events. First, an invited talk by Yael Kalai and Shafi Goldwasser (delivered by Yael) followed by panel on "cryptographic assumptions." Second, an invited talk by Yevgeniy Dodis. Finally, in addition to regular papers and invited events, the conference also featured a rump session.

We are greatly indebted to many people who were involved in making TCC 2016-A a success. First of all, a big thanks to the most important contributors: all the authors who submitted papers to the conference. Next, we would like to thank the PC members for their hard work, dedication, and diligence in reviewing the papers, verifying the correctness, and in-depth discussion. We are also thankful to the external reviewers for their volunteered hard work and investment in reviewing papers and answering questions, often under time pressure. For running the conference itself, we are very grateful to the general chairs, Ran Canetti and Iftach Haitner, as well as Galit Herzberg and the rest of the local Organizing Committee. Finally, we are thankful to the TCC Steering Committee as well as the entire thriving and vibrant TCC community.

January 2016

Eyal Kushilevitz
Tal Malkin

TCC 2016-A

The 13th Theory of Cryptography Conference

Suzanne Dellal Center, Tel Aviv, Israel
January 10–13, 2016

Sponsored by the *International Association for Cryptographic Research*

General Chairs

Ran Canetti   Tel Aviv University, Israel
          Boston University, USA
Iftach Haitner   Tel Aviv University, Israel

Program Chairs

Eyal Kushilevitz   Technion, Israel
Tal Malkin   Columbia University, USA

Program Commitee

Masayuki Abe   NTT, Japan
Amos Beimel   Ben-Gurion University, Israel
Nir Bitansky   MIT, USA
Andrej Bogdanov   Chinese University of Hong Kong, SAR China
Zvika Brakerski   Weizmann Institute of Science, Israel
Christina Brzuska   Hamburg University of Technology, Germany
Nishanth Chandran MSR India
Melissa Chase   MSR Redmond, USA
Dana Dachman-Soled University of Maryland, USA
Yuval Ishai   Technion, Israel
Jonathan Katz   University of Maryland, USA
Hugo Krawczyk   IBM Research, USA
Huijia Lin   UC Santa Barbara, USA
Claudio Orlandi   Aarhus University, Denmark
Omkant Pandey   Drexel University, USA
Valerio Pastro   Columbia University, USA
Leonid Reyzin   Boston University, USA
Guy Rothblum   Samsung Research America, USA
Gil Segev   Hebrew University, Israel
Adam Smith   Pennsylvania State University, USA
Vinod Vaikuntanathan MIT, USA
Ivan Visconti   University of Salerno, Italy
Brent Waters   UT Austin, USA
Vassilis Zikas   ETH, Switzerland

External Reviewers

Divesh Aggarwal
Prabhanjan Ananth
Daniel Apon
Benny Applebaum
Gilad Asharov
Nuttapong Attrapadung
Pablo Azar
Saikrishna
   Badrinarayanan
Allison Bishop
Elette Boyle
Ignacio Cascudo
David Cash
Binyi Chen
Yilei Chen
Mahdi Cheragchi
Kai-Min Chung
Michele Ciampi
Aloni Cohen
Sandro Coretti
Akshay Degwekar
Gregory Demay
Itai Dinur
Yevgeniy Dodis
Nico Döttling
Antonio Faonio
Sebastian Faust
Victoria Fehr
Dario Fiore
Nils Fleischhacker
Eiichiro Fujisaki
Juan Garay
Ran Gelles
Craig Gentry
Niv Gilboa
Alexander Golovnev
Sergey Gorbunov
Rishab Goyal
Jens Groth

Siyao Guo
Shai Halevi
Prahladh Harsha
Carmit Hazay
Brett Hemenway
Ryo Hiromasa
Justin Holmgren
Ai Ishida
Zahra Jafargholi
Abhishek Jain
Stanislaw Jarecki
Daniel Jost
Tomasz Kazana
Carmen Kempka
Dakshita Khurana
Susumu Kiyoshima
Saleet Klein
Ilan Komargodski
Venkata Koppula
Lucas Kowalczyk
Ranjit Kumaresan
Tancrède Lepoint
Feng-Hao Liu
Tianren Liu
Satya Lokam
Steve Lu
Anna Lysyanskaya
Vadim Lyubashevsky
Mohammad Mahmoody
Hemanta K. Maji
Christian Matt
Eric Miles
Arno Mittelbach
Pratyay Mukherjee
Moni Naor
Jesper Buus Nielsen
Ryo Nishimaki
Adam O'Neill
Miyako Ohkubo

Olya Ohrimenko
Omer Paneth
Sunoo Park
Anat Paskin-Cherniavsky
Giuseppe Persiano
Oxana Poburinnaya
Antigoni Polychroniadou
Tal Rabin
Silas Richelson
Mike Rosulek
Ron Rothblum
Yannis Rouselakis
Alessandra Scafuro
Karn Seth
Luisa Siniscalchi
John Steinberger
Stefano Tessaro
Aishwarya
   Thiruvengadam
Mehdi Tibouchi
Daniel Tschudi
Jalaj Upadhyay
Prashant Vasudevan
Muthu
   Venkitasubramaniam
Daniele Venturi
Dhinakaran
   Vinayagamurthy
Thomas Watson
Hoeteck Wee
Mor Weiss
Daniel Wichs
Keita Xagawa
Eylon Yogev
Ching-Hua Yu
Yu Yu
Mark Zhandry
Hong-Sheng Zhou

Contents – Part II

Codes and Interactive Proofs

Limitations of Obfuscation and Obfuscation-Avoiding Constructions

Contents – Part I

Public Key Encryption, Signatures, and VRF

Complexity of Cryptographic Primitives

Obfuscation-Based Cryptographic Constructions

Cryptographic Assumptions (Invited Talk followed by Panel)

Multiparty Computation

Zero Knowledge and PCP

Making the Best of a Leaky Situation: Zero-Knowledge PCPs from Leakage-Resilient Circuits

Yuval Ishai[1,2]([✉]), Mor Weiss[1], and Guang Yang[3]

[1] Department of Computer Science, Technion, Haifa, Israel
{yuvali,morw}@cs.technion.ac.il
[2] Department of Computer Science, UCLA, Los Angeles, CA, USA
[3] Institute for Interdisciplinary Information Sciences, Tsinghua University,
Beijing, China
guang.research@gmail.com

Abstract. A Probabilistically Checkable Proof (PCP) allows a randomized verifier, with oracle access to a purported proof, to probabilistically verify an input statement of the form "$x \in L$" by querying only few bits of the proof. A zero-knowledge PCP (ZKPCP) is a PCP with the additional guarantee that the view of any verifier querying a bounded number of proof bits can be efficiently simulated given the input x alone, where the simulated and actual views are statistically close.

Originating from the first ZKPCP construction of Kilian et al. [21], all previous constructions relied on locking schemes, an unconditionally secure oracle-based commitment primitive. The use of locking schemes makes the verifier *inherently* adaptive, namely, it needs to make at least two rounds of queries to the proof.

Motivated by the goal of constructing non-adaptively verifiable ZKPCPs, we suggest a new technique for compiling standard PCPs into ZKPCPs. Our approach is based on leakage-resilient circuits, which are circuits that withstand certain "side-channel" attacks, in the sense that these attacks reveal nothing about the (properly encoded) input, other than the output. We observe that the verifier's oracle queries constitute a side-channel attack on the wire-values of the circuit verifying membership in L, so a PCP constructed from a circuit resilient against such attacks would be ZK. However, a leakage-resilient circuit evaluates the desired function *only if* its input is properly encoded, i.e., has a specific structure, whereas by generating a "proof" from the wire-values of the circuit on an *ill-formed* "encoded" input, one can cause the verification to accept inputs $x \notin L$ *with probability 1*. We overcome this obstacle by constructing leakage-resilient circuits with the additional guarantee that ill-formed encoded inputs are detected. Using this approach, we obtain the following results:

– We construct the first *witness-indistinguishable* PCPs (WIPCP) for NP with non-adaptive verification. WIPCPs relax ZKPCPs by only requiring that different witnesses be indistinguishable. Our construction combines strong leakage-resilient circuits as above with the PCP

© International Association for Cryptologic Research 2016
E. Kushilevitz and T. Malkin (Eds.): TCC 2016-A, Part II, LNCS 9563, pp. 3–32, 2016.
DOI: 10.1007/978-3-662-49099-0_1

of Arora and Safra [2], in which queries correspond to side-channel attacks by shallow circuits, and with correlation bounds for shallow circuits due to Lovett and Srivinasan [22].

- Building on these WIPCPs, we construct non-adaptively verifiable *computational* ZKPCPs for NP in the common random string model, assuming that one-way functions exist.
- As an application of the above results, we construct *3-round* WI and ZK proofs for NP in a distributed setting in which the prover and the verifier interact with multiple servers of which t can be corrupted, and the total communication involving the verifier consists of $\mathsf{poly}\log(t)$ bits.

1   Introduction

In this work we study probabilistically checkable proofs with zero-knowledge properties, and establish a connection between such proofs and leakage-resilient circuits. Before describing our main results, we first give a short overview of these objects.

Probabilistically Checkable Proof (PCP) systems [1, 2] are proof systems that allow an efficient randomized verifier, with oracle access to a purported proof generated by an efficient prover (that is also given the witness), to probabilistically verify claims of the form "$x \in L$" (for an NP-language L) by probing only few bits of the proof. The verifier accepts the proof of a true claim with probability 1 (the *completeness* property), and rejects false claims with high probability (the probability that the verifier accepts a false claim is called *the soundness error*). The celebrated PCP theorem [1, 2, 8] asserts that any NP language admits a PCP system with soundness error $1/2$ in which the verifier reads only a *constant* number of proof bits (soundness can be amplified using repetition). Moreover, the verifier is *non-adaptive*, namely its queries are determined solely by his randomness (a verifier is *adaptive* if each of his queries may also depend on the oracle answers to previous queries).

A very different kind of proofs are zero-knowledge (ZK) proofs [14], namely proofs that carry no extra knowledge other than being convincing. Combining the advantages of ZK proofs and PCPs, a *zero-knowledge PCP* (ZKPCP) is defined similarly to a traditional PCP, except that the proof is also randomized and there is the additional guarantee that the view of any (possibly malicious) verifier who makes a bounded number of queries can be efficiently simulated up to a small statistical distance.

Previous ZKPCP constructions [17, 19, 21] are obtained from standard (i.e., non-ZK) PCPs in two steps. First, the standard PCP is transformed into a PCP with a weaker "honest-verifier" ZK guarantee (which is much easier to achieve than full-fledged ZK). Then, this "honest-verifier" ZKPCP is combined with an unconditionally secure oracle-based commitment primitive called a "locking scheme" [17, 21]. This transformation yields ZKPCPs for NP with statistical ZK against *query-bounded malicious* verifiers, namely ones who are only limited to asking at most $p(|x|)$ queries, for some *fixed* polynomial p that is much smaller than the proof length, but can be much bigger than the (polylogarithmic) number of queries asked by the honest verifier.

A common limitation of all previous ZKPCP constructions is that they require *adaptive* verification, even if the underlying non-ZK PCP can be non-adaptively verified. This raises the natural question of constructing PCPs that can be *non-adaptively* verified, and guarantee ZK against *malicious* verifiers. We note that the adaptivity of the verifier is inherent to any locking-scheme-based ZKPCP, since the unconditional security of locking schemes makes their opening inherently adaptive. Therefore, constructing ZKPCPs that can be verified non-adaptively requires a new approach towards ZKPCP construction. An additional advantage of eliminating the use of locking schemes is the possibility of constructing ZKPCPs preserving the proof length (which is important when these are used for cryptographic applications as described below), since locking schemes inherently incur a polynomial blow-up in the PCP length.

Motivated by these goals, we suggest a new approach for the construction of ZKPCPs. We apply *leakage-resilient circuit compilers* (LRCCs) to construct *witness-indistinguishable PCPs* (WIPCPs) for NP, a weaker variant of ZKPCPs in which the simulation is not required to be efficient. We then apply the so-called "FLS technique" [12] to convert these WIPCPs into *computational* ZKPCPs (CZKPCPs) in the common random string (CRS) model, based on the existence of one-way functions (OWFs). In such a CZKPCP, the view of any query-bounded PPT verifier can be *efficiently* simulated, in a way which is *computationally indistinguishable* from the actual view.

Informally, an LRCC compiles any circuit into a new circuit that operates on encoded inputs, and withstands side-channel attacks in the sense that these reveal nothing about the (properly encoded) input, other than what follows from the output. Works on LRCCs obtained information-theoretic security for different classes of leakage functions [10,11,15,18,23,25].

Other than the theoretical interest in this question, our study of PCPs with ZK properties is motivated by their usefulness for cryptographic applications. For instance, ZKPCPs are the underlying combinatorial building blocks of succinct zero-knowledge arguments, which have been the subject of a large body of recent work (see, e.g., [3–5] and references therein).

A more direct application of WIPCPs and ZKPCPs is for implementing efficiently verifiable zero-knowledge proofs in a distributed setting involving a prover, verifier, and multiple (potentially corrupted) servers. In this setting a prover can distribute a ZKPCP between the servers, allowing the verifier to efficiently verify the claim by polling a small random subset of the servers.[1] In this and similar situations, ZKPCPs that only offer security against an honest verifier are not sufficient for protecting against *colluding servers*. We use our non-adaptively verifiable WIPCPs and CZKPCPs for NP to construct *3-round* WI and CZK proofs for NP in this distributed setting, in which the total communication with the verifier is *sublinear* in the input length. The WI proofs are

---

[1] Unlike the ZKPCP model, the answers of malicious servers may depend on the identity of the verifier's queries, but this can be overcome using techniques of [19].

unconditional, whereas the CZK proofs are based on the existence of OWFs. This should be contrasted with standard sublinear ZK arguments, that require at least 4 rounds of interaction, and require the existence of collision resistant hash functions. We refer the reader to, e.g., [17] for additional discussion of ZKPCPs and their applications.

1.1   Our Results and Techniques

We now give a more detailed account of our results, and the underlying techniques.

FROM LRCCS AND PCPs TO WIPCPs. Let L be an NP-language with a corresponding NP-relation $\mathcal{R}_L$, and a boolean circuit C verifying $\mathcal{R}_L$. Recall that the prover P in a PCP system for $\mathcal{R}_L$ is given the input x and a witness y for the membership of x in L, and outputs a proof π that is obtained by applying some function f_P to x, y. For our purposes, it would be more convenient to think of f_P as a function of the *entire wire values* w of C, when evaluated on x, y. In a ZKPCP, few bits in the output of f_P should reveal essentially nothing about the wire values w, i.e., C should withstand "leakage" from f_P. In general, we cannot assume that C has this guarantee, but using an LRCC, C can be compiled into a circuit $\hat{C}$ with this property. Informally, an LRCC is associated with a function class $\mathcal{L}$ (the *leakage class*) and a (randomized) input encoding scheme E, and compiles a deterministic circuit C into a deterministic circuit $\hat{C}$, that emulates C, but operates on an encoded input. It is leakage-resilient in the following sense: for any input z for C, and any $\ell \in \mathcal{L}$, the output of ℓ on the wire values of $\hat{C}$, when evaluated on E(z), reveals nothing other than $C(z)$. This is formalized in the simulation-based paradigm (i.e., the wire-values of $\hat{C}$ can be efficiently simulated given only $C(z)$).

We establish a connection between ZKPCPs and LRCCs. Assume the existence of an LRCC associated with a leakage class $\mathcal{L}$, such that any restriction $f_P^{\mathcal{I}}$ of f_P to a "small" subset $\mathcal{I}$ of its outputs satisfies $f_P^{\mathcal{I}} \in \mathcal{L}$. Then the oracle answers to the queries of a query-bounded verifier V correspond to functions in $\mathcal{L}$, since for every possible set $\mathcal{I}$ of oracle queries, the answers are $f_P^{\mathcal{I}}(w)$. Therefore, if w is the wire values of a *leakage-resilient* circuit then the system is ZK. This gives a general method of transforming standard PCPs into ZKPCPs: P, V replace $C_x = C(x, \cdot)$ (i.e., C with x hard-wired into it) with $\hat{C}_x$; and P proves that $\hat{C}_x$ is satisfiable by generating the PCP π from the wire values of $\hat{C}_x$.

This transformation crucially relies on the fact that $\hat{C}_x$ emulates C_x (e.g., if $\hat{C}_x$ always outputs 1 then the resultant PCP system is not sound). However, in current constructions of LRCCs (e.g., [11,18,23]), this holds *only if the encoded input of $\hat{C}_x$ was honestly generated*. Moreover, there always exists a choice of an *ill-formed* "encoding" that satisfies $\hat{C}_x$ (i.e., causes it to output 1). In our case the *prover* generates the encoded input of $\hat{C}_x$ (the verifier does not know this input), so a malicious prover can pick an ill-formed "encoding" that satisfies $\hat{C}_x$, causing the verifier to accept *with probability 1*. Therefore, soundness requires that if C_x is not satisfiable, then there exists *no* satisfying input for $\hat{C}_x$ (either well- or ill-formed), a property which we call *SAT-respecting*. The main tool we use are *SAT-respecting* LRCCs, which we construct based on the LRCC of Faust et al. [11]. To describe our construction, we first need to delve deeper into their construction.

The LRCC of [11] transforms a circuit C into a circuit $\hat{C}$ that operates on encodings generated by a linear encoding scheme, and emulates the operations of C on these encodings. Leakage-resilience against functions in a restricted function class $\mathcal{L}$ is obtained by "refreshing" the encoded intermediate values of the computation after every operation, using encodings of 0. (The LRCCs of [18,23] operate essentially in the same way.) The input of $\hat{C}$ includes sufficiently many encodings of 0 to be used for the entire computation.[2] However, by providing $\hat{C}$ also with 1-encodings (i.e., encodings of 1), one can change the functionality emulated by $\hat{C}$. (In particular, if the encoding "refreshing" the output gate is a 1-encoding, the output is flipped.) This is not just an artifact of the construction, but rather is *essential* for their leakage-resilience argument. Concretely, to simulate the wire values of $\hat{C}$ *without knowing its input*, the simulator sometimes uses 1-encodings, which rules out the natural solution of verifying that the encodings used for "refreshing" are 0-encodings. We observe that if C were emulated twice, *it would suffice to know that at least one copy used only 0-encodings*, since then $\hat{C}$ is satisfiable only if the honestly-evaluated copy is satisfiable (i.e., C is satisfiable). At first, this may seem as no help at all, but it turns out that by emulating C twice, we can construct what we call a *relaxed* LRCC, which is similar to an LRCC, except that the simulator is *not* required to be efficient. Specifically, assume that before compiling C into $\hat{C}$, we would replace it with a circuit C' that computes C twice, and outputs the AND of both evaluations. Then $\hat{C'}$ would be relaxed leakage-resilient, since an unbounded simulator could simulate the wire values of $\hat{C'}$ by finding a satisfying input z_S for C, and honestly evaluating $\hat{C'}$ on a pair of encodings of z_S. Using a hybrid argument, we can prove that functions in $\mathcal{L}$ cannot distinguish the simulated wire values $\mathcal{W}_S$ from the actual wire values $\mathcal{W}_R$ of $\hat{C'}$ when evaluated on a satisfying input z_R. Indeed, we can first replace the input in the *first* copy from z_R to z_S (using the leakage-resilience of the LRCC of [11] to claim that functions in $\mathcal{L}$ cannot distinguish this hybrid distribution from $\mathcal{W}_R$), then do the same in the *second* copy. By replacing the inputs one at a time, we only need to use 1-encodings in a *single* copy.[3] However, holding two copies of the original circuit still does not guarantee that the evaluation in at least one of them uses only 0-encodings.

The natural solution would again be to add a sub-circuit verifying that the encodings used are 0-encodings, but this sub-circuit should hide the identity of

---

[2] Actually, [11] consider a model of *continuous* leakage, in which the circuit is invoked multiple times on different inputs, and maintains a secret state. Their construction uses tamper-proof hardware (called *opaque gates*) to generate the encodings of 0 used for refreshing. We consider the simpler model of *one-time* leakage on circuits that operated on *encoded* inputs [18,23], and as a result we can incorporate the necessary encodings (used for refreshing) into the encoded input.

[3] This technique is reminiscent of the "2-key trick" of [24], used to convert a CPA-secure encryption scheme into a CCA-secure one.

the "correctly evaluated" copy. This is because the hybrid argument described above first uses 1-encodings in the first copy (and 0-encodings in the second), and then uses 1-encodings in the second copy (and only 0-encodings in the first). Therefore, if functions in $\mathcal{L}$ could determine which copy uses only 0-encodings, they could also distinguish between the hybrids. Instead, we describe an "oblivious" checker $\mathcal{T}_0$, which at a high-level operates as follows. To check that *either* the first *or* the second copy use only 0-encodings, it checks that for every pair of encodings, one from the first copy, and one from the second, the product of the encoded values is 0. To guarantee that leakage on $\mathcal{T}_0$ reveals no information regarding *which* copy uses only 0-encodings, we use the LRCC of [11] to compile $\mathcal{T}_0$ into a leakage-resilient circuit $\hat{\mathcal{T}}_0$. This introduces the additional complication that now we must also verify the encodings used to "refresh" the computation in $\hat{\mathcal{T}}_0$ (otherwise 1-encodings may be used, potentially changing the functionality of $\hat{\mathcal{T}}_0$ and rendering it useless). However, since $\hat{\mathcal{T}}_0$ does not operate directly on the *inputs to* $\hat{C}'$ (it operates only on the encodings used for "refreshing"), we show that the "refreshing" encodings used in $\hat{\mathcal{T}}_0$ can be checked directly (by decoding the encoded values and verifying that they are 0). Additional technicalities arise since introducing these additional components prevents us from using the LRCC of [11] as a black box (see Sect. 3 for additional details on the analysis). Finally, we note that our circuit-compiler is *relaxed*-leakage-resilient because in all hybrids, we need the honestly-evaluated copy to be satisfied, so the simulator needs to find a satisfying input for C. This is also the reason that we get
· *WI*PCPs, and not *ZK*PCPs. If we had a SAT-respecting LRCC, the transformation described above would give a ZKPCP. However, we show that known LRCCs withstanding global leakage [11, 18, 23] cannot be transformed into SAT-respecting *non-relaxed* LRCCs (i.e., LRCCs with an *efficient* simulator), unless NP $\subseteq$ BPP. Intuitively, this is because these constructions admit a simulator which is *universal* in the sense that it simulates the wire values of the compiled circuit *without knowing the leakage function*, and the simulated values "fool" *all* functions in $\mathcal{L}$. Combining such a SAT-respecting LRCC with PCPs for NP (through the transformation described above) would give a BPP algorithm of deciding any NP-language.

CONSTRUCTING WIPCPs FOR NP. Recall that our general transformation described above relied on f_P being in the function class $\mathcal{L}$ that is associated with the SAT-respecting relaxed-LRCC. We observe that the PCP system of Arora and Safra [2] has the property that every "small" subset of proof bits can be generated using a low-depth circuit of polynomial size over the operations $\wedge, \vee, \neg, \oplus$, with "few" $\oplus$ gates. We use recent correlation bounds of Lovett and Srivinasan [22], which roughly state that such circuits have negligible correlation with the boolean function that counts the number of 1's modulo 3 in its input, to construct a SAT-respecting circuit compiler that is relaxed leakage-resilient with respect to this function class. Combining this relaxed LRCC with our general transformation, we prove the following, where NA-WIPCP denotes the class of all NP-languages that have a PCP system with a negligible soundness error, polynomial-length proofs, a non-adaptive honest verifier that queries poly-logarithmically many proof bits, and guarantee WI against (adaptive) malicious verifiers querying a fixed polynomial number of proof bits.

Theorem 1 (NA-WIPCPs for NP). NP = NA − WIPCP.

CONSTRUCTING CZKPCPs FOR NP. Using a general technique of Feige et al. [12], and assuming the existence of OWFs, we transform our WIPCP into a CZKPCP in the CRS model, in which the PCP prover and verifier both have access to a common random string. Concretely, we prove the following result, where NA-CZKPCP corresponds exactly to the class NA-WIPCP, except that the WI property is replaced with CZK in the CRS model.

Corollary 1 (NA-CZKPCPs for NP). *Assume that OWFs exist. Then* NP = NA − CZKPCP.

In Sect. 4 we describe a simple alternative approach for constructing CZKPCPs by applying a PCP on top of a standard non-interactive zero-knowledge (NIZK) proof. This should be contrasted with our main construction that only relies on a OWF.

2 Preliminaries

Let $\mathbb{F}$ be a finite field, and Σ be a finite alphabet (i.e., a set of symbols). In the following, function composition is denoted as $f \circ g$, where $(f \circ g)(x) := f(g(x))$. If F, G are families of functions then $F \circ G = \{f \circ g : f \in F, g \in G\}$. Vectors will be denoted by boldface letters (e.g., $\mathbf{a}$). If $\mathcal{D}$ is a distribution then $X \leftarrow \mathcal{D}$, or $X \in_R \mathcal{D}$, denotes sampling X according to the distribution $\mathcal{D}$. Given two distributions X, Y, $\mathsf{SD}(X, Y)$ denotes the statistical distance between X and Y. For a natural n, $\mathsf{negl}(n)$ denotes a function that is negligible in n. For a function family $\mathcal{L}$, we sometimes use the term "leakage family $\mathcal{L}$", or "leakage class $\mathcal{L}$". In the following, n usually denotes the input length, m usually denotes the output length, d, s denote depth and size, respectively (e.g., of circuits, as defined below), t is used to count $\oplus$ gates, and σ is a security parameter. We assume that standard cryptographic primitives (e.g., OWFs) are secure against non-uniform adversaries.

Definition 1 (Leakage-indistinguishability of distributions). *Let D, D' be finite sets, $\mathcal{L} = \{\ell : D \to D'\}$ be a family of leakage functions, and $\epsilon > 0$. We say that two distributions X, Y over D are $(\mathcal{L}, \epsilon)$-leakage-indistinguishable, if for any function $\ell \in \mathcal{L}$, $\mathsf{SD}(\ell(X), \ell(Y)) \leq \epsilon$.*

Remark 1. In case $\mathcal{L}$ consists of functions over different domains, we say that X, Y over D are $(\mathcal{L}, \epsilon)$-leakage-indistinguishable if $\mathsf{SD}(\ell(X), \ell(Y)) \leq \epsilon$ for every function $\ell \in \mathcal{L}$ *with domain D.*

ENCODING SCHEMES. An encoding scheme E over alphabet Σ is a pair (Enc, Dec) of algorithms, where the *encoding algorithm* Enc is a probabilistic polynomial-time (PPT) algorithm that given a message $x \in \Sigma^n$ outputs an encoding $\hat{x} \in \Sigma^{\hat{n}}$

for some $\hat{n} = \hat{n}(n)$; and the *decoding algorithm* Dec is a deterministic algorithm, that given an $\hat{x}$ of length $\hat{n}$ in the image of Enc, outputs an $x \in \Sigma^n$. Moreover, $\Pr[\mathsf{Dec}(\mathsf{Enc}(x)) = x] = 1$ for every $x \in \Sigma^n$. We say that E is *onto*, if Dec is defined for every $x \in \Sigma^{\hat{n}(n)}$.

An encoding scheme $\mathsf{E} = (\mathsf{Enc}, \mathsf{Dec})$ over $\mathbb{F}$ is *linear* if for every n, n divides $\hat{n}(n)$, and there exists a decoding vector $\mathbf{r}^{\hat{n}(n)} \in \mathbb{F}^{\hat{n}(n)/n}$ such that the following holds for every $x \in \mathbb{F}^n$. First, every encoding $\mathbf{y}$ in the support of $\mathsf{Enc}(x)$ can be partitioned into n equal-length parts $\mathbf{y} = (\mathbf{y}^1, ..., \mathbf{y}^n)$. Second, $\mathsf{Dec}(\mathbf{y}) = (\langle \mathbf{r}^{\hat{n}(n)}, \mathbf{y}^1 \rangle, ..., \langle \mathbf{r}^{\hat{n}(n)}, \mathbf{y}^n \rangle)$ (where "$\langle \cdot, \cdot \rangle$" denotes inner product). Given an encoding scheme $\mathsf{E} = (\mathsf{Enc}, \mathsf{Dec})$ over $\mathbb{F}$, and $n \in \mathbb{N}$, we say that a vector $\mathbf{v} \in \mathbb{F}^{\hat{n}(n)}$ is *well-formed* if $\mathbf{v} \in \mathsf{Enc}(0^n)$.

PARAMETERIZED ENCODING SCHEMES. We consider encoding schemes in which the encoding and decoding algorithms are given an additional input 1^σ, which is used as a security parameter. Concretely, the encoding length depends also on σ (and not only on n), i.e., $\hat{n} = \hat{n}(n, \sigma)$, and for every σ the resultant scheme is an encoding scheme (in particular, for every $x \in \Sigma^n$ *and every* $\sigma \in \mathbb{N}$, $\Pr[\mathsf{Dec}(\mathsf{Enc}(x, 1^\sigma), 1^\sigma) = x] = 1$). We call such schemes *parameterized*. A parameterized encoding scheme is *onto* if it is onto for every σ. It is linear if it is linear for every σ (in particular, there exist decoding vectors $\{\mathbf{r}^{\hat{n}(n,\sigma)}\}$). For $n, \sigma \in \mathbb{N}$, a vector $\mathbf{v} \in \mathbb{F}^{\hat{n}(n,\sigma)}$ is *well-formed* if $\mathbf{v} \in \mathsf{Enc}(0^n, 1^\sigma)$. We will only consider parameterized encoding schemes, and therefore when we say "encoding scheme" we mean a *parameterized* encoding scheme.

Definition 2 (Leakage-indistinguishability of functions and encodings). *Let $\mathcal{L}$ be a family of leakage functions, and $\epsilon > 0$. A randomized function $f : \Sigma^n \to \Sigma^m$ is $(\mathcal{L}, \epsilon)$-leakage-indistinguishable if for every $x, y \in \Sigma^n$, the distributions $f(x), f(y)$ are $(\mathcal{L}, \epsilon)$-leakage-indistinguishable.*

We say that an encoding scheme E is $(\mathcal{L}, \epsilon)$-leakage-indistinguishable if for every large enough $\sigma \in \mathbb{N}$, $\mathsf{Enc}(\cdot, 1^\sigma)$ is $(\mathcal{L}, \epsilon)$-leakage indistinguishable.

CIRCUITS. We consider arithmetic circuits C over the field $\mathbb{F}$ and the set $X = \{x_1, ..., x_n\}$ of variables. C is a directed acyclic graph whose vertices are called *gates* and whose edges are called *wires*. The wires of C are labled with functions over X. Every gate in C of in-degree 0 has out-degree 1 and is either labeled by a variable from X and is referred to as an *input gate*; or is labeled by a constant $\alpha \in \mathbb{F}$ and is referred to as a const_α *gate*. Following [11], all other gates are labeled by one of the following functions $+, -, \times, \mathsf{copy}$ or id, where $+, -, \times$ are the addition, subtraction, and multiplication operations of the field (i.e., the outcoming wire is labeled with the addition, subtraction, or product (respectively) of the labels of the incoming wires), and these vertices have fan-in 2 and fan-out 1; copy vertices have fan-in 1 and fan-out 2, where the labels of the outcoming edges carry the same function as the incoming edge; and id vertices have fan-in and fan-out 1, and the label of the outcoming edge is the same as the incoming edge. We write $C : \mathbb{F}^n \to \mathbb{F}^m$ to indicate that C is an arithmetic circuit over $\mathbb{F}$ with n inputs and m outputs. The *size* of a circuit C, denoted $|C|$, is the number of wires in C, together with input and output gates. Shallow (d, s)

denotes the class of all depth-d, size-s, arithmetic circuits over $\mathbb{F}$. Similarly, ShallowB (d, s) denotes the class of all depth-d, size-s, boolean circuits with $\wedge, \vee$ gates (replacing the $+, -, \times$ gates of arithmetic circuits), id, copy, const_0, and const_1 gates (with fan-in and fan-out as specified above), and $\neg$ gates with fan-in and fan-out 1. Somewhat abusing notation, we use the same notations to denote the *families of functions* computable by circuits in the respective class of circuits. AC^0 denotes all constant-depth and polynomial-sized boolean circuits over *unbounded fan-in and fan out* $\wedge, \vee, \neg, \mathrm{const}_0$ and const_1 gates.

Definition 3. *For $\mathbb{F} = \mathbb{F}_2$, a circuit $C : \mathbb{F}^n \to \mathbb{F}$ over $\mathbb{F}_2$ is satisfiable if there exists an $x \in \mathbb{F}^n$ such that $C(x) = 1$. For $\mathbb{F} \neq \mathbb{F}_2$, C is satisfiable if there exists an $x \in \mathbb{F}^n$ such that $C(x) = 0$.*

2.1 Circuit Compilers

We define the notion of a circuit compiler. Informally, it consists of an encoding scheme and a compiler algorithm, that compiles a given circuit into a circuit operating on encodings, and emulating the original circuit. Formally,

Definition 4 (Circuit compiler over $\mathbb{F}$). *A circuit compiler over $\mathbb{F}$ is a pair* (Comp, E) *of algorithms with the following syntax.*

- E $=$ (Enc, Dec) *is an encoding scheme, where* Enc *is a PPT encoding algorithm that given a vector $x \in \mathbb{F}^n$, and 1^σ, outputs a vector $\hat{x}$. We assume that $\hat{x} \in \mathbb{F}^{\hat{n}}$ for some $\hat{n} = \hat{n}(n, \sigma)$.*
- Comp *is a polynomial-time algorithm that given an arithmetic circuit C over $\mathbb{F}$ outputs an arithmetic circuit $\hat{C}$.*

We require that (Comp, E) *satisfy the following* correctness *requirement. For any arithmetic circuit C, and any input x for C, we have* $\Pr\left[\hat{C}(\hat{x}) = C(x)\right] = 1$, *where $\hat{x}$ is the output of* Enc $\left(x, 1^{|C|}\right)$.

A boolean circuit compiler is a circuit compiler over $\mathbb{F}_2$.

We consider circuit compilers that are also "sound", meaning that satisfying (possibly *ill formed*) inputs for the compiled circuit exist only if the original circuit is satisfiable.

Definition 5 (SAT-respecting circuit compiler). *A circuit compiler* (Comp, E) *is* SAT-respecting *if it satisfies the following* soundness *requirement for every circuit $C : \mathbb{F}^n \to \mathbb{F}$. If $\hat{C} = \mathsf{Comp}(C)$ is satisfiable then C is satisfiable, i.e., if $\hat{C}(\hat{x}^*) = 0$ for some $\hat{x}^* \in \mathbb{F}^{\hat{n}}$, then there exists an $x \in \mathbb{F}^n$ such that $C(x) = 0$. (For $\mathbb{F} = \mathbb{F}_2$, we require that if $\hat{C}$ outputs 1 on some input, then so does C.)*

2.2   Leakage-Resilient Circuit Compilers (LRCCs)

We consider circuit compilers whose outputs are *leakage resilient* for a class $\mathcal{L}$ of functions, in the following sense. For every "not too large" circuit C, and every input x for C, the wire values of the compiled circuit $\hat{C}$, when evaluated on a random encoding $\hat{x}$ of x, can be simulated given only the output of C; and functions in $\mathcal{L}$ cannot distinguish between the actual and simulated wire values.

Notation 2. *For a Circuit C, a leakage function $\ell : \mathbb{F}^{|C|} \to \mathbb{F}^m$ for some natural m, and an input x for C, $[C, x]$ denotes the wire values of C when evaluated on x, and $\ell[C, x]$ denotes the output of ℓ on $[C, x]$.*

Definition 6 (Relaxed LRCC). *Let $\mathbb{F}$ be a finite field. For a function class $\mathcal{L}$, $\epsilon(n) : \mathbb{N} \to \mathbb{R}^+$, and a size function $\mathsf{S}(n) : \mathbb{N} \to \mathbb{N}$, we say that $(\mathsf{Comp}, \mathsf{E})$ is $(\mathcal{L}, \epsilon(n), \mathsf{S}(n))$-relaxed leakage-resilient if there exists an algorithm Sim such that the following holds. For all sufficiently large n's, every arithmetic circuit C over $\mathbb{F}$ of input length n and size at most $\mathsf{S}(n)$, every $\ell \in \mathcal{L}$ of input length $\left|\hat{C}\right|$, and every $x \in \mathbb{F}^n$, we have $\mathsf{SD}\left(\ell\left[\mathsf{Sim}\left(C, C\left(x\right)\right)\right], \ell\left[\hat{C}, \hat{x}\right]\right) \leq \epsilon(|x|)$, where $\hat{x} \leftarrow \mathsf{Enc}\left(x, 1^{|C|}\right)$.*

Definition 6 is relaxed in the sense that (unlike [11, 18, 23]) Sim is not required to be efficient.

The error in Definitions 5 and 6 is defined with relation to the input length n. Both definitions can be naturally extended such that the compiler is also given a security parameter κ, and the error depends on κ (and possibly also n).

3   SAT-Respecting Relaxed LRCC

In this section we construct a SAT-respecting relaxed LRCC. We first describe a relaxed LRCC over any finite field $\mathbb{F} \neq \mathbb{F}_2$, then use its instantiation over $\mathbb{F}_3$ to construct a boolean relaxed LRCC (which we later use to construct WIPCPs and CZKPCPs). Our starting point is the circuit-compiler of Faust et al. [11], which we denote by $\left(\mathsf{Comp}^{\mathrm{FRRTV}}, \mathsf{E}^{\mathrm{FRRTV}}\right)$. They present a general circuit-compiler that guarantees correctness, and a stronger notion of leakage-resilience (informally, that the wire values of the compiled circuit can be *efficiently* simulated). However, the correctness of their construction relies on the assumption that the inputs to the compiled circuit are honestly encoded. Therefore, their construction is not SAT-respecting, since by using ill-formed encoded inputs one can cause the compiled circuit to output arbitrary values, *even if other than that the compiler was honestly applied to the original circuit*. We describe a method of generalizing their construction such that the circuit-compiler is also SAT-respecting. We first give a high-level overview of the compiler of [11].

GADGETS. On input a circuit C, our compiler, and that of $\mathsf{Comp}^{\mathrm{FRRTV}}$, replace every wire of C with a *bundle* of wires, and every gate in C with a *gadget*. More specifically, a bundle is a string of field elements, encoding a field element according to some encoding scheme E; and a gadget is a circuit which operates on bundles and emulates the operation of the corresponding gate in C. A gadget has both standard inputs, that represent the wires in the original circuit, and masking inputs, that are used to achieve privacy. More formally, a gadget emulates a specific boolean or arithmetic operation on the standard inputs, and outputs a bundle encoding the correct output. Every gadget G is associated with a set M_G of "well-formed" masking input bundles (e.g., in the circuit compiler of [11], M_G consists of sets of 0-encodings). For every standard input x, on input a bundle $\mathbf{x}$ encoding x, and *any* masking input bundles $\mathsf{m} \in M_G$, the output of the gadget G should be consistent with the operation on x. For example, if G computes the operation $\times$, then for every standard input $x = (x_1, x_2)$, for every bundle encoding $\mathbf{x} = (\mathbf{x}_1, \mathbf{x}_2)$ of x according to E, and for every masking input bundles $\mathsf{m} \in M_G$, $G(\mathbf{x}, \mathsf{m})$ is a bundle encoding $x_1 \times x_2$ according to E. Since all the encoding schemes that we consider are onto, we may think of the masking input bundles m as encoding some set mask of values, in which case we say that G takes $|\mathsf{mask}|$ masking inputs. The privacy of the internal computations in the gadget will be achieved when the masking input bundles of the gadget are uniformly distributed over M_G, *regardless* of the actual values encoded by the masking input bundles.

GADGET-BASED CIRCUIT-COMPILERS. $\hat{C} = \mathsf{Comp}^{\mathrm{FRRTV}}(C)$ is a circuit in which every gate is replaced with the corresponding gadget, and output gates are followed by decoding sub-circuits (computing the decoding function of E). Recall that the gadgets also have masking inputs. These are provided as part of the encoded input of $\hat{C}$, in the following way. $\mathsf{E}^{\mathrm{FRRTV}}$ uses an "inner" encoding scheme $\mathsf{E}^{\mathrm{in}} = \left(\mathsf{Enc}^{\mathrm{in}}, \mathsf{Dec}^{\mathrm{in}}\right)$, where $\mathsf{Enc}^{\mathrm{FRRTV}}$ uses $\mathsf{Enc}^{\mathrm{in}}$ to encode the inputs of C, concatenated with 0^κ for a "sufficiently large" κ (these 0-encodings will be the masking inputs to the gadgets); and $\mathsf{Dec}^{\mathrm{FRRTV}}$ uses $\mathsf{Dec}^{\mathrm{in}}$ to decode its input, and discards the last κ symbols.

3.1 The Construction

Let $C : \mathbb{F}^n \to \mathbb{F}$ be the circuit to be compiled. In the following, let $\mathsf{r} = \mathsf{r}(\sigma)$ denote the number of masking inputs used in a circuit compiled according to the compiler of [11]. Recall that our compiler, given a circuit C, generates two copies C_1, C_2 of C (that operate on two copies of the inputs); compiles C_1, C_2 into circuits $\hat{C}_1, \hat{C}_2$ using the circuit-compiler of [11]; generates the circuit $\hat{C}'$ that outputs the AND of $\hat{C}_1, \hat{C}_2$; generates a circuit $\mathcal{T}_0$ verifying that at least one of the copies $\hat{C}_1, \hat{C}_2$ uses well-formed masking inputs (i.e., its masking inputs are well-formed vectors); compiles $\mathcal{T}_0$ into $\hat{\mathcal{T}}_0$ using the circuit-compiler of [11]; and finally verifies "in the clear" that $\hat{\mathcal{T}}_0$ uses well-formed masking inputs. We now describe these ingredients in more detail.

Our first ingredient checks the validity of the masking inputs used in the compiled circuit $\hat{C}'$. If $\mathsf{m}^1, \mathsf{m}^2$ are masking inputs used in the first and second copies $\hat{C}_1, \hat{C}_2$ in $\hat{C}'$, respectively (i.e., these copies are given encodings of $\mathsf{m}^1, \mathsf{m}^2$), then we compute $v_{ij} = \mathsf{m}_i^1 \times \mathsf{m}_j^2$ for every $i, j \in [\mathsf{r}]$, and check that all the v_{ij}'s are zero. To make this check easier, we will use the following "binarization" sub-circuit, which outputs 1 if its input is 0, and outputs 0 on all other values.

Construction 3 ("Binarization" sub-circuit $\mathcal{T}$). $\mathcal{T} : \mathbb{F} \to \mathbb{F}$ *is defined as* $\mathcal{T}(z) = -\prod_{0 \neq a \in \mathbb{F}} (z - a)$, *computed using* $O(|\mathbb{F}|) \times$ *and constant gates arranged in* $O(\log |\mathbb{F}|)$ *layers.*

Observation 4. $\mathcal{T}(0) = 1$, *and for every* $0 \neq z \in \mathbb{F}$, $\mathcal{T}(z) = 0$.

The sub-circuit $\mathcal{T}_0$ described next checks the masking inputs $\mathsf{m}^1, \mathsf{m}^2$ used in the copies of $\hat{C}$, and outputs 1 if and only if one of $\mathsf{m}^1, \mathsf{m}^2$ is the all-zero string. It computes all products of the form $\mathsf{m}_i^1 \times \mathsf{m}_j^2$, then applies $\mathcal{T}$ to every product, and computes the products of all these outputs.

Construction 5 (Oblivious mask-checking sub-circuit $\mathcal{T}_0$). $\mathcal{T}_0 : \mathbb{F}^\mathsf{r} \times \mathbb{F}^\mathsf{r} \to \mathbb{F}$ *is defined as follows.* $\mathcal{T}_0(y, z) = \prod_{i,j \in [\mathsf{r}]} \mathcal{T}(y_i \times z_j)$, *computed using a multiplication tree of size* $O(\mathsf{r})$ *and depth* $O(\log \mathsf{r})$ *(on top of the multiplication trees used to compute* $\mathcal{T}$*).*

Observation 6. *Since the outputs of* $\mathcal{T}$ *are in* $\{0, 1\}$, $\mathcal{T}_0(y, z) = 1$ *if and only if for every* $i, j \in [\mathsf{r}]$, $\mathcal{T}(y_i, z_j) = 1$ *(which by Observation 4 happens if and only if* $y_i \times z_j = 0$*), otherwise it outputs 0.*

Our final ingredient is a sub-circuit $\mathcal{T}_V$ checking the masking inputs used in the compiled sub-circuit $\hat{\mathcal{T}}_0$. At a high level, $\mathcal{T}_V$ decodes every masking input; uses $\mathcal{T}$ to map the decoded values into $\{0, 1\}$ such that only 0 is mapped to 1; and multiplies all these values, to guarantee that all the masking inputs are well-formed. In the following, $\mathsf{r}_0 = \mathsf{r}_0(\sigma)$ denotes the number of masking inputs used in $\hat{\mathcal{T}}_0$.

Construction 7 (Non-oblivious mask-checking sub-circuit $\mathcal{T}_V$). *Let* n, σ, $\kappa \in \mathbb{N}$, $\hat{n} = \hat{n}(n + \kappa, \sigma)$, *and* $\left\{ \boldsymbol{d}^{\hat{n}} \right\}$ *be the decoding vectors of* E^{in}. *We define the decoding sub-circuit* $\mathcal{D}_V : \mathbb{F}^{\hat{n}} \to \mathbb{F}$ *corresponding to* $\boldsymbol{d}^{\hat{n}}$ *as follows:* $\mathcal{D}_V(\boldsymbol{v}) = \langle \boldsymbol{d}^{\hat{n}}, \boldsymbol{v} \rangle$, *where* $\langle \cdot, \cdot \rangle$ *denotes inner-product.* $\mathcal{D}_V$ *is computed using any correct decoding circuit with* $O(\hat{n})$ *gates arranged in* $O(\log \hat{n})$ *layers.*

We define $\mathcal{T}_V : \left(\mathbb{F}^{\hat{n}} \right)^{\mathsf{r}_0} \to \mathbb{F}$ *as follows: for* $\boldsymbol{R} = (\boldsymbol{r}_1, ..., \boldsymbol{r}_{\mathsf{r}_0})$ *where* $\boldsymbol{r}_i \in \mathbb{F}^{\hat{n}}$ *for every* $1 \leq i \leq \mathsf{r}_0$, $\mathcal{T}_V(\boldsymbol{R}) = \prod_{i \in [\mathsf{r}_0]} \mathcal{T}(\mathcal{D}_V(\boldsymbol{r}_i))$. $\mathcal{T}_V$ *is computed using* $O(\mathsf{r}_0) \times$ *gates, arranged in a tree of depth* $O(\log \mathsf{r}_0)$ *(on top of the sub-circuits* $\mathcal{T} \circ \mathcal{D}_V$*).*

Observation 8. *Let* $\boldsymbol{R} = (\boldsymbol{r}_1, ..., \boldsymbol{r}_{\mathsf{r}_0}) \in \left(\mathbb{F}^{\hat{n}} \right)^{\mathsf{r}_0}$, *then for every* $i \in [\mathsf{r}_0]$, $\mathcal{D}_V(\boldsymbol{r}_i) = v_i$, *where* v_i *is the value that* $\boldsymbol{r}_i$ *encodes. Since the outputs of* $\mathcal{T}$ *are in* $\{0, 1\}$, $\mathcal{T}(\mathcal{D}_V(\boldsymbol{r}_i)) = 1$ *if and only if* $v_i = 0$, *so* $\mathcal{T}_V = 1$ *if and only if all* $\boldsymbol{r}_i$*'s are well-formed, otherwise it outputs 0.*

Our circuit-compiler (Construction 9) uses the ingredients described above. Comp first compiles 2 copies C_1, C_2 of C, and $\mathcal{T}_0$, into $\hat{C}_1, \hat{C}_2, \hat{\mathcal{T}}_0$ (respectively), using the compiler of [11]. Then, it generates a flag bit indicating whether $\hat{C}_1, \hat{C}_2$

have the same output, and the masking inputs used in $\hat{C}_1, \hat{C}_2, \hat{T}_0$ are well-formed. If so, the output is that of $\hat{C}_1$, otherwise it is 1. (Recall that an arithmetic circuit is satisfied iff its output is 0.) The encodings scheme generates encoded inputs for both copies $\hat{C}_1, \hat{C}_2$, as well as sufficient masking inputs to be used in $\hat{C}_1, \hat{C}_2, \hat{T}_0$.

Construction 9 $((\mathcal{L}, \epsilon(n), \mathsf{S}(n))$-LRCC over $\mathbb{F})$. *The circuit compiler* $(\mathsf{Comp}, \mathsf{E} = (\mathsf{Enc}, \mathsf{Dec}))$ *is defined as follows. Let* $\mathsf{r} = \mathsf{r}(\sigma), \mathsf{r}_0 = \mathsf{r}_0(\sigma) : \mathbb{N} \to \mathbb{N}$ *be parameters whose value will be set later.*

Let $\mathsf{E}^{\mathrm{in}} = \left(\mathsf{Enc}^{\mathrm{in}}, \mathsf{Dec}^{\mathrm{in}}\right)$ *be a linear encoding scheme over* $\mathbb{F}$, *with encodings of length* $\hat{n}_{\mathrm{in}} = \hat{n}_{\mathrm{in}}(n, \sigma)$, *and decoding vectors* $\{\boldsymbol{d}^{\hat{n}_{\mathrm{in}}}\}$. *Then* $\mathsf{Enc}(x, 1^\sigma) = (\hat{x}_1, \hat{x}_2)$, *where* $\hat{x}_i \leftarrow \mathsf{Enc}^{\mathrm{in}}\left((x, 0^{\mathsf{r}+\mathsf{r}_0}), 1^\sigma\right)$; *and* $\mathsf{Dec}\left((\hat{x}_1, \hat{x}_2), 1^\sigma\right)$ *computes* $\mathsf{Dec}^{\mathrm{in}}(\hat{x}_1, 1^\sigma)$, *and discards the last* $\mathsf{r} + \mathsf{r}_0$ *symbols. We use* $\hat{n} = \hat{n}(n, \sigma)$ *to denote the length of encodings output by* Enc, *and* $\hat{n}_1 = \hat{n}_1(\sigma) := \hat{n}(1, \sigma)$. *(Notice that* $\hat{n}(n, \sigma) = 2\hat{n}_{\mathrm{in}}(n + \mathsf{r} + \mathsf{r}_0, \sigma)$.*) For* $(\hat{x}_1, \hat{x}_2) \leftarrow \mathsf{Enc}(x, 1^\sigma)$, *we denote* $\hat{x}_i = (\hat{x}_i^{\mathrm{in}}, \boldsymbol{R}_i, \boldsymbol{R}_i^0)$, *where* $\hat{x}_i^{\mathrm{in}}$ *is the encoding of* x, *and* $\boldsymbol{R}_i, \boldsymbol{R}_i^0$ *are encodings of* $0^{\mathsf{r}}, 0^{\mathsf{r}_0}$, *respectively. ($\boldsymbol{R}_2^0$ is not used in the construction, but it is part of $\hat{x}_2$ because the same internal encoding scheme $\mathsf{Enc}^{\mathrm{in}}$ is used to generate $\hat{x}_1, \hat{x}_2$.)*

Let $\left(\mathsf{Comp}^{\mathrm{FRRTV}}, \mathsf{E}^{\mathrm{FRRTV}}\right)$ *be the circuit compiler of [11]. Comp on input a circuit* $C : \mathbb{F}^n \to \mathbb{F}$, *outputs the circuit* $\hat{C} : \mathbb{F}^{\hat{n}(n, |C|)} \to \mathbb{F}$ *defined as follows.*

- *Let* C_1, C_2 *be two copies of* C, $\hat{C}_i = \mathsf{Comp}^{\mathrm{FRRTV}}(C_i)$ *for* $i = 1, 2$, *and* $\hat{T}_0 = \mathsf{Comp}^{\mathrm{FRRTV}}(T_0)$.
- *Let* $\mathsf{f}\left((\hat{x}_1^{\mathrm{in}}, \boldsymbol{R}_1, \boldsymbol{R}_1^0), (\hat{x}_2^{\mathrm{in}}, \boldsymbol{R}_2, \boldsymbol{R}_2^0)\right) := \mathcal{T}\left(\hat{C}_1(\hat{x}_1^{\mathrm{in}}, \boldsymbol{R}_1) - \hat{C}_2(\hat{x}_2^{\mathrm{in}}, \boldsymbol{R}_2)\right) \times \hat{T}_0\left((\boldsymbol{R}_1, \boldsymbol{R}_2), \boldsymbol{R}_1^0\right) \times \mathcal{T}_V\left(\boldsymbol{R}_1^0\right)$. *($\mathsf{f} = 1$ if $\hat{C}_1, \hat{C}_2$ have the same output, and in addition the masking inputs used in $\hat{T}_0$, and at least one of $\hat{C}_1, \hat{C}_2$, are well-formed. Otherwise, $\mathsf{f} = 0$.) Then:*

$$\hat{C}\left((\hat{x}_1^{\mathrm{in}}, \boldsymbol{R}_1, \boldsymbol{R}_1^0), (\hat{x}_2^{\mathrm{in}}, \boldsymbol{R}_2, \boldsymbol{R}_2^0)\right) = \left(1 - \mathsf{f}\left((\hat{x}_1^{\mathrm{in}}, \boldsymbol{R}_1, \boldsymbol{R}_1^0), (\hat{x}_2^{\mathrm{in}}, \boldsymbol{R}_2, \boldsymbol{R}_2^0)\right)\right)$$

$$+ \mathsf{f}\left((\hat{x}_1^{\mathrm{in}}, \boldsymbol{R}_1, \boldsymbol{R}_1^0), (\hat{x}_2^{\mathrm{in}}, \boldsymbol{R}_2, \boldsymbol{R}_2^0)\right) \cdot \hat{C}_1\left(\hat{x}_1^{\mathrm{in}}, \boldsymbol{R}_1, \boldsymbol{R}_1^0\right)$$

(Notice that the output is $\hat{C}_1\left(\hat{x}_1^{\mathrm{in}}, \boldsymbol{R}_1, \boldsymbol{R}_1^0\right)$ if $\mathsf{f} = 1$, otherwise it is 1.)

Let $\mathsf{r}^{\mathrm{FRRTV}}$ *denote the maximal number of masking inputs used in a gadget used by the compiler of [11], and* $\mathsf{S}_0(\mathsf{r})$ *denote the size of* $\mathcal{T}_0$. *Then* $\mathsf{r}(\sigma) = \sigma \cdot \mathsf{r}^{\mathrm{FRRTV}}$ *and* $\mathsf{r}_0(\sigma) = \sigma \cdot \mathsf{S}_0\left(\mathsf{r}^{\mathrm{FRRTV}}\right)$.

Next, we briefly analyze the properties of the construction. (The full analysis appears in the full version.)

SAT-RESPECTING. If the masking inputs of $\hat{T}_0$ are ill-formed, then $\mathcal{T}_V$ resets the flag, so the output is 1 (i.e., $\hat{C}$ is not satisfied). Conditioned on $\hat{T}_0$ having well-formed masking inputs, the correctness of the compiler of [11] (applied to $\hat{T}_0$)), guarantees that the flag is reset if the masking inputs of *both* $\hat{C}_1, \hat{C}_2$ are ill-formed. Finally, if at least one of $\hat{C}_1, \hat{C}_2$ has well-formed masking inputs, and $\hat{C}$ is satisfied (in particular, the flag is not reset), then there exists an $x \in \mathbb{F}^n$ that satisfies the correctly evaluated copy, and therefore also satisfies C. We note that the encoding scheme should be onto, otherwise computations in compiled circuits may *not* correspond to computations in the original circuits (since the "encoded" input may not correspond to a *valid* input for the original circuit).

RELAXED LEAKAGE-RESILIENCE. At a high level, on input $C : \mathbb{F}^n \to \mathbb{F}$, and $C(x)$ for $x \in \mathbb{F}^n$, Sim finds a $y \in \mathbb{F}^n$ such that $C(y) = C(x)$ (this is the reason that Sim is unbounded); generates $\hat{C} = \mathsf{Comp}(C)$ and $\hat{y} \leftarrow \mathsf{Enc}\left(y, 1^{|C|}\right)$; honestly evaluates $\hat{C}$ on $\hat{y}$; and outputs the wire values of $\hat{C}$. If E is leakage-indistinguishable for a leakage class which is "somewhat stronger" than $\mathcal{L}$, then for every $\ell \in \mathcal{L}$, $\mathsf{SD}\left(\ell\left[\hat{C}, \hat{x}\right], \ell\left[\hat{C}, \hat{y}\right]\right) \leq \epsilon(n)$, where $\hat{x} \leftarrow \mathsf{Enc}\left(x, 1^{|C|}\right)$. Informally, this follows from a hybrid argument, where we first replace the input of $\hat{C}_1$ from $\hat{x}$ to $\hat{y}$, and then do the same for $\hat{C}_2$. (This is also the reason that we do not explicitly verify that $\hat{C}_1, \hat{C}_2$ are evaluated on encodings of the same input.)

To show that each adjacent pair of hybrids is leakage-indistinguishable, we first use an argument similar to that of [11], where we first replace the bundles of $\hat{C}_1$ or $\hat{C}_2$ (depending on the pair of hybrids in question) that are external to the gadgets (i.e., bundles that correspond to wires of the original circuit C) with random encoding of the "correct" values; and then replacing the bundles internal to the gadgets of $\hat{C}_1$ (or $\hat{C}_2$) with simulated values. However, our compiled circuit $\hat{C}$ consists also of $\hat{\mathcal{T}}_0, \mathcal{T}_V$, so the analysis in our case is more complex, and in particular we cannot use the leakage-resilience analysis of [11] as a black box. To explain the difficulty in generating these wires values, we need to take a closer look at their leakage-resilience analysis.

Recall that the leakage-indistinguishability proof for every pair of adjacent hybrids contains in itself two series of hybrid arguments, one replacing external bundles, and the other replacing internal bundles. In the first case, leakage-indistinguishability is reduced to that of the underlying encoding scheme E^{in}, whereas in the second it is reduced to the leakage-indistinguishability of the actual and simulated wire values of a single gadget. Specifically, the leakage function ℓ^{in} in the reduction is given either an encoding of a single field element, or the wire values of a single gadget; uses its input to generate *all the wire values of the compiled circuit*; and then evaluates ℓ on these wire values. Thus, if originally we could withstand leakage from some function class $\mathcal{L}^{\mathrm{in}}$, and the additional wires can be generated by a function class $\mathcal{L}_R$, then after the reduction we can withstand leakage from any function class $\mathcal{L}$ such that $\mathcal{L} \circ \mathcal{L}_R \subseteq \mathcal{L}^{\mathrm{in}}$. In particular, if $\mathcal{L}^{\mathrm{in}}$ consists of functions computable by low-depth circuits, and computing the internal wires of $\hat{\mathcal{T}}_0, \mathcal{T}_V$ require deep circuits (consequently, $\mathcal{L}_R$ necessarily contains functions whose computation requires deep circuits), then we have no leakage-resilience. To overcome this, we show how to simulate these additional wires using shallow circuits. This is possible because (due to the way in which the hybrids are defined) the masking inputs in at least one copy are well-formed. Specifically, the structure of $\hat{\mathcal{T}}_0, \mathcal{T}_V$ guarantees that *conditioned on the masking inputs of $\hat{C}_2$ being well-formed*, these wire values can be computed by shallow circuits. When the masking inputs of $\hat{C}_2$ are *ill-formed*, we are guaranteed that the masking inputs of $\hat{C}_1$ are *well-formed*. Conditioned on this event, we show an *alternative* method of computing the internal wires of $\hat{T}_0, \mathcal{T}_V$, which can be done by shallow circuits. Thus, we get the following result.

Proposition 1 (SAT-respecting relaxed LRCC over $\mathbb{F}$). *Let $\mathcal{L}, \mathcal{L}_\mathsf{E}$ be families of functions, $\mathsf{S}(n) : \mathbb{N} \to \mathbb{N}$ be a size function, and $\epsilon(n) : \mathbb{N} \to \mathbb{R}^+$. Let $\mathsf{E}^{\mathrm{in}} = \left(\mathsf{Enc}^{\mathrm{in}}, \mathsf{Dec}^{\mathrm{in}}\right)$ be a linear, onto, $(\mathcal{L}_\mathsf{E}, \epsilon(n))$-leakage-indistinguishable encoding scheme with parameters $n = 1$, σ and $\hat{n} = \hat{n}(\sigma)$, such that $\mathcal{L}_\mathsf{E} = \mathcal{L} \circ \mathsf{Shallow}\left(7, O\left(\hat{n}^4\left(\mathsf{S}(n)\right) \cdot \mathsf{S}(n)\right)\right)$. Then there exists a SAT-respecting, $(\mathcal{L}, 8\epsilon(n) \cdot \mathsf{S}(n), \mathsf{S}(n))$-relaxed-LRCC over $\mathbb{F}$. Moreover, For every $C : \mathbb{F}^n \to \mathbb{F}$, the compiled circuit $\hat{C}$ has size $\left|\hat{C}\right| = O\left(|\mathbb{F}| \cdot \hat{n}^5\left(\mathsf{S}(n)\right) \cdot |C|^2\right)$.*

3.2 A SAT-Respecting Relaxed LRCC over $\mathbb{F}_2$

In this section we describe a relaxed LRCC over $\mathbb{F}_2$. Our starting point is the circuit-compiler of Construction 9 over the field $\mathbb{F}$, which we apply to an "arithmetic version" of the boolean circuit. At a high-level, we construct our circuit compiler over $\mathbb{F}_2$ as follows: we represent field elements of $\mathbb{F}$ using bit-strings; and operations $+, -, \times, \mathsf{id}, \mathsf{copy}, \mathsf{const}_\alpha, \alpha \in \mathbb{F}$ as functions over $\lceil\log|\mathbb{F}|\rceil$-bit strings. (For now, we assume that there exist gates operating on $\lceil\log|\mathbb{F}|\rceil$-bit strings and computing these operations.) We "translate" boolean circuits into arithmetic circuits with such operations, and apply the circuit-compiler of Construction 9 (where the field operations are implemented using the boolean operations described in Sect. 2) to the "translated" circuit. (We note that leakage-resilience deteriorates when an arithmetic compiler is transformed to a boolean one, but only by a constant factor in the depth and size of circuits computing the leakage functions.) Concretely, we set $\mathbb{F} = \mathbb{F}_3$.

FROM BOOLEAN CIRCUITS TO ARITHMETIC CIRCUITS. Our boolean circuit-compiler operates on *boolean circuits*, but employs an arithmetic circuit-compiler operating on *arithmetic circuits* over $\mathbb{F}$. Therefore, we first transform the boolean circuit into an equivalent arithmetic circuit in the natural manner (i.e., representing every bit operation as a polynomial over the arithmetic field).

The field elements of $\mathbb{F}$, and the arithmetic operations over $\mathbb{F}$ that are used by the arithmetic relaxed LRCC (Construction 9) will be represented using bit strings and boolean operations, respectively.

REPRESENTING FIELD ELEMENTS AS BIT STRINGS. We can use any 1:1 transformation $E_b : \mathbb{F}_3 \to \{0, 1\}^2$, such that every bit string is associated with a field element. This is required for the SAT-respecting property, to guarantee that whatever values are carried on the wires of the boolean circuit, they can be "translated" into wires of the arithmetic circuit over $\mathbb{F}_3$, and is achieved by defining a "reverse" mapping E_b^{-1}.

IMPLEMENTING FIELD OPERATIONS. The compiled arithmetic circuit uses the field operations $+, -, \times$, and also $\mathsf{copy}, \mathsf{id}$ and $\mathsf{const}_\alpha, \alpha \in \mathbb{F}_3$. These operations are represented using bit operations over bit strings generated by E_b. Specifically, we think of every field operation as a boolean function with 4 inputs (a pair of 2-bit strings representing the pair of input field elements) and 2 outputs (a 2-bit string representing the output field element). We stress that though an honest construction over bits uses only 3 of the 4 possible 2-bit strings encoding field elements (i.e., only the strings in the image of E_b as defined, for example, in Construction 11), the function representing a field operation in $\mathbb{F}_3$ should be defined to output the correct values on *all* 2-bit strings. The truth table of each output bit has constant size, and can be represented by a constant-size, depth-3 boolean circuit. copy, id and const$_\alpha$ gates are handled similarly. Therefore, the size (depth) of each gadget (and consequently, of the entire compiled circuit) increases by a constant multiplicative factor (specifically, by a factor of 3).

Notice that representing boolean circuits using arithmetic circuits introduces the following obstacle. For a satisfiable circuit $\hat{C}$, we are only guaranteed the existence of an $x \in \mathbb{F}^n$ satisfying the original *arithmetic* circuit, whereas for boolean circuits we require that $x \in \{0,1\}^n$. Therefore, we need an additional "input checker" sub-circuit that will guarantee that the inputs to the compiled circuit encode binary strings.

Definition 7 (Input-checker $\mathcal{T}^{\mathrm{in}}$). $\mathcal{T}^{\mathrm{in}} : \mathbb{F} \rightarrow \mathbb{F}$ *is defined as follows:* $\mathcal{T}^{\mathrm{in}}(z) = \mathcal{T}(z^2 - z)$.

Observation 10. *For every* $z \in \mathbb{F}_3$, $\mathcal{T}^{\mathrm{in}}(z) \in \{0,1\}$, *and* $\mathcal{T}^{\mathrm{in}}(z) = 1$ *if and only if* $z \in \{0,1\}$.

Construction 11 (SAT-respecting relaxed LRCC). *Let* $E_b : \mathbb{F}_3 \rightarrow \{0,1\}^2$ *such that* $E_b(0) = 00$, $E_b(1) = 01$, *and* $E_b(2) = 11$, *and let* $E_b^{-1} : \{0,1\}^2 \rightarrow \mathbb{F}_3$ *such that* $E_b^{-1}(00) = 0$, $E_b^{-1}(01) = E_b^{-1}(10) = 1$, *and* $E_b^{-1}(11) = 2$. *Let* T' *be an algorithm transforming boolean circuits into arithmetic circuits over* $\mathbb{F}_3$, *and* (Comp, E = (Enc, Dec)) *be the circuit compiler over* $\mathbb{F}_3$ *of Construction 9. The circuit compiler over* $\mathbb{F}_2$ *is* $\left(\mathsf{Comp}^b, \mathsf{E}^b = \left(\mathsf{Enc}^b, \mathsf{Dec}^b\right)\right)$, *where:*

- $\mathsf{Enc}^b = E_b \circ \mathsf{Enc}$ *and* $\mathsf{Dec}^b = \mathsf{Dec} \circ E_b^{-1}$
- Comp^b *on input* $C : \{0,1\}^n \rightarrow \{0,1\}$:

  - *Uses* T' *to transform* C *into an equivalent arithmetic circuit* $C' : \mathbb{F}_3^n \rightarrow \mathbb{F}_3$.
  - *Constructs the circuit* $C'' : \mathbb{F}_3^n \rightarrow \mathbb{F}_3$ *such that* $C''(x_1, ..., x_n) = 1 - (C'(x_1, ..., x_n) \times (\times_{i=1}^n \mathcal{T}^{\mathrm{in}}(x_i)))$. *(Notice that* $C''(x_1, ..., x_n)$ *outputs 0 if and only if* $C'(x_1, ..., x_n) = 1$ *and* $x_1, ..., x_n \in \{0,1\}$.)
  - *Computes* $\hat{C}'' = \mathsf{Comp}(C'')$.
  - *Replaces every gate in* $\hat{C}''$ *with a constant-size, depth-3 boolean circuit computing the truth table of the gate operation.* Comp^b *can use any correct circuit, as long as these circuits are used consistently (i.e., for every gate the same circuit is used to replace all appearances of the gate in* $\hat{C}''$).

- Denote the output of $\hat{C}''$ by $e \in \mathbb{F}_3$, represented by the string $(e_1, e_2) \in \{0,1\}^2$. Then Comp^b outputs the circuit $\hat{C}_b$ obtained from $\hat{C}''$ by applying a $\vee$ gate, followed by a $\neg$ gate, to the output of $\hat{C}''$. (This reduces the output string of $\hat{C}''$ to a single bit, and flips the output of $\hat{C}''$, which is required due to the negation added in step 2.)

We use $\hat{C}_{1,b}, \hat{C}_{2,b}, \hat{T}_{0,b}, \mathcal{T}_{V,b}$ to denote the components of $\hat{C}_b$ corresponding to $\hat{C}_1, \hat{C}_2, \hat{T}_0, \mathcal{T}_V$, respectively.

Observation 12. $\hat{C}_b(\hat{x}) \in \{0,1\}$ for every $\hat{x}$. Moreover, $\hat{C}_b(\hat{x}) = 1$ if and only if $\hat{C}''(\hat{x}) = 0$. If Comp is SAT-respecting, then this guarantees that $C''(x) = 0$ for some $x \in \mathbb{F}_3$. The definition of C'', and the correctness of T', guarantees that $x \in \{0,1\}^n$, and that $C'(x) = C(x) = 1$.

In the full version, we prove that if Construction 9 is a SAT-respecting relaxed-LRCC over $\mathbb{F}_3$, then so is Construction 11 (over $\mathbb{F}_2$), against a somewhat-weaker leakage family. The leakage family is weaker because relaxed leakage-resilience is proved by reduction to the relaxed leakage-resilience of Construction 9 (the leakage function in the reduction, given the wire values of the arithmetic compiled circuit, generate the internal wires emulating these operations using boolean operations). Formally, we obtained the following.

Proposition 2. Let $\mathcal{L}, \mathcal{L}_\mathsf{E}$ be families of functions, $\mathsf{S}(n) : \mathbb{N} \to \mathbb{N}$ be a size function, and $\epsilon(n) : \mathbb{N} \to \mathbb{R}^+$. Let E^{in} be a linear, onto encoding scheme over $\mathbb{F}_3$ with parameters $n = 1$, σ and $\hat{n} = \hat{n}(\sigma)$, that is $(\mathcal{L}_\mathsf{E}, \epsilon(n))$-leakage-indistinguishable, and $\mathcal{L}_\mathsf{E} = \mathcal{L} \circ \mathsf{ShallowB}\left(33, O\left(\hat{n}^5(\mathsf{S}(n)) \cdot \mathsf{S}(n)^2\right)\right)$. Then there exists a constant $c > 0$, and a SAT-respecting, $(\mathcal{L}, c \cdot \epsilon(n) \cdot \mathsf{S}(n), \mathsf{S}(n))$-relaxed-LRCC over $\mathbb{F}_2$. Moreover, $\left|\hat{C}_b\right| = O\left(\hat{n}^5(\mathsf{S}(n)) |C|^2\right)$.

Taking E^{in} to be the parity encoding in the previous proposition, and using a result of Håstad [16] that AC^0 circuits (i.e., constant-depth and polynomial-sized boolean circuits with unbounded fan-in $\wedge, \vee$ and $\neg$ gates) cannot distinguish parity encodings of 0 and 1, we obtain an LRCC against AC^0-leakage. (We note that the compiler can also be made to withstand leakage that outputs more than one bit, using a result of Dubrov and Ishai [9]. The details of this construction, and the proof of Corollary 2, are deferred to the full version.)

Corollary 2. There exists a SAT-respecting $\left(\mathsf{AC}^0, \mathsf{negl}(n), \mathsf{poly}(n)\right)$-relaxed-LRCC over $\mathbb{F}_2$.

3.3 Withstanding Leakage from AC^0 Circuits with $\oplus$ Gates

Recall that AC^0 denotes the class of constant-depth, polynomial-sized boolean circuits over unbounded fan-in and fan-out $\wedge, \vee, \neg$ gates. In this section we describe a SAT-respecting circuit-compiler withstanding leakage computed by AC^0 circuits, augmented with a sublinear number of $\oplus$ gates of unbounded fan-in and fan-out. Concretely, we use Construction 11, where the underlying arithmetic LRCC over $\mathbb{F}_3$ is instantiated with the encoding scheme E^{in} that maps an element $\gamma \in \mathbb{F}_3$ into a vector $v \in \{0,1\}^k$ (for some natural k), which is random subject to the constraint that the number of 1's in v is congruent to γ modulo 3. We show, by reduction to correlation bounds of [22], that AC^0 circuits, augmented with a sublinear number of $\oplus$ gates, have a negligible advantage in distinguishing between random encodings of 0 and 1 according to E^{in}. (This reduction is non-trivial and appears in Appendix A.) Using the leakage-indistinguishability of E^{in}, we prove the existence of a circuit compiler withstanding leakage from AC^0 circuits that have several output bits and are augmented with a sublinear number of $\oplus$ gates. (The proof appears in the full version.)

Theorem 13. *For input length parameter n, leakage length bound $\hat{n} = \hat{n}(n)$, size bound $s = s(n)$, output length bound $m = m(n)$, parity gate bound $t = t(n)$, and depth bound d, let $\mathcal{L}^m_{\hat{n},d,s,\oplus t} = \bigcup_{n \in \mathbb{N}} \mathcal{L}^{m(n)}_{\hat{n}(n),d,s(n),\oplus t(n)}$, where $\mathcal{L}^{m_0}_{\hat{n}_0,d_0,s_0,\oplus t_0}$ denotes the class of boolean circuits of input length $\hat{n}_0$ over $\neg$ gates and unbounded fan-in $\wedge, \vee, \oplus$ gates, whose depth, size, output length, and number of parity gates are bounded by d_0, s_0, m_0, t_0, respectively. Then for every positive constants d, c, polynomials m, t, and polynomial size bound $s' = s'(n)$, there exists a polynomial $l(n)$, such that there exists a SAT-respecting $\left(\mathcal{L}^m_{l,d,l^c,\oplus t}, 2^{-n^c}, s'(n)\right)$-relaxed LRCC over $\mathbb{F}_2$, which on input a circuit $C : \{0,1\}^n \to \{0,1\}$ of size $|C| \le s'(n)$ outputs a circuit $\hat{C}$ of size $|\hat{C}| \le l(n)$.*

4   WIPCPs and CZKPCPs

Given a relation $\mathcal{R} = \mathcal{R}(x,w)$, we let $L_{\mathcal{R}} := \{x : \exists w, (x,w) \in \mathcal{R}\}$. A *probabilistic proof system* (P,V) for an NP-relation $\mathcal{R} = \mathcal{R}(x,w)$ consists of a *PPT* prover P that on input (x,w) outputs a proof π (in standard probabilistically checkable proofs the prover is deterministic, but our constructions will crucially rely on the prover being probabilistic), and a probabilistic verifier V that given input x and oracle access to a proof π outputs either accept or reject. We say that V is *q-query-bounded* if V makes at most q queries to π.

WIPCPs. A probabilistic proof system is a WIPCP for an NP-relation $\mathcal{R} = \mathcal{R}(x,w)$ if it satisfies the following. First, when given $x \in L_{\mathcal{R}}$, and oracle access to an honestly generated proof, the verifier accepts with probability 1 (this is called *completeness*). Second, given $x \notin L_{\mathcal{R}}$, the verifier rejects except with some probability ϵ_S, *regardless* of its "proof" oracle (this is called ϵ_S-*soundness*). Thirdly, for every (possibly malicious, possibly adaptive) q^*-query bounded verifier V^*, every $x \in L_{\mathcal{R}}$, and every pair w_1, w_2 of witnesses for x, the view of V^* when verifying an honestly generated proof for (x, w_1) is ϵ_{ZK}-statistically close to its view when verifying an honestly generated proof for (x, w_2) (this is called $(\epsilon_{\mathrm{ZK}}, q^*)$-*WI*). A WIPCP is a *non-adaptive WIPCP (NA-WIPCP)* system for a relation $\mathcal{R} = \mathcal{R}(x,w)$, if the *honest* verifier is non-adaptive. In the following, we denote by $\mathsf{NA} - \mathsf{WIPCP}[r, q, q^*, \epsilon_S, \epsilon_{\mathrm{ZK}}, \ell]$ the class of NP-languages that admit an NP-relation $\mathcal{R}$ with a non-adaptive $(\epsilon_{\mathrm{ZK}}, q^*)$-WIPCP, in which the prover outputs proofs of length ℓ, the honest verifier tosses $O(r)$ coins, queries $O(q)$ proof bits, and rejects false claims except with probability at most ϵ_S. We use $\mathsf{PCP}[r, q, \epsilon, \ell]$ to denote the class of NP-languages admitting a standard (i.e., non-WI) PCP system with the same properties, and write $\mathcal{R} \in \mathsf{PCP}[r, q, \epsilon, \ell]$ to denote that $L_\mathcal{R} \in \mathsf{PCP}[r, q, \epsilon, \ell]$. We denote $\mathsf{NA} - \mathsf{WIPCP} := \mathsf{NA} - \mathsf{WIPCP}[\mathsf{poly}\log n, \mathsf{poly}\log n, \mathsf{poly}(n), \mathsf{negl}(n), \mathsf{negl}(n), \mathsf{poly}(n)]$.

We describe a transformation from PCPs to NA-WIPCPs, which can be applied to any PCP system in which the proof is obtained from the witness through an "easy" function (we formalize this notion below). Recall that a standard PCP π can be generated from the wire values $[C_\mathcal{R}, (x, w)]$ of the verification circuit $C_\mathcal{R}$ of the relation, on input x and witness w. If the function f taking $[C_\mathcal{R}, (x, w)]$ to π is in a function class $\mathcal{L}$, then the system can be made WI as follows. The prover and verifier both compile $C_\mathcal{R}(x, \cdot)$ (i.e., $C_\mathcal{R}$ with x hard-wired into it) into a SAT-respecting circuit $\hat{C}_\mathcal{R}$ that is relaxed leakage-resilient against $\mathcal{L}$. The prover then samples a random encoding $\hat{w}$ of w, and generates the PCP $\pi = f\left[\hat{C}_\mathcal{R}, \hat{w}\right]$. The verifier probabilistically verifies that $\hat{C}_\mathcal{R}$ is satisfiable by reading few symbols of π, which (if the verifier is non-adaptive) correspond to applying a leakage function from $\mathcal{L}$ to the wire values of $\hat{C}_\mathcal{R}$. This gives the following result. (The detailed construction, and the proof of Proposition 3, appear in the full version.)

Proposition 3. *Let n be a length parameter, $\epsilon_S, \epsilon_{\mathrm{ZK}} \in [0, 1]$, $\mathsf{S} = \mathsf{S}(n)$ be a size function, $q^* = q^*(n)$ be a query function, and $g(\cdot)$ be a polynomial. Let $\mathcal{L}$ be a family of leakage functions, such that:*

- *there is a SAT-respecting $(\mathcal{L}, \epsilon_{\mathrm{ZK}}, \mathsf{S})$-relaxed LRCC $(\mathsf{Comp}, \mathsf{E})$ satisfying $|\mathsf{Comp}(C)| \leq g(|C|)$;*
- *there is a $\mathsf{PCP}[r(n), q(n), \epsilon_S, \ell(n)]$ system for 3SAT, such that for every $(\varphi, W) \in$ 3SAT, every subset $\mathcal{Q}$ of q^* bits of an honestly-generated proof $\pi = \pi(\varphi, W)$ is computable from W by a function $f_{\varphi, \mathcal{Q}} \in \mathcal{L}$.*

Then for every NP-relation $\mathcal{R} = \mathcal{R}(x, w)$ with verification circuit $C^\mathcal{R}$ of size at most S, we have that $\mathcal{R} \in \mathsf{NA} - \mathsf{WIPCP}[r(t), q(t), q^*, \epsilon_S, 2\epsilon_{\mathrm{ZK}}, \ell(t)]$, where $t = O\left(g\left(|C^\mathcal{R}|\right)\right)$, and WI holds against non-adaptive verifiers.

In the full version we use techniques of [7] to generalize the WI property of Proposition 3 to *adaptive* verifiers, while increasing the statistical distance of the WI by a multiplicative factor of roughly ℓ^{q^*} (all other parameters remain unchanged). Then, we prove that the PCP system of [2] for 3SAT has the property that every proof bit is generated from the NP-witness by an AC^0 circuit, augmented with "few" $\oplus$ gates. Theorem 1 follows by combining these two results with Theorem 13.

CZKPCPs in the CRS model. A probabilistic proof system is a CZKPCP in the CRS model for an NP-relation $\mathcal{R} = \mathcal{R}(x, w)$ if the prover and verifier have access to a common random string s; correctness holds for *any* s; soundness holds for a uniformly random s; and there exists a PPT simulator Sim such that for every q^*-query bounded verifier V^*, and every $x \in L_{\mathcal{R}}$, Sim (x) is computationally indistinguishable from the joint distribution of a uniformly random s, and the view of V^* given s and oracle access to an honestly generated proof for x (this is called *computational ZK (CZK)*). Similar to NA-WIPCPs, a CZKPCP system is *non-adaptive (NA-CZKPCP)* if the honest verifier is non-adaptive. Applying the techniques of [12] to Proposition 3, we obtain a general transformation from NA-WIPCPs to NA-CZKPCPs, and Corollary 1 follows by using the NA-WIPCP of Theorem 1 (see the full version for details).

We note that a simple alternative construction of CZKPCP for NP can be obtained by applying a standard PCP on top of a standard NIZK proof [6,13]. Concretely, the CZKPCP prover generates a PCP for the NP-claim "there exists a NIZK for the claim $x \in L_{\mathcal{R}}$, relative to the CRS s, that would cause the NIZK-verifier to accept", where the witness is the NIZK proof string. Since the NIZK itself is CZK, the resultant PCP is also CZK. However, NIZK proofs for NP are not known to follow from the existence of one-way functions, and can currently be based only on much stronger assumptions such as the existence of trapdoor permutations [12].

THE (IM)POSSIBILITY OF SAT-RESPECTING NON-RELAXED LRCCs. Known constructions of LRCCs withstanding global leakage [11,18,23] guarantee a *universal* simulation property, in the sense that the simulator generates the simulated wire values *without* knowing the identity of the leakage function; and these values are guaranteed to be indistinguishable from the actual wire values, *for every leakage function in the leakage class*. Consequently, our construction (which is based on the LRCC of [11]), also guarantees this universal simulation property. Our general transformation from SAT-respecting relaxed LRCCs to WIPCPs can also be applied to a SAT-respecting *non-relaxed* LRCC, in which case we would get ZKPCP for all NP, with a *universal PPT* simulator that generates a simulated proof *without seeing the queries of the verifier*. This simulator can be used to decide the NP-language, so the existence of SAT-respecting LRCCs with a universal simulator would imply that NP $\subseteq$ BPP. (See the full version for additional details.) We note that our transformation of Sect. 4 does *not* require the LRCC simulator to be universal. However, the construction of (SAT-respecting) non-relaxed LRCCs with a non-universal simulator would require developing new techniques for constructing LRCCs.

4.1   Distributed ZK and WI Proofs

We use our WIPCPs and CZKPCPs to construct *3-round* distributed WI and CZK proofs (respectively) for NP in a distributed setting, in which the PPT prover P and verifier V are aided by m polynomial-time servers $S_1, ..., S_m$. We call such systems *m-distributed proof systems*. We note that P has input (x, w), V has input x, and the servers have no input. Our motivation for studying proofs in a distributed setting is to minimize the round complexity, and underlying assumptions, of sublinear ZK proofs. Concretely, it is known that assuming the existence of collision resistant hash functions, there exist 2-party 4-round sublinear ZK arguments for NP [17,20]. (Arguments guarantee soundness only against *bounded* malicious provers.) We show that in the distributed setting, there exist *3-round* sublinear CZK (respectively, WI) *proofs* for NP, *assuming the existence of OWFs* (respectively, *unconditional*). Thus, the distributed setting allows us to improve previous results in terms of round complexity, underlying assumptions, and soundness type.

DISTRIBUTED CZK\WI PROOF SYSTEMS. An m-distributed proof system is a (t, m)-distributed ZK proof system for an NP-relation $\mathcal{R}$ if it satisfies the following properties. First, if all parties are honest and $(x, w) \in \mathcal{R}$ then V accepts x with probability 1 (the *correctness* property). Second, if $x \notin L_\mathcal{R}$ then V rejects x except with negligible probability, even if the prover is corrupted and colludes with at most t corrupted servers (the *soundness* property). Thirdly, for every adversary $\mathcal{A}$ corrupting V and $t' \leq t$ servers there exists a PPT simulator Sim such that for every $x \in L_\mathcal{R}$, Sim (x) is computationally indistinguishable from the the view of $\mathcal{A}$ in the protocol execution, when it has input x. This notion can be naturally relaxed to WI, or CZK in the CRS model.

We use WIPCPs (respectively, CZKPCPs) to construct a *3-round* distributed-WI proof system (respectively, CZK proof system in the CRS model) which, at a high level, operates as follows. In the first round the prover distributes a WIPCP (respectively, a CZKPCP) between the servers, and in the second and third rounds the verifier and servers emulate the WIPCP (respectively, CZKPCP) verification procedure (the verifier sends the proof queries of the WIPCP or CZKPCP verifier, and the servers provide the corresponding proof bits). This overview is an over-simplification of the construction: the verification procedure of the WIPCP (respectively, CZKPCP) cannot be used as-is since it only guarantees soundness when the verification is performed with a proof *oracle*, whereas corrupted servers *can determine their answers after seeing the queries of the verifier*. We overcome this by using techniques of [19] (a more detailed description and analysis of these distributed proof systems appears in the full version). Thus, we obtain the following results.

Theorem 14 (Sublinear distributed WI proofs). *For every NP-relation $\mathcal{R}$, and polynomial $t(n)$, there exists a polynomial $m(n) > t(n)$ such that $\mathcal{R}$ has a 3-round sublinear (t, m)-distributed WI proof system, where n is the input length.*

Theorem 15 (Sublinear distributed CZK proofs in the CRS model). *Assume that OWFs exist. Then for every NP-relation $\mathcal{R}$, and polynomial $t(n)$, there exists a polynomial $m(n) > t(n)$ such that $\mathcal{R}$ has a 3-round sublinear (t, m)-distributed CZK proof system in the CRS model, where n is the input length.*

These constructions *crucially* rely on the *non-adaptivity* of the honest WIPCP (respectively, CZKPCP) verifier (otherwise we would need at least 4 rounds, since rounds cannot be compressed). Moreover, the verifier may collude with a subset of servers, so the PCP should be WI (respectively, CZK) against *malicious* verifiers.

Acknowledgements. We thank the anonymous TCC reviewers for helpful comments, and in particular for pointing out the simple construction of CZKPCP from PCP and NIZK. The first author was supported by ERC starting grant 259426, ISF grant 1709/14, and BSF grant 2012378. Research done in part while visiting the Simons Institute for the Theory of Computing, supported by the Simons Foundation and by the DIMACS/Simons Collaboration in Cryptography through NSF grant #CNS-1523467. Research also supported in part from a DARPA/ARL SAFEWARE award, NSF Frontier Award 1413955, NSF grants 1228984, 1136174, 1118096, and 1065276. This material is based upon work supported by the Defense Advanced Research Projects Agency through the ARL under Contract W911NF-15-C-0205. The views expressed are those of the author and do not reflect the official policy or position of the Department of Defense, the National Science Foundation, or the U.S. Government. The second author was supported by ERC starting grant 259426 and an IBM PhD Fellowship. The third author was supported by the National Basic Research Program of China Grant 2011CBA00300, 2011CBA00301, and the National Natural Science Foundation of China Grant 61033001, 61350110536, 61361136003.

A   A Leakage-Indistinguishable Encoding Scheme

In this section we define the encoding scheme that is used to prove Theorem 13, and use correlation bounds of [22] to show that it is leakage-indistinguishable against leakage computable by AC^0 circuits, augmented with few $\oplus$ gates.

Notation 16. *For $\gamma \in \{0, 1, 2\}$ and $n \in \mathbb{N}$, U_γ^n denotes the uniform distribution over $\{v \in \{0,1\}^{3n} : \#_1(v) \equiv \gamma \mod 3\}$; $\#_1(v)$ denotes the number of 1's in v; and $U_{1,2}^n$ denotes the uniform distribution over $\{v \in \{0,1\}^{3n} : \#_1(v) \not\equiv 0 \mod 3\}$.*

Definition 8. *We define an encoding scheme $\mathsf{E}_3 = (\mathsf{Enc}_3, \mathsf{Dec}_3)$ over $\mathbb{F}_3$ such that for every $e \in \mathbb{F}_3$, $\mathsf{Enc}_3(e, 1^n)$ is distributed according to U_e^n,[4] and $\mathsf{Dec}_3(v)$ returns $(\#_1(v) \mod 3)$. Notice that E_3 is linear, with decoding vectors $\{1^{3n}\}$, and consequently also onto.*

The leakage class we consider is "AC^0, augmented with few $\oplus$ gates":

Definition 9 ($\mathcal{L}_{n,d,s,\oplus t}^m$ leakage family). *Let $n \in \mathbb{N}$ be a length parameter, $d \in \mathbb{N}$ be a depth parameter, $s \in \mathbb{N}$ be a size parameter, and $t \in \mathbb{N}$ be a parity gate bound. The family $\mathcal{L}_{n,d,s,\oplus t}$ consists of all functions computable by a boolean circuit $C : \{0,1\}^n \to \{0,1\}$ of size at most s and depth d, with unbounded fan-in and fan-out $\wedge, \vee, \neg, \oplus$ gates, out of which at most t are $\oplus$ gates. The family $\mathcal{L}_{d,s,\oplus t}$ of functions is defined as $\mathcal{L}_{d,s,\oplus t} = \cup_{n \in \mathbb{N}} \mathcal{L}_{n,d,s,\oplus t}$.*

For a length parameter $m \in \mathbb{N}$, and a function $f : \{0,1\}^n \to \{0,1\}^m$, let $f_i(x_1, ..., x_n), i \in [m]$ denote the i'th output bit of f. We use the following notation: $\mathcal{L}_{n,d,s,\oplus t}^m = \{f : \{0,1\}^n \to \{0,1\}^m : \forall 1 \leq i \leq m, f_i \in \mathcal{L}_{n,d,s,\oplus t}\}$, and $\mathcal{L}_{d,s,\oplus t}^m := \cup_{n \in \mathbb{N}} \left(\mathcal{L}_{n,d,s,\oplus t}^m \right)$.

---

[4] Enc_3 can be computed efficiently by repeating the following procedure n^2 times. Pick $v \in \{0,1\}^n$ uniformly at random, compute $t := \#_1(v)$, and if $t = e$ then return v. If all iterations fail, return a fixed $v_e \in \{0,1\}^n$ such that $\#_1(v) = e$. Then the output of Enc_3 is statistically close to U_e^n.

We use a correlation bound of Lovett and Srinivasan [22, Theorem 6] which, informally, states that AC^0 circuits, augmented with "few" $\oplus$ gates, have negligible correlation with the boolean function MOD_3 where $\mathrm{MOD}_3(v) = 0$ if and only if $\#_1(v) \equiv 1 \mod 3$. (Their result is more general, but we state a weaker and simpler version that suffices for our needs.) We first define the notion of correlation.

Definition 10 (Correlation). *Let* $n \in \mathbb{N}$, $g, f : \{0,1\}^n \to \{0,1\}$, *and let* $\mathcal{D}$ *be a distribution over* $\{0,1\}^n$. *The correlation of* g *and* f *in relation to* $\mathcal{D}$ *is* $\mathsf{Corr}_{\mathcal{D}}(g, f) = 2 \left| \frac{1}{2} - \Pr_{x \leftarrow \mathcal{D}}[g(x) = f(x)] \right|$.
For a class $\mathcal{G}$ *of functions,* $\mathsf{Corr}_{\mathcal{D}}(\mathcal{G}, f) = \max_{g \in \mathcal{G}} \mathsf{Corr}_{\mathcal{D}}(g, f)$.

We are interested in correlations with the following function:

Notation 17 (MOD$_s$ function). *Let* $s \in \mathbb{N}$. *The function* $\mathrm{MOD}_s^n : \{0,1\}^{3n} \to \{0,1\}$ *is defined as* $\mathrm{MOD}_s(x) = 0$ *if and only if* $\sum_{i=1}^{3n} x_i \equiv 0 \mod s$. *We use* MOD_s *to denote the family of functions* $\cup_{n \in \mathbb{N}} \mathrm{MOD}_s^n$.

Theorem 18 ([22], Theorem 6 (rephrased)). *For every constant depth parameter* $d \in \mathbb{N}$ *there exist constants* $c, \epsilon \in (0,1)$, *such that for every constant* $l \in \mathbb{N}$ *there exists a minimal length parameter* $n_0 \in \mathbb{N}$ *such that for every* $n \geq n_0$, $\mathsf{Corr}_{\mathcal{D}_3^n}(\mathcal{L}_{3n,d,n^l,\oplus n^\epsilon}, \mathrm{MOD}_3^n) \leq 2^{-n^c}$, *where* $\mathcal{D}_3^n$ *is the distribution induced by the following process: first pick a random bit* $b \in_R \{0,1\}$; *if* $b = 0$ *pick* $x \in \{0,1\}^{3n}$ *according to the distribution* U_0^n, *otherwise pick* $x \in \{0,1\}^{3n}$ *according to* $U_{1,2}^n$.

Next, we use Theorem 18 to show that AC^0 circuits, augmented with "few" $\oplus$ gates, have a negligible advantage in distinguishing between random encodings of 0,1, and 2 according to the encoding scheme of Definition 8. Formally:

Corollary 3. *For every constant depth parameter* $d \in \mathbb{N}$ *there exist constants* $c, \epsilon \in (0,1)$, *such that for every constant* $l \in \mathbb{N}$ *there exists a minimal length parameter* $n_0 \in \mathbb{N}$ *such that for every* $n \geq n_0$ *the encoding scheme* $\mathsf{Enc}_3(\cdot, 1^n)$ *of Definition 8 is* $\left(\mathcal{L}_{3n,d,n^l,\oplus n^\epsilon}, 2^{-n^c}\right)$*-leakage-indistinguishable.*

We proceed to prove Corollary 3 in two steps. First, we show that Theorem 18 implies that AC^0 circuits, augmented with "few" $\oplus$ gates, cannot distinguish between random encodings of 0, and random encodings of either 1 or 2. Second, we show that this implies indistinguishability of encodings of every pair of values in $\{0, 1, 2\}$. The first step follows from the next lemma.

Lemma 1. *Let* $\epsilon \in (0,1)$, $n \in \mathbb{N}$, *and* $\mathcal{G}$ *be a class of functions from* $\{0,1\}^{3n}$ *to* $\{0,1\}$. *If* $\mathsf{Corr}_{\mathcal{D}_3^n}(\mathcal{G}, \mathrm{MOD}_3^n) \leq \epsilon$ *then* $U_0^n, U_{1,2}^n$ *are* $(\mathcal{G}, \epsilon)$-*leakage-indistinguishable, where* $\mathcal{D}_3^n$ *is the distribution defined in Theorem 18.*

Proof. Let $g \in \mathcal{G}$. We first establish the connection between the probability $p_g := \Pr_{x \leftarrow \mathcal{D}_3^n}[g(x) = \mathrm{MOD}_3^n(x)]$ that g computes MOD_3^n correctly, and the distinguishing advantage of g:

$$p_g = \Pr_{x \leftarrow \mathcal{D}_3^n} [g(x) = \mathrm{MOD}_3^n(x) \,|\, \mathrm{MOD}_3^n(x) = 0] \cdot \Pr_{x \leftarrow \mathcal{D}_3^n} [\mathrm{MOD}_3^n(x) = 0]$$

$$+ \Pr_{x \leftarrow \mathcal{D}_3^n} [g(x) = \mathrm{MOD}_3^n(x) \,|\, \mathrm{MOD}_3^n(x) = 1] \cdot \Pr_{x \leftarrow \mathcal{D}_3^n} [\mathrm{MOD}_3^n(x) = 1]$$

observing that for $x \leftarrow \mathcal{D}_3^n$, $\mathrm{MOD}_3^n(x)$ is 0 (or 1) with probability half, and that

$$\Pr_{x \leftarrow \mathcal{D}_3^n} [g(x) = \mathrm{MOD}_3^n(x) \,|\, \mathrm{MOD}_3^n(x) = 0] = \Pr_{x \leftarrow U_0^n} [g(x) = 0]$$

$$\Pr_{x \leftarrow \mathcal{D}_3^n} [g(x) = \mathrm{MOD}_3^n(x) \,|\, \mathrm{MOD}_3^n(x) = 1] = \Pr_{x \leftarrow U_{1,2}^n} [g(x) = 1]$$

we get:

$$p_g = \frac{1}{2} + \frac{1}{2} \left(\Pr_{x \leftarrow U_{1,2}^n} [g(x) = 1] - \Pr_{x \leftarrow U_0^n} [g(x) = 1] \right).$$

By the assumption of the lemma,

$$2 \left| \frac{1}{2} - p_g \right| = \mathsf{Corr}_{\mathcal{D}_3^n}(g, \mathrm{MOD}_3^n) \le \epsilon.$$

Therefore, we get:

$$\left| \Pr_{x \leftarrow U_{1,2}^n} [g(x) = 1] - \Pr_{x \leftarrow U_0^n} [g(x) = 1] \right| \le \epsilon.$$

$\square$

Next, we establish a connection between the distinguishing advantage of circuits between the following pairs of distributions: $U_0^{2n}, U_{1,2}^{2n}$ (over $6n$-bit vectors); $U_0^n, U_{1,2}^n$; and U_0^n, U_1^n (over $3n$-bit vectors).

Lemma 2. *Let $d, s, t \in \mathbb{N}$, and $c \in (0, 1)$ be a constant. If there exists an $n_0 \in \mathbb{N}$ such that for every $n \ge n_0$, $U_0^n, U_{1,2}^n$ are $(\mathcal{L}_{3n,d,s,\oplus t}, \epsilon)$-leakage-indistinguishable for $\epsilon = 2^{-n^c}$, and $U_0^{2n}, U_{1,2}^{2n}$ are $(\mathcal{L}_{6n,d+1,2s+1,\oplus 2t}, \epsilon)$-leakage-indistinguishable, then there exists an n_0' such that for every $n \ge n_0'$, U_0^n, U_1^n are $(\mathcal{L}_{3n,d,s,\oplus t}, \sqrt{7\epsilon})$-leakage-indistinguishable.*

In the following proofs, we use the following notation, and the following observation regarding the connection between U_1^n, U_2^n and $U_{1,2}^n$.

Notation 19. *Let $n \in \mathbb{N}$. For $\gamma \in \{0, 1, 2\}$, we use $\mathcal{S}_\gamma^n$ to denote $\mathsf{supp}(U_\gamma^n)$, $\mathcal{S}_{1,2}^n$ to denote $\mathsf{supp}(U_{1,2}^n)$, and k_γ^n to denote $|\mathcal{S}_\gamma^n|$.*

Observation 20. *For every $n \in \mathbb{N}$, and every function $g : \{0, 1\}^{3n} \to \{0, 1\}$, by the law of total probability, and since $\Pr_{x \leftarrow U_{1,2}^n} [x \in \mathcal{S}_1^n] = \Pr_{x \leftarrow U_{1,2}^n} [x \in \mathcal{S}_2^n] = \frac{1}{2}$,*

$$\Pr_{x \leftarrow U_{1,2}^n} [g(x) = 1] = \frac{1}{2} \left(\Pr_{x \leftarrow U_1^n} [g(x) = 1] + \Pr_{x \leftarrow U_2^n} [g(x) = 1] \right).$$

Proof (of Lemma 2). If the lemma does not hold, then there exist infinitely many n's, for each of which U_0^n, U_1^n are *not* $(\mathcal{L}_{3n,d,s,\oplus t}, \sqrt{7\epsilon})$-leakage-indistinguishable. Let $\epsilon' = \epsilon'(n) > \sqrt{7\epsilon}$ denote the maximal distinguishing advantage between U_0^n, U_1^n, let $\hat{D} = \left\{\hat{D}_n\right\}$ be a family of distinguishers obtaining this advantage, and let $\mathcal{N}$ be the infinite set of n's for which $\hat{D}$ obtains this advantage. For $\gamma \in \{0,1,2\}$, let $p_\gamma^n := \Pr_{x \leftarrow U_\gamma^n}\left[\hat{D}_n(x) = 1\right]$. Assume first that $p_0^n > p_1^n$ for infinitely many n's in $\mathcal{N}$. There are two possible cases: either for infinitely many n's in $\mathcal{N}$, $p_2^n \leq p_0^n$; or $p_2^n > p_0^n$ for infinitely many n's in $\mathcal{N}$. In the first case, $\hat{D}$ has advantage at least $\frac{\epsilon'}{2} > \frac{\sqrt{7\epsilon}}{2} > \frac{\sqrt{4\epsilon}}{2} \geq^{\epsilon \leq 1} \epsilon$ in distinguishing between $U_0^n, U_{1,2}^n$, for every n such that $p_0^n \geq p_2^n$ and $p_0^n \geq p_1^n + \epsilon'$. Indeed, using Observation 20,

$$\left|\Pr_{x \leftarrow U_0^n}\left[\hat{D}_n(x) = 1\right] - \Pr_{x \leftarrow U_{1,2}^n}\left[\hat{D}_n(x) = 1\right]\right| = \left|p_0^n - \frac{1}{2}\left(p_1^n + p_2^n\right)\right|$$

using the case assumption that $p_0^n \geq p_1^n, p_2^n$, this advantage is equal to:

$$\frac{1}{2}\left(p_0^n - p_1^n\right) + \frac{1}{2}\left(p_0^n - p_2^n\right) \geq \frac{1}{2}\left(p_0^n - p_1^n\right) \geq \frac{\epsilon'}{2}.$$

Therefore, only the second case remains, and Lemma 3 below shows that there exists an $\hat{n}_0 \in \mathbb{N}$ such that for every such n which is greater than $\hat{n}_0$, $U_0^{2n}, U_{1,2}^{2n}$ are distinguishable in $\mathcal{L}_{6n,d+1,2s+1,\oplus 2t}$ with advantage at least $\frac{(\epsilon')^2}{6} + E(n) > \frac{(\sqrt{7\epsilon})^2}{6} + E(n) = \epsilon + \frac{\epsilon + E(n)}{6}$, where $E(n) = O\left(2^{-3n}\right)$. Recall that $\epsilon = 2^{-n^c}$, so $E(n) = o(\epsilon)$, and let $n' \in \mathbb{N}$ such that for every $n \geq n'$, $|E(n)| \leq \epsilon$ (notice that $E(n)$ may be negative). Then for every $n \geq \max\{n', \hat{n}_0\}$ in $\mathcal{N}$ such that $p_2^n > p_0^n \geq p_1^n + \epsilon'$ (there are infinitely many such n's by the case assumption), $\epsilon + \frac{\epsilon + E(n)}{6} \geq \epsilon$, meaning that $U_0^{2n}, U_{1,2}^{2n}$ can be distinguished in $\mathcal{L}_{6n,d+1,2s+1,\oplus 2t}$ with advantage more than ϵ, a contradiction to the assumption of the lemma. Therefore, if $p_0^n \geq p_1^n + \epsilon'$ for infinitely many n's in $\mathcal{N}$, then U_0^n, U_1^n are $(\mathcal{L}_{3n,d,s,\oplus t}, \sqrt{7\epsilon})$-distinguishable only for finitely many n's.

Assume now that $p_0^n \geq p_1^n$ only for finitely many n's in $\mathcal{N}$, i.e., $p_1^n \geq p_0^n$ for infinitely many n's in $\mathcal{N}$. If for infinitely many n's in $\mathcal{N}$, $p_2^n \geq p_0^n$ and $p_1^n > p_0^n$, then the advantage of $\hat{D}_n$ in distinguishing between $U_0^n, U_{1,2}^n$ is at least

$$\left|p_0^n - \frac{p_1^n + p_2^n}{2}\right| = \frac{p_1^n - p_0^n}{2} + \frac{p_2^n - p_0^n}{2} \geq \frac{p_1^n - p_0^n}{2} \geq \frac{\epsilon'}{2}.$$

The second case, where $p_2^n < p_0^n < p_1^n$ for infinitely many n's, follows from Lemma 3 in the same manner as before. □

We now prove the lemma used in the proof of Lemma 2, for the case $p_2^n > p_0^n > p_1^n$ (or $p_1^n > p_0^n > p_2^n$) for infinitely many n's. Notice that Lemma 3 uses the distributions $U_0^{2n}, U_{1,2}^{2n}$ over $6n$-*bit vectors*, and distinguishers over $3n$-*bit vectors*.

Lemma 3. *Let $n, d, s, t \in \mathbb{N}$, $\epsilon > 0$, and $\{D_n \in \mathcal{L}_{3n,d,s,\oplus t}\}_{n\in\mathbb{N}}$. For $\gamma \in \{0,1,2\}$, denote $p_\gamma^n := \Pr_{x \leftarrow U_\gamma^n}[D_n(x) = 1]$. Then there exist error terms $E^+(n), E^-(n) = O(2^{-3n})$, and an $n_0 \in \mathbb{N}$, such that the following holds for every $n_0 \le n \in \mathbb{N}$. If $p_2^n > p_0^n > p_1^n$ and $p_0^n - p_1^n \ge \epsilon$, then $U_0^{2n}, U_{1,2}^{2n}$ are $\left(\mathcal{L}_{6n,d+1,2s+1,\oplus 2t}, \frac{\epsilon^2}{6} + E^+(n)\right)$-distinguishable; and if $p_2^n < p_0^n < p_1^n$ and $p_1^n - p_0^n \ge \epsilon$, then $U_0^{2n}, U_{1,2}^{2n}$ are $\left(\mathcal{L}_{6n,d+1,2s+1,\oplus 2t}, \frac{\epsilon^2}{6} + E^-(n)\right)$-distinguishable.*

Proof. Let D_n' be the distinguisher that interprets its input as a pair (x, y) of $3n$-bit vectors, and outputs $D_n(x) \wedge D_n(y)$. Notice that if $D_n \in \mathcal{L}_{3n,d,s,\oplus t}$, then $D_n' \in \mathcal{L}_{6n,d+1,2s+1,\oplus 2t}$. We now analyze the advantage of D_n' in distinguishing between $U_0^{2n}, U_{1,2}^{2n}$. Using Lemma 5, $\Pr_{(x,y) \leftarrow U_0^{2n}}[D_n'(x, y) = 1] = \frac{(p_0^n)^2 + 2p_1^n p_2^n}{3} + E_0(n) + E_0'(n) \cdot p_2^n$, where $E_0(n), E_0'(n)$ are error terms, and $|E_0(n)|, |E_0'(n)| = O(2^{-3n})$. Using Lemma 6, $\Pr_{(x,y) \leftarrow U_{1,2}^{2n}}[D_n'(x, y) = 1] = \frac{2p_0^n p_1^n + (p_1^n)^2 + 2p_0^n p_2^n + (p_2^n)^2}{6} + E_{1,2}(n) + E_{1,2}'(n) \cdot p_2^n + E_{1,2}''(n) \cdot (p_2^n)^2$, where $E_{1,2}(n), E_{1,2}'(n), E_{1,2}''(n)$ are error terms, and $|E_{1,2}(n)|, |E_{1,2}'(n)|, |E_{1,2}''(n)| = O(2^{-3n})$. Therefore,

$$\mathcal{E}_{D_n'} := \Pr_{x \leftarrow U_{1,2}^{2n}}[D_n'(x, y) = 1] - \Pr_{x \leftarrow U_0^{2n}}[D_n'(x, y) = 1]$$

$$= \frac{2p_0^n p_1^n + (p_1^n)^2 + 2p_0^n p_2^n + (p_2^n)^2 - 2(p_0^n)^2 - 4p_1^n p_2^n}{6}$$

$$+ E(n) + E'(n) \cdot p_2^n + E''(n) \cdot (p_2^n)^2$$

where $E(n), E'(n), E''(n)$ are error terms, and $|E(n)|, |E'(n)|, |E''(n)| = O(2^{-3n})$. Thinking of $\mathcal{E}_{D_n'}$ as a function of p_2^n, there exists an n_0 such that for every $n \ge n_0$, the minimal value of $\mathcal{E}_{D_n'}(p_2^n)$ is obtained when $p_2^n = \frac{2p_1^n - p_0^n - 3E'(n)}{1 + 6E''(n)} \approx 2p_1^n - p_0^n$. Let $n \ge n_0$, and assume first $p_2^n > p_0^n > p_1^n$ and $p_0^n - p_1^n \ge \epsilon$. Then $\frac{2p_1^n - p_0^n - 3E'(n)}{1 + 6E''(n)} \approx 2p_1^n - p_0^n < p_0$, and in the domain $z \ge \frac{2p_1^n - p_0^n - 3E'(n)}{1 + 6E''(n)}$, $\mathcal{E}_{D'}$ is monotonically increasing, so the minimal value of $\mathcal{E}_{D'}$ in this section is obtained when $p_2^n = p_0^n$ (since by the case assumption, $p_2^n \ge p_0^n$), in which case $\mathcal{E}_{D_n'}|_{p_2^n = p_0^n} = \frac{(p_0^n - p_1^n)^2}{6} + E(n) + E'(n) \cdot p_0^n + E''(n) \cdot (p_0^n)^2 \ge \frac{\epsilon^2}{6} + E(n) + E'(n) \cdot p_0^n + E''(n) \cdot (p_0^n)^2 =_{p_0^n \in (0,1)} \frac{\epsilon^2}{6} + E^+(n)$, where $E^+(n) = O(2^{-3n})$, so D_n' obtaining advantage $\delta^+ := \frac{\epsilon^2}{6} + E^+(n)$ in distinguishing between $U_0^{2n}, U_{1,2}^{2n}$, where $E^+(n) = O(2^{-3n})$.

Second, assume that $p_2^n < p_0^n < p_1^n$ and $p_1^n - p_0^n \ge \epsilon$. Then $\frac{2p_1^n - p_0^n - 3E'(n)}{1 + 3E''(n)} \approx 2p_1^n - p_0^n > p_0$. Since by the case assumption $p_2^n < p_0^n$ then in the domain $z \le \frac{2p_1^n - p_0^n - 3E'(n)}{1 + 3E''(n)}$ the function is monotonically decreasing, so the minimal advantage is obtained when $p_0^n = p_2^n$, and the rest of the analysis follows as in the previous case. $\qquad\square$

We now state and prove the lemmas that were used in the proof of Lemma 3. We will need the following result about the values of k_0^n, k_1^n, k_2^n. (The proof, which is by induction and uses Observation 20, appears in the full version.)

Lemma 4. *Let* $n \in \mathbb{N}$. *Then* $k_1^n = k_2^n = \frac{2^{3n} + (-1)^{n-1}}{3}$, *and* $k_0^n = \frac{2^{3n} + 2 \cdot (-1)^n}{3}$.

Lemma 5. *Let* $D_n', p_0^n, p_1^n, p_2^n$ *be as defined in the proof of Lemma 3. Then*
$$\Pr_{(x,y) \leftarrow U_0^{2n}} [D_n'(x,y) = 1] = \frac{(p_0^n)^2 + 2p_1^n p_2^n}{3} + E_0(n) + E_0'(n) \cdot p_2^n, \quad \text{where}$$
$E_0(n), E_0'(n)$ *are error terms, and* $|E_0(n)|, |E_0'(n)| = O\left(2^{-3n}\right)$.

Proof. Since
$$\mathcal{S}_0^{2n} = \left\{(x,y) : x, y \in \{0,1\}^{3n} \wedge (x, y \in \mathcal{S}_0^n \vee x \in \mathcal{S}_1^n, y \in \mathcal{S}_2^n \vee x \in \mathcal{S}_2^n, y \in \mathcal{S}_1^n)\right\}$$

then by the law of total probability, $\Pr_{(x,y) \leftarrow U_0^{2n}} [D_n'(x,y) = 1]$ is equal to:

$$\Pr_{(x,y) \leftarrow U_0^{2n}} [D_n'(x,y) = 1 | x, y \in \mathcal{S}_0^n] \cdot \Pr_{(x,y) \leftarrow U_0^{2n}} [x, y \in \mathcal{S}_0^n]$$

$$+ \Pr_{(x,y) \leftarrow U_0^{2n}} [D_n'(x,y) = 1 | x \in \mathcal{S}_1^n, y \in \mathcal{S}_2^n] \cdot \Pr_{(x,y) \leftarrow U_0^{2n}} [x \in \mathcal{S}_1^n, y \in \mathcal{S}_2^n]$$

$$+ \Pr_{(x,y) \leftarrow U_0^{2n}} [D_n'(x,y) = 1 | x \in \mathcal{S}_2^n, y \in \mathcal{S}_1^n] \cdot \Pr_{(x,y) \leftarrow U_0^{2n}} [x \in \mathcal{S}_2^n, y \in \mathcal{S}_1^n]$$

$$= \left(\Pr_{x \leftarrow U_0^n} [D(x) = 1]\right)^2 \cdot \frac{|\mathcal{S}_0^n|^2}{|\mathcal{S}_0^{2n}|} + 2 \Pr_{x \leftarrow U_1^n} [D(x) = 1] \cdot \Pr_{x \leftarrow U_2^n} [D(x) = 1] \cdot \frac{|\mathcal{S}_1^n| \cdot |\mathcal{S}_2^n|}{|\mathcal{S}_0^{2n}|}$$

If n is even, then by Lemma 4: $k_0^n = |\mathcal{S}_0^n| = \frac{2^{3n}+2}{3}$; $k_0^{2n} = |\mathcal{S}_0^{2n}| = \frac{2^{6n}+2}{3}$; and $k_1^n = |\mathcal{S}_1^n| = \frac{2^{3n}-1}{3}$. Therefore,

$$\frac{|\mathcal{S}_0^n|^2}{|\mathcal{S}_0^{2n}|} = \frac{\left(\frac{2^{3n}+2}{3}\right)^2}{\frac{2^{6n}+2}{3}} = \frac{1}{3} \cdot \frac{2^{6n} + 2^{3n+2} + 4}{2^{6n} + 2} = \frac{1}{3} \cdot \left(1 + \frac{2^{3n+2} + 2}{2^{6n} + 2}\right) = \frac{1}{3} + O\left(2^{-3n}\right)$$

$$\frac{|\mathcal{S}_1^n| \cdot |\mathcal{S}_2^n|}{|\mathcal{S}_0^{2n}|} = \frac{|\mathcal{S}_1^n|^2}{|\mathcal{S}_0^{2n}|} = \frac{\left(\frac{2^{3n}-1}{3}\right)^2}{\frac{2^{6n}+2}{3}} = \frac{1}{3} \cdot \frac{2^{6n} - 2^{3n+1} + 1}{2^{6n} + 2} = \frac{1}{3} - O\left(2^{-3n}\right)$$

Otherwise, n is odd, and by Lemma 4: $k_0^n = |\mathcal{S}_0^n| = \frac{2^{3n}-2}{3}$; $k_0^{2n} = |\mathcal{S}_0^{2n}| = \frac{2^{6n}+2}{3}$; and $k_1^n = |\mathcal{S}_1^n| = \frac{2^{3n}+1}{3}$. Similar calculations give:

$$\frac{|\mathcal{S}_0^n|^2}{|\mathcal{S}_0^{2n}|} = \frac{1}{3} - O\left(2^{-3n}\right), \quad \frac{|\mathcal{S}_1^n| \cdot |\mathcal{S}_2^n|}{|\mathcal{S}_0^{2n}|} = \frac{1}{3} + O\left(2^{-3n}\right)$$

Consequently,

$$\Pr_{(x,y) \leftarrow U_0^{2n}} [D_n'(x,y) = 1] = \frac{(p_0^n)^2 + 2p_1^n p_2^n}{3} + E_0(n) + E_0'(n) \cdot p_2^n$$

where E_0, E_0' are error terms, and $|E_0(n)|, |E_0'(n)| = O\left(2^{-3n}\right)$. $\qquad \square$

The proof of the following lemma is similar to the proof of Lemma 5, and appears in the full version.

Lemma 6. *Let* $D'_n, p_0^n, p_1^n, p_2^n$ *be as defined in the proof of Lemma 3. Then* $\Pr_{(x,y)\leftarrow U_{1,2}^{2n}} [D'_n(x,y) = 1] = \frac{2p_0^n p_1^n + (p_1^n)^2 + 2p_0^n p_2^n + (p_2^n)^2}{6} + E_{1,2}(n) + E'_{1,2}(n) \cdot p_2^n + E''_{1,2}(n) \cdot (p_2^n)^2$, *where* $E_{1,2}, E'_{1,2}, E''_{1,2}$ *are error terms, and* $|E_{1,2}(n)|, |E'_{1,2}(n)|, |E''_{1,2}(n)| = O(2^{-3n})$.

Next, we prove that if U_0^n, U_1^n are leakage-indistinguishable against some family of leakage functions, then E_3 is leakage indistinguishable against a slightly weaker family of leakage functions.

Lemma 7. *Let* $n, d, s, t \in \mathbb{N}$, *and* $\epsilon = \epsilon(n) > 0$. *If there exists an* $n_0 \in \mathbb{N}$ *such that for every* $n \geq n_0$, U_0^n, U_1^n *are* $(\mathcal{L}_{3n,d,s,\oplus t}, \epsilon)$-*leakage-indistinguishable, then for every* $n \geq n_0$, $\mathsf{E}_3(\cdot, 1^n)$ *is* $(\mathcal{L}_{3n,d-1,s-3n,\oplus t}, 2\epsilon)$-*leakage-indistinguishable.*

Proof. We show first that $\mathsf{Enc}_3(0, 1^n)$, $\mathsf{Enc}_3(2, 1^n)$ are $(\mathcal{L}_{3n,d-1,s-3n,\oplus t}, \epsilon)$-leakage-indistinguishable for every $n \geq n_0$. Otherwise, there exist infinitely many n's and for each a distinguisher $D_n \in \mathcal{L}_{3n,d-1,s-3n,\oplus t}$ that achieves advantage $\epsilon' > \epsilon$ in distinguishing between the distributions $\mathsf{Enc}_3(0, 1^n)$, $\mathsf{Enc}_3(2, 1^n)$. For every such n we define D'_n to apply negation gates on its inputs, and run D_n. Then $D'_n \in \mathcal{L}_{3n,d,s,\oplus t}$, and notice that since the encoding length is divisible by 3, and the transformation $v \to \bar{v}$ is 1:1 and onto (where $\bar{v}$ denotes the vector obtained by coordinate-wise negating v) then: if $v \leftarrow \mathsf{Enc}_3(0, 1^n)$ then $\bar{v} \leftarrow \mathsf{Enc}_3(0, 1^n)$; and if $v \leftarrow \mathsf{Enc}_3(1, 1^n)$ then $\bar{v} \leftarrow \mathsf{Enc}_3(2, 1^n)$. Therefore, for every such n, $|\Pr[D'_n(\mathsf{Enc}(0, 1^n)) = 1] - \Pr[D'_n(\mathsf{Enc}(1, 1^n)) = 1]| = |\Pr[D_n(\mathsf{Enc}(0, 1^n)) = 1] - \Pr[D_n(\mathsf{Enc}(2, 1^n)) = 1]| = \epsilon' > \epsilon$, contradicting the assumption of the lemma. Second, since for every $n \geq n_0$, $\mathsf{Enc}_3(0, 1^n)$, $\mathsf{Enc}_3(2, 1^n)$ are $(\mathcal{L}_{3n,d-1,s-3n,\oplus t}, \epsilon)$-leakage-indistinguishable, and $\mathsf{Enc}_3(0, 1^n)$, $\mathsf{Enc}_3(1, 1^n)$ are $(\mathcal{L}_{3n,d,s,\oplus t}, \epsilon)$-leakage-indistinguishable, then using the triangle inequality $\mathsf{Enc}_3(1, 1^n)$, $\mathsf{Enc}_3(2, 1^n)$ are $(\mathcal{L}_{3n,d-1,s-3n,\oplus t}, 2\epsilon)$-leakage-indistinguishable. □

We are finally ready to prove Corollary 3.

Proof. (of Corollary 3). Let $d' = d + 2$, let ϵ, c be the constants for which Theorem 18 holds for depth parameter d', and we set $c' = \frac{c}{2}$, and $\epsilon' = \frac{\epsilon}{2}$. Given l, let $l' = l + 1$, and let n_0 be the minimal length parameter for which Theorem 18 holds with parameters d', l'. Let n'_0 be such that for every $n \geq n'_0$, $2(n^l + 3n) + 1 \leq n^{l'}$, $2n^{\epsilon'} \leq n^{\epsilon}$, and $2\sqrt{7} \cdot 2^{-\frac{n^c}{2}} \leq 2^{-n^{c'}}$. Let n''_0 be the minimal length parameter whose existence is guaranteed in Lemma 2 for the length parameter $\max\{n_0, n'_0\}$, constant c, depth parameter $d + 2$, size parameter $s = n^l + 3n$, and parity gate bound $t = n^{\epsilon'}$. Let $\tilde{n}_0 = \max\{n_0, n'_0, n''_0\}$. We show that the corollary holds for minimal length parameter $\tilde{n}_0$ and constants c', ϵ'. Indeed, for every $n \geq \tilde{n}_0$ Theorem 18 guarantees that $\mathsf{Corr}_{\mathcal{D}_3^n}\left(\mathcal{L}_{3n,d+2,2(n^l+3n)+1,\oplus n^{\epsilon'}}, \mathsf{MOD}_3^n\right) \leq 2^{-n^c}$ (since $n \geq n_0$ and $n \geq n'_0$). By Lemma 1, this implies that for every $n \geq \tilde{n}_0$, $U_0^n, U_{1,2}^n$ are $\left(\mathcal{L}_{3n,d+1,n^l+3n,\oplus n^{\epsilon'}}, 2^{-n^c} \right)$-leakage-indistinguishable, and $U_0^{2n}, U_{1,2}^{2n}$ are $\left(\mathcal{L}_{6n,d+2,2(n^l+3n)+1,\oplus 2n^{\epsilon'}}, 2^{-n^c} \right)$-leakage-indistinguishable. By Lemma 2, for every $n \geq \tilde{n}_0$, U_0^n, U_1^n are $\left(\mathcal{L}_{3n,d+1,n^l+3n,\oplus n^{\epsilon'}}, \sqrt{7} \cdot 2^{-\frac{n^c}{2}} \right)$-leakage-indistinguishable (because $n \geq n_0''$). By Lemma 7, $\mathsf{E}_3\left(\cdot, 1^n \right)$ is $\left(\mathcal{L}_{3n,d,n^l,\oplus n^{\epsilon'}}, 2\sqrt{7} \cdot 2^{-\frac{n^c}{2}} \right)$-leakage-indistinguishable. Since $\tilde{n}_0 \geq n_0'$, $\mathsf{E}_3\left(\cdot, 1^n \right)$ is $\left(\mathcal{L}_{3n,d,n^l,\oplus n^{\epsilon'}}, 2^{-n^{c'}} \right)$-leakage-indistinguishable. $\qquad\square$

References

1. Arora, S., Lund, C., Motwani, R., Sudan, M., Szegedy, M.: Proof verification and hardness of approximation problems. In: Proceedings of the 33rd Annual IEEE Symposium on Foundations of Computer Science, FOCS 1992, pp. 14–23, Pittsburgh, Pennsylvania, USA, 24–27 October 1992
2. Arora, S., Safra, S.: Probabilistic checking of proofs: a new characterization of NP. In: Proceedings of the 33rd Annual IEEE Symposium on Foundations of Computer Science, FOCS 1992, pp. 2–13, Pittsburgh, Pennsylvania, USA, 24–27 October 1992
3. Ben-Sasson, E., Chiesa, A., Genkin, D., Tromer, E., Virza, M.: SNARKs for C: verifying program executions succinctly and in zero knowledge. In: Canetti, R., Garay, J.A. (eds.) CRYPTO 2013, Part II. LNCS, vol. 8043, pp. 90–108. Springer, Heidelberg (2013)
4. Ben-Sasson, E., Chiesa, A., Tromer, E., Virza, M.: Scalable zero knowledge via cycles of elliptic curves. In: Garay, J.A., Gennaro, R. (eds.) CRYPTO 2014, Part II. LNCS, vol. 8617, pp. 276–294. Springer, Heidelberg (2014)
5. Ben-Sasson, E., Chiesa, A., Tromer, E., Virza, M.: Succinct non-interactive zero knowledge for a Von Neumann architecture. In: Proceedings of the 23rd USENIX Security Symposium, pp. 781–796, San Diego, CA, USA, 20–22 August 2014
6. Blum, M., De Santis, A., Micali, S., Persiano, G.: Noninteractive zero-knowledge. SIAM J. Comput. **20**(6), 1084–1118 (1991)
7. Canetti, R., Damgård, I., Dziembowski, S., Ishai, Y., Malkin, T.: On adaptive vs. non-adaptive security of multiparty protocols. In: Pfitzmann, B. (ed.) EURO-CRYPT 2001. LNCS, vol. 2045, pp. 262–279. Springer, Heidelberg (2001)
8. Dinur, I.: The PCP theorem by gap amplification. In: Proceedings of the 38th Annual ACM Symposium on Theory of Computing, STOC 2006, pp. 241–250, Seattle, WA, USA, 21–23 May 2006
9. Dubrov, B., Ishai, Y.: On the randomness complexity of efficient sampling. In: Proceedings of the 38th Annual ACM Symposium on Theory of Computing, STOC 2006, pp. 711–720, Seattle, WA, USA, 21–23 May 2006
10. Dziembowski, S., Faust, S.: Leakage-resilient circuits without computational assumptions. In: Cramer, R. (ed.) TCC 2012. LNCS, vol. 7194, pp. 230–247. Springer, Heidelberg (2012)
11. Faust, S., Rabin, T., Reyzin, L., Tromer, E., Vaikuntanathan, V.: Protecting circuits from leakage: the computationally-bounded and noisy cases. In: Gilbert, H. (ed.) EUROCRYPT 2010. LNCS, vol. 6110, pp. 135–156. Springer, Heidelberg (2010)
12. Feige, U., Lapidot, D., Shamir, A.: Multiple non-interactive zero knowledge proofs based on a single random string (extended abstract). In: 31st Annual Symposium

on Foundations of Computer Science, vol. I, pp. 308–317. St. Louis, Missouri, USA, 22–24 October 1990

13. Goldreich, O.: The Foundations of Cryptography, vol. 1, Basic Techniques. Cambridge University Press (2001)

14. Goldwasser, S., Micali, S., Rackoff, C.: The knowledge complexity of interactive proof-systems (extended abstract). In: Proceedings of the 17th Annual ACM Symposium on Theory of Computing, STOC 1985, pp. 291–304, Providence, Rhode Island, USA, 6–8 May 1985

15. Goldwasser, S., Rothblum, G.N.: How to compute in the presence of leakage. In: Proceedings of the 53rd Annual IEEE Symposium on Foundations of Computer Science, FOCS 2012, pp. 31–40, New Brunswick, NJ, USA, 20–23 October 2012

16. Håstad, J.: Almost optimal lower bounds for small depth circuits. In: Proceedings of the 18th Annual ACM Symposium on Theory of Computing, STOC 1986, pp. 6–20, Berkeley, California, USA, 28–30 May 1986

17. Ishai, Y., Mahmoody, M., Sahai, A.: On efficient zero-knowledge PCPs. In: Cramer, R. (ed.) TCC 2012. LNCS, vol. 7194, pp. 151–168. Springer, Heidelberg (2012)

18. Ishai, Y., Sahai, A., Wagner, D.: Private circuits: securing hardware against probing attacks. In: Boneh, D. (ed.) CRYPTO 2003. LNCS, vol. 2729, pp. 463–481. Springer, Heidelberg (2003)

19. Ishai, Y., Weiss, M.: Probabilistically checkable proofs of proximity with zero-knowledge. In: Lindell, Y. (ed.) TCC 2014. LNCS, vol. 8349, pp. 121–145. Springer, Heidelberg (2014)

20. Kilian, J.: A note on efficient zero-knowledge proofs and arguments (extended abstract). In: Proceedings of the 24th Annual ACM Symposium on Theory of Computing, STOC 1992, pp. 723–732, Victoria, British Columbia, Canada, 4–6 May 1992

21. Kilian, J., Petrank, E., Tardos, G.: Probabilistically checkable proofs with zero knowledge. In: Proceedings of the 29th Annual ACM Symposium on Theory of Computing, STOC 1997, pp. 496–505. El Paso, Texas, USA, 4–6 May 1997

22. Lovett, S., Srinivasan, S.: Correlation bounds for poly-size AC^0 circuits with $n^{1-o(1)}$ symmetric gates. In: Goldberg, L.A., Jansen, K., Ravi, R., Rolim, J.D.P. (eds.) RANDOM 2011 and APPROX 2011. LNCS, vol. 6845, pp. 640–651. Springer, Heidelberg (2011)

23. Miles, E., Viola, E.: Shielding circuits with groups. In: Proceedings of the 45th Annual ACM Symposium on Theory of Computing, STOC 2013, pp. 251–260, Palo Alto, CA, USA, 1–4 June 2013

24. Naor, M.,Yung, M.: Public-key cryptosystems provably secure against chosen ciphertext attacks. In: Proceedings of the 22nd Annual ACM Symposium on Theory of Computing, pp. 427–437, Baltimore, Maryland, USA, 13–17 May 1990

25. Rothblum, G.N.: How to compute under AC^0 leakage without secure hardware. In: Safavi-Naini, R., Canetti, R. (eds.) CRYPTO 2012. LNCS, vol. 7417, pp. 552–569. Springer, Heidelberg (2012)

Quasi-Linear Size Zero Knowledge from Linear-Algebraic PCPs

Eli Ben-Sasson[2]([✉]), Alessandro Chiesa[3], Ariel Gabizon[2], and Madars Virza[1]

[1] MIT, Cambridge, USA
[2] Technion, Haifa, Israel
eli@cs.technion.ac.il
[3] UC Berkeley, Berkeley, USA

Abstract. The seminal result that every language having an interactive proof also has a zero-knowledge interactive proof assumes the existence of one-way functions. Ostrovsky and Wigderson [33] proved that this assumption is necessary: if one-way functions do not exist, then only languages in BPP have zero-knowledge interactive proofs.

Ben-Or et al. [9] proved that, nevertheless, every language having a multi-prover interactive proof also has a zero-knowledge multi-prover interactive proof, unconditionally. Their work led to, among many other things, a line of work studying zero knowledge without intractability assumptions. In this line of work, Kilian, Petrank, and Tardos [28] defined and constructed zero-knowledge probabilistically checkable proofs (PCPs).

While PCPs with quasilinear-size proof length, but without zero knowledge, are known, no such result is known for zero knowledge PCPs. In this work, we show how to construct "2-round" PCPs that are zero knowledge and of length $\tilde{O}(K)$ where K is the number of queries made by a malicious polynomial time verifier. Previous solutions required PCPs of length at least K^6 to maintain zero knowledge. In this model, which we call *duplex PCP* (DPCP), the verifier first receives an oracle string from the prover, then replies with a message, and then receives another oracle string from the prover; a malicious verifier can make up to K queries in total to both oracles.

Deviating from previous works, our constructions do not invoke the PCP Theorem as a blackbox but instead rely on certain algebraic properties of a specific family of PCPs. We show that if the PCP has a certain linear algebraic structure — which many central constructions can be shown to possess, including [2, 4, 15] — we can add the zero knowledge property at virtually no cost (up to additive lower order terms) while introducing only minor modifications in the algorithms of the prover and verifier. We believe that our linear-algebraic characterization of PCPs may be of independent interest, as it gives a simplified way to view previous well-studied PCP constructions.

© International Association for Cryptologic Research 2016
E. Kushilevitz and T. Malkin (Eds.): TCC 2016-A, Part II, LNCS 9563, pp. 33–64, 2016.
DOI: 10.1007/978-3-662-49099-0_2

1   Introduction

We continue the study of proof systems that provide soundness and zero knowledge, simultaneously and unconditionally (i.e., no intractability assumptions are needed to achieve the two), as we now explain.

Interactive Proofs. An *interactive proof* [6,20] for a language $\mathscr{L}$ is a pair of interactive algorithms (P, V), where P is known as the prover and V as the verifier, that satisfies the following: (i) (completeness) for every instance x in $\mathscr{L}$, $P(x)$ can make $V(x)$ accept with probability 1; (ii) (soundness) for every instance x not in $\mathscr{L}$, every prover $\tilde{P}$ can make $V(x)$ accept with at most a small probability ϵ. Shamir [35] showed the expressive power of interactive proofs by proving that **IP** = **PSPACE**, i.e., all and only languages in **PSPACE** have interactive proofs.

Zero Knowledge. An interactive proof is *zero knowledge* [20] if the verifier, even if malicious, cannot learn any information about an instance x in $\mathscr{L}$, by interacting with the prover, besides the fact x is in $\mathscr{L}$: for any efficient verifier $\tilde{V}$ there exists an efficient simulator S such that $S(x)$ is "indistinguishable" from the view of $\tilde{V}$ while interacting with $P(x)$. Depending on the choice of definition for indistinguishability, one gets different flavors of zero knowledge.

If indistinguishability is required to hold for efficient deciders only, then one gets *computational* zero knowledge; **CZK** denotes the corresponding complexity class. A seminal result in cryptography says that if one-way functions exist then **CZK** = **IP**, i.e., every language having an interactive proof also has a computational zero-knowledge interactive proof [8,20,23]. If indistinguishability is required to hold for all deciders, then one gets *statistical* zero knowledge; if instead the simulator's output and the verifier's view are the same distribution (and not merely close to each other), then one gets *perfect* zero knowledge. These stronger notions determine the corresponding complexity classes **SZK** and **PZK**, both of which are contained in **AM** ∩ **coAM**; of course, **PZK** ⊆ **SZK** ⊆ **CZK**.

Unfortunately, zero knowledge cannot be achieved unconditionally for non-trivial languages: Ostrovsky and Wigderson [33] proved that if one-way functions do not exist then **CZK** equals an average-case variant of **BPP**.

Other Types of Proof Systems. Due to the limitations of interactive proofs with respect to zero knowledge that holds unconditionally, researchers have explored other types of proof systems, as an alternative to interactive proofs.

– **MIP.** Ben-Or et al. [9] first studied statistical zero knowledge, and proved that it can be achieved in a new model, *multi-prover interactive proof* (MIPs), where the verifier interacts with multiple provers that are not allowed to communicate while interacting with the verifier (though they may share a random string before such an interaction begins). More precisely, Ben-Or et al. prove that every language having a multi-prover interactive proof also has a perfect zero-knowledge multi-prover interactive proof (again, without relying on intractability assumptions). The result of [9] was subsequently improved in a number of papers [5,19,29].

- **PCP.** Kilian et al. [28] study statistical zero knowledge in the model of *probabilistically checkable proofs* (PCPs) [2–4], where the verifier has oracle access to a string. Essentially, the oracle string can be thought of as a stateless prover: the answer to a query depends only on the query itself, but not any other queries that were previously made. Building on results implicit in [19], Kilian et al. showed two main theorems. First, every language in **NEXP** has a PCP that is statistical zero knowledge against verifiers that make at most any polynomial number of queries to the PCP. Second, every language in **NP** has, for every constant $c > 0$, a PCP that is statistically zero knowledge against verifiers that make at most $\mathsf{k}(n) := n^c$ queries to the PCP.

  Subsequent works [24–26,31] provided simplifications (giving alternative constructions or simplifying that of [28]) and limitations (showing that for languages in **NP** one cannot efficiently sample the oracle if one seeks statistical zero knowledge against verifiers that make at most a polynomial number of queries).

- **IPCP.** Goyal et al. [21] study statistical zero knowledge in the model of *interactive PCPs* (IPCPs) [27], where the verifier interacts with two provers of which one is restricted to be an oracle. Goyal et al. prove that every language in **NP** has a constant-round interactive PCP that is statistical zero knowledge against verifiers that make at most any polynomial number of queries to the PCP, and where both provers' strategies can be implemented efficiently as a function of the instance and the witness.

A Limitation of Prior Work. PCPs with quasilinear-size proof length, but without zero knowledge, are known: for every language $\mathscr{L}$ in **NTIME**$(T(n))$, there is a PCP with proof length $\tilde{O}(T(n))$ and query complexity $O(1)$ [14,15,17,32]. On the other hand, no such result for statistical zero knowledge PCPs is known: even when applied to PCPs of length $\tilde{O}(T(n))$, [28]'s result and followup improvements yields a proof length that is polynomial in $T(n) \cdot \mathsf{k}(n)$, where $\mathsf{k}(n)$, known as the *knowledge bound*, is a bound on the number of queries by any verifier (see Sect. 4.1 for further discussion). We thus ask the following question: are there statistical zero knowledge PCPs with proof length quasilinear in $T(n) + \mathsf{k}(n)$?

1.1 Our Contributions

We do not answer the above question in the PCP model, but we give a positive answer in a closely related model that can be thought of as a "2-round PCP", which we call *duplex PCP* (DPCP). At a high level, a DPCP works as follows: the prover first sends an oracle string π_0 to the verifier, just as in a PCP; then, the verifier sends a message ρ to the prover; finally, the prover answers with a second oracle string π_1; the verifier may query both oracles, and then accept or reject. In other words, a DPCP is merely a 2-round interactive proof in which the prover sends oracle strings rather than messages. We prove the following theorem:

Theorem 1 (see Theorem 4 for formal statement). *For every language $\mathscr{L}$ in* **NTIME**$(T) \cap$ **NP** *and polynomially-bounded knowledge bound* k *there exists a DPCP system satisfying the following:*

- *the proof length (in fact, also the prover running time) is quasilinear in* $n + T(n) + \mathsf{k}(n)$*;*
- *the query complexity is polynomial in* $\log(T(n) + \mathsf{k}(n))$*;*
- *the verifier running time is polynomial in* $n + \log(T(n) + \mathsf{k}(n))$*;*
- *perfect zero knowledge holds against any verifier that makes at most* $\mathsf{k}(n)$ *adaptive queries (in total to both oracles);*
- *the soundness error is* $1/2$ *(and can be reduced by repetition to* $2^{-\lambda}$ *while preserving perfect zero knowledge, provided that the number of queries does not exceed* $\mathsf{k}(n)$*).*

Moreover, similarly to the PCPs of [28], the DPCP system that we construct is in fact not only sound but is also a *proof of knowledge* [7]; however, in contrast to [28], the DPCP verifier is *non-adaptive*, in the sense that the query locations depend only on the verifier's random tape.

Perhaps the main difference between our construction and prior work is the techniques that we use. While previous works use the PCP Theorem as a black box, compiling a PCP into a zero knowledge PCP by using *locking schemes* [28], we use certain *algebraic* properties of a specific family of PCPs to guarantee zero knowledge. In comparison to the generic approach, we are more specific, but the addition of zero knowledge essentially comes "for free" when compared to the corresponding constructions without zero knowledge. (In contrast, [28] achieves a proof length of $\Omega(\mathsf{k}(n)^6 \cdot \mathsf{l}(n)^c)$, for some large enough c, when starting from a PCP with proof length $\mathsf{l}(n)$.)

DPCP vs IPCP. Duplex PCPs are an alternative to interactive PCPs that combine PCPs and interaction. In a DPCP, the verifier gets an oracle string from the prover, replies with a message, and then gets another oracle string from the prover; in an IPCP, the verifier gets an oracle string from the prover, and then engages in an interactive proof with him.

Both [21] and our work are similar in that both address aspects that we do not know how to address in the PCP model, and resort to studying alternative models, i.e., IPCP and DPCP respectively. The two works however give different flavors of results: [21] obtain IPCPs that are zero knowledge against verifiers that ask at most any polynomial number of queries $\mathsf{k}(n)$ but their oracle is of polynomial size in $\mathsf{k}(n)$ (actually, of exponential size but with a polynomial-size circuit describing it); on the other hand, our work obtains DPCPs that are zero knowledge against verifiers that ask at most a fixed polynomial number of queries $\mathsf{k}(n)$ and our oracles are of quasilinear size in $\mathsf{k}(n)$.

Finally, we note that our construction can be also cast as an IPCP, because the knowledge bound $\mathsf{k}(n)$ holds only for the first oracle, i.e., perfect zero knowledge is preserved even if the verifier reads the second oracle in full. This provides a result on a 2-round IPCP incomparable to [21]'s 4-round IPCP.

On the Minimal Computational Gap Between Prover and Verifier Needed for Zero Knowledge. IP and MIP systems assume a computational gap between prover and verifier. The prover is allowed (and often assumed) to be computationally unbounded and the verifier is polynomially bounded. An intriguing corollary of our theorem is that the computational gap between prover and verifier can be drastically reduced, to a mere polylogarithmic one. Namely, suppose that we wish to create zero-knowledge systems in which the verifier runs in time $\mathsf{tv}(n)$; in the model above, as long as $\mathsf{tp}(n) > \mathsf{tv}(n) \cdot (\log \mathsf{tv}(n))^c$ for an absolute constant c, then perfect zero knowledge with a small soundness error can be obtained under no intractability assumptions. (See Corollary 1 for a formal statement.)

2   Preliminaries

Functions and Distributions. We use $f \colon D \to R$ to denote a function with domain D and range R; given a subset $\tilde{D}$ of D, we use $f|_{\tilde{D}}$ to denote the restriction of f to $\tilde{D}$. Given a distribution $\mathcal{D}$, we write $x \leftarrow \mathcal{D}$ to denote that x is sampled according to $\mathcal{D}$.

Distances. A distance measure is a function $\Delta \colon \Sigma^n \times \Sigma^n \to [0,1]$ such that for all $x, y, z \in \Sigma^n$: (i) $\Delta(x,x) = 0$, (ii) $\Delta(x,y) = \Delta(y,x)$, and (iii) $\Delta(x,y) \leq \Delta(x,z) + \Delta(z,y)$. For example, the *relative Hamming distance* over alphabet Σ is a distance measure: $\Delta_{\Sigma}^{\mathrm{Ham}}(x,y) := |\{i \mid x_i \neq y_i\}|/n$. We extend Δ to distances of strings to sets: given $x \in \Sigma^n$ and $S \subseteq \Sigma^n$, we define $\Delta(x,S) := \min_{y \in S} \Delta(x,y)$ (or 1 if S is empty). We say that a string x is ϵ-close to another string y if $\Delta(x,y) \leq \epsilon$, and ϵ-far from y if $\Delta(x,y) > \epsilon$; similar terminology applies for a string x and a set S.

Fields and Polynomials. We denote by $\mathbb{F}$ a finite field, by $\mathbb{F}_q$ the field of size q, and by $\mathscr{F}$ the set of all finite fields. We denote by $\mathbb{F}[X_1, \ldots, X_m]$ the ring of polynomials in m variables over $\mathbb{F}$; given a polynomial P in $\mathbb{F}[X_1, \ldots, X_m]$, $\deg_{X_i}(P)$ is the degree of P in the variable X_i; the total degree of P is the sum of all of these individual degrees.

Linear Spaces. Given $n \in \mathbb{N}$, a subset S of $\mathbb{F}^n$ is an $\mathbb{F}$-linear space if $\alpha x + \beta y \in S$ for all $\alpha, \beta \in \mathbb{F}$ and $x, y \in S$.

Languages and Relations. We denote by $\mathscr{R}$ a relation consisting of pairs (x, w), where x is the *instance* and w is the *witness*. We denote by $\mathrm{Lan}(\mathscr{R})$ the language corresponding to $\mathscr{R}$, and by $\mathscr{R}|_{\mathsf{x}}$ the set of witnesses in $\mathscr{R}$ for x.

Complexity Classes. We write complexity classes in bold capital letters: **NP**, **PSPACE**, **NEXP**, and so on. We take a "relation-centric" point of view: we view **NTIME** as a class of relations rather than as the class of the corresponding languages; we thus may write things like "let $\mathscr{R}$ be in **NP**". If $\mathscr{R}$ is in **NTIME**(T), we fix an arbitrary machine $M_{\mathscr{R}}$ that decides $\mathscr{R}$ in time $T(n)$, i.e., $M_{\mathscr{R}}(\mathsf{x}, \mathsf{w})$ always halts after $T(|\mathsf{x}|)$ steps and $M_{\mathscr{R}}(\mathsf{x}, \mathsf{w}) = 1$ if and only if $(\mathbb{x}, \mathbb{w}) \in \mathscr{R}$; we then say that $M_{\mathscr{R}}$ decides $\mathscr{R}$ (or $\mathrm{Lan}(\mathscr{R})$). Throughout, we assume that $T(n) \geq n$.

Codes. An error correcting code C is a set of functions $w \colon H \to \Sigma$, where H, Σ are finite sets. The message length of C is $n := \log_{|\Sigma|} |C|$, its block length is $\ell := |H|$, its rate is $\rho := n/\ell$, its (minimum) distance is $d := \min\{\Delta(w, z) \mid w, z \in C, w \neq z\}$ when Δ is the (absolute) Hamming distance, and its (minimum) relative distance is $\delta := d/\ell$. Given a code family $\mathscr{C}$, we denote by $\mathrm{Rel}(\mathscr{C})$ the relation that naturally corresponds to $\mathscr{C}$, i.e., $\{(C, w) \mid C \in \mathscr{C}, w \in C\}$. A code C is linear if Σ is a finite field and C is a Σ-linear space in Σ^{ℓ}; we denote by $\dim(C)$ the dimension of C when viewed as a linear space. A code C is t-wise independent if, for every subset I of $[\ell]$ with cardinality t, the distribution of $w|_I$ (viewed as a string) for a random $w \in C$ equals the uniform distribution on Σ^t.

Random Shifts. We later use the following folklore claim about distance preservation for random shifts in linear spaces; for completeness, we include its short proof.

Claim. Let n be in $\mathbb{N}$, $\mathbb{F}$ a finite field, S an $\mathbb{F}$-linear space in $\mathbb{F}^n$, and $x, y \in \mathbb{F}^n$. If x is ϵ-far from S, then $\alpha x + y$ is $\epsilon/2$-far from S, with probability $1 - |\mathbb{F}|^{-1}$ over a random $\alpha \in \mathbb{F}$. (Distances are relative Hamming distances.)

Proof. Suppose, by way of contradiction, that there exist $\alpha_1, \alpha_2 \in \mathbb{F}$ and $y_1, y_2 \in S$ with $\alpha_1 \neq \alpha_2$ such that, for every $i \in \{1, 2\}$, $\alpha_i x + y$ is $\epsilon/2$ close to y_i. Then, by the triangle inequality, $z := y_1 - y_2$ is ϵ-close to $(\alpha_1 x + y) - (\alpha_2 x + y) = (\alpha_1 - \alpha_2)x$. We conclude that x is ϵ-close to $\frac{1}{\alpha_1 - \alpha_2} z \in S$, a contradiction.

2.1 Probabilistically Checkable Proofs

A *PCP system* [2–4] for a relation $\mathscr{R}$ is a tuple $\mathsf{PCP} = (P, V)$ that works as follows.

- The *prover* P is a probabilistic algorithm that, given as input an instance-witness pair $(\mathbb{x}, \mathbb{w})$ with $n := |\mathbb{x}|$, outputs a proof $\pi \colon D(n) \to \Sigma(n)$, where both $D(n)$ and $\Sigma(n)$ are finite sets.
- The *verifier* V is a probabilistic oracle algorithm that, given as input an instance $\mathbb{x}$ with $n := |\mathbb{x}|$ and with oracle access to a proof $\pi \colon D(n) \to \Sigma(n)$, queries π at a few locations and then outputs a bit.

The system PCP has (perfect) completeness and soundness error $\mathsf{e}(n)$ if the following two conditions hold. (Below, we explicitly denote the prover's and verifier's randomness as r_P and r_V.)

Completeness: For every instance-witness pair $(\mathbb{x}, \mathbb{w})$ in the relation $\mathscr{R}$,

$$\Pr_{r_P, r_V} \left[V^{P(\mathbb{x}, \mathbb{w}; r_P)}(\mathbb{x}; r_V) = 1 \right] = 1 \;.$$

Soundness: For every instance x not in the language $\mathrm{Lan}(\mathscr{R})$ and proof $\pi \colon D(n) \to \Sigma(n)$,

$$\Pr_{r_V}[V^\pi(\mathrm{x}; r_V) = 1] \le \mathsf{e}(n) .$$

A relation $\mathscr{R}$ belongs to the complexity class **PCP**$[\mathsf{a}, \mathsf{l}, \mathsf{q}, \mathsf{e}, \mathsf{tp}, \mathsf{tv}]$ if there is a PCP system for $\mathscr{R}$ in which:

- the answer alphabet (i.e., $\Sigma(n)$) is $\mathsf{a}(n)$,
- the proof length over that alphabet (i.e., $|D(n)|$) is at most $\mathsf{l}(n)$,
- the verifier queries the proof in at most $\mathsf{q}(n)$ locations,
- the soundness error is $\mathsf{e}(n)$,
- the prover runs in time $\mathsf{tp}(n)$, and
- the verifier runs in time $\mathsf{tv}(n)$.

Finally, we add the symbol na in the square brackets (i.e., we write **PCP**$[\ldots, \mathsf{na}]$) if the queries to the proof are non-adaptive (i.e., the queried locations only depend on the verifier's inputs).

2.2 Probabilistically Checkable Proofs of Proximity

A *PCPP system* [12,18] for a relation $\mathscr{R}$ is a tuple $\mathsf{PCPP} = (P, V)$ that works as follows.

- The *prover* P is a probabilistic algorithm that, given as input an instance-witness pair (x, w) with $n := |\mathrm{x}|$, outputs a proof $\pi \colon D(n) \to \Sigma(n)$, where both $D(n)$ and $\Sigma(n)$ are finite sets.
- The *verifier* V is a probabilistic oracle algorithm that, given as input an instance x with $n := |\mathrm{x}|$ and with oracle access to a witness w and proof $\pi \colon D(n) \to \Sigma(n)$, queries w and π at a few locations and then outputs a bit.

The system PCPP has (perfect) completeness, soundness error e, distance measure Δ, and proximity parameter d if the following two conditions hold. (Below, we explicitly denote the prover's and verifier's randomness as r_P and r_V.)

Completeness: For every instance-witness pair (x, w) in the relation $\mathscr{R}$,

$$\Pr_{r_P, r_V}\left[V^{(\mathrm{w}, P(\mathrm{x}, \mathrm{w}; r_P))}(\mathrm{x}; r_V) = 1\right] = 1 .$$

Soundness: For every instance-witness pair (x, w), perhaps not in the language, such that $\Delta(\mathrm{w}, \mathscr{R}|_{\mathrm{x}}) \ge \mathsf{d}(n)$ and proof $\pi \colon D(n) \to \Sigma(n)$,

$$\Pr_{r_V}\left[V^{(\mathrm{w}, \pi)}(\mathrm{x}; r_V) = 1\right] \le \mathsf{e}(n) .$$

A relation $\mathscr{R}$ belongs to the complexity class **PCPP**$[\mathsf{a}, \mathsf{l}, \mathsf{q}, \Delta, \mathsf{d}, \mathsf{e}, \mathsf{tp}, \mathsf{tv}]$ if there is a PCPP system for $\mathscr{R}$ in which:

- the answer alphabet (i.e., $\Sigma(n)$) is $\mathsf{a}(n)$,
- the proof length over that alphabet (i.e., $|D(n)|$) is at most $\mathsf{l}(n)$,

- the verifier queries the two oracles (codeword and proof) in at most $q(n)$ locations (in total),
- the distance measure is Δ,
- the proximity parameter is $d(n)$,
- the soundness error is $e(n)$,
- the prover runs in time $tp(n)$, and
- the verifier runs in time $tv(n)$.

Finally, we add the symbol na in the square brackets (i.e., we write $\mathbf{PCPP}[\ldots, \text{na}]$) if the queries to the oracles are non-adaptive (i.e., the queried locations only depend on the verifier's inputs).

2.3 Zero Knowledge PCPs

The notion of zero knowledge for PCPs was first considered in [19,28]. A PCP system $\mathsf{PCP} = (P, V)$ for a relation $\mathscr{R}$ has *perfect zero knowledge with knowledge bound* k if there exists an expected-polynomial-time probabilistic algorithm S such that, for every k-query polynomial-time probabilistic oracle algorithm $\tilde{V}$, the following two distribution families are identical:

$$\{S(\tilde{V}, \mathsf{x})\}_{(\mathsf{x},\mathsf{w}) \in \mathscr{R}} \quad \text{and} \quad \{\text{PCPView}(\tilde{V}, P, \mathsf{x}, \mathsf{w})\}_{(\mathsf{x},\mathsf{w}) \in \mathscr{R}} \ ,$$

where $\text{PCPView}(\tilde{V}, \pi, \mathsf{x}, \mathsf{w})$ is the view of $\tilde{V}$ in its execution when given input x and oracle access to $\pi := P(\mathsf{x}, \mathsf{w})$. The definition of statistical and computational zero knowledge (with knowledge bound k) are similar: rather than identical, the two distribution families are required to be statistically and computationally close (as $|\mathsf{x}|$ grows), respectively.

A relation $\mathscr{R}$ belongs to the complexity class $\mathbf{PCP}_{\text{pzk}}[a, l, q, e, tp, tv, k]$ if there exists a PCP system for $\mathscr{R}$ that (i) puts $\mathscr{R}$ in $\mathbf{PCP}[a, l, q, e, tp, tv]$, and (ii) has perfect zero knowledge with knowledge bound k; as for $\mathbf{PCP}$, we add the symbol na in the square brackets of $\mathbf{PCP}_{\text{pzk}}$ if the queries to the proof are non-adaptive. The complexity classes $\mathbf{PCP}_{\text{szk}}$ and $\mathbf{PCP}_{\text{czk}}$ are similarly defined for statistical and computational zero knowledge.

The KPT Result. Kilian, Petrank, and Tardos proved the following theorem:

Theorem 2 [28]. *For every polynomial time function* $T \colon \mathbb{N} \to \mathbb{N}$, *polynomial security function* $\lambda \colon \mathbb{N} \to \mathbb{N}$, *and polynomial knowledge bound function* $k \colon \mathbb{N} \to \mathbb{N}$,

$$\mathbf{NTIME}(T) \subseteq \mathbf{PCP}_{\text{szk}} \begin{bmatrix} a & = \mathbb{F}_{2^{\text{poly}(\lambda)}} \\ l & = \text{poly}(T, k) \\ q & = \text{poly}(\lambda) \\ e & = 2^{-\lambda} \\ tp & = \text{poly}(\lambda, T) \\ tv & = \text{poly}(\lambda, T, k) \\ k & \end{bmatrix} \ .$$

Remark 1. We make two remarks: (i) the symbol na does not appear above because [28]'s construction relies on adaptively querying the proof; (ii) inspection of [28]'s construction reveals that $l(n) \geq \text{poly}(T(n)) \cdot k(n)^6$.

2.4   Reed–Muller and Reed–Solomon Codes

We define Reed–Muller and Reed–Solomon codes, as well as their "vanishing" variants [15]; all of these are linear codes. We then state a theorem about PCPPs for certain families of RS codes.

RM Codes. Let $\mathbb{F}$ be a finite field, H, V subsets of $\mathbb{F}$, m a positive integer, and ϱ a constant in $(0, 1]$; ϱ is called the *fractional degree*. The Reed–Muller code with parameters $\mathbb{F}, H, m, \varrho$ is $\mathsf{RM}[\mathbb{F}, H, m, \varrho] := \{w \colon H^m \to \mathbb{F} \mid \max_{i \in [m]} \deg_{X_i}(w) < \varrho|H|\}$; its message length is $n = (\varrho|H|)^m$, block length is $\ell = |H|^m$, rate is $\rho = \varrho^m$, and relative distance is $\delta = 1 - \varrho$. The vanishing Reed–Muller code with parameters $\mathbb{F}, H, m, \varrho, V$ is $\mathsf{VRM}[\mathbb{F}, H, m, \varrho, V] := \{w \in \mathsf{RM}[\mathbb{F}, H, m, \varrho] \mid w(V^m) = \{0\}\}$; it is a subcode of $\mathsf{RM}[\mathbb{F}, H, m, \varrho]$.

RS Codes. Let $\mathbb{F}$ be a finite field, H, V subsets of $\mathbb{F}$, and ϱ a constant in $(0, 1]$. The Reed–Solomon code with parameters $\mathbb{F}, H, \varrho$ is $\mathsf{RS}[\mathbb{F}, H, \varrho] := \mathsf{RM}[\mathbb{F}, H, 1, \varrho]$. The vanishing Reed–Solomon code with parameters $\mathbb{F}, H, \varrho, V$ is $\mathsf{VRS}[\mathbb{F}, H, \varrho, V] := \{w \in \mathsf{RS}[\mathbb{F}, H, \varrho] \mid w(V) = \{0\}\}$.

Two RS Code Families and Their PCPPs. Given $\varrho \in (0, 1]$, we denote by: (i) $\mathcal{RS}_\varrho^*$ the set of Reed–Solomon codes $\mathsf{RS}[\mathbb{F}, H, \varrho]$ for which $\mathbb{F}$ has characteristic 2 and H is an $\mathbb{F}_2$-affine space; and (ii) $\mathcal{VRS}_\varrho^*$ the set of vanishing Reed–Solomon codes $\mathsf{VRS}[\mathbb{F}, H, \varrho, V]$ for which $\mathbb{F}$ has characteristic 2 and H is an $\mathbb{F}_2$-affine space. The following theorem is from [10,15] (the prover running time is shown in [10] and the other parameters in [15]).

Theorem 3. *For every security function* $\lambda \colon \mathbb{N} \to \mathbb{N}$, $\varrho \in (0, 1)$, *and* $s > 0$,

$$
\mathrm{Rel}(\mathcal{RS}_\varrho^*), \, \mathrm{Rel}(\mathcal{VRS}_\varrho^*) \in \mathbf{PCPP}
\begin{bmatrix}
\mathsf{a} & = \mathbb{F}_{2^{s+\log \ell}} \\
\mathsf{l} & = \tilde{O}(\ell) \\
\mathsf{q} & = \lambda \cdot \mathrm{polylog}(\ell) \\
\varDelta & = \varDelta_\mathsf{a}^{\mathrm{Ham}} \\
\mathsf{d} & = \varrho/2 \\
\mathsf{e} & = 2^{-\lambda} \\
\mathsf{tp} & = \mathrm{poly}(s) \cdot \tilde{O}(\ell) \\
\mathsf{tv} & = \lambda \cdot \mathrm{poly}(s + \log \ell) \\
\mathsf{na} &
\end{bmatrix}.
$$

We will also require the following folklore claim, whose correctness can be proved by induction on m:

Claim. Let $\mathbb{F}$ be a finite field, H, V subsets of $\mathbb{F}$ with $H \cap V = \emptyset$, m a positive integer, and t a positive integer not exceeding $|H| - |V|$. Then $\mathsf{VRM}[\mathbb{F}, H, m, \frac{|V| + t}{|H|}, V]$ is t-wise independent.

3   Duplex PCPs

We define duplex PCPs, and then define notions of zero knowledge for this model. Our main theorem is the construction of a duplex PCP with certain parameters; see Sect. 4. The difference between a PCP and a duplex PCP is that all provers (both honest and malicious) produce two proof oracles rather than one: the prover produces a proof π_0; then the verifier sends a message ρ to the prover; then the prover produces another proof π_1; finally the verifier queries both π_0 and π_1 and either accepts or rejects. (Thus, a PCP is a special case of a duplex PCP, but not vice versa.) More precisely, a *duplex PCP system* for a relation $\mathscr{R}$ is a tuple $\mathsf{DPCP} = (P, V)$ that works as follows.

- The *prover* P is a pair (P_0, P_1) of probabilistic algorithms, with shared randomness, where: (a) given as input an instance-witness pair (x, w) with $n := |\mathsf{x}|$, P_0 outputs a proof $\pi_0 \colon D_0(n) \to \Sigma(n)$; (b) given as input (x, w) and the verifier's message ρ (see below), P_1 outputs a proof $\pi_1 \colon D_1(n) \to \Sigma(n)$. Here $D_0(n), D_1(n), \Sigma(n)$ are finite sets.
- The *verifier* V is a pair (V_0, V_1) of probabilistic algorithms, with shared randomness, where: (a) given as input an instance x with $n := |\mathsf{x}|$, V_0 outputs a message ρ; (b) given as input x and with oracle access to proofs $\pi_0 \colon D_0(n) \to \Sigma(n)$ and $\pi_1 \colon D_1(n) \to \Sigma(n)$, V_1 queries π_0 and π_1 at a few locations and then outputs a bit.

The system DPCP has (perfect) completeness and soundness error $\mathsf{e}(n)$ if the following two conditions hold. (Below, we explicitly denote the prover's and verifier's randomness as r_P and r_V.)

Completeness: For every instance-witness pair (x, w) in the relation $\mathscr{R}$,

$$
\Pr_{r_P, r_V} \left[V_1^{\pi_0, \pi_1}(\mathsf{x}; r_V) = 1 \; \middle| \; \begin{array}{l} \pi_0 \leftarrow P_0(\mathsf{x}, \mathsf{w}; r_P) \\ \rho \leftarrow V_0(\mathsf{x}; r_V) \\ \pi_1 \leftarrow P_1(\mathsf{x}, \mathsf{w}, \rho; r_P) \end{array} \right] = 1 \;.
$$

Soundness: For every instance x not in the language $\mathrm{Lan}(\mathscr{R})$ and pair of algorithms $\tilde{P} = (\tilde{P}_0, \tilde{P}_1)$,

$$
\Pr_{r_V} \left[V_1^{\pi_0, \pi_1}(\mathsf{x}; r_V) = 1 \; \middle| \; \begin{array}{l} \pi_0 \leftarrow \tilde{P}_0 \\ \rho \leftarrow V_0(\mathsf{x}; r_V) \\ \pi_1 \leftarrow \tilde{P}_1(\rho) \end{array} \right] \leq \mathsf{e}(n) \;.
$$

A relation $\mathscr{R}$ belongs to the complexity class $\mathbf{DPCP}[\mathsf{a}, \mathsf{l}, \mathsf{q}, \mathsf{e}, \mathsf{tp}, \mathsf{tv}]$ if there is a DPCP system for $\mathscr{R}$ in which:

- the answer alphabet (i.e., $\Sigma(n)$) is $\mathsf{a}(n)$,
- the proof length over that alphabet (i.e., $(|D_0(n)| + |D_1(n)|)$) is at most $\mathsf{l}(n)$,
- the verifier queries the two proofs in at most $\mathsf{q}(n)$ locations (in total),
- the soundness error is $\mathsf{e}(n)$,
- the prover runs in time $\mathsf{tp}(n)$, and
- the verifier runs in time $\mathsf{tv}(n)$.

Finally, we add the symbol na in the square brackets (i.e., we write $\mathbf{DPCP}[\ldots, \mathsf{na}]$) if the queries to the proof are non-adaptive (i.e., the queried locations only depend on the verifier's inputs).

Zero Knowledge. A DPCP system $\mathsf{DPCP} = (P, V)$ for a relation $\mathscr{R}$ has *perfect zero knowledge with knowledge bound* k if there exists an expected-polynomial-time probabilistic algorithm S such that for every pair of polynomial-time probabilistic oracle algorithms $\tilde{V} := (\tilde{V}_0, \tilde{V}_1)$ the following two distribution families are identical:

$$\{S(\tilde{V}, \mathsf{x})\}_{(\mathsf{x},\mathsf{w})\in\mathscr{R}} \quad \text{and} \quad \{\mathrm{DPCPView}(\mathsf{k}, \tilde{V}, P, \mathsf{x}, \mathsf{w})\}_{(\mathsf{x},\mathsf{w})\in\mathscr{R}} \ ,$$

where $\mathrm{DPCPView}(\mathsf{k}, \tilde{V}, P, \mathsf{x}, \mathsf{w})$ is the view of $\tilde{V}_1$ in its execution when given input x and when allowed to make a total of $\mathsf{k}(n)$ adaptive queries to π_0, π_1, where $\pi_0 := P_0(\mathsf{x}, \mathsf{w})$ and $\pi_1 := P_1(\mathsf{x}, \mathsf{w}, \tilde{V}_0^{\pi_0}(\mathsf{x}))$. (As above, P_0, P_1 share the same randomness r_P; ditto for $\tilde{V}_0, \tilde{V}_1$.) The definition of statistical and computational zero knowledge (with knowledge bound k) are similar: rather than identical, the two distribution families are required to be statistically and computationally close (as $|\mathsf{x}|$ grows), respectively.

4 Main Theorem

The main result of this paper is the following.

Theorem 4. *For every polynomial time function* $T\colon \mathbb{N} \to \mathbb{N}$, *polynomial knowledge bound function* $\mathsf{k}\colon \mathbb{N} \to \mathbb{N}$,

$$\mathbf{NTIME}(T) \subseteq \mathbf{DPCP}_{\mathrm{pzk}} \begin{bmatrix} \mathsf{a} &= \mathbb{F}_{2^{O(\log(T+\mathsf{k}))}} \\ \mathsf{l} &= \tilde{O}(T + \mathsf{k}) \\ \mathsf{q} &= \mathrm{polylog}(T + \mathsf{k}) \\ \mathsf{e} &= \frac{1}{2} \\ \mathsf{tp} &= \mathrm{poly}(n) \cdot \tilde{O}(T + \mathsf{k}) \\ \mathsf{tv} &= \mathrm{poly}(n + \log(T + \mathsf{k})) \\ \mathsf{k} & \\ \mathsf{na} & \end{bmatrix}$$

A Corollary. The theorem above implies that, fixing T, the prover running time is merely quasilinear in the knowledge bound k, while the verifier running time increases only polylogarithmically in k. This leads to an intriguing corollary: a poly-logarithmic computational overhead of the prover over the verifier is all that is needed to maintain perfect zero knowledge in the duplex PCP model. We state this formally next.

Corollary 1. *For every polynomial time function* $T\colon \mathbb{N} \to \mathbb{N}$ *and relation* $\mathscr{R} \in \mathbf{NTIME}(T)$, *there is a constant* c *such that, for every function* $\mathsf{tv}\colon \mathbb{N} \to \mathbb{N}$ *with* $\mathsf{tv}(n) \geq n \cdot (\log T(n))^c$, *there is a DPCP system with:*

- *completeness 1 and soundness* $2^{-\mathsf{tv}(n)/\mathrm{polylog}(T(n))}$;
- *perfect zero knowledge;*
- *the verifier running time is* $\mathsf{tv}(n)$ *and prover running time is* $\mathsf{tp}(n) := \max\{T(n) \cdot (\log T(n))^c, \mathsf{tv}(n) \cdot (\log \mathsf{tv}(n))^c\}$.

The verifier has no limitations other than a bound on its running time (its query complexity can be as large as $\mathsf{tv}(n)$).

4.1   Proof Sketch

Let $\mathscr{R}$ be a relation in **NP**, and let (x, w) be an instance-witness pair in $\mathscr{R}$. The prover and verifier both know x, while the prover also knows w. The prover wishes to convince the verifier that he knows a witness w for x, in such a way that the verifier does not learn anything about w (beyond what can be inferred from the prover's claim).

The KPT Approach. We introduce our ideas by contrasting them with those of [28]. Suppose that the prover wishes to convince the verifier by sending him a PCP proof $\pi = \pi(w)$ such that any k values in π do not reveal anything about w. Loosely speaking, [28] (building on [19]) provide a probabilistic transformation that maps the PCP proof π to a new proof π', in which each bit of π is "hidden" amongst many bits of π'. The main tool employed in the transformation is a *locking scheme*, and its use imposes certain limitations: (i) the new proof π' is poly(k) larger than the original one (k^6 by inspection of [19,28]); (ii) zero knowledge holds only statistically, but not perfectly, because a malicious verifier can be "lucky" and obtain information on the bit of π being locked with fewer queries to π' than expected.

Our Approach (Ideally). We take a different approach: apply a "local" PCP to a "random" witness, as we now explain. Suppose that $\pi = \pi(w)$ is (t, k)-*local*, i.e., any k positions of the PCP proof π jointly depend on at most t positions of the witness w. Note that, even if π is (t, k)-local, a single bit of π can still leak information about w. So suppose further that the relation $\mathscr{R}$ is t-*randomizable*: given $(x, w) \in \mathscr{R}$, one can efficiently sample a witness w' from a t-wise independent subset of the set of witnesses for x. In such a case, the prover can produce a zero-knowledge PCP as follows: (1) sample a witness w' from the t-wise independent subset; then (2) send to the verifier the PCP proof $\pi = \pi(w')$. Indeed, the locality of π ensures that seeing any k indices of π reveals nothing about w, because these k indices are a function of t random bits. In sum, if we had a (t, k)-local PCP for a t-randomizable relation $\mathscr{R}$, then we could obtain a PCP for $\mathscr{R}$ that is zero knowledge against verifiers that ask at most k queries.

Our Approach (in Reality). Unfortunately, we do not know how to obtain local PCPs for randomizable relations. However, we are able to obtain "partially local" *duplex* PCPs for certain randomizable relations, and also show that **NTIME** can be efficiently reduced to these randomizable relations, as we now explain.

Our starting point are *algebraic PCPs*: certain PCPs that prove satisfiability of *algebraic problems* (APs) [34]. Numerous known PCP constructions can be viewed as algebraic PCPs. Informally, in this work we make two basic observations: (i) algebraic PCPs exist for certain randomizable relations; and (ii) an algebraic PCP proof can be split in two parts, one part is local, while the other part is not local but enjoys convenient linear algebraic properties that, nevertheless, enable us to hide information about the witness, in the duplex PCP model. (Recall that, in the duplex PCP model, the prover produces a proof π_0; then the verifier sends a message ρ to the prover; then the prover produces another proof π_1; finally the verifier queries both π_0 and π_1 and either accepts or rejects.)

In more detail, from a technical viewpoint, we proceed as follows. First, we introduce a family of constraint satisfaction problems (CSPs) called *linear algebraic CSPs*, and show that **NTIME** is efficiently reducible to *randomizable* linear algebraic CSPs. The reduction consists of two parts: we go through an intermediary that we call *group preserving algebraic problems* (GAPs), a special case of APs that we believe to be of independent interest for the study of algebraic PCPs. Second, we construct a duplex PCP system for randomizable linear algebraic CSPs that is zero knowledge against verifiers that ask at most a certain number of queries.

A Technical Piece: Zero-Knowledge Duplex PCPP for Low-Degreeness. Later sections address all of the above steps (see Sect. 4.2 for a roadmap of these), and for now we only sketch one of these steps. Above we mention that an algebraic PCP proof has two parts: a local part, and a non-local part. This latter part of the proof arises from a central component of many PCP proofs: a *PCP of proximity* (PCPP) [13,18] that facilitates low-degree testing. Informally, given a function $f\colon H \to \mathbb{F}$ and an integer d, a *PCPP for degree d* is a proof $\pi(f)$ that f is ϵ-close to an evaluation of a polynomial degree at most degree d. We explain how to transform a PCPP for low-degreeness into a duplex PCPP for low-degreeness that is zero knowledge against verifiers that make at most t queries.

The set C of functions $f\colon H \to \mathbb{F}$ that are evaluations of a polynomial of degree at most d is a subspace of $\mathbb{F}^{|H|}$. The basic idea is that, in order for the prover to convince the verifier that a function f is close to C, it suffices for the prover to convince the verifier that a *random offset* of f is close to C: one can verify that, for any $u\colon H \to \mathbb{F}$, if f is ϵ-far from C, then $\alpha f + u$ is $\epsilon/2$-far from C, with probability $1 - |\mathbb{F}|^{-1}$ over a random $\alpha \in \mathbb{F}$. Hence, we can let the duplex PCP work as follows: (i) the prover samples a witness w' from the t-wise independent subset, chooses a random $u \in C$, and sends $\pi_0 := (\mathsf{w}', u)$ to the verifier; (ii) the verifier sends to the prover a random $\alpha \in \mathbb{F}$; (iii) the prover sends $\pi_1 = (v, \pi(v))$ to the verifier, where $v := \alpha \mathsf{w}' + u$ and $\pi(v)$ is a PCPP for low-degreeness of v; (iv) the verifier runs the PCPP verifier on (v, π) to check that v is close to C, and then checks that $v_i = \alpha \mathsf{w}'_i + u_i$ for a few random indices i in $\{1, \ldots, |H|\}$.

Let us discuss the various properties of the duplex PCPP.

- COMPLETENESS: If $\mathsf{w} \in C$, then $\alpha \mathsf{w}' + u \in C$; therefore, the prover convinces the verifier.
- ZERO-KNOWLEDGE: If the verifier asks at most t queries, then he learns nothing about w because: $\pi_0 = (\mathsf{w}', u)$ contains w' sampled from a t-wise independent subset and u random in C; $\pi_1 = (v, \pi(v))$ is running the PCPP on a vector v that is random in C.
- SOUNDNESS: If v does equal $\alpha \cdot \mathsf{w} + u$, then the verifier rejects with high probability because v is far from C (and the PCPP verifier rejects π with high probability). If instead v does not equal $\alpha \cdot \mathsf{w} + u$, then the fact that v is

close to C does not prove anything about whether w is also close. So, in this case, we need to reason about the success probability of the verifier's linearity tests: if these pass with enough probability, then with high probability v is close to $\alpha w + u$, which again suffices for our purpose. Overall, soundness holds.

Next, we discuss how the technical sections are organized, and how they come together to yield our main theorem.

4.2   Roadmap of the Rest of the Paper

The rest of the paper is dedicated to turn the above intuition into a more formal proof. To do so, we introduce various intermediate steps, as follows.

- In Sect. 5, we introduce *linear algebraic CSPs* (a family of constraint satisfaction problems), and then describe how to obtain a *canonical PCP* for any linear algebraic CSP.
- In Sect. 6, we introduce *randomizable* linear algebraic CSPs, a subfamily of linear algebraic CSPs; then we show that, for every randomizable linear algebraic CSP, we can convert the CSP's canonical PCP into a corresponding zero-knowledge duplex PCP, incurring only little overheads.
- In Sect. 7, we show an efficient reduction from **NTIME** to randomizable linear algebraic CSPs; along the way, we introduce a family of algebraic problems, having special symmetry properties, that we believe to be of independent interest (e.g., for studying other questions about PCPs).

Combining (i) the efficient reduction from **NTIME** to randomizable linear algebraic CSPs together with (ii) the zero-knowledge duplex PCP for such problems yields Theorem 4. In Sect. 8 we provide details about how these components are combined.

5   Linear Algebraic CSPs and Their Canonical PCPs

We introduce *linear algebraic CSPs*, a family of constraint satisfaction problems; then we describe how to obtain a *canonical PCP* for any linear algebraic CSP.

5.1   Linear Algebraic Constraint Satisfaction Problems

A constraint satisfaction problem asks whether, for a given "local" function g, there exists an input α such that $g(\alpha)$ is an "accepting" output. For example, in the case of 3-SAT with n variables and m clauses, the function g maps $\{0,1\}^n$ to $\{0,1\}^m$, and $g(\alpha)$ indicates which clauses are satisfied by $\alpha \in \{0,1\}^n$; hence α yields an accepting output if (and only if) $g(\alpha) = 1^m$. Below we introduce a family of constraint satisfaction problems whose domain and range are linear-algebraic objects, namely, linear error correcting codes.

We begin by providing the notion of locality that we use for g; we also provide two other notions, one for the efficiency of computing a single coordinate of g's output, and another for measuring g's "pseudorandomness".

Definition 1. *Let* $g: \Sigma^n \to \Sigma^m$ *be a function. We say that* g *is:*

- *q-local if for every $j \in [m]$ there exists $I_j \subseteq [n]$ with $|I_j| \le q$ such that $g(\alpha)[j]$ (the j-th coordinate of $g(\alpha)$) depends only on $\alpha|_{I_j}$ (the restriction of α to I_j);*
- *c-efficient if there is a time c algorithm that, given j and $\alpha|_{I_j}$, computes the set I_j and value $g(\alpha)[j]$;*
- *(γ, ϵ)-sampling if $\Pr[\, I_j \cap I \ne \emptyset \mid j \leftarrow [m]\,] \le \gamma$ for every $I \subseteq [n]$ with $|I|/n \le \epsilon$.*

Next we introduce $\mathscr{R}_{\mathsf{LA}}$, the relation of **linear algebraic CSPs**:

Definition 2 ($\mathscr{R}_{\mathsf{LA}}$). *Given functions* $f: \mathbb{N} \to \mathscr{F}$, $\ell, q, c: \mathbb{N} \to \mathbb{N}$, *and* $\rho, \delta, \gamma, \epsilon: \mathbb{N} \to (0, 1]$, *the relation*

$$\mathscr{R}_{\mathsf{LA}}[f, \ell, \rho, \delta, q, c, \gamma, \epsilon]$$

consists of instance-witness pairs $(\mathbf{x}, \mathbf{w})$ *satisfying the following.*

- *The instance $\mathbf{x}$ is a tuple $(1^n, C_\circ, C_\bullet, g)$ where:*
  - *$C_\circ, C_\bullet$ are linear error correcting codes with block lengths $\ell_\circ(n), \ell_\bullet(n)$ at most $\ell(n)$, each with rate at most $\rho(n)$ and relative distance at least $\delta(n)$ over the same field $f(n)$;*
  - *$g: f(n)^{\ell_\circ(n)} \to f(n)^{\ell_\bullet(n)}$ is a $q(n)$-local, $c(n)$-efficient, $(\gamma(n), \epsilon(n))$-sampling function;*
  - *$C_\bullet \cup g(C_\circ)$ has relative distance at least $\delta(n)$ (though may not be a linear space).*
- *The witness $\mathbf{w}$ is a tuple $(\alpha_\circ, \alpha_\bullet)$ where $\alpha_\circ \in f(n)^{\ell_\circ(n)}$ and $\alpha_\bullet \in f(n)^{\ell_\bullet(n)}$.*
- *The instance $\mathbf{x}$ and witness $\mathbf{w}$ jointly satisfy the following: $\alpha_\circ \in C_\circ$, $\alpha_\bullet \in C_\bullet$, and $g(\alpha_\circ) = \alpha_\bullet$.*

We prove a simple claim about instances not in the language $\mathrm{Lan}(\mathscr{R}_{\mathsf{LA}})$, which we use several times later on.

Claim. For every instance $\mathbf{x} = (1^n, C_\circ, C_\bullet, g)$ not in the language $\mathrm{Lan}(\mathscr{R}_{\mathsf{LA}})$ and (candidate) witness $\tilde{\mathbf{w}} = (\tilde{\alpha}_\circ, \tilde{\alpha}_\bullet) \in f(n)^{\ell_\circ(n)} \times f(n)^{\ell_\bullet(n)}$ at least one of the following holds:

- at least one of $\tilde{\alpha}_\circ$ and $\tilde{\alpha}_\bullet$ is ϵ-far in relative Hamming distance from $C_\circ$ or $C_\bullet$, respectively; or
- there exist $\alpha_\circ \in C_\circ$ and $\alpha_\bullet \in C_\bullet$ such that $\tilde{\alpha}_\circ$ and $\tilde{\alpha}_\bullet$ are ϵ-close to $\alpha_\circ$ and $\alpha_\bullet$, respectively, but $g(\alpha_\circ) \ne \alpha_\bullet$.

Proof. If neither of the two cases hold, then there exist $\alpha_\circ \in C_\circ$ and $\alpha_\bullet \in C_\bullet$ such that $g(\alpha_\circ) = \alpha_\bullet$. But then $(\alpha_\circ, \alpha_\bullet)$ is a satisfying assignment for $\mathbf{x}$, contradicting our assumption that $\mathbf{x}$ is not in the language $\mathrm{Lan}(\mathscr{R}_{\mathsf{LA}})$.

Finally we need notation for referring to codes appearing in instances of $\mathscr{R}_{\mathsf{LA}}$:

Definition 3. *Given* $\mathscr{R} \subseteq \mathscr{R}_{\mathsf{LA}}$, *we denote by*

- *$\mathscr{C}_{\mathscr{R}, \circ}$ the set of codes C for which there is an instance $\mathbf{x} = (1^n, C_\circ, C_\bullet, g)$ in the relation $\mathscr{R}$ with $C = C_\circ$;*
- *$\mathscr{C}_{\mathscr{R}, \bullet}$ the set of codes C for which there is an instance $\mathbf{x} = (1^n, C_\circ, C_\bullet, g)$ in the relation $\mathscr{R}$ with $C = C_\bullet$.*

5.2 A Canonical PCP for Linear Algebraic CSPs

We show how to construct a "canonical" PCP system for instances in $\mathscr{R}_{\mathsf{LA}}$ (the relation of linear algebraic CSPs). At a high level, a canonical PCP proof for a $\mathscr{R}_{\mathsf{LA}}$-instance x consists of a witness $\mathsf{w} = (\alpha_\circ, \alpha_\bullet)$ concatenated with two PCPP proofs $\pi_\circ, \pi_\bullet$, showing that $\alpha_\circ, \alpha_\bullet$ are close to $C_\circ, C_\bullet$ respectively. The canonical PCP verifier first checks the two PCPP proofs and then checks that $g(\alpha_\circ)[j] = \alpha_\bullet[j]$ for a uniformly random $j \in [\ell_\bullet]$.

Definition 4. *Given (i) a relation $\mathscr{R} \subseteq \mathscr{R}_{\mathsf{LA}}$, (ii) a PCPP system $\mathsf{PCPP}_\circ = (P_\circ, V_\circ)$ for $\mathrm{Rel}(\mathscr{C}_{\mathscr{R},\circ})$, and (iii) a PCPP system $\mathsf{PCPP}_\bullet = (P_\bullet, V_\bullet)$ for $\mathrm{Rel}(\mathscr{C}_{\mathscr{R},\bullet})$, the canonical PCP system for the triple $(\mathscr{R}, \mathsf{PCPP}_\circ, \mathsf{PCPP}_\bullet)$ is the PCP system $\mathsf{PCP} = (P, V)$ constructed as follows.*

- **Prover.** *Given $(\mathsf{x}, \mathsf{w}) \in \mathscr{R}_{\mathsf{LA}}$, the PCP prover P outputs $\pi := (\mathsf{w}, \pi_\circ, \pi_\bullet)$ where $\pi_\circ := P_\circ(C_\circ, \alpha_\circ)$ and $\pi_\bullet := P_\bullet(C_\bullet, \alpha_\bullet)$. In other words, the PCP prover outputs a PCP proof that is the concatenation of the witness $\mathsf{w} = (\alpha_\circ, \alpha_\bullet)$ and a pair of PCPP proofs, the first proving that $\alpha_\circ \in C_\circ$ and the second proving that $\alpha_\bullet \in C_\bullet$.*
- **Verifier.** *Given x and oracle access to a PCP proof $\pi = (\mathsf{w}, \pi_\circ, \pi_\bullet)$, the PCP verifier V works as follows:*
  - *(proximity) check that $V_\circ^{(\alpha_\circ, \pi_\circ)}(C_\circ)$ and $V_\bullet^{(\alpha_\bullet, \pi_\bullet)}(C_\bullet)$ both accept;*
  - *(consistency) check that $g(\alpha_\circ)[j] = \alpha_\bullet[j]$ for a uniformly random $j \in [\ell_\bullet]$.*

The next lemma says that the above construction is a PCP system when $\mathscr{R}_{\mathsf{LA}}$'s parameters are sufficiently "good".

Lemma 1 ($\mathscr{R}_{\mathsf{LA}} \to \mathbf{PCP}$). *Suppose that $\mathscr{R}$ is a relation that satisfies the following conditions:*

(i) $\mathscr{R} \subseteq \mathscr{R}_{\mathsf{LA}}[f_1, \ell_1, \rho_1, \delta_1, q_1, c_1, \gamma_1, \epsilon_1]$ with $\epsilon_1 < \min\{\frac{\delta_1}{2}, \delta_1 - \gamma_1\}$;
(ii) $\mathrm{Rel}(\mathscr{C}_{\mathscr{R},\circ}), \mathrm{Rel}(\mathscr{C}_{\mathscr{R},\bullet}) \in \mathbf{PCPP}[\mathsf{a}_2, \mathsf{l}_2, \mathsf{q}_2, \Delta_{\mathsf{a}_2}^{\mathrm{Ham}}, \mathsf{d}_2, \mathsf{e}_2, \mathsf{tp}_2, \mathsf{tv}_2, \mathsf{na?}]$ with $\mathsf{a}_2 = f_1$ and $\mathsf{d}_2 \leq \epsilon_1$.

Then there is a canonical PCP system for a triple $(\mathscr{R}, \mathsf{PCPP}_\circ, \mathsf{PCPP}_\bullet)$ that yields

$$\mathscr{R} \in \mathbf{PCP} \begin{bmatrix} \mathsf{a} & = f_1 \ (= \mathsf{a}_2) \\ \mathsf{l} & = 2\mathsf{l}_2(\ell_1) + 2\ell_1 \\ \mathsf{q} & = 2\mathsf{q}_2(\ell_1) + q_1 + 1 \\ \mathsf{e} & = \max\{1 - \delta_1 + \gamma_1 + \epsilon_1, \mathsf{e}_2\} \\ \mathsf{tp} & = 2\mathsf{tp}_2(\ell_1) \\ \mathsf{tv} & = 2\mathsf{tv}_2(\ell_1) + c_1 + \log \ell_1 \\ \mathsf{na?} & \end{bmatrix} .$$

Above, $\mathsf{na?}$ denotes the fact that if the PCPP systems are non-adaptive so is the canonical PCP system.

Proof (Proof of Lemma 1). First, we show that the canonical PCP system satisfies completeness and soundness; afterwards, we discuss the efficiency parameters achieved by it.

Completeness. Consider an instance-witness pair (x, w) in the relation $\mathscr{R}$. Parse the instance x as $(1^n, C_\circ, C_\bullet, g)$ and the witness w as $(\alpha_\circ, \alpha_\bullet)$. Since $(x, w) \in \mathscr{R}$, we have that $\alpha_\circ \in C_\circ$, $\alpha_\bullet \in C_\bullet$, and $g(\alpha_\circ) = \alpha_\bullet$. Therefore, the PCP proof $(w, \pi_\circ, \pi_\bullet)$ generated by the PCP prover is accepted by the PCP verifier with probability 1: the PCPP verifiers $V_\circ^{(\alpha_\circ, \pi_\circ)}(C_\circ)$ and $V_\bullet^{(\alpha_\bullet, \pi_\bullet)}(C_\bullet)$ always accept and $g(\alpha_\circ)[j] = \alpha_\bullet[j]$ for every $j \in [\ell_\bullet]$.

Soundness. Consider an instance x not in the language $\mathrm{Lan}(\mathscr{R})$ and a PCP proof $\tilde{\pi} = (\tilde{w}, \tilde{\pi}_\circ, \tilde{\pi}_\bullet)$. Parse the instance x as $(1^n, C_\circ, C_\bullet, g)$ and the wintess $\tilde{w}$, inside $\tilde{\pi}$, as $(\tilde{\alpha}_\circ, \tilde{\alpha}_\bullet)$. We use Claim in Sect. 5.1 to prove that V accepts $\tilde{\pi}$ with probability at most $\max\{1 - \delta_1 + \gamma + \epsilon_1, e_2\}$, by considering the following three cases.

- *Case 1: $\tilde{\alpha}_\circ$ is ϵ_1-far in relative Hamming distance from $C_\circ$.* The canonical PCP verifier's proximity test fails, because $\Delta_a^{\mathrm{Ham}}(\tilde{\alpha}_\circ, C_\circ) \geq \epsilon_1 \geq d_2$, and so the PCPP verifier $V_\circ^{(\alpha_\circ, \tilde{\pi}_\circ)}(C_\circ)$ accepts with probability at most e_2.
- *Case 2: $\tilde{\alpha}_\bullet$ is ϵ_1-far in relative Hamming distance from $C_\bullet$.* This case is analogous to the previous one.
- *Case 3: there exist $\alpha_\circ \in C_\circ$ and $\alpha_\bullet \in C_\bullet$ with $\Delta_a^{\mathrm{Ham}}(\alpha_\circ, \tilde{\alpha}_\circ) \leq \epsilon_1$ and $\Delta_a^{\mathrm{Ham}}(\alpha_\bullet, \tilde{\alpha}_\bullet) \leq \epsilon_1$.*
  First, since ϵ_1 is less than $\delta_1/2$ (the unique decoding radius of $C_\circ$ and $C_\bullet$), the codewords $\alpha_\circ$ and $\alpha_\bullet$ are unique.
  Next, we claim that $\alpha_\bullet' := g(\alpha_\circ)$ and $g(\tilde{\alpha}_\circ)$ are γ_1-close. Indeed, since g is (γ_1, ϵ_1)-sampling, $\alpha_\circ$ and $\tilde{\alpha}_\circ$ differ in at most $\epsilon_1 \cdot \ell_\circ(n)$ positions, and so at most $\gamma_1 \cdot \ell_\bullet(n)$ positions of $g(\tilde{\alpha}_\circ)$ depend on an index where $\alpha_\circ$ and $\tilde{\alpha}_\circ$ differ.
  Next, we claim that $\Delta_a^{\mathrm{Ham}}(\alpha_\bullet, \alpha_\bullet') \geq \delta_1$. Indeed, we have that $\alpha_\bullet \neq \alpha_\bullet'$ because otherwise $(\alpha_\circ, \alpha_\bullet)$ would be a satisfying assignment for x (contradicting the assumption that $x \notin \mathrm{Lan}(\mathscr{R})$); moreover, we also have that $C_\bullet \cup g(C_\circ)$ has relative distance at least δ_1.
  We now use the triangle inequality, along with the above observations, to obtain that

  $$\delta_1 \Delta_a^{\mathrm{Ham}}(\alpha_\bullet, \alpha_\bullet') \leq \Delta_a^{\mathrm{Ham}}(\alpha_\bullet, \tilde{\alpha}_\bullet) + \Delta_a^{\mathrm{Ham}}(\tilde{\alpha}_\bullet, g(\tilde{\alpha}_\circ)) + \Delta_a^{\mathrm{Ham}}(g(\tilde{\alpha}_\circ), \alpha_\bullet')$$
  $$\leq \epsilon_1 + \Delta_a^{\mathrm{Ham}}(\tilde{\alpha}_\bullet, g(\tilde{\alpha}_\circ)) + \gamma_1 .$$

Thus, $\Delta_a^{\mathrm{Ham}}(\tilde{\alpha}_\bullet, g(\tilde{\alpha}_\circ)) \geq \delta_1 - (\gamma_1 + \epsilon_1)$, and so the canonical PCP verifier's consistency check passes with probability at most $1 - \delta_1 + \gamma_1 + \epsilon_1$.

We conclude that V accepts $\tilde{\pi}$ with probability at most $\max\{1 - \delta_1 + \gamma_1 + \epsilon_1, e_2\}$.

Other Parameters. The remaining parameters are straightforward to establish. The canonical PCP does not change the alphabet, so $a = f_1$ (which also equals a_2). The proof length, and the running times of the prover and verifier are the sum of the same measures of the canonical PCP's components: the PCP proof has $\mathsf{l} = 2\mathsf{l}_2(\ell_1) + 2\ell_1$ symbols, is produced in time $\mathsf{tp} = 2\mathsf{tp}_2(\ell_1)$, and is verified in time $\mathsf{tv} = 2\mathsf{tv}_2(\ell_1) + c_1 + O(1)$. The canonical PCP verifier makes $q_1 + 1$ queries on top of those made by the PCPP verifiers, so its query complexity is $\mathsf{q} = 2\mathsf{q}_2(\ell_1) + q_1 + 1$. The $q_1 + 1$ additional queries are non-adaptive; so if the PCPP verifiers are non-adaptive, so is the canonical PCP verifier.

6   Zero-Knowledge Duplex PCPs from Randomizable Linear Algebraic CSPs

We introduce *randomizable linear algebraic CSPs*, a subfamily of linear algebraic CSPs. Then we show that, for every randomizable linear algebraic CSP, we can convert the CSP's canonical PCP into a corresponding zero-knowledge duplex PCP, incurring only little overheads.

6.1   Randomizable Linear Algebraic CSPs

The definition below specifies the notion of randomizability for linear algebraic CSPs.

Definition 5 ($\mathscr{R}_{\mathsf{RLA}}$). *The relation $\mathscr{R}_{\mathsf{RLA}}[f, \ell, \rho, \delta, q, c, \gamma, \epsilon, t, r]$ is the sub-relation of $\mathscr{R}_{\mathsf{LA}}[f, \ell, \rho, \delta, q, c, \gamma, \epsilon]$ obtained by restricting it to instances that are t-randomizable in time r. An instance $\mathsf{x} = (1^n, C_\circ, C_\bullet, g)$ is $t(n)$-randomizable in time $r(n)$ if: (i) there exists a $t(n)$-wise independent subcode $C' \subseteq C_\circ$ such that if $(w_\circ, g(w_\circ))$ satisfies x, then, for every $w'_\circ$ in $C' + w_\circ := \{w' + w_\circ \mid w' \in C'\}$, the witness $(w'_\circ, g(w'_\circ))$ satisfies x; and (ii) one can sample, in time $r(n)$, three uniformly random elements in C', $C_\circ$ and $C_\bullet$ respectively.*

6.2   Construction of Zero-Knowledge Duplex PCPs

We construct a zero-knowledge duplex PCP system for randomizable linear algebraic CSPs. The duplex PCP system does little more than invoking, as a subroutine, the canonical PCP system for the linear algebraic CSP; hence, the efficiency of the duplex PCP and of the canonical PCP system are closely related. The construction demonstrates that "adding zero knowledge to an algebraic PCP" is cheap, provided that one moves from the PCP model to the (more general) duplex PCP model. More precisely, we prove the following theorem.

Theorem 5 ($\mathscr{R}_{\mathsf{RLA}} \to \mathbf{DPCP}_{\mathrm{pzk}}$). *Suppose that $\mathscr{R}$ is a relation that satisfies the following conditions:*

(i) *$\mathscr{R} \subseteq \mathscr{R}_{\mathsf{RLA}}[f_1, \ell_1, \rho_1, \delta_1, q_1, c_1, \gamma_1, \epsilon_1, t_1, r_1]$ with $\epsilon_1 < \min\{\frac{\delta_1}{2}, \delta_1 - \gamma_1\}$ and r_1 polynomially bounded;*

(ii) *$\mathrm{Rel}(\mathscr{C}_{\mathscr{R},\circ}), \mathrm{Rel}(\mathscr{C}_{\mathscr{R},\bullet}) \in \mathbf{PCPP}[a_2, \mathsf{l}_2, q_2, \Delta_{\mathsf{a}_2}^{\mathrm{Ham}}, \mathsf{d}_2, \mathsf{e}_2, \mathsf{tp}_2, \mathsf{tv}_2, \mathsf{na?}]$ with $a_2 = f_1$ and $\mathsf{d}_2 \le \epsilon_1/4$.*

Then there is a duplex PCP system for $\mathscr{R}$ that yields

$$\mathscr{R} \in \mathbf{DPCP}_{pzk} \begin{bmatrix} \mathsf{a} & = f_1 \ (= \mathsf{a}_2) \\ \mathsf{l} & = 2\mathsf{l}_2(\ell_1) + 6\ell_1 \\ \mathsf{q} & = 2\mathsf{q}_2(\ell_1) + q_1 + 7 \\ \mathsf{e} & = \max\{1 - \delta_1 + \gamma_1 + \epsilon_1\,,\, (1 - |f_1|^{-1}) \cdot \max\{\mathsf{e}_2, \epsilon_1/4\} + |f_1|^{-1}\} \\ \mathsf{tp} & = 2\mathsf{tp}_2(\ell_1) + (c_1 + 5)\ell_1 + r_1 \\ \mathsf{tv} & = 2\mathsf{tv}_2(\ell_1) + c_1 + \log \ell_1 \\ \mathsf{k} & = t_1/q_1 \\ \mathsf{na?} \end{bmatrix}.$$

Above, na? *denotes the fact that if the PCPP systems are non-adaptive so is the duplex PCP system.*

Proof. We prove the claim by constructing a suitable duplex PCP system $\mathsf{DPCP} = (P, V)$ for the relation $\mathscr{R}$. Recall that: the prover P is a pair of algorithms (P_0, P_1), and the verifier V is also a pair of algorithms (V_0, V_1); moreover, an instance x of $\mathscr{R}$ is of the form $(1^n, C_\circ, C_\bullet, g)$, while a witness w of $\mathscr{R}$ is of the form $(\alpha_\circ, \alpha_\bullet)$; finally, randomizability implies that there is a $t(n)$-wise independent subcode $C' \subseteq C_\circ$ such that if $(w_\circ, g(w_\circ))$ satisfies x then so does the witness $(w'_\circ, g(w'_\circ))$, for every $w'_\circ$ in $C' + w_\circ$.

We now describe the construction of the duplex PCP system $\mathsf{DPCP} = (P, V)$:

– $P_0(\mathsf{x}, \mathsf{w}) \to \pi_0$
  Sample uniformly random $v_\circ \in C_\circ, v_\bullet \in C_\bullet, u' \in C'$; compute $w_\circ := u' + \alpha_\circ$, $w_\bullet := g(w_\circ)$ and output $\pi_0 := (w_\circ \| v_\circ \| w_\bullet \| v_\bullet)$.

– $V_0(\mathsf{x}) \to \rho$
  Sample uniformly random $\rho_\circ, \rho_\bullet \in f_1$, and output $\rho := (\rho_\circ, \rho_\bullet)$.

– $P_1(\mathsf{x}, \mathsf{w}, \rho) \to \pi_1$
  Compute $z_\circ := \rho_\circ w_\circ + v_\circ$ and $z_\bullet := \rho_\bullet w_\bullet + v_\bullet$; compute $\pi_\circ := P_\circ(C_\circ, z_\circ)$ and $\pi_\bullet = P_\bullet(C_\bullet, z_\bullet)$; and output $\pi_1 := (z_\circ \| z_\bullet \| \pi_\circ \| \pi_\bullet)$. (Essentially, this step corresponds to running the canonical PCP prover with respect to a uniformly random pair $(z_\circ, z_\bullet)$ in $(C_\circ, C_\bullet)$.)

– $V_1^{\pi_0, \pi_1}(\mathsf{x}) \to b$
  Conduct the following tests (and reject if any of them fails):
  - (proximity) check that $V_\circ^{(z_\circ, \pi_\circ)}(C_\circ)$ and $V_\bullet^{(z_\bullet, \pi_\bullet)}(C_\bullet)$ both accept;
  - (consistency) check that $g(w_\circ)[j] = w_\bullet[j]$ for a random $j \in [\ell_\bullet]$;
  - (linearity) check that $z_\circ[i] = \rho_\circ w_\circ[i] + v_\circ[i]$ and $z_\bullet[k] = \rho_\bullet w_\bullet[k] + v_\bullet[k]$ for random $i \in [\ell_\circ(n)]$ and $k \in [\ell_\bullet(n)]$.
  (Essentially the first two steps correspond to running the canonical PCP verifier on modified inputs, while the third step consists of two linearity tests.)

Having described the duplex PCP system, we now show that it satisfies completeness, soundness and zero-knowledge; afterwards, we discuss the efficiency parameters achieved by it.

Completeness. Consider an instance-witness pair (x, w) in the relation $\mathscr{R}$. Since $(\mathsf{x}, \mathsf{w}) \in \mathscr{R}$, we have that $\alpha_\circ \in C_\circ$, $\alpha_\bullet \in C_\bullet$, and $g(\alpha_\circ) = \alpha_\bullet$. Since $w_\circ \in C' + \alpha_\circ$ and $\mathscr{R}$ is randomizable, we have that $(w_\circ, w_\bullet) := (w_\circ, g(w_\circ))$ satisfies x;

thus V_1's consistency check passes with probability 1. Since the codes $C_\circ$ and $C_\bullet$ are linear and $w_\circ, v_\circ \in C_\circ$, $w_\bullet, v_\bullet \in C_\bullet$, we have that $z_\circ := \rho_\circ w_\circ + v_\circ \in C_\circ$ and $z_\bullet := \rho_\bullet w_\bullet + v_\bullet \in C_\bullet$; thus the PCPP verifiers $V_\circ^{(z_\circ, \pi_\circ)}(C_\circ)$ and $V_\bullet^{(z_\bullet, \pi_\bullet)}(C_\bullet)$ accept with probability 1. Finally, by construction of $z_\circ$ and $z_\bullet$, V_1's linearity tests also accept with probability 1. We conclude that the duplex PCP system described above has perfect completeness.

Soundness. Consider an instance x not in the language $\mathrm{Lan}(\mathscr{R})$. Fix an arbitrary proof string $\tilde{\pi}_0 = (\tilde{w}_\circ \| \tilde{v}_\circ \| \tilde{w}_\bullet \| \tilde{v}_\bullet)$, and let the proof string $\tilde{\pi}_1 = (\tilde{z}_\circ \| \tilde{z}_\bullet \| \tilde{\pi}_\circ \| \tilde{\pi}_\bullet)$ depend arbitrarily on the verifier message $\rho = (\rho_\circ, \rho_\bullet)$. We use Claim in Sect. 5.1 with respect to the instance x and witness $(\tilde{w}_\circ, \tilde{w}_\bullet)$ and distinguish between three cases below.

– *Case 1: $\tilde{w}_\circ$ is ϵ_1-far in relative Hamming distance from $C_\circ$.*
  Claim in Sect. 2 implies that $z_\circ' := \rho_\circ \tilde{w}_\circ + \tilde{v}_\circ$ is $\epsilon_1/2$-far from $C_\circ$, with probability $1 - |f_1|^{-1}$ over a random choice of $\rho_\circ$. Let $\theta := \Delta_\mathsf{a}^{\mathrm{Ham}}(z_\circ', \tilde{z}_\circ)$ and $\eta := \Delta_\mathsf{a}^{\mathrm{Ham}}(\tilde{z}_\circ, C_\circ)$. By the triangle inequality, $\theta + \eta \geq \Delta_\mathsf{a}^{\mathrm{Ham}}(z_\circ', C_\circ) \geq \epsilon_1/2$; hence, at least one of the inequalities $\theta \geq \epsilon_1/4$ and $\eta \geq \epsilon_1/4$ holds. In the former case, V_1's first linearity test accepts with probability at most $1 - \epsilon_1/4$; in the latter case, the PCPP verifier $V_\circ^{(\tilde{z}_\circ, \tilde{\pi}_\circ)}(C_\circ)$ for V_1's first proximity test accepts with probability at most e_2, as $\Delta_\mathsf{a}^{\mathrm{Ham}}(\tilde{z}_\circ, C_\circ) \geq \epsilon_1/4 \geq \mathsf{d}_2$.
– *Case 2: $\tilde{w}_\bullet$ is ϵ_1-far in relative Hamming distance from $C_\bullet$.*
  This case is analogous to the previous one.
– *Case 3: there exist $w_\circ \in C_\circ$ and $w_\bullet \in C_\bullet$ with $\Delta_\mathsf{a}^{\mathrm{Ham}}(w_\circ, \tilde{w}_\circ) \leq \epsilon_1$ and $\Delta_\mathsf{a}^{\mathrm{Ham}}(w_\bullet, \tilde{w}_\bullet) \leq \epsilon_1$.*
  In this case we follow the very end of the soundness analysis in Lemma 1's proof, replacing $\tilde{\alpha}_\circ, \tilde{\alpha}_\bullet$ there with $\tilde{w}_\circ, \tilde{w}_\bullet$, and conclude that the verifier accepts with probability at most $1 - \delta_1 + \gamma_1 + \epsilon_1$.

Summing up, in the first case the verifier's acceptance probability is at most $(1 - |f_1|^{-1}) \cdot \max\{\mathsf{e}_2, \epsilon_1/4\} + |f_1|^{-1}$; similarly for the second case. In the third case the rejection probability is $1 - \delta_1 + \gamma_1 + \epsilon_1$, that of the canonical PCP consistency verifier. This completes the soundness analysis.

Zero Knowledge. We construct a simulator S that yields perfect zero knowledge with knowledge bound k. Consider an instance-witness pair (x, w) in the relation $\mathscr{R}$, and a malicious verifier $\tilde{V} = (\tilde{V}_0, \tilde{V}_1)$ making at most k adaptive queries. $S(\tilde{V}, x)$, the output of the simulator S, when given as input $\tilde{V}$ and x, has to be identically distributed to $\mathrm{DPCPView}(\mathsf{k}, \tilde{V}, P, x, w)$, which is the view of $\tilde{V}_1$ in its execution when given input x and when allowed to make a total of $\mathsf{k}(n)$ adaptive queries to π_0, π_1, where $\pi_0 := P_0(x, w)$ and $\pi_1 := P_1(x, w, \tilde{V}_0^{\pi_0}(x))$. In fact, we will prove a stronger statement: the output of the simulator continues to exactly match the view of the verifier, interacting with the honest prover, even if the verifier is allowed unbounded access to π_1, provided that $\tilde{V}$ makes at most k queries to π_0.

  We now discuss how S works. At a high level, S treats $\tilde{V}$ as a black box, running it once without rewinding; along the way, S samples suitable answers for each query (as discussed below); when $\tilde{V}$ halts, S outputs all the answers and $\tilde{V}$'s randomness (which together form the view of the verifier). The simulator S runs in strict polynomial time, without ever aborting. We now describe how S answers each query.

The simulator S maintains a proof string π^S that is initially unspecified at all locations; we write $\pi^S[i] = *$ if the i-th location of this proof string is unspecified. During the simulation, S adaptively specifies locations in π^S as a result of answering $\tilde{V}$'s queries; this specification process is definitive, in the sense that queries to locations that have been previously specified are answered consistently with the previously-specified value. We now discuss how S adaptively specifies locations in π^S. We distinguish between two parts of the simulation: before the point when $\tilde{V}$ sends his message ρ, and only queries to π_0 are possible; and afterwards, when queries to both π_0 and π_1 are possible.

- *Simulating answers to $\pi_0 = (w_\circ \| v_\circ \| w_\bullet \| v_\bullet)$, before $\tilde{V}$ outputs $\tilde{\rho} = (\tilde{\rho}_\circ, \tilde{\rho}_\bullet)$.*
  1. For a query $j \in [\ell_\circ]$ to $w_\circ[j]$: if unspecified, answer with a random field element. That is, if $w_\circ^S[j] = *$, then sample a random $\beta \in f_1$ and set $w_\circ^S := \beta$.
  2. For a query $j \in [\ell_\circ]$ to $v_\circ[j]$: if unspecified, answer with a random field element. That is, if $v_\circ^S[j] = *$, then sample a random $\gamma \in f_1$ and set $v_\circ^S[j] = \gamma$. Then check if there are any unspecified locations of $v_\circ^S$ that are determined by the linear constraint "$v_\circ^S \in C_\circ$" and the currently specified locations of $v_\circ^S$; if there are, set these accordingly.
  3. For a query $j \in [\ell_\bullet]$ to $w_\bullet[j]$: if unspecified, (i) compute the set $I_j \subseteq [\ell_\circ]$ of locations on which $g(w_\circ^S)[j]$ depends (see Definition 2); (ii) deduce $w_\bullet^S|_{I_j}$ by querying each $i \in I_j$ according to Step 1; and (iii) set $w_\bullet^S[j] := g(w_\circ^S|_{I_j})$.
  4. For a query $j \in [\ell_\bullet]$ to $v_\bullet[j]$: answer in an analogous way to the case of a query $j \in [\ell_\circ]$ to $v_\circ$.
- *Simulating answers to $\pi_0 = (w_\circ \| v_\circ \| w_\bullet \| v_\bullet)$ and $\pi_1 = (z_\circ \| z_\bullet \| \pi_\circ \| \pi_\bullet)$, after $\tilde{V}$ outputs $\tilde{\rho} = (\tilde{\rho}_\circ, \tilde{\rho}_\bullet)$.*
  5. After receiving $\tilde{\rho} = (\tilde{\rho}_\circ, \tilde{\rho}_\bullet)$, immediately do the following:
     (a) sample a random $z_\circ^S \in C_\circ$ under the constraint "$z_\circ^S[i] = \tilde{\rho}_\circ w_\circ^S[i] + v_\circ^S[i]$ for all i s.t. $w_\circ^S[i] \neq * \wedge v_\circ^S[i] \neq *$";
     (b) sample a random $z_\bullet^S \in C_\bullet$ under the analogous constraint;
     (c) compute $\pi_\circ^S := P_\circ(C_\circ, z_\circ^S)$;
     (d) compute $\pi_\bullet^S := P_\bullet(C_\bullet, z_\bullet^S)$.
  6. All queries to $z_\circ, z_\bullet, \pi_\circ, \pi_\bullet$ are answered according to the values specified in Step 5.
  7. For a query $j \in [\ell_\circ]$ to $w_\circ[j]$ or $v_\circ[j]$: if both are unspecified, answer with a random field element; otherwise, the one that is unspecified is determined according to the constraint $z_\circ^S[i] = \tilde{\rho}_\circ w_\circ^S[i] + v_\circ^S[i]$ (except that, if $\tilde{\rho}_\circ = 0$, then answer according to the constraint $z_\circ^S[i] = v_\circ^S[i]$ by setting $w^S[i]$ to be a random field element).
  8. For a query $j \in [\ell_\bullet]$ to $w_\bullet[j]$: answer analogously to Step 3, except that subqueries to $w_\circ[j]$ follow Step 7.
  9. For a query $j \in [\ell_\bullet]$ to $v_\bullet[j]$: compute $w_\bullet^S[j]$ as in Step 8 and set $v_\bullet^S[j] := \tilde{\rho}_\bullet w_\bullet^S[j] - z_\bullet^S[j]$.

We claim that the above simulation achieves perfect zero-knowledge, that is, $S(\tilde{V}, \mathbf{x})$ is identically distributed to $\mathrm{DPCPView}(\mathbf{k}, \tilde{V}, P, \mathbf{x}, \mathbf{w})$. We show that the distribution of answers provided by the simulation to $\tilde{V}$ is the same as the distribution of answers obtained by $\tilde{V}$ from the oracles provided by the honest prover. First, we discuss the answers to queries asked before $\tilde{V}$ sends $\tilde{\rho} = (\tilde{\rho}_\circ, \tilde{\rho}_\bullet)$, which can only be to the oracle $\pi_0 = (w_\circ \| v_\circ \| w_\bullet \| v_\bullet)$:

(i) In an honest proof, $v_\circ$ and $v_\bullet$ are random in $C_\circ$ and $C_\bullet$, respectively. The simulator answers a query to either of these by selecting a random field element and then propagating to other locations the linear constraints imposed by belonging to the linear code.

(ii) In an honest proof, $w_\circ$ is computed as $w_\circ := u' + \alpha_\circ$, where u' is random in C'. Any t values from a random codeword in C' are distributed identically to t random field elements, because C' is t-wise independent. The queries of $\tilde{V}$ determine at most $\mathsf{k} \cdot q = t$ locations of $w_\circ$. Hence, in an honest proof, $\tilde{V}$ gets uniformly random answers for its queries to $w_\circ$; this matches the simulated view where S answers $\tilde{V}$'s queries to $w_\circ$ with random fields elements.

(iii) In an honest proof, $w_\bullet$ is a deterministic function of $w_\circ$: $w_\bullet := g(w_\circ)$. As described above, the $\le t$ positions of $w_\circ$ determined by the verifier's questions are uniformly random in the honest proof, as well as in the simulated proof. Therefore the honest and the simulated views of $w_\bullet$ are identically distributed, as deterministic functions of identically distributed random variables.

Next, we discuss the answers to queries asked after $\tilde{V}$ sends $\tilde{\rho} = (\tilde{\rho}_\circ, \tilde{\rho}_\bullet)$; now $\tilde{V}$ can query both $\pi_0 = (w_\circ \| v_\circ \| w_\bullet \| v_\bullet)$ and $\pi_1 = (z_\circ \| z_\bullet \| \pi_\circ \| \pi_\bullet)$.

In an honest proof, answers to verifiers queries after sending $\tilde{\rho}$ are from an uniform distribution of $v_\circ \in C_\circ, v_\bullet \in C_\bullet, u' \in C'$ (and deterministic functions of those and $\alpha_\circ$), that is further conditioned on the answers given before sending $\tilde{\rho}$.

We conclude the discussion of the simulator by examining the time complexity of the simulation. Most steps of the simulation require (a) sampling a random field element and, possibly, (b) solving a linear system with a polynomial number of equations. The only expensive part of the simulation is Step 5, because it requires sampling random codewords in $C_\circ$ and $C_\bullet$, as well as computing PCPP proofs for these two codewords. Provided that r_1 is polynomially bounded, the entire simulation also runs in polynomial time in the instance size n. (The definition of zero knowledge in Sect. 3 prescribes, as typically done, a simulator that runs in expected probabilistic polynomial time; our simulator runs in strict probabilistic polynomial time.)

7   From NTIME to Randomizable Linear Algebraic CSPs

- $\mathscr{R}_{\mathsf{AP}}$ & $\mathscr{R}_{\mathsf{GAP}}$. In Sect. 7.1, we define *algebraic problems*, implicit in several influential works on PCPs and IP [2, 4, 5, 30] and explicitly defined in [22, 34, 37]. Afterward, we define *group-preserving algebraic problems*, a new "symmetric" variant of algebraic problems that not only are powerful enough to efficiently capture **NTIME** but are also naturally "randomizable", as discussed below.

- $\mathscr{R}_{\mathsf{AP}} \to \mathscr{R}_{\mathsf{LA}}$. In Sect. 7.2 (see Lemma 2), we show that algebraic problems are a sublanguage of linear algebraic CSPs. This observation shows that the techniques of this paper could potentially be applied to many PCP systems (e.g., those in [2,4,5,11,13–16,22,30,37] to name a few) and also provides a "warm up" for the next item.
- $\mathscr{R}_{\mathsf{GAP}} \to \mathscr{R}_{\mathsf{RLA}}$. In Sect. 7.3 (see Lemma 3), we show an efficient reduction from group-preserving algebraic problems to randomizable linear algebraic CSPs. In other words, the property of group preservation allows the corresponding linear algebraic CSPs to be randomizable.
- **NTIME** $\to \mathscr{R}_{\mathsf{GAP}}$. In Sect. 7.4 (see Lemma 4), we show an efficient reduction from **NTIME** to group-preserving algebraic problems.
- **NTIME** $\to \mathscr{R}_{\mathsf{RLA}}$. In Sect. 7.5 (see Theorem 6), we explain how to combine the above to obtain the efficient reduction from **NTIME** to randomizable linear algebraic CSPs.

7.1 Algebraic Problems and Group Preservation

The definition below of **algebraic problems** is essentially due to [34] (though the term "algebraic problem" is from [22]); variants of it appear in later works such as [10,14–16,22,36,37].

Definition 6 ($\mathscr{R}_{\mathsf{AP}}$). *Given functions $F\colon \mathbb{N} \to \mathscr{F}$, and $h, m, \eta, d, \sigma \colon \mathbb{N} \to \mathbb{N}$, the relation*

$$\mathscr{R}_{\mathsf{AP}}[F, h, m, \eta, d, \sigma]$$

consists of instance-witness pairs $(\mathbf{x}, \mathbf{w})$ satisfying the following.

- *The instance $\mathbf{x}$ is a tuple $(1^n, H, Q, \boldsymbol{N})$ where:*
  - *H is a subset of $F(n)$ with cardinality $h(n)$;*
  - *Q is a polynomial in $F(n)[X_1, \ldots, X_{m(n)}, Y_1, \ldots, Y_{\eta(n)}]$ such that (i) it has degree less than $h(n)$ in each variable X_i, (ii) it has total degree at most $d(n)$ when viewed as a polynomial in the variables $Y_1, \ldots, Y_{\eta(n)}$ with coefficients in $F(n)[X_1, \ldots, X_{m(n)}]$, (iii) it can be evaluated by an arithmetic circuit of size $\sigma(n)$;*
  - *$\boldsymbol{N} = (N_1, \ldots, N_{\eta(n)})$ and each $N_i \colon F(n)^{m(n)} \to F(n)^{m(n)}$ is an invertible affine function.*
- *The witness $\mathbf{w}$ is a polynomial A in $F(n)[X_1, \ldots, X_{m(n)}]$.*
- *The instance $\mathbf{x}$ and witness $\mathbf{w}$ jointly satisfy the following:*

$$\text{for every } \alpha \in H^{m(n)}, (Q \circ A \circ \boldsymbol{N})(\alpha) = 0 \tag{1}$$

where

$$(Q \circ A \circ N)(X) := Q(X_1, \ldots, X_{m(n)}, A(N_1(X_1, \ldots, X_{m(n)})), \ldots, A(N_{\eta(n)}(X_1, \ldots, X_{m(n)}))). \tag{2}$$

Next, we define **group-preserving algebraic problems**, a family of algebraic problems in which the set H is a subgroup of $F(n)$ and the neighbor functions act on the product group $H^{m(n)}$. The additional symmetry enables a reduction to randomizable linear algebraic CSPs, which give rise to zero knowledge duplex PCPs. We believe that group-preserving algebraic problems may find applications in the study of PCPs beyond their use in this paper.

Definition 7 ($\mathscr{R}_{\mathsf{GAP}}$). *The relation $\mathscr{R}_{\mathsf{GAP}}[F, h, m, \eta, d, \sigma]$ is the sub-relation of $\mathscr{R}_{\mathsf{AP}}[F, h, m, \eta, d, \sigma]$ obtained via restriction to instances that are group preserving. An instance $\mathsf{x} = (1^n, H, Q, \boldsymbol{N})$ is* group preserving *if: (i) H is an additive or a multiplicative subgroup of $F(n)$; (ii) each $N_i \colon F(n)^{m(n)} \to F(n)^{m(n)}$ in $\boldsymbol{N}$ can be identified with an element χ_i in $H^{m(n)}$ such that $N_i(x) = \chi_i \odot x$, where $\odot$ denotes the group operation of the product group $H^{m(n)}$.*

We also write $\mathscr{R}_{\mathsf{GAP}}[F, h, m, \eta, d, \sigma, +]$ to denote the further restriction to instances that are additively group preserving (i.e., H is an additive subgroup); similarly, we write $\mathscr{R}_{\mathsf{GAP}}[F, h, m, \eta, d, \sigma, \times]$ to denote the restriction to instances that are multiplicatively group preserving.

- The *degree* of x, denoted $|\mathsf{x}|_{\deg}$, is $\deg_{Y_1, \ldots, Y_{\eta(n)}}(Q)$, i.e., the total degree of Q viewed as a polynomial in the variables $Y_1, \ldots, Y_{\eta(n)}$ with coefficients in the ring $\mathbb{F}[X_1, \ldots, X_{m(n)}]$.
- The *circuit size* of x, denoted $|\mathsf{x}|_{\mathrm{circ}}$, is the circuit size of Q.

7.2 Algebraic Problems Naturally Reduce to Linear Algebraic CSPs

Lemma 2 ($\mathscr{R}_{\mathsf{AP}} \to \mathscr{R}_{\mathsf{LA}}$). *For every $F \colon \mathbb{N} \to \mathscr{F}$, $h, m, \eta, d, \sigma \colon \mathbb{N} \to \mathbb{N}$, $\epsilon \colon \mathbb{N} \to (0, 1)$, and $\mathscr{R} \subseteq \mathscr{R}_{\mathsf{AP}}[F, h, m, \eta, d, \sigma]$ there exist a relation $\mathscr{R}'$ and algorithms* inst, $\mathsf{wit}_1, \mathsf{wit}_2$ *satisfying the following conditions:*

- EFFICIENT REDUCTION. *For every instance x, letting $\mathsf{x}' := \mathsf{inst}(\mathsf{x})$:*
  - *for every witness w, if $(\mathsf{x}, \mathsf{w}) \in \mathscr{R}$ then $(\mathsf{x}', \mathsf{wit}_1(\mathsf{x}, \mathsf{w})) \in \mathscr{R}'$;*
  - *for every witness w', if $(\mathsf{x}', \mathsf{w}') \in \mathscr{R}'$ then $(\mathsf{x}, \mathsf{wit}_2(\mathsf{x}, \mathsf{w}')) \in \mathscr{R}$.*
  Moreover, inst runs in time $\mathrm{poly}(|\mathsf{x}|)$, wit_1 in time $\mathrm{poly}(|\mathsf{x}|) \cdot \tilde{O}(|\mathsf{w}| \cdot \eta \cdot \sigma)$, and wit_2 in time $\mathrm{poly}(|\mathsf{x}|) \cdot \tilde{O}(|\mathsf{w}'|)$.
- LINEAR ALGEBRAIC CSP. *The relation $\mathscr{R}'$ is a subset of*

$$
\mathscr{R}_{\mathsf{LA}} \begin{bmatrix} f = F \\ \ell = |F|^m \\ \rho = \left(\frac{hd}{|F|}\right)^m \\ \delta = 1 - \frac{hd}{|F|} \\ q = \eta \\ c = \sigma + \eta \\ \gamma = \eta\epsilon \\ \epsilon \end{bmatrix} .
$$

- RM CODES. *If $\mathsf{x} = (1^n, H, Q, \boldsymbol{N})$ then $\mathsf{inst}(\mathsf{x}) = (1^n, C_{\circ}, C_{\bullet}, g)$ with*
  - $C_{\circ} = \mathsf{RM}\left[F(n), F(n), m(n), \frac{h(n)}{|F(n)|}\right]$;
  - $C_{\bullet} = \mathsf{VRM}\left[F(n), F(n), m(n), \frac{h(n)d(n)}{|F(n)|}, H\right]$;
  - *g is the function that maps $F(n)[X_1, \ldots, X_{m(n)}]$ to $F(n)^{F(n)^{m(n)}}$ as follows: given A in $F(n)[X_1, \ldots, X_{m(n)}]$ and $\omega \in F(n)^{m(n)}$, the ω-th coordinate of $g(A)$ equals to $(Q \circ A \circ \boldsymbol{N})(\omega)$.*

Proof (Proof of Lemma 2). Let $\mathbf{x} = (1^n, H, Q, \mathbf{N})$ be an instance of $\mathcal{R}_{\mathsf{AP}}[F, h, m, \eta, d, \sigma]$, and construct $\mathbf{x}' := \mathsf{inst}(\mathbf{x}) = (1^n, C_\circ, C_\bullet, g)$ as above. We first argue that $\mathbf{x}'$ is an instance of $\mathcal{R}_{\mathsf{LA}}[f, \ell, \rho, \delta, q, c, \gamma, \epsilon]$.

First, $C_\circ$ and $C_\bullet$ are linear error correcting codes with block length at most $\ell := |F|^m$, rate at most $\rho := \max\{(\frac{h}{|F|})^m, (\frac{hd}{|F|})^m\}$, and relative distance at least $\delta := \min\{1 - \frac{h}{|F|}, 1 - \frac{hd}{|F|}\}$ over the same field F. (See Sect. 2.4.)

By construction, the function g is q-local with $q := \eta$ and c-efficient with $c := \sigma + \eta$; moreover, g is (γ, ϵ)-sampling with $\gamma := \eta\epsilon$, as we now explain. (See Definition 1 for definitions of these properties.) For every $\omega \in F^m$, I_ω denotes the set of indices in F^m that $g(\cdot)[\omega]$ depends on; for the g above, I_ω equals $\{N_1(\omega), \ldots, N_\eta(\omega)\}$. For every $\omega' \in F^m$ and $\omega \in F^m$, if $\omega' \in I_\omega$ then $\omega \in \{N_1^{-1}(\omega'), \ldots, N_\eta^{-1}(\omega')\}$. Hence, the number of ω's with $\omega' \in I_\omega$ is at most η, because each N_i is invertible. We deduce that $\Pr[I_\omega \cap I \neq \emptyset \mid \omega \leftarrow F^m] \leq (\eta \cdot |I|)/|F|^m \leq \eta\epsilon$.

Finally, $C_\bullet \cup g(C_\circ)$ has relative distance at least δ because it is a subset of $\mathsf{RM}[F, F, m, \frac{hd}{|F|}]$. This claim is immediate for $C_\bullet$; for $g(C_\circ)$, it follows from the fact that $Q \circ A \circ \mathbf{N}$ has, in each variable, a degree that is at most a multiplicative factor of d larger than the degree of A.

We conclude the proof by explaining how one obtains the two witness maps $\mathsf{wit}_1, \mathsf{wit}_2$. For wit_1, suppose that $\mathbf{w} = A \in F[X_1, \ldots, X_m]$ is a witness for $\mathbf{x}$; then one can verify that $\mathbf{w}' := (\alpha_\circ, \alpha_\bullet)$, where $\alpha_\circ := A$ and $\alpha_\bullet := Q \circ A \circ \mathbf{N}$, is a witness for $\mathbf{x}'$; $\alpha_\bullet$ can be efficiently obtained by first computing the evaluation of A on F^m (via an FFT), then computing the evaluation of $Q \circ A \circ \mathbf{N}$ on F^m (via point-to-point computation), and finally interpolating (via an inverse FFT). Conversely, for wit_2, suppose that $\mathbf{w}' = (\alpha_\circ, \alpha_\bullet)$ is a witness for $\mathbf{x}'$; then one can verify that $\mathbf{w} := \alpha_\circ$ is a witness for $\mathbf{x}$.

7.3 From Group-Preserving Algebraic Problems to Randomizable Linear Algebraic CSPs

Lemma 3 ($\mathcal{R}_{\mathsf{GAP}} \to \mathcal{R}_{\mathsf{RLA}}$). *For every $F: \mathbb{N} \to \mathscr{F}$, $h, m, \eta, d, \sigma, t: \mathbb{N} \to \mathbb{N}$, $\delta, \epsilon: \mathbb{N} \to (0, 1)$ with $|F| \geq \hat{h}$, where $\hat{h}$ denotes the smallest integral multiple of h that is greater than $\frac{(h+t)d}{1-\delta}$, and for any $\mathscr{R} \subseteq \mathcal{R}_{\mathsf{GAP}}[F, h, m, \eta, d, \sigma]$ there exist a relation $\mathscr{R}'$ and algorithms $\mathsf{inst}, \mathsf{wit}_1, \mathsf{wit}_2$ satisfying the following conditions:*

– EFFICIENT REDUCTION. *For every instance $\mathbf{x}$, letting $\mathbf{x}' := \mathsf{inst}(\mathbf{x})$:*
  - *for every witness $\mathbf{w}$, if $(\mathbf{x}, \mathbf{w}) \in \mathscr{R}$ then $(\mathbf{x}', \mathsf{wit}_1(\mathbf{x}, \mathbf{w})) \in \mathscr{R}'$;*
  - *for every witness $\mathbf{w}'$, if $(\mathbf{x}', \mathbf{w}') \in \mathscr{R}'$ then $(\mathbf{x}, \mathsf{wit}_2(\mathbf{x}, \mathbf{w}')) \in \mathscr{R}$.*
  Moreover, inst runs in time $\mathrm{poly}(|\mathbf{x}|)$, wit_1 in time $\mathrm{poly}(|\mathbf{x}|) \cdot \tilde{O}(|\mathbf{w}| \cdot \eta \cdot \sigma)$, and wit_2 in time $\mathrm{poly}(|\mathbf{x}|) \cdot \tilde{O}(|\mathbf{w}'|)$.

- RANDOMIZABLE LINEAR ALGEBRAIC CSP. *The relation $\mathcal{R}'$ is a subset of*

$$
\mathcal{R}_{\mathsf{RLA}} \begin{bmatrix} f = F \\ \ell = \hat{h}^m \\ \rho = (\frac{(h+t)d}{\hat{h}})^m \\ \delta = 1 - (\frac{(h+t)d}{\hat{h}}) \\ q = \eta \\ c = \sigma + \eta \\ \gamma = \eta\epsilon \\ \epsilon \\ t \\ r = \tilde{O}(\hat{h}^m) \end{bmatrix}.
$$

Proof (Proof of Lemma 3). Let $\mathsf{x} = (1^n, H, Q, \mathbf{N})$ be an instance of $\mathcal{R}_{\mathsf{GAP}}[F, h, m, \eta, d, \sigma]$. We construct an instance $\mathsf{x}' := \mathsf{inst}(\mathsf{x}) = (1^n, C_\circ, C_\bullet, g)$ of $\mathcal{R}_{\mathsf{RLA}}[f, \ell, \rho, \delta, q, c, \gamma, \epsilon, t, r]$ as follows.

Let $\hat{H}$ be a subset of F that is a union of cosets of H with $|\hat{H}| = \hat{h}$ and $\hat{H} \cap H = \emptyset$. (This can be done as follows: let S be a subset of the quotient group $F^\odot / H$ with cardinality $|S| = \hat{h}/h$ that does not include $1_\odot$, where $F^\odot$ denotes the additive or multiplicative group of F, depending on whether H is additive or multiplicative, and $1_\odot$ is the identity in H; then set $\hat{H} := \{x \odot y \mid x \in S, y \in H\}$.) Analogously to the proof of Lemma 2, we define:

- $C_\circ := \mathsf{RM}\left[F(n), \hat{H}, m(n), \frac{h(n)+t(n)}{\hat{h}(n)}\right]$;
- $C_\bullet := \mathsf{VRM}\left[F(n), \hat{H}, m(n), \frac{(h(n)+t(n))d(n)}{\hat{h}(n)}, H\right]$;
- g to be the function that maps $F(n)[X_1, \ldots, X_{m(n)}]$ to $F(n)^{\hat{H}^{m(n)}}$ as follows: given A in $F(n)[X_1, \ldots, X_{m(n)}]$ and $\omega \in \hat{H}^{m(n)}$, the ω-th coordinate of $g(A)$ equals to $(Q \circ A \circ \mathbf{N})(\omega)$. Note that g is well-defined, i.e., $g(A)$ is a function from $\hat{H}^{m(n)}$ to $F(n)$; this follows from the group preservation property of x (see Definition 7): for every $\omega \in \hat{H}^m$ and $i \in [\eta]$, it holds that $N_i(\omega) \subseteq \hat{H}^m$ because $\hat{H}$ is a union of cosets of H and N_i multiplies every coordinate of ω by an element of H.

We first argue that x' constructed above is an instance of $\mathcal{R}_{\mathsf{RLA}}[f, \ell, \rho, \delta, q, c, \gamma, \epsilon, t, r]$.

First, analogously to the proof of Lemma 2, we note that $C_\circ$ and $C_\bullet$ are linear error correcting codes with block length at most $\ell := \hat{h}^m$, rate at most $\rho := \max\{(\frac{h+t}{\hat{h}})^m, (\frac{(h+t)d}{\hat{h}})^m\}$, and relative distance at least $\delta := \min\{1 - \frac{h+t}{\hat{h}}, 1 - \frac{(h+t)d}{\hat{h}}\}$ over the same field F; also, we deduce that g is q-local with $q := \eta$, c-efficient with $c := \sigma + \eta$, and (γ, ϵ)-sampling with $\gamma := \eta\epsilon$.

Next, recalling Definition 5, x' is t-randomizable in time $r := \tilde{O}(\hat{h}^m)$ because: (i) $C' := \mathsf{VRM}[F(n), \hat{H}, m, \frac{h+t}{\hat{h}}, H]$ is a subcode of $C_\circ$ and it is t-wise independent due to Claim in Sect. 2.4 (C' satisfies the hypotheses because $H \cap \hat{H} = \emptyset$ and $\hat{h} - h \geq \frac{(h+t)d}{1-\delta} - h \geq t$); and (ii) one can sample random elements from C', $C_\circ$

and $C_\bullet$ in time $\tilde{O}(\hat{h}^m)$ by using the quasilinear FFT algorithms for multipoint evaluation and interpolation (sampling the random polynomial in necessary basis is easy for $C_\circ$; for vanishing Reed–Muller codes we rely on Alon's Combinatorial Nullstellensatz [1] as per Lemma 4.11 of [15]).

We conclude the proof by observing that necessary witness maps $\mathsf{wit}_1, \mathsf{wit}_2$ exist. Just as in Lemma 2, if $\mathsf{w} = A \in F(n)[X_1, \ldots, X_{m(n)}]$ is a witness for x then $\mathsf{wit}_1(\mathsf{x}, \mathsf{w})$ outputs $\mathsf{w}' := (A, Q \circ A \circ N)$, which is a witness for x'; conversely, if $\mathsf{w}' = (\alpha_\circ, \alpha_\bullet)$ is a witness for x' then $\mathsf{wit}_2(\mathsf{x}, \mathsf{w}')$ outputs $\mathsf{w} := \alpha_\circ$, which is a witness for x.

7.4   An Efficient Reduction from NTIME to Group-Preserving Algebraic Problems

The following lemma gives an efficient reduction from **NTIME** to group-preserving algebraic problems in which instances are over fields of characteristic 2 and preserve additive groups.

Lemma 4 (NTIME $\to \mathscr{R}_{\mathsf{GAP}}$). *For every* $h, m, T \colon \mathbb{N} \to \mathbb{N}$ *with* $h(n)^{m(n)} = \Omega(T(n) \log T(n))$ *and* $\mathscr{R} \in \mathbf{NTIME}(T)$ *there exist a relation* $\mathscr{R}'$ *and algorithms* $\mathsf{inst}, \mathsf{wit}_1, \mathsf{wit}_2$ *satisfying the following conditions:*

– EFFICIENT REDUCTION. *For every instance* x, *letting* $\mathsf{x}' := \mathsf{inst}(\mathsf{x})$:
  - *for every witness* w, *if* $(\mathsf{x}, \mathsf{w}) \in \mathscr{R}$ *then* $(\mathsf{x}', \mathsf{wit}_1(\mathsf{x}, \mathsf{w})) \in \mathscr{R}'$;
  - *for every witness* w', *if* $(\mathsf{x}', \mathsf{w}') \in \mathscr{R}'$ *then* $(\mathsf{x}, \mathsf{wit}_2(\mathsf{x}, \mathsf{w}')) \in \mathscr{R}$.

  Moreover, inst *runs in time* $\mathrm{poly}(n + \log h(n) + m(n))$ *and* $\mathsf{wit}_1, \mathsf{wit}_2$ *run in time* $\tilde{O}(T(n))$.
– GROUP PRESERVING ALGEBRAIC PROBLEM. *The relation* $\mathscr{R}'$ *is a subset of*

$$\mathscr{R}_{\mathsf{GAP}} \begin{bmatrix} F = \mathbb{F}_{2^{\log T + O(\log \log T)}} \\ h \\ m \\ \eta = \mathrm{polylog}(T) \\ d = O(1) \\ \sigma = \mathrm{poly}(n + \log T) \\ + \end{bmatrix}.$$

The proof appears in the full version.

7.5   Combining the Two Reductions

By combining Lemmas 3 and 4, we obtain the following theorem, which gives the reduction claimed at the beginning of this section.

Theorem 6 (NTIME $\to \mathscr{R}_{\mathsf{RLA}}$). *For every* $T, t \colon \mathbb{N} \to \mathbb{N}$, $\delta, \epsilon \colon \mathbb{N} \to (0, 1)$, *and* $\mathscr{R} \in \mathbf{NTIME}(T)$ *there exist a relation* $\mathscr{R}'$ *and algorithms* $\mathsf{inst}, \mathsf{wit}_1, \mathsf{wit}_2$ *satisfying the following conditions:*

- EFFICIENT REDUCTION. *For every instance* x, *letting* x′ := inst(x)*:*
  - *for every witness* w, *if* $(x, w) \in \mathscr{R}$ *then* $(x', wit_1(x, w)) \in \mathscr{R}'$;
  - *for every witness* w′, *if* $(x', w') \in \mathscr{R}'$ *then* $(x, wit_2(x, w')) \in \mathscr{R}$.

  Moreover, inst *runs in time* $poly(n + \log(\frac{T(n)+t(n)}{1-\delta(n)}))$ *and* wit_1, wit_2 *run in time*
  $poly(n) \cdot \tilde{O}(\frac{T(n)+t(n)}{1-\delta(n)})$.

- RANDOMIZABLE LINEAR ALGEBRAIC CSP. *The relation* $\mathscr{R}'$ *is a subset of*

$$
\mathscr{R}_{\mathsf{RLA}} \begin{bmatrix}
f = \mathbb{F}_{2^{\log(T+t)+O(\log\log(T+t))}} \\
\ell = \tilde{O}(\frac{T+t}{1-\delta}) \\
\rho = 1 - \delta \\
\delta \\
q = \text{polylog}(T) \\
c = \text{poly}(n + \log T) \\
\gamma = \text{polylog}(T) \cdot \epsilon \\
\epsilon \\
t \\
r = \tilde{O}(\frac{T+t}{1-\delta})
\end{bmatrix} .
$$

- AFFINE RS CODES OVER CHARACTERISTIC 2. *Both* $\mathscr{C}_{\mathscr{R}',\circ}$ *and* $\mathscr{C}_{\mathscr{R}',\bullet}$ *are subsets of* $\mathcal{RS}_\rho^* \cup \mathcal{VRS}_\rho^*$ *(see Sect. 2.4).*

Proof (Proof of Theorem 6). First, we invoke Lemma 4 with h, m, T such that $m(n) = 1$ and $h(n) = O(T(n) \log T(n))$; this yields a relation $\mathscr{R}^{(1)}$ and algorithms $inst^{(1)}, wit_1^{(1)}, wit_2^{(1)}$ such that: (i) $inst^{(1)}, wit_1^{(1)}, wit_2^{(1)}$ provide a reduction from $\mathscr{R} \in \mathbf{NTIME}(T)$ to $\mathscr{R}^{(1)}$, with $inst^{(1)}(x)$ running in time $poly(n + \log h(n) + m(n))$ and $wit_1^{(1)}(x, w), wit_2^{(1)}(x, w^{(1)})$ in time $\tilde{O}(T(n))$; and (ii) $\mathscr{R}^{(1)}$ is a subset of

$$
\mathscr{R}_{\mathsf{GAP}} \begin{bmatrix}
F = \mathbb{F}_{2^{\log T + O(\log\log T)}} \\
h = O(T(n) \log T(n)) \\
m = 1 \\
\eta = \text{polylog}(T) \\
d = O(1) \\
\sigma = \text{poly}(n + \log T) \\
+
\end{bmatrix} .
$$

Next, we invoke Lemma 3 on $\mathscr{R}^{(1)}$, using δ, ϵ, t from the theorem statement. Note that the conditions of the theorem are satisfied as $|F| \geq \frac{(h+t)d}{1-\delta} + h \geq \hat{h}$. Therefore this yields a relation $\mathscr{R}^{(2)}$ and algorithms $inst^{(2)}, wit_1^{(2)}, wit_2^{(2)}$ such that: (i) $inst^{(2)}, wit_1^{(2)}, wit_2^{(2)}$ provide a reduction from $\mathscr{R}^{(1)}$ to $\mathscr{R}^{(2)}$, with $inst^{(2)}(x^{(1)})$ running in time $poly(|x^{(1)}|)$, $wit_1^{(2)}(x^{(1)}, w^{(1)})$ in time $poly(|x^{(1)}|) \cdot \tilde{O}(|w^{(1)}| \cdot \eta \cdot \sigma)$ and $wit_2^{(2)}(x^{(1)}, w^{(2)})$ in time $poly(|x^{(1)}|) \cdot \tilde{O}(|w^{(2)}|)$; and (ii) $\mathscr{R}^{(2)}$ is a subset of

$$\mathscr{R}_{\mathsf{RLA}} \begin{bmatrix} f = F \\ \ell = O(\frac{h+t}{1-\delta}) \\ \rho = 1 - \delta \\ \delta \\ q = \eta \\ c = \sigma + \eta \\ \gamma = \eta\epsilon \\ \epsilon \\ t \\ r = \tilde{O}(\frac{h+t}{1-\delta}) \end{bmatrix} .$$

One can check that $\mathscr{R}^{(2)}$ achieves the parameters specified in the theorem statement.

The desired reduction from $\mathscr{R}$ to $\mathscr{R}^{(2)}$ is given by the algorithms $\mathsf{inst}(\mathbb{x}) := \mathsf{inst}^{(2)}(\mathsf{inst}^{(1)}(\mathbb{x}))$, $\mathsf{wit}_1(\mathbb{x}, \mathbb{w}) := \mathsf{wit}_1^{(2)}(\mathsf{inst}^{(1)}(\mathbb{x}), \mathsf{wit}_1^{(1)}(\mathbb{x}, \mathbb{w}))$, and $\mathsf{wit}_2(\mathbb{x}, \mathbb{w}') := \mathsf{wit}_2^{(1)}(\mathbb{x}, \mathsf{wit}_2^{(2)}(\mathsf{inst}^{(1)}(\mathbb{x}), \mathbb{w}'))$. One can verify that inst runs in time $\mathrm{poly}(n + \log(\frac{T(n)+t(n)}{1-\delta(n)}))$ and $\mathsf{wit}_1, \mathsf{wit}_2$ run in time $\mathrm{poly}(n) \cdot \tilde{O}(\frac{T(n)+t(n)}{1-\delta(n)})$.

8   Proof of Theorem 4

Proof (Proof of Theorem 4). We explain how to combine Theorem 6 and Lemma 5 (and Theorem 3) so to obtain Theorem 4.

Let $\mathscr{R}$ be a relation in **NTIME**(T); we need to construct a duplex PCP system for $\mathscr{R}$ with the claimed parameters. For now we focus on achieving soundness of $\frac{1}{2}$, and discuss the general case at the end of the proof.

We first reduce **NTIME** to randomizable linear algebraic CSPs: invoke Theorem 6 on $\mathscr{R}$ to obtain a relation $\mathscr{R}'$ and algorithms $\mathsf{inst}, \mathsf{wit}_1, \mathsf{wit}_2$ such that: (i) $\mathsf{inst}, \mathsf{wit}_1, \mathsf{wit}_2$ provide a reduction from $\mathscr{R}$ to $\mathscr{R}'$, with inst running in time $\mathrm{poly}(n + \log(T(n) + t_1(n)))$ and $\mathsf{wit}_1, \mathsf{wit}_2$ in time $\tilde{O}(T(n) + t_1(n))$; and (ii) $\mathscr{R}'$ is a subset of

$$\mathscr{R}_{\mathsf{RLA}} \begin{bmatrix} f_1 = \mathbb{F}_{2^{\log(T+t_1)+O(\log\log(T+t_1))}} \\ \ell_1 = \tilde{O}(T + t_1) \\ \rho_1 = 1 - \delta_1 \\ \delta_1 \\ q_1 = \mathrm{polylog}(T) \\ c_1 = \mathrm{poly}(n + \log T) \\ \gamma_1 = \mathrm{polylog}(T) \cdot \epsilon_1 \\ \epsilon_1 \\ t_1 \\ r_1 = \tilde{O}(\frac{T+t_1}{1-\delta_1}) \end{bmatrix} .$$

Above, as parameters of Theorem 6, we chose ϵ_1, δ_1 and t_1 as follows: ϵ_1 such that $\gamma_1 = \mathrm{polylog}(T) \cdot \epsilon_1 \leq \frac{2}{9}$, then $\delta_1 := 1 - \epsilon_1/4$, and $t_1 := \mathsf{k} \cdot q_1 = \mathsf{k} \cdot \mathrm{polylog}(T)$.

Next we obtain PCPP systems for the relations corresponding to codes appearing in instances of $\mathscr{R}'$. Theorem 6 guarantees that both $\mathscr{C}_{\mathscr{R}', \circ}$ and $\mathscr{C}_{\mathscr{R}', \bullet}$

are subsets of $\mathcal{RS}_\rho^* \cup \mathcal{VRS}_\rho^*$. We now invoke Theorem 3, choosing $\lambda = 2$ and s such that fields f_1 for $\mathscr{R}'$ and a_2 for the PCPPs match. That is, we chose $s = \tilde{O}(\log\log(T + t_1))$ and obtain:

$$\mathrm{Rel}(\mathscr{C}_{\mathscr{R}',\circ}),\ \mathrm{Rel}(\mathscr{C}_{\mathscr{R}',\bullet}) \in \mathbf{PCPP}\begin{bmatrix} a_2 &= \mathbb{F}_{2^{s+\log\ell_1}} \\ l_2 &= \tilde{O}(\ell_1) \\ q_2 &= \mathrm{polylog}(\ell_1) \\ \Delta_2 &= \Delta_a^{\mathrm{Ham}} \\ d_2 &= \rho_1/2 \\ e_2 &= 1/4 \\ tp_2 &= \mathrm{poly}(s)\cdot\tilde{O}(\ell_1) \\ tv_2 &= \mathrm{poly}(s+\log\ell_1) \\ na \end{bmatrix}.$$

Finally we invoke Theorem 5 for $\mathscr{R}'$ to obtain a duplex PCP system for $\mathscr{R}'$, supplying the PCPPs we just obtained from Theorem 3. Note that our choices satisfy the hypothesis of Theorem 5 is satisfied, as the two fields match, r_1 is polynomially bounded, and as we chose $\gamma_1, \epsilon_1 \leq \frac{2}{9}$, $\delta_1 \geq \frac{17}{18}$, we also have $\epsilon_1 < \min\{\frac{\delta_1}{2}, \delta_1 - \gamma_1\}$ and $d_2 \leq \epsilon_1/4$. This establishes our claim that:

$$\mathscr{R} \in \mathbf{DPCP}_{\mathrm{pzk}}\begin{bmatrix} a &= & & & \mathbb{F}_{2^{\log(T+t_1)+O(\log\log(T+t_1))}} \\ l &= & 2l_2(\ell_1)+6\ell_1 & = & \tilde{O}(T+t_1) \\ q &= & 2q_2(\ell_1)+q_1+7 & = & \mathrm{polylog}(T) \\ e &= & & & \frac{1}{2} \\ tp &= & \mathrm{inst}+\mathrm{wit}_1+(2tp_2(\ell_1)+(c_1+5)\ell_1+r_1) = & & \mathrm{poly}(n)\cdot\tilde{O}(T+k) \\ tv &= & \mathrm{inst}+(2tv_2(\ell_1)+c_1+\log\ell_1) & = & \mathrm{poly}(n+\log(T+k)) \\ k & & & & \\ na \end{bmatrix}.$$

The precise expression for soundness error is $e := \max\{1-\delta_1+\gamma_1+\epsilon_1,\ (1-|f_1|^{-1})\cdot \max\{e_2, \epsilon_1/4\}+|f_1|^{-1}\}$, but it is upper bounded by $\frac{1}{2}$, as for us $1-\delta_1+\gamma_1+\epsilon_1 \leq \frac{1}{2}$, $\max\{e_2, \epsilon_1/4\} = \frac{1}{4}$ and $|f_1| \geq 4$.

Acknowledgments. We thank Yuval Ishai and Mor Weiss for helpful discussions. The research leading to these results has received funding from: the European Community's Seventh Framework Programme (FP7/2007–2013) under grant agreement number 240258; the Israeli Science Foundation (grant 1501/14); and the Center for Science of Information (CSoI), an NSF Science and Technology Center, under grant agreement CCF-0939370.

References

1. Alon, N.: Combinatorial Nullstellensatz. Comb. Probab. Comput. **8**, 7–29 (1999)
2. Arora, S., Lund, C., Motwani, R., Sudan, M., Szegedy, M.: Proof verification and the hardness of approximation problems. JACM **45**, 501–555 (1998)
3. Arora, S., Safra, S.: Probabilistic checking of proofs: a new characterization of NP. JACM **45**, 70–122 (1998)
4. Babai, L., Fortnow, L., Levin, L.A., Szegedy, M.: Checking computations in polylogarithmic time. In: STOC 1991 (1991)

5. Babai, L., Fortnow, L., Lund, C.: Non-deterministic exponential time has two-prover interactive protocols. Comput. Complex. **1**, 3–40 (1991)
6. Babai, L., Moran, S.: Arthur-Merlin games: a randomized proof system, and a hierarchy of complexity class. J. Comput. Syst. Sci. **36**, 254–276 (1988)
7. Bellare, M., Goldreich, O.: On defining proofs of knowledge. In: Brickell, E.F. (ed.) CRYPTO 1992. LNCS, vol. 740, pp. 390–420. Springer, Heidelberg (1993)
8. Ben-Or, M., Goldreich, O., Goldwasser, S., Håstad, J., Kilian, J., Micali, S., Rogaway, P.: Everything provable is provable in zero-knowledge. In: Goldwasser, S. (ed.) CRYPTO 1988. LNCS, vol. 403, pp. 37–56. Springer, Heidelberg (1990)
9. Ben-Or, M., Goldwasser, S., Kilian, J., Wigderson, A.: Multi-prover interactive proofs: how to remove intractability assumptions. In: STOC 1988 (1988)
10. Ben-Sasson, E., Chiesa, A., Genkin, D., Tromer, E.: Fast reductions from RAMs to delegatable succinct constraint satisfaction problems. In: ITCS 2013 (2013)
11. Ben-Sasson, E., Chiesa, A., Genkin, D., Tromer, E.: On the concrete efficiency of probabilistically-checkable proofs. In: STOC 2013 (2013)
12. Ben-Sasson, E., Goldreich, O., Harsha, P., Sudan, M., Vadhan, S.: Robust PCPs of proximity, shorter PCPs and applications to coding. In: STOC 2004 (2004)
13. Ben-Sasson, E., Goldreich, O., Harsha, P., Sudan, M., Vadhan, S.: Short PCPs verifiable in polylogarithmic time. In: CCC 2005 (2005)
14. Ben-Sasson, E., Goldreich, O., Harsha, P., Sudan, M., Vadhan, S.: Robust PCPs of proximity, shorter PCPs, and applications to coding. SIAM J. Comput. **36**, 889–974 (2006)
15. Ben-Sasson, E., Sudan, M.: Short PCPs with polylog query complexity. SIAM J. Comput. **38**, 551–607 (2008)
16. Ben-Sasson, E., Viola, E.: Short PCPs with projection queries. In: Esparza, J., Fraigniaud, P., Husfeldt, T., Koutsoupias, E. (eds.) ICALP 2014. LNCS, vol. 8572, pp. 163–173. Springer, Heidelberg (2014)
17. Dinur, I.: The PCP theorem by gap amplification. JACM **54**, 12:1–12:44 (2007)
18. Dinur, I., Reingold, O.: Assignment testers: towards a combinatorial proof of the PCP theorem. In: FOCS 2004 (2004)
19. Dwork, C., Feige, U., Kilian, J., Naor, M., Safra, M.: Low communication 2-prover zero-knowledge proofs for NP. In: Brickell, E.F. (ed.) CRYPTO 1992. LNCS, vol. 740, pp. 215–227. Springer, Heidelberg (1993)
20. Goldwasser, S., Micali, S., Rackoff, C.: The knowledge complexity of interactive proof-systems. In: STOC 1985 (1985)
21. Goyal, V., Ishai, Y., Mahmoody, M., Sahai, A.: Interactive locking, zero-knowledge PCPs, and unconditional cryptography. In: Rabin, T. (ed.) CRYPTO 2010. LNCS, vol. 6223, pp. 173–190. Springer, Heidelberg (2010)
22. Harsha, P., Sudan, M.: Small PCPs with low query complexity. Comput. Complex. **9**, 157–201 (2000)
23. Impagliazzo, R., Yung, M.: Direct minimum knowledge computations. In: Pomerance, C. (ed.) CRYPTO 1987. LNCS, vol. 293, pp. 40–51. Springer, Heidelberg (1988)
24. Ishai, Y., Kushilevitz, E., Ostrovsky, R., Sahai, A.: Zero-knowledge proofs from secure multiparty computation. SIAM J. Comput. **39**, 1121–1152 (2009)
25. Ishai, Y., Mahmoody, M., Sahai, A.: On efficient zero-knowledge PCPs. In: Cramer, R. (ed.) TCC 2012. LNCS, vol. 7194, pp. 151–168. Springer, Heidelberg (2012)
26. Ishai, Y., Mahmoody, M., Sahai, A., Xiao, D.: On zero-knowledge PCPs: limitations, simplifications, and applications (2015). http://www.cs.virginia.edu/mohammad/files/papers/ZKPCPs-Full.pdf

27. Kalai, Y.T., Raz, R.: Interactive PCP. In: Aceto, L., Damgård, I., Goldberg, L.A., Halldórsson, M.M., Ingólfsdóttir, A., Walukiewicz, I. (eds.) ICALP 2008, Part II. LNCS, vol. 5126, pp. 536–547. Springer, Heidelberg (2008)

28. Kilian, J., Petrank, E., Tardos, G.: Probabilistically checkable proofs with zero knowledge. In: STOC 1997 (1997)

29. Lapidot, D., Shamir, A.: A one-round, two-prover, zero-knowledge protocol for NP. Combinatorica **15**, 204–214 (1995)

30. Lund, C., Fortnow, L., Karloff, H., Noam, N.: Algebraic methods for interactive proof systems. JACM **39**, 859–868 (1992)

31. Mahmoody, M., Xiao, D.: Languages with efficient zero-knowledge PCPs are in SZK. In: Sahai, A. (ed.) TCC 2013. LNCS, vol. 7785, pp. 297–314. Springer, Heidelberg (2013)

32. Mie, T.: Polylogarithmic two-round argument systems. J. Math. Cryptol. **2**, 343–363 (2008)

33. Ostrovsky, R., Wigderson, A.: One-way functions are essential for non-trivial zero-knowledge. In: ISTCS 1993 (1993)

34. Polishchuk, A., Spielman, D.A.: Nearly-linear size holographic proofs. In: STOC 1994 (1994)

35. Shamir, A.: IP = PSPACE. JACM **39**, 869–877 (1992)

36. Spielman, D.: Computationally efficient error-correcting codes and holographic proofs. Ph.D. thesis, Massachusetts Institute of Technology (1995)

37. Szegedy, M.: Many-valued logics and holographic proofs. In: Wiedermann, J., Van Emde Boas, P., Nielsen, M. (eds.) ICALP 1999. LNCS, vol. 1644, pp. 676–686. Springer, Heidelberg (1999)

From Private Simultaneous Messages to Zero-Information Arthur-Merlin Protocols and Back

Benny Applebaum[✉] and Pavel Raykov

School of Electrical Engineering, Tel-Aviv University, Tel Aviv, Israel
{bennyap,pavelraykov}@post.tau.ac.il

Abstract. Göös, Pitassi and Watson (ITCS, 2015) have recently introduced the notion of *Zero-Information Arthur-Merlin Protocols* (ZAM). In this model, which can be viewed as a private version of the standard Arthur-Merlin communication complexity game, Alice and Bob are holding a pair of inputs x and y respectively, and Merlin, the prover, attempts to convince them that some public function f evaluates to 1 on (x, y). In addition to standard completeness and soundness, Göös et al., require a "zero-knowledge" property which asserts that on each yes-input, the distribution of Merlin's proof leaks no information about the inputs (x, y) to an external observer.

In this paper, we relate this new notion to the well-studied model of *Private Simultaneous Messages* (PSM) that was originally suggested by Feige, Naor and Kilian (STOC, 1994). Roughly speaking, we show that the randomness complexity of ZAM corresponds to the communication complexity of PSM, and that the communication complexity of ZAM corresponds to the randomness complexity of PSM. This relation works in both directions where different variants of PSM are being used. Consequently, we derive better upper-bounds on the communication-complexity of ZAM for arbitrary functions. As a secondary contribution, we reveal new connections between different variants of PSM protocols which we believe to be of independent interest.

1   Introduction

In this paper we reveal an intimate connection between two seemingly unrelated models for non-interactive information-theoretic secure computation. We begin with some background.

Research supported by the European Union's Horizon 2020 Programme (ERC-StG-2014-2020) under grant agreement No. 639813 ERC-CLC, ISF grant 1155/11, GIF grant 1152/2011, and the Check Point Institute for Information Security. This work was done in part while the first author was visiting the Simons Institute for the Theory of Computing, supported by the Simons Foundation and by the DIMACS/Simons Collaboration in Cryptography through NSF grant CNS-1523467.

E. Kushilevitz and T. Malkin (Eds.): TCC 2016-A, Part II, LNCS 9563, pp. 65–82, 2016.
DOI: 10.1007/978-3-662-49099-0_3

1.1   Zero-Information Unambiguous Arthur-Merlin Communication Protocols

Consider a pair of computationally-unbounded (randomized) parties, Alice and Bob, each holding an n-bit input, x and y respectively, to some public function $f : \{0,1\}^n \times \{0,1\}^n \to \{0,1\}$. In our first model, a third party, Merlin, wishes to convince Alice and Bob that their joint input is mapped to 1 (i.e., (x,y) is in the language $f^{-1}(1)$). Merlin gets to see the parties' inputs (x,y) and their private randomness r_A and r_B, and is allowed to send a single message ("proof") p to both parties. Then, each party decides whether to accept the proof based on its input and its private randomness. We say that the protocol accepts p if both parties accept it. The protocol is required to satisfy natural properties of (perfect) completeness and soundness. Namely, if $(x,y) \in f^{-1}(1)$ then there is always a proof $p = p(x,y,r_A,r_B)$ that is accepted by both parties, whereas if $(x,y) \in f^{-1}(0)$ then, with probability $1 - \delta$ (over the coins of Alice and Bob), no such proof exists. As usual in communication-complexity games the goal is to minimize the communication complexity of the protocol, namely the length of the proof p.

This model, which is well studied in the communication complexity literature [BFS86, Kla03, Kla10], is viewed as the communication complexity analogue of AM protocols [BM88]. Recently, Göös et al. [GPW15] suggested a variant of this model which requires an additional "zero-knowledge" property defined as follows: For any 1-input $(x,y) \in f^{-1}(1)$, the proof sent by the honest prover provides no information on the inputs (x,y) to an external viewer. Formally, the random variable $p_{x,y} = p(x,y,r_A,r_B)$ induced by a random choice of r_A and r_B should be distributed according to some universal distribution D which is independent of the specific 1-input (x,y). Moreover, an additional *Unambiguity* property is required: any 1-input $(x,y) \in f^{-1}(1)$ and any pair of strings (r_A, r_B) uniquely determine a single accepting proof $p(x,y,r_A,r_B)$.

This modified version of AM protocols (denoted by ZAM) was originally presented in attempt to explain the lack of explicit nontrivial lower bounds for the communication required by AM protocols. Indeed, Göös et al., showed that any function $f : \{0,1\}^n \times \{0,1\}^n \to \{0,1\}$ admits a ZAM protocol with at most exponential communication complexity of $O(2^n)$. Since the transcript of a ZAM protocol carries no information on the inputs, the mere existence of such protocols forms a "barrier" against "information complexity" based arguments. This suggests that, at least in their standard form, such arguments cannot be used to prove lower bounds against AM protocols (even with Unambiguous completeness).

Regardless of the original motivation, one may view the ZAM model as a simple and natural information-theoretic analogue of (non-interactive) zero-knowledge proofs where instead of restricting the computational power of the verifier, we split it between two non-communicating parties (just like AM communication games are derived from the computational-complexity notion of AM protocols). As cryptographers, it is therefore natural to ask:

> How does the ZAM model relate to other more standard models of information-theoretic secure computation?

As we will later see, answering this question also allows us to make some (modest) progress in understanding the communication complexity of ZAM protocols.

1.2 Private Simultaneous Message Protocols

Another, much older, notion of information-theoretically secure communication game was suggested by Feige et al. [FKN94]. As in the previous model, there are three (computationally-unbounded) parties: Alice, Bob and a Referee. Here too, an input (x, y) to a public function $f : \{0, 1\}^n \times \{0, 1\}^n \rightarrow \{0, 1\}$ is split between Alice and Bob, which, in addition, share a common random string c. Alice (resp., Bob) should send to the referee a single message a (resp., b) such that the transcript (a, b) reveals $f(x, y)$ but nothing else. That is, we require two properties: (Correctness) There exists a decoder algorithm Dec which recovers $f(x, y)$ from (a, b) with high probability; and (Privacy) There exists a simulator Sim which, given the value $f(x, y)$, samples the joint distribution of the transcript (a, b) up to some small deviation error. (See Sect. 4 for formal definitions.)

Following [IK97], we refer to such a protocol as a private simultaneous messages (PSM) protocol. A PSM protocol for f can be alternatively viewed as a special type of *randomized encoding* of f [IK00, AIK04], where the output of f is encoded by the output of a randomized function $F((x, y), c)$ such that F can be written as $F((x, y), c) = (F_1(x, c), F_2(y, c))$. This is referred to as a "2-decomposable" encoding in [Ish13].

1.3 ZAM vs. PSM

Our goal will be to relate ZAM protocols to PSM protocols. Since the latter object is well studied and strongly "connected" to other information-theoretic notions (cf. [BIKK14]), such a connection will allow us to place the new ZAM in our well-explored world of information-theoretic cryptography.

Observe that ZAM and PSM share some syntactic similarities (illustrated in Fig. 1). In both cases, the input is shared between Alice and Bob and the third party holds no input. Furthermore, in both cases the communication pattern consists of a single message. On the other side, in ZAM the third party (Merlin) attempts to convince Alice and Bob that the joint input is mapped to 1, and so the communication goes from Merlin to Alice/Bob who generate the output (accept/reject). In contrast, in a PSM protocol, the messages are sent in the other direction: from Alice and Bob to the third party (the Referee) who ends up with the output. In addition, the privacy guarantee looks somewhat different. For ZAM, privacy is defined with respect to an external observer and only over 1-inputs, whereas soundness is defined with respect to the parties (Alice and Bob) who hold the input (x, y). (Indeed, an external observer cannot even tell whether the joint input (x, y) is a 0-input.) Accordingly, in the ZAM model, correctness and privacy are essentially two different concerns that involve different parties. In contrast, for PSM protocols privacy should hold with respect to the view of the receiver who should still be able to decode.

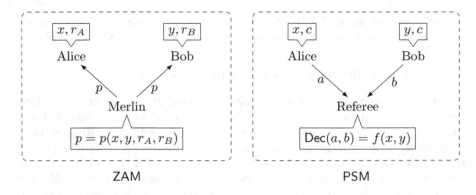

Fig. 1. Flow of messages

These differences seem to point to non-trivial gaps between these two notions. The picture becomes even more confusing when looking at existing constructions. On one hand, the general ZAM constructions presented by [GPW15, Theorem 6] (which use a reduction to Disjointness) seem more elementary than the simplest PSM protocols of [FKN94]. On the other hand, there are ZAM constructions which share common ingredients with existing PSM protocols. Concretely, the branching-program (BP) representation of the underlying function have been used both in the context of PSM [FKN94, IK97] and in the context of ZAM [GPW15, Theorem 1]. (It should be mentioned that there is a quadratic gap between the complexity of the two constructions.) Finally, both in ZAM and in PSM, it is known that any function $f : \{0,1\}^n \times \{0,1\}^n \rightarrow \{0,1\}$ admits a protocol with exponential complexity, but the best known lower-bound is only linear in n. Overall, it is not clear whether these relations are coincidental or point to a deeper connection.[1]

2   Our Results

We prove that ZAM protocols and PSM protocols are intimately related. Roughly speaking, we will show that the *inverse* of ZAM is PSM and vice versa. Therefore, the randomness complexity of ZAM essentially corresponds to the communication complexity of PSM and the communication complexity of ZAM essentially corresponds to the randomness complexity of PSM. This relation works in both directions where different variants of PSM are being used. We proceed with a formal statement of our results. See Fig. 2 for an overview of our transformations.

---

[1] The authors of [GPW15] seem to suggest that there is no formal connection between the two models. Indeed, they explicitly mention PSM as "a different model of private two-party computation, [...] where the best upper and lower bounds are also exponential and linear."

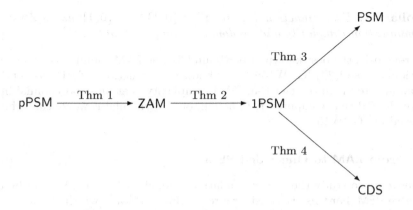

Fig. 2. Overview of the constructions

2.1 From Perfect PSM to ZAM

We begin by showing that a special form of *perfect* PSM protocols (referred to pPSM) yields ZAM protocols.

Theorem 1. *Let f be a function with a* pPSM*protocol that has communication complexity t and randomness complexity s. Then f has a $1/2$-sound* ZAM *scheme with randomness complexity of t and communication complexity of $s + 1$.*

A pPSM protocol is a PSM in which both correctness and privacy are required to be errorless (perfect), and, in addition, the encoding should satisfy some regularity properties.[2]

To prove the theorem, we use the combinatorial properties of the perfect encoding to define a new function $g(x, y, p) = (g_1(x, p), g_2(y, p))$ which, when restricted to a 1-input (x, y), forms a bijection from the randomness space to the output space, and when (x, y) is a 0-input the restricted function $g(x, y, \cdot)$ covers only half of the range. Given such a function, it is not hard to design a ZAM: Alice (resp., Bob) samples a random point r_A in the range of g_1 (resp., r_B in the range of g_2), and accepts a proof $p = (p_1, p_2)$ if p_1 is a preimage of r_A under g_1 (resp. p_2 is a preimage of r_B under g_2). It is not hard to verify that the protocol satisfies Unambiguous completeness, $1/2$-soundness and zero-information. (See Sect. 5.)

Although the notion of pPSM looks strong, we note that all known general PSM protocols are perfect. (See full version for details.) By plugging in the best known protocol from [BIKK14], we derive the following corollary.

---

[2] Essentially, the range of $F = (F_1, F_2)$ can be partitioned into two equal sets S_0 and S_1 and for every input (x, y) the function $F_{x,y}(c)$ that maps the randomness c to the transcript (a, b) forms a bijection from the randomness space to the set $S_{f(x)}$. In the context of randomized encoding, this notion was originally referred to as *perfect randomized encoding* [AIK04]. See Sect. 4 for formal definitions.

Corollary 1. *Every function* $f : \{0,1\}^n \times \{0,1\}^n \to \{0,1\}$ *has a* ZAM *with communication complexity and randomness complexity of* $O(2^{n/2})$.

Previously, the best known upper-bound for the ZAM complexity of a general function f was $O(2^n)$ [GPW15]. Using known constructions of BP-based pPSM, we can also re-prove the fact that ZAM complexity is at most polynomial in the size of the BP that computes f. (Though, our polynomial is worse than the one achieved by [GPW15].)

2.2 From ZAM to One-Sided PSM

We move on to study the converse relation. Namely, whether ZAM can be used to derive PSM. For this, we consider a relexation of PSM in which privacy should hold only with respect to 1-inputs. In the randomized encoding literature, this notion is referred to as *semi-private randomized encoding* [AIK04, AIK15]. In the context of PSM protocols we refer to this variant as 1PSM.

Theorem 2. *Let* $f : \{0,1\}^n \times \{0,1\}^n \to \{0,1\}$ *be a function with a* δ-*complete* ZAM *protocol that has communication complexity* ℓ *and randomness complexity* m. *Then, for all* $k \in \mathbb{N}$, *the following hold:*

1. f *has* $(2^{2n}\delta^k)$-*correct and* 0-*private* 1PSM *with communication complexity of* km *and* $2km$ *bits of shared randomness.*
2. f *has* $(2^{2n}\delta^k + 2^{-\ell k})$-*correct and* $(2^{-\ell k})$-*private* 1PSM *with communication complexity of* km *and* $2\ell k$ *bits of shared randomness.*

In particular, if the underlying ZAM protocol has a constant error (e.g., $\delta = 1/2$), we can get a 1PSM with an exponential small error of $\exp(-\Omega(n))$ at the expense of a linear overhead in the complexity, i.e., communication complexity and randomness complexity of $O(nm)$ and $O(\ell n)$, respectively.

Both parts of the theorem are proven by "inverting" the ZAM scheme. That is, as a common randomness Alice and Bob will take a proof p sampled according to the ZAM's accepting distribution. Since each proof forms a rectangle, Alice and Bob can locally sample a random point (r_A, r_B) from p's rectangle (Alice samples r_A and Bob samples r_B). The 1PSM's encoding functions output the sampled point (r_A, r_B). We show that if (x, y) is a 1-input then (r_A, r_B) is distributed uniformly, while in the case of the 0-input the sampled point belongs to some specific set Z that covers only a small fraction of the point space. Therefore, the 1PSM's decoder outputs 0 if the sampled point is in Z and 1, otherwise.

The difference between the two parts of Theorem 2 lies in the way that the common randomness is sampled. In the first part we sample p according to the exact ZAM's accepting distribution, whereas in the second part we compromise on imperfect sampling. This allows us to reduce the length of the shared randomness in 1PSM at the expense of introducing the sampling error in privacy and correctness. The proof of the theorem appears in Sect. 6.

2.3   From 1PSM to PSM and CDS

Theorem 2 shows that a ZAM protocol with low randomness complexity implies communication-efficient 1PSM protocol. However, the latter object is not well-studied and one may suspect that, for one-sided privacy, such low-communication 1PSM protocols may be easily achievable. The following theorem shows that this is unlikely by relating the worst-case communication complexity of 1PSM to the worst-case communication complexity of general PSM (here "worst case" ranges over all functions of given input length).

Theorem 3. *Assume that for all n, each function $f : \{0,1\}^n \times \{0,1\}^n \to \{0,1\}$ has a $\delta(n)$-correct $\varepsilon(n)$-private 1PSM protocol with communication complexity $t(n)$ and randomness complexity $s(n)$. Then, each f has a $[\delta(n) + \delta(t(n))]$-correct $\max(\varepsilon(n), \delta(n) + \varepsilon(t(n)))$-private PSM protocol with communication complexity $t(t(n))$ and randomness complexity $s(n) + s(t(n))$. In particular, if every such f has a 1PSM with polynomial communication and randomness, and negligible privacy and correctness errors, then every f has a PSM with polynomial communication and randomness, and negligible privacy and correctness errors.*

The existence of a PSM for an arbitrary function $f : \{0,1\}^n \times \{0,1\}^n \to \{0,1\}$ with polynomial communication and randomness and negligible privacy and correctness errors is considered to be an important open question in information-theoretic cryptography, and so constructing 1PSM with such parameters would be considered to be a major breakthrough. Together with Theorem 2, we conclude that it will be highly non-trivial to discover randomness-efficient ZAM protocols for general functions.

Finally, we observe that 1PSM protocols yield (almost) directly protocols for *Conditional Disclosure of Secrets* (CDS) [GIKM00]. In this model, Alice holds an input x and Bob holds an input y, and, in addition, both parties hold a common secret bit s. The referee, Carol, holds both x and y, but it does not know the secret s. Similarly to the PSM case, Alice and Bob use shared randomness to compute the messages m_1 and m_2 that are sent to Carol. The CDS requires that Carol can recover s from (m_1, m_2) iff $f(x,y) = 1$. Moving to the complement $\overline{f} = 1 - f$ of f, one can view the CDS model as a variant of 1PSM, in which the privacy leakage in case of 0-inputs is full, i.e., given the messages sent by Alice and Bob, one can recover their input (x, y). Indeed, it is not hard to prove the following observation (whose proof is deferred to the full version).

Theorem 4. *Let f be a function with a δ-complete and ε-private 1PSM that has communication complexity t and randomness complexity s. Then, the function $\overline{f} = 1 - f$ has a δ-complete and ε-private CDS scheme with communication complexity t and randomness complexity s.*

In the full version we also describe a direct transformation from ZAM to CDS which does not suffer from the overhead introduced in Theorem 2. We note that CDS protocols have recently found applications in Attribute-Based Encryption (see [GKW15]).

3   Preliminaries

For an integer $n \in \mathbb{N}$, let $[n] = \{1, \ldots, n\}$. The complement of a bit b is denoted by $\bar{b} = 1 - b$. For a set S, we let S^k be the set of all possible k-tuples with entries in S, and for a distribution D, we let D^k be the probability distribution over k-tuples such that each tuple's element is drawn according to D. We let $s \leftarrow_R S$ denote an element that is sampled uniformly at random from the finite set S. The uniform distribution over n-bit strings is denoted by U_n. For a boolean function $f : S \to \{0, 1\}$, we say that $x \in S$ is 0-input if $f(x) = 0$, and is 1-input if $f(x) = 1$. A subset R of a product set $A \times B$ is a *rectangle* if $R = A' \times B'$ for some $A' \subseteq X$ and $B' \subseteq Y$.

The statistical distance between two random variables, X and Y, denoted by $\Delta(X; Y)$ is defined by $\Delta(X; Y) := \frac{1}{2} \sum_z |\Pr[X = z] - \Pr[Y = z]|$. We will also use statistical distance for probability distributions, where for a probability distribution D the value $\Pr[D = z]$ is defined to be $D(z)$.

We write $\Delta_{x_1 \leftarrow D_1, \ldots, x_k \leftarrow D_k} (F(x_1, \ldots, x_k); G(x_1, \ldots, x_k))$ to denote the statistical distance between two distributions obtained as a result of sampling x_i's from D_i's and applying the functions F and G to $(x_1, \ldots, x_k)$, respectively. We use the following facts about the statistical distance. For every distributions X and Y and a function F (possibly randomized), we have that $\Delta(F(X), F(Y)) \leq \Delta(X, Y)$. In particular, for a boolean function F this implies that $\Pr[F(X) = 1] \leq \Pr[F(Y) = 1] + \Delta(X; Y)$.

For a sequence of probability distributions $(D_1, \ldots, D_k)$ and a probability vector $W = (w_1, \ldots, w_k)$ we let $Z = \sum w_i D_i$ denote the "mixture distribution" obtained by sampling an index $i \in [k]$ according to W and then outputting an element $z \leftarrow D_i$.

Lemma 1. *For any distribution $Z = \sum w_i D_i$ and probability distribution S, it holds that*

$$\Delta(S; M) \leq \sum_{i=1}^{k} w_i \, \Delta(S; D_i).$$

Proof. By the definition of statistical distance we can write $\Delta(S; Z)$ as

$$\frac{1}{2} \sum_z \left| S(z) - \sum_{i=1}^{k} w_i D_i(z) \right| = \frac{1}{2} \sum_z \left| \sum_{i=1}^{k} w_i (S(z) - D_i(z)) \right|$$

$$\leq \frac{1}{2} \sum_z \sum_{i=1}^{k} w_i \, |S(z) - D_i(z)|$$

$$= \frac{1}{2} \sum_{i=1}^{k} w_i \sum_z |S(z) - D_i(z)|$$

$$= \sum_{i=1}^{k} w_i \, \Delta(S; D_i).$$

$\square$

4 Definitions

4.1 PSM-Based Models

Definition 1 (PSM, 1PSM, pPSM). *Let* $f : \{0,1\}^n \times \{0,1\}^n \rightarrow \{0,1\}$ *be a boolean function. We say that a pair of (possibly randomized[3]) encoding algorithms* $F_1, F_2 : \{0,1\}^n \times \{0,1\}^s \rightarrow \{0,1\}^t$ *are* PSM *for* f *if they satisfy the following properties:*

δ-CORRECTNESS: *There exists a deterministic algorithm* Dec, *called decoder, such that for every input* (x,y) *we have that*

$$\Pr_{c \leftarrow_R \{0,1\}^s} [\mathsf{Dec}(F_1(x,c), F_2(y,c)) \neq f(x,y)] \leq \delta.$$

ε-PRIVACY: *There exists a randomized algorithm (simulator)* Sim *such that for any input* (x,y) *it holds that*

$$\Delta_{c \leftarrow_R \{0,1\}^s} (\mathsf{Sim}(f(x,y)); (F_1(x,c), F_2(y,c))) \leq \varepsilon,$$

where we write $\Delta_{x_1 \leftarrow D_1,\ldots,x_k \leftarrow D_k}(F(x_1,\ldots,x_k); G(x_1,\ldots,x_k))$ *to denote the statistical distance between two distributions obtained as a result of sampling* x_i*'s from* D_i*'s and applying the functions* F *and* G *to* $(x_1,\ldots,x_k)$*, respectively.*

If privacy holds only on 1-inputs then the protocol is referred to as 1PSM. *A* pPSM *protocol is a* PSM *which satisfies* 0-correctness, (standard) 0-privacy, and, in addition, satisfies the following properties:*

BALANCE: *There exists a* 0-private (perfectly private) simulator Sim *such that* $\mathsf{Sim}(U_1) \equiv U_{2t}$.

STRETCH-PRESERVATION: *We have that* $1 + s = 2t$, *i.e., the total output length equals to the randomness complexity plus a single bit.[4]*

The communication complexity of the PSM *(resp., 1PSM, pPSM) protocol is defined as the encoding length* t, *and the randomness complexity of the protocol is defined as the length* s *of the common randomness.*

Remark 1 (pPSM– combinatorial view). One can also formulate the pPSM definition combinatorially [AIK04]: For f's b-input (x,y), let $F_{xy}(c)$ denote the joint output of the encoding $(F_1(x,c), F_2(y,c))$. Let $S_b := \{F_{xy}(c) \mid c \in \{0,1\}^s, (x,y) \in f^{-1}(b)\}$ and let $R = \{0,1\}^t \times \{0,1\}^t$ denote the joint range of (F_1, F_2). Then, (F_1, F_2) is a pPSM of f if and only if (1) The 0-image S_0 and the 1-image S_1 are disjoint; (2) The union of S_0 and S_1 equals to the range R; and (3) for

---

[3] In the original paper [FKN94], the functions F_1, F_2 are deterministic. We extend this model by allowing Alice and Bob to use local randomness that is assumed to be available freely.

[4] Intuitively, this bit carries the outcome of the function.

all (x, y) the function F_{xy} is a bijection on $S_{f(x,y)}$. One can also consider a case when F_1 and F_2 have arbitrary ranges, i.e., $F_i : \{0,1\}^n \times \{0,1\}^s \to \{0,1\}^{t_i}$. In this case we say that (F_1, F_2) is a pPSM of f if the above conditions hold with respect to the joint range $R = \{0,1\}^{t_1} \times \{0,1\}^{t_2}$.

4.2   ZAM

Definition 2 (ZAM). *Let* $f : \{0,1\}^n \times \{0,1\}^n \to \{0,1\}$. *We say that a pair of deterministic boolean functions* $A, B : \{0,1\}^n \times \{0,1\}^m \times \{0,1\}^\ell \to \{0,1\}$ *is a* ZAM *for* f *if it satisfies the following properties:*

UNAMBIGUOUS COMPLETENESS: *For any 1-input* (x, y) *and any randomness* $(r_A, r_B) \in \{0,1\}^m \times \{0,1\}^m$ *there exists a unique* $p \in \{0,1\}^\ell$ *such that* $A(x, r_A, p) = 1 = B(y, r_B, p)$.

ZERO INFORMATION: *There exists a distribution* D *on the proof space* $\{0,1\}^\ell$ *such that for any 1-input* (x, y) *we have that*

$$\forall p \in \{0,1\}^\ell \; D(p) = \Pr_{r_A, r_B \leftarrow_R \{0,1\}^m}[A(x, r_A, p) = 1 = B(y, r_B, p)].$$

The distribution D *is called the* accepting distribution.

δ-SOUNDNESS: *For any 0-input* (x, y) *it holds that*

$$\Pr_{r_A, r_B \leftarrow_R \{0,1\}^m}[\exists p \in \{0,1\}^\ell : A(x, r_A, p) = 1 = B(y, r_B, p)] \le \delta.$$

The communication complexity (resp., randomness complexity) of the ZAM *protocol is defined as the length* ℓ *of the proof (resp., the length* m *of the local randomness).*

The Zero Information property asserts that for every accepting input (x, y) the distribution $D_{x,y}$, obtained by sampling r_A and r_B and outputting the (unique) proof p which is accepted by Alice and Bob, is identical to a single universal distribution D.

Following [GPW15], we sometimes refer to the proofs as "rectangles" because for each (x, y) a proof p naturally corresponds to a set of points $\{(r_A, r_B) : A(x, r_A, p) = 1 = B(y, r_B, p)\}$ which forms a rectangle in $\{0,1\}^m \times \{0,1\}^m$.

5   From pPSM to ZAM

In this section we construct a ZAM scheme from a pPSM protocol. By exploiting the combinatorial structure of pPSM, for each input (x, y) we construct a function h_{xy} that is a bijection if (x, y) is a 1-input and is two-to-one if (x, y) is a 0-input. In the constructed ZAM scheme Alice and Bob use their local randomness to sample a uniform point in h's range (Alice samples its x-coordinate r_A and Bob samples its y-coordinate r_B). Merlin's proof is the preimage p for the sampled point, i.e., a point p such that $h_{xy}(p) = (r_A, r_B)$. In order to accept the proof p, Alice and Bob verify that it is a preimage for the sampled point (r_A, r_B).

First, the constructed ZAM is unambiguously complete because h_{xy} is a bijection if (x, y) is a 1-input of f. Second, the constructed ZAM satisfies the zero-information property because the distribution of the accepted proofs is uniform. Third, the constructed ZAM is sound, because if (x, y) is a 0-input, then h_{xy} is two-to-one, implying that with probability at least $1/2$ no preimage can be found.

Theorem 1. *Let f be a function with a* pPSM *protocol that has communication complexity t and randomness complexity s. Then f has a $1/2$-sound* ZAM *scheme with randomness complexity of t and communication complexity of $s + 1$.*

Proof. Let $f : \{0,1\}^n \times \{0,1\}^n \to \{0,1\}$ be a function with a pPSM $F_1, F_2 : \{0,1\}^n \times \{0,1\}^s \to \{0,1\}^t$. We show that there exists a $1/2$-sound ZAM protocol for f with Alice's and Bob's local randomness spaces $\{0,1\}^m$ and proof space $\{0,1\}^\ell$, where $m = t$ and $\ell = 2t$.

First, we prove some auxiliary statement about pPSM. Let $g(x, y, c) := (F_1(x, c), F_2(y, c))$. For any (x, y), we define a new function $h_{xy} : \{0,1\}^s \times \{0,1\} \to \{0,1\}^t \times \{0,1\}^t$ as follows.

$$h_{xy}(c, b) := \begin{cases} g(x, y, c), \text{if } b = 0; \\ g(x_0, y_0, c), \text{if } b = 1 \text{ (where } (x_0, y_0) \text{ is a canonical } 0 - \text{input for} f). \end{cases}$$

The function h satisfies the following useful properties as follows from the combinatorial view of pPSM (Remark 1).

Fact 1. *If (x, y) is a 1-input for f, then the function h_{xy} is a bijection. Otherwise, if (x, y) is a 0-input for f, then the image of the function h_{xy} covers exactly half of the range $\{0,1\}^t \times \{0,1\}^t$.*

We now describe a ZAM protocol for f in which the local randomness of Alice and Bob is sampled from $\{0,1\}^t$, and the proof space is $\{0,1\}^s \times \{0,1\}$. Recall that (F_1, F_2) is a pPSM and therefore $s + 1 = 2t$ and $\{0,1\}^s \times \{0,1\} = \{0,1\}^{2t}$. The ZAM's accepting functions A, B are defined as follows:

$$A(x, m_1, (c, b)) = \begin{cases} 1, \text{if } (m_1 = F_1(x, c) \text{ and } b = 0) \text{ or} \\ \quad (m_1 = F_1(x_0, c) \text{ and } b = 1); \\ 0, \text{otherwise.} \end{cases}$$

$$B(y, m_2, (c, b)) = \begin{cases} 1, \text{if } (m_2 = F_2(y, c) \text{ and } b = 0) \text{ or} \\ \quad (m_2 = F_2(y_0, c) \text{ and } b = 1); \\ 0, \text{otherwise.} \end{cases}$$

Observe that the following equivalence holds.

Claim. $\forall x, y, c, b, m_1, m_2 \left[h_{xy}(c, b) = (m_1, m_2) \right] \Leftrightarrow \left[A(x, m_1, (c, b)) = 1 = B(y, m_2, (c, b)) \right].$

Now we verify that A, B is ZAM for f:

UNAMBIGUOUS COMPLETENESS: Consider any f's 1-input (x, y) and take any $(m_1, m_2) \in \{0, 1\}^t \times \{0, 1\}^t$. Since (x, y) is a 1-input for f, we have that h_{xy} is a bijection. This means that there exists a unique (c, b) such that $h_{xy}(c, b) = (m_1, m_2)$. By Claim 5, this proof (c, b) is the only proof which is accepted by both Alice and Bob when the randomness is set to m_1, m_2.

ZERO INFORMATION: We show that the accepting distribution is uniform, i.e., for any 1-input (x, y) and for any $p \in \{0, 1\}^s \times \{0, 1\}$ it holds that

$$\Pr_{r_A, r_B \leftarrow_R \{0,1\}^t} [A(x, r_A, p) = 1 = B(y, r_B, p)] = 2^{-2t}.$$

Take any 1-input (x, y). Since (x, y) is a 1-input for f, we have that h_{xy} is a bijection. Hence, there exists a unique $(m_1^*, m_2^*) \in \{0, 1\}^n \times \{0, 1\}^n$ such that $h_{xy}(c, b) = (m_1^*, m_2^*)$. By Claim 5, this means that Alice and Bob accept only this (m_1^*, m_2^*). Hence, for all proofs p we have that

$$\Pr_{r_A, r_B \leftarrow_R \{0,1\}^t} [A(x, r_A, p) = 1 = B(y, r_B, p)] =$$
$$\Pr_{r_A, r_B \leftarrow_R \{0,1\}^t} [r_A = m_1^*, r_B = m_2^*] = 2^{-2t}.$$

1/2-SOUNDNESS: Fix some 0-input (x, y), and recall that the image H of h_{xy} covers exactly half of the range $\{0, 1\}^t \times \{0, 1\}^t$, i.e., $|H| = \left| \{0, 1\}^t \times \{0, 1\}^t \right| / 2$. It follows that, with probability $1/2$, the randomness of Alice and Bob (m_1, m_2) chosen randomly from $\{0, 1\}^t \times \{0, 1\}^t$ lands outside H. In this case, the set $h_{xy}^{-1}(m_1, m_2)$ is empty and so there is no proof (c, b) that will be accepted.

$\square$

6   From ZAM to 1PSM

In this section we construct 1PSM protocols from a ZAM scheme and prove Theorem 2 (restated here for convenience).

Theorem 2. *Let* $f : \{0, 1\}^n \times \{0, 1\}^n \to \{0, 1\}$ *be a function with a δ-complete* ZAM *protocol that has communication complexity ℓ and randomness complexity m. Then, for all $k \in \mathbb{N}$, the following hold:*

1. *f has $(2^{2n} \delta^k)$-correct and 0-private 1PSM with communication complexity of km and $2km$ bits of shared randomness.*
2. *f has $(2^{2n} \delta^k + 2^{-\ell k})$-correct and $(2^{-\ell k})$-private 1PSM with communication complexity of km and $2\ell k$ bits of shared randomness.*

Proof. Let $f : \{0,1\}^n \times \{0,1\}^n \to \{0,1\}$ be a function with a δ-sound ZAM protocol (A, B) with Alice's and Bob's local randomness spaces $\{0,1\}^m$ and the proof space $\{0,1\}^\ell$. Fix some integer k. We start by constructing the first 1PSM protocol.

We first define some additional notation and prove auxiliary claims. For a pair of inputs (x, y) let

$$E_{xy} := \{(r_A, r_B) \in \{0,1\}^m \times \{0,1\}^m \mid \exists p : A(x, r_A, p) = 1 = B(y, r_B, p)\}$$

and $Z := \bigcup_{(x,y) \in f^{-1}(0)} E_{xy}^k.$

Claim. $|Z| \leq 2^{2n}(\delta 2^{2m})^k.$

Proof. By the soundness property of ZAM, we have that $|E_{xy}| \leq \delta 2^{2m}$ for any 0-input (x, y). Hence, each $|E_{xy}^k| \leq (\delta 2^{2m})^k$. We conclude that

$$|Z| = \left| \bigcup_{(x,y) \in f^{-1}(0)} E_{xy}^k \right| \leq \sum_{(x,y) \in f^{-1}(0)} |E_{xy}^k| \leq 2^{2n}(\delta 2^{2m})^k = \delta^k 2^{2n+2mk}.$$

$\square$

Let $\mathcal{A}_p^x := \{r_A \in \{0,1\}^m \mid A(x, r_A, p) = 1\}$ and $\mathcal{B}_p^y := \{r_B \in \{0,1\}^m \mid B(y, r_B, p) = 1\}.$

Claim. Let D_{ACC} be the accepting distribution of ZAM. Then, for any 1-input (x, y) and $p \in \{0,1\}^\ell$ we have that $D_{\mathrm{ACC}}(p) = 2^{-2m}|\mathcal{A}_p^x||\mathcal{B}_p^y|.$

Proof. By definition

$$D_{\mathrm{ACC}}(p) = \frac{|\{(r_A, r_B) \in \{0,1\}^m \times \{0,1\}^m \mid A(x, r_A, p) = 1 = B(y, r_B, p)\}|}{|\{0,1\}^m| \cdot |\{0,1\}^m|}.$$

In order to derive the claim, it remains to notice that since every proof forms a "rectangle" [GPW15], we have that

$$\{(r_A, r_B) \in \{0,1\}^m \times \{0,1\}^m \mid A(x, r_A, p) = 1 = B(y, r_B, p)\} = \mathcal{A}_p^x \times \mathcal{B}_p^y.$$

$\square$

We can now describe the encoding algorithms G_1 and G_2 and the decoder Dec. First, G_1 and G_2 use the shared randomness to sample a proof p according to the accepting distribution. Then G_1 and G_2 sample (private) randomness that can lead to the acceptance of p on their input (x, y), i.e., G_1 computes $a \leftarrow_R \mathcal{A}_p^x$ and G_2 computes $b \leftarrow_R \mathcal{B}_p^y$. We have that if $f(x, y) = 1$ then (a, b) is distributed uniformly, while if $f(x, y) = 0$ then (a, b) is sampled from the set Z. The task of the decoder is to verify whether it is likely that a point has been sampled from Z or uniformly. This is achieved by repeating the protocol k times. Below is the formal description of the algorithms G_1, G_2 and decoder.

- **Shared Randomness.** The common randomness $c \in \{0,1\}^{k \cdot 2m}$ is used for sampling k independent samples $(p_1, \ldots, p_k)$ from D_{ACC}. (Each such sample can be obtained by sampling $r = (r_A, r_B) \leftarrow_R \{0,1\}^{2m}$ and outputting the unique proof p that corresponds to r and to some fixed 1-input (x_0, y_0).)
- **Encoders.** The encoder $G_1(x, c)$ outputs $(a_1, \ldots, a_k) \leftarrow_R \mathcal{A}_{p_1}^x \times \cdots \times \mathcal{A}_{p_k}^x$ and the encoder G_2 outputs $(b_1, \ldots, b_k) \leftarrow_R \mathcal{B}_{p_1}^y \times \cdots \times \mathcal{B}_{p_k}^x$.
- **Decoder.** $\mathsf{Dec}((a_1, \ldots, a_k), (b_1, \ldots, b_k))$
  If $((a_1, b_1), \ldots, (a_k, b_k)) \in Z$ then output 0, otherwise output 1.

Let us verify that the proposed protocol is a 1PSM for f.

$(2^{2n}\delta^k)$-**Correctness.** Since that the decoder never errs on 0-inputs, it suffices to analyze the probability that some 1-input (x, y) is incorrectly decoded to 0. Fix some 1-input (x, y). Below we will show that the message $s = ((a_1, b_1), \ldots, (a_k, b_k))$ generated by the encoders G_1 and G_2 is uniformly distributed over the set $(\{0,1\}^m \times \{0,1\}^m)^k$. Hence, the probability that s lands in Z (and decoded incorrectly to 0) is exactly $\frac{|Z|}{|(\{0,1\}^m \times \{0,1\}^m)^k|}$, which, by Claim 6, is upper-bounded by $2^{2n}\delta^k$.

It is left to show that s is uniformly distributed. To see this, consider the marginalization of (a_i, b_i)'s probability distribution: For a fixed (r_A, r_B) we have that

$$\Pr[(a_i, b_i) = (r_A, r_B)] = \sum_{p \in \{0,1\}^\ell} \Pr[(a_i, b_i) = (r_A, r_B) \mid p_i = p] \Pr[p_i = p].$$

Because of the unambiguous completeness property of ZAM, we have that there exists a single p^* such that $(r_A, r_B) \in \mathcal{A}_{p^*}^x \times \mathcal{B}_{p^*}^y$. Hence, all probabilities $\Pr[(a_i, b_i) = (r_A, r_B) \mid p_i = p]$ are zero, if $p \neq p^*$. This implies that

$$\Pr[(a_i, b_i) = (r_A, r_B)] = \Pr[(a_i, b_i) = (r_A, r_B) \mid p_i = p^*] \Pr[p_i = p^*].$$

We have that $\Pr[p_i = p] = D_{\mathrm{ACC}}(p) = 2^{-2m}|\mathcal{A}_p^x||\mathcal{B}_p^y|$ (due to Claim 6), and $\Pr[(a_i, b_i) = (r_A, r_B) \mid p_i = p^*]$ is $\frac{1}{|\mathcal{A}_p^x| \cdot |\mathcal{B}_p^y|}$ by the construction of the encoding functions. Hence, $\Pr[(a_i, b_i) = (r_A, r_B)] = 2^{-2m}$. Because all pairs (a_i, b_i) are sampled independently, we get that the combined tuple $s = ((a_1, b_1), \ldots, (a_k, b_k))$ is sampled uniformly from $(\{0,1\}^m \times \{0,1\}^m)^k$, as required.

Privacy for 1-inputs. As shown above, if (x, y) is a 1-input, then s is uniformly distributed over $(\{0,1\}^m \times \{0,1\}^m)^k$. Hence, the simulator for proving the privacy property of PSM can be defined as a uniform sampler from $(\{0,1\}^m \times \{0,1\}^m)^k$.

The Second Protocol. The second item of the theorem is proved by using the first protocol, except that the point $\boldsymbol{p} = (p_1, \ldots, p_k)$ is sampled from a different distribution D'. For a parameter t, the distribution D' is simply the distribution D_{ACC}^k discretized into $2^{-(\ell k + t)}$-size intervals. Such D' can be sampled using only $\ell k + t$ random bits. Moreover, for each point $\boldsymbol{p}$, the difference between $D_{\mathrm{ACC}}^k(\boldsymbol{p})$ and $D'(\boldsymbol{p})$ is at most $2^{-(\ell k + t)}$. Since the support of D_{ACC}^k is of size at most $2^{\ell k}$, it follows that $\Delta(S(U_{\ell k + t}); D_{\mathrm{ACC}}^k) \leq 2^{-(\ell k + t)} \cdot 2^{\ell k} = 2^{-t}$. As a result, we introduce an additional error of 2^{-t} in both privacy and correctness. By setting t to ℓk, we derive the second 1PSM protocol. $\qquad\square$

7 From 1PSM to PSM

In this section we show how to upgrade a 1PSM protocol into a PSM protocol. We assume that we have a way of constructing 1PSM for all functions. Our main idea is to reduce a construction of a PSM scheme for f to two 1PSM schemes. The first 1PSM scheme computes the function f, and the second 1PSM scheme computes the function $\overline{\mathsf{Dec}_f}$, i.e., the complement of the decoder Dec_f of the first scheme. We show how to combine the two schemes such that the first scheme protects the privacy of 1-inputs and the second scheme protects the privacy of 0-inputs.

Theorem 3. *Assume that for all n, each function $f : \{0,1\}^n \times \{0,1\}^n \to \{0,1\}$ has a $\delta(n)$-correct $\varepsilon(n)$-private 1PSM protocol with communication complexity $t(n)$ and randomness complexity $s(n)$. Then, each f has a $[\delta(n) + \delta(t(n))]$-correct $\max(\varepsilon(n), \delta(n) + \varepsilon(t(n)))$-private PSM protocol with communication complexity $t(t(n))$ and randomness complexity $s(n) + s(t(n))$. In particular, if every such f has a 1PSM with polynomial communication and randomness, and negligible privacy and correctness errors, then every f has a PSM with polynomial communication and randomness, and negligible privacy and correctness errors.*

Proof. Let $f : \{0,1\}^n \times \{0,1\}^n \to \{0,1\}$. Let $F_1, F_2 : \{0,1\}^n \times \{0,1\}^{s(n)} \to \{0,1\}^{t(n)}$ be a $\delta(n)$-correct and $\varepsilon(n)$-private on 1 inputs 1PSM for f with decoder Dec_f and simulator Sim_f. Define a function $g : \{0,1\}^{t(n)} \times \{0,1\}^{t(n)} \to \{0,1\}$ to be $1 - \mathsf{Dec}_f(m_1, m_2)$. Let $G_1, G_2 : \{0,1\}^{t(n)} \times \{0,1\}^{s(t(n))} \to \{0,1\}^{t(t(n))}$ be a $\delta(t(n))$-correct and $\varepsilon(t(n))$-private on 1 inputs 1PSM for g with decoder Dec_g and simulator Sim_g.

We construct a (standard) PSM for f as follows. Let $\{0,1\}^u = \{0,1\}^{s(n)} \times \{0,1\}^{s(t(n))}$ be the space of shared randomness, let $\{0,1\}^v = \{0,1\}^{t(t(n))}$ be the output space and define the encoding functions $H_1, H_2 : \{0,1\}^n \times \{0,1\}^u \to \{0,1\}^v$, by

$$H_1(x, (c,r)) = G_1(F_1(x,c), r) \quad \text{and} \quad H_2(y, (c,r)) = G_2(F_2(y,c), r).$$

We show that H_1, H_2 satisfy the security properties of PSM:

$\delta(n) + \delta(t(n))$-CORRECTNESS: On an input (e_1, e_2) define the decoding algorithm Dec to output $1 - \text{Dec}_g(e_1, e_2)$. The decoding algorithm Dec works correctly whenever both Dec_g and Dec_f succeed. Hence, the error probability for decoding can be bounded as follows:

$$\Pr_{(c,r) \leftarrow_R \{0,1\}^u}[\text{Dec}(H_1(x, (c,r)), H_2(y, (c,r))) \neq f(x,y)]$$

$$= \Pr_{(c,r) \leftarrow_R \{0,1\}^u}[1 - \text{Dec}_g(G_1(F_1(x,c), r)), G_2(F_2(y,c), r))) \neq f(x,y)]$$

$$\leq \Pr_{c \leftarrow_R \{0,1\}^{s(n)}}[1 - (1 - (\text{Dec}_f(F_1(x,c), F_2(y,c)))) \neq f(x,y)] + \delta(t(n))$$

$$= \Pr_{c \leftarrow_R \{0,1\}^{s(n)}}[\text{Dec}_f(F_1(x,c), F_2(y,c)) \neq f(x,y)] + \delta(t(n))$$

$$\leq \delta(n) + \delta(t(n)).$$

ε-PRIVACY: We define the simulator Sim as follows: on 0-inputs it outputs Sim_g and on 1-inputs it computes $\text{Sim}_f = (m_1, m_2)$, randomly samples r from $\{0,1\}^{s(t(n))}$, and outputs $(G_1(m_1, r), G_2(m_2, r))$. We verify that the simulator truthfully simulates the randomized encoding (H_1, H_2) with deviation error of at most ε.

We begin with the case where (x, y) is a 0-input for f. For any c, let L_c denote the distribution of the random variable $(G_1(F_1(x,c), r), G_2(F_2(y,c), r))$ where $r \leftarrow_R \{0,1\}^{s(t(n))}$. Let M denote the "mixture distribution" which is defined by first sampling $c \leftarrow_R \{0,1\}^{s(n)}$ and then outputting a random sample from L_c, that is, the distribution $M = \sum_{c \in \{0,1\}^{s(n)}} \Pr[U_{s(n)} = c] L_c$. Due to Lemma 1, we have that

$$\Delta(\text{Sim}_g; M) \leq \sum_{c \in \{0,1\}^{s(n)}} \Pr[U_{s(n)} = c] \Delta(\text{Sim}_g; L_c).$$

Let C denote a subset of $c \in \{0,1\}^{s(n)}$ such that $(F_1(x,c), F_2(y,c))$ is a 1-input for g. The set C satisfies the following two properties: (1) $\forall c \in C \Delta(\text{Sim}_g; L_c) \leq \varepsilon(t(n))$ and (2) $|C|/2^{s(n)} \geq 1 - \delta(n)$. The property (1) holds because G_1, G_2 is private on 1-inputs of g. The property (2) holds because Dec_f decodes correctly with the probability at least $1 - \delta(n)$. After splitting the mixture sum in two, we have that

$$\sum_{c \in \{0,1\}^{s(n)}} \Pr[U_{s(n)} = c] \Delta(\text{Sim}_g; L_c) = \sum_{c \in C} 2^{-s(n)} \Delta(\text{Sim}_g; L_c)$$

$$+ \sum_{c \notin C} 2^{-s(n)} \Delta(\text{Sim}_g; L_c).$$

Because of the properties of C, we have that the first sum is upperbounded by $\varepsilon(t(n))$ and the second one is upperbounded by $\delta(n)$. This implies that $\Delta(\text{Sim}_g; M) \leq \delta(n) + \varepsilon(t(n))$.

We move on to the case where (x, y) is a 1-input. Then

$$\mathop{\Delta}_{c \leftarrow_R \{0,1\}^{s(n)}} (\mathsf{Sim}_f \; ; \; (F_1(x, c), F_2(y, c))) \leq \varepsilon(n).$$

Consider the randomized procedure G which, given (m_1, m_2), samples $r \leftarrow_R \{0,1\}^{s(t(n))}$ and outputs the pair $(G_1(m_1, r), G_2(m_2, r))$. Applying G to the above distributions we get:

$$\mathop{\Delta}_{(c,r) \leftarrow_R \{0,1\}^u} (G(\mathsf{Sim}_f; r); \; G(F_1(x, c), F_2(y, c); r)) \leq \varepsilon(n). \qquad (1)$$

Recall that, for a random $r \leftarrow_R \{0,1\}^{s(t(n))}$, it holds that $G(\mathsf{Sim}_f; r) \equiv \mathsf{Sim}(1)$, and for every r, $G(F_1(x, c), F_2(y, c); r) = (H_1(x, (c, r)), H_2(y, (c, r)))$. Hence, Eq. 1 can be written as

$$\mathop{\Delta}_{(c,r) \leftarrow_R \{0,1\}^u} (\mathsf{Sim}(1); \; (H_1(x, (c, r)), H_2(y, (c, r)))) \leq \varepsilon(n).$$

Since $\varepsilon(n) \leq \max(\varepsilon(n), \delta(n) + \varepsilon(t(n)))$, the theorem follows.

□

References

[AIK04] Applebaum, B., Ishai, Y., Kushilevitz, E.: Cryptography in NC^0. In: Proceedings of 45th Symposium on Foundations of Computer Science (FOCS 2004), pp. 166–175. IEEE Computer Society, Rome, Italy, 17–19 October 2004

[AIK15] Applebaum, B., Ishai, Y., Kushilevitz, E.: Minimizing locality of one-way functions via semi-private randomized encodings. Electron. Colloq. Comput. Complex. (ECCC) **22**, 45 (2015)

[BFS86] Babai, L., Frankl, P., Simon, J.: Complexity classes in communication complexity theory (preliminary version). In: 27th Annual Symposium on Foundations of Computer Science, pp. 337–347. IEEE Computer Society, Toronto, Canada, 27–29 October 1986

[BIKK14] Beimel, A., Ishai, Y., Kumaresan, R., Kushilevitz, E.: On the cryptographic complexity of the worst functions. In: Lindell, Y. (ed.) TCC 2014. LNCS, vol. 8349, pp. 317–342. Springer, Heidelberg (2014)

[BM88] Babai, L., Moran, S.: Arthur-merlin games: a randomized proof system, and a hierarchy of complexity classes. J. Comput. Syst. Sci. **36**(2), 254–276 (1988)

[FKN94] Feige, U., Kilian, J., Naor, M.: A minimal model for secure computation (extended abstract). In: Leighton, F.T., Goodrich, M.T. (eds.) Proceedings of the Twenty-Sixth Annual ACM Symposium on Theory of Computing, pp. 554–563. ACM, Montréal, Québec, Canada, 23–25 May 1994

[GIKM00] Gertner, Y., Ishai, Y., Kushilevitz, E., Malkin, T.: Protecting data privacy in private information retrieval schemes. J. Comput. Syst. Sci. **60**(3), 592–629 (2000)

[GKW15] Gay, R., Kerenidis, I., Wee, H.: Communication complexity of conditional disclosure of secrets and attribute-based encryption. In: Gennaro, R., Robshaw, M. (eds.) CRYPTO 2015, Part II. LNCS, vol. 9216, pp. 485–502. Springer, Heidelberg (2015)

[GPW15] Göös, M., Pitassi, T., Watson, T.: Zero-information protocols and unambiguity in arthur-merlin communication. In: Roughgarden, T. (ed.) Proceedings of the 2015 Conference on Innovations in Theoretical Computer Science, ITCS 2015, pp. 113–122. ACM, Rehovot, Israel, 11–13 January 2015

[IK97] Ishai, Y., Kushilevitz, E.: Private simultaneous messages protocols with applications. In: Proceedings of the 5th Israeli Symposium on Theory of Computing and Systems, pp. 174–183, June 1997

[IK00] Ishai, Y., Kushilevitz, E.: Randomizing polynomials: a new representation with applications to round-efficient secure computation. In: 41st Annual Symposium on Foundations of Computer Science, FOCS 2000, pp. 294–304. IEEE Computer Society, Redondo Beach, California, USA, 12–14 November 2000

[Ish13] Ishai, Y.: Randomization techniques for secure computation. In: Prabhakaran, M., Sahai, A., (eds.) Secure Multi-Party Computation, vol. 10 of Cryptology and Information Security Series, pp. 222–248. IOS Press (2013)

[Kla03] Klauck, H.: Rectangle size bounds and threshold covers in communication complexity. In: 18th Annual IEEE Conference on Computational Complexity (Complexity 2003), pp. 118–134. IEEE Computer Society, Aarhus, Denmark, 7–10 July 2003

[Kla10] Klauck, H.: A strong direct product theorem for disjointness. In: Schulman, L.J. (ed.) Proceedings of the 42nd ACM Symposium on Theory of Computing, STOC 2010, pp. 77–86. ACM, Cambridge, Massachusetts, USA, 5–8 June 2010

A Transform for NIZK Almost as Efficient and General as the Fiat-Shamir Transform Without Programmable Random Oracles

Michele Ciampi[1]([✉]), Giuseppe Persiano[2], Luisa Siniscalchi[1], and Ivan Visconti[1]

[1] DIEM, University of Salerno, Salerno, Italy
{mciampi,lsiniscalchi,visconti}@unisa.it
[2] DISA-MIS, University of Salerno, Salerno, Italy
giuper@gmail.com

Abstract. The Fiat-Shamir (FS) transform is a popular technique for obtaining practical zero-knowledge argument systems. The FS transform uses a hash function to generate, without any further overhead, non-interactive zero-knowledge (NIZK) argument systems from public-coin honest-verifier zero-knowledge (public-coin HVZK) proof systems. In the proof of zero knowledge, the hash function is modeled as a *programmable* random oracle (PRO).

In TCC 2015, Lindell embarked on the challenging task of obtaining a similar transform with improved heuristic security. Lindell showed that, for several interesting and practical languages, there exists an efficient transform in the *non-programmable* random oracle (NPRO) model that also uses a common reference string (CRS). A major contribution of Lindell's transform is that zero knowledge is proved without random oracles and this is an important step towards achieving efficient NIZK arguments in the CRS model without random oracles.

In this work, we analyze the efficiency and generality of Lindell's transform and notice a significant gap when compared with the FS transform. We then propose a new transform that aims at filling this gap. Indeed our transform is almost as efficient as the FS transform and can be applied to a broad class of public-coin HVZK proof systems. Our transform requires a CRS and an NPRO in the proof of soundness, similarly to Lindell's transform.

1 Introduction

Non-interactive zero-knowledge (NIZK) proofs[1] introduced in [5,6,24] are widely used in Cryptography. Such proofs allow a prover to convince a verifier with just one message about the membership of an instance x in a language L without leaking any additional information. NIZK proofs are not possible without a setup

---

[1] When discussing informally we will use the word proof to mean both an unconditionally sound proof and a computationally sound proof (i.e., an argument). Only in the more formal part of the paper we will make a distinction between arguments and proofs.

© International Association for Cryptologic Research 2016
E. Kushilevitz and T. Malkin (Eds.): TCC 2016-A, Part II, LNCS 9563, pp. 83–111, 2016.
DOI: 10.1007/978-3-662-49099-0_4

assumption and the one proposed initially in [5] is the existence of a *Common Reference String* (CRS) received as input both by the prover and the verifier. The CRS model has been the standard setup for NIZK in the last 25 years. Another setup that has been proposed in literature is the existence of registered public keys in [2,13,21].

Starting with the breakthrough of [29,30] we know that NIZK proofs in the CRS model exist for any NP language with the additional appealing feature of using just one CRS for any polynomial number of proofs. Moreover NIZK proofs and their stronger variations [23,39,48] have been shown to be not only interesting for their original goal of being a non-interactive version of classic zero-knowledge (ZK) proofs [36,37], but also because they are powerful building blocks in many applications (e.g., for CCA encryption [45], ZAPs [27,28]).

Efficient NIZK. Generic constructions of NIZK proofs are rather inefficient since they require to first compute an NP reduction and then to apply the NIZK proof for a given NP-complete language to the instance output by the reduction. A significant progress in efficiency has been proposed in [40] where several techniques have been proposed to obtain efficient NIZK proofs that can be used in bilinear groups.

The most popular use of NIZK proofs in real-world scenarios consists in taking an efficient *interactive* public-coin honest-verifier zero-knowledge (HVZK) proof system and in making it a NIZK argument through the so called *Fiat-Shamir (FS) transform* [31]. The FS transform replaces the verifier by calls to a hash function on input the transcript so far. In the random oracle [3] (RO) model the hash function can only be evaluated through calls to an oracle that answers as a random function. The security proof allows the simulator for HVZK to program the RO (i.e., the simulator decides how to answer to a query) and this allows to convert the entire transcript of a public-coin HVZK proof into a single message that is indistinguishable from the single message computed by a honest NIZK prover. The efficiency of the FS transform led to many practical applications. The transform is also a method to obtain signatures of knowledge, as discussed in [14].

The main disadvantage of the FS transform is the fact that the random oracle methodology has been proved to be unsound both in general [7] and both for the specific case [4,35] of turning identification schemes into signatures as considered in [31]. Nevertheless, the examples of constructions proved secure in the RO model and insecure for any concrete hash function are seemingly artificial while no natural construction has been successfully attacked yet. Therefore the RO methodology remains widely used in practice.

The FS transform applied to 3-round HVZK proofs is one of the major uses of the RO model for real-world protocols, therefore any progress in this research direction (either on the security of the transform, or on its efficiency, or on its generality) is of extreme interest.

In [38] Groth showed an efficient transform for NIZK where soundness is proved requiring a programmable RO while no random oracle is needed to prove zero knowledge.

Efficient NIZK with Designated/Registered Verifiers. A first attempt to get efficient NIZK arguments from some restricted class of 3-round public-coin HVZK proofs without ROs was done by [21] (the proof of soundness required complexity leveraging) and later on by [13] that achieved a weaker form of soundness in the registered public-key model. The limitation of this model is that a NIZK proof can be verified only by a designated verifier (i.e., the proof requires a secret known to the verifier). Moreover there is an inconvenient preliminary registration phase where the verifier has to register her public key.

Lindell's Transform. Very recently, in [43], Lindell proposed a very interesting transform that can be seen as an attempt towards obtaining efficient constructions without random oracles. Starting from a Σ-protocol for a language L (i.e., a special type of 3-round public-coin HVZK proof used already in several efficient constructions of zero knowledge [1, 10, 19, 25, 44, 46, 49, 51, 54]), Lindell shows how to construct an efficient NIZK[2] argument system for L in the CRS model. Two are the major advantages of Lindell's transform with respect to the FS transform. First, in Lindell's transform the proof of ZK does not need the existence of a random oracle and this allows to avoid some issues due to protocol composition [52]. We remark that the proof of ZK for Lindell's transform needs a CRS but this is unavoidable as one-round ZK in the plain model is possible only for trivial languages. Second, the soundness of Lindell's transform can be proved by relying on a *non-programmable random oracle* (NPRO). An NPRO is a RO that in the protocol and in the security proofs can be used only as a black box and can not be programmed by a simulator or by the adversary of a reduction. This is a considerable advantage compared to the FS transform since replacing a RO by an NPRO is a step towards removing completely the need of ROs in a cryptographic construction. Indeed the work of Lindell goes precisely in the direction of solving a major open problem in Cryptography: obtaining an efficient RO-free transform for NIZK arguments to be used in place of the FS transform.

The main drawback of Lindell's transform is that it requires extra computation on top of the one needed to run the Σ-protocol for the language L. In contrast, the FS transform does not incur into any overhead on top of a 3-round public-coin HVZK proof for L. In addition, since 3-round public-coin HVZK proofs are potentially less demanding than Σ-protocols, we have that requiring a Σ-protocol as starting protocol for a transform instead of a public-coin HVZK proof may already result in an efficiency loss.

Lindell's transform is based on a primitive named *dual-mode* (DM) commitment scheme (DMCS). A DMCS is based on a membership-hard language Λ and each specific commitment takes as input an instance ρ of Λ and has the following property: if $\rho \notin \Lambda$, the DM commitment is perfectly binding; on the other hand, if $\rho \in \Lambda$, the DM commitment can be arbitrarily equivocated if a

---

[2] Lindell's NIZK argument is a not an argument of knowledge in contrast to the NIZK argument obtained through an FS transform.

witness for $\rho \in \Lambda$ is known. Moreover, the two modes are indistinguishable[3]. Lindell showed that DMCSs can be constructed efficiently from Σ-protocols for membership-hard languages and also provided a concrete example based on the language of Diffie-Hellman tuples (DH). Then, Lindell's transform shows how to combine DM commitments and Σ-protocols along with a hash function[4] to obtain an efficient NIZK argument.

1.1 Our Results

In this paper, we continue the study of generic and efficient transforms from 3-round public-coin HVZK proofs to NIZK arguments.

We start by studying the generality and efficiency of Lindell's transform in terms of the Σ-protocol used for instantiating the DMCS (and in turn instantiating the CRS) and the Σ-protocol to which the transform is applied. As a result, we point out a significant gap in generality and efficiency of Lindell's transform compared to the FS transform.

Then we show an improved transform that is based on weaker requirements. Specifically, our transform only requires computational HVZK and optimal soundness instead of perfect special HVZK[5] and special soundness. More interestingly and surprisingly despite being based on weaker requirements, our transform is also significantly more efficient than Lindell's transform and very close to the efficiency of the FS transform. We next discuss our contributions in more details.

The Classes of Σ-protocols Needed in [43]. Lindell defines Σ-protocols as 3-round public-coin proofs that enjoy *perfect* special HVZK and special soundness. The former property means that the simulator on input any valid statement x and challenge e can compute (a, z) such that the triple (a, e, z) is perfectly indistinguishable from an accepting transcript where the verifier sends e as challenge. Special soundness instead means that from any two accepting transcripts (a, e, z) and (a, e', z') for the same statement x that share the first message but have different challenges $e \neq e'$, one can efficiently compute a witness w for $x \in L$. Lindell in [42] shows a construction of a DMCS from any (defined as above) Σ-protocol for a membership-hard language[6].

The Efficiency of Lindell's Transform. Lindell's transform uses a DMCS derived from a Σ-protocol $\Pi_\Lambda = (\mathcal{P}_\Lambda, \mathcal{V}_\Lambda)$ for language Λ whose commitment algorithm com works by running the simulator of Π_Λ. The CRS contains an instance ρ of Λ along with the description of a hash function h. The argument produced by

---

[3] A similar notion was introduced in [11,12] and a scheme with similar features was proposed in [22].

[4] In the proof of soundness this function will be modeled as an NPRO.

[5] The latest version of Lindell's transform [42] works by assuming just perfect special HVZK instead of *strong* perfect special HVZK needed in [43].

[6] The construction in [43] needs an additional property that however is enjoyed by classic Σ-protocol s as we discuss in Appendix A.

the NIZK $\Pi = (\mathcal{P}, \mathcal{V})$ for $x \in L$ starting from a Σ-protocol $\Pi_L = (\mathcal{P}_L, \mathcal{V}_L)$ for L is computed as a tuple (a', e, z, r) where $a' = \mathsf{com}(a, r)$, $e = h(x|a')$, and z is the 3rd round of Π_L answering to the challenge e and having a as first round. The verifier checks that a' is a commitment of a with randomness r, that e is the output of $h(x|a')$ and that (a, e, z) is accepted by $\mathcal{V}_L$.

As an example, in [43] Lindell discussed the use of the Σ-protocol for the language DH for which the transform produces a very efficient NIZK proof; indeed the additional cost is of only 8 modular exponentiations: 4 to be executed by the prover and 4 by the verifier.

In this work we notice however that there is a caveat when analyzing the efficiency of Lindell's transform. The caveat is due to the message space of the DMCS. Indeed, once the CRS is fixed the max length of a message that can be committed to with only one execution of com is limited to the challenge length l_Λ of Π_Λ. Therefore in case the first round a of Π_L is much longer than l_Λ, the transform of Lindell requires multiple executions of com therefore suffering of a clear efficiency loss.

We show indeed in Tables 2 and 3 that Lindell's transform can generate in the resulting NIZK argument a blow up of the computations compared to what $\mathcal{P}_L$ and $\mathcal{V}_L$ actually do, and therefore compared to the FS transform.

Our Transform. In this paper, we present a different transform that is closer to the FS transform both on generality and on efficiency.

Our transform can be used to obtain a NIZK for any language L with a 3-round HVZK proofs enjoying optimal soundness (i.e., a weaker soundness requirement compared to special soundness). The CRS can be instantiated based on any membership-hard language Λ with a 3-round HVZK proofs enjoying optimal soundness. More specifically, we do not require perfect HVZK nor special HVZK for the involved Σ-protocols. Moreover, instead of special soundness, we will just require that, for any false statement and any first round message a, there is at most one challenge c that can be answered correctly. This is clearly a weaker requirement than special soundness and was already used by [44].

Essentially we just need that both protocols Π_L and Π_Λ are 3-round public-coin HVZK proofs with optimal soundness. Our transform produces a NIZK argument $\Pi = (\mathcal{P}, \mathcal{V})$ that does not require multiple executions of Π_L and Π_Λ and, therefore, it remains efficient under any scenario without suffering of the previously discussed issue about challenge spaces in Lindell's transform.

Techniques. We start by considering the FS transform in the NPRO model and by noticing that, as already claimed and proved in [53], if the original 3-round public-coin HVZK proof is witness indistinguishable (WI)[7], then the transformed protocol is still WI, and of course the proof of WI is RO free.

Notice that as in [43], $\mathcal{P}$ and $\mathcal{V}$ need a common hash function (modeled as an NPRO in the soundness proof) to run the protocol and this can be enforced through a setup (i.e., a non-programmable CRS [47], or a global hash function

---

[7] We use WI both to mean witness indistinguishable and witness indistinguishability.

[9]). The use of the FS transform in the NPRO model is not sufficient for our purposes. Indeed we want generality and the HVZK proof might not be witness indistinguishable. Moreover we should make a witness available to the simulator. We solve this problem by using the OR composition of 3-round perfect HVZK proofs proposed in [18]. We will let the prover $\mathcal{P}$ for NIZK to prove that either $x \in L \vee \rho \in \Lambda$. We notice that in [18] the proposed OR composition is proved to guarantee WI only when applied to two instances of the same language having a public-coin *perfect* HVZK proof. We can avoid this limitation using a generalization discussed already in [32,33] that allows the OR composition different protocols for different languages relying on *computational HVZK* only.

1.2 Comparison

Here we compare the computational effort, both for the prover and the verifier, required to execute Lindell's NIZK argument, our NIZK argument and the FS one. The properties of the three transforms are summarized in Table 1. The cost for the prover can be found in Table 2, while the one for the verifier can be found in Table 3. The comparison of the computational effort is performed with respect to three Σ-protocols[8]. Roughly speaking, in the comparisons, we consider the CRS to contain an instance of the the language DH of Diffie-Hellman triples with respect to 1024-bit prime p_{CRS} and consider two Σ-protocols: the one to prove that a triples is Diffie-Hellman[9] with respect to a prime p, for which we consider the cases in which p is 1024-bit and 2048-bit long[10], and the Σ-protocol for graph isomorphism (GI). For the Σ-protocol for graph isomorphism, we count only the modular exponentiations and do not count other operations (e.g., random selection of a permutation and generation of the adjacency matrix of permuted graphs) since they are extremely efficient and clearly dominated by the cost of modular exponentiations. A detailed description of the Σ-protocols and of the way we measure the computational effort is found in Sect. 6.

The tables give evidence of the fact that while Lindell's transform on some specific cases can replace the FS transform by paying a small overhead, in other cases there is a significant loss in performance. Our transform instead remains very close to the FS transform both when considering the amount of computation and when considering the generality of the protocols that can be given as input to the transform.

Which Protocols can be Given in Input to the Transform? We stress that our transform allows for additional proof systems to be used for instantiating the

---

[8] We consider the same Σ-protocol discussed in [43] and in addition we consider the one for Graph Isomorphism since it has the special property of having a very long first round that can be computed very efficiently.

[9] See Sect. 6 for a formal definition of the polynomial relation and the respective Σ-protocol s.

[10] Clearly, in case p is such that $|p| < |p_{\text{CRS}}|$, then Lindell's transform has a slightly smaller number of exponentiations with respect to the number of exponentiations that we count in the tables.

Table 1. Requirements for the proofs in input to the three transforms.

Transform	$HVZK$ for Λ	$HVZK$ for L	Soundness	Model
Lindell [42]	Special + perfect	Special + Perfect	Special	NPRO + CRS
This paper	Computational	Computational	Optimal	NPRO + CRS
FS	/	Computational	Classic	PRO

Table 2. Efficiency of the three transforms: modular exponentiations for the prover.

Transform	DH		GI
	$|p| = 1024$	$|p| = 2048$	n vertices
Lindell [42]	$2\ \mod p + 12\ \mod p_{\mathrm{CRS}}$	$2\ \mod p + 20\ \mod p_{\mathrm{CRS}}$	$4n^2\ \mod p_{\mathrm{CRS}}$
This paper	$2\ \mod p + 4\ \mod p_{\mathrm{CRS}}$	$2\ \mod p + 4\ \mod p_{\mathrm{CRS}}$	$4\ \mod p_{\mathrm{CRS}}$
FS	$2\ \mod p$	$2\ \mod p$	/

Table 3. Efficiency of the three transforms: modular exponentiations for the verifier.

Transform	DH		GI
	$|p| = 1024$	$|p| = 2048$	n vertices
Lindell [42]	$4\ \mod p + 12\ \mod p_{\mathrm{CRS}}$	$4\ \mod p + 20\ \mod p_{\mathrm{CRS}}$	$4n^2\ \mod p_{\mathrm{CRS}}$
This paper	$4\ \mod p + 4\ \mod p_{\mathrm{CRS}}$	$4\ \mod p + 4\ \mod p_{\mathrm{CRS}}$	$4\ \mod p_{\mathrm{CRS}}$
FS	$4\ \mod p$	$4\ \mod p$	/

CRS and for obtaining a NIZK argument system. This is not only a theoretical progress. Indeed there exist efficient constructions such as the one of [51] that is a variation of the one of [44]. The construction of [51] is an efficient 3-round HVZK proof system with optimal soundness for a language L and is not a Σ-protocol for the corresponding relation $\mathcal{R}_L$. For further details, see Appendix B.

2   HVZK Proof Systems and Σ-Protocols

We denote the security parameter by n and use "|" as concatenation operator (i.e., if a and b are two strings then by $a|b$ we denote the concatenation of a and b). For a finite set S, $x \leftarrow S$ denotes the algorithm that chooses x from S with uniform distribution.

A polynomial-time relation $\mathcal{R}$ (or *polynomial relation*, in short) is a subset of $\{0,1\}^* \times \{0,1\}^*$ such that membership of (x,w) in $\mathcal{R}$ can be decided in time polynomial in $|x|$. For $(x,w) \in \mathcal{R}$, we call x the *instance* and w a *witness* for x. For a polynomial-time relation $\mathcal{R}$, we define the NP-language $L_{\mathcal{R}}$ as $L_{\mathcal{R}} = \{x | \exists w : (x,w) \in \mathcal{R}\}$. We will model a random oracle as a random function $\mathcal{O} : \{0,1\}^* \to \{0,1\}^n$. Analogously, unless otherwise specified, for an NP-language L we denote by $\mathcal{R}_L$ the corresponding polynomial-time relation (that is, $\mathcal{R}_L$ is such that $L = L_{\mathcal{R}_L}$).

We remark that for simplicity we will omit the modulus in modular arithmetic calculations.

For two interactive machines A and B, we denote by $\langle A(\alpha), B(\beta)\rangle(\gamma)$ the distribution of B's output after running on private input β with A using private input α, both running on common input γ. Typically, one of the two machines receives the security parameter 1^n as input.

Definition 1. *A pair of PPT interactive machines $(\mathcal{P}_L, \mathcal{V}_L)$ constitutes a proof system (resp., an argument system) for NP-language L, if the following conditions hold:*

- *Completeness. For every $x \in L$ and w such that $(x, w) \in \mathcal{R}_L$, it holds:*

$$\text{Prob}\left[\,\langle \mathcal{P}_L(w, 1^n), \mathcal{V}_L\rangle(x) = 1\,\right] = 1.$$

- *Soundness. For every interactive (resp., PPT interactive) machine $\mathcal{P}_L^\star$, there exists a negligible function ν such that for every $x \notin L$ and every z:*

$$\text{Prob}\left[\,\langle \mathcal{P}_L^\star(z, 1^n), \mathcal{V}_L\rangle(x) = 1\,\right] \leq \nu(n).$$

An interactive protocol $\Pi_L = (\mathcal{P}_L, \mathcal{V}_L)$ is *public coin* if, at every round, $\mathcal{V}_L$ simply tosses a predetermined number of coins (random challenge) and sends the outcome to the prover.

In a 3-round public-coin protocol $\Pi_L = (\mathcal{P}_L, \mathcal{V}_L)$ for an NP-language L, $\mathcal{P}_L$ and $\mathcal{V}_L$ receive the common input x and, additionally, $\mathcal{P}_L$ receives security parameter 1^n in unary and w such that $(x, w) \in \mathcal{R}_L$ as private input. The interaction, with challenge length l, proceeds as follows:

The 3-round public-coin protocol Π_L:

1. $\mathcal{P}_L$, on input $1^n, x$ and w, computes message a and sends it to $\mathcal{V}_L$.
2. $\mathcal{V}_L$ chooses a random challenge $e \leftarrow \{0, 1\}^l$ and sends it to $\mathcal{P}_L$.
3. $\mathcal{P}_L$, on input x, w, e, and the randomness used to compute a, computes message z and sends it to $\mathcal{V}_L$.
4. $\mathcal{V}_L$ decides to accept or reject based on its view (i.e., (x, a, e, z)).

A triple (a, e, z) of messages exchanged during the execution of a 3-round proof (resp., argument) system is called a *3-round transcript*. We say that a 3-round transcript (a, e, z) is an *accepting transcript* for x if the argument system Π_L instructs $\mathcal{V}_L$ to accept based on the values (x, a, e, z). Two accepting 3-rounds transcripts (a, e, z) and (a', e', z') for an instance x constitute a *collision* if $a = a'$ and $e \neq e'$.

Definition 2. *A 3-round proof or argument system $\Pi_L = (\mathcal{P}_L, \mathcal{V}_L)$ for NP-language L is* Honest-Verifier Zero Knowledge (HVZK) *if there exists a PPT simulator algorithm* Sim *that takes as input security parameter 1^n and instance $x \in L$ and outputs an accepting transcript for x. Moreover, the distribution of the output of the simulator on input x is computationally indistinguishable from the distribution of the honest transcript obtained when $\mathcal{V}_L$ and $\mathcal{P}_L$ run Π_L on common input x and any private input w such that $(x, w) \in \mathcal{R}_L$.*

If the transcripts are identically distributed we say that Π_L is perfect HVZK.

Definition 3. *A 3-round public-coin proof system $\Pi_L = (\mathcal{P}_L, \mathcal{V}_L)$ for language L with challenge length l enjoys* optimal soundness *if for every $x \notin L$ and for every first-round message a there is at most one challenge $e \in \{0,1\}^l$ for which there exists a third-round message z such that (a, e, z) is accepting for x.*

Note that any 3-round public-coin optimally sound proof system with challenge length l has soundness error 2^{-l} [44].

Definition 4. *A 3-round public-coin proof system $\Pi_L = (\mathcal{P}_L, \mathcal{V}_L)$ with challenge length l is a Σ-protocol for an* NP-*language L if it enjoys the following properties:*

- *Completeness. If $(x, w) \in \mathcal{R}_L$ then all honest 3-round transcripts for (x, w) are accepting.*
- *Special Soundness. There exists an efficient algorithm mathsf Extract that, on input x and a collision for x, outputs a witness w such that $(x, w) \in \mathcal{R}_L$.*
- *Special Honest Verifier Zero Knowledge (special HVZK). There exists a PPT simulator algorithm* Sim *that takes as input security parameter 1^n, $x \in L$ and $e \in \{0,1\}^l$ and outputs an accepting transcript for x where e is the challenge. Moreover for all l-bit strings e, the distribution of the output of the simulator on input (x, e) is perfect indistinguishable from the distribution of the 3-round honest transcript obtained when $\mathcal{V}_L$ sends e as challenge and $\mathcal{P}_L$ runs on common input x and any private input w such that $(x, w) \in \mathcal{R}_L$.*

Sometimes, we will abuse notion and say that a proof system or Σ-protocol is for a polynomial relation $\mathcal{R}$ instead of referring to NP-language $L_{\mathcal{R}}$.

It is easy to see that Σ-protocols enjoy optimal soundness. The converse, however, is not true. See Appendix B for an example of an optimal-sound 3-round public-coin proof system that does not enjoy special soundness (and is special perfect HVZK).

In order not to overburden the descriptions of protocols and simulators, we will omit the specification of the security parameter when it is clear from the context.

2.1   3-Round Public-Coin HVZK Proofs and WI

Following [33], for an NP-language L, we define $\hat{L}$ to be the input language that includes both L and all false instances that are well formed and can be used by an adversarial prover in order to prove a false statement. More formally, $L \subseteq \hat{L}$ and membership in $\hat{L}$ can be tested in polynomial time. We implicitly assume that a verifier executes the protocol only if the common input $x \in \hat{L}$; otherwise, it rejects immediately.

Definition 5. *A 3-round public-coin proof system $\Pi = (\mathcal{P}_L, \mathcal{V}_L)$ is* Witness Indistinguishable (WI) *for polynomial relation $\mathcal{R}$ if, for every malicious verifier $\mathcal{V}_L^\star$, there exists a negligible function ν such that for all x, w, w' with $(x, w) \in \mathcal{R}$ and $(x, w') \in \mathcal{R}$, it holds that:*

$$|\text{Prob}\,[\,\langle \mathcal{P}_L(w, 1^n), \mathcal{V}_L^\star \rangle(x) = 1\,] - \text{Prob}\,[\,\langle \mathcal{P}_L(w', 1^n), \mathcal{V}_L^\star \rangle(x) = 1\,]| \leq \nu(n).$$

The notion of a perfect *WI 3-round proof system is obtained by requiring that $\nu(n) = 0$.*

Sometimes we abuse the above definition and say that a proof system is WI for a NP-language L instead of referring to the associated polynomial relation $\mathcal{R}_L$.

We recall the following result.

Theorem 1 ([18]). *Every 3-round public-coin proof system with perfect HVZK for an NP-language L is perfect WI for $\mathcal{R}_L$.*

2.2 Challenge Lengths of 3-Round HVZK Proofs

Challenge-Length Amplification. The challenge of a 3-round public-coin proof system with HVZK and optimal soundness can be extended through parallel repetition.

Lemma 1. *Let Π_L be a 3-round public-coin proof system with optimal soundness for NP-language L that enjoys perfect HVZK and has challenge length l. The protocol Π_L^k consisting of k parallel instances of Π_L is a 3-round public-coin proof system for relation L that enjoys perfect HVZK, has optimal soundness and has challenge length $k \cdot l$.*

Proof. The HVZK it is preserved by Π_L^k for the same arguments of [18]. About the optimal soundness of Π_L^k, it is simple to see that if the protocol Π_L^k in not optimal sound then also Π_L is not optimal sound.

A similar lemma can be proved for a Σ-protocol (as in [15,16,32]) for which HVZK is not perfect.

Challenge-Length Reduction. We now show that starting from any 3-round public-coin proof system that enjoys HVZK and has optimal soundness with challenge length l, one can construct a 3-round public-coin proof system that still enjoys HVZK, has optimal soundness but works with a shorter challenge. Moreover perfect HVZK is preserved. A similar transformation was shown in [20] for the case of Σ-protocol that are special perfect HVZK.

Lemma 2. *Let Π_L be a HVZK 3-round public-coin proof system for L with optimal soundness and challenge length l. Then for every $l' < l$, there exists a 3-round public-coin proof system Π_L' for L with HVZK and optimal soundness and challenge length l'. Protocol Π_L' has the same efficiency as Π_L and, moreover, if Π_L is perfect HVZK so is Π_L'.*

Proof. Following is a description of Π_L'.
Common input: instance x for an NP-language L.
Private input of $\mathcal{P}_L'$: w s.t. $(x, w) \in \mathcal{R}_L$.
The protocol Π_L':

1. $\mathcal{P}_L'$ computes $a \leftarrow \mathcal{P}_L(x, w)$ and sends it to $\mathcal{V}_L'$;
2. $\mathcal{V}_L'$ randomly chooses challenge $e \leftarrow \{0, 1\}^{l'}$ and sends it to $\mathcal{P}_L'$;
3. $\mathcal{P}_L'$ randomly chooses $pad \leftarrow \{0, 1\}^{(l-l')}$, sets $e' = e|pad$, computes $z \leftarrow \mathcal{P}_L(x, w, a, e')$ and sends $z' = (z, pad)$ to $\mathcal{V}_L'$;
4. $\mathcal{V}_L'$ outputs the output of $\mathcal{V}_L(x, a, e|pad, z)$.

Completeness follows directly from the completeness of Π.

HVZK. We can consider the simulator Sim', that on input x runs as follows:

1. run $(a, e', z) \leftarrow \mathsf{Sim}(x)$;
2. set *pad* equal to the last $l - l'$ bits of e', and set e equal to the fist l' bits of e';
3. output $(a, e, (z, pad))$.

This concludes the proof.

Optimal soundness follows directly from the optimal soundness of Π.

The following theorem follows from Lemmas 1 and 2,

Theorem 2. *Suppose* NP-*language L admits a HVZK 3-round public-coin proof system Π_L that has optimal soundness and challenge length l. Then for any $l' > 0$ there exists HVZK 3-round public-coin proof system Π'_L that has optimal soundness and challenge length l'. If $l' \leq l$ then Π'_L is as efficient as Π_L. Otherwise the communication and computation complexities of Π'_L are at most l'/l times the ones of Π_L. Moreover, perfect HVZK is preserved.*

2.3   3-Round Public-Coin HVZK Proofs for or Composition of Statements

In this section we recall the construction of [18] that starts from a HVZK 3-round public-coin proof system Π_L for an NP-language L and constructs a HVZK 3-round public-coin proof system $\Pi_{L \vee L}$ for the "OR" language of L; that is the NP-language $L \vee L = \{(x_0, x_1) : x_0 \in L \vee x_1 \in L\}$. Below we give the descriptions of the prover $\mathcal{P}_{L \vee L}$ and of the verifier $\mathcal{V}_{L \vee L}$ of $\Pi_{L \vee L}$. In the description, we let Sim denote the simulator for Π_L and l denote the challenge length of Π_L. We also let $b \in \{0, 1\}$ be such that w is a witness for $x_b \in L$; that is, $(x_b, w) \in \mathcal{R}_L$.

Common input: instances x_0, x_1 for an NP-language L.
Private input of $\mathcal{P}_{L \vee L}$: w s.t $(x_0, x_1, w) \in \hat{\mathcal{R}}_{L \vee L}$.
The protocol $\Pi_{L \vee L}$:

1. $\mathcal{P}_{L \vee L}$ computes $a_b \leftarrow \mathcal{P}_L(x_b, w)$, $(a_{1-b}, e_{1-b}, z_{1-b}) \leftarrow \mathsf{Sim}(x_{1-b})$ and sends (a_0, a_1) to $\mathcal{V}_{L \vee L}$.
2. $\mathcal{V}_{L \vee L}$ chooses at random challenge $e \leftarrow \{0, 1\}^l$ and sends e to $\mathcal{P}_{L \vee L}$.
3. $\mathcal{P}_{L \vee L}$ sets $e_b = e \oplus e_{1-b}$, computes $z_b \leftarrow \mathcal{P}_L(x_b, w, a_b, e_b)$ and outputs $\Big((e_0, e_1), (z_0, z_1)\Big)$.
4. $\mathcal{V}_{L \vee L}\Big((x_0, x_1), (a_0, a_1), e, ((e_0, e_1), (z_0, z_1))\Big)$. $\mathcal{V}_{L \vee L}$ accepts if and only if $e = e_0 \oplus e_1$ and $\mathcal{V}_L(x_0, a_0, e_0, z_0) = 1$ and $\mathcal{V}_L(x_1, a_1, e_1, z_1) = 1$.

Theorem 3 ([18,33]). *If Π_L is a HVZK 3-round public-coin proof system with optimal soundness for* NP-*language L then $\Pi_{L \vee L}$ is a HVZK 3-round public-coin proof system with optimal soundness for* NP-*language $L \vee L$ and is WI for polynomial-time relation*

$$\mathcal{R}_{L \vee L} = \Big\{((x_0, x_1), w) : \Big((x_0, w) \in \mathcal{R}_L \wedge x_1 \in L\Big) \vee \Big((x_1, w) \in \mathcal{R}_L \wedge x_0 \in L\Big)\Big\}.$$

Moreover if Π_L is perfect HVZK then $\Pi_{L \vee L}$ is perfect WI for polynomial-time relation

$$\hat{\mathcal{R}}_{L \vee L} = \left\{ ((x_0, x_1), w) : \left((x_0, w) \in \mathcal{R}_L \wedge x_1 \in \hat{L} \right) \vee \left((x_1, w) \in \mathcal{R}_L \wedge x_0 \in \hat{L} \right) \right\}.$$

We remark that results of [18,33] are known to hold for Σ-protocol s, but in the proof of WI they use only HVZK. Therefore their results also hold starting from a HVZK 3-round public-coin proof system with optimal soundness (and not necessarily special soundness) that we consider in the above theorem. Indeed we observe that $\Pi_{L \vee L}$ has optimal soundness for the following reason. Suppose that $\Pi_{L \vee L}$ does not enjoy optimal soundness. This means that for a false instance and the same first round (a_0, a_1) there are two accepting conversation, namely:

$$\left((a_0, a_1), e, ((e_0, e_1), (z_0, z_1)) \right), \left((a_0, a_1), e', ((e'_0, e'_1), (z'_0, z'_1)) \right)$$

with $e \neq e'$. Then it must be the case that for some $b = 0$ or $b = 1$, $e_b \neq e'_b$ and then (a_b, e_b, z_b) (a_b, e'_b, z'_b) are two accepting transcripts with the same first round for the protocol Π_L, and thus the optimal soundness of Π_L is violated.

It is possible to extend the above construction to handle two different NP-languages L_0, L_1 that admit HVZK 3-round public-coin proof system with optimal soundness. Indeed by Theorem 2, we can assume, without loss of generality, that L_0 and L_1 have 3-round public-coin proof systems Π_{L_0} and Π_{L_1} with the same challenge length. Assuming that L_0 and L_1 have 3-round public-coin proof systems Π_{L_0} and Π_{L_1} that are HVZK and have optimal soundness with the same challenge length. We can apply the same construction outlined above to obtain a 3-round public-coin proof system $\Pi_{L_0 \vee L_1}$ that enjoys HVZK and has optimal soundness for relation

$$\hat{\mathcal{R}}_{L_0 \vee L_1} = \left\{ ((x_0, x_1), w) : \left((x_0, w) \in \mathcal{R}_{L_0} \wedge x_1 \in \hat{L}_1 \right) \vee \left((x_1, w) \in \mathcal{R}_{L_1} \wedge x_0 \in \hat{L}_0 \right) \right\}.$$

We have the following theorem.

Theorem 4. *If Π_{L_0} and Π_{L_1} are HVZK 3-round public-coin proof systems with optimal soundness for NP-languages L_0 and L_1 then $\Pi_{L_0 \vee L_1}$ is a HVZK 3-round public-coin proof system with optimal soundness for the for NP-language $L_0 \vee L_1 = \{(x_0, x_1) : x_0 \in L_0 \vee x_1 \in L_1\}$ and is WI for polynomial-time relation*

$$\mathcal{R}_{L_0 \vee L_1} = \left\{ ((x_0, x_1), w) : \left((x_0, w) \in \mathcal{R}_{L_0} \wedge x_1 \in L_1 \right) \vee \left((x_1, w) \in \mathcal{R}_{L_1} \wedge x_0 \in L_0 \right) \right\}.$$

Moreover, if Π_{L_0} and Π_{L_1} are perfect then $\Pi_{L_0 \vee L_1}$ is perfect WI for polynomial-time relation $\hat{\mathcal{R}}_{L \vee L}$.

3   Non-Interactive Argument Systems

Part of the definitions of this section are taken from [43].

Definition 6. *A non-interactive argument system for an NP-language L consists of three PPT machines $(\mathcal{CRS}, \mathcal{P}, \mathcal{V})$, that have the following properties:*

– *Completeness: for all* $(x, w) \in \mathcal{R}_L$, *it holds that:*

$$\text{Prob}\,[\,\sigma \leftarrow \mathcal{CRS}(1^n); \mathcal{V}(\sigma, x, \mathcal{P}(\sigma, x, w)) = 1\,] = 1.$$

– *Adaptive Soundness: for every PPT function* $f : \{0,1\}^{poly(n)} \rightarrow \{0,1\}^n \setminus L$ *for all PPT prover* $\mathcal{P}^\star$, *there exists a negligible function* ν, *such that for all* n:

$$\text{Prob}\left[\,\sigma \leftarrow \mathcal{CRS}(1^n); \mathcal{V}^{\mathcal{O}}(\sigma, f(\sigma), \mathcal{P}^{\star \mathcal{O}}(\sigma)) = 1\,\right] \leq \nu(n)$$

where $\mathcal{O} : \{0,1\}^* \rightarrow \{0,1\}^n$ *is a random function.*

Definition 7. *A non-interactive argument system is* adaptive unbounded zero knowledge *(NIZK) for an* NP-*language L if there exists a probabilistic PPT simulator S such that for every PPT function*

$$f : \{0,1\}^{\texttt{poly(n)}} \rightarrow \left(\{0,1\}^n \times \{0,1\}^{\texttt{poly(n)}}\right) \cap \mathcal{R}_L,$$

for every polynomial $p(\cdot)$ *and for every PPT malicious verifier* $\mathcal{V}^\star$, *there exists a negligible function* ν *such that,*

$$\left|\text{Prob}\,[\,\mathcal{V}^\star\left(R_f(\mathcal{P}^f(n,p))\right) = 1\,] - \text{Prob}\,[\,\mathcal{V}^\star\left(S_f(n,p)\right) = 1\,]\right| \leq \nu(n)$$

where f_1 *and* f_2 *denote the first and second output of* f, *respectively, and* $R_f(\mathcal{P}^f(n,p))$ *and* $S_f(n,p)$ *denote the output from the following experiments.*

Real proofs $R_f(\mathcal{P}^f(n,p))$:

– $\sigma \leftarrow \mathcal{CRS}(1^n)$ *a common reference string is sampled.*
– *For* $i = 1, \ldots, p(n)$ *(initially* $\boldsymbol{x}$ *and* $\boldsymbol{\pi}$ *are empty):*
  • $x_i \leftarrow f_1(\sigma, \boldsymbol{x}, \boldsymbol{\pi})$: *the next statement* x_i *to be proven is chosen.*
  • $\pi_i \leftarrow \mathcal{P}(\sigma, f_1(\sigma, \boldsymbol{x}, \boldsymbol{\pi}), f_2(\sigma, \boldsymbol{x}, \boldsymbol{\pi}))$: *the ith proof is generated.*
  • *set* $\boldsymbol{x} = x_1 \ldots x_i$ *and* $\boldsymbol{\pi} = \pi_1 \ldots \pi_i$.
– *output* $(\sigma, \boldsymbol{x}, \boldsymbol{\pi})$.

Simulation $S_f(n,p)$:

– $\sigma \leftarrow S(1^n)$ *a common reference string is sampled.*
– *For* $i = 1, \ldots, p(n)$ *(initially* $\boldsymbol{x}$ *and* $\boldsymbol{\pi}$ *are empty):*
  • $x_i \leftarrow f_1(\sigma, \boldsymbol{x}, \boldsymbol{\pi})$: *the next statement* x_i *to be proven is chosen.*
  • $\pi_i \leftarrow S(x_i)$: *simulator S generates a simulated proof* π_i *that* $x_i \in L$.
  • *set* $\boldsymbol{x} = x_1 \ldots x_i$ *and* $\boldsymbol{\pi} = \pi_1 \ldots \pi_i$.
– *output* $(\sigma, \boldsymbol{x}, \boldsymbol{\pi})$.

Definition 8. *A non-interactive argument system is* adaptive unbounded witness indistinguishable *(NIWI) for an* NP-*language L if for every PPT adversary* $\mathcal{V}^\star$, *for every PPT function*

$$f : \{0,1\}^{\texttt{poly(n)}} \rightarrow \left(\{0,1\}^n \times \{0,1\}^{\texttt{poly(n)}} \times \{0,1\}^{\texttt{poly(n)}}\right) \cap \mathcal{R}_L^\wedge,$$

and for every polynomial $p(\cdot)$, there exists a negligible function ν such that

$$\left| \mathrm{Prob}\left[\mathcal{V}^\star(R_0^{\mathcal{P},f}(n,p)) = 1 \right] - \mathrm{Prob}\left[\mathcal{V}^\star(R_1^{\mathcal{P},f}(n,p)) = 1 \right] \right| \leq \nu(n),$$

where $\mathcal{R}_L^\wedge = \{(x, w^0, w^1) : (x, w^0) \in \mathcal{R}_L \wedge (x, w^1) \in \mathcal{R}_L\}$ and $R_b^{\mathcal{P},f}$ is the following experiment. $R_b^{\mathcal{P},f}(n,p)$:

- $\sigma \leftarrow \mathcal{CRS}(1^n)$.
- For $i = 1, \ldots, p(n)$ (initially $\boldsymbol{x}$ and $\boldsymbol{\pi}$ are empty):
  - $(x_i, w_i^0, w_i^1) \leftarrow f(\sigma, \boldsymbol{x}, \boldsymbol{\pi})$:
    statement x_i to be proven and witnesses w_i^0, w_i^1 for x_i are generated.
  - $\pi_i \leftarrow \mathcal{P}(\sigma, x_i, w_i^b)$: the ith proof is generated.
  - set $\boldsymbol{x} = x_1 \ldots x_i$ and $\boldsymbol{\pi} = \pi_1 \ldots \pi_i$.
- output $(\sigma, \boldsymbol{x}, \boldsymbol{\pi})$.

4   NIWI Argument Systems from 3-Round HVZK Proofs

In this section we discuss the FS transform in the NPRO model in order to obtain a NIWI argument system $\Pi = (\mathcal{P}, \mathcal{V})$ for a polynomial relation $\mathcal{R}_L$. We start from a 3-round public-coin WI HVZK proof system with optimal soundness $\Pi_L = (\mathcal{P}_L, \mathcal{V}_L)$ for L. $\mathcal{P}$ and $\mathcal{V}$ have access to an NPRO $H : \{0,1\}^* \to \{0,1\}^n$. We describe Π below and we assume that the challenge length of Π_L is the security parameter n.

Common input: instance x for NP-language L.
Private input to $\mathcal{P}$: w s.t. $(x, w) \in \mathcal{R}_L$.
Common reference string: $\mathcal{CRS}$ samples a key s for a hash function family H and sets $\sigma = s$.

1. $\mathcal{P} \to \mathcal{V}$: The prover $\mathcal{P}$ executes the following steps:
   1.1. $a \leftarrow \mathcal{P}_L(x, w)$;
   1.2. $e \leftarrow H_s(x, a)$;
   1.3. $z \leftarrow \mathcal{P}_L(x, w, a, e)$;
   1.4. send $\pi = (a, e, z)$ to $\mathcal{V}$.
2. $\mathcal{V}$'s output: $\mathcal{V}$ outputs 1 if and only if $\mathcal{V}_L(x, a, e, z) = 1$ and $e = H_s(x, a)$.

The following theorem was proved by Yung and Zhao in [53] (see Claim 1, page 4). For completeness, we provide a proof of the claim below.

Theorem 5 ([53]). *Let Π_L be a 3-round public-coin WI proof system for the polynomial relation $\mathcal{R}_L$. Then Π is adaptive WI for $\mathcal{R}_L$ in the CRS model.*

Proof We show that Π is adaptive WI for $\mathcal{R}_L$ through the following hybrids.

1. $\mathcal{H}_1$ is the experiment $R_0^{\mathcal{P},f}(n,p)$ (Definition 8), where $\mathcal{P}$ for $j = 1, \ldots, p(n)$ executes Π and outputs π_j using the first of the two witnesses given in output by f.

2. $\mathcal{H}_i$ (with $i > 0$) differs from $\mathcal{H}_1$ in the first i interactions, where $\mathcal{P}$ executes Π using the second witness given in output by f. Namely: $\mathcal{P}$ on input (x_j, w_j^1) executes Π and outputs π_j using w_j^1 for all $j : 1 \leq j < i$. Instead, for the interactions $i \leq j < p(n) + 1$, $\mathcal{P}$ on input (x_j, w_j^0) executes Π using w_j^0 as a witness and outputs π_j.
3. $\mathcal{H}_{p(n)+1}$ is the experiment $R_1^{\mathcal{P}, f}(n, p)$ (Definition 8), where $\mathcal{P}$ for $j = 1, \ldots, p(n)$ executes Π and outputs π_j using the second witness given in output by f.

$\mathcal{H}_i \approx \mathcal{H}_{i+1}$: Suppose there exists a malicious adversary $\mathcal{V}^\star$ that distinguishes between the experiments $\mathcal{H}_i$ and $\mathcal{H}_{i+1}$ with $1 \leq i \leq p(n)$, then we can show that there exists an adversary $\mathcal{A}$ that breaks the WI property of Π_L. The reduction works as follows.

1. For $j = 1, \ldots, i - 1$, $\mathcal{A}$ on input (x_j, w_j^1) executes Π using w_j^1 to obtain π_j.
2. For $j = i$, $\mathcal{A}$ interacts with the WI challenger of Π_L as follows:
   (a) $\mathcal{A}$ has on input (x_j, w_j^0, w_j^1) and sends it to the challenger of WI;
   (b) the challenger computes and sends the first message a_j to $\mathcal{A}$;
   (c) $\mathcal{A}$ computes $e_j = H_s(a_j)$ and sends it to the challenger of WI;
   (d) the challenger computes and sends z_j to $\mathcal{A}$;
   (e) $\mathcal{A}$ sends $\pi_j = (a_j, e_j, z_j)$ to $\mathcal{V}^\star$;
   (f) $\mathcal{A}$ adds to $\boldsymbol{x}$ the theorem x_j and to $\boldsymbol{\pi}$ the proof π_j.
3. $\forall j = i + 1, \ldots, p(n)$ $\mathcal{A}$ on input (x_j, w_j^0) executes Π using w_j^0 to obtain π_j.
4. Set $\boldsymbol{x} = x_1, \ldots, x_{p(n)}$ and $\boldsymbol{\pi} = \pi_1, \ldots, \pi_{p(n)}$.

$\mathcal{A}$ sends $\boldsymbol{x}$ and $\boldsymbol{\pi}$ to $\mathcal{V}^\star$ and outputs what $\mathcal{V}^\star$ outputs.

We now observe that if the challenger of WI has used the first witness we are in $\mathcal{H}_i$ otherwise we are in $\mathcal{H}_{i+i}$. It follows that $R_0^{\mathcal{P}, f}(n, p) \equiv \mathcal{H}_1 \approx \cdots \approx \mathcal{H}_{p(n)} \approx \mathcal{H}_{p(n)+1} \equiv R_1^{\mathcal{P}, f}(n, p)$ to conclude the proof.

Adaptive Soundness. To prove soundness we follow [43] and use the fact that, for every function g, with a sufficiently large co-domain, relation $\mathcal{R} = \{(x, g(x))\}$ is evasive [8] in the NPRO model. A relation $\mathcal{R}$ is *evasive* if, given access to a random oracle $\mathcal{O}$, it is infeasible to find a string x so that the pair $(x, \mathcal{O}(x)) \in \mathcal{R}$.

Theorem 6 *Let Π_L be a 3-round public-coin proof system with optimal soundness for the NP-language L, and let H be a non programmable random oracle. Then, Π is a non-interactive argument system with (adaptive) soundness for L in the NPRO model.*

Proof Completeness of Π follows from the completeness of Π_L. Let $\mathcal{O}$ be an NPRO. In order to prove the soundness of Π we use the fact that for any function g, the relation $\mathcal{R} = \{(x, g(x))\}$ is evasive. We define the function g s.t. $g(x, a) = e$, where there exists z such that the transcript (a, e, z) is accepting for the instance x. If $x \notin L$ by the optimal soundness property we have that for every a there is a single e for which there is some z so that (a, e, z) is accepting. Therefore g is a function, as required and it follows that the relation $\mathcal{R} = \{((x, a), g(x, a))\}$ is evasive. Suppose that there exist a polynomial function f and a malicious prover

$\mathcal{P}^\star$ such that $\mathcal{P}^\star$ proves a false statement (i.e., $\mathcal{V}^\mathcal{O}(\sigma, f(\sigma), \mathcal{P}^{\star\mathcal{O}}(\sigma)) = 1$, where $\sigma \leftarrow \mathcal{CRS}(1^n)$) with non-negligible probability, then there is an adversary $\mathcal{A}$ that finds (x, a) s.t. $\mathcal{O}(x, a) = g(x, a)$ with non-negligible probability. The adversary $\mathcal{A}$ works as follows. First, it runs $\sigma \leftarrow \mathcal{CRS}(1^n)$. Then it runs $(x, a, e, z) \leftarrow \mathcal{P}^\star(\sigma)$. Finally it outputs $(x, \mathcal{O}(x, a))$. From the contradicting assumption we know that $\mathcal{V}^\mathcal{O}(\sigma, f(\sigma), (a, e, z)) = 1$ with non-negligible probability. This implies that the transcript $(a, \mathcal{O}(x, a), z)$ is accepting with non-negligible probability. Since $x \notin L$ there exists only one e for which $(a, \mathcal{O}(x, a), z)$ is accepting. Therefore we have that with non-negligible probability it holds that $\mathcal{O}(x, a) = e$ (i.e., $\mathcal{O}(x, a) = g(x, a)$) and this contradicts the fact that any function g is evasive for an NPRO.

5 Our Transform: NIZK from HVZK

From the previous section we know that if we have a 3-round HVZK proof system with optimal soundness $\Pi_{L\vee\Lambda} = (\mathcal{P}_{L\vee\Lambda}, \mathcal{V}_{L\vee\Lambda})$ for polynomial relation

$$\hat{\mathcal{R}}_{L\vee\Lambda} = \{((x, \rho), w) : ((x, w) \in \mathcal{R}_L \wedge \rho \in \hat{\Lambda}) \vee ((\rho, \omega) \in \mathcal{R}_\Lambda \wedge x \in \hat{L})\}$$

that is also WI for polynomial relation

$$\mathcal{R}_{L\vee\Lambda} = \{((x, \rho), w) : ((x, w) \in \mathcal{R}_L \wedge \rho \in \Lambda) \vee ((\rho, \omega) \in \mathcal{R}_\Lambda \wedge x \in L)\}$$

we can apply the FS transform to make it non-interactive still preserving WI and soundness. To run the protocol a common hash function is needed and such a function is modeled as an NPRO in the proof of soundness.

Here we make use of the above result in order to transform a 3-round HVZK proof system with optimal soundness for an NP-language L into a NIZK argument for L in the CRS model using an NPRO in the proof of soundness. The transformed NIZK argument $\Pi = (\mathcal{P}, \mathcal{V})$ is described below.

Common input: instance x for an NP-language L.

Private input of $\mathcal{P}$: w s.t $(x, w) \in \mathcal{R}_L$.

Common reference string: $\mathcal{CRS}$ on input 1^n runs $\rho \leftarrow S_\Lambda(1, 1^n)$ where Λ is an membership-hard language and samples a key s for a hash function family H. Then it sets $\sigma = (\rho, s)$.

$\mathcal{P} \rightarrow \mathcal{V}$: $\mathcal{P}$ executes the following steps:

1. $a \leftarrow \mathcal{P}_{L\vee\Lambda}((x, \rho), w)$;
2. $e \leftarrow H_s(x, a)$;
3. $z \leftarrow \mathcal{P}_{L\vee\Lambda}((x, \rho), w, a, e)$;
4. send $\pi = (a, e, z)$ to $\mathcal{V}$.

$\mathcal{V}'s$ output: $\mathcal{V}$ accepts if and only if $\mathcal{V}_{L\vee\Lambda}((x, \rho), a, e, z) = 1$ and $e = H_s(x, a)$.

In our construction we suppose that the challenge length of Π_Λ is n, where n denotes the security parameter. Therefore to use the OR composition of [18] we need to consider a 3-round public-coin proof system with HVZK and optimal soundness Π_L for $\mathcal{R}_L$ that has challenge length n and therefore soundness error 2^{-n}). This is not a problem because we can use Theorem 2 to transform every 3-round public-coin proof system with HVZK and optimal soundness with challenge n' (where $n' \neq n$) to another one with challenge length n. More precisely, if $n' > n$ we can use Lemma 2 to reduce n' to n almost for free. If $n' < n$ we need to use Lemma 1, therefore we have to run multiple executions of Π_L to apply the OR composition of [18]. Notice that this potential computational effort is implicit also for the FS transform and for Lindell's transform. Indeed if the original 3-round public-coin proof system with HVZK and optimal soundness has just a one-bit (or in general a short) challenge then clearly the resulting NIZK is not sound. Therefore the parallel repetition of the 3-round public-coin proof system with HVZK and optimal soundness is required before applying the transform in order to reduce the soundness error (see Sect. 2.2).

Theorem 7. *Let $\Pi_{L\vee\Lambda}$ be a 3-round public-coin proof system for polynomial relation $\hat{\mathcal{R}}_{L\vee\Lambda}$ that is WI for polynomial relation $\mathcal{R}_{L\vee\Lambda}$. Then Π is zero knowledge for $\mathcal{R}_L$ in the CRS model.*

Proof. The simulator S works as follows:

1. S on input 1^n, runs $(\rho, \omega) \leftarrow S_\Lambda(0, 1^n)$; samples a key s for a hash function and sets $\sigma = \{\rho, s\}$ and outputs σ.
2. S on input σ, ω and x_i (for every $i = 1, \ldots, p(n)$) computes $a \leftarrow \mathcal{P}_{L\vee\Lambda}((x_i, \rho), \omega)$, $e \leftarrow H_s(x_i, a)$ and $z \leftarrow \mathcal{P}_{L\vee\Lambda}((x_i, \rho), \omega, a, e)$. It outputs $\pi_i = (a, e, z)$.

We show that the output of S is computationally indistinguishable from a real transcript given in output by $\mathcal{P}$ in a real execution of Π through the following hybrids games.

1. $\mathcal{H}_0$ is the experiment $R_f(\mathcal{P}^f(n, p))$ (Definition 7).
2. $\mathcal{H}_1$ differs from $\mathcal{H}_0$ in the way that ρ is generated. Indeed in $\mathcal{H}_1$ we have that σ is computed by running $S_\Lambda(0, 1^n)$. The second output ω of S_Λ is not used. Clearly $\mathcal{H}_0$ and $\mathcal{H}_1$ are indistinguishable otherwise the membership-hard property of Λ would be contradicted. More details on this reduction will be given below.
3. $\mathcal{H}_2$ differs from $\mathcal{H}_1$ just on the witness used by $\mathcal{P}_{L\vee\Lambda}$. Indeed now ω is used as witness. The WI property of $\Pi_{L\vee\Lambda}$ guarantees that $\mathcal{H}_2$ can not be distinguished from $\mathcal{H}_1$. More details on this reduction will be given below. Notice that $\mathcal{H}_2$ corresponds to the simulation.

$\mathcal{H}_0 \approx \mathcal{H}_1$: If there exists a malicious verifier $\mathcal{V}^\star$ that distinguishes between $\mathcal{H}_0$ and $\mathcal{H}_1$, then there exists an adversary $\mathcal{A}$ that breaks the membership-hard property of Λ. The reduction works as follows.

1. $\mathcal{A}$ queries the challenger of S_Λ that sends back ρ.
2. $\mathcal{A}$ samples a key s for a hash function family H and sets $\sigma = \{\rho, s\}$.
3. $\mathcal{A}$ on input $(x_i, w_i) \in \mathcal{R}_L$ for $i = 1, \ldots, p(n)$ computes the following steps:
   3.1. compute $a_i \leftarrow \mathcal{P}_{L\vee\Lambda}((x_i, \rho), w_i)$;
   3.2. compute $e_i \leftarrow H_s(x_i, a_i)$;

     3.3. compute $z_i \leftarrow \mathcal{P}_{L \vee \Lambda}((x_i, \rho), w_i, a_i, e_i)$;

     3.4. set $\pi_i = (a_i, e_i, z_i)$;

     3.5. set $\boldsymbol{x} = x_1, \ldots, x_i$ and $\boldsymbol{\pi} = \pi_1, \ldots, \pi_i$.

4. $\mathcal{A}$ sends $\sigma, \boldsymbol{x}, \boldsymbol{\pi}$ to $\mathcal{V}^\star$.

5. $\mathcal{A}$ outputs the output of $\mathcal{V}^\star$.

We now observe that if the challenger of a sampling algorithm S_Λ sends $\rho \notin \Lambda$ we are in $\mathcal{H}_0$ otherwise we are in $\mathcal{H}_1$. This implies that $\mathcal{H}_0 \approx \mathcal{H}_1$.

$\mathcal{H}_1 \approx \mathcal{H}_2$: If there exists a distinguisher $\mathcal{V}^\star$ that distinguishes between $\mathcal{H}_1$ and $\mathcal{H}_2$, then there exists an adversary $\mathcal{A}$ against the adaptive NIWI property of $\Pi_{L \vee \Lambda}$, therefore contradicting Theorem 5. The reduction works as follows.

1. $\mathcal{A}$ runs $(\rho, \omega) \leftarrow S_\Lambda(0, 1^n)$, samples a key s for a hash function and sets $\sigma = \{\rho, s\}$.

2. $\mathcal{A}$ has on input a PPT function $f = (f_1, f_2)$ and defines $f' = (f_1', f_2')$ as follows: $f'(\sigma, \boldsymbol{t}, \boldsymbol{\pi})$ on input a CRS σ, a vector of theorems $\boldsymbol{t} = (x_1, \rho), \ldots, (x_{p(n)}, \rho)$ and a vector of proofs $\boldsymbol{\pi} = \pi_1, \ldots, \pi_{p(n)}$ returns $(f_1(\sigma, \boldsymbol{x}, \boldsymbol{\pi}), \rho), (f_2(\sigma, \boldsymbol{x}, \boldsymbol{\pi}), \omega)$.

3. $\mathcal{A}$ interacts with the challenger of adaptive NIWI, using f', in order to obtain x_i, $\pi_i = \{a_i, e_i, z_i\}$, for $i = 1, \ldots, p(n)$.

4. $\mathcal{A}$ sets $\boldsymbol{x} = x_1, \ldots, x_{p(n)}$ and $\boldsymbol{\pi} = \pi_1, \ldots, \pi_{p(n)}$.

5. $\mathcal{A}$ sends $\sigma, \boldsymbol{x}, \boldsymbol{\pi}$ to $\mathcal{V}^\star$ and outputs the output of $\mathcal{V}^\star$.

We now observe that if the challenger of NIWI chooses the first witness w_i we are in $\mathcal{H}_1$ otherwise we are in $\mathcal{H}_2$. This implies that $\mathcal{H}_1 \approx \mathcal{H}_2$. We can thus conclude that $\mathcal{H}_0 \approx \mathcal{H}_1 \approx \mathcal{H}_2$ and therefore the output of S is computational indistinguishable from a real transcript.

Theorem 8. *Let $\Pi_{L \vee \Lambda}$ be a 3-round public-coin HVZK proof system with optimal soundness for relation $\mathcal{R}_{L \vee \Lambda}$, and WI for relation $\tilde{\mathcal{R}}_{L \vee \Lambda}$, and let H be an NPRO. Then, Π is a non-interactive argument system with adaptive soundness for the relation $\mathcal{R}_L$ in the CRS model using the NPRO model for soundness.*

Proof. The completeness of Π follows from the completeness of $\Pi_{L \vee \Lambda}$. In order to prove adaptive soundness we notice that an adversarial prover proving a false statement $x \in L$ can be directly reduced to an adversarial prover proving a false statement for $\Pi_{L \vee \Lambda}$ in the NPRO model. This contradicts Theorem 6. Indeed the only subtlety that is worthy to note is that when the adversarial prover runs the protocol, we have that the statement "$\rho \in \Lambda$" stored in the CRS is false, therefore if also the instance "$x \in L$" proved by the prover is false then the OR composition of the two statements is also false.

6   Details on Some Σ-Protocols

First of all we need to briefly introduce two Σ-protocols, one to prove that a tuple is a DH tuple ($\Pi_{\mathcal{DH}}$ [41]), and the other one to prove that two graphs are isomorphic ($\Pi_{\mathcal{GH}}$ [34]). Our comparison assumes that the CRS is a DH

tuple $((G_{\text{CRS}}, q_{\text{CRS}}, p_{\text{CRS}}, g_{\text{CRS}}), A_{\text{CRS}}, B_{\text{CRS}}, C_{\text{CRS}})$ with p_{CRS} and q_{CRS} primes such that $p_{\text{CRS}} = 2q_{\text{CRS}} + 1$ and $|p_{\text{CRS}}| = 1024$. We distinguish two cases. In the first one the prover wants to prove that a tuple $((G, q, p, g), A, B, C)$ is a DH tuple, and in the other one the prover tries to convince the verifier that two graphs G_0 and G_1 with n vertices each are isomorphic.

A Σ-protocol for Diffie-Hellman tuples. We consider the following polynomial-time relation $\mathcal{R}_{\mathcal{DH}} = \{(((G, q, g), A = g^r, B = h, C = h^r), r) : B^r = C\}$ over cyclic groups G_q of prime-order q. Typically, G is the subgroup of quadratic residues of $\mathbb{Z}_p$ for prime $p = 2q + 1$. We next briefly describe Σ-protocol $\Pi_{\mathcal{DH}} = (\mathcal{P}_{\mathcal{DH}}, \mathcal{V}_{\mathcal{DH}})$ for $\mathcal{R}_{\mathcal{DH}}$.

Common input: instance x and language DH.
Private input of $\mathcal{P}_{\mathcal{DH}}$: r.
The protocol $\Pi_{\mathcal{DH}}$:

1. $\mathcal{P}_{\mathcal{DH}}$ picks $t \in \mathbb{Z}_q$ at random, computes and sends $a = g^t$, $b = h^t$ to $\mathcal{V}_{\mathcal{DH}}$;
2. $\mathcal{V}_{\mathcal{DH}}$ chooses a random challenge $e \in \mathbb{Z}_q$ and sends it to $\mathcal{P}_{\mathcal{DH}}$;
3. $\mathcal{P}_{\mathcal{DH}}$ computes and sends $z = t + er$ to $\mathcal{V}_{\mathcal{DH}}$;
4. $\mathcal{V}_{\mathcal{DH}}$ checks $g^z = a \cdot A^e$ AND $h^z = b \cdot C^e$ accepts if and only if it is the case.

We show the special HVZK simulator Sim for $\Pi_{\mathcal{DH}}$. Sim, on input x and a challenge e of length $|q| - 1$ executes the following steps:

1. randomly chooses $z \in \mathbb{Z}_q$;
2. computes $a = g^z \cdot A^{-e}$;
3. computes $b = h^z \cdot C^{-e}$.

Graph Isomorphism. We show a Σ-protocol $\Pi_{\mathcal{GH}} = (\mathcal{P}_{\mathcal{GH}}, \mathcal{V}_{\mathcal{GH}})$ to prove that two graphs are isomorphic. Given two graphs G_0 and G_1, prover $\mathcal{P}_{\mathcal{GH}}$ wants to convince verifier $\mathcal{V}_{\mathcal{GH}}$ that he knows a permutation ϕ such that $\phi(G_0) = G_1$.

Common input: theorem $x = (G_0, G_1)$.
Private input of $\mathcal{P}_{\mathcal{GH}}$: ϕ.
The protocol $\Pi_{\mathcal{GH}}$:

1. $\mathcal{P}_{\mathcal{GH}}$ randomly chooses a permutation ψ and a bit $b \in \{0, 1\}$, computes and sends $P = \psi(G_b)$;
2. $\mathcal{V}_{\mathcal{GH}}$ chooses and sends a random bit $b' \in \{0, 1\}$ $\mathcal{P}_{\mathcal{GH}}$;
3. $\mathcal{P}_{\mathcal{GH}}$ sends the permutation τ to $\mathcal{V}_{\mathcal{GH}}$, where

$$\tau = \begin{cases} \psi & if \ b = b' \\ \psi\phi^{-1} & if \ b = 0, b' = 1 \\ \psi\phi & if \ b = 1, b' = 0 \end{cases}$$

4. $\mathcal{V}_{\mathcal{GH}}$ accepts if and only if $P = \tau(G_{b'})$.

Computational Effort: Two Cases. We show a summary of the comparison among our transform and Lindell's transform in Tables 2 and 3. The cost is measured by considering the computations in terms of number of exponentiations made by $\mathcal{P}$ and of $\mathcal{V}$. In our comparison we consider that a CRS contains a DH tuple $((G_{\text{CRS}}, q_{\text{CRS}}, p_{\text{CRS}}, g_{\text{CRS}}), A_{\text{CRS}}, B_{\text{CRS}}, C_{\text{CRS}})$ with $|p_{\text{CRS}}| = n = 1024$, with security parameter n (therefore $|q_{\text{CRS}}| = 1023$). We consider two cases. In the first one we use the NIZK argument to prove that a tuple $((G, q, p, g), A, B, C)$ is a DH tuple; in particular we take in account two sub-cases: when $p = 1024$ and when $p = 2048$. In the second case we use the NIZK argument to prove the isomorphism between two graphs G_0 and G_1, and we assume that $k = n^2$ bits are needed to represent a graph with n vertices. We stress that Lindell's transform needs to commit the first round of a Σ-protocol (plus the instance to be proved, but for our comparison we ignore that the instance has to be committed) associated to the language that we take into account (the language of the DH tuples or the language of the isomorphic graphs). Therefore, using the described CRS, to commit to a string of 1023 bit, 4 exponentiations are required. This is a consequence of the fact that the commitment is made by executing the simulator associated with $\Pi_{\mathcal{DH}}$ (with $|q_{\text{CRS}}| = 1023$).

Case 1: proving that a tuple is a DH tuple.

- [43]. When the instance to be proved is $((G, q, p, g), A, B, C)$ with $p = 1024$, the prover $\mathcal{P}$ needs to compute $a = g^t$, $b = h^t$ (as describe before) and needs to commit to them. The total size of a and b is 2048 bits, therefore to commit to 2048 bits we need to execute the DM commitment 3 times. This implies that the prover needs to compute $3 \cdot 4$ exponentiations mod p_{CRS} and 2 exponentiations mod p. The verifier $\mathcal{V}$ needs to checks if open of the DM commitments was correct, and also needs to compute $g^z = a \cdot A^e p$ and $h^z = b \cdot C^e$. For this reason the verifier needs to compute $3 \cdot 4$ exponentiations mod p_{CRS} plus 4 exponentiations mod p. With the same arguments we can count the amount of exponentiations needed to prove that the instance is a DH tuple with $p = 2048$.
- Our transform. When $|p| = 1024$ (resp., $|p| = 2048$) the prover need to run the simulator Sim of $\Pi_{\mathcal{DH}}$ with the instance $((G_{\text{CRS}}, q_{\text{CRS}}, p_{\text{CRS}}, g_{\text{CRS}}), A_{\text{CRS}}, B_{\text{CRS}}, C_{\text{CRS}})$ (this costs 4 exponentiations), also we need to compute $a = g^t$, $b = h^t$. The total number of exponentiations is 6 (2 exponentiations mod p, and 4 exponentiations mod p_{CRS}). The verifier needs to perform two times the verifier's algorithm for $\Pi_{\mathcal{DH}}$, one with the instance $((G_{\text{CRS}}, q_{\text{CRS}}, p_{\text{CRS}}, g_{\text{CRS}}), A_{\text{CRS}}, B_{\text{CRS}}, C_{\text{CRS}})$, the other one with the instance $((G, q, p, g), A, B, C)$, for a total amount of 4 exponentiations mod p_{CRS}, and 4 exponentiations mod p.

Case 2: Graph isomorphism.

- [43]. We consider that the instance to be proved is composed by two graphs (G_0, G_1). Also we assume that to represent one graph with n vertices $k = n^2$ bits are necessary. In this case we remark that because the security parameter is $n = 1024$ we need to execute n times the protocol $\Pi_{\mathcal{GH}}$ described before.

For the described assumptions we have that the first round of $\Pi_{\mathcal{GH}}$ is $P = \sigma(G_b)$ and $|P| = n^2$. Therefore the prover needs to run n executions of the DM commitment function to commit to P, where each of them costs 4 exponentiations. Also we need to execute n iteration of this process, for a total amount of $4n^2$ exponentiations mod p_{CRS}. Even in this case the verifier needs to checks if all opens with respect to the n commitments are correctly computed for a total amount of $4n^2$ exponentiations mod p_{CRS}.

- Our transform. In this case the prover $\mathcal{P}$ computes only 2 exponentiations mod p to compute the first round of $\Pi_{\mathcal{DH}}$. The verifier runs the verifier's algorithm of $\Pi_{\mathcal{DH}}$ that costs 4 exponentiations mod p.

Acknowledgments. We thank Alessandra Scafuro and Berry Schoenmakers for various useful discussions on Σ-protocols. An updated version of this work appears in [17].

A   Dual Mode Commitments and the Need for *Strong* Σ-protocols

The following definition of a dual-mode commitment scheme (DMCS, in short) is from [43].

Definition 9 ([43]). *A dual-mode commitment scheme (DMCS) is a tuple of PPT algorithms* (GenCRS, Com, Scom) *such that:*

- GenCRS(1^n) *outputs a common reference string, denoted by ρ.*
- (GenCRS, Com): *when $\rho \leftarrow$ GenCRS(1^n) and $m \in \{0,1\}^n$, algorithm* Com$_\rho(m; r)$ *with randomness r is a non-interactive perfectly-binding commitment scheme.*
- (Com, Scom): *For every PPT adversary $\mathcal{A}$ and every polynomial $p(\cdot)$, the output of the following two experiments is computationally indistinguishable:*

$Real_{\text{Com},\mathcal{A}}(1^n)$	$Simulation_{\text{Scom}}(1^n)$
- $\rho \leftarrow$ GenCRS(1^n)	- $\rho \leftarrow$ Scom(1^n)
- For $i = 1, \ldots, p(n)$:	- For $i = 1, \ldots, p(n)$:
1. $m_i \leftarrow \mathcal{A}(\rho, \boldsymbol{c}, \boldsymbol{r})$	1. $c_i \leftarrow$ Scom
2. $r_i \leftarrow \{0,1\}^{\text{poly}(n)}$	2. $m_i \leftarrow \mathcal{A}(\rho, \boldsymbol{c}, \boldsymbol{r})$
3. $c_i = $ Com$_\rho(m_i; r_i)$	3. $r_i \leftarrow$ Scom(m_i)
4. Set $\boldsymbol{c} = c_1, \ldots, c_i$ and $\boldsymbol{r} = r_1, \ldots, r_i$	4. Set $\boldsymbol{c} = c_1, \ldots, c_i$ and $\boldsymbol{r} = r_1, \ldots, r_i$
- Output $\mathcal{A}(\rho, m_1, r_1, \ldots, m_{p(n)}, r_{p(n)})$	- Output $\mathcal{A}(\rho, m_1, r_1, \ldots, m_{p(n)}, r_{p(n)})$

Membership-Hard Languages with Efficient Sampling. Lindell defines a membership-hard language Λ as a language such that one can efficiently sampleboth instances that belong to the language and instances that do not belong to the language. Still distinguishing among these two types of instances is hard. This is formalized through a sampling algorithm S_Λ that on input a bit b outputs an instance $\rho \in \Lambda$ along with a witness ω when $b = 0$, and outputs an instance $\rho \notin \Lambda$ otherwise. No polynomial-time distinguisher on input ρ can guess b with probability non-negligibly better than $1/2$. Let S_Λ^ρ denote the instance part of the output (i.e., without the witness when b is 0).

Definition 10 ([43]). *Let Λ be a language. We say that Λ is membership-hard with efficient sampling if there exists a PPT sampler S_Λ such that for every PPT distinguisher $\mathcal{D}$ there exists a negligible function μ such that:*

$$|\mathrm{Prob}\,[\,\mathcal{D}(S_\Lambda^\rho(0, 1^n), 1^n) = 1\,] - \mathrm{Prob}\,[\,\mathcal{D}(S_\Lambda(1, 1^n), 1^n) = 1\,]\,| \leq \mu(n).$$

There are several popular membership-hard languages in literature. We will in particular consider the one considered by Lindell in [43]: the language DH of Diffie-Hellman triples.

Lindell's construction of a DMCS from Σ-protocols. Let us describe Lindell's construction of a DMCS from any membership-hard language Λ admitting a Σ-protocol $\Pi_\Lambda = (\mathcal{P}_\Lambda, \mathcal{V}_\Lambda)$ with simulator Sim_Λ for perfect special HVZK.

Regular ρ generation: Run sampler S_Λ for Λ with input $(1, 1^n)$ and receive back ρ (recall that $\rho \notin \Lambda$).
Commitment: To commit to a value $m \in \{0, 1\}^n$ with randomness r, Com sets $e = m$, runs $\mathsf{Sim}_\Lambda(\rho, e)$ with randomness r and obtains (a, z). The output of Com is the commitment $c = a$ and the decommitment information (e, r).
Decommitment: To decommit, provide e, z and the receiver checks that $\mathcal{V}_\Lambda(\rho, a, e, z) = 1$.
Simulator Scom:
  – On input 1^n, Scom runs the sampler S_Λ with input $(0, 1^n)$, and receives back (ρ, ω) (recall that $\rho \in \Lambda$ and ω is a witness to this fact). Then, Scom computes $a = \mathcal{P}_\Lambda(\rho, \omega)$, sets $c = a$ and outputs (c, ρ).
  – On input $m \in \{0, 1\}^n$, Scom sets $e = m$ and outputs $z = \mathcal{P}_\Lambda(\rho, \omega, a, e)$.

A.1   A Subtlety in Lindell's Construction: The Need of Strong Σ-protocols

We now discuss a subtlety in the construction of a DMCS from any Σ-protocol for a membership-hard language given in [43]. We stress that the content of this section does not apply when considering [42].

We observe that the construction of a DMCS from any Σ-protocol for a membership-hard language given in [43] works when the Σ-protocol is equipped with a simulator such that when the simulator gets as randomness the 3rd round of the prover, then the simulator is able to output the *same* first round of the prover. This special property has been investigated in [26] where it was called *strong* perfect special HVZK. In more details, a Σ-protocol is strong perfect special HVZK if it admits a simulator Sim that on input any challenge e outputs a transcript (a, e, z) that is perfectly indistinguishable from the distribution of the transcript generated by the prover when the challenge is e, but in addition it is required that the transcript is computed by sampling the 3rd round uniformly at random. The strong perfect special HVZK property is formalized below.

Definition 11 ([26]). *The special perfect HVZK property is strong if there exists a PPT simulator* Sim *for the special perfect HVZK property that on input $x \in L_\mathcal{R}$ and a challenge "e" works by sampling the 3rd round "z" uniformly at random and then computing the 1st round "a" deterministically from "x, e" and "z".*

Lindell's construction of a DMCS showed in [43] requires a simulator for strong perfect special HVZK.
A Σ-protocol , Π_{DH} *for DH.* Now we show an artificial but useful example that shows a Σ-protocol with a simulator Sim for perfect special HVZK that however does not works if strong perfect special HVZK is desired.

The most widely used Σ-protocol $\Pi_{DH} = (\mathcal{P}_{DH}, \mathcal{V}_{DH})$ for the language DH consists in running in parallel two instances of a Σ-protocol for $DLog$ each proving knowledge a discrete logarithm. The two instances are linked together by having the verifier send the same challenge and expecting to receive the same third-round message. Schnorr's protocol [50] constitutes a natural choice for a Σ-protocol for $DLog$.

Consider instead instantiating the Σ-protocol for DH with the following Σ-protocol $\Pi_{DLog} = (\mathcal{P}_{DLog}, \mathcal{V}_{DLog})$ for proving knowledge of the discrete logarithm w of x with base g. $\mathcal{P}_{DLog}$ first selects another random group element x' along with its discrete logarithm w' to the base g and then sends x' to $\mathcal{V}_{DLog}$. Then $\mathcal{P}_{DLog}$ and $\mathcal{V}_{DLog}$ run two instances of Schnorr's Σ-protocol using the same challenge so that $\mathcal{P}_{DLog}$ proves to $\mathcal{V}_{DLog}$ knowledge of both w and w'. Clearly, Π_{DLog} is a Σ-protocol for $DLog$ (this comes from the fact that the AND of two Σ-protocols is still a Σ-protocol and from the fact that knowledge of a pair (w, w') implies knowledge of w) and, consequently, Π_{DH} instantiated with Π_{DLog} is a Σ-protocol for DH. Moreover notice that Π_{DLog} admits a simulator $\mathsf{Sim}^\star_{DLog}$ for perfect HVZK that uses the simulator of Schnorr's protocol to compute the transcript of the first instance, while it uses the prover of Schnorr's protocol for producing the transcript associated to x', after having selected x' along with a witness w' when the protocol starts. We now provide a formal description of this Σ-protocol.

More precisely we show a Σ-protocol $\Pi_{DLog} = (\mathcal{P}_{DLog}, \mathcal{V}_{DLog})$ for relation $\mathcal{R}_{DLog} = \{((\mathcal{G}, g, q, x), w) : x = g^w\}$ that is special perfect HVZK and such that there exists a simulator for special perfect HVZK that does not satisfy the requirement of *strong* perfect special HVZK of Π_{DLog} (see Definition 11).

Common Input: $(\mathcal{G}, g, q, x)$ and relation $\mathcal{R}_{DLog}$.
Input of $\mathcal{P}_{DLog}$: w t.c $((\mathcal{G}, g, q, x), w) \in \mathcal{R}_{DLog}$.
The protocol Π_{DLog}:

1. $\mathcal{P}_{DLog}$ chooses r_0, r_1, w_1 at random from $\mathcal{Z}_q$, and g_1 at random from $\mathcal{G}$. Then it computes $(a_0, a_1) = (g^{r_0}, g_1^{r_1})$, and $x_1 = g_1^{w_1}$. $\mathcal{P}_{DLog}$ sends (a_0, g_1, x_1, a_1) to $\mathcal{V}_{DLog}$.

2. $\mathcal{V}_{DLog}$ chooses a random challenge $e \leftarrow \{0,1\}^l$ (where $2^l < q$) and sends e to $\mathcal{P}_{DLog}$.
3. $\mathcal{P}_{DLog}$ computes $z_0 = r_0 + ew$ and $z_1 = r_1 + ew_1$ it sends (z_0, z_1) to $\mathcal{V}_{DLog}$.
4. $\mathcal{V}_{DLog}$ checks $g^{z_0} = a_0 x^e$ and $g_1^{z_1} = a_1 x_1^e$ accepts if and only if it is the case.

Special HVZK The simulator Sim of Π_{DLog} on input the theorem $(\mathcal{G}, g, q, x)$ and challenge e works as follows:

1. pick z_0, r_1, w_1 at random from $\mathcal{Z}_q$ and g_1 at random from $\mathcal{G}$.
2. compute $a_0 = g^{z_0} x^{-e}$ and $a_1 = g_1^{r_1}$.
3. compute $x_1 = g_1^{w_1}$ and $z_1 = r_1 + ew_1$.
4. return $(a_0, g_1, x_1, a_1, z_0, z_1)$.

Completeness. In order to see that completeness holds, observe that when $\mathcal{P}_{DLog}$ runs the protocol honestly we have:

$$g^{z_0} = g^{r_0 + we} = g^{r_0} \cdot g^{we} = a_0 \cdot x^e \quad \text{and} \quad g_1^{z_1} = g_1^{r_1 + w_1 e} = g_1^{r_1} \cdot g_1^{w_1 e} = a_1 \cdot x_1^e.$$

Special Soundness. Let $(a_0, g_1, x_1, a_1, e, z_0, z_1)$ $(a_0, g_1, x_1, a_1, e', z_0', z_1')$ be a collision. We have that $g^{z_0} = a_0 x^e$ and $g^{z_0'} = a_0 x^{e'}$, and thus we have $g^{z_0 - z_0'} = x^{e - e'}$ that implies that $x = g^{\frac{z_0 - z_0'}{e - e'}}$, therefore $w = \frac{z_0 - z_0'}{e - e'}$.

Special Perfect HVZK. We now check that the transcript returned by Sim, on input the theorem $(\mathcal{G}, g, q, x)$ and challenge e, is identically distributed w.r.t. the transcript obtained from the interaction between $\mathcal{P}_{DLog}$ and $\mathcal{V}_{DLog}$, when the challenge is e. The transcript differs only in the computation of a_0 and z_0. In the case of the $\mathcal{P}_{DLog}$ $a_0 = g^{r_0}$ where r_0 is chosen uniformly at random and $z_0 = r_0 + ew$. Instead, Sim chooses z_0 uniformly at random and $r_0 = z_0 - ew$, therefore clearly Sim and $\mathcal{P}_{DLog}$ produce a_0 and z_0 with the same distribution.

Π_{DH} *does not produce a DMCS.* We observe that Lindell's construction of a DMCS from any Σ-protocol for a membership-hard language [43] does not seem to work when Π_{DH} is used as Σ-protocol. Indeed consider the steps of experiments $Real_{Com, \mathcal{A}}(1^n)$ and $Simulation_{Scom}(1^n)$ in which $\mathcal{A}$ obtains as input (ρ, c, r) and consider iteration with $i = 2$ of the loop.

In $Real_{Com, \mathcal{A}}(1^n)$, $\mathcal{A}$'s view includes (m_1, r_1, c_1) and thus $\mathcal{A}$ can check that indeed c_1 is the output of $Com(m_1; r_1)$. This means that in the above construction, c_1 is the first component of the pair given in output by $Sim_\Lambda(\rho, e)$ when running with randomness r_1, and this is precisely the way in which c_1 was produced in Step 3 when $i = 1$. Therefore the check of $\mathcal{A}$ succeeds in $Real_{Com, \mathcal{A}}(1^n)$.

In $Simulation_{Scom}(1^n)$, $\mathcal{A}$'s view includes (m_1, r_1, c_1) and thus $\mathcal{A}$ can still perform the check that c_1 is the output of $Com(m_1; r_1)$ by running $Sim_\Lambda(\rho, e)$ with randomness r_1. However, in this case it is *not* true that c_1 is computed by running $Com(m_1; r_1)$. Indeed, in the execution of $Simulation_{Scom}(1^n)$, c_1 is computed by running $c_1 \leftarrow Scom$ and then r_1 is computed by running $r_1 \leftarrow Scom(m_1)$. In the above construction Scom computes c_1 and r_1 as the 1st and 3rd messages that are computed by $\mathcal{P}_\Lambda$ when the challenge is m_1. Therefore whenever the 3rd round r_1 computed by $\mathcal{P}_\Lambda$ does not correspond to a randomness that can be given as input to $\mathsf{Sim}_\Lambda(\rho, m_1)$ to get the same c_1 computed by $\mathcal{P}_\Lambda$, we have that the check of $\mathcal{A}$ fails.

By noticing that the 3rd round r_1 of $\mathcal{P}_{DH}$ in Π_{DH} does not give any information about the random instance x' of $DLog$ that $\mathcal{P}'_{DH}$ would compute and that would be part of c_1, we have that there exists a simulator for DH, using internally $\mathsf{Sim}^\star_{DLog}$, that on input (ρ, m_1) and running with randomness r_1 computes c_1 only with negligible probability and thus the above $\mathcal{A}$ is a successful distinguisher of experiments $Real_{\mathsf{Com}, \mathcal{A}}(1^n)$ and $Simulation_{\mathsf{Scom}}(1^n)$.

B   An Optimal-Sound (and Not Special Sound) 3-Round Perfect Special HVZK Proof

In this section we show a 3-round public-coin perfect special HVZK proof system that is optimal sound and not special sound. First of all we briefly describe the Σ-protocol of [44] to prove that, given a commitment and a message m, m is committed in com. Then we show the protocol of [51] that is a modification of [44] and given a commitment com and a value Ψ, allows to prove that the discrete logarithm of Ψ is committed in com.

In order to describe the protocol of [44] and [51] we consider two prime p and q s.t. $p = 2q + 1$, a group of order $\mathcal{G}$ of order q such that the DDH assumption is hard. Also we consider two random elements, g and h, taken from $\mathcal{G}$. We next describe Σ-protocol $\Pi_{Com} = (\mathcal{P}_{Com}, \mathcal{V}_{Com})$ of [44] for relation

$$\mathcal{R}_{Com} = \left\{ \left(\left((\mathcal{G}, q, g, h), v, \mathsf{com} = (\hat{g}, \hat{h}) \right), w \right) : \hat{g} = g^w, \hat{h} = h^{w+v} \right\}.$$

Common Input: $(\mathcal{G}, g, v, h, \mathsf{com} = (\hat{g}, \hat{h}), q)$ and relation $\mathcal{R}_{Com}$.
Input of $\mathcal{P}_{Com}$: w s.t. $((\mathcal{G}, v, g, h, \mathsf{com} = (\hat{g}, \hat{h}), q), w) \in \mathcal{R}_{Com}$.
The protocol Π_{Com}:

1. The prover $\mathcal{P}_{Com}$ chooses r from $\mathcal{Z}_q$ and sends $(\tilde{g} = g^r, \tilde{h} = h^r)$ to $\mathcal{V}_{Com}$;
2. The verifier $\mathcal{V}_{Com}$ chooses a random challenge $e \leftarrow \mathcal{Z}_q$ and sends e to $\mathcal{P}_{Com}$;
3. $\mathcal{P}_{Com}$ sends $z = ew + r$ to $\mathcal{V}_{Com}$;
4. $\mathcal{V}_{Com}$ checks that $\hat{g}^e \tilde{g} = g^z$ and $\left(\frac{\hat{h}}{h^v}\right)^e \tilde{h} = h^z$ accepts if and only if the checks are successful.

In [51] a similar protocol was used to prove that com is a commitment of the discrete logarithm of a value $\Psi \in \mathcal{G}$ with $h^\psi = \Psi$. Formally the protocol is for the NP language

$$L = \left\{ \left(\Psi = h^\psi, \mathsf{com} = (\hat{g} = g^w, \hat{h} = h^{w+\psi}) \right) : g, h \leftarrow \mathcal{G}, \psi \in \mathbb{Z}_q, w \in \mathbb{Z}_q \right\}$$

and for the corresponding relation

$$\mathcal{R}_L = \left\{ \left((\Psi = h^\psi, \mathsf{com} = (\hat{g} = g^w, \hat{h} = h^{w+\psi})), (w, \psi) \right) : g, h \leftarrow \mathcal{G}, \psi \in \mathbb{Z}_q, w \in \mathbb{Z}_q \right\}$$

The protocol follows Π_{Com} with the differences that the common input is $(\mathcal{G}, q, g, \Psi = h^\psi, h, \mathsf{com} = (\hat{g}, \hat{h})$ and that the verifier decide whether to accept or not checking if it holds that $\hat{g}^e \widetilde{g} = g^z$ and $\left(\frac{\hat{h}}{\Psi}\right)^e \widetilde{h} = h^z$. While this protocol preserves the perfect special HVZK property, it is not a proof of knowledge for $\mathcal{R}_L$ and neither special sound even though it still enjoys optimal soundness. We now proceed more formally.

Optimal soundness. We now consider an instance that is not in the NP language L, and show that, once the first round of the protocol is fixed, there exists only one challenge e s.t. the prover can answer successfully computing the third round z of the protocol. Consider the instance $\left(\Psi = h^\psi, \mathsf{com} = (\hat{g} = g^w, \hat{h} = h^{w+\psi\prime})\right) \notin L$ (with $\psi \neq \psi\prime$). Assume by contradiction that given the fist round of the protocol $(\widetilde{g}, \widetilde{h})$ there exist two distinct challenges e_0 and e_1 for which the prover can make the verifier accept with answers z_0, z_1 respectively. In the end we prove that $\psi = \psi\prime$.

Proof Since the verifier accepts, it must be that for all $i \in \{0, 1\}$, the following checks are successful: $\hat{g}^{e_i} \widetilde{g} = g^{z_i}$ and $\left(\frac{\hat{h}}{\Psi}\right)^{e_i} \widetilde{h} = h^{z_i}$. It follows that $\hat{g}^{e_0 - e_1} = g^{z_0 - z_1}$ and $\left(\frac{\hat{h}}{\Psi}\right)^{e_0 - e_1} = h^{z_0 - z_1}$. Suppose that $h = g^\omega$, we get

$$g^{w\omega(e_0 - e_1)} = \hat{g}^{(e_0 - e_1)\omega} = g^{(z_0 - z_1)\omega} = h^{(z_0 - z_1)} = (\hat{h} \cdot \Psi^{-1})^{e_0 - e_1}$$
$$= h^{z_0 - z_1} = g^{\omega(w + \psi\prime - \psi)(e_0 - e_1)}.$$

Therefore, if $e_0 \neq e_1$ we get the contradiction that $\psi = \psi\prime$.

The Protocol is not Special Sound for $\mathcal{R}_L$. To argue that the protocol of [51] is not special sound, we note that in order to compute a commitment of the discrete logarithm of Ψ, knowledge of this discrete logarithm is not necessary since it is possible to compute $\mathsf{com} = (\hat{g}, h^w \cdot \Psi)$ with $w \in \mathbb{Z}_q$. Indeed, notice that the discrete logarithm ψ of Ψ is never used in the proof. Formally, we suppose that the protocol is special sound for the polynomial relation $\mathcal{R}_L$ and then construct an adversary $\mathcal{A}$ that, given $Y = g^y \in \mathcal{G}$, returns the discrete logarithm y of Y.

We have shown that there exist 3-round public-coin proof systems that are optimal sound and not special sound. It also easy to observe that special soundness implies optimal soundness. Indeed, consider an NP-Language L and a corresponding relation $\mathcal{R}_L$. All Σ-protocols for $\mathcal{R}_L$ must also be 3-round HVZK proofs for L with optimal soundness. If not, than the violation of optimal soundness ($\mathcal{P}^*$ for a false statement can generate (a, c, z) and (a, c', z') with c' different from c and both accepting) implies directly also a violation of special soundness.

References

1. Almeida, J.B., Bangerter, E., Barbosa, M., Krenn, S., Sadeghi, A.-R., Schneider, T.: A certifying compiler for zero-knowledge proofs of knowledge based on sigma-protocols. In: Gritzalis, D., Preneel, B., Theoharidou, M. (eds.) ESORICS 2010. LNCS, vol. 6345, pp. 151–167. Springer, Heidelberg (2010)

2. Barak, B., Canetti, R., Nielsen, J.B., Pass, R.: Universally composable protocols with relaxed set-up assumptions. In: 45th Symposium on Foundations of Computer Science (FOCS 2004), Rome, Italy, 17–19 October 2004

3. Bellare, M., Rogaway, P.: Random oracles are practical: a paradigm for designing efficient protocols. In: CCS 1993, Proceedings of the 1st ACM Conference on Computer and Communications Security, Fairfax, Virginia, USA, pp. 62–73, 3–5 November 1993

4. Bitansky, N., Dachman-Soled, D., Garg, S., Jain, A., Kalai, Y.T., López-Alt, A., Wichs, D.: Why "fiat-shamir for proofs" lacks a proof. In: Sahai, A. (ed.) TCC 2013. LNCS, vol. 7785, pp. 182–201. Springer, Heidelberg (2013)

5. Blum, M., De Santis, A., Micali, S., Persiano, G.: Noninteractive zero-knowledge. SIAM J. Comput. 20(6), 1084–1118 (1991)

6. Blum, M., Feldman, P., Micali, S.: Non-interactive zero-knowledge and its applications. In: Proceedings of the 20th Annual ACM Symposium on Theory of Computing, Chicago, Illinois, USA, pp. 103–112, 2–4 May 1988

7. Canetti, R., Goldreich, O., Halevi, S.: The random oracle methodology, revisited. In: Proceedings of the Thirtieth Annual ACM Symposium on the Theory of Computing, Dallas, Texas, USA, pp. 209–218, 23–26 May 1998

8. Canetti, R., Goldreich, O., Halevi, S.: The random oracle methodology, revisited. J. ACM 51(4), 557–594 (2004)

9. Canetti, R., Lin, H., Paneth, O.: Public-coin concurrent zero-knowledge in the global hash model. In: Sahai, A. (ed.) TCC 2013. LNCS, vol. 7785, pp. 80–99. Springer, Heidelberg (2013)

10. Catalano, D., Dodis, Y., Visconti, I.: Mercurial commitments: minimal assumptions and efficient constructions. In: Halevi, S., Rabin, T. (eds.) TCC 2006. LNCS, vol. 3876, pp. 120–144. Springer, Heidelberg (2006)

11. Catalano, D., Visconti, I.: Hybrid trapdoor commitments and their applications. In: Caires, L., Italiano, G.F., Monteiro, L., Palamidessi, C., Yung, M. (eds.) ICALP 2005. LNCS, vol. 3580, pp. 298–310. Springer, Heidelberg (2005)

12. Catalano, D., Visconti, I.: Hybrid commitments and their applications to zero-knowledge proof systems. Theor. Comput. Sci. 374(1–3), 229–260 (2007)

13. Chaidos, P., Groth, J.: Making sigma-protocols non-interactive without random oracles. In: Katz, J. (ed.) PKC 2015. LNCS, vol. 9020, pp. 650–670. Springer, Heidelberg (2015)

14. Chase, M., Lysyanskaya, A.: On signatures of knowledge. In: Dwork, C. (ed.) CRYPTO 2006. LNCS, vol. 4117, pp. 78–96. Springer, Heidelberg (2006)

15. Ciampi, M., Persiano, G., Scafuro, A., Siniscalchi, L., Visconti, I.: Improved OR composition of Sigma-protocols. IACR Cryptology ePrint Archive 2015, 810 (2015). http://eprint.iacr.org/2015/810

16. Ciampi, M., Persiano, G., Scafuro, A., Siniscalchi, L., Visconti, I.: Improved OR composition of sigma-protocols. In: Theory of Cryptography - 13th Theory of Cryptography Conference, TCC 2016-A, Tel Aviv, Israel, 10–13 January 2016

17. Ciampi, M., Persiano, G., Siniscalchi, L., Visconti, I.: A transform for NIZK almost as efficient and general as the Fiat-Shamir transform without programmable random oracles. IACR Cryptology ePrint Archive, 770 (2015). http://eprint.iacr.org/2015/770

18. Cramer, R., Damgård, I.B., Schoenmakers, B.: Proof of partial knowledge and simplified design of witness hiding protocols. In: Desmedt, Y.G. (ed.) CRYPTO 1994. LNCS, vol. 839, pp. 174–187. Springer, Heidelberg (1994)

19. Damgård, I.B.: Efficient concurrent zero-knowledge in the auxiliary string model. In: Preneel, B. (ed.) EUROCRYPT 2000. LNCS, vol. 1807, pp. 418–430. Springer, Heidelberg (2000)

20. Damgård, I.: On Σ-protocol (2010). http://www.cs.au.dk/ivan/Sigma.pdf

21. Damgård, I.B., Fazio, N., Nicolosi, A.: Non-interactive zero-knowledge from homomorphic encryption. In: Halevi, S., Rabin, T. (eds.) TCC 2006. LNCS, vol. 3876, pp. 41–59. Springer, Heidelberg (2006)

22. Damgård, I., Groth, J.: Non-interactive and reusable non-malleable commitment schemes. In: Proceedings of the 35th Annual ACM Symposium on Theory of Computing, San Diego, CA, USA, pp. 426–437, 9–11 June 2003

23. De Santis, A., Di Crescenzo, G., Ostrovsky, R., Persiano, G., Sahai, A.: Robust non-interactive zero knowledge. In: Kilian, J. (ed.) CRYPTO 2001. LNCS, vol. 2139, pp. 566–598. Springer, Heidelberg (2001)

24. De Santis, A., Micali, S., Persiano, G.: Non-interactive zero-knowledge proof systems. In: Advances in Cryptology - CRYPTO 1987, A Conference on the Theory and Applications of Cryptographic Techniques, Proceedings, Santa Barbara, California, USA, pp. 52–72, 16–20 August 1987

25. Di Crescenzo, G., Visconti, I.: Concurrent zero knowledge in the public-key model. In: Caires, L., Italiano, G.F., Monteiro, L., Palamidessi, C., Yung, M. (eds.) ICALP 2005. LNCS, vol. 3580, pp. 816–827. Springer, Heidelberg (2005)

26. Dodis, Y.: G22.3220-001/g63.2180 Advanced Cryptography - Lecture 3 (Fall 2009)

27. Dwork, C., Naor, M.: Zaps and their applications. In: 41st Annual Symposium on Foundations of Computer Science, FOCS 2000, Redondo Beach, California, USA, pp. 283–293, 12–14 November 2000

28. Dwork, C., Naor, M.: Zaps and their applications. SIAM J. Comput. **36**(6), 1513–1543 (2007)

29. Feige, U., Lapidot, D., Shamir, A.: Multiple non-interactive zero knowledge proofs based on a single random string. In: 31st Annual Symposium on Foundations of Computer Science, St. Louis, Missouri, USA, vol. I, pp. 308–317, 22–24 October 1990

30. Feige, U., Lapidot, D., Shamir, A.: Multiple non-interactive zero knowledge proofs under general assumptions. SIAM J. Comput. **29**(1), 1–28 (1999)

31. Fiat, A., Shamir, A.: How to prove yourself: practical solutions to identification and signature problems. In: Odlyzko, A.M. (ed.) CRYPTO 1986. LNCS, vol. 263, pp. 186–194. Springer, Heidelberg (1987)

32. Garay, J.A., MacKenzie, P., Yang, K.: Strengthening zero-knowledge protocols using signatures. J. Cryptology **19**(2), 169–209 (2006)

33. Garay, J.A., MacKenzie, P.D., Yang, K.: Strengthening zero-knowledge protocols. In: Biham, E. (ed.) EUROCRYPT 2003. LNCS, vol. 2656, pp. 177–194. Springer, Heidelberg (2003)

34. Goldreich, O., Micali, S., Wigderson, A.: Proofs that yield nothing but their validity and a methodology of cryptographic protocol design. In: 27th Annual Symposium on Foundations of Computer Science, Toronto, Canada, pp. 174–187, 27–29 October 1986

35. Goldwasser, S., Kalai, Y.T.: On the (in)security of the fiat-shamir paradigm. In: 44th Symposium on Foundations of Computer Science (FOCS 2003), Proceedings, Cambridge, MA, USA, pp. 102–113, 11–14 October 2003

36. Goldwasser, S., Micali, S., Rackoff, C.: The knowledge complexity of interactive proof-systems. In: Proceedings of the 17th Annual ACM Symposium on Theory of Computing, Providence, Rhode Island, USA, pp. 291–304, 6–8 May 1985

37. Goldwasser, S., Micali, S., Rackoff, C.: The knowledge complexity of interactive proof systems. SIAM J. Comput. **18**(1), 186–208 (1989)
38. Groth, J.: Honest verifier zero-knowledge arguments applied. Dissertation Series DS-04-3, BRICS. PhD thesis, xii+119 (2004)
39. Groth, J., Ostrovsky, R., Sahai, A.: Perfect non-interactive zero knowledge for NP. In: Vaudenay, S. (ed.) EUROCRYPT 2006. LNCS, vol. 4004, pp. 339–358. Springer, Heidelberg (2006)
40. Groth, J., Sahai, A.: Efficient non-interactive proof systems for bilinear groups. In: Smart, N.P. (ed.) EUROCRYPT 2008. LNCS, vol. 4965, pp. 415–432. Springer, Heidelberg (2008)
41. Lindell, Y.: An efficient transform from Sigma Protocols to NIZK with a CRS andnon-programmable random oracle. Cryptology ePrint Archive, Report 2014/710 (2014). http://eprint.iacr.org/2014/710/20150906:203011
42. Lindell, Y.: An efficient transform from Sigma Protocols to NIZK with a CRS and non-programmable random oracle. Cryptology ePrint Archive, Report 2014/710 (2014). http://eprint.iacr.org/2014/710/20150906:203011
43. Lindell, Y.: An efficient transform from sigma protocols to NIZK with a CRS and non-programmable random oracle. In: Dodis, Y., Nielsen, J.B. (eds.) TCC 2015, Part I. LNCS, vol. 9014, pp. 93–109. Springer, Heidelberg (2015)
44. Micciancio, D., Petrank, E.: Simulatable commitments and efficient concurrent zero-knowledge. In: Biham, Eli (ed.) EUROCRYPT 2003. LNCS, vol. 2656, pp. 140–159. Springer, Heidelberg (2003)
45. Naor, M., Yung, M.: Public-key cryptosystems provably secure against chosen ciphertext attacks. In: Proceedings of the 22nd Annual ACM Symposium on Theory of Computing, Baltimore, Maryland, USA, pp. 427–437, 13–17 May 1990
46. Ostrovsky, R., Pandey, O., Visconti, I.: Efficiency preserving transformations for concurrent non-malleable zero knowledge. In: Micciancio, D. (ed.) TCC 2010. LNCS, vol. 5978, pp. 535–552. Springer, Heidelberg (2010)
47. Pass, R.: On deniability in the common reference string and random oracle model. In: Boneh, D. (ed.) CRYPTO 2003. LNCS, vol. 2729, pp. 316–337. Springer, Heidelberg (2003)
48. Sahai, A.: Non-malleable non-interactive zero knowledge and adaptive chosen-ciphertext security. In: 40th Annual Symposium on Foundations of Computer Science, FOCS 1999, New York, NY, USA, pp. 543–553, 17–18 October 1999
49. Scafuro, A., Visconti, I.: On round-optimal zero knowledge in the bare public-key model. In: Pointcheval, D., Johansson, T. (eds.) EUROCRYPT 2012. LNCS, vol. 7237, pp. 153–171. Springer, Heidelberg (2012)
50. Schnorr, C.-P.: Efficient Identification and Signatures for Smart Cards. In: Brassard, G. (ed.) CRYPTO 1989. LNCS, vol. 435, pp. 239–252. Springer, Heidelberg (1990)
51. Visconti, I.: Efficient zero knowledge on the internet. In: Bugliesi, M., Preneel, B., Sassone, V., Wegener, I. (eds.) ICALP 2006. LNCS, vol. 4052, pp. 22–33. Springer, Heidelberg (2006)
52. Wee, H.: Zero knowledge in the random oracle model, revisited. In: Matsui, M. (ed.) ASIACRYPT 2009. LNCS, vol. 5912, pp. 417–434. Springer, Heidelberg (2009)
53. Yung, M., Zhao, Y.: Interactive zero-knowledge with restricted random oracles. In: Halevi, S., Rabin, T. (eds.) TCC 2006. LNCS, vol. 3876, pp. 21–40. Springer, Heidelberg (2006)
54. Yung, M., Zhao, Y.: Generic and practical resettable zero-knowledge in the bare public-key model. In: Naor, M. (ed.) EUROCRYPT 2007. LNCS, vol. 4515, pp. 129–147. Springer, Heidelberg (2007)

Improved OR-Composition of Sigma-Protocols

Michele Ciampi[1]([⊠]), Giuseppe Persiano[2], Alessandra Scafuro[3],
Luisa Siniscalchi[1], and Ivan Visconti[1]

[1] DIEM, University of Salerno, Salerno, Italy
{mciampi,lsiniscalchi,visconti}@unisa.it
[2] DISA-MIS, University of Salerno, Salerno, Italy
giuper@gmail.com
[3] Boston University and Northeastern University, Boston, USA
scafuro@bu.edu

In [18] Cramer, Damgård and Schoenmakers (CDS) devise an OR-composition technique for Σ-protocols that allows to construct highly-efficient proofs for compound statements. Since then, such technique has found countless applications as building block for designing efficient protocols.

Unfortunately, the CDS OR-composition technique works only if *both* statements are fixed before the proof starts. This limitation restricts its usability in those protocols where the theorems to be proved are defined at different stages of the protocol, but, in order to save rounds of communication, the proof must start even if not all theorems are available. Many round-optimal protocols ([21, 30,41,44]) crucially need such property to achieve round-optimality, and, due to the inapplicability of CDS's technique, are currently implemented using proof systems that requires expensive NP reductions, but that allow the proof to start even if no statement is defined (a.k.a., LS proofs from Lapidot-Shamir [31]).

In this paper we show an improved OR-composition technique for Σ-protocols, that requires only one statement to be fixed when the proof starts, while the other statement can be defined in the last round. This seemingly weaker property is sufficient for the applications, where typically one of the theorems is fixed before the proof starts. Concretely, we show how our new OR-composition technique can directly improve the round complexity of the efficient perfect quasi-polynomial time simulatable argument system of Pass [38] (from four to three rounds) and of efficient resettable WI arguments (from five to four rounds).

1 Introduction

Witness-Indistinguishable (WI) Proofs. WI[1] proofs are fundamental for the design of cryptographic protocols, particularly when they are also proofs of knowledge (PoK). In a WIPoK the prover $\mathcal{P}$ proves knowledge of a witness certifying the veracity of a statement $x \in L$ to a verifier $\mathcal{V}$. WIPoKs can be used directly in some applications (e.g., in identification schemes) or can be a building block for stronger security notions (e.g., for zero-knowledge proofs using the FLS [24] paradigm or for round-optimal secure computation [30]).

---

[1] We will use WI to mean both "witness indistinguishability" and "witness indistinguishable".

© International Association for Cryptologic Research 2016
E. Kushilevitz and T. Malkin (Eds.): TCC 2016-A, Part II, LNCS 9563, pp. 112–141, 2016.
DOI: 10.1007/978-3-662-49099-0_5

Round complexity of cryptographic protocols has been extensively studied both for its practical relevance and for its natural and conceptual interest. Regarding WIPoKs, we know from Blum's protocol [5] that 3-round WIPoKs exist for all NP languages under the sole assumptions that one-way permutations exist. This result is obtained by designing a WIPoK for the language of Hamiltonian graphs and then by leveraging on the NP-completeness of the language of Hamiltonian graphs. Under stronger cryptographic assumptions, 2-round WI proofs, called ZAPs, and non-interactive WI (NIWI) proofs have been shown in [4,23,28]. Neither ZAPs nor NIWI proofs are PoKs.

Since NPreductions are extremely expensive, several practical interactive PoKs have been designed for languages that are used in real-world cryptographic protocols (e.g., for proving knowledge of a discrete logarithm (DLog)). The study of such ad-hoc protocols mainly concentrates on a standardized form of a 3-round PoK referred to as Σ-protocol [19,42].

Σ-protocols. A Σ-protocol for an NPlanguage L with witness relation R_L is a 3-round proof system jointly run by a prover $\mathcal{P}$ and a verifier $\mathcal{V}$ in which $\mathcal{P}$ proves knowledge of a witness w for $x \in L$. In a Σ-protocol the only message sent by $\mathcal{V}$ is a random string. Such proof systems have two very useful properties: special soundness, which is a strong form of proof of knowledge, and special honest-verifier zero knowledge (SHVZK). The latter property basically says the following: if the challenge is known in advance, then by just knowing also the theorem, it is possible to generate an accepting transcript without using the witness. This is formalized through the existence of a special simulator, called the SHVZK simulator that, on input *a theorem* x and a challenge c, will output (a, z) such that (a, c, z) is an accepting 3-message transcript for x and is indistinguishable from the transcript produced by the honest prover when the challenge is c. Blum's protocol for Graph Hamiltonicity is an example of a Σ-protocol. Another popular example of Σ-protocols is Schnorr's protocol [42] for proving knowledge of a discrete logarithm.

The security provided by the SHVZK property is clearly insufficient as it gives no immediate guarantees against verifiers who deviates from the protocol. Despite of this, the success of Σ-protocols and their impact in various constructions [1,2,6,9–12,14,15,20,22,25,27,32,33,36,37,40,41,43] is a fact. This is due to a breakthrough of Cramer et al. [18] that adds WI to the security of Σ-protocol.

OR Composition of Σ-Protocols. Let L be a language that admits a Σ-protocol Π_L. In [18] it is shown how to use Π_L and its properties to construct a new Σ-protocol, Π_L^{OR}, for proving the OR composition of theorems in L *avoiding* the NPreduction by crucially exploiting the honest-verifier zero-knowledge (HVZK[2]) property of Π_L. The rationale behind the transformation can be informally explained as follows. The prover wishes to prove a statement of the form $((x_0 \in L) \vee (x_1 \in L))$. The naïve idea of simply running Π_L twice in parallel

---

[2] HVZK requires the existence of a simulator that by receiving in input the theorem gives in output an accepting triple (a, c, z). Clearly HVZK is implied by SHVZK.

would not work because the prover knows only one of the witnesses, say w_b, and cannot compute two accepting transcripts without knowing w_{1-b}. However, due to the HVZK property, the prover can generate an accepting transcript for $x_{1-b} \in L$ even without knowing w_{1-b}, by running the HVZK simulator Sim associated with Π_L. Indeed, Sim "only" needs in input the theorem x_{1-b} and will output the entire transcript, challenge included. The trick is then to generate the challenges for the two executions of Π_L, in such a way that the prover can control the challenge of exactly one of them (but not both), and set it to the value generated by Sim. Note that, if running the algorithm of Sim is as efficient as running the algorithm of $\mathcal{P}$, then the composed protocol is efficient. We stress that this OR-composition technique preserves SHVZK and will refer to it as the CDS-OR technique.

A very interesting property of this transformation, besides the fact that it does not need NPreduction, is that if Sim is a simulator for perfect HVZK then Π_L^{OR} is WI (this was shown in [18]). This result was further extended by Garay et al. [25] that noted that the CDS-OR technique can be used also for Σ-protocols that are computational HVZK. In this case the relation proved is slightly different, namely, starting with a relation $\mathcal{R}_L$ and instances x_0 and x_1, the resulting Π_L^{OR} protocol is computational WI for the relation

$$\mathcal{R}_L^{OR} = \{((x_0, x_1), w) : ((x_0, w) \in \mathcal{R}_L \wedge (x_1 \in L)) \vee ((x_1, w) \in \mathcal{R}_L \wedge (x_0 \in L))\}.$$

Input-Delayed Proofs. Often in cryptographic protocols there is a preamble phase that has the purpose of establishing, at least in part, a statement to be proven with a WI proof. In such cases, since one of the statements is fully specified only when the preamble is completed, the WI proof can start only after the preamble ends. Hence, the overall round complexity of protocols that follow this paradigm amounts to the sum of the round complexity of the preamble and of the WI proof.

In [31], Lapidot and Shamir (and later on Feige et al. in [24]) show a 3-round proof of knowledge for Hamiltonian Graphs which has the special property that a prover can compute the first round of the proof, *without* knowing the theorem to be proved (that is, the graph) but only needs to know its size (that is, the number of vertices). Such a 3-round protocol is a Σ-protocol (and thus satisfies the SHVZK property) and is a WI proof. We will refer to this protocol as LS. Also, we will call *input delayed* a Σ-protocol where the prover computes the first message without knowledge of the statement to be proved.

The input-delayed property directly improves the round complexity of all the cryptographic protocols that follow the paradigm described above. The reason is that now the WI proof can start even if the preamble that generates the statement is not completed yet. It is worthy to note that in many applications the preamble serves as a mean to generate some trapdoor theorem, that is used only in the security proof. The "honest" theorem instead is typically known already at the beginning of the protocol. This technique has been used extensively and, most notably, it led to the celebrated FLS paradigm that upgrades any WI proof system into a zero-knowledge (ZK) proof system.

The input-delayed property of LS has been instrumental to provide round-efficient constructions from general assumptions, such as: 4-round (optimal) secure 2PC where only one player gets the output (5 rounds when both players get the output) [30], 4-round resettable WI arguments [41,44], 4-round (optimal) resettable ZK for NP in the BPK model [41,44].

Despite being so influential to achieve round efficiency for cryptographic protocols, the power of LS unfortunately vanishes as soon as practical constructions are desired. Indeed, similarly to Blum's protocol, LS is crucially based on specific properties of Hamiltonian graphs. Thus, when used to prove more natural languages, which is the case of most of the applications using WI proofs, it requires to perform rather inefficient NP reductions.

Efficient Protocols and Limits of the CDS-OR Technique. A natural question is what happens if we want to avoid the NP reduction and we try to use the CDS-OR technique to construct input-delayed adaptive WI proofs. A bit more specifically, we know that there exist Σ-protocols that are input delayed. Schnorr's protocol [42] for DLog is such an example since the first message can be computed without knowing the instance, but only a group generator. Thus the question is what happens if we apply the CDS-OR technique to an input-delayed Σ-protocol. Do we obtain a WI Σ-protocol that is input delayed as well?

Unfortunately, the answer is negative. The CDS-OR technique does *not* preserve the input-delayed property, not even when used to compose two Σ-protocols that are both input delayed. To see why, recall that the CDS-OR composition technique when applied to Σ-protocol Π_L for language L requires the prover to compute two accepting transcripts, one of which is computed by running the HVZK simulator Sim. Recall that Sim needs in input the theorem to be proved. Hence, to prove knowledge of a witness for the compound theorem $(x_0 \in L \vee x_1 \in L)$, the prover, who knows one witness, say w_b, needs to know also x_{1-b} already at the first round to be able to run the simulator. Thus, in the CDS-OR technique the prover can successfully complete the protocol if and only if *both*[3] instances are specified already at the first round.

Because of this missing feature, the CDS-OR technique has limited power in allowing one to obtain round-efficient/optimal cryptographic protocols, compared to the number of rounds obtained by using LS. As such, in some cases when focusing on efficient constructions, the *best* round-complexity that we can achieve using efficient Σ-protocols and avoiding NP reductions needs at least one additional round, therefore requiring at least 5-round if one wants to match the previously mentioned applications (e.g., 5-round resettable ZK for NP in the BPK model [41,44] and 5-round resettable WI [41,44]) argument systems.

Additionally, we note that the CDS-OR technique is the bottleneck in the round-complexity of the 4-round straight-line perfect simulatable in quasi-polynomial time argument shown by Pass in [38]. This argument uses quasi-polynomial time simulation and, potentially, it would only need three rounds as any Σ-protocol. The additional first round is required precisely to define the

---

[3] Note that the WI property requires that the prover would be able to prove any of the two theorems, and thus potentially use the simulator on either x_0 or x_1.

trapdoor theorem. Hence, the following natural question arises: *Given a language L with an input-delayed Σ-protocol Π_L, is it possible to design an efficient Witness Indistinguishable Σ-protocol Π_{OR}^L for proving knowledge of a witness certifying that $(x_0 \in L \vee x_1 \in L)$ that does not require knowledge of both x_0 and x_1 to play the first round?*

1.1 Our Contribution

In this paper we answer the above question positively for a large class of Σ-protocols that includes *all* Σ-protocols used in efficient constructions. Specifically, we propose a new OR-composition technique for Σ-protocols that relaxes the need of having both instances fixed before the Σ-protocol starts. Our technique allows the composition of Σ-protocols for different languages and leads to improved round complexity in previous efficient constructions based on CDS-OR technique. Namely, we describe the following two results that we obtain by making use of our new OR-composition technique:

- Efficient 3-round straight-line perfect quasi-polynomial time simulatable argument system for a large class of useful languages. The previous construction required four rounds [38].
- Efficient 4-round rWI argument system. Previous constructions required five rounds [41,44].

Our new technique can also be used to replace LS towards obtaining efficient round-optimal resettable zero-knowledge arguments in the BPK model (using the constructions of [41,44]), round-optimal secure two-party computation (using the construction of [30]) and 4-round non-malleable commitments (using the construction of [26]).

Finally, we provide a precise classification of the Σ-protocols that can be used in our new OR-composition technique. In the following paragraphs we first provide a high-level description our OR-composition technique, then we discuss the applications in more details.

1.2 Our Techniques

Overview. We start by defining the setting we are considering. Let L_0 and L_1 be any pair of languages admitting Σ-protocols Π_0 and Π_1. We want to construct a Σ-protocol Π_L^{OR} for the language $L = L_0 \vee L_1$. An instance of L is a pair (x_0, x_1) and we want only x_0 to be specified before Π_L^{OR} starts while x_1 is specified only upon the last round of the protocol[4]. We assume that Π_1 is an *input-delayed* Σ-protocol and thus the first prover message of Π_1 can be computed without knowing x_1. As mentioned earlier this property is satisfied by popular Σ-protocols such as the ones for Discrete Log, Diffie-Hellman triples, and of course, LS itself.

---

[4] Like LS, we will just need the size of x_1 to be known when Π_L^{OR} starts.

Now, recall that the problem with the CDS-OR technique was that a prover needs to run Sim to compute the first round of the protocol, and this necessarily requires knowledge of *both* theorems before the protocol starts. We want instead that the prover uses only knowledge of x_0.

We solve this problem by introducing a new OR-composition technique that does not require the prover to run Sim on x_1 already in the first round. Instead, our technique allows the prover to wait and take action only in the third round when x_1 is finally defined.

Our starting point is the well known fact that given any Σ-protocol there exists an instance-dependent trapdoor commitment (IDTC) scheme where the witness for the membership of the instance in the language can be used as a trapdoor to open a committed message as any desired message, as in [20]. Our next observation is that, instead of having the prover send the first round for protocol Π_1 in the clear, we can have him send a commitment to it, and such commitment can be computed using an instance-dependent trapdoor commitment based on Π_0 with respect to instance x_0. Recall that this is possible, as in our setting we assume that Π_1 is an input-delayed Σ-protocol, so the prover can honestly compute the first message of Π_1 without knowing x_1. Therefore, the first round of our Π_L^{OR} protocol, is simply an IDTC of a honest Π_1's first round.

Later on, upon receiving the challenge c from the verifier, and after the theorem x_1 is defined, the prover computes the third round as follows. If she has received a witness for x_0, then she will run Sim on input (x_1, c) to compute an accepting transcript of Π_1 for x_1. Then, using the witness w_0 she will equivocate the commitment sent in the first round, according to the message output by Sim. Otherwise, if she has received a witness for x_1 then she does not need to equivocate: she will honestly open the commitment, and honestly compute the third message of Π_1. Therefore, the third round of Π_L^{OR}, simply consists of an opening of the IDTC together with the third message of Π_1.

Now note that this idea works only if we have a special IDTC scheme that has the following strong trapdoor property: a sender can equivocate even a commitment that has been computed honestly. Unfortunately, this property is not satisfied in general by any trapdoor commitment based on Σ-protocols, but only for some. This would restrict the class of Σ-protocols that we can use as L_0 in our technique. For example, this class would not contain Blum's protocol.

Our next contribution is the construction of IDTC schemes that satisfy this strong trapdoor property, for a large class of Σ-protocols. Towards this goal, we define the notion of a t-IDTC scheme which are IDTCs for which the ability to open a commitment in t ways implies knowledge of a witness for the instance associated with the commitment. Next, we construct 2-IDTC and 3-IDTC schemes based on two different classes of Σ-protocols, the union of which includes all the Σ-protocols that are commonly used in cryptographic protocols. Finally, we provide a general OR-composition technique for any pair of languages L_0 and L_1 such that L_0 has a t-IDTC scheme and L_1 has an input-delayed Σ-protocol.

t-Instance-Dependent Trapdoor Commitment Scheme.

For integer $t \geq 2$, a t-IDTC scheme for a polynomial-time relation $\mathcal{R}$ admitting Σ-protocol $\Pi_{\mathcal{R}}$

is a triple (TCom, TDec, TFake) where TCom, TDec are the honest commitment/decommitment procedures and TFake is the equivocation procedure that, given a witness for an instance x, equivocates any commitment with respect to x computed by TCom. The crucial differences between a t-IDTC scheme and a regular trapdoor commitment scheme are: (a) the trapdoor property is strong in the sense that knowledge of the trapdoor (that is, the witness of the instance x) allows to equivocate even commitments that have been honestly computed; (b) the binding property is relaxed: in a t-IDTC scheme, the sender can open the same commitment in $t - 1$ different ways, even without the trapdoor. This relaxation allows us to build an IDTC scheme from a wider class of Σ-protocols, which will cover all the Σ-protocols that have been used in literature.

Constructing a 2-IDTC Scheme. A 2-IDTC scheme can be directly constructed from any Σ-protocol Π_0 that has the following property: even if the first message a_0 was computed by the SHVZK simulator Sim, an accepting z_0 can be efficiently computed, for every challenge c_0, by using knowledge of the witness and of the randomness used by Sim to produce a_0. We call the Σ-protocols that satisfy this property, *chameleon* Σ-protocols, and we denote by $\mathsf{P}_{\mathsf{sim}}$ the special prover strategy that can answer any challenge even starting from a simulated a_0.

More precisely, given a chameleon Σ-protocol Π_0 for a language L_0, one can construct a 2-IDTC scheme as follows. Let $x_0 \in L_0$. To commit to a message m, the sender runs $\mathsf{Sim}(x_0, m; r_0)$ and obtains a_0, z_0. The commitment is the value a_0. The opening is the pair m, z_0. The commitment is accepted iff (x_0, a_0, m, z_0) is accepting. To equivocate a_0, as a message m', run the special prover algorithm $\mathsf{P}_{\mathsf{sim}}((x_0, m, r_0), w_0, m')$ and obtain an accepting z_0.

Constructing a 3-IDTC Scheme. We now discuss a different committing strategy that works for Σ-protocols in which the simulated first message a_0 can only be continued for the challenge specified by Sim, even if a witness is made available. Blum's protocol for Hamiltonicity is an example of such Σ-protocol.

To commit to m, the sender sends a pair (a_0, a_0') where, with probability $1/2$, a_0 is obtained by running $\mathsf{Sim}(x_0, m)$ while a_0' is computed by running the prover of Π_0, and with probability $1/2$ the above order is inverted. One can think of a commitment as composed of two threads: a *simulated* thread and a *honest* thread. To open the commitment, the prover sends m and z^*, and the verifier accepts the decommitment if m, z^* are accepting for one of the threads; namely, the verifier checks that either (a_0, m, z^*) or (a_0', m, z^*) is accepting for $x_0 \in L_0$. To equivocate (a_0, a_0') to a message m', the sender simply continues the thread of the honest prover, using m' as challenge and computes z^* using the witness. Clearly, a malicious sender can open in two different ways even when $x_0 \notin L$. Nevertheless, three openings allow the extraction of the witness for x_0.

When our OR-composition technique is instantiated with a 3-IDTC scheme we have that the resulting protocol is still WI since no power is added to the verifier. However the protocol is *not* a Σ-protocol since the special-soundness property is not guaranteed. The reason is that, in a 3-IDTC scheme the sender can open the commitment in two different ways even without having the trapdoor (i.e., the witness for $x_0 \in L_0$). Therefore, for any challenge c sent by $\mathcal{V}$, the fact that the commitment of a_1 can be opened in two ways gives a malicious prover $\mathcal{P}^*$ two chances (a_1, c, z_1) and (a_1', c, z_1') to successfully complete the protocol for a false statement x_1. Nevertheless, this extra freedom does not hurt soundness as both openings (i.e., a_1 and a_1') are fixed in advance, and thus when x_1 is not an instance of the language there exist only two challenges c' and c'' that would allow $\mathcal{P}^*$ to succeed. When the challenge is long enough the success probability of $\mathcal{P}^*$ is therefore negligible.

Our construction when starting from a 3-IDTC scheme is 3-special sound (i.e., answering to 3 challenges allows one to compute a witness efficiently), and therefore it is a proof of knowledge when the challenge is long enough.

1.3 Discussion

What Really Matters. Our new OR-composition technique works only when the theorem that has not been defined yet (i.e., x_1), admits an input-delayed Σ-protocol). We stress that this is not a limitation for the applications that we have in mind. In fact, in all *efficient* protocols that make use of input-delayed proofs that we are aware of, the preamble has always the purpose of generating the trapdoor theorem. In practical scenarios[5] L_1 usually corresponds to DLog or DDH. The fact that we can not have Blum's Σ-protocol for L_1 when L_1 is the language of Hamiltonian graphs, is therefore not relevant as the actual language of interest is L_0.

Comparison with the CDS-OR Technique. We remark that even in the extremely simplified case where:

1. the two instances x_0 and x_1 are for the same language L,
2. L admits an *input-delayed* Σ-protocol Π_L which is also special HVZK,
3. Π_L is chameleon and thus one can compute the first message using Sim and then continue with the prover to answer to arbitrary challenges,
4. the prover knows in advance the witness w and instance x_b for which she will be able to honestly complete the protocol,

the CDS-OR technique fails in obtaining a Σ-protocol (or a WIPoK) for the OR composition of instances of L if any one of the instances is not known when the protocol starts.

Beyond Schnorr's Protocol. The works of Cramer [16], Cramer and Damgård [17], and Maurer [34,35] showed that a protocol (referred to as the *Pre-Image Protocol*) for proving knowledge of a pre-image of a group homomorphism unifies and generalizes a large number of protocols in the literature. Classic Σ-protocols, such as Schnorr's protocol [42] and the Guillou-Quisquater protocol [29], are particular cases of this abstraction. We show that the *Pre-Image Protocol* is a chameleon Σ-protocol and can thus be used in our construction.

---

[5] These are the only scenarios of interest for our work since if practicality is not desired than one can just rely on the LS Σ-protocol and use NPreductions.

What Is In and What Is Out. As mentioned previously, the Σ-protocol for L_1 can be any *input-delayed* Σ-protocol. We now discuss which Σ-protocols can be used to instantiate L_0 in our OR transform. For this purpose, we identify four classes of Σ-protocols and we prove that any Σ-protocol that falls in any of the first three classes can be used in our OR transform (by instantiating either a 2-IDTC, or a 3-IDTC scheme).

We also identify a class of Σ-protocols that is not suitable for any of our techniques. Luckily, we have no example of natural Σ-protocols that fall in this class, and in order to prove the separation we had to construct a very contrived scheme. The four classes are listed below.

- *(Class 1)* Σ-protocols that *are* Chameleon and *do not* require the witness to compute the first round. This class of Σ-protocols can be used to construct both 2-IDTC and 3-IDTC schemes.
- *(Class 2)* Σ-protocols that *are* Chameleonand *require* the prover to use the witness already to compute the first round. This class of Σ-protocols can be used to construct a 2-IDTC scheme.
- *(Class 3)* Σ-protocols that *are not* Chameleon but *do not require* the prover to use the witness in the first round. This class of Σ-protocols can be used to construct a 3-IDTC scheme.
- *(Class 4)* Σ-protocols that *are not* Chameleon and *require* the witness to be used already in the first round. This class of Σ-protocols can not be used in our techniques.

The Input-Delayed Features. We stress here that our techniques allow to start and complete an efficient OR composition of two Σ-protocols (with the discussed restrictions) provided that one instance is known and another one will be known later. Having a witness for the first or the second instance always allows $\mathcal{P}$ to convince $\mathcal{V}$. This contrasts with the CDS-OR technique where knowing a witness for x_0 would block $\mathcal{P}$ immediately since $\mathcal{P}$ would need immediately x_1 to continue, but x_1 will not be available until the third round.

1.4   Applications

Our new OR-composition technique does not provide the full power of LS because it needs one theorem to be known before the protocol starts. However, as we show below, this seemingly weaker property suffices to improve the round-complexity of some of the previous constructions based on the CDS-OR technique. Such constructions aim to efficiently[6] transform a Σ-protocol for a relation $\mathcal{R}$ into a *round-efficient* argument with more appealing features.

Efficient 3-Round Straight-Line Perfect Quasi-Polynomial Time Simulatable Argument System. We achieve this result directly, using the construction of Pass [38] and replacing the CDS-OR technique with our technique. As a

---

[6] By *efficiently* we mean that no NPreduction is needed and only a constant number of modular exponentiations are added. We do not discuss the practicality of the resulting constructions.

result the first round of the verifier of [38] can be postponed, therefore reducing the round complexity from four to three rounds. Our construction works for all languages admitting a perfect chameleon Σ-protocol.

Efficient 4-Round Resettable WI Arguments. It is well known [8] how to transform a Σ-protocol into a resettable WI protocol: the verifier commits to the challenge c using a perfectly hiding commitment scheme and sends it to the prover in the first round; the prover then computes its messages with randomness derived by applying a pseudo-random function (PRF) on the commitment received. Soundness follows directly from the soundness of the Σ-protocol due to the perfect hiding of the commitment. WI follows from the fact that the protocol is zero knowledge against a stand-alone verifier and thus concurrent WI. Then the use of the PRF and the fact that all messages of the verifier are committed in advance upgrades concurrent WI to resettable WI. This approach, however, generates a 5-round protocol.

Achieving the same result *efficiently*, namely, avoiding NP reductions, in only four rounds is non-trivial. The reason is that if we attempt to replace the 2-round perfectly hiding commitment with a non-interactive commitment, we lose the unconditional soundness property, and then it is not clear how to argue about computational soundness. More specifically, black-box extraction of the witness is not possible (black-box extraction and resettable WI can not coexist) and the adversarial prover could try to maul the commitment of the verifier and adaptively generate the first round of the Σ-protocol. In fact, even allowing complexity-leveraging arguments (and thus, straight-line extraction), constructing a 4-round WI argument system that avoids NP reductions and adds only a few modular exponentiations to the underlying Σ-protocol has remained so far an open problem.

We solve this problem by using our new OR-composition technique. We have the verifier commit to the challenge in the first round, but then later, instead of sending the decommitment, she will directly send the challenge and prove that either the challenge is the correct opening of the commitment or she solved some hard puzzle (in our construction, computing the Discrete Log of a random group element chosen by the prover). The puzzle is sent by the prover in the second round and it will be solved by the reduction in super-polynomial time in the proof of soundness.

This trick has been proposed in literature in various forms [21, 38] and we are using the form used in [21] where the puzzle is sent only in the second round. [21] must use the LS transform and therefore needs NP-reduction. As explained earlier, going through LS *was* necessary as the CDS-OR transform can be applied only if both statements are fixed at the beginning.

Our new OR transform solves precisely this problem, and it allows the verifier to start the proof before the puzzle is defined, and this proof can be done efficiently without NP reductions.

Resettable WI follows from the CGGM transformation and the WI property of the proof generated by the prover. The groups used for the commitment of the challenge and for the puzzle sent by the prover, will be chosen appropriately so that the hardness of computing discrete logarithms are different and guarantee that our reductions work (i.e., we make use of complexity leveraging).

Further Applications. Our new OR-composition technique can find various other applications. Indeed, wherever there is a round-efficient (but otherwise inefficient) construction based on the use of LS without a corresponding efficient construction with the *same* round complexity, then our technique constitutes a powerful tool towards achieving computationally efficient and round-efficient constructions. For instance, the 4-round (optimal) resettable ZK argument systems in the BPK model provided in [41,44], consists (roughly) of the parallel execution of a (resettable) WI protocol from the prover to the verifier, where the prover proves that either $x \in L$ or he knows the secret key associated to the public identity of the verifier, and a 3-round (resettably-sound) WI protocol from the verifier to the prover, where $\mathcal{V}$ proves knowledge of the secret key associate to its public key, or knowledge of the solution of a puzzle computed by the prover. When instantiated with efficient Σ-protocols, such construction requires 5-rounds, where the additional round, from the prover to the verifier, is used to send the puzzle necessary for the verifier to start a proof using the CDS-OR technique. We observe that this setting closely resembles the setting of the 4-round resettable WI (rWI) protocol that we provide in this paper. As such, one could directly instantiate the proof provided by the prover of the BPK model, with our 4-round rWI protocol, and have the verifier just prove knowledge of its secret keys, thus avoiding the need of the additional first round.

Our OR-composition technique could also be useful in replacing the use of LS in the 4-round non-malleable commitment scheme of [26], and in the round-optimal secure two-party computation protocol of [30].

1.5 Open Problems

Our OR-composition technique relaxes the requirement of CDS-OR of requiring knowledge of *all* instances already at the beginning of the protocol. However still our result does not match the power of LS where *no* theorem is required for the protocol to start. An immediate open question is whether one can improve our OR transform so that the first round can be run without the knowledge of any theorem. Perhaps a first step in this direction would be to answer a related relaxed question, which is to design an OR transform for proving (still preserving WI) knowledge of 1 out of n theorems and that requires knowledge of (at least some) theorems only after the second round. It would also be interesting to extend our technique in order to make it applicable to *all* Σ-protocols.

2 Definitions

In this section we set-up our notation and give some useful definitions. More definitions can be found in the full version.

We denote the security parameter by λ. If A is a probabilistic algorithm then $A(x)$ denotes the probability distribution of the output of A when it receives x as input. By $A(x; R)$ instead we denote the output of A on input x when coin tosses R are used as randomness.

A *polynomial-time relation* $\mathcal{R}$ (or, simply, a *relation*) is a subset of $\{0,1\}^\star \times \{0,1\}^\star$ for which membership of (x, w) to $\mathcal{R}$ can be decided in time polynomial in $|x|$. We define the NP-language $L_\mathcal{R}$ as $L_\mathcal{R} = \{x | \exists w : (x, w) \in \mathcal{R}\}$. If $(x, w) \in \mathcal{R}$, we say that w is a *witness* for *instance* x. Following [25], we define $\hat{L}_\mathcal{R}$ to be the input language that includes both $L_\mathcal{R}$ and all well formed instances that do not have a witness. More formally, $L_\mathcal{R} \subseteq \hat{L}_\mathcal{R}$ and membership in $\hat{L}_\mathcal{R}$ can be tested in polynomial time. We implicitly assume that the verifier of a protocol for relation $\mathcal{R}$ executes the protocol only if the common input x belongs to $\hat{L}_\mathcal{R}$ and rejects immediately common inputs not in $\hat{L}_\mathcal{R}$.

Number-Theoretic Assumptions. We define *group generator* algorithms to be probabilistic polynomial-time algorithms that take as input security parameter 1^λ and output $(\mathcal{G}, q, g)$, where $\mathcal{G}$ is (the description of) a cyclic group of order q and g is a generator of $\mathcal{G}$. We assume that membership in $\mathcal{G}$ and its group operations can be performed in time polynomial in the length of q and that there is an efficient procedure to randomly select elements from $\mathcal{G}$. Moreover, with a slight abuse of notation, we will use $\mathcal{G}$ to denote the group and its description.

We consider the sub-exponential versions of the DLog and of the DDH assumptions that posit the hardness of the computation of discrete logarithms and of breaking the Decisional Diffie-Hellman assumption with respect to the group generator algorithm IG that, on input λ, randomly selects a λ-bit prime q such that $p = 2q + 1$ is also prime and outputs the order q group $\mathcal{G}$ of the quadratic residues modulo p along with a random generator g of $\mathcal{G}$. The strong versions of the two assumptions posit the hardness of the same problems even if p (and q) and generator g are chosen adversarially.

3 Σ-Protocols

We consider *3-move protocols* Π for a polynomial-time relation $\mathcal{R}$. Protocol Π is played by a prover $\mathcal{P}$ and a verifier $\mathcal{V}$ that receive a common input x. $\mathcal{P}$ receives as an additional private input a witness w for x and the security parameter 1^λ in unary. The protocol Π has the following form:

1. $\mathcal{P}$ runs algorithm P_1 on common input x, private input w, security parameter 1^λ and randomness R obtaining $a = \mathsf{P}_1(x, w, 1^\lambda; R)$ and sends a to $\mathcal{V}$.
2. $\mathcal{V}$, after receiving a from $\mathcal{P}$, chooses a random *challenge* $c \leftarrow \{0,1\}^l$ and sends c to $\mathcal{P}$.
3. $\mathcal{P}$ runs algorithm P_2 on input x, w, R, c and sends $z \leftarrow \mathsf{P}_2(x, w, R, c)$ to $\mathcal{V}$.
4. $\mathcal{V}$ outputs $\mathsf{V}(x, a, c, z)$ (i.e., $\mathcal{V}$'s decision to accept ($b = 1$) or reject ($b = 0$)).

We call $(\mathsf{P}_1, \mathsf{P}_2, \mathsf{V})$ the algorithms *associated* with Π and l the challenge length such that, wlog, the challenge space $\{0,1\}^l$ is composed of 2^l different challenges.

The triple (a, c, z) of messages exchanged is called a *3-move transcript*. A 3-move transcript is *honest* if a, z correspond to the messages computed running the honest algorithms, respectively, of P_1 and P_2, and c is a random string, in $\{0, 1\}^l$. A 3-move transcript (a, c, z) is *accepting* for x if and only if $V(x, a, c, z) = 1$. Two accepting 3-move transcripts (a, c, z) and (a', c', z') for an instance x constitute a *collision* if $a = a'$ and $c \neq c'$.

Definition 1 (Σ-protocol [18]). *A 3-move protocol Π with challenge length l is a Σ-protocol for a relation $\mathcal{R}$ if it enjoys the following properties:*

1. **Completeness.** *If $(x, w) \in \mathcal{R}$ then all honest 3-move transcripts for (x, w) are accepting.*
2. **Special Soundness.** *There exists an efficient algorithm* Extract *that, on input x and a collision for x, outputs a witness w such that $(x, w) \in \mathcal{R}$.*
3. **Special Honest-Verifier Zero Knowledge (SHVZK).** *There exists a PPT simulator algorithm* Sim *that takes as input $x \in L_{\mathcal{R}}$, security parameter 1^λ and $c \in \{0, 1\}^l$ and outputs an accepting transcript for x where c is the challenge. Moreover, for all l-bit strings c, the distribution of the output of the simulator on input (x, c) is computationally indistinguishable from the distribution of the 3-move honest transcript obtained when $\mathcal{V}$ sends c as challenge and $\mathcal{P}$ runs on common input x and any private input w such that $(x, w) \in \mathcal{R}$. We say that Π is* Perfect *when the two distributions are identical.*

Not to overburden the descriptions of protocols and simulators, we will omit the specification of the security parameter when it is clear from the context.

In the rest of the paper, we will call a 3-move protocol that enjoys Completeness, Special Soundness and Honest-Verifier Zero Knowledge (HVZK[7]) a $\tilde{\Sigma}$-*protocol*. The next theorem shows that SHVZK can be added to a 3-move protocol with HVZK without any significant penalty in terms of efficiency.

Theorem 1 [19]. *Suppose relation $\mathcal{R}$ admits a 3-move protocol Π' that is HVZK (resp., perfect HVZK). Then $\mathcal{R}$ admits a 3-move protocol Π that is SHVZK (resp., perfect SHVZK) and has the same efficiency.*

Proof. Let l be the challenge length of Π', let (P'_1, P'_2, V') be the algorithms associated with Π' and let Sim$'$ be the simulator for Π'. Consider the following algorithms.

1. P_1, on input $(x, w) \in \mathcal{R}$, security parameter 1^λ and randomness R_1, parses R_1 as (r_1, c'') where $|c''| = l$, computes $a' \leftarrow P'_1(x, w, 1^\lambda; r_1)$, and outputs $a = (a', c'')$.
2. P_2, on input $(x, w) \in \mathcal{R}$, R_1 and randomness R_2 parses R_1 as (r_1, c''), c, sets $c' = c \oplus c''$, computes $z' \leftarrow P'_2(x, w, r_1, c'; R_2)$, and sends it to $\mathcal{V}$.
3. V, on input x, $a = (a', c'')$, c and z', returns the output of $V'(x, a', c \oplus c'', z')$ to decide whether to accept or not.

---

[7] Recall that HVZK requires the existence of a simulator that generates a full transcript. This is a seemingly weaker requirement than SHVZK where the challenge is an input for the simulator.

Consider the following PPT simulator Sim that, on input an instance x and a challenge c, runs Sim$'$ on input x and obtains (a', c', z'). Then Sim sets $c'' = c \oplus c'$ and $a = (a', c'')$ and outputs (a, c, z'). It is easy to see that if Sim$'$ is a HVZK (resp. perfect HVZK) simulator for Π' then Sim is a SHVZK (resp. perfect SHVZK) simulator for Π.

We will use the definition of proof of knowledge given in [3, 19].

Theorem 2 [19]. *Let Π be a Σ-protocol for a relation $\mathcal{R}$ with challenge length l. Then Π is a proof of knowledge with knowledge error 2^{-l}.*

Definition 2 (Input-Delayed Σ-protocol). *A Σ-protocol $\Pi = (\mathcal{P}, \mathcal{V})$ with $\mathcal{P}$ running PPT algorithms $(\mathsf{P}_1, \mathsf{P}_2)$ is an input-delayed Σ-protocol if P_1 takes as input only the length of the common instance and P_2 takes as input the common instance x, the witness w, the randomness R_1 used by P_1 and the challenge c received from the verifier.*

Definition 3 (Witness-Delayed Σ-protocol). *A Σ-protocol $\Pi = (\mathcal{P}, \mathcal{V})$ for a relation $\mathcal{R}$ with associated algorithms $(\mathsf{P}_1, \mathsf{P}_2, \mathsf{V})$ is a witness-delayed Σ-protocol if P_1 takes as input only the common instance x.*

In a *Chamelon Σ-protocol*, the prover can compute the first message by using the simulator and thus knowing only the input but not the witness. Once the challenge has been received, the prover can compute the last message (thus completing the interaction) by using the witness w (which is thus used only to compute the last message) and the coin tosses used by the simulator to compute the first message.

Definition 4 (Chameleon Σ-protocol). *A Σ-protocol Π for polynomial-time relation $\mathcal{R}$ is a* Chameleon *Σ-protocol if there exists an SHVZK simulator Sim and an algorithm $\mathsf{P}_{\mathsf{sim}}$ satisfying the following property:*

Delayed Indistinguishability*: for all pairs of challenges c_0 and c_1 and for all $(x, w) \in \mathcal{R}$, the following two distributions*

$$\left\{ R \leftarrow \{0,1\}^{|x|^d}; (a, z_0) \leftarrow \mathsf{Sim}(x, c_0; R); z_1 \leftarrow \mathsf{P}_{\mathsf{sim}}((x, c_0, R), w, c_1) : \right.$$
$$\left. (x, a, c_1, z_1) \right\}$$

and

$$\left\{ (a, z_1) \leftarrow \mathsf{Sim}(x, c_1) : (x, a, c_1, z_1) \right\}$$

are indistinguishable, where Sim is the Special HVZK simulator and d is such that Sim, on input an λ-bit instance, uses at most λ^d random coin tosses. If the two distributions above are identical then we say that delayed indistinguishability is perfect, and Π is a Perfect Chameleon *Σ-protocol.*

We remark that a chameleon Σ-protocol Π has two modes of operations: the standard mode when $\mathcal{P}$ runs P_1 and P_2, and a *delayed* mode when $\mathcal{P}$ uses Sim and $\mathsf{P}_{\mathsf{sim}}$. Moreover, observe that since Sim is a simulator for Π, it follows from the delayed-indistinguishability property that, for all challenges c and $\tilde{c}$ and common inputs x, distribution

$$\{R \leftarrow \{0,1\}^{|x|^d}; (a, \tilde{z}) \leftarrow \mathsf{Sim}(x, \tilde{c}; R); z \leftarrow \mathsf{P}_{\mathsf{sim}}((x, \tilde{c}, R), w, c) : (a, c, z)\}$$

is indistinguishable from

$$\{R \leftarrow \{0,1\}^{|x|^d}; a \leftarrow \mathsf{P}_1(x, w; R); z \leftarrow \mathsf{P}_2(x, w, R, c) : (a, c, z)\}.$$

That is, the two modes of operations of Π are indistinguishable. This property make us able to claim that if Π is WI when a WI challenger interacts with an adversary using $(\mathsf{P}_1, \mathsf{P}_2)$, then Π is WI even when the pair $(\mathsf{Sim}, \mathsf{P}_{\mathsf{sim}})$ is used. Finally, we observe that Chameleon Σ-protocols do exist and Schnorr's protocol [42] is one example. When considering the algorithms associated to a Chameleon Σ-protocol, we will add $\mathsf{P}_{\mathsf{sim}}$.

3.1 Σ-protocols and Witness Indistinguishability

Definition 5. *A 3-move protocol $\Pi = (\mathcal{P}, \mathcal{V})$ is* Witness Indistinguishable (WI) *for a relation $\mathcal{R}$ if, for every malicious verifier $\mathcal{V}^\star$, there exists a negligible function ν such that for all x, w, w' such that $(x, w) \in \mathcal{R}_L$ and $(x, w') \in \mathcal{R}_L$*

$$\left| \mathrm{Prob}\left[\langle \mathcal{P}(w, 1^\lambda), \mathcal{V}^\star \rangle(x) = 1 \right] - \mathrm{Prob}\left[\langle \mathcal{P}(w', 1^\lambda), \mathcal{V}^\star \rangle(x) = 1 \right] \right| \leq \nu(\lambda).$$

The notion of a *perfect* WI 3-move protocol is obtained by requiring the two distributions to be identical. We start by recalling the following result.

Theorem 3 [18]. *Every Perfect $\tilde{\Sigma}$-protocol[8] is Perfect WI.*

For completeness, in the full version we show a $\tilde{\Sigma}$-*protocol* that it is not WI.

3.2   Or Composition of $\tilde{\Sigma}$-protocols: the CDS-OR Transform

In this section we describe the CDS-OR [18] transform in details. Let Π be a $\tilde{\Sigma}$-*protocol* for polynomial-time relation $\mathcal{R}$ with challenge length l, associated algorithms $(\mathsf{P}_1, \mathsf{P}_2, \mathsf{V})$ and HVZK simulator Sim. The CDS-OR transform constructs a $\tilde{\Sigma}$-protocol Π_{OR} with associated algorithms $(\mathsf{P}_1^{\mathsf{OR}}, \mathsf{P}_2^{\mathsf{OR}}, \mathsf{V}_\Sigma^{\mathsf{OR}})$ for the relation

$$\mathcal{R}_{\mathsf{OR}} = \left\{ ((x_0, x_1), w) : \left((x_0, w) \in \mathcal{R} \wedge x_1 \in \hat{L}_\mathcal{R} \right) \text{ OR } \left((x_1, w) \in \mathcal{R} \wedge x_0 \in \hat{L}_\mathcal{R} \right) \right\}.$$

---

[8] We remind the reader that we call a 3-move protocol that enjoys Completeness, Special Soundness and Honest-Verifier Zero Knowledge (HVZK) a $\tilde{\Sigma}$-protocol.

Protocol 1. *CDS-OR Transform.*

Common input: (x_0, x_1).

$\mathcal{P}$'s private input: (b, w) *with* $b \in \{0, 1\}$ *and* $(x_b, w) \in \mathcal{R}$.

$\mathsf{P}_1^{\mathsf{OR}}((x_0, x_1), (b, w); R_1)$. Set $a_b = \mathsf{P}_1(x_b, w; R_1)$. Compute $(a_{1-b}, c_{1-b}, z_{1-b}) \leftarrow \mathsf{Sim}(x_{1-b})$. Output (a_0, a_1).

$P_2^{\mathsf{OR}}((x_0, x_1), (b, w), c, R_1)$. Set $c_b = c \oplus c_{1-b}$. Compute $z_b \leftarrow \mathsf{P}_2(x_b, w, c_b, R_1)$. Output $((c_0, c_1), (z_0, z_1))$.

$\mathsf{V}_\Sigma^{\mathsf{OR}}((x_0, x_1), (a_0, a_1), c, ((c_0, c_1), (z_0, z_1)))$. $\mathsf{V}_\Sigma^{\mathsf{OR}}$ accepts if and only if $c = c_0 \oplus c_1$ and $\mathsf{V}(x_0, a_0, c_0, z_0) = 1$ and $\mathsf{V}(x_1, a_1, c_1, z_1) = 1$.

Theorem 4 [18, 25]. *If Π is a $\tilde{\Sigma}$-protocol for $\mathcal{R}$ then Π_{OR} is a $\tilde{\Sigma}$-protocol for $\mathcal{R}_{\mathsf{OR}}$ and is WI for relation*

$$\mathcal{R}'_{\mathsf{OR}} = \{((x_0, x_1), w) : ((x_0, w) \in \mathcal{R} \wedge x_1 \in L_\mathcal{R}) \text{ OR } ((x_1, w) \in \mathcal{R} \wedge x_0 \in L_\mathcal{R})\}.$$

Moreover, if Π is a Perfect $\tilde{\Sigma}$-protocol for $\mathcal{R}$ then Π^{OR} is WI for $\mathcal{R}_{\mathsf{OR}}$.

It is possible to extend the above construction to handle two different relations $\mathcal{R}_0$ and $\mathcal{R}_1$ that admit $\tilde{\Sigma}$-protocols. Indeed we can assume, wlog, that $\mathcal{R}_0$ and $\mathcal{R}_1$ have $\tilde{\Sigma}$-protocols Π_0 and Π_1 with the same challenge length (details are available in the full version). Hence, the construction outlined above can be used to construct $\tilde{\Sigma}$-*protocol* $\Pi_{\mathsf{OR}}^{\mathcal{R}_0, \mathcal{R}_1}$ for relation

$$\mathcal{R}_{\mathsf{OR}} = \left\{((x_0, x_1), w) : \left((x_0, w) \in \mathcal{R}_0 \wedge x_1 \in \hat{L}_{\mathcal{R}_1}\right) \text{ OR } \left((x_1, w) \in \mathcal{R}_1 \wedge x_0 \in \hat{L}_{\mathcal{R}_0}\right)\right\}.$$

Theorem 5. *If Π_0 and Π_1 are $\tilde{\Sigma}$-protocols for $\mathcal{R}_0$ and $\mathcal{R}_1$, respectively, then $\Pi_{\mathsf{OR}}^{\mathcal{R}_0, \mathcal{R}_1}$ is a $\tilde{\Sigma}$-protocol for relation $\mathcal{R}_{\mathsf{OR}}$ and is WI for relation*

$$\mathcal{R}'_{\mathsf{OR}} = \{((x_0, x_1), w) : ((x_0, w) \in \mathcal{R}_0 \wedge x_1 \in L_{\mathcal{R}_1}) \text{ OR } ((x_1, w) \in \mathcal{R}_1 \wedge x_0 \in L_{\mathcal{R}_0})\}.$$

Moreover, if Π_0 and Π_1 are Perfect $\tilde{\Sigma}$-protocols for $\mathcal{R}_0$ and $\mathcal{R}_1$ then Π^{OR} is WI for $\mathcal{R}_{\mathsf{OR}}$.

We remark that if Π_0 and Π_1 are Σ-protocols then the CDS-OR transform yields a Σ-protocol for $\mathcal{R}_{\mathsf{OR}}$ and Theorems 4 and 5 still hold.

4 t-Instance-Dependent Trapdoor Commitment Schemes

In this section, for integer $t \geq 2$, we define the notion of a t-Instance-Dependent Trapdoor Commitmentscheme associated with a polynomial-time relation $\mathcal{R}$ and show constructions for $t = 2$ and $t = 3$.

Definition 6 (t-Instance-Dependent Trapdoor Commitment Scheme). *Let $t \geq 2$ be an integer and let $\mathcal{R}$ be a polynomial-time relation. A t-Instance-Dependent Trapdoor Commitment (a t-IDTC, in short) scheme for $\mathcal{R}$ with message space M is a triple of PPT algorithms $(\mathsf{TCom}, \mathsf{TDec}, \mathsf{TFake})$ where*

TCom *is the randomized* commitment *algorithm that takes as input security para-meter* 1^λ, *an instance* $x \in \hat{L}_\mathcal{R}$ *(with* $|x| = \texttt{poly}(\lambda)$*) and a message* $m \in M$ *and outputs* commitment com, *decommitment* dec, *and* auxiliary information rand; TDec *is the* verification *algorithm that takes as input* $(x, \texttt{com}, \texttt{dec}, m)$ *and decides whether* m *is the decommitment of* com; TFake *is the randomized* equivocation *algorithm that takes as input* $(x, w) \in \mathcal{R}$, *messages* m_1 *and* m_2 *in* M, *commitment* com *of* m_1 *with respect to instance* x *and associated auxiliary information* rand *and produces decommitment information* $\texttt{dec}_2$ *such that* TDec, *on input* $(x, \texttt{com}, \texttt{dec}_2, m_2)$, *outputs* 1.

A t-*Instance-Dependent Trapdoor Commitment enjoys:*

- **Correctness**: *for all* $x \in \hat{L}_\mathcal{R}$, *all* $m \in M$, *it holds that*

$$\text{Prob}\left[(\texttt{com}, \texttt{dec}, \texttt{rand}) \leftarrow \textsf{TCom}(1^\lambda, x, m) : \textsf{TDec}(x, \texttt{com}, \texttt{dec}, m) = 1 \right] = 1.$$

- t-**Special Extract**: *there exists an efficient algorithm* ExtractTCom *that, on input* x, *commitment* com, *pairs* $(\texttt{dec}_i, m_i)_{i=1}^t$ *of openings and messages such that*
  - *for* $1 \le i < j \le t$ *we have that* $m_i \ne m_j$;
  - $\textsf{TDec}(x, \texttt{com}, \texttt{dec}_i, m_i) = 1$, *for* $i = 1, \dots, t$;
  outputs w *such that* $(x, w) \in \mathcal{R}$.

- **Hiding (resp., Perfect Hiding)**: *for every PPT (resp., unbounded) adversary* $\mathcal{A}$ *there exists a negligible function* ν *(resp.,* $\nu(\cdot) = 0$*) such that, for all* $x \in L_\mathcal{R}$ *and all* $m_0, m_1 \in M$, *it holds that*

$$\text{Prob}\big[b \leftarrow \{0, 1\}; (\texttt{com}, \texttt{dec}, \texttt{rand}) \leftarrow \textsf{TCom}(1^\lambda, x, m_b) :$$
$$b = \mathcal{A}(x, \texttt{com}, m_0, m_1)\big] \le \frac{1}{2} + \nu(\lambda).$$

- **Trapdoorness**: *the following two families of probability distributions are indistinguishable:*

$$\{(\texttt{com}, \texttt{dec}_1, \texttt{rand}) \leftarrow \textsf{TCom}(1^\lambda, x, m_1);$$
$$\texttt{dec}_2 \leftarrow \textsf{TFake}(x, w, m_1, m_2, \texttt{com}, \texttt{rand}) : (\texttt{com}, \texttt{dec}_2)\}$$

  and $\{(\texttt{com}, \texttt{dec}_2, \texttt{rand}) \leftarrow \textsf{TCom}(1^\lambda, x, m_2) : (\texttt{com}, \texttt{dec}_2)\}$ *over all families* $\{(x, w, m_1, m_2)\}$ *such that* $(x, w) \in \mathcal{R}$ *and* $m_1, m_2 \in M$.
  The perfect trapdoorness *property requires the two probability distributions to coincide for all* (x, w, m_1, m_2) *such that* $(x, w) \in \mathcal{R}$ *and* $m_1, m_2 \in M$.

Constructing a 2-IDTC scheme from a Chameleon Σ-*protocol.* Let $\Pi = (\mathcal{P}, \mathcal{V})$ with associated algorithms $(\textsf{P}_1, \textsf{P}_2, \textsf{V}, \textsf{P}_{\textsf{sim}})$ be a Chameleon Σ-protocol for polynomial-time relation $\mathcal{R}$ with a security parameter 1^λ. Let l be the challenge length of Π and let Sim be a SHVZK simulator associated to Π. We construct a t-IDTC scheme $(\textsf{TCom}_\Pi, \textsf{TDec}_\Pi, \textsf{TFake}_\Pi)$ for $\mathcal{R}$ with messages space $M = \{0, 1\}^l$ for $x \in \hat{L}_R$ as follows.

Protocol 2. 2-IDTC scheme from Chameleon Σ-protocol Π.

- $\mathsf{TCom}_\Pi(1^\lambda, x, m_1)$: On input x and $m_1 \in M$, pick randomness R and compute $(a, z) \leftarrow \mathsf{Sim}(x, m_1; R)$. Output $\mathsf{com} = a$, $\mathsf{dec} = z$ and $\mathsf{rand} = R$;
- $\mathsf{TDec}_\Pi(x, \mathsf{com}, \mathsf{dec}, m_1)$: On input x, com, dec and m_1, run $b = \mathsf{V}(x, \mathsf{com}, m_1, \mathsf{dec})$ and accept m_1 as the decommitted message iff $b = 1$.
- TFake_Π: On input $(x, w) \in \mathcal{R}$, messages $m_1, m_2 \in M$, for m_2 and rand for com, output $z = \mathsf{P}_{\mathsf{sim}}((x, m_1, \mathsf{rand}), w, m_2)$.

Theorem 6. If Π is a Chameleon Σ-protocol for $\mathcal{R}$ then Protocol 2 is a 2-IDTC scheme for $\mathcal{R}$. Moreover, if Π is Perfect then so is Protocol 2.

Proof. Correctness follows directly from the Completeness property of Π.

2-Special-Extract. Suppose com is a commitment with respect to instance x and let dec_1 and dec_2 be two openings of com as messages $m_1 \neq m_2$, respectively. Then, triplets $(\mathsf{com}, m_1, \mathsf{dec}_1)$ and $(\mathsf{com}, m_2, \mathsf{dec}_2)$ are accepting transcripts for Π on common input x with the same first round; that is, they constitute a collision for Π. Therefore, we define algorithm $\mathsf{ExtractTCom}$ to be the algorithm that runs algorithm $\mathsf{Extract}$ (that exists by the special soundness of Π) on input the collision. $\mathsf{ExtractTCom}$ returns the witness for x computed by $\mathsf{Extract}$.

(Perfect) Trapdoorness. It follows from the Perfect Delayed-Indistinguishability property of Π as well as the (perfect) Hiding property.

Constructing a 3-IDTC Scheme. Let $\mathcal{R}$ be a polynomial-time relation as above admitting a witness-delayed Σ-protocol Π with associated algorithms $(\mathsf{P}_1, \mathsf{P}_2, \mathsf{V})$ and security parameter 1^λ. Let l denote the challenge length of Π. We construct a 3-IDTC scheme for message space $M = \{0, 1\}^l$ for $x \in L_R$, as follows.

Protocol 3. 3-IDTC scheme.

- TCom_Π: On input 1^λ, x and $m_1 \in M$, pick randomness R and compute $(a_0, z) \leftarrow \mathsf{Sim}(x, m_1)$ and $a_1 \leftarrow \mathsf{P}_1(x; R)$. Let $\mathsf{com}_0 = a_0$ and $\mathsf{com}_1 = a_1$. Output $\mathsf{com} = (\mathsf{com}_b, \mathsf{com}_{1-b})$ for a randomly selected bit b, $\mathsf{dec} = z$ and $\mathsf{rand} = R$.
- TDec_Π: On input x, $\mathsf{com} = (\mathsf{com}_0, \mathsf{com}_1)$, dec and m_1, accept m_1 if and only if either $\mathsf{V}(x, \mathsf{com}_0, m_1, \mathsf{dec}) = 1$ or $\mathsf{V}(x, \mathsf{com}_1, m_1, \mathsf{dec}) = 1$.
- TFake_Π: On input $(x, w) \in \mathcal{R}$, messages $m_1, m_2 \in M$, commitment com for m_1 and rand for com, output $z \leftarrow \mathsf{P}_2(x, w, \mathsf{rand}, m_2)$.

Theorem 7. If Π is a witness-delayed Σ-protocol for $\mathcal{R}$, with the associated algorithms $(\mathsf{P}_1, \mathsf{P}_2, \mathsf{V})$, then Protocol 3 is a 3-IDTC scheme for $\mathcal{R}$. Moreover, if Π is Perfect then so is Protocol 3.

Proof. Correctness follows from the completeness of Π.

3-Special Extract. It follows from the special soundness of Π. Assume that the committer generates 3 accepting openings dec_1, dec_2 and dec_3, for distinct messages m_1, m_2 and m_3, for the same commitment com computed w.r.t. x. In this case, we have three accepting transcript for Π and therefore at least two of them must share the same first message, i.e., it is a collision. Thus we can run the extractor Extract for Π on the collision and obtain a witness for x.

Trapdoorness. It follows from the SHVZK property of Π. We prove this property via hybrid arguments.

The first hybrid, $\mathcal{H}_1$ is the real execution, where a honest prover commits to a message following the honest commitment and decommitment procedure, without using the trapdoor. More formally, in the hybrid $\mathcal{H}_1$ the prover performs the following steps:

- On input x and $m_1, m_2 \in M$, the prover selects random coin tosses R and computes $(a_0, z) \leftarrow \mathsf{Sim}(x, m_2)$, $a_1 \leftarrow \mathsf{P}_1(x; R)$. It picks $b \leftarrow \{0, 1\}$ and sends com $= (a_b, a_{1-b})$, dec $= z$, m_2.

The second hybrid $\mathcal{H}_2$ is equal to $\mathcal{H}_1$ with the difference that a_0 is computed using the algorithm P_1 and z using P_2. Formally:

- On input x and $m_1, m_2 \in M$, the prover selects random coin tosses $R = (r_1, r_2)$ and computes $a_0 \leftarrow \mathsf{P}_1(x; r_1)$, $z \leftarrow \mathsf{P}_2(x, w, r_1, m_2)$ and $a_1 \leftarrow \mathsf{P}_1(x; r_2)$. It picks $b \leftarrow \{0, 1\}$ and sends com $= (a_b, a_{1-b})$, dec $= z$, m_2.

Due to the SHVZK property of Π, $\mathcal{H}_1$ is indistinguishable from $\mathcal{H}_2$. Now we consider the hybrid $\mathcal{H}_3$ in which a_1 is computed using $\mathsf{Sim}(x, m_2)$. Formally:

- On input x and $m_1, m_2 \in M$, the prover selects random coin tosses R and computes $a_0 \leftarrow \mathsf{P}_1(x; R)$, $z \leftarrow \mathsf{P}_2(x, w, R, m_2)$ and $(a_1, \overline{z}) \leftarrow \mathsf{Sim}(x, m_1)$. It picks $b \leftarrow \{0, 1\}$ and sends com $= (a_b, a_{1-b})$, dec $= z$, m_2.

Even in this case, we can claim that $\mathcal{H}_3$ is indistinguishable from $\mathcal{H}_2$ because of the SHVZK of Π. The proof ends with the observation that $\mathcal{H}_3$ is the experiment in which a sender commits to a message m_1 and opens to m_2 using the trapdoor.

If Π is a perfect SHVZK protocol, then the sequence of hybrids produces identical distributions.

5 Our New OR-Composition Technique

In this section we formally describe our new OR transform. Let $\mathcal{R}_0$ be a relation admitting a t-IDTC scheme, $I = (\mathsf{TCom}_{\Pi_0}, \mathsf{TDec}_{\Pi_0}, \mathsf{TFake}_{\Pi_0})$, with $t = 2$ or $t = 3$, and $\mathcal{R}_1$ a relation admitting an input-delayed Σ-protocol Π_1 with associated algorithms $(\mathsf{P}_1^1, \mathsf{P}_2^1, \mathsf{V}^1)$ and simulator Sim^1. We show a Σ-protocol Π^{OR} for the OR relation:

$$\mathcal{R}_{\mathsf{OR}} = \{((x_0, x_1), w) : ((x_0, w) \in \mathcal{R}_0 \wedge x_1 \in \hat{L}_{\mathcal{R}_1}) \text{ OR } ((x_1, w) \in \mathcal{R}_1 \wedge x_0 \in \hat{L}_{\mathcal{R}_0})\}.$$

We denote by $(\mathsf{P}_1^{\mathsf{OR}}, \mathsf{P}_2^{\mathsf{OR}}, \mathsf{V}^{\mathsf{OR}})$ the algorithms associated with Π^{OR}. We assume that the initial common input is x_0. The other input x_1 and the witness w for (x_0, x_1) are made available to the prover only after the challenge has been received. We let $b \in \{0,1\}$ be such that $(x_b, w) \in \mathcal{R}_b$ and assume that the message space of the t-IDTC scheme I includes all possible first-round messages of Π_1. Note that for the constructions of the t-IDTC scheme we provide, the message space coincides with the set of challenges of the underlying Σ-protocol and, in the full version we show that the challenge length of a Σ-protocol can be easily expanded/reduced.

We remind that prover algorithm $\mathsf{P}_2^{\mathsf{OR}}$ receives as further input the randomness $(R_1, \mathbf{rand}_1)$ used by $\mathsf{P}_1^{\mathsf{OR}}$ to produce the first-round message.

Protocol 4. Protocol Π^{OR} for $\mathcal{R}_{\mathsf{OR}}$.
   Common input: $(x_0, 1^\lambda)$, where 1^λ is the security parameter.

1. $\mathsf{P}_1^{\mathsf{OR}}(x_0, 1^\lambda)$. Pick random R_1 and compute $a_1 \leftarrow \mathsf{P}_1^1(1^\lambda; R_1)$. Then commit to a_1 by running $(\mathbf{com}, \mathbf{dec}_1, \mathbf{rand}_1) \leftarrow \mathsf{TCom}_{\Pi_0}(1^\lambda, x_0, a_1)$. Output $\mathbf{com}$.
2. $\mathsf{P}_2^{\mathsf{OR}}((x_0, x_1), c, (w, b), (\mathbf{rand}_1, R_1))$ (with $(x_b, w) \in \mathcal{R}_b$).
   If $b = 1$, compute $z_1 \leftarrow \mathsf{P}_2^1(x_1, w, R_1, c)$ and output $(\mathbf{dec}_1, a_1, z_1)$.
   If $b = 0$, compute $(a_2, z_2) \leftarrow \mathsf{Sim}^1(x_1, c)$, $\mathbf{dec}_2 \leftarrow \mathsf{TFake}_{\Pi_0}(x_0, w, a_1, a_2, \mathbf{com}, \mathbf{rand}_1)$ and output $(\mathbf{dec}_2, a_2, z_2)$.
3. V^{OR}, on input (x_0, x_1), $\mathbf{com}$, c, and $(\mathbf{dec}, a, z)$ received from Π^{OR}, outputs 1 iff
$$\mathsf{TDec}_{\Pi_0}(x_0, \mathbf{com}, \mathbf{dec}, a) = 1 \text{ and } \mathsf{V}^1(x_1, a, c, z) = 1;$$

Theorem 8. If $\mathcal{R}_0$ admits a 2-IDTC (resp., 3-IDTC) scheme and if $\mathcal{R}_1$ admits an input-delayed Σ-protocol, then Π^{OR} is a Σ-protocol (resp., is a 3-round public-coin SHVZK PoK) for relation $\mathcal{R}_{\mathsf{OR}}$.

Proof. Completeness follows by inspection. We next prove the properties of Protocol 4 when instantiated with a 2-IDTC and 3-IDTC schemes.

Proof for the construction based on the 2-IDTC scheme. Special Soundness. It follows from the special soundness of the underlying Σ-protocol Π_1 and the 2-Special Extract of the 2-IDTC scheme. More formally, consider a collision $(\mathbf{com}, c, (\mathbf{dec}, a, z))$ and $(\mathbf{com}, c', (\mathbf{dec}', a', z'))$ for input (x_0, x_1). We observe that:

- if $a = a'$ then (a, c, z) and (a', c', z') is a collision for Π_1 for input x_1; then we can obtain a witness w_1 for x_1 by the Special Soundness property of Π_1;
- if $a \neq a'$, then $\mathbf{dec}$ and $\mathbf{dec}'$ are two openings of $\mathbf{com}$ with respect to x_0 for messages $a \neq a'$; then we can obtain a witness w_0 by the 2-Special Extract of the 2-IDTC scheme.

SHVZK Property. Consider simulator $\mathsf{Sim}^{\mathsf{OR}}$ that, on input (x_0, x_1) and challenge c, sets $(a, c, z) \leftarrow \mathsf{Sim}_1(x_1, c)$ and $(\mathbf{com}, \mathbf{dec}) \leftarrow \mathsf{TCom}_{x_0}(a)$, and outputs $(\mathbf{com}, c, (\mathbf{dec}, a, z))$. Next, we show that the transcript generated by $\mathsf{Sim}^{\mathsf{OR}}$ is indistinguishable from the one generated by a honest prover.

Let us first consider the case in which the prover of Π^{OR} receives a witness for x_1. In this case, if we sample a random distribution $(\mathbf{com}, c, (\mathbf{dec}, a, z))$ of Π^{OR} on input (x_0, x_1) constrained to c being the challenge we have that (a, c, z)

has the same distribution as in random transcript of Π_1 on input x_1 constrained to c being the challenge; moreover, $(\mathsf{com}, \mathsf{dec})$ is a pair of commitment and decommitment of a with respect to x_0. By the property of Sim_1, this distribution is indistinguishable from (a, c, z) computed as $\mathsf{Sim}_1(x_1, c)$ which is exactly as in the output $\mathsf{Sim}^{\mathsf{OR}}$.

Let us now consider the case in which the prover of Π^{OR} receives a witness for x_0. If we sample a random distribution $(\mathsf{com}, c, (\mathsf{dec}, a, z))$ of Π^{OR} on input (x_0, x_1) constrained to c being the challenge we have that (a, c, z) are distributed exactly as in the output of $\mathsf{Sim}^{\mathsf{OR}}$ (that is by running Sim_1 on input x_1 and c). In addition, in the output of $\mathsf{Sim}^{\mathsf{OR}}$, $(\mathsf{com}, \mathsf{dec})$ are commitment and decommitment of a whereas in the view of Π^{OR} they are computed by means of TFake algorithm. However, the two distributions are indistinguishable by the trapdoorness of the Instance-Dependent Trapdoor Commitment.

Proof for the construction based on the 3-IDTC *scheme.* 3-*Special Soundness.* This property ensures that there exists an efficient algorithm that, given three accepting transcripts, (a, c_0, z_0), (a, c_1, z_1), (a, c_2, z_2) with $c_i \neq c_j$ for $1 \leq i < j \leq 3$, for the same common input, outputs a witness for x.

Consider three accepting transcripts for Π^{OR} and input (x_0, x_1): $(\mathsf{com}, c_1, (\mathsf{dec}_1, a_1, z_1))$, $(\mathsf{com}, c_2, (\mathsf{dec}_2, a_2, z_2))$ and $(\mathsf{com}, c_3, (\mathsf{dec}_3, a_3, z_3))$.

We observe that:

- if $a_i = a_j$ for some $i \neq j$ then (a_i, c_i, z_i) and (a_j, c_j, z_j) is a collision for Π_1 for input x_1; thus we can obtain a witness w_1 for x_1 by the Special Soundness property of Π_1;
- if $a_i \neq a_j$ for all $i \neq j$, then dec_1 and dec_2 and dec_3 are three openings of the same com with respect to x_0 for messages a_1, a_2 and a_3; then we can obtain a witness w_0 for x_0 by the 3-Special Extract of the 3-IDTC scheme.

We stress that having a long enough challenge, 3-special soundness implies the proof of knowledge property.

SHVZK Property. This is similar to the proof for the construction based on 2-IDTC.

5.1   Witness Indistinguishability of Our Transform

In this section we discuss the *adaptive WI* property of Π^{OR}. Roughly speaking, adaptive WI means that in the WI experiment the adversary $\mathcal{A}$ is not forced to choose *both* theorems x_0 and x_1 at the onset of the experiment. Rather, she can choose theorem x_1 and witnesses w_0, w_1 adaptively, *after* seeing the first message of Π^{OR} played by the prover on input x_0. After x_1, w_0, w_1 have been selected by $\mathcal{A}$, the experiment randomly selects $b \leftarrow \{0, 1\}$. The prover then receives x_1 and w_b and proceeds to complete the protocol. The adversary wins the game if she can guess b with probability non-negligibly greater than $1/2$. More formally, we consider adaptive WIfor polynomial-time relation

$$\mathcal{R}^p_{\mathsf{OR}} = \left\{ ((x_0, x_1), w) : \left((x_0, w) \in \mathcal{R}_0 \wedge x_1 \in \hat{L}_{\mathcal{R}_1} \right) \text{ OR } \left((x_1, w) \in \mathcal{R}_1 \wedge x_0 \in \hat{L}_{\mathcal{R}_0} \right) \right\}$$

and for the weaker relation

$$\mathcal{R}^c_{OR} = \Big\{((x_0, x_1), w) : \big((x_0, w) \in \mathcal{R}_0 \wedge x_1 \in L_{\mathcal{R}_1}\big) \text{ OR } \big((x_1, w) \in \mathcal{R}_1 \wedge x_0 \in L_{\mathcal{R}_0}\big)\Big\}.$$

The adaptive WI experiment, $\mathsf{ExpWI}^\delta_{\mathcal{A}}(x_0, \lambda, \mathsf{aux})$ with $\delta \in \{c, p\}$, is parameterized by PPT adversary $\mathcal{A}$ and has three inputs: instance x_0, security parameter λ, and auxiliary information aux for $\mathcal{A}$.

$\mathsf{ExpWI}^\delta_{\mathcal{A}}(x_0, \lambda, \mathsf{aux})$:

1. $a = \mathsf{P}^{OR}_1(x_0, 1^\lambda; R_1)$, for random coin tosses R_1;
2. $\mathcal{A}(x_0, a, \mathsf{aux})$ outputs $((x_1, w_0, w_1), c, \mathsf{state})$
   such that $((x_0, x_1), w_0), ((x_0, x_1), w_1) \in \mathcal{R}^\delta_{OR}$;
3. $b \leftarrow \{0, 1\}$;
4. $z \leftarrow \mathsf{P}^{OR}_2((x_0, x_1), w_b, R_1, c)$;
5. $b' \leftarrow \mathcal{A}(z, \mathsf{state})$;
6. If $b = b'$ then output 1 else output 0.

We set $\mathsf{Adv}^\delta_{\mathcal{A}}(x_0, \lambda, \mathsf{aux}) = \Big| \mathrm{Prob} \Big[\mathsf{ExpWI}^\delta_{\mathcal{A}}(x_0, \lambda, \mathsf{aux}) = 1 \Big] - \frac{1}{2} \Big|$.

Definition 7. Π^{OR} *is Adaptive Witness Indistinguishable (resp., Adaptive Perfect Witness Indistinguishable) if for every adversary $\mathcal{A}$ there exists a negligible function ν such that for all aux and x_0 it holds that $\mathsf{Adv}^c_{\mathcal{A}}(x_0, \lambda, \mathsf{aux}) \leq \nu(\lambda)$ (resp., $\mathsf{Adv}^p_{\mathcal{A}}(x_0, \lambda, \mathsf{aux}) = 0$).*

Next, in Theorem 9, we prove the Adaptive Perfect WI of Π^{OR} when both Π_0 and Π_1 are perfect SHVZK. When one of Π_0 and Π_1 is not perfect, we would like to prove that Π^{OR} is Adaptive WI. In Theorem 10 we prove a weaker form of Adaptive WI in which the adversary is restricted in his choice of witnesses (w_0, w_1) for relation $\mathcal{R}^c_{OR}$. We leave open the problem of an OR-composition technique that gives Adaptive WI when the Σ-protocol composed are not both perfect SHVZK.

Theorem 9. *If Π_0 and Π_1 are perfect SHVZK then Π^{OR} is Adaptive Perfect Witness Indistinguishable.*

Proof. The proof considers the following three cases:

Case 1. $(x_0, w_0) \in \mathcal{R}_0$ and $(x_1, w_1) \in \mathcal{R}_1$;
Case 2. $(x_0, w_0) \in \mathcal{R}_0$ and $(x_0, w_1) \in \mathcal{R}_0$;
Case 3. $(x_1, w_0) \in \mathcal{R}_1$ and $(x_1, w_1) \in \mathcal{R}_1$.

For each case we present a sequence of hybrids and prove that pairs of consecutive hybrids are perfectly indistinguishable.

Case 1. The first hybrid experiment $\mathcal{H}_1(x_0, \lambda, \mathsf{aux})$ is the original experiment $\mathsf{ExpWI}^p_{\mathcal{A}}(x_0, \lambda, \mathsf{aux})$ in which $b = 1$ (and thus $\mathcal{P}$ uses witness w_1). That is,

- In Step 1 of $\mathsf{ExpWI}_{\mathcal{A}}^P(x_0, \lambda, \mathsf{aux})$, the following steps are executed:
  1. $a = \mathsf{P}_1^1(1^\lambda; R_1)$, for random coin tosses R_1;
  2. $(\mathsf{com}, \mathsf{dec}, \mathsf{rand}) \leftarrow \mathsf{TCom}_{\Pi_0}(x_0, 1^\lambda, a)$ and outputs com.
- In Step 4 of $\mathsf{ExpWI}_{\mathcal{A}}^P(x_0, \lambda, \mathsf{aux})$, the following steps are executed:
  1. set $a' = a$;
  2. $z \leftarrow \mathsf{P}_2^1(x_1, w_1, c, R_1)$;
  3. set $\mathsf{dec}' = \mathsf{dec}$;
  4. output (dec', a', z).

The second hybrid experiment $\mathcal{H}_2(x_0, \lambda, \mathsf{aux})$ differs from $\mathcal{H}_1(x_0, \lambda, \mathsf{aux})$ in the way a' and dec' are computed. More specifically,

- Step 1 of $\mathsf{ExpWI}_{\mathcal{A}}^P(x_0, \lambda, \mathsf{aux})$ stays the same.
  1. $a = \mathsf{P}_1^1(1^\lambda; R_1)$, for random coin tosses R_1;
  2. $(\mathsf{com}, \mathsf{dec}, \mathsf{rand}) \leftarrow \mathsf{TCom}_{\Pi_0}(x_0, 1^\lambda, a)$ and outputs com.
- In Step 4 of $\mathsf{ExpWI}_{\mathcal{A}}^P(x_0, \lambda, \mathsf{aux})$, the following steps are executed:
  1. $a' = \mathsf{P}_1^1(1^\lambda; R_1')$, for random coin tosses R_1';
  2. $z \leftarrow \mathsf{P}_2^1(x_1, w_1, c, R_1')$;
  3. $\mathsf{dec}' \leftarrow \mathsf{TFake}_{\Pi_0}(x_0, w_0, a, a', \mathsf{com}, \mathsf{rand})$;
  4. (dec', a', z).

The trapdoorness of the instance-dependent trapdoor commitment scheme based on Π_0 guarantees that $\mathcal{H}_1(x_0, \lambda, \mathsf{aux})$ and $\mathcal{H}_2(x_0, \lambda, \mathsf{aux})$ are perfectly indistinguishable for all λ.

The third hybrid experiment $\mathcal{H}_3(x_0, \lambda, \mathsf{aux})$ differs from $\mathcal{H}_2(x_0, \lambda, \mathsf{aux})$ in the way a' and z are computed. More specifically,

- Step 1 of $\mathsf{ExpWI}_{\mathcal{A}}^P(x_0, \lambda, \mathsf{aux})$ stays the same.
  1. $a = \mathsf{P}_1^1(1^\lambda; R_1)$, for random coin tosses R_1;
  2. $(\mathsf{com}, \mathsf{dec}, \mathsf{rand}) \leftarrow \mathsf{TCom}_{\Pi_0}(x_0, 1^\lambda, a)$ and outputs com.
- In Step 4 of $\mathsf{ExpWI}_{\mathcal{A}}^P(x_0, \lambda, \mathsf{aux})$, the following steps are executed:
  1. $(a', z) \leftarrow \mathsf{Sim}^1(x_1, c)$;
  2. $\mathsf{dec}' \leftarrow \mathsf{TFake}_{\Pi_0}(x_0, w_0, a, a', \mathsf{com}, \mathsf{rand})$;
  3. (dec', a', z).

By the perfect SHVZK of Π_1, we have that $\mathcal{H}_2(x_0, \lambda, \mathsf{aux})$ and $\mathcal{H}_3(x_0, \lambda, \mathsf{aux})$ are perfectly indistinguishable for all λ. The proof ends with the observation that $\mathcal{H}_3(x_0, \lambda, \mathsf{aux})$ is exactly experiment $\mathsf{ExpWI}_{\mathcal{A}}^P(x_0, \lambda, \mathsf{aux})$ when $b = 0$.

Case 2. The first hybrid experiment $\mathcal{H}_1(x_0, \lambda, \mathsf{aux})$ is again the original experiment $\mathsf{ExpWI}_{\mathcal{A}}^P(x_0, \lambda, \mathsf{aux})$ in which $b = 1$ (and thus $\mathcal{P}$ uses witness w_1). The second hybrid experiment $\mathcal{H}_2(x_0, \lambda, \mathsf{aux})$ differs from $\mathcal{H}_1(x_0, \lambda, \mathsf{aux})$ in the way TFake is executed (namely, using as input w_0 instead of w_1). More specifically,

- Step 1 of $\mathsf{ExpWI}_{\mathcal{A}}^P(x_0, \lambda, \mathsf{aux})$ stays the same.
  1. $a = \mathsf{P}_1^1(1^\lambda; R_1)$, for random coin tosses R_1;
  2. $(\mathsf{com}, \mathsf{dec}, \mathsf{rand}) \leftarrow \mathsf{TCom}_{\Pi_0}(x_0, 1^\lambda, a)$ and outputs com.
- In Step 4 of $\mathsf{ExpWI}_{\mathcal{A}}^P(x_0, \lambda, \mathsf{aux})$, the following steps are executed:

1. $(a', z) = \mathsf{Sim}^1(x_1, c)$;
2. $\mathsf{dec}' \leftarrow \mathsf{TFake}_{\varPi_0}(x_0, w_0, a, a', \mathsf{com}, \mathsf{rand})$;
3. (dec', a', z).

The trapdoorness of the instance-dependent trapdoor commitment scheme based on $\varPi_0$ implies that $\mathcal{H}_1(x_0, \lambda\mathsf{aux})$ is perfectly indistinguishable from $\mathcal{H}_2(x_0, \lambda\mathsf{aux})$ for all λ. The proof ends with the observation that $\mathcal{H}_2(x_0, \lambda, \mathsf{aux})$ is exactly experiment $\mathsf{ExpWI}^p_{\mathcal{A}}(x_0, \lambda, \mathsf{aux})$ when $b = 0$.

Case 3. The first hybrid experiment $\mathcal{H}_1(x_0, \lambda, \mathsf{aux})$ is again the original experiment $\mathsf{ExpWI}^p_{\mathcal{A}}(x_0, \mathsf{aux})$ in which $b = 1$ (and thus $\mathcal{P}$ uses witness w_1). The second hybrid experiment $\mathcal{H}_2(x_0, \lambda, \mathsf{aux})$ differs from $\mathcal{H}_1(x_0, \lambda, \mathsf{aux})$ in the way z is computed (using as input w_1 instead of w_0 when P_2 is executed). More specifically,

- In Step 1 of $\mathsf{ExpWI}^p_{\mathcal{A}}(x_0, \lambda, \mathsf{aux})$, the following steps are executed:
  1. $a = \mathsf{P}^1_1(1^\lambda; R_1)$, for random coin tosses R_1;
  2. $(\mathsf{com}, \mathsf{dec}, \mathsf{rand}) \leftarrow \mathsf{TCom}_{\varPi_0}(x_0, 1^\lambda, a)$ and outputs com.
- In Step 4 of $\mathsf{ExpWI}^p_{\mathcal{A}}(x_0, \lambda, \mathsf{aux})$, the following steps are executed:
  1. $z \leftarrow \mathsf{P}^1_2(x_1, w_0, c, R_1)$;
  2. output (dec, a, z)

The Perfect WI property of $\varPi_1$ implies that $\mathcal{H}_1(x_0, \lambda, \mathsf{aux})$ is perfectly indistinguishable from $\mathcal{H}_2(x_0, \lambda, \mathsf{aux})$. The proof ends with the observation that $\mathcal{H}_2(x_0, \lambda, \mathsf{aux})$ is exactly the experiment $\mathsf{ExpWI}^p_{\mathcal{A}}(x_0, \lambda, \mathsf{aux})$ when $b = 0$.

Next we consider the computational case in which one of $\varPi_0$ and $\varPi_1$ is not Perfect SHVZK (but they are still both SHVZK).

Theorem 10. *If $\varPi_0$ and $\varPi_1$ are SHVZK then $\varPi^{\mathsf{OR}}$ is Adaptive Witness Indistinguishable with respect to adversaries that output (x_1, w_0, w_1) such that at least one of w_0 and w_1 is a witness for $x_1 \in L_{\mathcal{R}_1}$.*

Proof. We prove this theorem by considering the following two cases:
(1) $(x_0, w_0) \in \mathcal{R}_0$ and $(x_1, w_1) \in \mathcal{R}_1$;
(2) $(x_1, w_0) \in \mathcal{R}_1$ and $(x_1, w_1) \in \mathcal{R}_1$.

Case 1. In this case the proof follows closely the one of Case 1 of Theorem 9, with the difference that hybrids here are only computationally indistinguishable.

Case 2. In this case we show that there exists $\mathcal{A}'$ for Case 1 that has the same success probability of $\mathcal{A}$. Suppose indeed that both w_0 and w_1 are witnesses for x_1 and that $\mathcal{A}$ breaks the adaptive WI property of $\varPi^{\mathsf{OR}}$. Then, by definition of $\mathcal{R}^c_{\mathsf{OR}}$ and by Definition 7, there exists $\mathcal{A}'$ that has in his description a witness w_2 for x_0. Indeed, the output of $\mathcal{A}$ interacting with $\mathcal{P}((x_0, x_1), w_2)$ would necessarily be distinguishable from the output of the interaction with either $\mathcal{P}((x_0, x_1), w_0)$ or $\mathcal{P}((x_0, x_1), w_1)$. Therefore $\mathcal{A}'$ would contradict Case 1 and thus there exists no successful $\mathcal{A}$ for Case 2.

6 Applications

In this section, we describe the application of our new OR-composition technique for constructing a 3-round straight-line perfect quasi-polynomial time simulatable argument system. In the full version we also show an efficient 4-round resettable WI argument system and an efficient 4-round resettable zero knowledge with concurrent soundness argument system in the BPK model.

A 3-Round Efficient Perfect Quasi-Polynomial Time Simulatable Argument System. In [38], Pass introduced relaxed notions of zero knowledge and knowledge extraction in which the simulator and the extractor are allowed to run in quasi-polynomial time. Allowing the simulator to run in quasi-polynomial time typically dispenses with the need of rewinding the verifier; that is, the simulator is *straight-line*. In [38], Pass first describes the following 2-round perfect ZK argument for any language L: the verifier $\mathcal{V}$ sends a value $Y = f(y)$ for a randomly chosen y where f is a sub-exponentially hard OWF and the first round of a ZAP protocol. The prover $\mathcal{P}$ then sends a commitment to $(y'|w')$ and uses the second round of the ZAP to prove that either $y' = f^{-1}(y)$ or w' is a witness for $x \in L$. If language L admits a Σ-protocol Π_L then the above construction can be implemented as an efficient 4-round argument with quasi-polynomial time simulation: the function f is concretely instantiated to be an exponentiation in a group in which the Discrete Log problem is hard and the ZAP is replaced with the CDS-OR composition of Π_L and Schnorr's Σ-protocol for the Discrete Log.

Note that Schnorr's Σ-protocol is input delayed and thus we can use it as Σ-protocol Π_1 in our OR transform in conjunction with any Chameleon Σ-protocol Π_0. One drawback of reducing to 3 rounds the result of [38] is that we can use only a perfect Σ-protocol since the goal is to obtain perfect WI in 3 rounds.

Simulation in Quasi-Polynomial Time. Since the verifier in an interactive argument is often modeled as a PPT machine, the classical zero-knowledge definition requires that the simulator runs also in (expected) polynomial time. In [38], the simulator is allowed to run in time $\lambda^{\texttt{poly}(\log(\lambda))}$. Loosely speaking, we say that an interactive argument is $\lambda^{\texttt{poly}(\log(\lambda))}$-perfectly simulatable if for any adversarial verifier there exists a simulator running in time $\lambda^{\texttt{poly}(\log(\lambda))}$, where λ is the size of the statement being proved, whose output is identically distributed to the output of the adversarial verifier.

Definition 8 (One-way functions for sub-exponential circuits [38]). *A function $f : \{0,1\}^* \rightarrow \{0,1\}^*$ is called one-way for sub-exponential circuits if there exists a constant α such that the following two condition holds:*

- *there exist a deterministic polynomial-time algorithm that on input y outputs $f(y)$;*
- *for every probabilistic algorithm $\mathcal{A}$ with running time bounded by 2^{λ^α}, all sufficiently large λ's, and every auxiliary input $z \in \{0,1\}^{\texttt{poly}(\lambda)}$*

$$\text{Prob}\left[\, y \xleftarrow{R} \{0,1\}^* : \mathcal{A}(f(y), z) \in f^{-1}(f(y))\,\right] < \frac{1}{\texttt{poly}(2^{\lambda^\alpha})}.$$

Now we define straight-line $T(\lambda)$-perfectly simulatable interactive arguments.

For our result we consider a one-way functions for sub-exponential circuits that is also one-to-one.

Definition 9 (straight-line $T(\lambda)$ simulatability, Definition 31 of [39]). *Let $T(\lambda)$ be a class of functions that is closed under composition with any polynomial. We say that an interactive argument (proof) $(\mathcal{P}, \mathcal{V})$ for the language $L \in NP$, with the witness relation $\mathcal{R}_L$, is straight-line $T(\lambda)$-simulatable if for every PPT machine $\mathcal{V}^*$ there exists a probabilistic simulator S with running time bounded by $T(\lambda)$ such that the following two ensembles are computationally indistinguishable (when the distinguish gap is a function in $\lambda = |x|$)*

- *$\{(\langle \mathcal{P}(w), \mathcal{V}^*(z)\rangle(x))\}_{z\in\{0,1\}^*, x\in L}$ for arbitrary w s.t. $(x, w) \in \mathcal{R}_L$*
- *$\{(\langle S, \mathcal{V}^*(z)\rangle(x))\}_{z\in\{0,1\}^*, x\in L}$*

We note that the above definition is very restrictive. In fact, the simulator is supposed to act as a cheating prover, with its only advantage being the possibility of running in time $T(\lambda)$, instead of in polynomial time. Trivially, it do not exist a straight-line $T(\lambda)$-simulatable proof for non-trivial languages (this should be contrasted with straight-line simulatable interactive arguments, which instead do exist).

For any NP-language L we consider the perfect chameleon Σ-protocol Π_L for the relation R_L. Also we consider the Schnorr Σ-protocol Π_{DLOG} the following relation $\mathsf{DLOG} = \{((\mathcal{G}, q, g, Y), y) : g^y = Y\}$ with the associated NP-language L_{DLOG}, over groups $\mathcal{G}$ of prime-order q, and use our OR-composition technique to obtain a new Σ-protocol $\Pi^{\mathsf{OR}} = (\mathcal{P}^{\mathsf{OR}}, \mathcal{V}^{\mathsf{OR}})$ for the relation

$$\mathcal{R}_{\mathsf{OR}} = \Big\{((x_L, x_{\mathsf{DLOG}}), w) : ((x_L, w) \in \mathcal{R}_L \land x_{\mathsf{DLOG}} \in \hat{L}_{\mathsf{DLOG}})\mathsf{OR}$$

$$((x_{\mathsf{DLOG}}, w) \in \mathsf{DLOG} \land x_L \in \hat{L}_{\mathcal{R}_L})\Big\}$$

with challenge length $l = \lambda$ and associated algorithms $\mathsf{P}_1^{\mathsf{OR}}$, $\mathsf{P}_2^{\mathsf{OR}}$ and V^{OR}.

Let f be a sub-exponentially hard one-to-one one-way function implemented using DLog as described before, with the only change that for some constant α, f is one-way w.r.t circuits of size 2^{λ^α}. Let $L \in NP$ and $k = \frac{1}{\alpha} + 1$. Our 3-round straight-line quasi-polynomial time simulatable argument system for $x \in L$ is the following.

Protocol 5. A 3-round straight-line quasi-polynomial time simulatable argument system.

Common input: *An instance x of a language $L \in NP$ with witness relation R_L with a perfect chameleon Σ-protocol, and 1^λ as security parameter.*

Private input: *$\mathcal{P}$ has w as a private input, s.t. $(x, w) \in \mathcal{R}_L$.*

Round 1. $\mathcal{P} \to \mathcal{V}$:

1. *On input a randomness R_1, $\mathcal{P}$ uniformly chooses (p, q, g) where $p = 2q + 1$ is a safe prime and g is a generator of a group $\mathcal{G}_q$ of size q. We remark that (p, q, g) are parameters selected so that the function $f(y) = g^y$ is a one-to-one one-way function for some constant α w.r.t circuits of size 2^{λ^α}.*
2. *$\mathcal{P}$ computes $a \leftarrow \mathsf{P}_1^{\mathsf{OR}}((x, 1^{\lambda^\alpha}); R_1)$.*
3. *$\mathcal{P}$ sends (p, q, g) and a to $\mathcal{V}$.*

Round 2. $\mathcal{V} \to \mathcal{P}$:

1. *$\mathcal{V}$ chooses $y \leftarrow \mathbb{Z}_q$ and computes $Y = g^y$.*
2. *$\mathcal{V}$ chooses $c \leftarrow \{0, 1\}^l$.*
3. *$\mathcal{V}$ sends c and Y to $\mathcal{P}$.*

Round 3. $\mathcal{P} \to \mathcal{V}$:

1. *$\mathcal{P}$ computes $z \leftarrow \mathsf{P}_2^{\mathsf{OR}}((x, ((p, q, g), Y)), w, c, R_1)$.*
2. *$\mathcal{P}$ sends z to $\mathcal{V}$.*
3. *$\mathcal{V}$ accepts if and only if $\mathsf{V}^{\mathsf{OR}}((x, ((p, q, g), Y)), a, c, z) = 1$.*

We remark that we are using the same assumption of [7] that allows the adversary of DLog to generate the DLog parameters while the challenger selects the random element of the group.

Theorem 11. *If $\varPi^{\mathsf{OR}}$ is a perfect Σ-protocol for OR composition of $\mathcal{R}_L$ and DLOG, then Protocol 5 is a 3-round straight-line perfectly $\lambda^{O(\log^k \lambda)}$-simulatable argument of knowledge.*

Proof. Completeness follows directly from the completeness of $\varPi^{\mathsf{OR}}$.

Soundness/Knowledge Extraction. We show that $\varPi$ is an argument of knowledge; this directly implies soundness. The claim follows from the fact that the argument system $\varPi^{\mathsf{OR}}$ used is a proof of knowledge when the challenge is long enough. and from the fact that a PPT adversary only finds a pre-image to Y (for f) with negligible probability. More formally, we construct a polynomial-time extractor E for every polynomial-time $\mathcal{P}^\star$ for protocol $\varPi$. E internally incorporates $\mathcal{P}^\star$ and each time $\varPi^{\mathsf{OR}}$ proves a new theorem it proceeds as follows. E invokes the extractor E^{OR} for $\varPi^{\mathsf{OR}}$. E outputs whatever E^{OR} outputs. By the proof knowledge property of $\varPi^{\mathsf{OR}}$, the output of E will either be a witness w for the statement proved, or the pre-image of Y. If E outputs w, we are done. Otherwise, if it outputs y with non-negligible probability, then we can construct a reduction that breaks the DLog assumption (still in the form proposed by [7]).

Quasi-Polynomial Time Perfect Simulation. Consider a straight-line simulator Sim that computes the first round as the honest prover. This is possible because $\varPi^{\mathsf{OR}}$ does not need any witness to computes the first round. After the simulator receives Y it checks that Y has a pre-image. Sim thereafter performs an exhaustive search to find a pre-image y of a value Y for the function f. To perform this task Sim tries all possible values $y' \in \{0,1\}^{\log^k \lambda}$ and checks if $f(y') = Y$. This thus takes time $\texttt{poly}(2^{\log^k \lambda})$, since the time it takes to evaluate the function f is a polynomial in λ. After having found a value y such that $f(y) = Y$, Sim uses y as witness to complete the execution of Π^{OR} (instead of using a real witness for x, as the honest prover would do). Clearly the running time of Sim is bounded by $\lambda^{O(\log^k \lambda)}$. We proceed to show that the output of the simulator is identically distributed to the output of any adversarial verifier in a real execution with an honest prover. Note that the only difference between a real execution and a simulated execution is in the choice of the witness used in the last stage of the protocol. Therefore, from the adaptive WI property of Π^{OR} we have that the output of the simulated execution is identically distributed to the output of the real execution.

Acknowledgments. We thank Berry Schoenmakers for various useful discussions on Σ-protocols.

The work of the third author was supported by the MACS project under NSF Frontier grant CNS-1414119 and by NSF grant 1012798. This work was done in part while the third author was visiting the Simons Institute for the Theory of Computing, supported by the Simons Foundation and by the DIMACS/Simons Collaboration in Cryptography through NSF grant CNS-1523467.

For the full version of this work see [13].

References

1. Abe, M., Okamoto, T., Suzuki, K.: Message recovery signature schemes from sigma-protocols. IEICE Trans. **96-A**(1), 92–100 (2013)
2. Bellare, M., Fischlin, M., Goldwasser, S., Micali, S.: Identification protocols secure against reset attacks. In: Pfitzmann, B. (ed.) EUROCRYPT 2001. LNCS, vol. 2045, pp. 495–511. Springer, Heidelberg (2001)
3. Bellare, M., Goldreich, O.: On defining proofs of knowledge. In: Brickell, E.F. (ed.) CRYPTO 1992. LNCS, vol. 740, pp. 390–420. Springer, Heidelberg (1993)
4. Bitansky, N., Paneth, O.: ZAPs and non-interactive witness indistinguishability from indistinguishability obfuscation. In: Dodis, Y., Nielsen, J.B. (eds.) TCC 2015, Part II. LNCS, vol. 9015, pp. 401–427. Springer, Heidelberg (2015)
5. Blum, M.: How to prove a theorem so no one else can claim it. In: International Congress of Mathematicians, p. 1444 (1986)
6. Blundo, C., Persiano, G., Sadeghi, A.-R., Visconti, I.: Improved security notions and protocols for non-transferable identification. In: Jajodia, S., Lopez, J. (eds.) ESORICS 2008. LNCS, vol. 5283, pp. 364–378. Springer, Heidelberg (2008)
7. Canetti, R., Dakdouk, R.R.: Extractable perfectly one-way functions. In: Aceto, L., Damgård, I., Goldberg, L.A., Halldórsson, M.M., Ingólfsdóttir, A., Walukiewicz, I. (eds.) ICALP 2008, Part II. LNCS, vol. 5126, pp. 449–460. Springer, Heidelberg (2008)
8. Canetti, R., Goldreich, O., Goldwasser, S., Micali, S.: Resettable zero-knowledge (extended abstract). In: STOC, pp. 235–244 (2000)
9. Catalano, D., Dodis, Y., Visconti, I.: Mercurial commitments: minimal assumptions and efficient constructions. In: Halevi, S., Rabin, T. (eds.) TCC 2006. LNCS, vol. 3876, pp. 120–144. Springer, Heidelberg (2006)

10. Catalano, D., Visconti, I.: Hybrid trapdoor commitments and their applications. In: Caires, L., Italiano, G.F., Monteiro, L., Palamidessi, C., Yung, M. (eds.) ICALP 2005. LNCS, vol. 3580, pp. 298–310. Springer, Heidelberg (2005)
11. Catalano, D., Visconti, I.: Hybrid commitments and their applications to zero-knowledge proof systems. Theor. Comput. Sci. **374**(1–3), 229–260 (2007)
12. Chaidos, P., Groth, J.: Making sigma-protocols non-interactive without random oracles. PKC **2015**, 650–670 (2015)
13. Ciampi, M., Persiano, G., Scafuro, A., Siniscalchi, L., Visconti, I.: Improved OR composition of Sigma-protocols. IACR Cryptology ePrint Archive 2015, vol. 810 (2015). http://eprint.iacr.org/2015/810
14. Ciampi, M., Persiano, G., Siniscalchi, L., Visconti, I.: A transform for NIZK almost as efficient and general as the Fiat-Shamir transform without programmable random oracles. IACR Cryptology ePrint Archive, vol. 770 (2015). http://eprint.iacr.org/2015/770
15. Ciampi, M., Persiano, G., Siniscalchi, L., Visconti, I.: A transform for NIZK almost as efficient and general as the Fiat-Shamir transform without programmable random oracles. In: Theory of Cryptography - 13th Theory of Cryptography Conference, TCC 2016-A, Tel Aviv, Israel, 10–13 January 2016
16. Cramer, R.: Modular design of secure yet practical cryptographic protocols. Ph.D. thesis, University of Amsterdam (1996)
17. Cramer, R., Damgård, I.B.: Zero-knowledge proofs for finite field arithmetic or: can zero-knowledge be for free? In: Krawczyk, H. (ed.) CRYPTO 1998. LNCS, vol. 1462, pp. 424–441. Springer, Heidelberg (1998)
18. Cramer, R., Damgård, I.B., Schoenmakers, B.: Proof of partial knowledge and simplified design of witness hiding protocols. In: Desmedt, Y.G. (ed.) CRYPTO 1994. LNCS, vol. 839, pp. 174–187. Springer, Heidelberg (1994)
19. Damgård, I.: On Σ-protocol (2010). http://www.cs.au.dk/~ivan/Sigma.pdf
20. Damgård, I., Groth, J.: Non-interactive and reusable non-malleable commitment schemes. In: STOC 2003, pp. 426–437 (2003)
21. Di Crescenzo, G., Persiano, G., Visconti, I.: Constant-round resettable zero knowledge with concurrent soundness in the bare public-key model. In: Franklin, M. (ed.) CRYPTO 2004. LNCS, vol. 3152, pp. 237–253. Springer, Heidelberg (2004)
22. Di Crescenzo, G., Visconti, I.: Concurrent zero knowledge in the public-key model. In: Caires, L., Italiano, G.F., Monteiro, L., Palamidessi, C., Yung, M. (eds.) ICALP 2005. LNCS, vol. 3580, pp. 816–827. Springer, Heidelberg (2005)
23. Dwork, C., Naor, M.: Zaps and their applications. FOCS **2000**, 283–293 (2000)
24. Feige, U., Lapidot, D., Shamir, A.: Multiple non-interactive zero knowledge proofs based on a single random string (extended abstract). In: FOCS 1990, pp. 308–317. IEEE Computer Society (1990)
25. Garay, J.A., MacKenzie, P., Yang, K.: Strengthening zero-knowledge protocols using signatures. J. Cryptology **19**(2), 169–209 (2006)
26. Goyal, V., Richelson, S., Rosen, A., Vald, M.: An algebraic approach to non-malleability. In: 55th FOCS 2014, pp. 41–50, Philadelphia, PA, USA. IEEE Computer Society, 18–21 October 2014
27. Groth, J., Kohlweiss, M.: One-out-of-many proofs: or how to leak a secret and spend a coin. In: Oswald, E., Fischlin, M. (eds.) EUROCRYPT 2015. LNCS, vol. 9057, pp. 253–280. Springer, Heidelberg (2015)
28. Groth, J., Ostrovsky, R., Sahai, A.: Perfect Non-interactive Zero Knowledge for NP. In: Vaudenay, S. (ed.) EUROCRYPT 2006. LNCS, vol. 4004, pp. 339–358. Springer, Heidelberg (2006)

29. Guillou, L.C., Quisquater, J.-J.: A practical zero-knowledge protocol fitted to security microprocessor minimizing both transmission and memory. In: Günther, C.G. (ed.) EUROCRYPT 1988. LNCS, vol. 330, pp. 123–128. Springer, Heidelberg (1988)

30. Katz, J., Ostrovsky, R.: Round-optimal secure two-party computation. In: Franklin, M. (ed.) CRYPTO 2004. LNCS, vol. 3152, pp. 335–354. Springer, Heidelberg (2004)

31. Lapidot, D., Shamir, A.: Publicly Verifiable Non-interactive Zero-Knowledge Proofs. In: Menezes, A., Vanstone, S.A. (eds.) CRYPTO 1990. LNCS, vol. 537, pp. 353–365. Springer, Heidelberg (1991)

32. Lindell, Y.: An efficient transform from sigma protocols to NIZK with a CRS and non-programmable random oracle. In: Dodis, Y., Nielsen, J.B. (eds.) TCC 2015, Part I. LNCS, vol. 9014, pp. 93–109. Springer, Heidelberg (2015)

33. Lindell, Y., Pinkas, B.: An efficient protocol for secure two-party computation in the presence of malicious adversaries. J. Cryptology $28(2)$, 312–350 (2015)

34. Maurer, U.: Zero-knowledge proofs of knowledge for group homomorphisms. Des. Codes Crypt. 77, 663–676 (2015)

35. Maurer, U.: Unifying zero-knowledge proofs of knowledge. In: Preneel, B. (ed.) AFRICACRYPT 2009. LNCS, vol. 5580, pp. 272–286. Springer, Heidelberg (2009)

36. Ostrovsky, R., Pandey, O., Visconti, I.: Efficiency preserving transformations for concurrent non-malleable zero knowledge. In: Micciancio, D. (ed.) TCC 2010. LNCS, vol. 5978, pp. 535–552. Springer, Heidelberg (2010)

37. Ostrovsky, R., Rao, V., Visconti, I.: On selective-opening attacks against encryption schemes. In: Abdalla, M., De Prisco, R. (eds.) SCN 2014. LNCS, vol. 8642, pp. 578–597. Springer, Heidelberg (2014)

38. Pass, R.: Simulation in quasi-polynomial time, and its application to protocol composition. In: Biham, E. (ed.) EUROCRYPT 2003. LNCS, vol. 2656, pp. 160–176. Springer, Heidelberg (2003)

39. Pass, R.: Alternative Variants of Zero-Knowledge Proofs. Master's thesis, Kungliga Tekniska Högskolan, licentiate Thesis Stockholm, Sweden (2004)

40. Pointcheval, D., Stern, J.: Security proofs for signature schemes. In: Maurer, U.M. (ed.) EUROCRYPT 1996. LNCS, vol. 1070, pp. 387–398. Springer, Heidelberg (1996)

41. Scafuro, A., Visconti, I.: On round-optimal zero knowledge in the bare public-key model. In: Pointcheval, D., Johansson, T. (eds.) EUROCRYPT 2012. LNCS, vol. 7237, pp. 153–171. Springer, Heidelberg (2012)

42. Schnorr, C.-P.: Efficient identification and signatures for smart cards. In: Brassard, G. (ed.) CRYPTO 1989. LNCS, vol. 435, pp. 239–252. Springer, Heidelberg (1990)

43. Visconti, I.: Efficient zero knowledge on the internet. In: Bugliesi, M., Preneel, B., Sassone, V., Wegener, I. (eds.) ICALP 2006. LNCS, vol. 4052, pp. 22–33. Springer, Heidelberg (2006)

44. Yung, M., Zhao, Y.: Generic and practical resettable zero-knowledge in the bare public-key model. In: Naor, M. (ed.) EUROCRYPT 2007. LNCS, vol. 4515, pp. 129–147. Springer, Heidelberg (2007)

Oblivious RAM

Onion ORAM: A Constant Bandwidth Blowup Oblivious RAM

Srinivas Devadas[1], Marten van Dijk[2], Christopher W. Fletcher[1(✉)],
Ling Ren[1(✉)], Elaine Shi[3], and Daniel Wichs[4]

[1] Massachusetts Institute of Technology, Cambridge, USA
{devadas,cwfletch,renling}@mit.edu
[2] University of Connecticut, Storrs, USA
vandijk@engr.uconn.edu
[3] Cornell University, Ithaca, USA
elaine@cs.cornell.edu
[4] Northeastern University, Boston, USA
wichs@ccs.neu.edu

Abstract. We present Onion ORAM, an Oblivious RAM (ORAM) with constant worst-case bandwidth blowup that leverages poly-logarithmic server computation to circumvent the logarithmic lower bound on ORAM bandwidth blowup. Our construction does not require fully homomorphic encryption, but employs an additively homomorphic encryption scheme such as the Damgård-Jurik cryptosystem, or alternatively a BGV-style somewhat homomorphic encryption scheme without bootstrapping. At the core of our construction is an ORAM scheme that has "shallow circuit depth" over the entire history of ORAM accesses. We also propose novel techniques to achieve security against a malicious server, without resorting to expensive and non-standard techniques such as SNARKs. To the best of our knowledge, Onion ORAM is the first concrete instantiation of a constant bandwidth blowup ORAM under standard assumptions (even for the semi-honest setting).

1 Introduction

Oblivious RAM (ORAM), initially proposed by Goldreich and Ostrovsky [19,20,36], is a cryptographic primitive that allows a *client* to store private data on an *untrusted server* and maintain *obliviousness* while accessing that data — i.e., guarantee that the server or any other observer learns nothing about the data or the client's access pattern (the sequence of addresses or operations) to that data. Since its initial proposal, ORAM has been studied in theory [21,25,39,41,45,49], or in various application settings including secure outsourced storage [8,29,32, 42,43,50], secure processors [10–12,31,38,40,51] and secure multi-party computation [13,14,24,28,47,48].

1.1 Server Computation in ORAM

The ORAM model considered historically, starting with the work of Goldreich and Ostrovsky [19,20,36], assumed that the server acts as a simple storage device that

© International Association for Cryptologic Research 2016
E. Kushilevitz and T. Malkin (Eds.): TCC 2016-A, Part II, LNCS 9563, pp. 145–174, 2016.
DOI: 10.1007/978-3-662-49099-0_6

allows the client to read and write data to it, but does not perform any computation otherwise. However, in many scenarios investigated by subsequent works [8,32,42,50] (e.g., the setting of remote oblivious file servers), the untrusted server has significant computational power, possibly even much greater than that of the client. Therefore, it is natural to extend the ORAM model to allow for server computation, and to distinguish between the amount of computation performed by the server and the amount of communication with the client.

Indeed, many recent ORAM schemes have implicitly or explicitly leveraged some amount of server computation to either reduce bandwidth cost [1,7,13,14, 29,32,39,43,52], or reduce the number of online roundtrips [49]. We remark that some prior works [1,32] call themselves oblivious storage (or oblivious outsourced storage) to distinguish from the standard ORAM model where there is no server computation. We will simply apply the term ORAM to both models, and refer to ORAM *with/without server computation* to distinguish between the two.

At first, many works implicitly used server computation in ORAM constructions [13,14,32,39,43,49,52], without making a clear definitional distinction from standard ORAM. Apon et al. were the first to observe that such a distinction is warranted [1], not only for the extra rigor, but also because the definition renders the important Goldreich-Ostrovsky ORAM lower bound [20] inapplicable to the server computation setting — as we discuss below.

1.2 Attempts to "Break" the Goldreich-Ostrovsky Lower Bound

Traditionally, ORAM constructions are evaluated by their *bandwidth, client storage* and *server storage*. Bandwidth is the amount of communication (in bits) between client/server to serve a client request, including the communication in the background to maintain the ORAM (i.e., ORAM evictions). We also define bandwidth blowup to be bandwidth measured in the number of blocks (i.e., blowup compared to a normal RAM). Client storage is the amount of trusted local memory required at the client side to manage the ORAM protocol and server storage is the amount of storage needed at the server to store all data blocks.

In their seminal work [20], Goldreich and Ostrovsky showed that an ORAM of N blocks must incur a $O(\log N)$ lower bound in bandwidth blowup, under $O(1)$ blocks of client storage. If we allow the server to perform computation, however, the Goldreich-Ostrovsky lower bound no longer applies with respect to client-server bandwidth [1]. The reason is that the Goldreich-Ostrovsky bound is in terms of the *number of operations* that must be performed. With server computation, though the number of operations is still subject to the bound, most operations can be performed on the server-side without client intervention, making it possible to break the bound in terms of bandwidth between client and server. Since historically bandwidth has been the most important metric and the bottleneck for ORAM, breaking the bound in terms of bandwidth constitutes a significant advance.

However, it turns out that this is not easy. Indeed, two prior works [1,32] have made endeavors towards this direction using homomorphic encryption.

Path-PIR [32] leverages additively homomorphic encryption (AHE) to improve ORAM online bandwidth, but its overall bandwidth blowup is still poly-logarithmic. On the other hand, Apon et al. [1] showed that using a fully homomorphic encryption (FHE) scheme with *constant ciphertext expansion*, one can construct an ORAM scheme with constant bandwidth blowup. The main idea is that, instead of having the client move data around on the server "manually" by reading and writing to the server, the client can instruct the server to perform ORAM request and eviction operations under an FHE scheme without revealing any data and its movement. While this is a very promising direction, it suffers from the following drawbacks:

- First, ORAM keeps access patterns private by continuously shuffling memory as data is accessed. This means the ORAM circuit depth that has to be evaluated under FHE depends on the number of ORAM accesses made and can grow unbounded (which we say to mean any polynomial amount in N). Therefore, Apon et al. [1] needs FHE bootstrapping, which not only requires circular security but also incurs a large performance penalty in practice.[1]
- Second, with the server performing homomorphic operations on encrypted data, achieving malicious security is difficult. Consequently, most existing works either only guarantee semi-honest security [32,52], or leveraged powerful tools such as SNARKs to ensure malicious security [1]. However, SNARKs not only require non-standard assumptions [18], but also incur prohibitive cost in practice.

1.3   Our Contributions

With the above observation, the goal of this work is to construct constant bandwidth blowup ORAM schemes from *standard assumptions* that have *practical efficiency* and *verifiability in the malicious setting*. Specifically, we give proofs by construction for the following theorems. Let B be the block size in bits and N the number of blocks in the ORAM.

Theorem 1 (Semi-honest Security Construction). *Under the Decisional Composite Residuosity assumption (DCR) or Learning With Errors (LWE) assumption, there exists an ORAM scheme with semi-honest security, $O(B)$ bandwidth, $O(BN)$ server storage and $O(B)$ client storage. To achieve negligible in N probability of ORAM failure and success from best known attacks, our schemes require poly-logarithmic in N block size and server computation.*

We use negligible in N security following prior ORAM work but also give asymptotics needed for exact exponential security in Sect. 6.

---

[1] While bootstrapping performance has been made asymptotically efficient by recent works [17], the cost in practice is still substantial, on the order of tens of seconds to minutes (amortized), whereas other homomorphic operations are on the order of milliseconds to seconds [22].

Looking at the big picture, our DCR-based scheme is the first demonstration of a constant bandwidth blowup ORAM using any additively homomorphic encryption scheme (AHE), as opposed to FHE. Our LWE-based scheme (detailed in the online version [9]) is the first time ORAM has been combined with SWHE/FHE in a way that does not require Gentry's bootstrapping procedure.

Our next goal is to extend our semi-honest constructions to the malicious setting. In Sect. 5, we will introduce the concept of "abstract server-computation ORAM" which both of our constructions satisfy. Then, we can achieve malicious security due to the following theorem:

Theorem 2 (Malicious Security Construction). *With the additional assumption of collision-resistant hash functions, any "abstract server-computation ORAM" scheme with semi-honest security can be compiled into a "verified server-computation ORAM" scheme which has malicious security.*

We stress that these are the *only* required assumptions. We do not need the circular security common in FHE schemes and do not rely on SNARKs for malicious security. We defer formal definitions of server-computation ORAM and malicious security to Appendix A.

Main Ideas. The key technical contributions enabling the above results are:

- (Sect. 3) An ORAM that, when combined with server computation, has *shallow circuit depth*, i.e., $O(\log N)$ over the entire history of all ORAM accesses. This is a necessity for our constructions based on AHE or SWHE, and removes the need for FHE (Gentry's bootstrapping operations). We view this technique as an important step towards practical constant bandwidth blowup ORAM schemes.
- (Sect. 5) A novel technique that combines a cut and choose-like idea with an error-correcting code to amplify soundness.

Table 1 summarizes our contributions and compares our schemes with some of the state-of-the-art ORAM constructions.

Practical Efficiency. To show how our results translate to practice, Sect. 6.4 compares our semi-honest AHE-based construction against Path PIR [32] and Circuit ORAM [47]—the best prior schemes with and without server computation that match our scheme in client/server storage. The top order bit is that as block size increases, our construction's bandwidth approaches $2B$. When all three schemes use an 8 MB block size (a proxy for modern image file size), Onion ORAM improves over Circuit ORAM and Path-PIR's bandwidth (in bits) by **35×** and **22×**, respectively. For larger block sizes, our improvement increases. We note that in many cases, block size is an application constraint: for applications asking for a large block size (e.g., image sharing), all ORAM schemes will use that block size.

1.4   Related Work

Recent non-server-computation ORAMs are approaching the Goldreich-Ostrovsky lower bound under $O(1)$ blocks of client storage. Goodrich et al. [21]

Table 1. Our Contribution. N is the number of blocks. The optimal block size is the data block size needed to achieve the stated bandwidth, and is measured in bits. All schemes have $O(B)$ client storage and $O(BN)$ server storage (both asymptotically optimal) and negligible failure probability in N. Computation measures the number of two-input plaintext gates evaluated per ORAM access. "M" stands for malicious security, and "SH" stands for semi-honest. We set parameters for AHE/SWHE (the Damgård-Jurik and Ring-LWE cryptosystems [3,6], respectively) to get super-poly in N defense to best known attacks [26,27]. For derivation of parameters for the SWHE schemes, see the extended version [9].

Scheme	Optimal block size B	Bandwidth	Server computation	Client computation	Security
Circuit ORAM [47]	$\Omega(\log^2 N)$	$\omega(B \log N)$	N/A	N/A	M
Path-PIR [32]	$\omega(\log^5 N)$	$O(B \log N)$	$\widetilde{\omega}(B \log^5 N)$	$\widetilde{O}(B \log^4 N)$	SH
AHE Onion ORAM	$\widetilde{\Omega}(\log^5 N)$	$O(B)$	$\widetilde{\omega}(B \log^4 N)$	$\widetilde{O}(B \log^4 N)$	SH
	$\widetilde{\omega}(\log^6 N)$	$O(B)$	$\widetilde{\omega}(B \log^4 N)$	$\widetilde{O}(B \log^4 N)$	M
SWHE Onion ORAM	$\widetilde{\omega}(\log^2 N)$	$O(B)$	$\widetilde{\omega}(B \log^2 N)$	$\widetilde{\omega}(B)$	SH
	$\widetilde{\omega}(\log^4 N)$	$O(B)$	$\widetilde{\omega}(B \log^2 N)$	$\widetilde{\omega}(B + \log^2 N)$	M

and Kushilevitz et al. [25] demonstrated $O(\log^2 N)$ and $O(\log^2 N / \log \log N)$ bandwidth blowup schemes, respectively. Recently, Wang et al. constructed Circuit ORAM [47], which achieves $\omega(\log N)$ bandwidth blowup.

Many state-of-the-art ORAM schemes or implementations make use of server computation. For example, the SSS construction [42,43], Burst ORAM [8] and Ring ORAM [39] assumed the server is able to perform matrix multiplication or XOR operations. Path-PIR [32] and subsequent work [7,52] increased the allowed computation to additively homomorphic encryption. Apon et al. [1] and Gentry et al. [13,14] further augmented ORAM with Fully Homomorphic Encryption (FHE). Williams and Sion rely on server computation to achieve a single online roundtrip [49]. We remark that the techniques of Gentry et al. [13] and Wang et al. [46], for improving data structure performance on top of ORAM, can be combined with our techniques.

Recent works on Garbled RAM [15,30] can also be seen as generalizing the notion of server-computation ORAM. However, existing Garbled RAM constructions incur $\mathsf{poly}(\lambda) \cdot \mathsf{polylog}(N)$ client work and bandwidth blowup, and therefore Garbled RAM does not give a server-computation RAM with constant bandwidth blowup. Reusable Garbled RAM [16] achieves constant client work and bandwidth blowup, but known reusable garbled RAM constructions rely on non-standard assumptions (indistinguishability obfuscation, or more) and are prohibitive in practice.

The mechanics of running our shallow depth ORAM over a homomorphic encryption scheme are similar to those used to evaluate encrypted branching programs [23]. (One may think of our contribution as formulating ORAM as a shallow enough circuit so that the techniques of [23] apply.)

2   Overview of Techniques

In our schemes, the client "guides" the server to perform ORAM accesses and evictions homomorphically by sending the server some "helper values". With these helper values, the server's main job will be to run a sub-routine called the "*homomorphic select*" operation (select operation for short), which can be implemented using either AHE or SWHE – resulting in two different constructions. We can achieve constant bandwidth blowup because helper value size is independent of data block size: when the block size sufficiently large, sending helper values does not affect the asymptotic bandwidth blowup. We now explain these ideas along with pitfalls and solutions in more detail. For the rest of the section, we focus on the AHE-based scheme but note that the story with SWHE is very similar.

Building Block: Homomorphic Select Operation. The select operation, which resembles techniques from private information retrieval (PIR) [27], takes as input m plaintext data blocks $\mathsf{pt}_1, \ldots, \mathsf{pt}_m$ and encrypted helper values which represent a user-chosen index i^*. The output is an encryption of block pt_{i^*}. Obviously, the helper values should not reveal i^*.

Our ORAM protocol will need select operations to be performed over the *outputs* of prior select operations. For this, we require a sequence of AHE schemes $\mathcal{E}_\ell$ with plaintext space $\mathbb{L}_\ell$ and ciphertext space $\mathbb{L}_{\ell+1}$ where $\mathbb{L}_{\ell+1}$ is again in the plaintext space of $\mathcal{E}_{\ell+1}$. Each scheme $\mathcal{E}_\ell$ is additively homomorphic meaning $\mathcal{E}_\ell(x) \oplus \mathcal{E}_\ell(y) = \mathcal{E}_\ell(x+y)$. We denote an ℓ-layer onion encryption of a message x by $\mathcal{E}^\ell(x) := \mathcal{E}_\ell(\mathcal{E}_{\ell-1}(\ldots \mathcal{E}_1(x)))$.

Suppose the inputs to a select operation are encrypted with ℓ layers of onion encryption, i.e., $\mathsf{ct}_i = \mathcal{E}^\ell(\mathsf{pt}_i)$. To select block i^*, the client sends an encrypted select vector (select vector for short), $\mathcal{E}_{\ell+1}(b_1), \ldots, \mathcal{E}_{\ell+1}(b_m)$ where $b_{i^*} = 1$ and $b_i = 0$ for all other $i \neq i^*$. Using this select vector, the server can homomorphically compute $\mathsf{ct}^* = \bigoplus_i \mathcal{E}_{\ell+1}(b_i) \cdot \mathsf{ct}_i = \mathcal{E}_{\ell+1}(\sum_i b_i \cdot \mathsf{ct}_i) = \mathcal{E}_{\ell+1}(\mathsf{ct}_{i^*}) = \mathcal{E}^{\ell+1}(\mathsf{pt}_{i^*})$. The result is the selected data block pt_{i^*}, with $\ell+1$ layers of onion encryption. Notice that the result has one more layer than the input.

All ORAM Operations can be Implemented Using Homomorphic Select Operations. In our schemes, for each ORAM operation, the client read/writes per-block metadata and creates a select vector(s) based on that metadata. The client then sends the encrypted select vector(s) to the server, who does the heavy work of performing actual computation over block contents.

Specifically, we will build on top of tree-based ORAMs [41,45], a standard type of ORAM without server computation. Metadata for each block includes its logical address and the path it is mapped to. To request a data block, the client first reads the logic addresses of all blocks along the read path. After this step, the client knows which block to select and can run the homomorphic select protocol with the server. ORAM eviction operations require that the client sends encrypted select vectors to indicate how blocks should percolate down the ORAM tree. As explained above, each select operation adds an encryption layer to the selected block.

Achieving Constant Bandwidth Blowup. To get constant bandwidth blowup, we must ensure that select vector bandwidth is smaller than the data block size. For this, we need several techniques. First, we will split each plaintext data block into C chunks $\mathsf{pt}_i = (\mathsf{pt}_i[1], \ldots, \mathsf{pt}_i[C])$, where each chunk is encrypted separately, i.e., $\mathsf{ct}_i = (\mathsf{ct}_i[1], \ldots, \mathsf{ct}_i[C])$ where $\mathsf{ct}_i[j]$ is an encryption of $\mathsf{pt}_i[j]$. Crucially, each select vector can be reused for all the C chunks. By increasing C, we can increase the data block size to decrease the relative bandwidth of select vectors.

Second, we require that each encryption layer adds a small *additive* ciphertext expansion (even a constant multiplicative expansion would be too large). Fortunately, we do have well established additively homomorphic encryption schemes that meet this requirement, such as the Damgård-Jurik cryptosystem [6]. Third, the "depth" of the homomorphic select operations has to be bounded and shallow. This requirement is the most technically challenging to satisfy, and we will now discuss it in more detail.

Bounding the Select Operation Depth. We address this issue by constructing a new tree-based ORAM, which we call a *"bounded feedback ORAM"*.[2] By "feedback", we refer to the situation where during an eviction some block a gets stuck in its current bucket b. When this happens, an eviction into b needs select operations that take both incoming blocks and block a as input, resulting in an extra layer on bucket b (on top of the layers bucket b already has). The result is that buckets will accumulate layers (with AHE) or ciphertext noise (with SWHE) on each eviction, which grows unbounded over time.

Our bounded feedback ORAM breaks the feedback loop by guaranteeing that bucket b will be empty at public times, which allows upstream blocks to move into b without feedback from blocks already in b. It turns out that breaking this feedback is not trivial: in all existing tree-based ORAM schemes [39, 41, 45, 47], blocks can get stuck in buckets during evictions which means there is no guarantee on when buckets are empty.[3] We remark that cutting feedback is equivalent to our claim of shallow circuit depth in Sect. 1.3: Without cutting feedback, the depth of the ORAM circuit keeps growing with the number of ORAM accesses.

Techniques for Malicious Security. We are also interested in achieving malicious security, i.e., enforcing honest behaviors of the server, while avoiding SNARKs. Our idea is to rely on probabilistic checking, and to leverage an error-correcting code to amplify the probability of detection. As mentioned before, each block is divided into C chunks. We will have the client randomly sample security parameter $\lambda \ll C$ chunks per block (the same random choice for all blocks), referred to as *verification chunks*, and use standard memory checking

---

[2] Previous versions of this report used the term *"steady progress"* which has been cited in subsequent works, but we feel bounded feedback is more accurate.

[3] We remark that some hierarchical ORAM schemes (e.g., [20]) also have bounded feedback, but achieve worse results in different respects relative our construction (e.g., worse server storage, deeper select circuits), when combined with server computation.

to ensure their authenticity and freshness. On each step, the server will perform homomorphic select operations on all C chunks in a block, and the client will perform the same homomorphic select operations on the λ verification chunks. In this way, whenever the server returns the client some encrypted block, the client can check whether the λ corresponding chunks match the verification chunks.

Unfortunately, the above scheme does not guarantee negligible failure of detection. For example, the server can simply tamper with a random chunk and hope that it's not one of the verification chunks. Clearly, the server succeeds with non-negligible probability. The fix is to leverage an error-correcting code to encode the original C chunks of each block into $C' = 2C$ chunks, and ensure that as long as $\frac{3}{4}C'$ chunks are correct, the block can be correctly decoded. Therefore, the server knows *a priori* that it will have to tamper with at least $\frac{1}{4}C'$ chunks to cause any damage at all, in which case it will get caught except with negligible probability.

3   Bounded Feedback ORAM

We now present the bounded feedback ORAM, a traditional ORAM scheme without server computation, to illustrate its important features. All notation used throughout the rest of the paper is summarized in Table 2.

3.1   Bounded Feedback ORAM Basics

We build on the tree-based ORAM framework of Shi et al. [41], which organizes server storage as a binary tree of nodes. The binary tree has $L + 1$ levels, where the root is at level 0 and the leaves are at level L. Each node in the binary tree is called a bucket and can contain up to Z data blocks. The leaves are numbered

Table 2. ORAM parameters and notations.

Notation	Meaning
N	Number of real data blocks in ORAM
B	Data block size in bits
C	The number of chunks in each data block
B_C	Chunk size in bits ($B = C \cdot B_C$)
L	Depth of the ORAM tree
Z	Maximum number of real blocks per bucket
A	Eviction frequency (larger means less frequent)
$\mathcal{P}(l)$	The path from the root to leaf l
$\mathcal{P}(l, i)$	The i-th bucket (towards the root) on $\mathcal{P}(l)$
G	Eviction counter
S	The set of chunk indices corresponding to verification chunks

$0, 1, \ldots, 2^L - 1$ in the natural manner. Pseudo-code for our algorithm is given in Fig. 1 and described below.

Note that many parts of our algorithm refer to *paths* down the tree where a path is a contiguous sequence of buckets from the root to a leaf. For a leaf bucket l, we refer to the path to l as path l or $\mathcal{P}(l)$. $\mathcal{P}(l, k)$ denotes the bucket at level $k \in [0..L]$ on $\mathcal{P}(l)$. Specifically, $\mathcal{P}(l, 0)$ denotes the root, and $\mathcal{P}(l, L)$ denotes the leaf bucket on $\mathcal{P}(l)$.

Main Invariant. Like all tree-based ORAMs, each block is associated with a random path and we say that each block can only live in a bucket along that path at any time. In a local position map, the client stores the path associated to each block.

Recursion. To avoid incurring a large amount of client storage, the position map should be recursively stored in other smaller ORAMs [41]. When the data block size is $\Omega(\log^2 N)$ for an N element ORAM—which will be the case for all of our final parameterizations—the asymptotic costs of recursion (in terms of server storage or bandwidth blowup) are insignificant relative to the main ORAM [44]. Thus, for the remainder of the paper, we no longer consider the bandwidth cost of recursion.

Metadata. To enable all ORAM operations, each block of data in the ORAM tree is stored alongside its address and leaf label (the path the block is mapped to). This metadata is encrypted using a semantically secure encryption scheme.

ORAM Request. Requesting a block with address a (ReadPath in Fig. 1) is similar to most tree-based ORAMs: look up the position map to obtain the path block a is currently mapped to, read all blocks on that path to find block a, invalidate block a, remap it to a new random path and add it to the root bucket. This involves decrypting the address metadata of every block on the path (Line 13) and setting one address to $\bot$ (Line 15). All addresses must be then re-encrypted to hide which block was invalidated.

ORAM Eviction. The goal of eviction is to percolate blocks towards the leaves to avoid bucket overflows and it is this procedure where we differ from existing tree-based ORAMs [13,39,41,45,47]. We now describe our eviction procedure in detail.

3.2   New Triplet Eviction Procedure

We combine techniques from [13,39,41] to design a novel eviction procedure (Evict in Fig. 1) that enables us to break select operation feedback.

Triplet Eviction on a Path. Similar to other Tree ORAMs, eviction is performed along a path. To perform an eviction: For every bucket $\mathcal{P}(l_e, k)$ (k from 0 to L, i.e., from root to leaf), we move blocks from $\mathcal{P}(l_e, k)$ to its two children. Specifically, each block in $\mathcal{P}(l_e, k)$ moves to either the left or right child bucket depending on which move keeps the block on the path to its leaf (this can be

```
 1: function Access(a, op, data′)
 2:     l′ ← UniformRandom(0, 2^L − 1)
 3:     l ← PositionMap[a]
 4:     PositionMap[a] ← l′
 5:     data ← ReadPath(l, a)
 6:     if op = read then
 7:         return data to client
 8:     if op = write then
 9:         data ← data′
10:     P(l, 0, cnt) ← (a, l′, data)
11:     Evict()

12: function ReadPath(l, a)
13:     Read all blocks on path P(l)
14:     Select and return the block with address a
15:     Invalidate the block with address a

16: function Evict( )
17:     Persistent variables cnt and G, initialized to 0
18:     cnt ← cnt + 1  mod A
19:     if cnt =? 0 then
20:         l_e ← bitreverse(G)
21:         EvictAlongPath(l_e)
22:         G ← G + 1  mod 2^L

23: function EvictAlongPath(l_e)
24:     for k ← 0 to L − 1 do
25:         Read all blocks in P(l_e, k) and its two children
26:         Move all blocks in P(l_e, k) to its two children
27:                          ▷ P(l_e, k) is empty at this point (Observation 1)
```

Fig. 1. Bounded Feedback ORAM (no server computation). Note that our construction differs from the original tree ORAM [41] only in the Evict procedure. We split Evict into EvictAlongPath to simplify the presentation later.

determined by comparing the block's leaf label to l_e). We call this process a bucket-triplet eviction.

In each of these bucket-triplet evictions, we call $P(l_e, k)$ the *source bucket*, the child bucket also on $P(l_e)$ the *destination bucket*, and the other child the *sibling bucket*. A crucial change that we make to the eviction procedure of the original binary-tree ORAM [41] is that we move *all* the blocks in the source bucket to its two children.

Eviction Frequency and Order. For every A (a parameter proposed in [39], which we will set later) ORAM requests, we select the next path to evict based on the reverse lexicographical order of paths (proposed in [13] and illustrated in Fig. 2). The reverse lexicographical order eviction most evenly and *deterministically* spreads out the eviction on all paths in the tree. Specifically, a bucket at level k will get evicted *exactly* every $A \cdot 2^k$ ORAM requests.

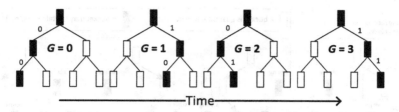

Fig. 2. The reverse lexicographical eviction order. **Black** buckets indicate those on each eviction path and G is the eviction count from Fig. 1. As indicated in Fig. 1, the eviction paths corresponding to $G = 4$ and $G = 0$ are equal: the exact eviction sequence shown above cycles forever. We mark the eviction path edges as 0/1 (goto left child = 0, right child = 1) to illustrate that the eviction path equals G in reverse binary representation.

Setting Parameters for Bounded Feedback. As mentioned, we require that during a bucket-triplet eviction, *all* blocks in the source bucket move to the two child buckets. The last step to achieve bounded feedback is to show that child buckets will have enough room to receive the incoming blocks, i.e., no child bucket should ever overflow except with negligible probability. (If any bucket overflows, we have experienced ORAM failure.) We guarantee this property by setting the bucket size Z and the eviction frequency A properly. According to the following lemma, if we simply set $Z = A = \Theta(\lambda)$, the probability that a bucket overflows is $2^{-\Theta(\lambda)}$, exponentially small.

Lemma 1 (No Bucket Overflows). *If $Z \geq A$ and $N \leq A \cdot 2^{L-1}$, the probability that a bucket overflows after an eviction operation is bounded by $e^{-\frac{(2Z-A)^2}{6A}}$.*

The proof of Lemma 1 relies on a careful analysis of the stochastic process stipulated by the reverse lexicographic ordering of eviction, and boils down to a Chernoff bound. We defer the full proof to Appendix B.1. Now, Lemma 1 with $Z = A = \Theta(\lambda)$ immediately implies the following key observation.

Observation 1 (Empty Source Bucket). *After a bucket-triplet eviction, the source bucket is empty.*

Furthermore, straightforwardly from the definition of reverse lexicographical order, we have,

Observation 2. *In reverse-lexicographic order eviction, each bucket rotates between the following roles in the following order: source, sibling, and destination.*

These observations together guarantee that buckets are empty at public and pre-determined times, as illustrated in Fig. 3.

Towards Bounded Feedback. The above two observations are the keys to achieving bounded feedback. An empty source bucket b will be a sibling bucket the next time it is involved in a triplet eviction. So select operations that move

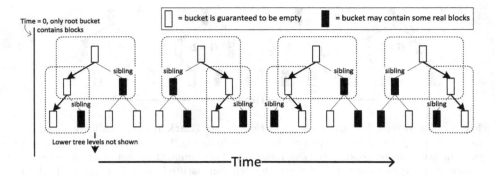

Fig. 3. ORAM tree state immediately after each of a sequence of four evictions. After an eviction, the buckets on the eviction path (excluding the leaves) are guaranteed to be empty. Further, at the start of each eviction, each sibling bucket for that eviction is guaranteed to be empty. **Notations:** Assume the ORAM tree has more levels (not shown for simplicity). The eviction path is marked with arrows. The dotted boxes indicate bucket triplets during each eviction.

blocks into b do not get feedback from b itself. Thus, the number of encryption layers (with AHE) or ciphertext noise (SWHE) becomes a function of previous levels in the tree only, which we can tightly bound later in Lemma 2 in Sect. 4.3.

Constant Server Storage Blowup. We note that under our parameter setting $N \leq A \cdot 2^{L-1}$ and $Z = A$, our bounded feedback ORAM's server storage is $O(2^{L+1} \cdot Z \cdot B) = O(BN)$, a constant blowup.

4   Semi-honest Onion ORAM with an Additively Homomorphic Encryption

In this section, we describe how to leverage an AHE scheme with additive ciphertext expansion to transform our bounded feedback ORAM into our semi-honest secure Onion ORAM scheme. First, we detail the homomorphic select operation that we introduced in Sect. 2.

4.1   Additively Homomorphic Select Sub-protocol

Suppose the client wishes to select the i^*-th block from m blocks denoted $\mathsf{ct}_1, \ldots, \mathsf{ct}_m$, each with $\ell_1, \ldots, \ell_m$ layers of encryption respectively. The sub-protocol works as follows:

1. Let $\ell := \max(\ell_1, \ldots, \ell_m)$. The client creates and sends to the server the following encrypted select vector $\langle \mathcal{E}_{\ell+1}(b_1), \mathcal{E}_{\ell+1}(b_2), \ldots \mathcal{E}_{\ell+1}(b_m) \rangle$, where $b_{i^*} = 1$ and $b_i = 0$ for $i \neq i^*$.
2. The server "lifts" each block to ℓ-layer ciphertexts, simply by continually re-encrypting a block until it has ℓ layers $\mathsf{ct}'_i[j] = \mathcal{E}_\ell(\mathcal{E}_{\ell-1}(\ldots \mathcal{E}_{\ell_i}(\mathsf{ct}_i[j])))$.

3. The server evaluates the homomorphic select operation on the lifted blocks:
$\mathsf{ct}_{out}[j] := \bigoplus_i (\mathcal{E}_{\ell+1}(b_i) \otimes \mathsf{ct}_i'[j]) = \mathcal{E}_{\ell+1}(\mathsf{ct}_{i*}')$. The outcome is the selected block ct_{i*} with $\ell + 1$ layers of encryption.

As mentioned in Sect. 2, we divide each block into C chunks. Each chunk is encrypted separately. All C chunks share the same select vector—therefore, encrypting each element in the select vector only incurs the chunk size (instead of the block size).

We stress again that every time a homomorphic select operation is performed, the output block gains an extra layer of encryption, on top of $\ell = \max(\ell_1, \ldots, \ell_m)$ onion layers. This poses the challenge of bounding onion encryption layers, which we address in Sect. 4.3.

4.2 Detailed Protocol

We now describe the detailed protocol. Recall that each block is tagged with the following metadata: the block's logical address and the leaf it is mapped to, and that the size of the metadata is independent of the block size.

Initialization. The client runs a key generation routine for all layers of encryption, and gives all public keys to the server.

Read Path. ReadPath(l, a) from Sect. 3.1 can be done with the following steps:

1. Client downloads and decrypts the addresses of all blocks on path l, locates the block of interest a, and creates a corresponding select vector $\boldsymbol{b} \in \{0, 1\}^{Z(L+1)}$.
2. Client and server run the homomorphic select sub-protocol with client's input being encryptions of each element in $\boldsymbol{b}$ and server's input being all encrypted blocks on path l. The outcome of the sub-protocol—block a—is sent to the client.
3. Client re-encrypts and writes back the addresses of all blocks on path l, with block a now invalidated. This removes block a from the path without revealing its location. Then, the client re-encrypts block a (possibly modified) under 1 layer, and appends it to the root bucket.

Eviction. To perform EvictAlongPath(l_e), do the following for each level k from 0 to $L - 1$:

1. Client downloads all the metadata (addresses and leaf labels) of the bucket triplet. Based on the metadata, the client determines each block's location after the bucket-triplet eviction.
2. For each slot to be written in the two child buckets:
   - Client creates a corresponding select vector $\boldsymbol{b} \in \{0, 1\}^{2Z}$.
   - Client and server run the homomorphic select sub-protocol with the client's input being encryptions of each element in $\boldsymbol{b}$, and the server's input being the child bucket (being written to) and its parent bucket. Note that if the child bucket is empty due to Observation 1 (which is public information to the server), it conceptually has zero encryption layers.

- Server overwrites the slot with the outcome of the homomorphic select sub-protocol.

4.3 Bounding Layers

Given the above protocol, we bound layers with the following lemma:

Lemma 2. *Any block at level $k \in [0..L]$ has at most $2k + 1$ encryption layers.*

The proof of Lemma 2 is deferred to Appendix B.2. The key intuition for the proof is that due to the reverse-lexicographic eviction order, each bucket will be written to exactly twice (i.e., be a destination or sibling bucket) before being emptied (as a source bucket). Also in Appendix B.2, we introduce a further optimization called the "copy-to-sibling" optimization, which yields a tighter bound: blocks at level $k \in [0..L]$ will have only $k + 1$ layers.

Eviction Post-processing—Peel off Layers in Leaf. The proof only applies to non-leaf buckets: blocks can stay inside a leaf bucket for an unbounded amount of time. Therefore, we need the following post-processing step for leaf nodes. After EvictAlongPath(l_e), the client downloads all blocks from the leaf node, peels off the encryption layers, and writes them back to the leaves as layer-$\Theta(L)$ re-encrypted ciphertexts (meeting the same layer bound as other levels). Since the client performs an eviction every A ORAM requests, and each leaf bucket has size $Z = A$, this incurs only $O(1)$ amortized bandwidth blowup.

4.4 Remarks on Cryptosystem Requirements

Let L' be the layer bound (derived in Sect. 4.3). For efficiency (in bandwidth for the overall protocol) we require the output of an arbitrary select operation performed during an ORAM request (note that $\ell = L'$ in this case) to be a constant times larger than the block size B. Since $L' = \omega(1)$, this implies we need additive blowup per encryption layer, independent of L'. One cryptosystem that satisfies the above requirement, for appropriate parameters, is the Damgård-Jurik cryptosystem (Sect. 6.2). We use this scheme to derive final parameters for the AHE construction in Sect. 6.

5 Security Against Fully Malicious Server

So far, we have seen an ORAM scheme that achieves security against an *honest-but-curious* server who follows the protocol correctly. We now show how to extend this to get a scheme that is secure against a fully malicious server who can deviate arbitrarily from the protocol.

5.1   Abstract Server-Computation ORAM

We start by describing several abstract properties of the Onion ORAM scheme from the previous section. We will call any server-computation ORAM scheme satisfying these properties an *abstract server-computation ORAM*.

Data Blocks and Metadata. The server storage consists of two types of data: *data blocks* and *metadata*. The server performs computation on data blocks, but never on metadata. The client reads and writes the metadata directly, so the metadata can be encrypted under any semantically secure encryption scheme.

Operations on Data Blocks. Following the notations in Sect. 2, each plaintext data block is divided into C chunks, and each chunk is separately encrypted $\mathsf{ct}_i = (\mathsf{ct}_i[1], \ldots, \mathsf{ct}_i[C])$. The client operates on the data blocks either by: (1) directly reading/writing an encrypted data block, or (2) instructing the server to apply a function f to form a new data block ct_i, where $\mathsf{ct}_i[j]$ only depends on the j-th chunk of other data blocks, i.e., $\mathsf{ct}_i[j] = f(\mathsf{ct}_1[j], \ldots, \mathsf{ct}_m[j])$ for all $j \in [1..C]$.

It is easy to check that the two Onion ORAM schemes are instances of the above abstraction. The metadata consists of the encrypted addresses and leaf labels of each data block, as well as additional space needed to implement ORAM recursion. The data blocks are encrypted under either a layered AHE scheme or a SWHE scheme. Function f is a "homomorphic select operation", and is applied to each chunk.

5.2   Semi-honest to Malicious Compiler

We now describe a generic compiler that takes any "abstract server-computation ORAM" that satisfies honest-but-curious security and compiles it into a "verified server-computation ORAM" which is secure in the fully malicious setting.

Verifying Metadata. We can use standard "memory checking" [2] schemes based on Merkle trees [33] to ensure that the client always gets the correct metadata, or aborts if the malicious server ever sends an incorrect value. A generic use of Merkle tree would add an $O(\log N)$ multiplicative overhead to the process of accessing metadata [29], which is good enough for us. This $O(\log N)$ overhead can also be avoided by aligning the Merkle tree with the ORAM tree [38], or using generic authenticated data structures [34]. In any case, verifying metadata is basically free in Onion ORAM.

Challenge of Verifying Data Blocks. Unfortunately, we cannot rely on standard memory checking to protect the encrypted data blocks when the client doesn't read/write them directly but rather instructs the server to compute on them. The problem is that a malicious server that learns some information about the client's access pattern based on *whether the client aborts or not*.

Consider Onion ORAM for example. The malicious server wants to learn if, during the homomorphic select operation of a ORAM request, the location being selected is i. The server can perform the operation correctly except that it would replace the ciphertext at position i with some incorrect value. In this case, if the location being selected was indeed i then the client will abort since the data it receives will be incorrect, but otherwise the client will accept. This violates ORAM's privacy requirement.

A more general way to see the problem is to notice that the client's abort decision above depends on the decrypted value, which depends on the secret key of the homomorphic encryption scheme. Therefore, we can no longer rely on the semantic security of the encryption scheme if the abort decision is revealed to the server. To fix this problem, we need to ensure that the client's abort decision only depends on ciphertext and not on the plaintext data.

Verifying Data Blocks. For our solution, the client selects a random subset S consisting of λ chunk positions. This set S is kept secret from the server. The subset of chunks in positions $\{j : j \in S\}$ of every encrypted data block are treated as additional metadata, which we call the "verification chunks". Verification chunks are encrypted and memory checked in the same way as the other metadata. Whenever the client instructs the server to update an encrypted data block, the client performs the same operation himself on the verification chunks. Then, when the client reads an encrypted data block from the server, he can check the chunks in S against the ciphertexts of verification chunks. This check ensures that the server cannot modify too many chunks without getting caught. To ensure that this check is sufficient, we apply an error-correcting code which guarantees that the server has to modify a large fraction of chunks to affect the plaintext. In more detail:

- Every plaintext data block $\mathsf{pt} = (\mathsf{pt}[1], \ldots, \mathsf{pt}[C])$ is first encoded via an error-correcting code into a codeword block $\mathsf{pt\_ecc} = \mathsf{ECC}(\mathsf{pt}) = (\mathsf{pt\_ecc}[1], \ldots, \mathsf{pt\_ecc}[C'])$. The error-correcting code ECC has a rate $C/C' = \alpha < 1$ and can efficiently recover the plaintext block if at most a δ-fraction of the codeword chunks are erroneous. For concreteness, we can use a Reed-Solomon code, and set $\alpha = \frac{1}{2}, \delta = (1 - \alpha)/2 = \frac{1}{4}$. The client then uses the "abstract server-computation ORAM" over the codeword blocks $\mathsf{pt\_ecc}$ (instead of pt).
- During initialization, the client selects a secret random set $S = \{s_1, \ldots, s_\lambda\} \subseteq [C']$. Each ciphertext data block ct_i has verification chunks $\mathsf{verCh}_i = (\mathsf{verCh}_i[1], \ldots, \mathsf{verCh}_i[\lambda])$. We ensure the invariant that, during an honest execution, $\mathsf{verCh}_i[j] = \mathsf{ct}_i[s_j]$ for $j \in [1..\lambda]$.
- The client uses a memory checking scheme to ensure the authenticity and freshness of the metadata including the verification chunks. If the client detects a violation in metadata at any point, the client aborts (we call this abort_0).
- Whenever the client directly updates or instructs the server to apply the aforementioned function f on an encrypted data block ct_i, it also updates or applies the same function f on the corresponding verification chunks $\mathsf{verCh}_i[j]$ for $j \in [1..\lambda]$, which possibly involves reading other verification chunks that are input to f.
- When the client reads an encrypted data block ct_i, it also reads verCh_i and checks that $\mathsf{verCh}_i[j] = \mathsf{ct}_i[s_j]$ for each $j \in [1..\lambda]$ and aborts if this is not the

case (we call this abort_1). Otherwise the client decrypts ct_i to get $\mathsf{pt\_ecc}_i$ and performs error-correction to recover pt_i. If the error-correction fails, the client aborts (we call this abort_2).

If the client ever aborts during any operation with $\mathsf{abort}_0, \mathsf{abort}_1$ or abort_2, it refuses to perform any future operations. This completes the compiler which gives us Theorem 2.

Security Intuition. Notice that in the above scheme, the decision whether abort_1 occurs does not depend on any secret state of the abstract server-computation ORAM scheme, and therefore can be revealed to the server without sacrificing privacy. We will argue that, if abort_1 does not occur, then the client retrieves the correct data (so abort_2 will not occur) with overwhelming probability. Intuitively, the only way that a malicious server can cause the client to either retrieve the incorrect data or trigger abort_2 without triggering abort_1 is to modify at least a δ (by default, $\delta = 1/4$) fraction of the chunks in an encrypted data block, but avoid modifying any of the λ chunks corresponding to the secret set S. This happens with probability at most $(1 - \delta)^\lambda$ over the random choice of S, which is negligible. The complete proof is given in Appendix B.3.

6 Optimizations and Analysis

In this section we present two optimizations, an asymptotic analysis and a concrete (with constants) analysis for our AHE-based protocol.

6.1 Optimizations

Hierarchical Select Operation and Sorting Networks. For simplicity, we have discussed select operations as inner products between the data vector and the coefficient vector. As an optimization, we may use the Lipmaa construction [27] to implement select hierarchically as a tree of d-to-1 select operations for a constant d (say $d = 2$). In that case, for a given 1 out of Z selection, $\boldsymbol{b}^{\mathsf{hier}} \in \{0, 1\}^{\log Z}$. Eviction along a path requires $O(\log N)$ bucket-triplet operations, each of which is a Z-to-Z permutation. To implement an arbitrary Z-to-Z permutation, we can use the Beneš sorting network, which consists of a total of $O(Z \log Z)$ 2-to-1 select operations per triplet.

At the same time, both the hierarchical select and the Beneš network add $\Theta(\log Z)$ layers to the output as opposed to a single layer. Clearly, this makes the layer bound from Lemma 2 increase to $\Theta(\log Z \log N)$. But we can set related parameters larger to compensate.

Permuted Buckets. Observe that on a request operation, the client and the server need to run a homomorphic select protocol among $O(\lambda \log N)$ blocks. We can reduce this number to $O(\lambda)$ using the permuted bucket technique from Ring ORAM [39] (similar ideas were used in hierarchical ORAMs [20]). Instead of reading all slots along the tree path during each read, we can randomly permute blocks in each bucket and only read/remove a block at a random looking slot (out of $Z = \Theta(\lambda)$ slots) per bucket. Each random-looking location will either contain the block of interest or a dummy block. We must ensure that no bucket runs out of dummies before the next eviction refills that bucket's dummies. Given our reverse-lexicographic eviction order, a simple Chernoff bound shows that adding $\Theta(A) = \Theta(\lambda)$ dummies, which increases bucket size by a constant factor, is sufficient to ensure that dummies do not run out except with probability $2^{-\Theta(\lambda)}$. We do not permute the root bucket since it will require additional techniques (and does not give much benefit). Therefore, a read path selects among $O(Z + \log N) = O(\lambda + \log N) = O(\lambda)$ blocks.

6.2   Damgård-Jurik Cryptosystem

We implement our AHE-based protocol over the Damgård-Jurik cryptosystem [6], a generalization of Paillier's cryptosystem [37]. Both schemes are based on the hardness of the decisional composite residuosity assumption. In this system, the public key $\mathsf{pk} = n = pq$ is an RSA modulus (p and q are two large, random primes) and the secret key $\mathsf{sk} = \mathsf{lcm}(p-1, q-1)$. In the terminology from our onion encryptions, $\mathsf{sk}_i, \mathsf{pk}_i = \mathcal{G}_i()$ for $i \geq 0$.

We denote the integers mod n as $\mathbb{Z}_n$. The plaintext space for the i-th layer of the Damgård-Jurik cryptosystem encryption, $\mathbb{L}_i$, is $\mathbb{Z}_{n^{s_0+i}}$ for some user specified choice of s_0. The ciphertext space for this layer is $\mathbb{Z}_{n^{s_0+i+1}}$. Thus, we clearly have the property that ciphertexts are valid plaintexts in the next layer. An interesting property that immediately follows is that if $s_0 = \Theta(i)$, then $|\mathbb{L}_i|/|\mathbb{L}_0|$ is a constant. In other words, by setting s_0 appropriately the ciphertext blowup after i layers of encryption is a constant.

We further have that $\oplus$ (the primitive for homomorphic addition) is integer multiplication and $\otimes$ (for scalar multiplication) is modular exponentiation. If these operations are performed on ciphertexts in $\mathbb{L}_i$, operations are mod $\mathbb{Z}_{n^{s_0+i}}$.

6.3   Asymptotic Analysis

We first perform the asymptotic analysis for exact exponential security. The results for negligible in N security in Table 1 is derived by setting $\lambda = \omega(\log N)$ and $\gamma = \Theta(\log^3 N)$ according to best known attacks [27].

Semi-honest Case

Chunk Size. The Damgård-Jurik cryptosystem encrypts a message of length γs_0 bits to a ciphertext of length $\gamma(s_0+1)$ bits, where γ is a parameter dependent on the security parameter λ, and s_0 is a user-chosen parameter. Using Beneš network, each ciphertext chunk accumulates $O(\log \lambda \log N)$ layers of encryption at the maximum. Suppose the plaintext chunk size is $B_c := \gamma s_0$, then at the maximum onion layer, the ciphertext size would be $\gamma(s_0 + O(\log \lambda \log N))$. Therefore, to ensure constant ciphertext expansion at all layers, it suffices to set $s_0 := \Omega(\log \lambda \log N)$ and chunk size $B_c := \Omega(\gamma \log \lambda \log N)$. This means ciphertext chunks and homomorphic select vectors are also $\Omega(\gamma \log \lambda \log N)$ bits.

Then we want our block size to be asymptotically larger than the select vectors at each step of our protocol (other metadata are much smaller).

Size of Select Vectors. Each read requires $O(\log \lambda)$ encrypted coefficients of $O(B_c)$ bits each. Eviction along a path requires $O(\log N)$ Beneš network (bucket-triplet operations), a total of $O(\lambda \log \lambda \log N)$ encrypted coefficients. Also recall that one eviction happens per $A = \Theta(\lambda)$ accesses. Therefore, the select vector size per ORAM access (amortized) is dominated by evictions, and is $\Theta(B_c \log \lambda \log N)$ bits.

Setting the Block Size. Clearly, if we set the block size to be $B :=$ $\Omega(B_c \log \lambda \log N)$, the cost of homomorphic select vectors could be asymptotically absorbed, thereby achieving constant bandwidth blowup. Since the chunk size $B_c = \Omega(\gamma \log \lambda \log N)$, we have $B = \Omega(\gamma \log^2 \lambda \log^2 N)$ bits.

Server Computation. The bottleneck of server computation is to homomorphically multiple a block with a encrypted select coefficient. In Damgård-Jurik, this is a modular exponentiation operation, which has $\widetilde{O}(\gamma^2)$ computational complexity for γ-bit ciphertexts. This means the *per-bit* computational overhead is $\widetilde{O}(\gamma)$. The server needs to perform this operation on $O(\lambda)$ blocks of size B, and therefore has a computational overhead of $\widetilde{O}(\gamma)O(B\lambda)$.

Client Computation. Client needs to decrypt $O(\log \lambda \log N)$ layers to get the plaintext block, and therefore has a computational overhead of $\widetilde{O}(\gamma)O(B \log \lambda \log N)$.

Malicious Case

Setting the Block Size. The main difference from semi-honest case is that on a read, the client must additionally download $\Theta(\lambda)$ verification chunks from each of the $\Theta(\lambda)$ blocks (assuming permuted buckets). Select vector size stays the same, and the error-correcting code increases block size by only a constant factor. Thus, the block size we need to achieve constant bandwidth over the entire protocol is $B = \Omega(B_c \lambda^2) = \Omega(\gamma \lambda^2 \log \lambda \log N)$.

Client Computation. Another difference is that the client now needs to emulate the server's homomorphic select operation on the verification chunks. But a simple analysis will show that the bottleneck of client computation is still onion decryption, and therefore remains the same asymptotically.

6.4 Concrete Analysis (Semi-honest Case Only)

Figure 4 shows bandwidth as a function of block size for our optimized semi-honest construction, taking into account all constant factors (including the extra bandwidth cost to recursively look up the position map). Other scheme variants in this paper have the same general trend. We compare to Path PIR and Circuit ORAM, the most bandwidth-efficient constructions with/without server computation that match our server/client storage asymptotics.

Takeaway. The high order bit is that as block size increases, Onion ORAM's bandwidth approaches $2B$. Note that $2B$ is the inherent lower bound in bandwidth since every ORAM access must at least the block of interest from the server and send it back after possibly modifying it. Given an 8 MB block size, which is approximately the size of an image file, we improve in bandwidth over Circuit ORAM by **35×** and improve over Path PIR by **22×**. For very large block sizes, our improvement continues to increase but Circuit ORAM and Path PIR improve less dramatically because their asymptotic bandwidth blowup has a $\log N$ factor. Note that for sufficiently small block sizes, both Path PIR and Circuit ORAM beat our bandwidth because our select vector bandwidth dominates. Yet, this crossover point is around 128 KB, which is reasonable in many settings.

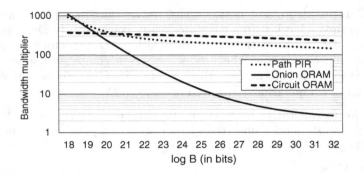

Fig. 4. Plots the bandwidth multiplier (i.e., the hidden constant for $O(B)$) for semi-honest Onion ORAM and two prior proposals. We fix the ORAM capacity to $NB = 2^{50}$ and give each scheme the same block size across different block sizes (hence as B increases, N decreases).

Constant Factor Optimization: Less Frequent Leaf Post-processing. In the above evaluation, we apply an additional constant factor optimization. Since $Z = A = \Theta(\lambda)$, we must send and receive one additional data block (amortized) per ORAM request to post-process leaf buckets during evictions (Sect. 4.3). To save bandwidth, we can perform this post-processing on a particular leaf bucket every p evictions to that leaf (p is a free variable). The consequence is that the number of layers that accumulate on leaf buckets increases by p which makes each ORAM read path more expensive by the corresponding amount. In practice, $p \geq 8$ yields the best bandwidth.

Parameterization Details. For both schemes, we set acceptable ORAM failure probability to 2^{-80} which results in $Z = A \approx 300$ for Onion ORAM, $Z = 120$ for Path PIR [41] and a stash size (stored on the server) of 50 blocks for Circuit ORAM [47]. For Onion ORAM and Path PIR we set $\gamma = 2048$ bits. For Circuit ORAM, we use the reverse lexicographic eviction order as described in that work, which gives 2 evictions per access and $Z = 2$. For Path PIR, we set the eviction frequency $v = 2$ [41].

6.5 Other Optimizations and Remarks

De-Amortization. We remark that it is easy to de-amortize the above algorithm so that the worst-case bandwidth equals amortized bandwidth and overall bandwidth doesn't increase. First, it is trivial to de-amortize the leaf bucket post-processing (Sect. 4.3) over the A read path operations because $A = Z$ and post-processing doesn't change the underlying plaintext contents of that bucket. Second, the standard de-amortization trick of Williams et al. [50] can be applied directly to our EvictAlongPath operation. We remark that it is easy to de-amortize evictions over the next A read operations because moving blocks from buckets (possibly on the eviction path) to the root bucket does not impact our eviction algorithm.

Online Roundtrips. The standard recursion technique [44] uses a small block size for position map ORAMs (to save bandwidth) and requires $O(\log N)$ roundtrips. In Onion ORAM, the block in the main ORAM is large $B = \Omega(\lambda \log N)$. We can use Onion ORAM with the same large block size for position map ORAMs. This achieves a constant number of recursive levels if N is polynomial in λ, and therefore maintains the constant bandwidth blowup.

7   Conclusion and Open Problems

This paper proposes *Onion ORAM*, the first concrete ORAM scheme with optimal asymptotics in worst-case bandwidth blowup, server storage and client storage in the single-server setting. We have shown that FHE or SWHE are not necessary in constructing constant bandwidth ORAMs, which instead can be constructed using only an additively homomorphic scheme such as the Damgård-Jurik cryptosystem. Yet combining SWHE with Onion ORAM improves the computational efficiency of the scheme. We further extend Onion ORAM to be secure in the fully malicious setting using standard assumptions. Due to the known efficiency of SWHE schemes like BGV, we think of our work as an important step towards *practical* constant bandwidth blowup ORAM schemes.

We do note that while our block size is poly-logarithmic, the exponent is rather large (especially for our malicious construction). Subsequent to our proposal of Onion ORAM, Moataz et al. [35] combined our bounded feedback ORAM with an optimized merge procedure for evictions which reduces server computation and block size for the semi-honest construction. We applaud this effort and argue that semi-honest constant bandwidth blowup ORAM is practical (or nearly practical). We leave tightening up poly-logarithmic factors for our malicious security construction as future work.

Beyond tightening parameters, an open problem is whether constant bandwidth blowup ORAMs can be constructed from non-homomorphic encryption schemes. The computational complexity of the Damgård-Jurik cryptosystem (which relies on modular exponentiation for homomorphic operations), or even more efficient SWHE schemes may be a bottleneck in practice. Can we construct constant bandwidth ORAM using simple computation such as XOR and any semantically secure encryption scheme with small ciphertext blowup? A partial result in this direction comes from Burst ORAM [8]: simple computation on ciphertexts (mod 2 XOR) enables a family of schemes (e.g., [39]) to achieve constant online bandwidth blowup on a *request*. Whether similar ideas can lead to constant bandwidth blowup on eviction is unclear.

Acknowledgements. We thank Vinod Vaikuntanathan for helpful discussion on this work.

A   Definitions of Server-Computation ORAM

We directly adopt the definitions and notations used by Apon et al. [1] who are the first to define server-computation ORAM as a reactive two-party protocol between the client and the server, and define its security in the Universal Composability model [5]. We use the notation

$$((c\_out, c\_state), (s\_out, s\_state)) \leftarrow \mathsf{protocol}((c\_in, c\_state), (s\_in, s\_state))$$

to denote a (stateful) protocol between a client and server, where c_in and c_out are the client's input and output; s_in and s_out are the server's input and output; and c_state and s_state are the client and server's states before and after the protocol.

We now define the notion of a *server-computation ORAM*, where a client outsources the storage of data to a server, and performs subsequent read and write operations on the data.

Definition 1 (Server-Computation   ORAM). *A   server-computation ORAM scheme consists of the following interactive protocols between a client and a server.*

$((\bot, z), (\bot, Z)) \leftarrow \mathsf{Setup}(1^\lambda, (D, \bot), (\bot, \bot))$: An interactive protocol where the client's input is a memory array $D[1..n]$ where each memory *block* has bit-length β; and the server's input is $\bot$. At the end of the Setup protocol, the client has secret state z, and server's state is Z (which typically encodes the memory array D).

$((\mathsf{data}, z'), (\bot, Z')) \leftarrow \mathsf{Access}((\mathsf{op}, z), (\bot, Z))$: To access data, the client starts in state z, with an input op where $\mathsf{op} := (\mathsf{read}, ind)$ or $\mathsf{op} := (\mathsf{write}, ind, \mathsf{data})$; the server starts in state Z, and has no input. In a correct execution of the protocol, the client's output data is the current value of the memory D at location ind (for writes, the output is the old value of $D[ind]$ before the write takes place). The client and server also update their states to z' and Z' respectively. The client outputs $\mathsf{data} := \bot$ if the protocol execution aborted.

We say that a server-computation ORAM scheme is correct, if for any initial memory $D \in \{0,1\}^{\beta n}$, for any operation sequence $\mathsf{op}_1, \mathsf{op}_2, \ldots, \mathsf{op}_m$ where $m = \mathrm{poly}(\lambda)$, an $\mathsf{op} := (\mathsf{read}, ind)$ operation would always return the last value written to the logical location ind (except with negligible probability).

A.1   Security Definition

We adopt a standard simulation-based definition of secure computation [4], requiring that a real-world execution "simulate" an ideal-world (reactive) functionality $\mathcal{F}$.

Ideal World. We define an ideal functionality $\mathcal{F}$ that maintains an up-to-date version of the data D on behalf of the client, and answers the client's access queries.

- *Setup.* An environment $\mathcal{Z}$ gives an initial database D to the client. The client sends D to an ideal functionality $\mathcal{F}$. $\mathcal{F}$ notifies the ideal-world adversary $\mathcal{S}$ of the fact that the setup operation occurred as well as the size of the database $N = |D|$, but not of the data contents D. The ideal-world adversary $\mathcal{S}$ says ok or abort to $\mathcal{F}$. $\mathcal{F}$ then says ok or $\bot$ to the client accordingly.
- *Access.* In each time step, the environment $\mathcal{Z}$ specifies an operation $\mathsf{op} := (\mathsf{read}, ind)$ or $\mathsf{op} := (\mathsf{write}, ind, \mathsf{data})$ as the client's input. The client sends op to $\mathcal{F}$. $\mathcal{F}$ notifies the ideal-world adversary $\mathcal{S}$ (without revealing to $\mathcal{S}$ the operation op). If $\mathcal{S}$ says ok to $\mathcal{F}$, $\mathcal{F}$ sends $D[ind]$ to the client, and updates $D[ind] := \mathsf{data}$ accordingly if this is a write operation. The client then forwards $D[ind]$ to the environment $\mathcal{Z}$. If $\mathcal{S}$ says abort to $\mathcal{F}$, $\mathcal{F}$ sends $\bot$ to the client.

Real World. In the real world, an environment $\mathcal{Z}$ gives an honest client a database D. The honest client runs the Setup protocol with the server $\mathcal{A}$. Then at each time step, $\mathcal{Z}$ specifies an input $\mathsf{op} := (\mathsf{read}, ind)$ or $\mathsf{op} := (\mathsf{write}, ind, \mathsf{data})$ to the client. The client then runs the Access protocol with the server. The environment $\mathcal{Z}$ gets the view of the adversary $\mathcal{A}$ after every operation. The client outputs to the environment the data fetched or $\bot$ (indicating abort).

Definition 2 (Simulation-Based Security: Privacy + Verifiability). *We say that a protocol $\Pi_{\mathcal{F}}$ securely computes the ideal functionality $\mathcal{F}$ if for any probabilistic polynomial-time real-world adversary (i.e., server) $\mathcal{A}$, there exists an ideal-world adversary $\mathcal{S}$, such that for all non-uniform, polynomial-time environment $\mathcal{Z}$, there exists a negligible function negl such that*

$$|\Pr\left[\mathrm{REAL}_{\Pi_{\mathcal{F}}, \mathcal{A}, \mathcal{Z}}(\lambda) = 1\right] - \Pr\left[\mathrm{IDEAL}_{\mathcal{F}, \mathcal{S}, \mathcal{Z}}(\lambda) = 1\right]| \leq \mathsf{negl}(\lambda)$$

At an intuitive level, our definition captures the privacy and verifiability requirements for an honest client (the client is never malicious in our setting), in the presence of a malicious server. The definition simultaneously captures *privacy* and *verifiability*. Privacy ensures that the server cannot observe the data contents or the access pattern. Verifiability ensures that the client is guaranteed to read the correct data from the server — if the server cheats, the client can detect it and abort the protocol.

B   Proofs

B.1   Bounded Feedback ORAM: Bounding Overflows

We now give formal proofs to show that buckets do not overflow in bounded feedback ORAM except with negligible probability.

Proof. (of Lemma 1). First of all, notice that when $Z \geq A$, the root bucket will never overflow. So we will only consider non-root buckets. Let b be a non-root bucket, and $Y(b)$ be the number of blocks in it after an eviction operation. We will first assume all buckets have infinite capacity and show that $E[Y(b)] \leq A/2$, i.e., the expected number of blocks in a non-root bucket after an eviction operation is no more than $A/2$ at any time. Then, we bound the overflow probability given a finite capacity.

If b is a leaf bucket, each of the N blocks in the system has a probability of 2^{-L} to be mapped to b independently. Thus $E[Y(b)] \leq N \cdot 2^{-L} \leq A/2$.

If b is a non-leaf (and non-root) bucket, we define two variables m_1 and m_2: the last EvictAlongPath operation where b is on the eviction path is the m_1-th EvictAlongPath operation, and the EvictAlongPath operation where b is a sibling bucket is the m_2-th EvictAlongPath operation. If $m_1 > m_2$, then $Y(b) = 0$, because b becomes empty when it is the source bucket in the m_1-th EvictAlongPath operation. (Recall that buckets have infinite capacity so this outcome is guaranteed.) If $m_1 < m_2$, there will be some blocks in b and we now analyze what blocks will end up in b. We time-stamp the blocks as follows. When a block is accessed and remapped, it gets time stamp m^*, which is the number of EvictAlongPath operations that have happened. Blocks with $m^* \leq m_1$ will not be in b as they will go to either the left child or the right child of b. Blocks with $m^* > m_2$ will not be in b as the last eviction operation that touches b (m_2-th) has already passed. Therefore, only blocks with time stamp $m_1 < m^* \leq m_2$ can be in b. There are at most $d = A|m_1 - m_2|$ such blocks. Such a block goes to b if and only if it is mapped to a path containing b. Thus, each block goes to b independently with a probability of 2^{-i}, where i is the level of b. The deterministic order of EvictAlongPath makes it easy to see[4] that $|m_1 - m_2| = 2^{i-1}$. Therefore, $E[Y(b)] \leq d \cdot 2^{-i} = A/2$ for any non-leaf bucket as well.

Now that we have independence and the bound on expectation, a simple Chernoff bound completes the proof.

B.2   Onion ORAM: Bounding Layers of Encryption

To bound the layers of onion encryption, we consider the following abstraction. Suppose all buckets in the tree have a layer associated with it.

– The root bucket contains layer-1 ciphertexts.

---

[4] One way to see this is that a bucket b at level i will be on the evicted path every 2^i EvictAlongPath operations, and its sibling will be on the evicted path halfway in that period.

- For a bucket known to be empty, we define bucket.layer := 0.
- Each bucket-triplet operation moves data from parent to child buckets. After the operation, child.layer := max{parent.layer, child.layer} + 1.

Recall that we use the following terminology. The bucket being evicted from is called the *source*, its child bucket on the eviction path is called the *destination*, and its other child forking off the path is called the *sibling*.

Proof. (of Lemma 2). We prove by induction.

Base case. The lemma holds obviously for the root bucket.

Inductive step. Suppose that this holds for all levels $\ell < k$. We now show that this holds for level k. Let bucket denote a bucket at level k. We focus on this particular bucket, and examine bucket.layer after each bucket-triplet operation that involves bucket. It suffices to show that after each bucket-triplet operation involving bucket, it must be that bucket.layer $\leq 2k + 1$. If a bucket-triplet operation involves bucket as a source, we call it a *source operation* (from the perspective of bucket). Similarly, if a bucket-triplet operation involves bucket as a destination or sibling, we call it a *destination operation* or a *sibling operation* respectively.

Based on Observation 1,

$$\text{bucket.layer} = 0 \qquad \text{(after each source operation)}$$

Since a sibling operation must be preceded by a source operation (if there is any preceding operation), bucket must be empty at the beginning of each sibling operation. By induction hypothesis, after each sibling operation, it must be that

$$\text{bucket.layer} \leq 2(k - 1) + 1 + 1 = 2k \qquad \text{(after each sibling operation)}$$

Since a destination operation must be preceded by a sibling operation (if there is any preceding operation), from the above we know that at the beginning of a destination operation bucket.layer must be bounded by $2k$. Now, by induction hypothesis, it holds that

$$\text{bucket.layer} \leq 2k + 1 \qquad \text{(after each destination operation)}$$

Finally, our post-processing on leaves where the client peels of the onion layers extends this lemma to all levels including leaves.

Copy-to-Sibling Optimization and a Tighter Layer Bound. An immediate implication of Observation 1 plus Observation 2 is that whenever a source evicts into a sibling, the sibling bucket is empty to start with because it was a source bucket in the last operation it was involved in. This motivates the following optimization: the server can simply copy blocks from the source bucket into the sibling. The client would read the metadata corresponding to blocks in the source bucket, invalidate blocks that do not belong to the sibling, before writing the (re-encrypted) metadata to the sibling.

This copy-to-sibling optimization avoids accumulating an extra onion layer upon writes into a sibling bucket. With this optimization and using a similar inductive proof, it is not hard to show a bucket at level k in the tree have at most $k + 1$ layers.

B.3   Malicious Security Proof

The Simulator. To simulate the setup protocol with some data of size N, the simulator chooses a dummy database D' of size N consisting of all 0s. It then follows the honest setup procedure on behalf of the client with database D'. To simulate each access operation, the simulator follows the honest protocol for reading a dummy index, say, $ind' = 0$, on behalf of the client.

During each operation, if the client protocol that's being executed by the simulator aborts then the simulator sends abort to $\mathcal{F}$ and stops responding to future commands on behalf of the client, else it gives ok to $\mathcal{F}$.

Sequence of Hybrids. We now follow a sequence of hybrid games to show that the real world and the simulation are indistinguishable:

$$|\Pr\left[\text{REAL}_{\Pi_{\mathcal{F}},\mathcal{A},\mathcal{Z}}(\lambda) = 1\right] - \Pr\left[\text{IDEAL}_{\mathcal{F},\mathcal{S},\mathcal{Z}}(\lambda) = 1\right]| \leq \mathsf{negl}(\lambda)$$

Game 0. Let this be the real game $\text{REAL}_{\Pi_{\mathcal{F}},\mathcal{A},\mathcal{Z}}$ with an adversarial server $\mathcal{A}$ and an environment $\mathcal{Z}$.

Game 1. In this game, the client also keeps a local copy of the correct metadata and data-blocks (in plaintext) that should be stored on the server. Whenever the client reads any (encrypted) metadata from the server during any operation, if the memory checking does not abort, then instead of decrypting the read metadata, the client simply uses the locally stored plaintext copy.

The only difference between Game 0 and Game 1 occurs if in Game 0 the memory checking does not abort, but the client retrieves the incorrect encrypted metadata, which happens with negligible probability by the security of memory checking. Therefore Game 0 and Game 1 are indistinguishable.

Game 2. In this game the client doesn't store the correct values of verCh_i with the encrypted metadata on the server, but instead replaces these with dummy values. The client still stores the correct values of verCh_i in the plaintext metadata stored locally, which it uses to do all of the actual computations.

Game 1 and Game 2 are indistinguishable by the CPA security of the symmetric-key encryption scheme used to encrypt metadata. We only need CPA security since, in Games 1 and 2, the client never decrypts any of the metadata ciphertexts.

Game 3. In this game, whenever the client reads an encrypted data block ct_i from the server, if abort$_1$ does not occur, instead of decrypting and decoding the encrypted data-block, the client simply uses local copy of the plaintext data-block.

The only difference between Game 2 and Game 3 occurs if at some point in time the client reads an encrypted data block ct_i from the server such that at least a δ fraction of the ciphertext chunks $\{\mathsf{ct}_i[j]\}$ in the block have been modified (so that decoding either fails with abort$_2$ or returns an incorrect value) but none of the chunks in locations $i \in S$ have been modified (so that abort$_1$ does not occur).

We claim that Game 2 and Game 3 are statistically indistinguishable, with statistical distance at most $q(1 - \delta)^\lambda$, where q is the total number of operations performed by the client. To see this, note that in both games the set S is initially completely random and unknown to the adversarial server. In each operation i that the client reads an encrypted data-block, the server can choose some set $S_i' \subseteq [C']$ of positions in which the ciphertext chunks are modified, and if $S_i' \cap S = \emptyset$ the server learns this information about the set S and the game continues, else the client aborts and the game stops. The server never gets any other information about S throughout the game. The games 2 and 3 only diverge if at some point the adversarial server guesses a set S_i' of size $|S_i'| \geq \delta C'$ such that $S \cap S_i' = \emptyset$. We call this the "bad event". Notice that the sets S_i' can be thought of as being chosen non-adaptively at the beginning of the game prior to the adversary learning any knowledge about S (this is because we know in advance that the server will learn $S_i' \cap S = \emptyset$ for all i prior to the game ending). Therefore, the probability that the bad event happens in the j'th operation is

$$\Pr_S[S_j' \cap S = \emptyset] \leq \binom{(1 - \delta)C'}{\lambda} / \binom{C'}{\lambda} \leq (1 - \delta)^\lambda$$

where $S \subseteq [C']$ is a random subset of size $|S| = \lambda$. By the union bound, the probability that the bad event happens during some operation $j \in \{1, \ldots, q\}$ is at most $q(1 - \delta)^\lambda$.

Game' 3. In this game, the client runs the setup procedure using the dummy database D' (as in the simulation) instead of the one given by the environment. Furthermore, for each access operation, the client just runs a dummy operation consisting of a read with the index $ind' = 0$ instead of the operation chosen by the environment. (We also introduce an ideal functionality $\mathcal{F}$ in this world which is given the correct database D at setup and the correct access operations as chosen by the environment. Whenever the client doesn't abort, it forwards the outputs of $\mathcal{F}$ to the environment.)

Games 3 and Game' 3 are indistinguishable by the semi-honest Onion ORAM scheme. In particular, in both games whenever the client doesn't abort, the client reads the correct metadata and data blocks as when interacting with an honest server, and therefore follows the same protocols as when interacting with an honest server. Furthermore, the decision whether or not the client aborts in these games (with abort$_0$ or abort$_1$; there is no more abort$_2$) only depends on the secret set S and the internal state of the memory checking scheme, but is independent of any of the secret state or decryption keys of the underlying semi-honest Onion ORAM scheme. Therefore, the view of the adversarial server in these games can be simulated given the view of the honest server.

Game' 2,1,0. We define Game' i for $i = 0, 1, 2$ the same way as Game i except that the client uses the dummy database D' and the dummy operations (reads with index $idx' = 0$) instead of those specified by the environment.

The arguments that Game' $i + 1$ and Game' i are indistinguishable as the same as those for Game $i + 1$ and Game i. Finally, we notice that Game 0 is the ideal game $\text{IDEAL}_{\mathcal{F},\mathcal{S},\mathcal{Z}}$ with the simulator $\mathcal{S}$.

Putting everything together, we see that the real and ideal games $\text{REAL}_{\Pi_{\mathcal{F}},\mathcal{A},\mathcal{Z}}$ and $\text{IDEAL}_{\mathcal{F},\mathcal{S},\mathcal{Z}}$ are indistinguishable as we wanted to show.

References

1. Apon, D., Katz, J., Shi, E., Thiruvengadam, A.: Verifiable oblivious storage. In: Krawczyk, H. (ed.) PKC 2014. LNCS, vol. 8383, pp. 131–148. Springer, Heidelberg (2014)
2. Blum, M., Evans, W.S., Gemmell, P., Kannan, S., Naor, M.: Checking the correctness of memories. In: FOCS (1991)
3. Brakerski, Z., Vaikuntanathan, V.: Fully homomorphic encryption from ring-LWE and security for key dependent messages. In: Rogaway, P. (ed.) CRYPTO 2011. LNCS, vol. 6841, pp. 505–524. Springer, Heidelberg (2011)
4. Canetti, R.: Security and composition of multiparty cryptographic protocols. J. Cryptol. **13**, 143–202 (2000)
5. Canetti, R.: Universally composable security: a new paradigm for cryptographic protocols. In: FOCS (2001)
6. Damgard, I., Jurik, M.: A generalisation, a simplification and some applications of Paillier's probabilistic public-key system. In: Kim, K. (ed.) PKC 2001. LNCS, vol. 1992, pp. 119–136. Springer, Heidelberg (2001)
7. Dautrich, J., Ravishankar, C.: Combining ORAM with PIR to minimize bandwidth costs. In: CODASPY (2015)
8. Dautrich, J., Stefanov, E., Shi, E.: Burst ORAM: Minimizing ORAM response times for bursty access patterns. In: USENIX Security (2014)
9. Devadas, S., van Dijk, M., Fletcher, C.W., Ren, L., Shi, E., Wichs, D.: Onion ORAM: a constant bandwidth blowup oblivious RAM. Cryptology ePrint Archive, Report 2015/005 (2015)
10. Fletcher, C., Ren, L., Kwon, A., van Dijk, M., Devadas, S.: Freecursive ORAM: [nearly] free recursion and integrity verification for position-based oblivious RAM. In: ASPLOS (2015)
11. Fletcher, C., Ren, L., Kwon, A., Van Dijk, M., Stefanov, E., Serpanos, D., Devadas, S.: A low-latency, low-area hardware oblivious RAM controller. In: FCCM (2015)
12. Fletcher, C., van Dijk, M., Devadas, S.: Secure processor architecture for encrypted computation on untrusted programs. In: STC (2012)
13. Gentry, C., Goldman, K.A., Halevi, S., Julta, C., Raykova, M., Wichs, D.: Optimizing ORAM and using it efficiently for secure computation. In: De Cristofaro, E., Wright, M. (eds.) PETS 2013. LNCS, vol. 7981, pp. 1–18. Springer, Heidelberg (2013)
14. Gentry, C., Halevi, S., Jutla, C., Raykova, M.: Private database access with he-over-oram architecture. Cryptology ePrint Archive, Report 2014/345
15. Gentry, C., Halevi, S., Lu, S., Ostrovsky, R., Raykova, M., Wichs, D.: Garbled RAM revisited. In: Nguyen, P.Q., Oswald, E. (eds.) EUROCRYPT 2014. LNCS, vol. 8441, pp. 405–422. Springer, Heidelberg (2014)

16. Gentry, C., Halevi, S., Raykova, M., Wichs, D.: Outsourcing private RAM computation. In: FOCS (2014)

17. Gentry, C., Halevi, S., Smart, N.P.: Better bootstrapping in fully homomorphic encryption. In: Fischlin, M., Buchmann, J., Manulis, M. (eds.) Public Key Cryptography – PKC 2012. LNCS, vol. 7293. Springer, Heidelberg (2012)

18. Gentry, C., Wichs, D.: Separating succinct non-interactive arguments from all falsifiable assumptions. In: STOC (2011)

19. Goldreich, O.: Towards a theory of software protection and simulation on Oblivious RAMs. In: STOC (1987)

20. Goldreich, O., Ostrovsky, R.: Software protection and simulation on oblivious RAMs. J. ACM **43**, 431–473 (1996)

21. Goodrich, M.T., Mitzenmacher, M., Ohrimenko, O., Tamassia, R.: Privacy-preserving group data access via stateless oblivious RAM simulation. In: SODA (2012)

22. Halevi, S., Shoup, V.: Bootstrapping for HElib. In: Oswald, E., Fischlin, M. (eds.) EUROCRYPT 2015. LNCS, vol. 9056, pp. 641–670. Springer, Heidelberg (2015)

23. Ishai, Y., Paskin, A.: Evaluating branching programs on encrypted data. In: Vadhan, S.P. (ed.) TCC 2007. LNCS, vol. 4392, pp. 575–594. Springer, Heidelberg (2007)

24. Keller, M., Scholl, P.: Efficient, Oblivious data structures for MPC. Cryptology ePrint Archive, Report 2014/137 (2014)

25. Kushilevitz, E., Lu, S., Ostrovsky, R.: On the (in) security of hash-based oblivious RAM and a new balancing scheme. In: SODA (2012)

26. Lindner, R., Peikert, C.: Better key sizes (and attacks) for LWE-based encryption. In: Kiayias, A. (ed.) CT-RSA 2011. LNCS, vol. 6558, pp. 319–339. Springer, Heidelberg (2011)

27. Lipmaa, H.: An oblivious transfer protocol with log-squared communication. In: Zhou, J., López, J., Deng, R.H., Bao, F. (eds.) ISC 2005. LNCS, vol. 3650, pp. 314–328. Springer, Heidelberg (2005)

28. Liu, Y. Huang, E. Shi, J. Katz, and M. Hicks. Automating efficient RAM-model secure computation. In: Oakland (2014)

29. Lorch, J.R., Parno, B., Mickens, J. W., Raykova, M., Schiffman, J.: Shroud: ensuring private access to large-scale data in the data center. In: FAST (2013)

30. Lu, S., Ostrovsky, R.: How to garble RAM programs? In: Johansson, T., Nguyen, P.Q. (eds.) EUROCRYPT 2013. LNCS, vol. 7881, pp. 719–734. Springer, Heidelberg (2013)

31. Maas, M., Love, E., Stefanov, E., Tiwari, M., Shi, E., Asanovic, K., Kubiatowicz, J., Song, D.: Phantom: practical oblivious computation in a secure processor. In: CCS (2013)

32. Mayberry, T., Blass, E.-O., Chan, A. H.: Efficient private file retrieval by combining ORAM and PIR. In: NDSS (2014)

33. Merkle, R.C.: Protocols for public key cryptography. In: Oakland (1980)

34. Miller, A., Hicks, M., Katz, J., Shi, E.: Authenticated data structures, generically. In: POPL (2014)

35. Moataz, T., Mayberry, T., Blass, E.-O.: Constant communication oblivious RAM. Cryptology ePrint Archive, Report 2015/570 (2015)

36. Ostrovsky, R.: Efficient computation on oblivious RAMs. In: STOC (1990)

37. Paillier, P.: Public-key cryptosystems based on composite degree residuosity classes. In: Stern, J. (ed.) EUROCRYPT 1999. LNCS, vol. 1592. Springer, Heidelberg (1999)

38. Ren, L., Fletcher, C., Yu, X., van Dijk, M., Devadas, S.: Integrity verification for path oblivious-RAM. In: HPEC (2013)
39. Ren, L., Fletcher, C.W., Kwon, A., Stefanov, E., Shi, E., Dijk, M.V., Devadas, S.: Constants count: practical improvements to oblivious RAM. In: USENIX Security (2015)
40. Ren, L., Yu, X., Fletcher, C., van Dijk, M., Devadas, S.: Design space exploration and optimization of path oblivious RAM in secure processors. In: ISCA (2013)
41. Shi, E., Chan, T.-H.H., Stefanov, E., Li, M.: Oblivious RAM with $O((logN)^3)$ Worst-Case Cost. In: Lee, D.H., Wang, X. (eds.) ASIACRYPT 2011. LNCS, vol. 7073, pp. 197–214. Springer, Heidelberg (2011)
42. Stefanov, E., Shi, E.: Oblivistore: high performance oblivious cloud storage. In: S&P (2013)
43. Stefanov, E., Shi, E., Song, D.: Towards practical oblivious RAM. In: NDSS (2012)
44. Stefanov, E., van Dijk, M., Shi, E., Chan, T.-H.H., Fletcher, C., Ren, L., Yu, X., Devadas, S.: Path ORAM: an extremely simple oblivious RAM protocol. Cryptology ePrint Archive, Report 2013/280
45. Stefanov, E., van Dijk, M., Shi, E., Fletcher, C., Ren, L., Yu, X., Devadas, S.: Path ORAM: an extremely simple oblivious RAM protocol. In: CCS (2013)
46. Wang, X., Nayak, K., Liu, C., Shi, E., Stefanov, E., Huang, Y.: Oblivious data structures. In: IACR (2014)
47. Wang, X.S., Chan, T.-H.H., Shi, E.: Circuit ORAM: On tightness of the Goldreich-Ostrovsky lower bound. Cryptology ePrint Archive, Report 2014/672
48. Wang, X.S., Huang, Y., Chan, T.-H.H., Shelat, A., Shi, E.: Scoram: oblivious ram for secure computation. In: CCS (2014)
49. Williams, P., Sion, R.: Single round access privacy on outsourced storage. In: CCS (2012)
50. Williams, P., Sion, R., Tomescu, A.: Privatefs: a parallel oblivious file system. In: CCS (2012)
51. Yu, X., Fletcher, C.W., Ren, L., van Dijk, M., Devadas, S.: Generalized external interaction with tamper-resistant hardware with bounded information leakage. In: CCSW (2013)
52. Zhang, J., Ma, Q., Zhang, W., Qiao, D.: Kt-oram: a bandwidth-efficient ORAM built on k-ary tree of pir nodes. Cryptology ePrint Archive, Report 2014/624 (2014)

Oblivious Parallel RAM and Applications

Elette Boyle[1]([✉]), Kai-Min Chung[2], and Rafael Pass[3]

[1] IDC Herzliya, Herzliya, Israel
eboyle@alum.mit.edu
[2] Academica Sinica, Taipei, Taiwan
kmchung@iis.sinica.edu.tw
[3] Cornell University, Ithaca, USA
rafael@cs.cornell.edu

Abstract. We initiate the study of cryptography for *parallel RAM (PRAM)* programs. The PRAM model captures modern multi-core architectures and cluster computing models, where several processors execute in parallel and make accesses to shared memory, and provides the "best of both" circuit and RAM models, supporting both cheap random access and parallelism.

We propose and attain the notion of *Oblivious PRAM*. We present a compiler taking any PRAM into one whose distribution of memory accesses is statistically independent of the data (with negligible error), while only incurring a polylogarithmic slowdown (in both total and *parallel* complexity). We discuss applications of such a compiler, building upon recent advances relying on Oblivious (sequential) RAM (Goldreich Ostrovsky JACM'12). In particular, we demonstrate the construction of a *garbled PRAM* compiler based on an OPRAM compiler and secure identity-based encryption.

E. Boyle—The research of the first author has received funding from the European Union's Tenth Framework Programme (FP10/ 2010-2016) under grant agreement no. 259426 ERC-CaC, and ISF grant 1709/14. Supported by the ERC under the EU's Seventh Framework Programme (FP/2007-2013) ERC Grant Agreement n. 307952.

K.-M. Chung—supported in part by Ministry of Science and Technology, Taiwan, under Grant no. MOST 103-2221-E-001-022-MY3.

R. Pass—Work supported in part by a Microsoft Faculty Fellowship, Google Faculty Award, NSF Award CNS-1217821, NSF Award CCF-1214844, AFOSR Award FA9550-15-1-0262 and DARPA and AFRL under contract FA8750-11-2-0211. The views and conclusions contained in this document are those of the authors and should not be interpreted as representing the official policies, either expressed or implied, of the Defense Advanced Research Projects Agency or the US Government.

R. Pass—This work was done in part while the authors were visiting the Simons Institute for the Theory of Computing, supported by the Simons Foundation and by the DIMACS/Simons Collaboration in Cryptography through NSF grant #CNS-1523467.

© International Association for Cryptologic Research 2016
E. Kushilevitz and T. Malkin (Eds.): TCC 2016-A, Part II, LNCS 9563, pp. 175–204, 2016.
DOI: 10.1007/978-3-662-49099-0_7

1   Introduction

Completeness results in cryptography provide general transformations from arbitrary functionalities described in a particular computational model, to solutions for executing the functionality securely within a desired adversarial model. Classic results, stemming from [Yao82, GMW87], modeled computation as *boolean circuits*, and showed how to emulate the circuit securely gate by gate.

As the complexity of modern computing tasks scales at tremendous rates, it has become clear that the circuit model is not appropriate: Converting "lightweight," optimized programs first into a circuit in order to obtain security is not a viable option. Large effort has recently been focused on enabling direct support of functionalities modeled as Turing machines or random-access machines (RAM) (e.g., [OS97, GKK+12, LO13, GKP+13, GHRW14, GHL+14, GLOS15, CHJV15, BGL+15, KLW15]). This approach avoids several sources of expensive overhead in converting modern programs into circuit representations. However, it actually introduces a different dimension of inefficiency. RAM (and single-tape Turing) machines do not support *parallelism*: thus, even if an insecure program can be heavily parallelized, its secure version will be inherently *sequential*.

Modern computing architectures are better captured by the notion of a *Parallel RAM (PRAM)*. In the PRAM model of computation, several (polynomially many) CPUs are simultaneously running, accessing the same shared "external" memory. Note that PRAM CPUs can model physical processors within a single multicore system, as well as distinct computing entities within a distributed computing environment. We consider an expressive model where the number of active CPUs may vary over time (as long as the pattern of activation is fixed a priori). In this sense, PRAMs capture the "best of both" RAM and the circuit models: A RAM program handles random access but is entirely sequential, circuits handle parallelism with variable number of parallel resources (i.e., the circuit width), but not random access; variable CPU PRAMs capture both random access and variable parallel resources. We thus put forth the challenge of designing cryptographic primitives that directly support PRAM computations, while preserving computational resources (total computational complexity and parallel time) up to poly logarithmic, while using the same number of parallel processors.

Oblivious Parallel RAM (OPRAM). A core step toward this goal is to ensure that secret information is not leaked via the *memory access patterns* of the resulting program execution.

A machine is said to be *memory oblivious*, or simply *oblivious*, if the sequences of memory accesses made by the machine on two inputs with the same running time are identically (or close to identically) distributed. In the late 1970s, Pippenger and Fischer [PF79] showed that any Turing Machine Π can be compiled into an oblivious one Π' (where "memory accesses" correspond to the movement of the head on the tape) with only a logarithmic slowdown in running-time. Roughly ten years later, Goldreich and Ostrovsky [Gol87, GO96] proposed the notion of Oblivious RAM (ORAM), and showed a similar transformation result with polylogarithmic slowdown. In recent years, ORAM compilers have become a central tool in developing cryptography for RAM programs, and a great deal of research has gone toward improving both the asymptotic and concrete efficiency of ORAM compilers (e.g., [Ajt10, DMN11, GMOT11, KLO12, CP13, CLP14, GGH+13, SvDS+13, CLP14, WHC+14, RFK+14, WCS14]). However, for all such compilers, the resulting program is inherently sequential.

In this work, we propose the notion of *Oblivious Parallel RAM (OPRAM)*. We present the first OPRAM compiler, converting any PRAM into an oblivious PRAM, while only inducing a polylogarithmic slowdown to both the total *and parallel* complexities of the program.

Theorem 1 (OPRAM – Informally Stated). *There exists an OPRAM compiler with $O(\log(m)\log^3(n))$ worst-case overhead in total and parallel computation, and $f(n)$ memory overhead for any $f \in \omega(1)$, where n is the memory size and m is an upper-bound on the number of CPUs in the PRAM.*

We emphasize that applying even the most highly optimized ORAM compiler to an m-processor PRAM program inherently inflicts $\Omega(m\log(n))$ overhead in the parallel runtime, in comparison to our $O(\log(m)\mathsf{polylog}(n))$. When restricted to single-CPU programs, our construction incurs slightly greater logarithmic overhead than the best optimized ORAM compilers (achieving $O(\log n)$ overhead for optimal block sizes); we leave as an interesting open question how to optimize parameters. (As we will elaborate on shortly, some very interesting results towards addressing this has been obtained in the follow-up work of [CLT15].)

1.1   Applications of OPRAM

ORAM lies at the base of a wide range of applications. In many cases, we can *directly* replace the underlying ORAM with an OPRAM to enable *parallelism* within the corresponding secure application. For others, simply replacing ORAM with OPRAM does not suffice; nevertheless, in this paper, we demontrate one application (garbling of PRAM programs) where they can be overcome; follow-up works show further applications (secure computation and obfuscation).

Direct Applications of OPRAM. We briefly describe some direct applications of OPRAM.

Improved/Parallelized Outsourced Data. Standard ORAM has been shown to yield effective, practical solutions for securely outsourcing data storage to an untrusted server (e.g., the ObliviStore system of [SS13]). Efficient OPRAM compilers will enable these systems to support secure efficient *parallel* accesses to outsourced data. For example, OPRAM procedures securely aggregate parallel data requests and resolve conflicts client-side, minimizing expensive client-server communications (as was explored in [WST12], at a smaller scale). As network latency is a major bottleneck in ORAM implementations, such parallelization may yield significant improvements in efficiency.

Multi-client Outsourced Data. In a similar vein, use of OPRAM further enables secure access and manipulation of outsourced shared data by multiple (mutually trusting) clients. Here, each client can simply act as an independent CPU, and will execute the OPRAM-compiled program corresponding to the parallel concatenation of their independent tasks.

Secure Multi-processor Architecture. Much recent work has gone toward implementing secure hardware architectures by using ORAM to prevent information leakage via access patterns of the secure processor to the potentially insecure memory (e.g., the Ascend project of [FDD12]). Relying instead on OPRAM opens the door to achieving secure hardware in the multi-processor setting.

Garbled PRAM (GPRAM). Garbled circuits [Yao82] allow a user to convert a circuit C and input x into garbled versions $\tilde{C}$ and $\tilde{x}$, in such a way that $\tilde{C}$ can be evaluated on $\tilde{x}$ to reveal the output $C(x)$, but without revealing further information on C or x. Garbling schemes have found countless applications in cryptography, ranging from delegation of computation to secure multi-party protocols (see below). It was recently shown (using ORAM) how to directly garble RAM programs [GHL+14, GLOS15], where the cost of evaluating a garbled program $\tilde{P}$ scales with its RAM (and not circuit) complexity.

In the full version of this paper, we show how to employ any OPRAM compiler to attain a *garbled PRAM (GPRAM)*, where the time to generate and evaluate the garbled PRAM program $\tilde{P}$ scales with the *parallel* time complexity of P. Our construction is based on one of the construction of [GHL+14] and extends it using some of the techniques developed for our OPRAM. Plugging in our (unconditional) OPRAM construction, we obtain:

Theorem 2 (Garbled PRAM – Informally Stated). *Assuming identity-based encryption, there exists a secure garbled PRAM scheme with total and parallel overhead* $\mathsf{poly}(\kappa) \cdot \mathsf{polylog}(n)$*, where κ is the security parameter of the IBE and n is the size of the garbled data.*

Secure Two-Party and Multi-party Computation of PRAMs. Secure multi-party computation (MPC) enables mutually distrusting parties to jointly evaluate functions on their secret inputs, without revealing information on the inputs beyond the desired function output. ORAM has become a central tool in achieving efficient MPC protocols for securely evaluating RAM programs. By instead relying on OPRAM, these protocols can leverage parallelizability of the evaluated programs.

Our garbled PRAM construction mentioned above yields constant-round secure protocols where the time to execute the protocol scales with the parallel time of the program being evaluated. In a companion paper [BCP15], we further demonstrates how to use OPRAM to obtain efficient protocols for securely evaluating PRAMs in the multi-party setting; see [BCP15] for further details.

Obfuscation for PRAMs. In a follow-up work, Chen et al. [CCC+15] rely on our specific OPRAM construction (and show that it satisfies an additional "puncturability" property) to achieve obfuscation for PRAMs.

1.2   Technical Overview

Begin by considering the simplest idea toward memory obliviousness: Suppose data is stored in random(-looking) shuffled order, and for each data query i, the lookup is performed to its *permuted* location, $\sigma(i)$. One can see this provides some level of hiding, but clearly does not suffice for general programs. The problem with the simple solution is in *correlated lookups* over time—as soon as item i is queried again, this collision will be directly revealed. Indeed, hiding correlated lookups while maintaining efficiency is perhaps the core challenge in building oblivious RAMs. In order to bypass this problem, ORAM compilers heavily depend on the ability of the CPU to *move data around*, and to *update its secret state* after each memory access.

However, in the parallel setting, we find ourselves back at square one. Suppose in some time step, a group of processors all wish to access data item i. Having all processors attempt to perform the lookup directly within a standard ORAM construction corresponds to running the ORAM several times *without moving data or updating state*. This immediately breaks security in all existing ORAM compiler constructions. On the other hand, we cannot afford for the CPUs to "take turns," accessing and updating the data sequentially.

In this overview, we discuss our techniques for overcoming this and further challenges. We describe our solution somewhat abstractly, building on a sequential ORAM compiler with a tree-based structure as introduced by Shi *et al.* [SCSL11]. In our formal construction and analysis, we rely on the specific tree-based ORAM compiler of Chung and Pass [CP13] that enjoys a particularly clean description and analysis.

Tree-Based ORAM Compilers. We begin by roughly describing the structure of tree-based ORAMs, originating in the work of [SCSL11]. At a high level, data is stored in the structure of a binary tree, where each node of the tree corresponds to a fixed-size bucket that may hold a collection of data items. Each memory cell addr in the original database is associated with a random *path* (equivalently, leaf) within a binary tree, as specified by a position map $\mathsf{path}_{addr} = Pos(addr)$.

The schemes maintain three invariants: (1) The content of memory cell addr will be found in one of the buckets *along the path* path_{addr}. (2) Given the view of the adversary (i.e., memory accesses) up to any point in time, the current mapping Pos appears uniformly random. And, (3) with overwhelming probability, no node in the binary tree will ever "overflow," in the sense that its corresponding memory bucket is instructed to store more items than its fixed capacity.

These invariants are maintained by the following general steps:

1. Lookup: To access a memory item addr, the CPU accesses all buckets down the path path_{addr}, and removes it where found.
2. Data "put-back": At the conclusion of the access, the memory item addr is assigned a *freshly random* path $Pos(addr) \leftarrow \mathsf{path}'_{addr}$, and is returned to the *root node* of the tree.
3. Data flush: To ensure the root (and any other bucket) does not overflow, data is "flushed" down the tree via some procedure. For example, in [SCSL11], the

flush takes place by selecting and emptying two random buckets from each level into their appropriate children; in [CP13], it takes place by choosing an independent path in the tree and pushing data items down this path as far as they will go (see Fig. 1 in Sect. 2.2).

Extending to Parallel RAMs. We must address the following problems with attempting to access a tree-based ORAM in *parallel*.

– **Parallel Memory Lookups:** As discussed, a core challenge is in hiding correlations in parallel CPU accesses. In tree-based ORAMs, if CPUs access different data items in a time step, they will access different paths in the tree, whereas if they attempt to simultaneously access the same data item, they will each access the same path in the tree, blatantly revealing a collision.

To solve this problem, before each lookup we insert a *CPU-coordination* phase. We observe that in tree-based ORAM schemes, this problem only manifests when CPUs access *exactly* the same item, otherwise items are associated with independent leaf nodes, and there are no bad correlations. We thus resolve this issue by letting the CPUs check—through an oblivious aggregation operation—whether two (or more) of them wish to access the same data item; if so, a representative is selected (the CPU with the smallest id) to actually perform the memory access, and all the others merely perform "dummy" lookups. Finally, the representative CPU needs to communicate the read value back to all the other CPUs that wanted to access the same data item; this is done using an oblivious multi-cast operation.

The challenge is in doing so without introducing too much overhead— namely, allowing only (per-CPU) memory, computation, and parallel time *polylogarithmic* in both the database size and the number of CPUs—and that itself retains memory obliviousness.

– **Parallel "Put-backs":** After a memory cell is accessed, the (possibly updated) data is assigned a fresh random path and is reinserted to the tree structure. To maintain the required invariants, the item must be inserted somewhere along its new path, *without revealing* any information about the path. In tree-based ORAMs, this is done by reinserting at the root node of the tree. However, this single node can hold only a small bounded number of elements (corresponding to the fixed bucket size), whereas the number of processors m—each with an item to reinsert—may be significantly larger.

To overcome this problem, instead of returning data items to the root, we directly insert them into level $\log m$ of the tree, while ensuring that they are placed into the correct bucket along their assigned path. Note that level $\log m$ contains m buckets, and since the m items are each assigned to random leaves, each bucket will in expectation be assigned exactly 1 item.

The challenge in this step is specifying how the m CPUs can insert elements into the tree while maintaining *memory obliviousness*. For example, if each CPU simply inserts their own item into its assigned node, we immediately leak information about its destination leaf node. To resolve this issue, we have the CPUs obliviously route items between each other, so that eventually the ith CPU holds the items to be insert to the ith node, and all CPUs finally perform either a real or a dummy write to their corresponding node.

– **Preventing Overflows:** To ensure that no new overflows are introduced after inserting m items, we now flush m times instead of once, and all these m flushes are done in parallel: each CPU simply performs an independent flush. These parallel flushes may lead to conflicts in nodes accessed (e.g., each flush operation will likely access the root node). As before, we resolve this issue by having the CPUs elect some representative to perform the appropriate operations for each accessed node; note, however, that this step is required only for correctness, and not for security.

Our construction takes a modular approach. We first specify and analyze our compiler within a simplified setting, where oblivious communication between CPUs is "for free." We then show how to efficiently instantiate the required CPU communication procedures oblivious routing, oblivious aggregation, and oblivious multi-cast, and describe the final compiler making use of these procedures. In this extended abstract, we defer the first step to Appendix 3.1, and focus on the remaining steps.

1.3   Related Work

Restricted cases of parallelism in Oblivious RAM have appeared in a handful of prior works. It was observed by Williams, Sion, and Tomescu [WST12] in their PrivateFS work that existing ORAM compilers can support parallelization across data accesses up to the "size of the top level,"[1] (in particular, at most $\log n$), when coordinated through a central trusted entity. We remark that central coordination is not available in the PRAM model. Goodrich and Mitzenmacher [GM11] showed that parallel programs in MapReduce format can be made oblivious by simply replacing the "shuffle" phase (in which data items with a given key are routed to the corresponding CPU) with a fixed-topology sorting network. The goal of improving the parallel overhead of ORAM was studied by Lorch et al. [LPM+13], but does not support compilation of PRAMs without first sequentializing.

Follow-up Work. As mentioned above, our OPRAM compiler has been used in the recent works of Boyle, Chung, and Pass [BCP15] and Chen et al. [CCC+15] to obtain secure multi-party computation for PRAM, and indistinguishability obfuscation for PRAM, respectively. A different follow-up work by Nayak et al. [NWI+15] provides targeted optimizations and an implementation for secure computation of specific parallel tasks.

Very recently, an exciting follow-up work of Chen, Lin, and Tessaro [CLT15] builds upon our techniques to obtain two new construction: an OPRAM compiler whose overhead in expectation matches that of the best current sequential ORAM [SvDS+13]; and, a general transformation taking *any* generic ORAM

---

[1] E.g., for tree-based ORAMs, the size of the root bucket.

compiler to an OPRAM compiler with $\log n$ overhead in expectation. Their OPRAM constructions, however, only apply to the special case of PRAM with a *fixed* number of processors being activated at every step (whereas our notion of a PRAM requires handling also a variable number of processors[2]); for the case of variable CPU PRAMs, the results of [CLT15] incurr an additional multlicative overhead of m in terms of computational complexity, and thus the bounds obtained are incomparable.

2   Preliminaries

2.1   Parallel RAM (PRAM) Programs

We consider the most general case of Concurrent Read Concurrent Write (CRCW) PRAMs. An m-processor CRCW *parallel random-access machine (PRAM)* with memory size n consists of numbered processors $CPU_1, \ldots, CPU_m$, each with local memory registers of size $\log n$, which operate synchronously in parallel and can make access to shared "external" memory of size n.

A PRAM program Π (given m, n, and some input x stored in shared memory) provides CPU-specific execution instructions, which can access the shared data via commands Access(r, v), where $r \in [n]$ is an index to a memory location, and v is a word (of size $\log n$) or $\perp$. Each Access(r, v) instruction is executed as:

1. **Read** from shared memory cell address r; denote value by v_{old}.
2. **Write** value $v \neq \perp$ to address r (if $v = \perp$, then take no action).
3. **Return** v_{old}.

In the case that two or more processors simultaneously initiate Access(r, v_i) with the same address r, then all requesting processors receive the previously existing memory value v_{old}, and the memory is rewritten with the value v_i corresponding to the lowest-numbered CPU i for which $v_i \neq \perp$.

We more generally support PRAM programs with a dynamic number of processors (i.e., m_i processors required for each time step i of the computation), as long as this sequence of processor numbers $m_1, m_2, \ldots$ is public information. The complexity of our OPRAM solution will scale with the number of required processors in each round, instead of the maximum number of required processors.

The *(parallel) time complexity* of a PRAM program Π is the maximum number of time steps taken by any processor to evaluate Π, where each Access execution is charged as a single step. The PRAM complexity of a function f is defined as the minimal parallel time complexity of any PRAM program which evaluates f. We remark that the PRAM complexity of any function f is bounded above by its circuit depth complexity.

---

[2] As previously mentioned, dealing with a variable number of processors is needed to capture standard circuit models of computation, where the circuit topology may be of varying width.

Remark 1 (CPU-to-CPU Communication). It will be sometimes convenient notationally to assume that CPUs may communicate directly amongst themselves. When the identities of sending and receiving CPUs is known a priori (which will always be the case in our constructions), such communication can be emulated in the standard PRAM model with constant overhead by communicating *through memory*. That is, each action "CPU1 sends message m to CPU2" is implemented in two time steps: First, CPU1 writes m into a special designated memory location addr_{CPU1}; in the following time step, CPU2 performs a read access to addr_{CPU1} to learn the value m.

2.2 Tree-Based ORAM

Concretely, our solution relies on the ORAM due to Chung and Pass [CP13], which in turn closely follows the tree-based ORAM construction of Shi *et al.* [SCSL11]. We now recall the [CP13] construction in greater detail, in order to introduce notation for the remainder of the paper.

The [CP13] construction (as with [SCSL11]) proceeds by first presenting an intermediate solution achieving obliviousness, but in which the CPU must maintain a large number of registers (specifically, providing a means for securely storing n data items requiring CPU state size $\tilde{\Theta}(n/\alpha)$, where $\alpha > 1$ is any constant). Then, this solution is recursively applied $\log_\alpha n$ times to store the resulting CPU state, until finally reaching a CPU state size $\mathsf{polylog}(n)$, while only blowing up the computational overhead by a factor $\log_\alpha n$. The overall compiler is fully specified by describing one level of this recursion.

Step 1: Basic ORAM with $O(n)$ Registers. The compiler $ORAM$ on input $n \in \mathbb{N}$ and a program Π with memory size n outputs a program Π' that is identical to Π but each $\mathsf{Read}(r)$ or $\mathsf{Write}(r, val)$ is replaced by corresponding commands $\mathsf{ORead}(r)$, $\mathsf{OWrite}(r, val)$ to be specified shortly. Π' has the same registers as Π and additionally has n/α registers used to store a *position* map Pos plus a polylogarithmic number of additional *work* registers used by ORead and OWrite. In its external memory, Π' will maintain a complete binary tree Γ of depth $\ell = \log(n/\alpha)$; we index nodes in the tree by a binary string of length at most ℓ, where the root is indexed by the empty string λ, and each node indexed by γ has left and right children indexed $\gamma 0$ and $\gamma 1$, respectively. Each memory cell r will be associated with a random leaf *pos* in the tree, specified by the position map Pos; as we shall see shortly, the memory cell r will be stored at one of the nodes on the path from the root λ to the leaf *pos*. To ensure that the position map is smaller than the memory size, we assign a *block* of α consecutive memory cells to the same leaf; thus memory cell r corresponding to block $b = \lfloor r/\alpha \rfloor$ will be associated with leaf $pos = \mathsf{Pos}(b)$.

Each node in the tree is associated with a *bucket* which stores (at most) K tuples (b, pos, v), where v is the content of block b and pos is the leaf associated with the block b, and $K \in \omega(\log n) \cap \mathsf{polylog}(n)$ is a parameter that will determine the security of the ORAM (thus each bucket stores $K(\alpha + 2)$ words). We assume that all registers and memory cells are initialized with a special symbol $\bot$.

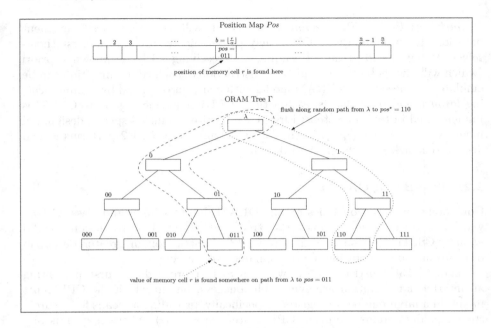

Fig. 1. Illustration of the basic [CP13] ORAM construction.

The following is a specification of the $\mathsf{ORead}(r)$ procedure:

Fetch: Let $b = \lfloor r/\alpha \rfloor$ be the block containing memory cell r (in the original database), and let $i = r \mod \alpha$ be r's component within the block b. We first look up the position of the block b using the position map: $pos = \mathsf{Pos}(b)$; if $\mathsf{Pos}(b) = \perp$, set $pos \leftarrow [n/\alpha]$ to be a uniformly random leaf.

   Next, traverse the data tree from the root to the leaf pos, making exactly one read and one write operation for the memory bucket associated with each of the nodes along the path. More precisely, we read the content once, and then we either write it back (unchanged), or we simply "erase it" (writing $\perp$) so as to implement the following task: search for a tuple of the form (b, pos, v) for the desired b, pos in any of the nodes during the traversal; if such a tuple is found, remove it from its place in the tree and set v to the found value, and otherwise take $v = \perp$. Finally, return the ith component of v as the output of the $\mathsf{ORead}(r)$ operation.

Update Position Map: Pick a uniformly random leak $pos' \leftarrow [n/\alpha]$ and let $\mathsf{Pos}(b) = pos'$.

Put Back: Add the tuple (b, pos', v) to the root λ of the tree. If there is not enough space left in the bucket, abort outputting overflow.

Flush: Pick a uniformly random leaf $pos^* \leftarrow [n/\alpha]$ and traverse the tree from the roof to the leaf pos^*, making exactly one read and one write operation for every memory cell associated with the nodes along the path so as to implement the following task: "push down" each tuple (b'', pos'', v'') read in the nodes traversed so far as possible along the path to pos^* while ensuring that the tuple is still on the path to its associated leaf pos'' (that is, the tuple ends up in the node $\gamma = $ longest common prefix of pos'' and pos^*.) Note that this operation can be performed trivially as long as the CPU has sufficiently many work registers to load two whole buckets into memory; since the bucket size is polylogarithmic, this is possible. If at any point some bucket is about to overflow, abort outputting overflow.

OWrite(r, v) proceeds identically in the same steps as ORead(r), except that in the "Put Back" steps, we add the tuple (b, pos', v'), where v' is the string v but the ith component is set to v (instead of adding the tuple (b, pos', v) as in ORead). (Note that, just as ORead, OWrite also outputs the ordinal memory content of the memory cell r; this feature will be useful in the "full-fledged" construction.)

The Full-fledged Construction: ORAM with Polylog Registers. The full-fledged construction of the CP ORAM proceeds as above, except that instead of storing the position map in registers in the CPU, we now recursively store them in another ORAM (which only needs to operate on n/α memory cells, but still using buckets that store K tuples). Recall that each invocation of ORead and OWrite requires reading one position in the position map and updating its value to a random leaf; that is, we need to perform a *single* recursive OWrite call (recall that OWrite updates the value in a memory cell, and returns the old value) to emulate the position map.

At the base of the recursion, when the position map is of constant size, we use the trivial ORAM construction which simply stores the position map in the CPU registers.

Theorem 3 ([CP13]). *The compiler ORAM described above is a secure Oblivious RAM compiler with* polylog(n) *worst-case computation overhead and* $\omega(\log n)$ *memory overhead, where n is the database memory size.*

2.3 Sorting Networks

Our protocol will employ an *n-wire sorting network*, which can be used to sort values on n wires via a fixed topology of comparisons. A sorting network consists of a sequence of *layers*, each layer in turn consisting of one or more comparator gates, which take two wires as input, and swap the values when in unsorted order. Formally, given input values $x = (x_1, \ldots, x_n)$ (which we assume to be integers wlog), a comparator operation compare(i, j, x) for $i < j$ returns x' where $x = x'$ if $x_i \leq x_j$, and otherwise, swaps these values as $x'_i = x_j$ and $x'_j = x_i$ (whereas $x'_k = x_k$ for all $k \neq i, j$). Formally, a layer in the sorting network is a set $L = \{(i_1, j_1), \ldots, (i_k, j_k)\}$ of pairwise-disjoint pairs of distinct indices of $[n]$. A d-depth sorting network is a list $SN = (L_1, \ldots, L_d)$ of layers, with the property that for any input vector x, the final output will be in sorted order $x_i \leq x_{i+1} \ \forall i < n$.

Ajtai, Komlós, and Szemerédi demonstrated a sorting network with depth logarithmic in n.

Theorem 4 ([AKS83]). *There exists an n-wire sorting network of depth $O(\log n)$ and size $O(n \log n)$.*

While the AKS sorting network is asymptotically optimal, in practical scenarios one may wish to use the simpler alternative construction due to Batcher [Bat68] which achieves significantly smaller linear constants.

3   Oblivious PRAM

The definition of an Oblivious PRAM (OPRAM) compiler mirrors that of standard ORAM, with the exception that the compiler takes as input and produces as output a *parallel* RAM program. Namely, denote the sequence of shared memory cell accesses made during an execution of a PRAM program Π on input (m, n, x) as $\tilde{\Pi}(m, n, x)$. And, denote by $\mathsf{ActivationPatterns}(\Pi, m, n., x)$ the (public) CPU activation patterns (i.e., number of active CPUs per timestep) of program Π on input (m, n, x). We present a definition of an OPRAM compiler following Chung and Pass [CP13], which in turn follows Goldreich [Gol87].

Definition 1 (Oblivious Parallel RAM). *A polynomial-time algorithm O is an Oblivious Parallel RAM (OPRAM) compiler with computational overhead $\mathsf{comp}(\cdot, \cdot)$ and memory overhead $\mathsf{mem}(\cdot, \cdot)$, if O given $m, n \in \mathbb{N}$ and a deterministic m-processor PRAM program Π with memory size n, outputs an m-processor program Π' with memory size $\mathsf{mem}(m, n) \cdot n$ such that for any input x, the parallel running time of $\Pi'(m, n, x)$ is bounded by $\mathsf{comp}(m, n) \cdot T$, where T is the parallel runtime of $\Pi(m, n, x)$, and there exists a negligible function μ such that the following properties hold:*

- **Correctness:** *For any $m, n \in \mathbb{N}$ and any string $x \in \{0, 1\}^*$, with probability at least $1 - \mu(n)$, it holds that $\Pi(m, n, x) = \Pi'(m, n, x)$.*
- **Obliviousness:** *For any two PRAM programs Π_1, Π_2, any $m, n \in \mathbb{N}$, and any two inputs $x_1, x_2 \in \{0, 1\}^*$, if $|\Pi_1(m, n, x_1)| = |\Pi_2(m, n, x_2)|$ and $\mathsf{ActivationPatterns}(\Pi_1, m, n, x_1)) = \mathsf{ActivationPatterns}(\Pi_2, m, n, x_2)$, then $\tilde{\Pi}'_1(m, n, x_1)$ is μ-close to $\tilde{\Pi}'_2(m, n, x_2)$ in statistical distance, where $\Pi'_i \leftarrow O(m, n, \Pi_i)$ for $i \in \{1, 2\}$.*

We remark that not all m processors may be active in every time step of a PRAM program Π, and thus its total computation cost may be significantly less than $m \cdot T$. We wish to consider OPRAM compilers that also preserve the processor activation structure (and thus total computation complexity) of the original program up to polylogarithmic overhead. Of course, we cannot hope to do so if the processor activation patterns themselves reveal information about the secret data. We thus consider PRAMs Π whose activation schedules $(m_1, \ldots, m_T)$ are a-priori fixed and public.

Definition 2 (Activation-Preserving). *An OPRAM compiler O with computation overhead $\mathsf{comp}(\cdot, \cdot)$ is said to be* activation preserving *if given $m, n \in \mathbb{N}$ and a deterministic PRAM program Π with memory size n and fixed (public) activation schedule $(m_1, \ldots, m_T)$ for $m_i \leq m$, the program Π' output by O has activation schedule $\left((m_1)_{i=1}^t, (m_2)_{i=1}^t, \ldots, (m_T)_{i=1}^t\right)$, where $t = \mathsf{comp}(m, n)$.*

It will additionally be useful in applications (e.g., our construction of garbled PRAMs, and the MPC for PRAMs of [BCP15]) that the resulting oblivious PRAM is *collision free*.

Definition 3 (Collision-Free). *An OPRAM compiler O is said to be* collision free *if given $m, n \in \mathbb{N}$ and a deterministic PRAM program Π with memory size n, the program Π' output by O has the property that no two processors ever access the same data address in the same timestep.*

We now present our main result, which we construct and prove in the following subsections.

Theorem 5 (Main Theorem: OPRAM). *There exists an activation-preserving, collision-free OPRAM compiler with $O(\log(m) \log^3(n))$ worst-case computational overhead and $f(n)$ memory overhead, for any $f \in \omega(1)$, where n is the memory size and m is the number of CPUs.*

3.1 Rudimentary Solution: Requiring Large Bandwidth

We first provide a solution for a simplified case, where we are not concerned with minimizing communication between CPUs or the size of required CPU local memory. In such setting, communicating and aggregating information between all CPUs is "for free."

Our compiler Heavy-O, on input $m, n \in \mathbb{N}$, fixed integer constant $\alpha > 1$, and m-processor PRAM program Π with memory size n, outputs a program Π' identical to Π, but with each $\mathsf{Access}(r, v)$ operation replaced by the modified procedure Heavy-OPAccess as defined in Fig. 2. (Here, "broadcast" means to send the specified message to all other processors).

Note that Heavy-OPAccess operates recursively for $t = 0, \ldots, \lceil \log_\alpha n \rceil$. This corresponds analogously to the recursion in the [SCSL11, CP13] ORAM, where in each step the size of the required "secure database memory" drops by a constant factor α. We additionally utilize a space optimization due to Gentry *et al.* [GGH+13] that applies to [CP13], where the ORAM tree used for storing data of size n' has depth $\log n'/K$ (and thus n'/K leaves instead of n'), where K is the bucket size. This enables the overall memory overhead to drop from $\omega(\log n)$ (i.e., K) to $\omega(1)$ with minimal changes to the analysis.

Lemma 1. *For any $n, m \in \mathbb{N}$, The compiler Heavy-O is a secure Oblivious PRAM compiler with parallel time overhead $O(\log^3 n)$ and memory overhead $\omega(1)$, assuming each CPU has $\tilde{\Omega}(m)$ local memory.*

Heavy-OPAccess$(t, (r_i, v_i))$: The Large Bandwidth Case
To be executed by $CPU_1, \ldots, CPU_m$ w.r.t. (recursive) database size $n_t := n/(\alpha^t)$, bucket size K.

Input: Each CPU_i holds: recursion level t, instruction pair (r_i, v_i) with $r_i \in [n_t]$, global parameter α.

Each CPU_i performs the following steps, in parallel

0. Exit Case: If $t \geq \log_\alpha n$, return 0.
   This corresponds to requesting the (trivial) position map for a block within a single-leaf tree.

1. Conflict Resolution
   (a) Broadcast the instruction pair (r_i, v_i) to all CPUs.
   (b) Let $b_i = \lfloor r_i/\alpha \rfloor$. Locally aggregate incoming instructions to block b_i as $\bar{v}_i = \bar{v}_i[1] \cdots \bar{v}_i[\alpha]$, resolving write conflicts (i.e., $\forall s \in [\alpha]$, take $\bar{v}_i[s] \leftarrow v_j$ for minimal j such that $r_j = b_i \alpha + s$).
   Denote by $\text{rep}(b_i) := \min\{j : \lfloor r_j/\alpha \rfloor = b_i\}$ the smallest index j of *any* CPU whose r_j is in this block b_i. (CPU $\text{rep}(b_i)$ will actually access b_i, while others perform dummy accesses).

2. Recursive Access to Position Map   (Define $L_t := 2n_t/K$, number of leaves in t'th tree).
   If $i = \text{rep}(b_i)$: Sample fresh leaf id $\ell_i' \leftarrow [L_t]$. Recurse as $\ell_i \leftarrow$ Heavy-OPAccess$(t + 1, (b_i, \ell_i'))$ to read the current value ℓ_i of $\text{Pos}(b_i)$ and rewrite it with ℓ_i'.
   Else: Recursively initiate *dummy* access $x \leftarrow$ Heavy-OPAccess$(t+1, (1, \bot))$ at arbitrary address (say 1); ignore the read value x. Sample fresh random leaf id $\ell_i \leftarrow [L_t]$ for a dummy lookup.

3. Look Up Current Memory Values
   Read the memory contents of all buckets down the path to leaf node ℓ_i defined in the previous step, copying all buckets into local memory.
   If $i = \text{rep}(b_i)$: locate and store target block triple (b_i, v_i^{old}, ℓ_i). Update $\bar{v}$ from Step 1 with existing data: $\forall s \in [\alpha]$, replace any non-written cell values $\bar{v}_i[s] = \emptyset$ with $\bar{v}_i[s] \leftarrow v_i^{old}[s]$. $\bar{v}_i$ now stores the entire data block to be rewritten for block b_i.

4. Remove Old Data from ORAM Database
   (a) If $i = \text{rep}(b_i)$: Broadcast (b_i, ℓ_i) to all CPUs. Otherwise: broadcast $(\bot, \ell_i)$.
   (b) Initiate UpdateBuckets$\big(n_t, (\text{remove-}b_i, \ell_i), \{(\text{remove-}b_j, \ell_j)\}_{j \in [m] \setminus \{i\}}\big)$, as in Figure 3.

5. Insert New Data into Database *in Parallel*
   (a) If $i = \text{rep}(b_i)$: Broadcast $(b_i, \bar{v}_i, \ell_i')$, with updated value $\bar{v}_i$ and target leaf ℓ_i'.
   (b) Let $\text{lev}^* := \lfloor \log(\min\{m, L_t\}) \rfloor$ be the ORAM tree level with number of buckets equal to number of CPUs (the level where data will be inserted). Locally aggregate all incoming instructions whose path ℓ_j' has lev^*-bit prefix i: $\text{Insert}_i := \{(b_j, \bar{v}_j, \ell_j') : (\ell_j')^{(\text{lev}^*)} = i\}$.
   (c) Access memory bucket i (at level lev^*) and rewrite contents, inserting data items Insert_i. If bucket i exceeds its capacity, abort with overflow.

6. Flush the ORAM Database
   (a) Sample a random leaf node $\ell_i^{\text{flush}} \leftarrow [L_t]$ along which to flush. Broadcast ℓ_i^{flush}.
   (b) If $i \leq L_t$: Initiate UpdateBuckets$\big(n_t, (\text{flush}, \ell_i^{\text{flush}}), \{(\text{flush}, \ell_j^{\text{flush}})\}_{j \in [m] \setminus \{i\}}\big)$, in Figure 3.
   Recall that flush means to "push" each encountered triple (b, ℓ, v) down to the lowest point at which his chosen flush path and ℓ agree.

7. Update CPUs
   If $i = \text{rep}(b_i)$: broadcast the *old* value v_i^{old} of block b_i to all CPUs.

Fig. 2. Pseudocode for oblivious parallel data access procedure Heavy-OPAccess (where we are temporarily not concerned with per-round bandwidth/memory).

UpdateBuckets $\left(n_t, (\mathsf{mycommand}, \mathsf{mypath}), \{(\mathsf{command}_j, \mathsf{path}_j)\}_{j \in [m] \setminus \{i\}}\right)$

Let $\mathsf{path}^{(0)}, \ldots, \mathsf{path}^{(\log L_t)}$ denote the bit prefixes of length 0 (i.e., $\emptyset$) to $\log(L_t)$ of path.

For each tree level $\mathsf{lev} = 0$ to $\log L_t$, each CPU i does the following at bucket $\mathsf{mypath}^{(\mathsf{lev})}$:

1. Define $\mathsf{CPUs}(\mathsf{mypath}^{(\mathsf{lev})}) := \{i\} \cup \{j : \mathsf{path}_j^{(\mathsf{lev})} = \mathsf{mypath}^{(\mathsf{lev})}\}$ to be the set of CPUs requesting changes to bucket $\mathsf{mypath}^{(\mathsf{lev})}$. Let $\mathsf{bucket\text{-}rep}(\mathsf{mypath}^{(\mathsf{lev})})$ denote the *minimal* index in the set.

2. If $i \neq \mathsf{bucket\text{-}rep}(\mathsf{mypath}^{(\mathsf{lev})})$, do nothing. Otherwise:

   Case 1: $\mathsf{mycommand} = \mathsf{remove}\text{-}b_i$.

   Interpret each $\mathsf{command}_j = \mathsf{remove}\text{-}b_j$ as a target block id b_j to be removed. Access memory bucket $\mathsf{mypath}^{(\mathsf{lev})}$ and rewrite contents, removing any block b_j for which $j \in \mathsf{CPUs}(\mathsf{mypath}^{(\mathsf{lev})})$.

   Case 2: $\mathsf{mycommand} = \mathsf{flush}$.

   Define $\mathsf{Flush} \subset \{L, R\}$ as $\{v : \exists\, \mathsf{path}_j \text{ s.t. } \mathsf{path}_j^{(\mathsf{lev}+1)} = \mathsf{mypath}^{(\mathsf{lev})} || v\}$, associating $L \equiv 0,\ R \equiv 1$. This determines whether data will be flushed left and/or right from this bucket.

   Access memory bucket $\mathsf{mypath}^{(\mathsf{lev})}$; denote its collection of stored data blocks b by ThisBucket. Partition ThisBucket $=$ ThisBucket-L $\cup$ ThisBucket-R into those blocks whose associated leaves continue to the left or right (i.e., ThisBucket-L $:= \{b_j \in \mathsf{ThisBucket} : \bar{\ell}_j^{(\mathsf{lev}+1)} = \mathsf{mypath}^{(\mathsf{lev})} || 0\}$, and similar for 1).

   – If $L \in \mathsf{Flush}$, then set ThisBucket $\leftarrow$ ThisBucket $\setminus$ ThisBucket-L, access memory bucket $\mathsf{mypath}^{(\mathsf{lev})} || 0$, and insert data items ThisBucket-L into it.

   – If $R \in \mathsf{Flush}$, then set ThisBucket $\leftarrow$ ThisBucket $\setminus$ ThisBucket-R, access memory bucket $\mathsf{mypath}^{(\mathsf{lev})} || 1$, and insert data items ThisBucket-R into it.

   Rewrite the contents of bucket $\mathsf{mypath}^{(\mathsf{lev})}$ with updated value of ThisBucket. If any bucket exceeds its capacity, abort with **overflow**.

Fig. 3. Procedure for combining CPUs' instructions for buckets and implementing them by a single representative CPU. (Used for correctness, not security). See Fig. 4 for a sample illustration.

We will address the desired claims of correctness, security, and complexity of the Heavy-O compiler by induction on the number of levels of recursion. Namely, for $t^* \in [\log_\alpha n]$, denote by Heavy-O_{t^*} the compiler that acts on memory size $n/(\alpha^{t^*})$ by executing Heavy-O only on recursion levels $t = t^*, (t^* + 1), \ldots, \lceil \log_\alpha n \rceil$. For each such t^*, we define the following property.

Level-t^* Heavy OPRAM: We say that Heavy-O_{t^*} is a *valid level-t^* heavy OPRAM* if the partial-recursion compiler Heavy-O_{t^*} is a secure Oblivious PRAM compiler for memory size $n/(\alpha^{t^*})$ with parallel time overhead $O(\log^2 n \cdot \log(n/\alpha^{t^*}))$ and memory overhead $\omega(1)$, assuming each CPU has $\tilde{\Omega}(m)$ local memory.

Then Lemma 1 follows directly from the following two claims.

Claim. Heavy-$O_{\log_\alpha n}$ is valid level-$(\log_\alpha n)$ heavy OPRAM.

Proof. Note that Heavy-$O_{\log_\alpha n}$, acting on trivial size-1 memory, corresponds directly to the exit case (Step 0) of Heavy-OPAccess in Fig. 2. Namely, correctness, security, and the required efficiency trivially hold, since there is a single data item in a fixed location to access.

Claim. Suppose Heavy-O_t is a valid level-t heavy OPRAM for $t > 0$. Then Heavy-O_{t-1} is a valid level-$(t - 1)$ heavy OPRAM.

Proof. We first analyze the correctness, security, and complexity overhead of Heavy-O_{t-1} conditioned on never reaching the event overflow (which may occur in Step 5(c), or within the call to UpdateBuckets). Then, we prove that the probability of overflow is negligible in n.

Correctness (w/o overflow). Consider the state of the memory (of the CPUs and server) in each step of Heavy-OPAccess, assuming no overflow. In Step 1, each CPU learns the instruction pairs of all other CPUs; thus all CPUs agree on single representative $\mathsf{rep}(b_i)$ for each requested block b_i, and a correct aggregation of all instructions to be performed on this block. Step 2 is a recursive execution of Heavy-OPAccess. By the inductive hypothesis, this access successfully returns the correct value ℓ_i of $\mathsf{Pos}(b_i)$ for each b_i queried, and rewrites it with the freshly sampled value ℓ_i' when specified (i.e., for each $\mathsf{rep}(b_i)$ access; the dummy accesses are read-only). We are thus guaranteed that each $\mathsf{rep}(b_i)$ will find the desired block b_i in Step 3 when accessing the memory buckets in the path down the tree to leaf ℓ_i (as we assume no overflow was encountered), and so will learn the current stored data value v_{old}.

In Step 4, each CPU learns the target block b_i and associated leaf ℓ_i of every representative CPU $\mathsf{rep}(b_i)$. By construction, each requested block b_i appears in some bucket B in the tree along his path, and there will necessarily be some CPU assigned as $\mathsf{bucket\text{-}rep}(B)$ in UpdateBuckets, who will then successfully remove

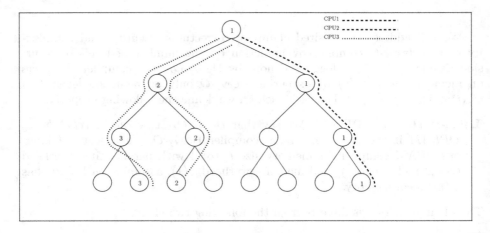

Fig. 4. UpdateBuckets sample illustration. Here, CPUs 1-3 each wish to modify nodes along their paths as drawn; for each overlapping node, the CPU with lowest id receives and implements the aggregated commands for the node.

the block b_i from B. At this point, none of the requested blocks b_i appear in the tree.

In Step 5, the CPUs insert each block b_i (with updated data value v_i) into the ORAM data tree at level $\min\{\log_\alpha n/\alpha^t, \lfloor\log_2(m)\rfloor\}$ along the path to its (new) leaf ℓ'_i.

Finally, the flushing procedure in Step 6 maintains the necessary property that each block b_i appears along the path to $\mathsf{Pos}(b_i)$, and in Step 7 all CPUs learn the collection of all queried values v_{old} (in particular, including the value they initially requested).

Thus, assuming no overflow, correctness holds.

Obliviousness (w/o overflow). Consider the access patterns to server-side memory in each step of Heavy-OPAccess, assuming no overflow. Step 1 is performed locally without communication to the server. Step 2 is a recursive execution of Heavy-OPAccess, which thus yields access patterns independent of the vector of queried data locations (up to statistical distance negligible in n), by the induction hypothesis. In Step 3, each CPU accesses the buckets along a single path down the tree, where representative CPUs $\mathsf{rep}(b_i)$ access along the path given by $\mathsf{Pos}(b_i)$ (for *distinct* b_i), and non-representative CPUs each access down an independent, random path. Since the adversarial view so far has been independent of the values of $\mathsf{Pos}(b_i)$, conditioned on this view all CPU's paths are independent and random.

In Step 4, all data access patterns are publicly determinable based on the accesses in the previous step (that is, the complication in Step 4 is to ensure correctness without access collisions, but is not needed for security). In Step 5, each CPU i accesses his corresponding bucket i in the tree. In the flushing procedure of Step 6, each CPU selects an independent, random path down the tree, and the communication patterns to the server reveal no information beyond the identities of these paths. Finally, Step 7 is performed locally without communication to the server.

Thus, assuming no overflow, obliviousness holds.

Protocol Complexity (w/o overflow). First note that the server-side memory storage requirement is simply that of the [CP13] ORAM construction, together with the $\log(2n_t/K)$ tree-depth memory optimization of [GHL+14]; namely, $f(n)$ memory overhead suffices for any $f \in \omega(1)$.

Consider the local memory required per CPU. Each CPU must be able to store: $O(\log n)$-size requests from each CPU (due to the broadcasts in Steps 1(a), 4(a), 5(a), and 7); and the data contents of at most 3 memory buckets (due to the flushing procedure in UpdateBuckets). Overall, this yields a per-CPU local memory requirement of $\tilde{\Omega}(m)$ (where $\tilde{\Omega}$ notation hides $\log n$ factors).

Consider the parallel complexity of the OPRAM-compiled program $\Pi' \leftarrow$ Heavy-$O(m, n, \Pi)$. For each parallel memory access in the underlying program Π, the processors perform: Conflict resolution (1 local communication round), Read/writing the position map (which has parallel complexity $O(\log^2 n \cdot \log(n/\alpha^t))$ by the inductive hypothesis), Looking up current memory values (sequential steps = depth of level-$(t-1)$ ORAM tree $\in O(\log(n/\alpha^{t-1}))$),

Removing old data from the ORAM tree (1 local communication round, plus depth of the ORAM tree $\in O(\log(n/\alpha^{t-1}))$ sequential steps), Inserting the new data in parallel (1 local communication round, plus 1 communication round to the server), Flushing the ORAM database (1 local communication round, and 2× the depth of the ORAM tree rounds of communication with the server, since each bucket along a flush path is accessed once to receive new data items and once to flush its own data items down), and Updating CPUs with the read values (1 local communication round). Altogether, this yields parallel complexity overhead $O(\log^2 n \cdot \log(n/\alpha^{t-1}))$.

It remains to address the probability of encountering overflow.

Claim. There exists a negligible function μ such that for any deterministic m-processor PRAM program Π, any database size n, and any input x, the probability that the Heavy-O-compiled program $\Pi'(m, n, x)$ outputs overflow is bounded by $\mu(n)$.

Proof. We consider separately the probability of overflow in each of the level-t recursive ORAM trees. Since there are $\lceil \log n \rceil$ of them, the claim follows by a straightforward union bound.

Taking inspiration from [CP13], we analyze the ORAM-compiled execution via an abstract dart game. The game consists of black and white darts. In each round of the game, m black darts are thrown, followed by m white darts. Each dart independently hits the bullseye with probability $p = 1/m$. The game continues until exactly K darts have hit the bullseye (recall $K \in \omega(\log n)$ is the bucket size), or after the end of the Tth round for some fixed polynomial bound $T = T(n)$, whichever comes first. The game is "won" (which will correspond to overflow in a particular bucket) if K darts hit the bullseye, and all of them are black.

Let us analyze the probability of winning in the above dart game.

Subclaim 1: With overwhelming probability in n, no more than $K/2$ darts hit the bullseye in any round. In any single round, associate with each of the $2 \cdot m$ darts thrown an indicator variable X_i for whether the dart strikes the target. The X_i are independent random variables each equal to 1 with probability $p = 1/m$. Thus, the probability that more than $K/2$ of the darts hit the target is bounded (via a Chernoff tail bound[3]) by

$$\Pr\left[\sum_{i=1}^{2m} X_i > K/2\right] \leq e^{\frac{2(K/4-1)^2}{2+(K/4-1)}} \leq e^{-\Omega(K)} \leq e^{-\omega(\log n)}.$$

Since there are at most $T = \mathsf{poly}(n)$ distinct rounds of the game, the subclaim follows by a union bound.

Subclaim 2: Conditioned on no round having more than $K/2$ bullseyes, the probability of winning the game is negligible in d. Fix an arbitrary such winning

---

[3] Explicit Chernoff bound used: for $X = X_1 + \cdots X_{2m}$ (X_i independent) and mean μ, then for any $\delta > 0$, it holds that $\Pr[X > (1+\delta)\mu] \leq e^{-\delta^2 \mu/(2+\delta)}$.

sequence s, which terminates sometime during some round r of the game. By assumption, the final partial round r contains no more than $K/2$ bullseyes. For the remaining $K/2$ bullseyes in rounds 1 through $r - 1$, we are in a situation mirroring that of [CP13]: for each such winning sequence s, there exist $2^{K/2} - 1$ distinct other "losing" sequences s' that each occur with the same probability, where any non-empty subset of black darts hitting the bullseye are replaced with their corresponding white darts. Further, every two distinct winning sequences s_1, s_2 yield disjoint sets of losing sequences, and all such constructed sequences have the property that no round has more than $K/2$ bullseyes (since this number of total bullseyes per round is preserved). Thus, conditioned on having no round with more than $K/2$ bullseyes, the probability of winning the game is bounded above by $2^{-K/2} \in e^{-\omega(\log n)}$.

We now relate the dart game to the analysis of our OPRAM compiler.

We analyze the memory buckets at the nodes in the t-th recursive ORAM tree, via three sub-cases.

Case 1: Nodes in level $\mathsf{lev} < \log m$. Since data items are inserted to the tree in parallel directly at level $\log m$, these nodes do not receive data, and thus will not overflow.

Case 2: Consider any internal node (i.e., a node that is not a leaf) γ in the tree at level $\log m \leq \mathsf{lev} < \log(L_t)$. (Recall $L_t := 2n_t/K$ is the number of leaves in the t'th tree when applying the [GHL+14] optimization). Note that when $m > L_t$, this case is vacuous. For purposes of analysis, consider the contents of γ as split into two parts: γ_L containing the data blocks whose leaf path continues to the left from γ (i.e., leaf $\gamma||0||\cdot$), and γ_R containing the data blocks whose leaf path continues right (i.e., $\gamma||1||\cdot$). For the bucket of node γ to overflow, there must be K tuples in it. In particular, either γ_L or γ_R must have $K/2$ tuples.

For each parallel memory access in $\Pi(m, n, x)$, in the t-th recursive ORAM tree for which $n_t \geq m/K$, (at most) m data items are inserted, and then m independent paths in the tree are flushed. By definition, an inserted data item will enter our bucket γ_L (respectively, γ_R) only if its associated leaf has the prefix $\gamma||0$ (resp., $\gamma||1$); we will assume the worst case in which all such data items arrive directly to the bucket. On the other hand, the bucket γ_L (resp., γ_R) will be completely emptied after any flush whose path contains this same prefix $\gamma||0$ (resp., $\gamma||1$). Since all leaves for inserted data items and data flushes are chosen randomly and independently, these events correspond directly to the black and white darts in the game above. Namely, the probability that a randomly chosen path will have the specific prefix $\gamma||0$ of length lev is $2^{-\mathsf{lev}} \leq 1/m$ (since we consider $\mathsf{lev} \geq \log m$); this corresponds to the probability of a dart hitting the bullseye. The bucket can only overflow if $K/2$ "black darts" (inserts) hit the bullseye without any "white dart" (flush) hitting the bullseye in between. By the analysis above, we proved that for any sequence of $K/2$ bullseye hits, the probability that all $K/2$ of them are black is bounded above by $2^{-K/4}$, which is negligible in n. However, since there is a fixed polynomial number $T = \mathsf{poly}(n)$ of parallel memory accesses in the execution of $\Pi(m, n, x)$ (corresponding to the number of "rounds" in the dart game), and in particular, $T(2m) \in \mathsf{poly}(n)$ total darts thrown, the probability that the sequence of bullseyes contains $K/2$

sequential blacks *anywhere* in the sequence is bounded via a direct union bound by $(T2m)2^{-K/4} \in e^{-\omega(\log n)}$, as desired.

Case 3: Consider any leaf node γ. This analysis follows the same argument as in [CP13] (with slightly tweaked parameters from the [GHL+14] tree-depth optimization). We refer the reader to the full version of this work for details.

Thus, the total probability of overflow is negligible in n, and the theorem follows.

3.2 Oblivious Routing, Aggregation, and Multi-cast

Oblivious Parallel Insertion (Oblivious Routing). Recall during the memory "put-back" phase, each CPU must insert its data item into the bucket at level $\log m$ of the tree lying along a freshly sampled random path, while *hiding* the path.

We solve this problem by delivering data items to their target locations via a *fixed-topology routing network*. Namely, the m processors $CPU_1, \ldots, CPU_m$ will first write the relevant m data items msg_i (and their corresponding destination addresses addr_i) to memory in fixed order, and then rearrange them in $\log m$ sequential rounds to the proper locations via the routing network. At the conclusion of the routing procedure, each node j will hold all messages msg_i for which $\mathsf{addr}_i = j$.

For simplicity, assume $m = 2^\ell$ for some $\ell \in \mathbb{N}$. The routing network has depth ℓ; in each level $t = 1, \ldots, \ell$, each node communicates with the corresponding node whose id agrees in all bit locations except for the tth (corresponding to his tth neighbor in the $\log m$-dimensional boolean hypercube). These nodes exchange messages according to the tth bit of their destination addresses addr_i. This is formally described in Fig. 5. After the tth round, each message msg_i is held by a party whose id agrees with the destination address addr_i in the first t bits. Thus, at the conclusion of ℓ rounds, all messages are properly delivered.

We demonstrate the case $m = 8 = 2^3$ below: first, CPUs exchange information along the depicted communication network in 3 sequential rounds (left); then, each CPU i inserts his resulting collection of items directly into node i of level 3 of the data tree (right).

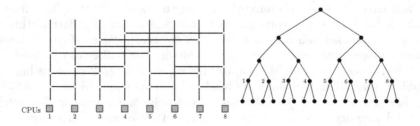

Parallel Insertion Routing Protocol Route$(m, (\mathsf{msg}_i, \mathsf{addr}_i))$

Input: CPU_i holds: message msg_i with target destination addr_i, and global threshold K.

Output: CPU_i holds $\{\mathsf{msg}_j : \mathsf{addr}_j = i\}$.

Let $\mathsf{lev}^* = \log m$ (assumed $\in \mathbb{N}$ for simplicity). Each CPU_i performs the following.

Initialize $M_{i,0} \leftarrow \mathsf{msg}_i$. For $t = 1, \ldots, \mathsf{lev}^*$:

  1. Perform the following symmetric message exchange with $CPU_{i \oplus 2^t}$:
$$M_{i,t+1} \leftarrow \{\mathsf{msg}_j \in M_{i,t} \cup M_{i \oplus 2^t, t} : (\mathsf{addr}_j)_t = (i)_t\}.$$
  2. If $|M_{i,t+1}| > K$ (i.e., memory overflow), then CPU_i aborts.

Fig. 5. Fixed-topology routing network for delivering m messages originally held by m processors to their corresponding destination addresses within $[m]$.

In the full version, we show that if the destination addresses addr_i are uniformly sampled, then with overwhelming probability no node will ever need to hold too many (the threshold K will be set to $\omega(\log n)$) messages at any point during the routing network execution:

Lemma 2 (Routing Network). *If L messages begin with target destination addresses addr_i distributed independently and uniformly over $[L]$ in the L-to-L node routing network in Fig. 5, then with probability bounded by $1 - (L \log L) 2^{-K}$, no intermediate node will ever hold greater than K messages at any point during the course of the protocol execution.*

Oblivious Aggregation. To perform the "CPU-coordination" phase, the CPUs efficiently identify a single representative and *aggregate* relevant CPU instructions; then, at the conclusion, the representative CPU must be able to *multi-cast* the resulting information to all relevant requesting CPUs. Most importantly, these procedures must be done *in an oblivious fashion*. We discuss oblivious aggregation first.

Formally, we want to achieve the following aggregation goal, with communication patterns independent of the inputs, using only $O(\log(m)\mathsf{polylog}(n))$ local memory and communication per CPU, in only $O(\log(m))$ sequential time steps. An illustrative example to keep in mind is where $\mathsf{key}_i = b_i$, $\mathsf{data}_i = v_i$, and Agg is the process that combines instructions to data items within the same data block, resolving conflicts as necessary.

Oblivious aggregation

Input: Each CPU $i \in [m]$ holds $(\mathsf{key}_i, \mathsf{data}_i)$. Let $\mathsf{K} = \bigcup \{\mathsf{key}_i\}$ denote the set of distinct keys. We assume that any (subset of) data associated with the same key can be aggregated by an aggregation function Agg to a short digest of size at most $\mathsf{poly}(\ell, \log m)$, where $\ell = |\mathsf{data}_i|$.

Goal: Each CPU i outputs out_i such that the following holds.
  – For every $\mathsf{key} \in \mathsf{K}$, there exists unique agent i with $\mathsf{key}_i = \mathsf{key}$ s.t. $\mathsf{out}_i = (\mathsf{rep}, \mathsf{key}, \mathsf{agg}_{\mathsf{key}})$, where $\mathsf{agg}_{\mathsf{key}} = \mathsf{Agg}(\{\mathsf{data}_j : \mathsf{key}_j = \mathsf{key}\})$.
  – For every remaining agent i, $\mathsf{out}_i = (\bot, \bot)$.

At a high level, we achieve this via the following steps. (1) First, the CPUs sort their data list with respect to the corresponding key values. This can be achieved via an implementation of a $\log(m)$-depth sorting network, and provides the useful guarantee that all data pertaining to the same key are necessarily held by an block of adjacent CPUs. (2) Second, we pass data among CPUs in a sequence of $\log(m)$ steps such that at the conclusion the "left-most" (i.e., lowest indexed) CPU in each key-block will learn the aggregation of *all* data pertaining to this key. Explicitly, in each step i, each CPU sends all held information to the CPU 2^i to the "left" of him, and simultaneously accepts any received information pertaining to his key. (3) Third, each CPU will learn whether he is the "left-most" representative in each key-block, by simply checking whether his left-hand neighbor holds the same key. From here, the CPUs have succeeded in aggregating information for each key at a single representative CPU; (4) in the fourth step, they now reverse the original sorting procedure to return this aggregated information to one of the CPUs who originally requested it.

Lemma 3 (Space-Efficient Oblivious Aggregation). *Suppose m processors initiate protocol* OblivAgg *w.r.t. aggregator* Agg, *on respective inputs $\{(\text{key}_i, \text{data}_i)\}_{i \in [m]}$, each of size ℓ. Then at the conclusion of execution, each processor $i \in [m]$ outputs a triple $(\text{rep}'_i, \text{key}'_i, \text{data}'_i)$ such that the following properties hold (where asymptotics are w.r.t. m):*

1. *The protocol terminates in $O(\log m)$ rounds.*
2. *The local memory and computation required per processor is $O(\log m + \ell)$.*
3. *(Correctness). For every key* key *$\in \bigcup\{\text{key}_i\}$, there exists a unique processor i with output* $\text{key}'_i = \text{key}$. *For each such processor, it further holds that* $\text{key}'_i = \text{key}_i$, $\text{rep}'_i = $ "rep", *and* $\text{data}'_i = \text{Agg}(\{\text{data}_j : \text{key}_j = \text{key}_i\})$. *For every remaining processor, the output tuple is* $(\perp, \perp)$.
4. *(Obliviousness). The inter-CPU communication patterns are independent of the inputs* $(\text{key}_i, \text{data}_i)$.

A full description of our Oblivious Aggregation procedure OblivAgg is given in Fig. 6. We defer the proof of Lemma 3 to the full version of this work and provide only a high-level sketch.

Proof Sketch of Lemma 3. Property (1): The parallel complexity of OblivAgg comes from Steps 1 and 4, which execute a sorting network and require $O(\log m)$ communication rounds.

Property (2): At any given time, a processor must only store and/or communicate a constant number of CPU id's (size $\log m$) and data items (size ℓ), yielding total $O(\log m + \ell)$.

Property (3): To show that the Aggregate Left phase in Step 2 is correct, it is proved (by induction) that for each pair of CPU indices $i < j$ with the same key, CPU_i will learn CPU_j's data after a number of rounds equal to the highest index in which the bit representations of i and j disagree.

Property (4): Both sorting network and aggregate-to-left have fixed communication topologies; thus the induced inter-CPU communications are independent of the initial CPU inputs.

Oblivious Multicasting. Our goal for Oblivious Multicasting is dual to that of the previous section: Namely, a subset of CPUs must deliver information to (unknown) collections of other CPUs who request it. This is abstractly modeled as follows, where key_i denotes which data item is requested by each CPU i.

Oblivious Multicasting

Input: Each CPU i holds $(\text{key}_i, \text{data}_i)$ with the following promise. Let $\mathsf{K} = \bigcup\{\text{key}_i\}$ denote the set of distinct keys. For every $\text{key} \in \mathsf{K}$, there exists a unique agent i with $\text{key}_i = \text{key}$ such that $\text{data}_i \neq \bot$; let data_key denote such data_i.

Goal: Each agent i outputs $\text{out}_i = (\text{key}_i, \text{data}_{\text{key}_i})$.

Oblivious Multicast can be solved in an analogous manner. We refer the reader to the full version of this work for the OblivMCast construction.

3.3   Putting Things Together

We now combine the so-called "Heavy-OPAccess" structure of our OPRAM formalized in Sect. 3.1 (Fig. 2) within the simplified "free CPU communication" setting, together with the (oblivious) Route, OblivAgg, and OblivMCast procedures constructed in the previous subsection. For simplicity, we describe the case in which the number of CPUs m is fixed; however, it can be modified in a straightforward fashion to the more general case (as long as the activation schedule of CPUs is a-priori fixed and public).

Recall the steps in Heavy-OPAccess where large memory/bandwidth are required.

- In Step 1, each CPU_i broadcasts (r_i, v_i) to all CPUs. Let $b_i = \lfloor r_i/\alpha \rfloor$. This is used to aggregate instructions to each b_i and determine its representative CPU $\text{rep}(b_i)$.
- In Step 4, each CPU_i broadcasts (b_i, ℓ_i) or $(\bot, \ell_i)$. This is used to aggregate instructions to each buckets along path ℓ_i about which blocks b_i's to be removed.
- In Step 5, each (representative) CPU_i broadcasts $(b_i, \bar{v}_i, \ell'_i)$. This is used to aggregate blocks to be inserted to each bucket in appropriate level of the tree.
- In Step 6, each CPU_i broadcasts ℓ_i^flush. This is used to aggregate information about which buckets the flush operation should perform.
- In Step 7, each (representative) $CPU_{\text{rep}(b)}$ broadcasts the old value v_old of block b to all CPUs, so that each CPU receives desired information.

We will use oblivious aggregation procedure to replace broadcasts in Step 1, 4, and 6; the parallel insertion procedure to replace broadcasts in Step 5, and finally the oblivious multicast procedure to replace broadcasts in Step 7.

Let us first consider the aggregation steps. For Step 1, to invoke the oblivious aggregation procedure, we set $\text{key}_i = b_i$ and $\text{data}_i = (r_i \bmod \alpha, v_i)$, and define the output of $\text{Agg}(\{(u_i, v_i)\})$ to be a vector $\bar{v} = \bar{v}[1] \cdots \bar{v}[\alpha]$ of read/write

---

Oblivious Aggregation Procedure OblivAgg (w.r.t. Agg)

Input: Each CPU $i \in [m]$ holds a pair $(\text{key}_i, \text{data}_i)$.

Output: Each CPU $i \in [m]$ outputs a triple $(\text{rep}_i, \text{key}_i, \text{aggdata}_i)$ corresponding to either $(\text{dummy}, \bot, \bot)$ or with $\text{aggdata}_i = \text{Agg}(\{\text{data}_j : \text{key}_j = \text{key}_i\})$, as further specified in Section 3.2.

1. **Sort on** key_i. Each CPU_i initializes a triple $(\text{sourceid}_i, \text{keytemp}_i, \text{datatemp}_i) \leftarrow (i, \text{key}_i, \text{data}_i)$.

   For each layer $L_1, \ldots, L_d$ in the sorting network:
   - Let $L_\ell = ((i_1, j_1), \ldots, (i_{m/2}, j_{m/2}))$ be the comparators in the current layer ℓ.
   - In *parallel*, for each $t \in [m/2]$, the corresponding pair of CPUs (CPU_{i_t}, CPU_{j_t}) perform the following pairwise sort w.r.t. key:

   If $\text{keytemp}_{j_t} < \text{keytemp}_{i_t}$, then

   swap $(\text{sourceid}_{i_t}, \text{keytemp}_{i_t}, \text{datatemp}_{i_t}) \leftrightarrow (\text{sourceid}_{j_t}, \text{keytemp}_{j_t}, \text{datatemp}_{j_t})$.

2. **Aggregate to left.** For $t = 0, 1, \ldots, \log m$:
   - (Pass to left). Each CPU_i for $i > 2^t$ sends his current pair $(\text{keytemp}_i, \text{datatemp}_i)$ to CPU_{i-2^t}.
   - (Aggregate). Each CPU_i for $i < m - 2^t$ receiving a pair $(\text{keytemp}_j, \text{datatemp}_j)$ will aggregate it into own pair if the keys match. That is, if $\text{keytemp}_i = \text{keytemp}_j$, then set $\text{datatemp}_i \leftarrow \text{Agg}(\text{datatemp}_i, \text{datatemp}_j)$. In both cases, the received pair is then erased.

   The left-most CPU_i with $\text{keytemp}_i = \text{key}$ now has $\text{Agg}(\{\text{datatemp}_j : \text{keytemp}_j = \text{key}\})$.

3. **Identify representatives.** For each value key_j, the left-most CPU i currently holding $\text{keytemp}_i = \text{key}_j$ will identify himself as (temporary) representative.
   - Each CPU_i for $i < m$: send keytemp_i to right-hand neighbor, CPU_{i+1}.
   - Each CPU_i for $i > 1$: If the received value keytemp_{i-1} matches his own keytemp_i, then set $\text{rep}_i \leftarrow$ "dummy" and zero out $\text{keytemp}_i \leftarrow \bot, \text{datatemp}_i \leftarrow \bot$. Otherwise, set $\text{rep}_i \leftarrow$ "rep". $(CPU_1$ always sets $\text{rep}_1 \leftarrow$ "rep").

4. **Reverse sort (i.e., sort on** sourceid_i**).** Return aggregated data to a requesting CPU.

   For each layer $L_1, \ldots, L_d$ in the sorting network:
   - Let $L_\ell = ((i_1, j_1), \ldots, (i_{m/2}, j_{m/2}))$ be the comparators in the current layer ℓ.
   - Each CPU_i initializes $\text{idtemp} \leftarrow \text{sourceid}_i$. In *parallel*, for each $t \in [m/2]$, the corresponding pair of CPUs (CPU_{i_t}, CPU_{j_t}) perform the following pairwise sort w.r.t. sourceid:

   If $\text{idtemp}_{j_t} < \text{idtemp}_{i_t}$, then

   swap $(\text{idtemp}_{i_t}, \text{rep}_{i_t}, \text{keytemp}_{i_t}, \text{datatemp}_{i_t}) \leftrightarrow$

   $(\text{idtemp}_{j_t}, \text{rep}_{j_t}, \text{keytemp}_{j_t}, \text{datatemp}_{j_t})$.

   At the conclusion, each CPU_i holds a tuple $(\text{idtemp}_i, \text{rep}_i, \text{keytemp}_i, \text{datatemp}_i)$ with $\text{idtemp}_i = i$ and $\text{keytemp}_i = \text{key}_i$.

5. **Output.** Each CPU_i outputs the triple $(\text{rep}_i, \text{key}_i, \text{datatemp}_i)$.

Fig. 6. Space-efficient oblivious data aggregation procedure.

instructions to each memory cell in the block, where conflicts are resolved by writing the value specified by the smallest CPU: i.e., $\forall s \in [\alpha]$, take $\bar{v}[s] \leftarrow v_j$ for minimal j such that $u_j = s$ and $v_j \neq \bot$. By the functionality of OblivAgg, at the conclusion of OblivAgg, each block b_i is assigned to a unique representative (not necessarily the smallest CPU), who holds the aggregation of all instructions on this block.

Both Step 4 and 6 invoke UpdateBuckets to update buckets along m random paths. In our rudimentary solution, the paths (along with instructions) are broadcast among CPUs, and the buckets are updated level by level. At each level, each update bucket is assigned to a representative CPU with minimal index, who performs aggregated instructions to update the bucket. Here, to avoid broadcasts, we invoke the oblivious aggregation procedure per level as follows.

- In Step 4, each CPU i holds a path ℓ_i and a block b_i (or $\perp$) to be removed. Also note that the buckets along the path ℓ_i are stored locally by each CPU i, after the read operation in the previous step (Step 3). At each level $\text{lev} \in [\log n]$, we invoke the oblivious aggregation procedure with $\text{key}_i = \ell_i^{(\text{lev})}$ (the lev-bits prefix of ℓ_i) and $\text{data}_i = b_i$ if b_i is in the bucket of node $\ell_i^{(\text{lev})}$, and $\text{data}_i = \perp$ otherwise. We simply define $\text{Agg}(\{\text{data}_i\}) = \{b : \exists \text{data}_i = b\}$ to be the union of blocks (to be removed from this bucket). Since $\text{data}_i \neq \perp$ only when data_i is in the bucket, the output size of Agg is upper bounded by the bucket size K. By the functionality of OblivAgg, at the conclusion of OblivAgg, each bucket $\ell_i^{(\text{lev})}$ is assigned to a unique representative (not necessarily the smallest CPU) with aggregated instruction on the bucket. Then the representative CPUs can update the corresponding buckets accordingly.
- In Step 6, each CPU i samples a path ℓ_i^{flush} to be flushed and the instructions to each bucket are simply left and right flushes. At each level $\text{lev} \in [\log n]$, we invoke the oblivious aggregation procedure with $\text{key}_i = \ell_i^{\text{flush}\,(\text{lev})}$ and $\text{data}_i = L$ (resp., R) if the $(\text{lev}+1)$-st bit of ℓ_i^{flush} is 0 (resp., 1). The aggregation function Agg is again the union function. Since there are only two possible instructions, the output has $O(1)$ length. By the functionality of OblivAgg, at the conclusion of OblivAgg, each bucket $\ell_i^{\text{flush}(\text{lev})}$ is assigned to a unique representative (not necessarily the smallest CPU) with aggregated instruction on the bucket. To update a bucket $\ell_i^{\text{flush}(\text{lev})}$, the representative CPU loads the bucket and its two children (if needed) into local memory from the server, performs the flush operation(s) locally, and writes the buckets back.

Note that since we update m *random* paths, we do not need to hide the access pattern, and thus the dummy CPUs do not need to perform dummy operations during UpdateBuckets. A formal description of full-fledged UpdateBuckets can be found in Fig. 7.

For Step 5, we rely on the parallel insertion procedure of Sect. 3.2, which routes blocks to proper destinations within the relevant level of the server-held data tree in parallel using a simple oblivious routing network. The procedure is invoked with $\text{msg}_i = b_i$ and $\text{addr}_i = \ell_i'$.

Finally, in Step 7, each representative CPU $\text{rep}(b)$ holds information of the block b, and each dummy CPU i wants to learn the value of a block b_i. To do so, we invoke the oblivious multicast procedure with $\text{key}_i = b_i$ and $\text{data}_i = v_i^{old}$ for representative CPUs and $\text{data}_i = \perp$ for dummy CPUs. By the functionality of OblivMCast, at the conclusion of OblivMCast, each CPU receives the value of the block it originally wished to learn.

The Final Compiler. For convenience, we summarize the complete protocol. Our OPRAM compiler O, on input $m, n_t \in \mathbb{N}$ and a m-processor PRAM program Π with memory size n_t (which in recursion level t will be $n_t = n/\alpha^t$), will output a program Π' that is identical to Π, but where each $\mathsf{Access}(r, v)$ operation is replaced by a sequence of operations defined by subroutine $\mathsf{OPAccess}(r, v)$, which we will construct over the following subsections. The $\mathsf{OPAccess}$ procedure begins with m CPUs, each with a requested data cell r_i (within some α-block b_i) and some action to be taken (either $\perp$ to denote read, or v_i to denote rewriting cell r_i with value v_i).

1. **Conflict Resolution:** Run $\mathsf{OblivAgg}$ on inputs $\{(b_i, v_i)\}_{i \in [m]}$ to select a unique representative $\mathsf{rep}(b_i)$ for each queried block b_i and aggregate all CPU instructions for this b_i (denoted $\bar{v}_i$).
2. **Recursive Access to Position Map:** Each representative CPU $\mathsf{rep}(b_i)$ samples a fresh random leaf id $\ell'_i \leftarrow [n_t]$ in the tree and performs a (recursive)

$\mathsf{UpdateBuckets}\,(m, (\mathsf{command}_i, \mathsf{path}_i))$

Let $\mathsf{path}^{(1)}, \mathsf{path}^{(2)}, \ldots, \mathsf{path}^{(\log n)}$ denote the bit prefixes of length 1 to $\log n$ of path.

For each level $\mathsf{lev} = 1, \ldots, \log n$ of the tree:

1. The CPUs invoke the oblivious aggregation procedure $\mathsf{OblivAgg}$ as follows.
   Case 1: $\mathsf{command}_i = \mathsf{remove}\text{-}b_i$.
   Each CPU i sets $\mathsf{key}_i = \mathsf{path}_i^{(\mathsf{lev})}$ and $\mathsf{data}_i = b_i$ if b_i is in the bucket of node $\ell_i^{(\mathsf{lev})}$, and $\mathsf{data}_i = \perp$ otherwise. Use the union function $\mathsf{Agg}(\{\mathsf{data}_i\}) = \{b : \exists \mathsf{data}_i = b\}$ as the aggregation function.
   Case 2: $\mathsf{command}_i = \mathsf{flush}$.
   Each CPU i sets $\mathsf{key}_i = \mathsf{path}_i^{(\mathsf{lev})}$ and $\mathsf{data}_i = L$ (resp., R) if the $(\mathsf{lev}+1)$-st bit of path_i is 0 (resp., 1). Use the union function as the aggregation function.
   At the conclusion of the protocol, each bucket $\mathsf{path}_i^{(\mathsf{lev})}$ is assigned to a representative CPU $\mathsf{bucket\text{-}rep}(\mathsf{path}_i^{(\mathsf{lev})})$ with aggregated commands $\mathsf{agg\text{-}command}_i$.
2. Each representative CPU performs the updates:
   If $i \neq \mathsf{bucket\text{-}rep}(\mathsf{path}_i^{(\mathsf{lev})})$, do nothing. Otherwise:
   Case 1: $\mathsf{command}_i = \mathsf{remove}\text{-}b_i$.
   Remove all blocks $b \in \mathsf{agg\text{-}command}_i$ in the bucket $\mathsf{path}_i^{(\mathsf{lev})}$ by accessing memory bucket $\mathsf{path}_i^{(\mathsf{lev})}$ and rewriting contents.
   Case 2: $\mathsf{command}_i = \mathsf{flush}$.
   Access memory buckets $\mathsf{path}_i^{(\mathsf{lev})}, \mathsf{path}_i^{(\mathsf{lev})} || 0, \mathsf{path}_i^{(\mathsf{lev})} || 1$, perform flush operation locally according to $\mathsf{agg\text{-}command}_i \subset \{L, R\}$, and write the contents back. Specifically, denote the collection of stored data blocks b in $\mathsf{path}_i^{(\mathsf{lev})}$ by $\mathsf{ThisBucket}$. Partition $\mathsf{ThisBucket} = \mathsf{ThisBucket\text{-}L} \cup \mathsf{ThisBucket\text{-}R}$ into those blocks whose associated leaves continue to the left or right (i.e., $\{b_j \in \mathsf{ThisBucket} : \bar{\ell}_j^{(\mathsf{lev}+1)} = \mathsf{mypath}^{(\mathsf{lev})} || 0\}$, and similar for 1).
   - If $L \in \mathsf{agg\text{-}command}_i$, then set $\mathsf{ThisBucket} \leftarrow \mathsf{ThisBucket} \setminus \mathsf{ThisBucket\text{-}L}$, and insert data items $\mathsf{ThisBucket\text{-}L}$ into bucket $\mathsf{path}_i^{(\mathsf{lev})} || 0$.
   - If $R \in \mathsf{agg\text{-}command}_i$, then set $\mathsf{ThisBucket} \leftarrow \mathsf{ThisBucket} \setminus \mathsf{ThisBucket\text{-}R}$, and insert data items $\mathsf{ThisBucket\text{-}L}$ into bucket $\mathsf{path}_i^{(\mathsf{lev})} || 0$.

Fig. 7. A space-efficient implementation of the $\mathsf{UpdateBuckets}$ procedure.

Read/Write access command on the position map database $\ell_i \leftarrow \mathsf{OPAccess}(t+1, (b_i, \ell_i'))$ to fetch the current position map value ℓ for block b_i and rewrite it with the newly sampled value ℓ_i'. Each dummy CPU performs an arbitrary dummy access (e.g., garbage $\leftarrow \mathsf{OPAccess}(t+1, (1, \emptyset))$).

3. **Look Up Current Memory Values:** Each CPU $\mathsf{rep}(b_i)$ fetches memory from the database nodes down the path to leaf ℓ_i; when b_i is found, it copies its value v_i into local memory. Each dummy CPU chooses a random path and make analogous dummy data fetches along it, ignoring all read values. (Recall that simultaneous data *reads* do not yield conflicts).

4. **Remove Old Data:** For each level in the tree,
   - Aggregate instructions across CPUs accessing the same "buckets" of memory (corresponding to nodes of the tree) on the server side. Each representative CPU $\mathsf{rep}(b)$ begins with the instruction of "remove block b if it occurs" and dummy CPUs hold the empty instruction. (Aggregation is as before, but at bucket level instead of the block level).
   - For each bucket to be modified, the CPU with the *smallest* id from those who wish to modify it executes the aggregated block-removal instructions for the bucket. Note that this aggregation step is purely for correctness and not security.

5. **Insert Updated Data into Database in *Parallel*:** Run Route on inputs $\{(m, (\mathsf{msg}_i, \mathsf{addr}_i))\}_{i \in [m]}$, where for each $\mathsf{rep}(b_i)$, $\mathsf{msg}_i = (b_i, \bar{v}_i, \ell_i')$ (i.e., updated block data) and $\mathsf{addr}_i = [\ell_i']_{\log m}$ (i.e., level-$\log m$-truncation of ℓ_i'), and for each dummy CPU, $\mathsf{msg}_i, \mathsf{addr}_i = \emptyset$.

6. **Flush the ORAM Database:** In parallel, each CPU initiates an independent flush of the ORAM tree. (Recall that this corresponds to selecting a random path down the tree, and pushing all data blocks in this path as far as they will go). To implement the simultaneous flush commands, as before, commands are aggregated across CPUs for each bucket to be modified, and the CPU with the smallest id performs the corresponding aggregated set of commands. (For example, all CPUs will wish to access the root node in their flush; the aggregation of all corresponding commands to the root node data will be executed by the lowest-numbered CPU who wishes to access this bucket, in this case CPU 1).

7. **Return Output:** Run OblivMCast on inputs $\{(b_i, v_i)\}_{i \in [m]}$ (where for dummy CPUs, $b_i, \bar{v}_i := \emptyset$) to communicate the *original* (pre-updated) value of each data block b_i to the subset of CPUs that originally requested it.

A few remarks regarding our construction.

Remark 2 (Truncating OPRAM for Fixed m). In the case that the number of CPUs m is fixed and known a priori, the OPRAM construction can be directly trimmed in two places.

Trimming Tops of Recursive Data Trees: Note that data items are always inserted into the OPRAM trees at level $\log m$, and flushed down from this level. Thus, the top levels in the ORAM tree are *never utilized*. In such case, the data buckets in the corresponding tops of the trees, from the root node to level $\log m$ for this bound, can simply be removed without affecting the OPRAM.

Truncating Recursion: In the t-th level of recursion, the corresponding database size shrinks to $n_t = n/\alpha^t$. In recursion level $\log_\alpha n/m$ (i.e., where $n_t = m$), we can then achieve oblivious data accesses via local CPU communication (storing each block $i \in [n_t] = [m]$ locally at CPU i, and running OblivAgg, OblivMCast directly) without needing any tree lookups or further recursion.

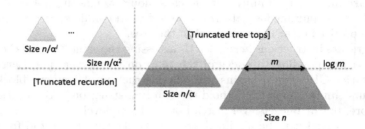

Remark 3 (Collision-Freeness). In the compiler above, CPUs only access the same memory address simultaneously in the (read-only) memory lookup in Step 3. However, a simple tweak to the protocol, replacing the direct memory lookups with an appropriate aggregation and multicast step (formally, the procedure UpdateBuckets as described in the appendix), yields collision freeness.

References

[Ajt10] Ajtai, M.: Oblivious rams without cryptogarphic assumptions. In: STOC, pp. 181–190 (2010)

[AKS83] Ajtai, M., Komlós, J., Szemerédi, E.: An 0(n log n) sorting network. In: Proceedings of the Fifteenth Annual ACM Symposium on Theory of Computing, STOC 1983, pp. 1–9 (1983)

[Bat68] Batcher, K.E.: Sorting networks and their applications. In: Proceedings of the Spring Joint Computer Conference, AFIPS 1968 (Spring), New York, NY, USA, 30 April–2 May 1968, pp. 307–314. ACM (1968)

[BCP15] Boyle, E., Chung, K.-M., Pass, R.: Large-scale secure computation: multi-party computation for (parallel) RAM programs. In: Gennaro, R., Robshaw, M. (eds.) CRYPTO 2015. LNCS, vol. 9216, pp. 742–762. Springer, Heidelberg (2015)

[BGL+15] Bitansky, N., Garg, S., Lin, H., Pass, R., Telang, S.: Succinct randomized encodings and their applications. In: Proceedings of the Forty-Seventh Annual ACM on Symposium on Theory of Computing, STOC 2015, pp. 439–448 (2015)

[CCC+15] Chen, Y.-C., Chow, S.S.M., Chung, K.-M., Lai, R.W.F., Lin, W.-K., Zhou, H.-S.: Computation-trace indistinguishability obfuscation and its applications. Cryptology ePrint Archive, Report 2015/406 (2015)

[CHJV15] Canetti, R., Holmgren, J., Jain, A., Vaikuntanathan, V.: Succinct garbling and indistinguishability obfuscation for RAM programs. In: Proceedings of the Forty-Seventh Annual ACM on Symposium on Theory of Computing, STOC 2015, pp. 429–437 (2015)

[CLP14] Chung, K.-M., Liu, Z., Pass, R.: Statistically-secure ORAM with $\tilde{O}(\log^2 n)$ overhead. In: Sarkar, P., Iwata, T. (eds.) ASIACRYPT 2014, Part II. LNCS, vol. 8874, pp. 62–81. Springer, Heidelberg (2014)

[CLT15] Chen, B., Lin, H., Tessaro, S.: Oblivious parallel RAM: improved efficiency and generic constructions. Cryptology ePrint Archive (2015)

[CP13] Chung, K.-M., Pass, R.: A simple ORAM. Cryptology ePrint Archive, Report 2013/243 (2013)

[DMN11] Damgård, I., Meldgaard, S., Nielsen, J.B.: Perfectly secure oblivious RAM without random oracles. In: Ishai, Y. (ed.) TCC 2011. LNCS, vol. 6597, pp. 144–163. Springer, Heidelberg (2011)

[FDD12] Fletcher, C.W., van Dijk, M., Devadas, S.: A secure processor architecture for encrypted computation on untrusted programs. In: Proceedings of the Seventh ACM Workshop on Scalable Trusted Computing, STC 2012, pp. 3–8 (2012)

[GGH+13] Gentry, C., Goldman, K.A., Halevi, S., Julta, C., Raykova, M., Wichs, D.: Optimizing ORAM and using it efficiently for secure computation. In: De Cristofaro, E., Wright, M. (eds.) PETS 2013. LNCS, vol. 7981, pp. 1–18. Springer, Heidelberg (2013)

[GHL+14] Gentry, C., Halevi, S., Lu, S., Ostrovsky, R., Raykova, M., Wichs, D.: Garbled RAM revisited. In: Nguyen, P.Q., Oswald, E. (eds.) EUROCRYPT 2014. LNCS, vol. 8441, pp. 405–422. Springer, Heidelberg (2014)

[GHRW14] Gentry, C., Halevi, S., Raykova, M., Wichs, D.: Outsourcing private RAM computation. In: Symposium on Foundations of Computer Science, FOCS 2014, pp. 404–413 (2014)

[GKK+12] Gordon, S.D., Katz, J., Kolesnikov, V., Krell, F., Malkin, T., Raykova, M., Vahlis, Y.: Secure two-party computation in sublinear (amortized) time. In: The ACM Conference on Computer and Communications Security, CCS 2012, Raleigh, NC, USA, 16–18 October 2012, pp. 513–524 (2012)

[GKP+13] Goldwasser, S., Kalai, Y.T., Popa, R.A., Vaikuntanathan, V., Zeldovich, N.: How to run turing machines on encrypted data. In: Canetti, R., Garay, J.A. (eds.) CRYPTO 2013, Part II. LNCS, vol. 8043, pp. 536–553. Springer, Heidelberg (2013)

[GLOS15] Garg, S., Steve, L., Ostrovsky, R., Scafuro, A.: Garbled RAM from one-way functions. In: Proceedings of the Forty-Seventh Annual ACM on Symposium on Theory of Computing, STOC 2015, pp. 449–458 (2015)

[GM11] Goodrich, M.T., Mitzenmacher, M.: Privacy-preserving access of outsourced data via oblivious RAM simulation. In: Aceto, L., Henzinger, M., Sgall, J. (eds.) ICALP 2011, Part II. LNCS, vol. 6756, pp. 576–587. Springer, Heidelberg (2011)

[GMOT11] Goodrich, M.T., Mitzenmacher, M., Ohrimenko, O., Tamassia, R.: Oblivious ram simulation with efficient worst-case access overhead. In: CCSW, pp. 95–100 (2011)

[GMW87] Goldreich, O., Micali, S., Wigderson, A.: How to play any mental game or a completeness theorem for protocols with honest majority. In: STOC, pp. 218–229 (1987)

[GO96] Goldreich, O., Ostrovsky, R.: Software protection and simulation on oblivious RAMs. J. ACM **43**(3), 431–473 (1996)

[Gol87] Goldreich, O.: Towards a theory of software protection and simulation by oblivious RAMs. In: STOC, pp. 182–194 (1987)

[KLO12] Kushilevitz, E., Lu, S., Ostrovsky, R.: On the (in)security of hash-based oblivious ram and a new balancing scheme. In: SODA, pp. 143–156 (2012)

[KLW15] Koppula, V., Lewko, A.B., Waters, B.: Indistinguishability obfuscation for turing machines with unbounded memory. In: Proceedings of the Forty-Seventh Annual ACM on Symposium on Theory of Computing, STOC, pp. 419–428 (2015)

[LO13] Lu, S., Ostrovsky, R.: Distributed oblivious RAM for secure two-party computation. In: Sahai, A. (ed.) TCC 2013. LNCS, vol. 7785, pp. 377–396. Springer, Heidelberg (2013)

[LPM+13] Lorch, J.R., Parno, B., Mickens, J.W., Raykova, M., Schiffman, J.: Shroud: ensuring private access to large-scale data in the data center. In: FAST, pp. 199–214 (2013)

[NWI+15] Nayak, K., Wang, X.S., Ioannidis, S., Weinsberg, U., Taft, N., Shi, E.: GraphSC: parallel secure computation made easy. In: IEEE Symposium on Security and Privacy (S&P) (2015)

[OS97] Ostrovsky, R., Shoup, V.: Private information storage (extended abstract). In: STOC, pp. 294–303 (1997)

[PF79] Pippenger, N., Fischer, M.J.: Relations among complexity measures. J. ACM **26**(2), 361–381 (1979)

[RFK+14] Ren, L., Fletcher, C.W., Kwon, A., Stefanov, E., Shi, E., van Dijk, M., Devadas, S.: Ring ORAM: closing the gap between small and large client storage oblivious RAM. IACR Cryptology ePrint Archive 2014:997 (2014)

[SCSL11] Shi, E., Chan, T.-H.H., Stefanov, E., Li, M.: Oblivious RAM with $O((\log N)^3)$ worst-case cost. In: Wang, X., Lee, D.H. (eds.) ASIACRYPT 2011. LNCS, vol. 7073, pp. 197–214. Springer, Heidelberg (2011)

[SS13] Stefanov, E., Shi, E.: ObliviStore: high performance oblivious cloud storage. In: IEEE Symposium on Security and Privacy, pp. 253–267 (2013)

[SvDS+13] Stefanov, E., van Dijk, M., Shi, E., Fletcher, C.W., Ren, L., Yu, X., Devadas, S.: Path ORAM: an extremely simple oblivious RAM protocol. In: ACM Conference on Computer and Communications Security, pp. 299–310 (2013)

[WCS14] Wang, X.S., Hubert Chan, T.-H., Shi, E.: Circuit ORAM: on tightness of the goldreich-ostrovsky lower bound. IACR Cryptology ePrint Archive 2014:672 (2014)

[WHC+14] Wang, X.S., Huang, Y., Hubert Chan, T.-H., Shelat, A., Shi, E.: SCORAM: oblivious RAM for secure computation. In: Proceedings of the 2014 ACM SIGSAC Conference on Computer and Communications Security, pp. 191–202 (2014)

[WST12] Williams, P., Sion, R., Tomescu, A.: PrivateFS: a parallel oblivious file system. In: Proceedings of the 2012 ACM Conference on Computer and Communications Security, CCS 2012, pp. 977–988 (2012)

[Yao82] Yao, A.C.-C.: Protocols for secure computations (extended abstract). In: 23rd Annual Symposium on Foundations of Computer Science (FOCS), pp. 160–164 (1982)

Oblivious Parallel RAM: Improved Efficiency and Generic Constructions

Binyi Chen[✉], Huijia Lin, and Stefano Tessaro

Department of Computer Science, University of California, Santa Barbara, USA
{binyichen,rachel.lin,tessaro}@cs.ucsb.edu

Abstract. Oblivious RAM (ORAM) garbles read/write operations by a client (to access a remote storage server or a random-access memory) so that an adversary observing the garbled access sequence cannot infer any information about the original operations, other than their overall number. This paper considers the natural setting of Oblivious *Parallel* RAM (OPRAM) recently introduced by Boyle, Chung, and Pass (TCC 2016A), where m clients simultaneously access in *parallel* the storage server. The clients are additionally connected via point-to-point links to coordinate their accesses. However, this additional inter-client communication must also remain oblivious.

The main contribution of this paper is twofold: We construct the first OPRAM scheme that (nearly) matches the storage and server-client communication complexities of the most efficient single-client ORAM schemes. Our scheme is based on an extension of Path-ORAM by Stefanov et al. [18]. Moreover, we present a *generic* transformation turning any (single-client) ORAM scheme into an OPRAM scheme.

1 Introduction

This paper considers the problem of hiding *access patterns* when reading from and writing to an untrusted memory or storage server. This is a fundamental problem in both in the context of software protection, as well as for secure outsourcing to a third-party storage provider.

The basic cryptographic method to hide access patterns is *Oblivious RAM* (ORAM) [8,9]. It compiles logical access sequences (from a client) into garbled ones (to a storage space, or *server*) so that a curious observer seeing the latter only (as well as the server contents) cannot infer anything *other than the overall number of logical accesses*—we say that such garbled access sequences are *oblivious*. Since its proposal, ORAM and its applications have been extensively studied (cf e.g. [1,3–7,9,11–21,24–28,30]). The state-of-the-art constructions [16,26] have a $\tilde{O}(\log^2 N)$ computation (and communication) overhead (per logical access),[1] where N is the size of the storage, i.e., the number of *data* bocks (of a certain bit size) it can store.

---

[1] The ORAM scheme of [16] has only $O(\log^2 N/\log\log N)$ overhead, while that of [26] has $O(\log^2 N)$ overhead. However, the latter construction is simpler and achieves better practical efficiency [26].

© International Association for Cryptologic Research 2016
E. Kushilevitz and T. Malkin (Eds.): TCC 2016-A, Part II, LNCS 9563, pp. 205–234, 2016.
DOI: 10.1007/978-3-662-49099-0_8

Parallel Oblivious Accesses. Existing ORAM schemes only support a *single* client, and in particular do not deal with *parallel accesses from multiple clients.* However, enabling such parallelism is important, e.g., to achieve scalable cloud storage services for multiple users, or to secure multi-processor architectures. To overcome this barrier, a few systems-oriented works [15,24,30] suggested to either use a trusted proxy shared by multiple clients to act as the "sole client" of ORAM, or to adapt known ORAM schemes (such as [7,9,29]) to support a limited, $O(\log N)$, number of parallel accesses.

Recently, Boyle, Chung, and Pass (BCP) [2] proposed the notion of *Oblivious Parallel RAM (OPRAM)*, which compiles *synchronous parallel* logical access sequences by m clients into, *parallel*, garbled sequences and inter-client messages, which together still reveal no information other than the total number of logical accesses. They also provided the first – and so far, the only – OPRAM scheme. Their construction is simple and elegant, but, has a server-client communication overhead of $\omega(\log^3 N)$—a factor of $\tilde{\Omega}(\log N)$ higher than state-of-the-art ORAM schemes [16,26]. Their approach seems not to extend directly to use the techniques behind existing communication-efficient ORAM schemes.

Hence, the natural question that arises is: *"Can we design an OPRAM scheme with the same per-client efficiency as the state-of-the-art ORAM schemes?"*

Our Contributions, in a Nutshell. Our first contribution answers this question affirmatively. In particular, we prove:

> **Theorem 1 (Informal):** *There is an OPRAM scheme with $O(\log^2 N)$ (amortized) server-client communication overhead, and constant storage overhead.*

Going beyond, an even more fundamental question concerns the basic relation between ORAM and OPRAM. We show that the two problems are related at a far more generic level:

> **Theorem 2 (Informal):** *There is a generic transformation that turns any ORAM scheme into an OPRAM scheme, with additional $O(\log N)$ (amortized) server-client communication overhead with respect to the original ORAM scheme.*

While the above results are in the amortized case, we note that in the worst case, the above complexity statements are true with O replaced by ω. Moreover, our OPRAM schemes all require client-to-client communication. Their inter-client communication is $\omega(\log N) \log m(\log m + \log N)B$ bits. We note that this also is an improvement by a factor $O(\log N)$ over BCP.

We stress that our approach is substantially different from that of BCP: One key idea is the use of partitioning, i.e., the fact that each client is responsible for a designated portion of the server storage. This eliminates much of the coordination necessary in BCP. Next, we move to explaining the high-level ideas behind our constructions in greater detail.

1.1 Subtree-OPRAM

We provide an overview of our scheme Subtree-OPRAM. Our construction of an m-client OPRAM scheme can be seen as consisting of two steps.

(1) First, we construct an ORAM scheme, called *Subtree-ORAM*, that enables a *single* client to batch-process m logical accesses at a time *in parallel*. Our Subtree-ORAM scheme is a generalization of Path-ORAM [26] to the setting with large client memory and parallel processing. We believe that this generalization is of independent interest.

(2) In a second step, we exploit the batch-processing structure of Subtree-ORAM to adapt it to the multiple-client setting, and derive our Subtree-OPRAM scheme by distributing its computation across multiple clients.

In the following, we explain all of this in more detail.

Review of Path-ORAM. Let us first give an overview of the tree-based ORAM approach by Shi et al. [23]. In particular, we review Path-ORAM [26], as it will serve as our starting point. (A more detailed review is given in Appendix B.)

To implement a storage space for N *data blocks*, basic (i.e., non-recursive) Path-ORAM organizes the storage space (virtually) as a complete binary tree with depth $O(\log N)$, where each node is a "bucket" that contains a fixed number $Z = O(1)$ of encrypted blocks (some of which may be dummies). To hide access patterns, each data block is assigned to a random path ℓ (from a leaf ℓ to the root, and we use ℓ to identify both the leaf and the associated path interchangeably) and stored in *some* bucket on path ℓ; after each access, the assignment is "refreshed" to a new random path ℓ'. The client keeps track of the current path assigned to each block using a *position map*. The client also keeps an additional (small) memory for overflowing blocks, called the *stash*. For each logical access to a certain block with address $\mathsf{a} \in [N]$, Path-ORAM takes the two following steps:

(1) Fetching a path. Retrieve the path ℓ currently associated with block a in the position map, and find block a on the path or in the local *stash*. Then, assign the block a to a new random path ℓ' and update the position map accordingly.

(2) Flushing along a path. Iterate over every block a' in the fetched path ℓ and in the stash (this includes the block a we just retrieved and possibly updated, and which was assigned to the new path ℓ'), and re-insert each block a' into the lowest possible bucket on ℓ that is also on the path assigned to a' according to the position map. If no suitable place is found (as each bucket can only contain at most Z blocks), the block is placed into the stash. The contents of the path are re-encrypted when being written back to the server (including dummy blocks).

The analysis of Path-ORAM [26] shows that the stash size is bounded by $\omega(\log N)$ with probability roughly $poly(\lambda)2^{-\omega(\log N)}$. To avoid keeping a large position map, Path-ORAM *recursively* stores the position map at the server. The final scheme has a recursion depth of $O(\log N)$—each logical access is translated to $O(\log N)$ actual accesses, each consisting of retrieving a path. Overall, the communication overhead is $O(\log^2 N)$. Also, the overall storage complexity at the server can be kept to $O(N)$ despite the recursion.

Subtree-ORAM. As our first contribution, we generalize Path-ORAM to process $m \geq 1$ logical accesses at a time. As the recursion step in Path-ORAM is rather generic, we focus on the non-recursive scheme, ignoring the costs of storing the position map.

The natural approach to achieve this is to retrieve a *subtree* of m paths, i.e., for every m logical accesses to blocks $a_1, \ldots, a_m$, we can do the following:

(1) Fetching subtree. Retrieve the subtree ST composed of the paths $\ell_1, \ldots, \ell_m$ assigned to the m blocks and find the blocks of interest in the subtree or in the stash, and possibly update their values.

(2) Path-by-path flushing. Execute the flushing procedure from Path-ORAM on the m paths in ST sequentially as in Path-ORAM, with each a_i assigned to a new random path ℓ_i'.

Unfortunately, there are *two* problems with this approach. First, if $a_1, \ldots, a_m$ are not all distinct, the accesses are not oblivious, as the same path would be retrieved multiple times. To avoid this, the final Subtree-ORAM scheme perform some pre-processing: For accesses to the same block, replace all but the first one with $\perp$ in the logical sequence to obtain $a_1', \ldots, a_m'$, and for each repetition $a_i' = \perp$, assign random path to be retrieved from the server — this is called a *fake read*.[2]

The second drawback is that repeating the flushing procedure of Path-ORAM m times in Step 2 is inherently sequential. To use Subtree-ORAM within Subtree-OPRAM below, we instead target a parallelizable flushing procedure. To this end, we introduce the following new flushing procedure, which we refer to as *subtree flushing*:

(2) Subtree flushing: Iterate over every block in ST and in the stash and place each block into the lowest node in the *entire subtree* ST that is still on its assigned path, and not yet full. The order in which blocks are processed can be arbitrary, and the process can be parallelized (subject to maintaining the size constraint of each node).

Security and correctness of Subtree-ORAM follow similar arguments as Path-ORAM. Furthermore, we bound the stash size of Subtree-ORAM by generalizing aspects of the analysis of Path-ORAM – we believe this to be of independent interest.

Subtree-OPRAM. Our end goal is to design an interactive protocol that enables m clients to access (and possibly alter) blocks $a_1, \ldots, a_m$ in parallel,

---

[2] Note that this random path may well collide with one of the other paths. Still, the key point is that it is chosen *independently* of the actual blocks. The use of such fake read has appeared in many previous works, such as, [2,24].

where client C_i is requesting in particular block a_i; both the access patterns to the server, as well as inter-client communication, must be oblivious.

We can think of our Subtree-OPRAM protocol as having the m clients collectively emulate the single Subtree-ORAM client. To this end, we use inter-client oblivious communication protocols based on tools developed in [2] to let clients interact with each other. Here, we focus our description on how to "distribute" Step 1 and Step 2 of Subtree-ORAM for the special cases that the requested blocks $a_1, \ldots, a_m$ are distinct. (Handling colliding requests in an oblivious way will require extra work.) For simplicity, we assume that all clients have access to the position map and all messages are implicitly encrypted (with a key shared by all clients). In particular, everything is re-encrypted before being written to the server.

Assume for simplicity that $m = 2^l$. We can think of the server storage in Subtree-OPRAM in terms of a tree of buckets, as in Path-ORAM and Subtree-ORAM. However, we remove the top l levels, effectively turning the tree into a forest of m trees $T_1, \ldots, T_m$; client C_i manages all read/write from/to T_i, and all blocks assigned to (a path in) T_i that do not fit in one of the buckets on the server remain in a local stash managed locally by C_i. More precisely:

(1) In parallel, each C_i finds the path ℓ_i assigned to a_i (using the position map) and delegates the job of reading path ℓ_i to the client C_j responsible for the tree T_j containing ℓ_i, to which it sends a request. Each C_j retrieves all paths for which it has received a request (again in parallel), which form a subtree ST_j of T_j; it then finds the blocks of interest in ST_j and its local stash, and sends them back to the respective clients who requested them.

(2) Each C_i assigns a_i a new path ℓ_i', and delegates the job of writing back (B_i, ℓ_i') to the client C_j responsible for the tree T_j containing ℓ_i'. To ensure obliviousness, the clients achieve this by running collectively the oblivious routing protocol of [2], which hides the destination of messages. Next, each C_i runs the subtree-flushing procedure locally on the retrieved subtree ST_i and its own stash, and finally writes the entire subtree ST_i back.

We will show that the m clients indeed collectively emulate the execution of the single client of Subtree-ORAM. In particular, parallel flushing on the individual subtrees emulates the effect of a global flushing over the union of these subtrees, but keeping the top l levels of the tree locally at the clients; also, the *union* of the stashes of all clients contains exactly the contents of the stash of the Subtree-ORAM client, as well as the contents of the top of the tree. This gives a bound on the overall sizes of the stashes.

In expectation, each client reads and writes one path per round, and thus the amortized client-server communication overhead is $O(\log N)$, and the final recursive Subtree-OPRAM has amortized overhead of $O(\log^2 N)$, with overwhelming probability. In fact, we prove that the worst-case overhead is not much higher, and is of the order of $\omega(\log^2 N)$, e.g., $O((\log^2 N) \cdot \log \log N)$, much smaller than BCP's $\omega(\log^3 N)$. We improve over BCP also in terms of inter-client communication complexity by a factor of $\log N$.

1.2   The Generic Transformation

Subtree-OPRAM is tailored at achieving the same overhead as Path-ORAM, and not surprisingly, the former heavily relies on the latter. Our second contribution is a generic transformation that converts *any* ORAM scheme into an OPRAM protocol. When applied to Path-ORAM, the resulting scheme is less efficient than Subtree-OPRAM – still, the main benefit here is *generality*.

Our approach generalizes ideas from partition-based ORAM [25]. Specifically, we split the server storage into m partitions each storing (roughly) N/m blocks, and let the m clients run each a copy of the basic ORAM algorithm (call them $\mathcal{O}_1, \ldots, \mathcal{O}_m$). Each client C_i thus manages the i-th partition independently using $\mathcal{O}_i$. Every block a is randomly assigned to one of the m partitions $P \in [m]$, and it is re-assigned to a new random partition after each access. The current assignment of blocks to the m partitions is recorded in a *partition map*, which we assume (for now) to be accessible by all clients. (In the end, it will be shared using recursion techniques.) Then, when m clients request the m blocks $a_1, \ldots, a_m$ in parallel, the clients simply find the respective partitions $P_1, \ldots, P_m$ containing these blocks, and let the corresponding clients retrieve the desired blocks and delete them from their partitions (if a block is accessed for multiple times, then "fake reads" are performed to a random partition). The actual access pattern *so far* is oblivious since all P_i's are random, and the basic ORAM scheme ensures that retrieving blocks from each partition is done obliviously.

However, writing these blocks back to new random partitions without revealing their destinations turns out to be non-trivial, *even if we can deliver the blocks obliviously to the clients responsible for the new partitions*. Indeed, naively invoking the corresponding ORAM copies to insert would reveal how many blocks are assigned to each partition. To hide this information, in our protocol each client inserts the *same* number κ of blocks to its partition, and keeps a queue of blocks to be inserted. We use a stochastic analysis to show that for any $R = \omega(\log \lambda)$, it is sufficient to insert *only $\kappa = 2$ blocks to each partition each time (and in particular, perform fake "insertions" if less than 2 blocks need to be inserted)*, and at most R "overflowing" blocks ever remain in the queue (except with negligible probability).

A challenge we have not addressed is how to use an ORAM for a partition of size $O(N/m)$ to store the blocks associated with it in an efficient way, i.e., without using the whole space of $[N]$ addresses. We will solve this by using an appropriate ORAM-based oblivious dictionary data structure.

As the expected number of read and write operations each client performs is 3 (one read and two writes), the non-recursive version has the same (amortized) computation and communication overhead as the underlying ORAM scheme. To obtain the final OPRAM scheme, we apply recursive techniques to outsource the partition map to the server.

Notation. Throughout this paper, we let $[n]$ denote the set $\{1, 2, ..., n\}$. We denote by $\Delta(X, Y)$ the statistical distance between distributions (or random variables) X and Y, i.e. $\Delta(X, Y) = \sum_x |\Pr[X = x] - \Pr[Y = x]|$. Also, we say that a function μ is *negligible* if for every polynomial p there exists a sufficiently large integer n_0, such that $\mu(n) \leq 1/p(n)$ for all $n > n_0$.

2   Oblivious (Parallel) RAM

We start by reviewing the notion of Oblivious RAM and its parallel extensions. We present definitions different from (yet essentially equivalent to) the ones by Goldreich and Ostrovsky [8,9] and BCP [2], considering clients and servers, instead of RAM compilers, which we consider to lead to more compact and natural descriptions, and are more in line with the applied ORAM literature.

Basic ORAM Setting. The basic ORAM setting considers two parties, a *client* and a *server*. The server $\mathcal{S}(M, B)$ has a large storage space consisting of M cells, each of size B bits, whereas the client has a much smaller memory. The client can access the storage space at the server using read and write commands, denoted as $\mathsf{Acc}(\mathsf{read}, \mathsf{a}, \perp)$ and $\mathsf{Acc}(\mathsf{write}, \mathsf{a}, v)$, where $\mathsf{a} \in [M]$ and $v \in \{0,1\}^B$. (We assume that all cells on the server are initialized to some fixed string, i.e., 0^B.) Both operations return the current value stored in cell a, in particular for the latter operation this is the value before the cell is overwritten with v.

An oblivious RAM (ORAM) scheme consists of an *ORAM client* $\mathcal{O}$ (or simply, an ORAM $\mathcal{O}$), which is a stateful interactive PPT machine which on initial input the security parameter λ, block size B, and storage size N, processes *logical* commands $\mathsf{Acc}(\mathsf{op}_i, \mathsf{a}_i, v_i)$, $\mathsf{op}_i \in \{\mathsf{read}, \mathsf{write}\}$, $\mathsf{a}_i \in [N]$, $v_i \in \{0,1\}^B \cup \{\perp\}$, by interacting with a server $\mathcal{S}(M, B)$ (for values $M = M(N)$ and $B = B(B, \lambda)$ explicitly defined by the scheme), via sequence of *actual* (read/write) accesses $\mathsf{Acc}(\overline{\mathsf{op}}_{i,1}, \overline{a}_{i,1}, \overline{v}_{i,1}), \ldots, \mathsf{Acc}(\overline{\mathsf{op}}_{i,q_i}, \overline{a}_{i,q_i}, \overline{v}_{i,q_i})$, and finally outputs a value val_i and updates its local state depending on the answers of these accesses.

An ORAM scheme hides the sequence of logical commands from an untrusted (honest-but-curious) server, who observes the actual sequence of accesses. The actual values written to the server can be hidden using semantically-secure encryption. Indeed, all known ORAM solutions have server cells hold each the encryption of a block, i.e., in general one has $B = B + O(\lambda)$. For this reason, we abstract away from the usage of encryption by dealing only with access-pattern security and tacitly assuming that all cells are going to be stored encrypted in the final scheme with a semantically secure encryption scheme, and that every write access to the server will be in form of a fresh re-encryption of the value. In this case, it makes sense to think of $B = B$, and an adversary who cannot see the value written to/read from the server.

We defer a definition of security and correctness for single-client ORAM in Appendix A, and here rather focus on generalizing above to the multi-client setting.

Multi-Client Setting. We now consider the setting of *oblivious parallel ORAM* (or OPRAM for short) with m clients. An m-client OPRAM is a set[3] of stateful

---

[3] For notational simplicity, we give definitions for the case where the number of clients m is fixed and independent of the security parameter. However, one can easily extend these definitions to the case where $m = m(\lambda)$ with some (straightforward) notational effort.

interactive PPT machines $\mathcal{PO} = \{\mathcal{O}_i\}_{i \in [m]}$ which all on initial input the security parameter λ, the storage size parameter N, and the block size B, proceed in rounds, interacting with the server $\mathcal{S}(M(N), B)$ (where M is a parameter of the scheme[4]) and with each other through point-to-point connections. At each round r the following steps happen: First, every client $\mathcal{O}_i$ receives as input a logical operation $\mathsf{Acc}(\mathsf{op}_{i,r}, \mathsf{a}_{i,r}, v_{i,r})$ where $\mathsf{op}_{i,r} \in \{\mathsf{read}, \mathsf{write}\}$, $\mathsf{a}_{i,r} \in [N]$ and $v_{i,r} \in \{0,1\}^B \cup \{\bot\}$. Then, the clients engage in an interactive protocol where at any time each client $\mathcal{O}_i$ can (1) Send messages to other clients, and (2) Perform one or more accesses to the server $\mathcal{S}(M, B)$. Finally, every $\mathcal{O}_i$ outputs some value $\mathsf{val}_{i,r}$.

Correctness and Obliviousness. We assume without loss of generality than the honest-but-curious adversary learns only the *access and communication patterns*. To this end, let us fix a sequence of logical access operations that are issued to the m clients in T successive rounds. First off, for all $i \in [m]$, we denote by
$$\boldsymbol{y}_i = \left(\mathsf{Acc}(\mathsf{op}_{i,r}, \mathsf{a}_{i,r}, v_{i,r})\right)_{r \in [T]}$$
the sequence of logical operations issued to $\mathcal{O}_i$ in the T rounds, and let $\boldsymbol{y} = (\boldsymbol{y}_1, \ldots, \boldsymbol{y}_m)$.

Now, for an execution of an OPRAM scheme $\mathcal{PO}$ for logical sequence of accesses $\boldsymbol{y}$ as above, we let ACP_i be the round-i communication pattern, i.e., the transcript of the communication among clients and between each client and the server in round $i \in [T]$, except that actual contents of the messages sent among clients, as well as the values v_i in server accesses by the clients, are removed. We define
$$\mathsf{ACP}_{\mathcal{PO}}(\lambda, N, B, \boldsymbol{y}) = (\mathsf{ACP}_1, \ldots, \mathsf{ACP}_T).$$

Finally, we also denote the outputs client i as $\boldsymbol{val}_i = (val_{i,1}, \ldots, val_{i,T})$ and
$$\mathsf{Out}_{\mathcal{PO}}(\lambda, N, B, \boldsymbol{y}) = (\boldsymbol{val}_1, \ldots, \boldsymbol{val}_m).$$

The outputs $\boldsymbol{z} = \mathsf{Out}_{\mathcal{PO}}(\lambda, N, B, \boldsymbol{y})$ of $\mathcal{PO}$ are correct w.r.t. the parallel accesses sequence $\boldsymbol{y}$, if it satisfies that for each command $\mathsf{Acc}(\mathsf{op}_{i,t}, \mathsf{a}_{i,t}, v_{i,t})$ in $\boldsymbol{y}$, the corresponding output $val_{i,t}$ in $\boldsymbol{z}$ is either the most recently written value on address a_i, or $\bot$ if a_i has not yet been written. Moreover, we assume that if two write operations occur in the same round for the same address, issued by clients $\mathcal{O}_i$ and $\mathcal{O}_j$, for $i < j$, then the value written by $\mathcal{O}_i$ is the one that takes effect. Let $\mathsf{Correct}$ be the predicate that on input $(\boldsymbol{y}, \boldsymbol{z})$ returns whether $\boldsymbol{z}$ is correct w.r.t. $\boldsymbol{y}$.

Definition 1 (Correctness and Security). *An OPRAM scheme $\mathcal{PO}$ achieves correctness and obliviousness if or all $N, B, T = poly(\lambda)$, there exists a negligible function μ such that, for every λ, every two parallel sequences $\boldsymbol{y}$ and $\boldsymbol{y}'$ of the same length $T(\lambda)$, the following are satisfied:*

(i) *Correctness.* $\Pr\left[\mathsf{Correct}\left(\boldsymbol{y}, \mathsf{Out}_{\mathcal{PO}}(\lambda, N, B, \boldsymbol{y})\right) = 1\right] \geq 1 - \mu(\lambda)$.
(ii) *Obliviousness.* $\Delta\left(\mathsf{ACP}_{\mathcal{PO}}(\lambda, N, B, \boldsymbol{y}), \mathsf{ACP}_{\mathcal{PO}}(\lambda, N, B, \boldsymbol{y}')\right) \leq \mu(\lambda)$.

---

[4] As in the single-client case above, we simply assume that server blocks and logical blocks have the same size for simplicity, as we only consider the unencrypted case.

Usually, the values λ, N, B are understood from the context, and we thus often use $\mathsf{ACP}(\boldsymbol{y}) = \mathsf{ACP}_{\mathcal{PO}}(\lambda, N, B, \boldsymbol{y})$ for notational simplicity.

OPRAM Complexity. The *server-communication* overhead and *inter-client communication* overhead of an OPRAM scheme $\mathcal{PO}$ are respectively the number of bits sent/received per client to/from the server, and to/from other clients, per logical access command, divided by the block size B. Finally, the server storage overhead of $\mathcal{PO}$ is the number of blocks stored at the server divided by N, and client storage overhead is the number of blocks stored at each client after each parallel access.

3   OPRAM with $O(\log^2 N)$ Server Communication Overhead

In this section, we present our first OPRAM scheme, called Subtree-OPRAM.

Theorem 1 (Subtree-OPRAM). *For every m, there is a m-client OPRAM scheme with the following properties: Let λ, N, and B denote the security parameter, the size of the logical space, and block size satisfying $B \geq 2 \log N$.*

- *Client Storage Overhead. Every client keeps a local stash consisting of $R = (\omega(\log \lambda) + O(\log m)) \log N$ blocks.*
- *Server Storage Overhead. $O(1)$.*
- *Server Communication Overhead. The amortized overhead is $O(\log^2 N)$ and the worst case overhead is $\omega(\log \lambda \log N) + O(\log^2 N)$ with overwhelming probability.*
- *Inter-Client Communication Overhead. The amortized and worst-case overheads are both $\omega(\log \lambda) \log m(\log m + \log N)$ with overwhelming probability.*

In particular, when the security parameter λ is set to N, the server communication complexity is $\omega(\log^2 N)$ in the worst case, and $O(\log^2 N)$ amortized.

To prove the theorem, as discussed in the introduction, we first present a single-client ORAM scheme, Subtree-ORAM, that supports parallel accesses in Sect. 3.1, and then adapt it to the multiple-client setting to obtain Subtree-OPRAM in Sect. 3.3. We analyze these two schemes in Appendixes C and D. Additional helper protocols needed by Subtree-OPRAM are given in Sect. 3.2.

3.1   Subtree-ORAM

In this section, we describe the non-recursive version of Subtree-ORAM, where the client keeps a large position map of size $O(N \log N)$; the same recursive technique as in Path-ORAM can be applied to reduce the client memory size.

The Subtree-ORAM client, $\mathsf{ST\text{-}\mathcal{O}}$, keeps a logical space of N blocks of size B using $M(N) = O(N)$ blocks on the server. The server storage space is organized (virtually) as a complete binary tree $\mathcal{T}$ of depth $D = \log N$ (we assume for simplicity that N is a power of two), where each node is a *bucket* capable of storing Z blocks. In particular, we associate leaves (and paths leading to them from the root) with elements of $[2^D] = [N]$. Additionally, ST-$\mathcal{O}$ locally maintains a position map pos.map and a stash stash of size respectively $O(N \log N)$ bits and $R(\lambda) \in \omega(\log \lambda)$ blocks.

In each iteration r, the Subtree-ORAM client ST-$\mathcal{O}$ processes a batch of m logical access operations $\{\mathsf{Acc}(\mathsf{op}_i, \mathsf{a}_i, v_i)\}_{i \in [m]}$ as follows:

1. **Pre-process.** Remove repetitive block accesses by producing a new m-component vector Q as follows: The i-th entry is set to $Q_i = (\mathsf{op}_i, \mathsf{a}_i)$ if the following condition holds, otherwise $Q_i = \bot$.
   - *Either*, there are (one or many) write requests to block a_i, and the i-th operation $\mathsf{Acc}(\mathsf{op}_i, \mathsf{a}_i, v_i)$ is the one with the minimal index among them.
   - *Or*, there are only read requests to block a_i, and the i-th operation $\mathsf{Acc}(\mathsf{op}_i, \mathsf{a}_i, v_i)$ is the one with the minimal index among them.
2. **Read paths in parallel.** Determine a set $S = \{\ell_1, \ldots, \ell_m\}$ of m paths to read, where each path is of one of the following two types:
   - **Real-read.** For each $Q_i = (\mathsf{op}_i, \mathsf{a}_i) \neq \bot$, set $\ell_i = \mathsf{pos.map}(\mathsf{a}_i)$ and immediately refresh $\mathsf{pos.map}(\mathsf{a}_i)$ to $\ell_i' \xleftarrow{\$} [N]$.
   - **Fake-read.** For each entry $Q_i = \bot$, sample a random path $\ell_i \xleftarrow{\$} [N]$.
   Then, retrieve all paths in S from the server, forming a subtree $\mathcal{T}_S$ of buckets with (at most) Z decrypted blocks in them.
3. **Post-process.** Answer each logical access $\mathsf{Acc}(\mathsf{op}_i, \mathsf{a}_i, v_i)$ as follows: Find block a_i in subtree $\mathcal{T}_S$ or stash, and returns the value of the block. Next, for each $Q_i \neq \bot$ if the corresponding logical access is a write operation $\mathsf{Acc}(\mathsf{write}, \mathsf{a}_i, v_i \neq \bot)$, update block a_i to value v_i.
4. **Flush subtree and write-back.** Let $\mathcal{T}_{\mathsf{real}}$ be the subtree consisting of only real-read paths in $\mathcal{T}_S$. Before (re-encrypting and) writing $\mathcal{T}_S$ back to the server, re-arrange the contents of $\mathcal{T}_{\mathsf{real}}$ and stash to fit as many blocks from stash into the subtree as follows:

   Subtree-flushing. Move all blocks in $\mathcal{T}_{\mathsf{real}}$ and stash to a temporary set Λ. Traverse through all blocks in Λ in an *arbitrary* order: Insert each block with address a from Λ, either into the lowest non-full bucket in $\mathcal{T}_{\mathsf{real}}$ that lies on the path pos.map(a) (if such bucket exists), or into stash. If at any point, the stash contains more than R blocks, output overflow and abort.

In Appendix C, we briefly discuss the analysis of Subtree-ORAM, noting the bulk of it (proving that the overflow probability is small) is deferred to the full version for lack of space.

3.2 Oblivious Inter-client Communication Protocols

Subtree-OPRAM, which we introduce in the next section, will use as components a few oblivious inter-client communication sub-protocols which will allow to emulate Subtree-ORAM in a distributed fashion. These are variants of similar protocols proposed in [2]. Their communication patterns are *statically fixed*, independent of inputs (and thus are oblivious in a very strong sense), and the communication and computation complexities of each protocol participant is small, i.e., roughly polylog(m) where m is the number of participants. We only describe the interfaces of these protocols; their implementations are based on $\log(m)$-depth sorting networks, and we refer the reader to [2] for further low-level details.

Oblivious Aggregation. Our first component protocol is used to aggregate data held by multiple users, and is parameterized by an *aggregation function* agg which can combine an arbitrary number of data items $d_1, d_2, \ldots$ (from a given data set) into an element $\mathsf{agg}(d_1, d_2, \ldots)$. The function agg is associative, i.e., $\mathsf{agg}(\mathsf{agg}(d_1, d_2, \ldots, d_k), d_{k+1}, \ldots d_{k+r})$ and $\mathsf{agg}(d_1, d_2, \ldots, d_k, \mathsf{agg}(d_{k+1}, \ldots d_{k+r}))$ both give us the same value as $\mathsf{agg}(d_1, \ldots, d_{k+r})$. Each party $i \in [m]$ starts the protocol with an input pair consisting of a pair (key_i, d_i). At the end of the execution, each party i obtains an output with one of two forms: (1) (rep, d^*), where d^* is the output of the aggregation function applied to $\{d_j : \mathsf{key}_j = \mathsf{key}_i\}$, or (2) $(\bot, \bot)$. Moreover, for every key which appears among the $\{\mathsf{key}_i\}_{i \in [m]}$, there exists exactly one party i with $\mathsf{key}_i = \mathsf{key}$ receiving an output of type (1). We refer to each such party as the *representative* for key_i.

An aggregation protocol with fixed communication patterns, called OblivAgg, is given in [2]. When the bit length of the data items and of the key values is at most ℓ bits, the protocol from [2] proceeds in $O(\log m)$ rounds, and in each round, every client sends $O(1)$ messages of size $O(\log m + \ell)$ bits.

Oblivious Routing. Another protocol we will use is the Oblivious Routing protocol OblivRoute from [2]. This m-party sub-protocol allows each party to send a message to another party; since the communication patterns are fixed, the recipients of the messages are hidden from an observer.

Protocol OblivRoute:

- Input of party i: (id_i, m_i) where m_i is the message of client i and id_i is the index of the recipient of the messages.
- Output of party i: $\{(\mathsf{id}_j, m_j) \mid \mathsf{id}_j = i\}$ the set of messages sent to party i.

We note that the implementation of OblivRoute is tailored at the case where each id_i is drawn independently and uniformly at random from $[m]$. (And this will be the case of our application below.) For a parameter $K \geq 0$, their protocol proceeds in $O(\log m)$ rounds, and in every round, a client sends a message of size $O(K \cdot (\ell + \log m))$ bits to another client, where ℓ is the size of the inputs. Then, the probability that the protocol aborts is roughly $O(m \log m 2^{-K})$, and thus one can set $K = \omega(\log \lambda)$ for this probability to be negligible in λ, or $K = \omega(\log N)$ in our ORAM applications where N becomes the security parameter.

Oblivious Election. We will need a variant of the above OblivAgg protocol with stronger guarantees. In particular, we need a protocol OblivElect that allows

m parties with requests $\{(\mathsf{op}_i, \mathsf{a}_i)\}_{i \in [m]}$ to elect a unique representative party for each unique address that appears among the m requests. This representative will be the party with the smallest identity $i \in [m]$ wanting to write to that address (if it exists), or otherwise the one with the smallest identity wanting to read from it. Formally, the protocol provides the following interface.

Protocol OblivElect:

- Input of party i: $(\mathsf{op}_i, \mathsf{a}_i)$, where $\mathsf{op}_i \in \{\mathsf{read}, \mathsf{write}\}$ and $\mathsf{a}_i \in [N]$.
- Output of party i: a value $o_i = \{\mathsf{rep}, \bot\}$, which is defined as follows. For each address a, define $S_a = \{i \mid \mathsf{a}_i = a\}$ and $W_a = \{i \mid \mathsf{a}_i = a \wedge \mathsf{op}_i = \mathsf{write}\}$, and let $i^*(a) = \min(W_a)$ if W_a is non-empty, or $i^*(a) = \min(S_a)$ otherwise. Then, we let $o_i = \mathsf{rep}$ if and only if $i = i^*(\mathsf{a}_i)$, and $o_i = \bot$ otherwise.

OblivElect can be implemented by modifying OblivAgg. At the high level, OblivAgg proceeds as follows (we refer to [2] for further details):

- Initially, every client i inputs a pair $(\mathsf{key}_i, \mathsf{d}_i)$, and these inputs are re-shuffled across clients and sorted according to the first component. That is, at the end of the first phase, any two clients $j < j'$ are going to hold a triple $(i(j), \mathsf{key}_{i(j)}, \mathsf{d}_{i(j)})$ and $(i(j'), \mathsf{key}_{i(j')}, \mathsf{d}_{i(j')})$, respectively, such that $\mathsf{key}_{i(j)} \leq \mathsf{key}_{i(j')}$ and $i(j) \neq i(j')$. This is achieved via a sorting network, where each client i initially holds $(i, \mathsf{key}_i, \mathsf{d}_i)$, and then such triples are swapped between pairs of clients (defined by the sorting network), according the key values.
- This guarantees that for every key which was initially input by $m' \geq 1$ clients, at the end of the first phase there exist m' consecutive clients $j, j + 1, \ldots, j + m' - 1$ (for some j) holding triples with $\mathsf{key}_{i(j)} = \cdots = \mathsf{key}_{i(j+m'-1)} = \mathsf{key}$. Then, client j is going to aggregate $\mathsf{d}_{i(j)}, \ldots, \mathsf{d}_{i(j+m'-1)}$, and the final representative for key is client $i(j)$. The aggregate information is sent back to the representatives by using once again a sorting network, sorting with respect to the $i(j)$'s.

We can easily modify OblivAgg to achieve OblivElect as follows. We run OblivAgg with client i inputting $\mathsf{key}_i = \mathsf{a}_i$ and $\mathsf{d}_i = (\mathsf{op}_i, i)$. However, the sorting network is not going to sort *solely* according to the key value, but *also* according to the associated d entry. In particular, we say that $(\mathsf{a}, \mathsf{op}, i) < (\mathsf{a}', \mathsf{op}', i')$ iff (1) $\mathsf{a} < \mathsf{a}'$, or (2) $\mathsf{a} = \mathsf{a}'$, $\mathsf{op} = \mathsf{write}$ and $\mathsf{op}' = \mathsf{read}$, or (3) $\mathsf{a} = \mathsf{a}'$, $\mathsf{op} = \mathsf{op}'$, and $i < i'$. The sorting now will ensure that the left-most client j holding a value for some $\mathsf{key} = \mathsf{a}$ will be such that $i(j)$ is our intended representative.

The complexity of OblivElect is the same as that of OblivAgg, setting $\ell = O(\log m + \log N)$. Thus we have $O(\log m)$ rounds, where each client sends $O(1)$ messages of size $O(\log m + \log N)$ bits.

Oblivious Multicasting. The oblivious multicast protocol OblivMCast is a m-party subprotocol that allows a subset of the parties, called the senders, to multicast values to others, called the receivers. More precisely:

Protocol OblivMCast:

– **Input of party i:** Input is either $(a_i, v_i \neq \perp)$ (where $a_i \in [N]$) indicating that party i is a sender with value v_i indexed by address a_i, or $(a_i, \perp)$ indicating that it is a receiver fetching the value indexed by a_i. For every possible a, there is at most one party with $a_i = a$ and $v_i \neq \perp$.

– **Output of party i:** If party i is a sender, its output is v_i. If party i is a receiver, its output is v_j, the value sent by party j with index $a_j = a_i$.

The protocol is in essence the reversal of our OblivElect protocol above. It can be built using similar techniques, achieving round complexity $O(\log m)$, and every client sends in each round $O(1)$ messages of size $O(B + \log N + \log m)$ bits, where B is the bit size of the values v_i.

3.3   Subtree-OPRAM

Non-Recursive Subtree-OPRAM. We first describe the non-recursive version of Subtree-OPRAM, where multiple clients share access to a global position map, which can be eliminated using recursive techniques as we explain further below. (Due to the constraints of coordinating access to the same items in OPRAM, our recursive techniques are somewhat more involved than in the basic ORAM case.)

Let m be the number of clients; assume for simplicity that it is a power of 2, i.e., $\log(m)$ is an integer. The Subtree-OPRAM protocol $\mathsf{ST\text{-}\mathcal{PO}} = \{\mathcal{O}_i\}_{i \in [m]}$, on common input (λ, N, B, m), organizes the server storage as a forest of m complete binary trees $\mathcal{T}_1, \ldots, \mathcal{T}_m$, each of depth $\log N - \log(m)$, where every node in each tree is a bucket of $Z = O(1)$ blocks of B bits. In other words, the union of $\mathcal{T}_i$ is the complete tree $\mathcal{T}$ in Subtree-ORAM, but with the top $\log(m)$ levels removed. Again, we identify paths with leaves in the tree, and we say that a path ℓ "belongs to" $\mathcal{T}_i$, if the leaf ℓ is in $\mathcal{T}_i$. Each client $\mathcal{O}_i$ is responsible for managing the portion of the storage space corresponding to $\mathcal{T}_i$, meaning that it reads/writes all paths belonging to $\mathcal{T}_i$, and maintains a local stash stash_i for storing all "overflowing" blocks whose assigned path belongs to $\mathcal{T}_i$. The Subtree-ORAM analysis will carry over, and imply that the size of each local stash is bounded by any function $R(\lambda, m) \in \omega(\log \lambda) + O(\log m)$, where the extra $O(\log m)$ is to store blocks that in the original Subtree-ORAM scheme would have belonged to the upper $\log(m)$ levels. The clients also share a global size-N position map $\mathsf{pos.map}$. (Recall that we are looking at the non-recursive version here.)

Recall that the m clients share a secret key for a semantically secure encryption scheme. In each iteration, each client i processes a logical access request $\mathsf{Acc}(op_i, addr_i, v_i)$. The m clients then proceed in parallel to process jointly the m logical requests from this iteration:

1. **Select block representatives.** The m clients run sub-protocol OblivElect, where client i uses input $(\mathsf{op}_i, \mathsf{a}_i)$ and receives either output rep or $\perp$; in the former case client i knows it is the *representative* for accessing block a_i.[5]

2. **Forward read-path requests.** Each client i determines the path ℓ_i it wants to fetch, and there are two possibilities:
   - **Real read.** If it is a representative, set path $\ell_i = \mathsf{pos.map}(\mathsf{a}_i)$ and $\mathsf{a}'_i = \mathsf{a}_i$, and immediately refresh $\mathsf{pos.map}(\mathsf{a}_i)$ to $\ell'_i \xleftarrow{\$} [N]$;
   - **Fake read.** If it is not a representative for a_i choose a random path $\ell_i \xleftarrow{\$} [N]$ and set $\mathsf{a}'_i = \perp$.

   If path ℓ_i belongs to tree $\mathcal{T}_j$, client i sends an encrypted message $(i, \mathsf{a}'_i, \ell_i)$ to client j.

3. **Read paths.** Each client $j \in [m]$ retrieves collects a set S_j of all paths contained in the messages $\{(i, \mathsf{a}'_i, \ell_i)\}$ received in the previous step, and then proceeds as follows:
   (1) Retrieve all paths in S_j, which form a subtree denoted $\mathcal{T}_{S_j}$.
   (2) For each $i \in [m]$ such that a request $(i, \perp, \ell_i)$ was received, send the encryption of a dummy block $\perp$ to client i
   (3) For each $i \in [m]$ such that a request $(i, \mathsf{a}'_i \neq \perp, \ell_i)$ was received, find block a'_i in $\mathcal{T}_{S_j}$ or in stash, *delete it*, and send the encryption of the value $\overline{v}_i$ of the block to client i.

4. **Answer client requests and update.** At the end of the previous step, each client holds a value $\overline{v}_i$ which is $\neq \perp$ if and only if i is the representative for a_i. Next, the m clients run sub-protocol OblivMCast to allow each representative to multicast the value it holds to other clients requesting the same block: Concretely, each client i uses input $(\mathsf{a}_i, \overline{v}_i)$ (recall a non-representative has $\overline{v}_i = \perp$) and receives output $\overline{v}'_i$, which is guaranteed to be the value of block a_i it requests. Each client i answers its logical request with $\overline{v}'_i$.
   Next, each representative i that has a write operation $\mathsf{Acc}(\mathsf{write}, \mathsf{a}_i, v_i)$ locally updates the value of block a_i to $\overline{v}_i = v_i$.

5. **Re-route blocks with newly assigned paths.** Each representative i send its block $(\mathsf{a}_i, \overline{v}_i)$ to the appropriate client for insertion according to the newly assigned path ℓ'_i (Step 1) as follows: Let j_i be the tree that path ℓ'_i belongs to; the m clients run sub-protocol OblivRoute where each representative i uses input $(j_i, (\ell'_i, \mathsf{a}_i, \overline{v}_i))$, and other clients use input $(j_i, \perp)$ for a randomly drawn $j_i \xleftarrow{\$} [m]$. [6]
   As the output of OblivRoute, each client j receives a set of blocks $\{(\ell'_i, \mathsf{a}_i, \overline{v}_i)\}$ whose path ℓ'_i belong to $\mathcal{T}_j$; it stores each $(\mathsf{a}_i, \overline{v}_i)$ in its local stash stash_j.

6. **Flush subtree and write-back.** For each client j, let $\mathcal{T}_{\mathrm{real}_j}$ be the subtree consisting of only real-read paths in $\mathcal{T}_{S_j}$. Before writing subtree $\mathcal{T}_{S_j}$ back to the server (re-encrypting all of its contents), client j runs the Subtree

---

[5] Note that the representatives are chosen consistently with how repetition is removed in Subtree-ORAM.

[6] Note that the destination addresses of OblivRoute here are all uniformly chosen, and thus we can use the implementation from [2].

Flushing Procedure on $\mathcal{T}_{\mathrm{real}_j}$ and stash_j (recall that if at any point, stash_j contains more than R blocks, the procedure output overflow).

Recursive Version. We can apply recursion to eliminate the use of the shared global position map in the above scheme. Observe that in each iteration, each client read/write the position map at most once in Step 2. In other words, the m clients, in order to answer a batch of m accesses, one per client, to a logical space of size $N \times B$ bits, clients need to first make a batch of at most m accesses, one per client, to the position map of size $N \times \log N$ bits. Since $B \geq \alpha \log N$ for some constant $\alpha > 1$ (for simplicity, N is a power of two), by recursively storing the shared position map to the server in $O(\log N)$ trees, the clients no longer need to share any position map. At the end of recursion, the size of the position map decreases to $O(1)$ and can be stored in the local memory of say, the first client. Other clients can access and update this position map using oblivious sub-protocols OblivAgg and OblivMCast.

This high-level strategy goes through almost identically as in Path-ORAM, except from the following caveat. Recall that in Step 2 of Subtree-OPRAM, if a client i is a representative, then it reads entry $\ell_i = \mathsf{pos.map}(a_i)$ of the position map and updates it to a new random address ℓ_i', and otherwise, it does not access the position map. Since $B \geq \alpha \log N$, the entire position map fits into a logical space of N/α blocks, where the block with address $\tilde{a}$ contains α position map entries, $\mathsf{pos.map}(\alpha\tilde{a}+1)\|\cdots\|\mathsf{pos.map}(\alpha(\tilde{a}+1))$. This means, when applying recursion and storing the position map at the server, client i needs to make the following logical access:

$$\mathsf{Acc}(\widetilde{\mathsf{op}}_i, \tilde{a}_i, \tilde{v}_i) = \begin{cases} \mathsf{Acc}(\mathsf{write}, \lfloor a_i/\alpha \rfloor, \ell_i') & \text{if } i \text{ is a representative} \\ \mathsf{Acc}(\mathsf{read}, 0, \bot) & \text{otherwise} \end{cases}$$

We assume without loss of generality above that clients who are not representatives simply make a read access to the block with address 0. By construction, different representatives i and j access different entries in the position map $a_i \neq a_j$. However, it is possible that two representatives i and j need to access the same logical address $\tilde{a} = \tilde{a}_i = \tilde{a}_j$, in order to update different entries of position map located in the same block $\tilde{a}$—call this a *write-collision*; since each block contains at most α position map entries, there are at most α write collisions for the same logical address. Recall that in Subtree-OPRAM, when multiple clients write to the same logical address, only the write operation with the smallest index is executed. Hence, naively applying recursion on Subtree-OPRAM means when write-collision occurs, only one position map entry would be updated.

This problem can be addressed by slightly modifying the interface of Subtree-OPRAM, so that, under the constraint that there are at most α writes to different parts of the same block, all writes are executed. In recursion, the modified scheme is invoked, to ensure that position maps are updated correctly, whereas at the top level, the original Subtree-OPRAM is used. To accommodate α write collisions, the only change appears in Step 1: In Subtree-OPRAM, the sub-protocol OblivElect is used, which ensures that for each address a, only the minimal indexed write is executed. We now modify this step to run the sub-protocol OblivAgg (with appropriate key, data and aggregate function specified shortly), so that, a unique representative is elected for each a, who receives all the write requests to that a, and executing all of them (note that while the write request are for the same block, they will concern different portions of the block corresponding to distinct position map entries, and thus "executing all of them" has a well-defined meaning):

1. **Select block representatives, modified.** The m clients run sub-protocol OblivAgg, where client i uses input $(\text{key}_i = a_i, d_i = v_i)$, and aggregate function $\text{agg}(d_1, d_2, \cdots) = d_1 || d_2, \cdots = V$. OblivAgg ensures that for each address a_i, a unique client j accessing that address a_i receives output (rep, V_i), and all other clients receive output $(\perp, \perp)$. In the former case, client j knows it is the *representative* for accessing block a_i, and V_i determines the new value of the block v_i.

The rest of the protocol proceeds identically as before. Since there are at most α write collision for each address, the length of the output of agg is bounded by $\ell = \alpha B$. Thus the protocol proceeds in $O(\log m)$ rounds, where in each round every client sends $O(1)$ messages of size $O(\log N + \log m + B)$ bits.

4 Generic OPRAM Scheme

In this section, we generalize the ideas from Subtree-OPRAM to obtain a generic transformation transforming an arbitrary single-client ORAM to an OPRAM scheme, incurring only in a $O(\log N)$ factor of efficiency loss. Overall, we are going to prove the following general theorem.

Theorem 2 (Generic-OPRAM). *There exists a generic transformation that turns any ORAM scheme $\mathcal{O}$ into an m-client OPRAM scheme Generic-OPRAM such that, for any $R = \omega(\log \lambda)$, the following are satisfied, as long as the block length satisfied $B \geq 2\log m$, and moreover $N/m \geq R$:*

- *Server Communication Overhead. The amortized communication overhead is $O(\log N \cdot \alpha(N/m))$ and the worst-case communication overhead is $O((\log N + \omega(\log \lambda)) \cdot \alpha(N/m))$, where $\alpha(N')$ is the communication overhead of ORAM scheme $\mathcal{O}$ with logical address space $[N']$.*
- *Inter-Client Communication Overhead. The amortized and worst-case overheads are both $\omega(\log \lambda) \log m(\log m + \log N)$ with overwhelming probability.*
- *Server and Client Storage. The sever stores $O(m \cdot M(N/m))$ blocks, where $M(N')$ is the number of blocks stored by $\mathcal{O}$ for logical address space N'. Moreover, the client's local storage overhead is $R + \text{polylog}(N)$.*

Our presentation will avoid taking the detour of introducing a single-client ORAM scheme allowing for parallel processing of batches of m access operations, as we have done above with Subtree-OPRAM. A direct description of Generic-OPRAM is conceptually simpler. Before we turn to discussing Generic-OPRAM, however, we discuss a basic building block behind our protocol.

4.1   Oblivious Dictionaries

In our construction below, every client will be responsible for a partition holding roughly N/m blocks. One of the challenges is to store these blocks obliviously using space which is roughly equivalent to that of storing N/m blocks. Ideally, we want to implement this using an ORAM with logical address space for N/m blocks, as this would result in constant storage overhead when the ORAM has also constant overhead. In particular, the elements assigned to a certain partition have addresses spread around the whole of $[N]$, and we have to map them efficiently to be stored into some block in $[N/m]$ in a way which is (a) storage efficient for the client, and (b) only requires accessing a small (i.e., constant) number of blocks to fetch or insert a new block. We going to solve this via an oblivious data structure implementing a dictionary interface and able to store roughly N/m blocks into a not-much-larger amount of memory.

The Data Structure. We want an oblivious implementation $\mathcal{OD}$ of a dictionary data structure holding at most n pairs (a, v), where v corresponds to a data block in our ORAM scheme, and $\mathsf{a} \in [N]$. (For our purposes, think of $n \approx N/m$.) At any point in time, $\mathcal{OD}$ stores at most one pair (a, v) for every a. It allows us to perform two operations:

- $\mathcal{OD}(\mathcal{I}, \mathsf{a}, v)$ inserts an item (a, v), where $\mathsf{a} \in [N]$, if the data structure contains less than n elements. Otherwise, if n elements are stored, it does not add an element, and returns an error symbol $\bot$.
- $\mathcal{OD}(\mathcal{R}\&\mathcal{D}, \mathsf{a})$ retrieves and deletes an item (a, v) stored in the data structure (if it exists), returning v, and otherwise returns an error $\bot$ if the element is not contained.

Moreover, $\mathcal{OD}$ enables two additional "dummy" operations $\mathcal{OD}(\mathcal{R}\&\mathcal{D}, \bot)$ and $\mathcal{OD}(\mathcal{I}, \bot, \bot)$ which are meant to have no effect on the data structure. Informally, for security, we demant that the access patterns resulting from any two equally long sequences of operations of type $\mathcal{OD}(\mathcal{I}, *, *)$ and $\mathcal{OD}(\mathcal{R}\&\mathcal{D}, *)$ are (statistically) indistinguishable.[7]

The Implementation. We can easily obtain the above $\mathcal{OD}$ data structure using for instance any Cuckoo-hashing based dictionary data structure with constant worst-case access complexity.[8]

Theorem 3 (Efficient Cuckoo-Hashing Based Dictionary [10]). *There exists an implementation of a dictionary data structure holding at most n blocks with the following properties: (1) It stores $n' = O(n)$ blocks in the memory. (2) Every insert, delete, and lookup operation, requires $c = O(1)$ accesses to blocks in memory. (3) The client stores* polylog(n) *blocks in local memory.*

---

[7] In fact, for our purposes, we could leak which operations are of which type, but it will be easy enough to achieve this even stronger notion.

[8] We think of a data structure as being in a simliar model as our ORAM scheme, namely consisting of a client interface, using a small amount of local memory, and the actual data being stored externally on the server.

(4) The failure probability is negligible (in n) for any poly(n)-long sequence of lookups, insertions, and deletions which guarantees that at most n elements are ever stored in the data structure.

From any ORAM scheme $\mathcal{O}$ with address space n', it is easy to implement the oblivious data-structure $\mathcal{OD}$: The client simply implements the dictionary data structure from Theorem 3 on top of the ORAM's logical address space, and uses additional polylog(n) local memory for managing this data structure. Dummy accesses can be performed by simply issuing c arbitrary read requests to the ORAM storage. We omit a formal analysis of this construction, which is immediate.

4.2   The Generic OPRAM Protocol

We finally show how to obtain our main generic construction of an oblivious parallel RAM: The server storage consists of m partitions, and the i-th client manages the i-th partition. In particular, client i runs the oblivious dictionary scheme $\mathcal{OD}$ presented above (we refer to its interface as $\mathcal{OD}_i$) on the i-th partition. Here, we assume that the clients have access to the partition map, mapping each address $\mathsf{a} \in [N]$ to some partition partition[a]. (We will discuss in the analysis how to eliminate this sharing using recursion.) Besides, Generic-OPRAM further takes care of the communication among clients using the algorithms OblivElect, OblivMCast, OblivRoute from Sect. 3.2.

We postpone a complexity, correctness, and security analysis to Appendix E, as well as a discussion of the recursion version.

Data Structures. The non-recursive version of Generic-OPRAM keeps a *partition map* with N entries that maps block addresses a to their currently assigned partition partition[a], and that can be accessed by all clients obviously (i.e., access to the partition map are secret). Every client additionally keeps a *stash* SS_i which contains at most R items to be inserted into $\mathcal{OD}_i$. For our analysis to work out, we need $R = \omega(\log \lambda)$. Also let $\kappa \geq 2$ be a constant.

Generic OPRAM Protocol. In each iteration, given the logical access requests $(\mathsf{Acc}(op_i, \mathrm{addr}_i, v_i))_{i \in [m]}$ input to the client, the m clients go through the following steps (all messages are tacitly encrypted with fresh random coins):

1. **Select block representatives.** Run OblivElect between clients with inputs $(\mathsf{a}_i, \mathrm{op}_i)_{i \in [m]}$. In the end, each client i knows whether it has been selected as the representative to get the block value a_i, or not.
2. **Query blocks.** Clients do one of two things:
   - **Real requests.** Each representative client i gets the partition index $p_i = $ partition[a_i], and sends a request a_i to client p_i. Moreover, it reassigns partition[a_i] $\overset{\$}{\leftarrow} [m]$.
   - **Fake requests.** Every non-representative client i generates a random $q_i \overset{\$}{\leftarrow} [m]$ and sends a request $\bot$ to client q_i.

3. **Retrieve the blocks.** Each client $p \in [m]$ processes the received requests according to some random ordering: For each request $\mathsf{a}_i \neq \perp$ received from client i, client p executes $\mathcal{OD}_p(\mathcal{R\&D}, \mathsf{a}_i)$ and denote the retrieved block value $\overline{v}_i$. If $\overline{v}_i = \perp$, then there must be some entry $(\mathsf{a}_i, \overline{v}'_i)$ in the SS_p. Then, client p deletes this entry, and sets $\overline{v}_i = \overline{v}'_i$. Finally, it sends $\overline{v}_i$ back to i. For every $\perp$ request received from some client i, client p executes the fake read access $\mathcal{OD}_p(\mathcal{R\&D}, \perp)$, and returns $\overline{v}_i = \perp$ to i.

4. **Representatives inform.** At the end of the previous step, each client holds a value $\overline{v}_i$ which is $\neq \perp$ if and only if i is the representative for a_i. Next, the m clients run sub-protocol OblivMCast to allow each representative to multicast the value it holds to other clients requesting the same block: Concretely, each client i uses input $(\mathsf{a}_i, \overline{v}_i)$ (recall a non-representative has $\overline{v}_i = \perp$) and receives output $\overline{v}'_i$, which is guaranteed to be the value of block a_i it requests. Each client i answers its logical request with $\overline{v}'_i$.

5. **Send updated values.** For each representative i such that $\mathsf{Acc}(op_i, \mathsf{a}_i, v_i)$ is a *write* command, let $\mathsf{id}_i = \mathsf{partition}[\mathsf{a}_i]$ and $\mathsf{msg}_i = (\mathsf{a}_i, v_i)$. Otherwise, if it is *not* a write command (but still, i a representative), it sets $\mathsf{msg}_i = (\mathsf{a}_i, \overline{v}_i)$ instead. Non-representative clients set $\mathsf{msg}_i = \perp$ and $\mathsf{id}_i \xleftarrow{\$} [m]$. Then, the clients run OblivRoute with respective inputs $(\mathsf{id}_i, \mathsf{msg}_i)$.

6. **Write back.** Each client $p \in [m]$ adds all pairs (a, v) received through OblivRoute to SS_p. Then, client p picks the first κ elements from SS_p, and for each such element (a, v), executes $\mathcal{OD}_i(\mathcal{I}, \mathsf{a}_i, v)$. If $\kappa' < \kappa$ elements are in SS_i, then the last $\kappa - \kappa'$ insertions are dummy insertions $\mathcal{OD}_i(\mathcal{I}, \perp, \perp)$. Anytime when stash SS_i needs to store more than R blocks or the partition holds more then $2N/m + R$ blocks, output "overflow" and halt.

Acknowledgments. The authors wish to thank Elette Boyle, Kai-Min Chung, and Mariana Raykova for insightful discussions.

Binyi Chen was partially supported by NSF grants CNS-1423566 and CNS-1514526, and a gift from the Gareatis Foundation. Huijia Lin was partially supported by NSF grants CNS-1528178 and CNS-1514526. Stefano Tessaro was partially supported by NSF grants CNS-1423566, CNS-1528178, and the Glen and Susanne Culler Chair. This work was done in part while the authors were visiting the Simons Institute for the Theory of Computing, supported by the Simons Foundation and by the DIMACS/Simons Collaboration in Cryptography through NSF grant CNS-1523467.

A   Correctness and Obliviousness of ORAM

For an access sequence $\boldsymbol{y}$ we let $\mathsf{AP}_i = \mathsf{AP}_i(\boldsymbol{y})$ be the access pattern of its i-th operation – i.e., the sequence of pairs $(\overline{op}_{i,1}, \overline{a}_{i,1}), \dots, (\overline{op}_{i,q_i}, \overline{a}_{i,q_i})$ describing the client's server accesses (*without* the actual values) when processing the i-th operation – and denote by val_i the answer of this operation. Then, we let

$$\mathsf{Out}_{\mathcal{O}}(\lambda, N, B, \boldsymbol{y}) = (\mathsf{val}_1, \mathsf{val}_2, \dots, \mathsf{val}_T) \,, \quad \mathsf{AP}_{\mathcal{O}}(\lambda, N, B, \boldsymbol{y}) = (\mathsf{AP}_1, \dots, \mathsf{AP}_T).$$

We say that the sequence of outputs $\boldsymbol{z} = \mathsf{Out}_{\mathcal{O}}(\lambda, N, B, \boldsymbol{y})$ of $\mathcal{O}$ is correct w.r.t. the sequence of logical accesses $\boldsymbol{y}$, if for each logical command $\mathsf{Acc}(op_i, \mathsf{a}_i, v_i)$ in

y, the corresponding output val_i in z is *either* the most recently written value on address a_i, *or* $\perp$ if a_i has not yet been written to. Let Correct be the predicate that on input (y, z) returns whether z is correct w.r.t. y.

Definition 2 (ORAM Correctness and Security). *An ORAM $\mathcal{O}$ achieves correctness and obliviousness if for all $N, T, B = poly(\lambda)$, there exists a negligible function μ, such that, for every λ, every two sequences y and y' of $T(\lambda)$ access operations, the following are satisfied:*

1. **Correctness:** $\Pr[\text{Correct}(y, \text{Out}_{\mathcal{O}}(\lambda, N, B, y)) = 1] \geq 1 - \mu(\lambda)$.
2. **Obliviousness:** $\Delta(\text{AP}_{\mathcal{O}}(\lambda, N, B, y), \text{AP}_{\mathcal{O}}(\lambda, N, B, y')) \leq \mu(\lambda)$.

We note that the above definition considers *statistical* obliviousness. This is generally achieved by tree-based ORAM schemes, but it can be relaxed to computational obliviousness, where the statistical distance is replaced by the best distinguishing advantage of a PPT distinguisher.

B   Review of Path-ORAM

In this section, we review the Path-ORAM scheme in detail, as it is used as a starting point for Subtree-ORAM and Subtree-OPRAM.

Overview. Path-ORAM is a tree-based ORAM that works for the single client setting. To implement a logical storage space for N data blocks, Path-ORAM organizes the storage space (virtually) as a complete binary tree with depth $D = \lceil \log N \rceil$. Each node of the tree is a bucket capable of storing Z blocks of size B (bits). Here Z is a constant, thus the server storage overhead is $O(1)$. To hide the logical access patterns, each data block a is assigned to a random path ℓ from root to leaf in the tree and stored at some node of the path; in order to hide the repetitive accesses to the same block, the assignment is updated to a new independent random path after each access.

In [26], Path-ORAM is constructed in two steps; first, a non-recursive version is proposed and analyzed, in which the client keeps a local position map with $N \log N$ bits; then the position map is recursively outsourced to the server, reducing the client storage to only $polylog(N)$ bits. Below we describe the non-recursive version first, and then show how to apply the recursive transformation.

Non-Recursive Version. The client maintains a *position map* that maps each block a to a path pos.map(a). Since each path can be specified using D (the depth of the tree) bits, the size of the position map is $ND = N\lceil \log N \rceil$ bits. Additionally, the client keeps a small local storage *stash* used for storing blocks that do not fit in the assigned path (due to limited space at each tree node). The capacity of the stash is bounded by $R = R(\lambda)$ for any function $R(\lambda) = \omega(\log \lambda)$, except with negligible probability in λ.

Given the i-th logical access Acc(op_i, a_i, v_i), Path-ORAM proceeds in two phases:

- **Phase 1: Processing the query.** Path-ORAM retrieves the path $\ell_i = $ pos.map(a_i) assigned to block a_i, and finds the block a_i on the path or in the stash. After returning the block value and potentially updating the block, Path-ORAM re-assigns block a to a new independent random path $\ell' \xleftarrow{\$} [N]$ and updates pos.map(a_i) $= \ell'_i$. It then moves the block to the stash.
- **Phase 2: Flushing and write-back.** Before re-encrypting and writing the path back to the server, in order to avoid the stash from "overflowing", Path-ORAM re-arranges path ℓ_i, to fit as many blocks from the stash into the path. More specifically, for each block a_j in the stash and on the path, Path-ORAM places it at the lowest non-full node p_j that intersects with its assigned path $\ell_j = $ pos.map(a_j). If no such node is found, the block remains in the stash. If at any point, the stash contains more than R blocks, Path-ORAM outputs "overflow" and aborts.

Recursive Version. In the above non-recursive version, the client keeps a large $N \log N$-bit position map. To reduce the client storage, Path-ORAM recursively outsources the position map to the server by adding extra $O(\log N)$ trees. More specifically, if the cell size $B \geq \alpha \lceil \log N \rceil$ for some integer $\alpha > 1$, the position map can be stored in $\lceil \frac{N}{\alpha} \rceil$ cells. This means, to answer an access to a logical storage space of size N, the non-recursive version only needs to make a query to another logical storage space (i.e. the position map) of size $\lceil \frac{N}{\alpha} \rceil$. Therefore, if the client further outsources the position map to the server, its local storage would be reduced to $\lceil \frac{N}{\alpha^2} \rceil$. This idea can be applied recursively until the client storage becomes polylog(N). In the final scheme, at the server, besides tree T_0 that stores data blocks, there are additional trees $rT_1, rT_2, ..., rT_l$ for position map queries, where $l = \lceil \log_\alpha N \rceil$. Tree rT_i has size $\tilde{O}(\lceil \frac{N}{\alpha^i} \rceil B)$ bits, and maintains the position map corresponding to tree rT_{i-1} which contains $\lceil \frac{N}{\alpha^{i-1}} \rceil$ cells. The position map corresponding to tree rT_l is stored in local storage. Now, to access a block a, the client needs to query pos.map(a) in T_0 by looking up the position in tree rT_1. In order to query the position map corresponding to tree rT_1, similarly, the client looks up in rT_2, so on and so forth. Finally the position map value of tree rT_l is stored in local storage.

Complexity. The storage overhead is $O(1)$ both in the non-recursive version and the recursive version. In the non-recursive version, for each logical access, Path-ORAM reads and writes a path with $\log N$ nodes, each of which contains Z cells, therefore the communication overhead is $O(\log N)$ per access. The computation overhead is $O(\log^2 N)$, since the flushing procedure takes time $O(\log^2 N)$ per access. After the recursive transformation is applied, to answer each logical access, the client needs to query $l = \lceil \log_\alpha N \rceil$ number of trees, and hence the communication/computation overhead blow by a factor of $\log N$.

C   Analysis of Subtree-ORAM

In the full version, we show that the overflow probability of Subtree-ORAM is negligible given any sequence of logical access requests. In particular, we prove the following proposition, which generalizes the analysis of Path-ORAM.

Proposition 1. *Fix the stash size to any $R(\lambda) \in \omega(\log \lambda)$. For every polynomial m, N, T, B, there exists a negligible function μ, such that, for every λ, and sequence $\mathbf{y}$ of T batches of m access requests, the probability that Subtree-ORAM outputs* overflow *is at most $\mu(\lambda)$.*

From Proposition 1, it is easy to show that Subtree-ORAM satisfies correctness and obliviousness.

- CORRECTNESS: Since the stash overflows with negligible probability, and Subtree-ORAM maintains the *block-path invariance* (as in Path-ORAM) – at any moment, each block can be found either on the path currently assigned to it or in the stash; by construction, Subtree-ORAM answers logical accesses correctly according to the correctness condition of ORAM.
- OBLIVIOUSNESS: Conditioned on no overflowing: (1) In each iteration, Subtree-ORAM always reads m independent and random paths from the server. (2) After each iteration, every requested block is assigned to a new random path, which is hidden from the adversary (as in Path-ORAM). Thus the construction is oblivious.

D   Analysis of Subtree-OPRAM

In this section, we give a high-level overview of why Subtree-OPRAM is correct and satisfies obliviousness. Also we discuss below the complexity of the protocol.

We discuss correctness and obliviousness for the non-recursive version only. The same properties are then also easily shown to be true for the recursive version. The first key observation is that the m clients of Subtree-OPRAM can be seen as collectively emulating the operations of the single client of Subtree-ORAM. Compare an execution of Subtree-OPRAM with a sequence $\mathbf{y}$ of T batches of m parallel logical accesses with the execution of Subtree-ORAM with the same sequence.[9] Then, we observe the following:

1. The ORAM tree T of Subtree-ORAM is stored in parts in Subtree-OPRAM: All but the top $\log m$ levels is stored as the m-tree forest $T_1, \cdots, T_m$ at the server, while the top $\log m$ levels are stored in a distributed way by the individual clients in their respective stashes — namely, if a block is stored at client j, its assigned path belongs to T_j. Therefore, the union of $\{\mathsf{stash}_i\}$ contains the same blocks as the stash of Subtree-ORAM, as well as all blocks in the top of the tree T of Subtree-ORAM.

---

[9] We are being somewhat informal here – one would have to define precisely what it means to "compare" in terms of executing both protocols with the same random choices. As it is somewhat tedious, we keep this on a more informal high level, hoping to convey the main ideas.

2. Subtree-OPRAM answers a batch of m requests (in each iteration) as Subtree-ORAM does: The m clients of Subtree-OPRAM choose a representative for each requested block (in Step 1) with the exactly same rule Subtree-ORAM uses to remove repetitive accesses, and to only keep one access per block. Later (in Step 4), Subtree-OPRAM first answers requests using the most recently written value from previous iterations and then executes the write operations; in particular, due to the way representatives are chosen, the write operation with the minimal index always takes effect.

3. Subtree-OPRAM maintains the *block-path invariant* as Subtree-ORAM. This is because each time a block is assigned to a new path ℓ', it is sent (using the OblivRoute sub-protocol) to client j managing the tree $\mathcal{T}_j$ the path ℓ' belongs to. Therefore, at any moment, a block is either on its assigned path or in the local stash of the client responsible for the tree its assigned path belongs to.

4. Subtree-OPRAM emulates the flushing procedure of Subtree-ORAM: Recall that Subtree-ORAM flushes along the subtree $\mathcal{T}_{real}$ of paths assigned to all requested blocks. Removing the top $\log m$ levels of $\mathcal{T}_{real}$ gives a set of m subtrees $\mathcal{T}_{real_1}, \cdots, \mathcal{T}_{real_m}$. Note that $\mathcal{T}_{real_i}$ is exactly the subtree that client i in Subtree-OPRAM performs flushing on (in Step 6). Indeed, by the design of the subtree flushing procedure, blocks that land in different subtrees $\mathcal{T}_{real_i} \neq \mathcal{T}_{real_j}$ can be operated on independently. Moreover, blocks that would land in the top $\log m$ levels of $\mathcal{T}_{real}$ or stash in Subtree-ORAM are naturally divided into the m local stashes according to which tree $\mathcal{T}_j$ their assigned path belongs to.

Correctness and Stash Analysis. By the above, if we fix any sequence $\boldsymbol{y}$ of parallel accesses, and consider the executions of (non-recursive) Subtree-OPRAM and (non-recursive) Subtree-ORAM with the same input sequence $\boldsymbol{y}$, since Subtree-ORAM answers every request correctly as long as it does not overflow, so does Subtree-OPRAM.

To argue that Subtree-OPRAM only overflows with negligible probability, recall that by Proposition 1, when the stash size of Subtree-ORAM is set to any $R'(\lambda) \in \omega(\log(\lambda))$, the probability of overflowing is negligible. We can thus bound the size of each local stash stash_i in Subtree-OPRAM, using the bound on the stash size of Subtree-ORAM. As noted above, after each iteration, the local stash stash_i of client i stores two types of blocks:

1. Blocks in the stash of Subtree-ORAM with an assigned path belonging to $\mathcal{T}_i$, and

2. Blocks in the top $\log m$ levels of the ORAM tree $\mathcal{T}$ of Subtree-ORAM, again with an assigned path belonging to $\mathcal{T}_i$.

By Proposition 1, the number of blocks of the first type is bounded by $\omega(\log \lambda)$ with overwhelming probability. Moreover, it is easy to see that the number of blocks of the second type is bounded by $O(\log m)$. Therefore, the size of stash_i is bounded by any $R(\lambda, m) \in \omega(\log \lambda) + O(\log m)$ with overwhelming probability. This is summarized by the following lemma.

Lemma 1. *Fix the stash size to any $R(\lambda, m) \in \omega(\log \lambda) + O(\log m)$. For every polynomial m, N, T, B, there is a negligible function μ, such that, for every λ, and sequence $\boldsymbol{y}$ of $T(\lambda)$ accesses, the probability that any client of the non-recursive Subtree-OPRAM outputs* overflow *is at most $\mu(\lambda)$.*

Complexity. The storage overhead of Subtree-OPRAM is the same as that of Path-ORAM, which is $O(1)$. The only contents stored at each client are the stashes, one per recursion level. Since each stash is of size $R(\lambda, m)B$, and the recursion depth is bounded by $O(\log N)$, the total client storage overhead is $O(\log N)R(\lambda, m) \in \omega(\log N \log \lambda) + O(\log N \log m)$.

Next, we analyze the communication and computation overheads (per client per access) of the recursive Subtree-OPRAM. In each iteration, to process m logical accesses (one per client), the m clients first recursively look up the position maps for $O(\log N)$ times using the non-recursive Subtree-OPRAM, and then process their requests using again the non-recursive Subtree-OPRAM. Fix any client i, we analyze its communication and computation complexities as follows:

- SERVER COMMUNICATION OVERHEAD: In each invocation of non-recursive Subtree-OPRAM, client i reads/writes a subtree of paths delegated to it by other clients. Since these paths are all chosen at random, in expectation client i read/write only 1 path in each invocation. Furthermore, across all $O(\log N)$ invocations of non-recursive Subtree-OPRAM, the probability that client i is delegated to read/write $\omega(\log \lambda) + O(\log N)$ paths is negligible. (Consider tossing $O(\log N) \times m$ balls (read/write path requests) randomly into m bins (clients); the probability that any bin has more than $O(\log N) + \omega(\log \lambda)$ balls is negligible in λ.) Since each path contains $O(\log N)$ blocks, the server communication overhead is bounded by $\omega(\log \lambda \log N) + O(\log^2 N)$ in the worst case, with overwhelming probability.

- INTER-CLIENT COMMUNICATION OVERHEAD: In each invocation of the non-recursive Subtree-OPRAM protocol, client i communicates with other clients in two ways: (1) using the oblivious sub-protocols (Steps 1, 4 and 5) and (2) sending the requests for reading certain block and path $(i, \mathsf{a}'_i, \ell_i)$ (Step 2) and sending back the retrieved block (Step 3). The maximum communicating complexity of the oblivious sub-protocols is $O(K \log m(\log m + \log N + B))$ bits, where K is in $\omega(\log \lambda)$. Therefore, across $O(\log N)$ invocations of non-recursive Subtree-OPRAM, the first type of inter-client communication involves sending/receiving at most $O(\log N K \log m(\log m + \log N + B))$ bits. On the other hand, by a similar argument as above, across $O(\log N)$ recursive invocations, with overwhelming probability, each client receives at most $O(\log N) + \omega(\log \lambda)$ requests of form $(i, \mathsf{a}'_i, \ell_i)$, and hence the second type of communication involves sending/receiving $(\omega(\log \lambda) + O(\log N)) \times O(\log m + \log N + B)$ bits with overwhelming probability. Thus, in total, the inter-client communication is $\omega(\log \lambda) \log m \log N(\log m + \log N + B)$ bits. Since $B \geq \alpha \log N$ for an $\alpha > 1$, the inter-client communication overhead is $\omega(\log \lambda) \log m(\log m + \log N)$.

Finally, we observe that when considering the communication overhead averaged over a sufficiently large number T of parallel accesses, the server communication overhead is bounded by $O(\log^2 N)$ with overwhelming probability. The inter-client communication complexity stays the same.

Obliviousness. The obliviousness of recursive Subtree-OPRAM follows from that of the non-recursive version. Conditioned on that the stash does not over-flow, the latter follows from three observations: (i) In each iteration, the paths $\{\ell_i\}$ read/write from/to the server (in Steps 3 and 6) are all independent and random, (ii) the communication between different clients is either through one of the oblivious sub-protocols (in Steps 1, 4, and 5), which has fixed communication pattern, or depends on the random paths $\{\ell_i\}$ (in Steps 2 and 3), and (iii) the new assignment of paths $\{\ell_i'\}$ to blocks accessed are hidden using OblivRoute (in Step 5). Combining these observations, we conclude that the access and communication patterns of Subtree-OPRAM is oblivious of the logical access pattern.

E   Analysis of Generic-OPRAM

Recursive Version. The above protocol assumes that every client has (private) access to the partition map to be accessed and updated throughout the execution of the protocol. This is of course not realistic. But similar to the case of the position map in Subtree-OPRAM, we can use $O(\log N)$-deep recursion. For this to work, we need block size to be at least, say, $B = 2 \log m$, since each entry in the partition map can be represented by $\log m$ bits.

Complexity Analysis. We now analyze the complexity of the Generic-OPRAM. We assume that $\mathcal{OD}_i$ is implemented from some ORAM scheme $\mathcal{O}_i$ which has communication overhead $\alpha(N')$ when using address space N', and that the same scheme stores $M(N')$ blocks on the server for the same address space. We make some assumptions in the following that appear reasonable, namely that $\alpha(O(N')) = O(\alpha(N'))$ and $M(O(N')) = O(M(N'))$ (this is true because these functions are polynomial). Moreover, we can also assume also that $M(N'/c) \leq M(N')/c$ for any constant c. (Note that the scheme is meaningful without these assumptions, but the resulting complexity statement would be somewhat more cumbersome.)

– **Server and Client Storage.** Let us start with the non-recursive case. Note that each partition needs enough blocks to implement a dynamic data structure to store $2N/m + R$ blocks. This will require an ORAM for $N' = O(N/m + R)$ blocks, which thus requires $O(M(N/m + R))$ blocks. Thus, the overall server storage complexity is of $O(m \cdot M(N/m + R))$ blocks. If $M(N') = O(N')$, in particular this implies that the overall storage com-plexity is $O(N + mR)$, and thus linear if $m \cdot R \in O(N)$.
  For the recursive case, note that the storage space is going to at least halve after each recursion level by our assumption on M. So if we assume that $N/m > R$, we see that the storage complexity remains $O(m \cdot M(N/m + R))$. Every client needs to store R blocks, and moreover, it needs $\text{polylog}(N)$ mem-ory for implementing $\mathcal{OD}$ and the underlying ORAM scheme $\mathcal{O}$, which we assume to have $\text{polylog}(N)$ client storage overhead.

- **Server Communication.** In contrast to Subtree-OPRAM above, a generic construction does not necessarily allow us to parallelize accesses to the data structure. The number of $\mathcal{OD}_i(\mathcal{R\&D}, \cdot)$ operations a client performs can thus vary in each round, but we can apply the same analysis as for Subtree-OPRAM above. Namely, given we are using $\log N$ levels of recursion, the per-client server communication is $O(\log N \cdot \alpha(N/m + R))$ in the amortized case, and $O((\log N + \omega(\log \lambda)) \cdot \alpha(N/m + R))$ in the worst case.
- **Inter-Client Communication.** The analysis is the same as the one for Subtree-OPRAM.

Correctness. We analyze our scheme and show that it is indeed a valid OPRAM scheme.

Lemma 2. *Generic-OPRAM satisfies correctness, and in particular only overflows with negligible probability, as long as $\mathcal{OD}$ is also correct and only fails with negligible probability.*

We omit part of the correctness proof, and restrict ourselves to the more involved part of the analysis, proving that none of the stashes SS_i ever overflows, and that none of the partition is supposed to hold more than $2N/m + R$ elements. Conditioned on no overflows, correctness can then be verified by inspection.

The final result on the overflow probability summarized by the following lemma.

Lemma 3. *For every constant $\kappa \geq 2$, every $T = T(\lambda)$, and every logical access sequence $\mathbf{y}$ of T batches of m parallel logical instructions,*

$$\Pr[The\,protocol\,outputs\,``\mathsf{overflow}"] \leq T \cdot m \cdot e^{-\Theta(R)},$$

where the randomness is taken over the partition assignment, and the constant in the exponent depends on κ only.

We split the proof of the lemma into two propositions – the first pertaining to SS_i overflowing, the second to partition load. We stress that the proof first proposition relies on some interesting (and non-elementary) fact from basic queueing theory to ensure that a constant outflow of (at most) two blocks is sufficient to avoid an overflow.

Proposition 2. *For every constant $\kappa \geq 2$, every $T = T(\lambda)$, and every logical access sequence $\mathbf{y}$ of T batches of m logical instructions,*

$$\Pr[One\,of\,the\,stashes\,\mathsf{SS}_i\,overflows] \leq T \cdot m \cdot e^{-\Theta(R)},$$

where the randomness is taken over the partition assignment, and the constant in the exponent depends on κ only.

Proof. We prove the lemma for $\kappa = 2$. It will be clear that the bound only improve for larger $\kappa > 2$. Let us look at what happens with one particular stash SS_i for some $i \in [m]$ over time, and compute the probability that it ever contains more than R elements. We model this via the following process:

Single-bin process: In each iteration, m balls are thrown into one out of m bins independently, and each one lands in the single bin we are looking at with probability $1/m$. Then, $\kappa = 2$ balls are taken out of the bin (if the bin contains at least $\kappa = 2$ balls), and otherwise the bin is emptied.

Note that in the actual protocol execution, less than m balls may be thrown into bins at each round because of possible repetition patterns, but it is clear that by always assuming that up to potential m balls can be thrown in the bin can only increase the probability of overflowing, and thus this will be assumed without loss of generality.

To analyze the probability that the bin overflows at some point in time (i.e., it contains more than R balls), we use the stochastic process proposed in Example 23 of [22], with $a = 0, b = +\infty$. There, it is shown that the number of balls in the bin at iteration T is distributed as the random variable

$$X_T = \max_{0 \leq i \leq T} Z_i$$

where $Z_0 = 0$ and

$$Z_i = \sum_{j \leq i} (V_j - U_j),$$

with V_j denoting the number of balls going to the bin in iteration j and U_j denoting the *potential* number of balls taken out from the bin in iteration j. Here $U_j = 2$ for every j, thus

$$Z_i = T_i - 2i,$$

where $T_i = \sum_{j \leq i} V_j$. Note that V_j can be seen as the sum of m independent Bernoulli random variables, each being one with probability $1/m$. Therefore, T_i is the sum of $m \cdot i$ Bernoulli random variables with expected value i. We want to show now that with very high probability, $T_i \leq 2i + R$. We can simply use the Chernoff bound, and consider two cases. First, if $R/i \geq 1$, then

$$\Pr\left[T_i \geq 2i + R\right] \leq \Pr[T_i \geq i \cdot (1 + R/i)] \leq e^{-\frac{\varepsilon^2}{2+\varepsilon} i},$$

where $\varepsilon = R/i$. Note that

$$\frac{\varepsilon^2}{2+\varepsilon} i = R \cdot \frac{1}{1 + 2i/R} \geq R/3.$$

Thus $\Pr\left[T_i \geq 2i + R\right] \leq e^{-R/3}$. The second case is that $R/i \leq 1$. Then,

$$\Pr\left[T_i \geq 2i + R\right] \leq \Pr[T_i \geq i \cdot (1 + 1)] \leq e^{-i/3} \leq e^{-R/3}.$$

Therefore, by the union bound, the probability that there exists some i such that $Z_i \geq R$ is at most $T \cdot e^{-R/3}$, i.e., $X_T \leq R$, except with probability $T \cdot e^{-R/3}$. To conclude, once again by the union bound, we obtain the bound on the probability that one of the m stashes overflows. □

We also need to analyze the probability that too many elements are assigned to one partition, as otherwise our protocol would also fail.

Lemma 4. *For a given partition map* partition : $[N] \to [m]$, *denote by L the maximum numbers of addresses* a $\in [N]$ *assigned to the same partition p. Then, for any sequence of T batches of m operations and any $R \geq 2$, the probability that any point in time, $L \geq 2N/m + R$ ist at most $T \cdot m \cdot e^{-R/2}$.*

Proof. Take the partition map contents at some fixed point in time, and fix some partition $i \in [m]$. The entire contants of the partition map are N independent random variables, and each one of them is equal to i with probability $1/m$. Let L^i be the number of addresses assigned to this given i, and let $L = \max_i L^i$. Note that L^i is a sum of Bernoulli random variables with expectation N/m. We can then use the Chernoff bound to see that

$$\Pr\left[L^i \geq 2N/m + R\right] = \Pr\left[S \geq N/m(1 + 1 + \varepsilon)\right] \leq e^{-R/2},$$

for $\varepsilon = Rm/N$. By the union bound,

$$\Pr\left[L \geq N/m + R\right] \leq m \cdot e^{-R/2}.$$

And finally, note that there are at most T different "assignments" of position maps due to the structure of the protocol, and thus the overall bound on the probability follows – once again – by the union bound. □

Obliviousness. Generic-OPRAM also satisfies the obliviousness property. The formal proof (which we omit) relies on the obliviousness of the $\mathcal{OD}_i$'s and the fact that whenever processing a batch of m logical accesses, the above scheme accesses first m randomly chosen partitions, and moreover, in the second phase, each partition is accessed exactly twice.

References

1. Boneh, D., Mazieres, D., Popa, R.: Remote oblivious storage: making oblivious ram practical. MIT Tech-report: MIT-CSAIL-TR-2011-018 (2011)
2. Boyle, E., Chung, K.-M., Pass, R.: Oblivious parallel RAM. In: Kushilevitz, E., Malkin, T. (eds.), TCC 2016A, LNCS (2016, To appear). http://eprint.iacr.org/2014/594
3. Chung, K.-M., Liu, Z., Pass, R.: Statistically-secure ORAM with $\tilde{O}(\log^2 n)$ overhead. In: Sarkar, P., Iwata, T. (eds.) ASIACRYPT 2014, Part II. LNCS, vol. 8874, pp. 62–81. Springer, Heidelberg (2014)
4. Chung, K.-M., Pass, R.: A simple ORAM. Cryptology ePrint Archive, Report 2013/243 (2013). http://eprint.iacr.org/2013/243
5. Fletcher, C.W., van Dijk, M., Devadas, S.: Towards an interpreter for efficient encrypted computation. In: Proceedings of the 2012 ACM Workshop on Cloud Computing Security, CCSW 2012, Raleigh, NC, USA, October 19, 2012, pp. 83–94 (2012)

6. Gentry, C., Goldman, K.A., Halevi, S., Julta, C., Raykova, M., Wichs, D.: Optimizing ORAM and using it efficiently for secure computation. In: De Cristofaro, E., Wright, M. (eds.) PETS 2013. LNCS, vol. 7981, pp. 1–18. Springer, Heidelberg (2013)
7. Goldreich, O.: Towards a theory of software protection. In: Odlyzko, A.M. (ed.) CRYPTO 1986. LNCS, vol. 263, pp. 426–439. Springer, Heidelberg (1987)
8. Goldreich, O.: Towards a theory of software protection and simulation by oblivious RAMs. In: Aho, A. (ed.) 19th ACM STOC, pp. 182–194. ACM Press, New York City, New York, USA (25–27 May 1987)
9. Goldreich, O., Ostrovsky, R.: Software protection and simulation on oblivious RAMs. J. ACM 43(3), 431–473 (1996)
10. Goodrich, M.T., Hirschberg, D.S., Mitzenmacher, M., Thaler, J.: Cache-oblivious dictionaries and multimaps with negligible failure probability. In: Even, G., Rawitz, D. (eds.) MedAlg 2012. LNCS, vol. 7659, pp. 203–218. Springer, Heidelberg (2012)
11. Goodrich, M.T., Mitzenmacher, M.: Privacy-preserving access of outsourced data via oblivious RAM simulation. In: Aceto, L., Henzinger, M., Sgall, J. (eds.) ICALP 2011, Part II. LNCS, vol. 6756, pp. 576–587. Springer, Heidelberg (2011)
12. Goodrich, M.T., Mitzenmacher, M., Ohrimenko, O., Tamassia, R.: Oblivious RAM simulation with efficient worst-case access overhead. In: Proceedings of the 3rd ACM Cloud Computing Security Workshop, CCSW 2011, Chicago, IL, USA, October 21, 2011, pp. 95–100 (2011)
13. Goodrich, M.T., Mitzenmacher, M., Ohrimenko, O., Tamassia, R.: Oblivious storage with low I/O overhead. CoRR, abs/1110.1851 (2011)
14. Goodrich, M.T., Mitzenmacher, M., Ohrimenko, O., Tamassia, R.: Privacy-preserving group data access via stateless oblivious RAM simulation. In: Rabani, Y. (ed.) 23rd SODA, pp. 157–167. ACM-SIAM, Kyoto, Japan (17–19 January 2012)
15. Dautrich, J.L. Jr., Stefanov, E., Shi, E.: Burst ORAM: minimizing ORAM response times for bursty access patterns. In: Proceedings of the 23rd USENIX Security Symposium, San Diego, CA, USA, August 20–22, 2014, pp. 749–764 (2014)
16. Kushilevitz, E., Lu, S., Ostrovsky, R.: On the (in)security of hash-based oblivious RAM and a new balancing scheme. In: Rabani, Y. (ed.) 23rd SODA, pp. 143–156. ACM-SIAM, Kyoto, Japan (17–19 January 2012)
17. Lorch, J.R., Parno, B., Mickens, J.W., Raykova, M., Schiffman, J.: Shroud: ensuring private access to large-scale data in the data center. In: Proceedings of the 11th USENIX Conference on File and Storage Technologies, FAST 2013, San Jose, CA, USA, February 12–15, 2013, pp. 199–214 (2013)
18. Maas, M., Love, E., Stefanov, E., Tiwari, M., Shi, E., Asanovic, K., Kubiatowicz, J., Song, D.: PHANTOM: practical oblivious computation in a secure processor. In: Sadeghi, A.-R., Gligor, V.D., Yung, M. (eds.) ACM CCS 2013, pp. 311–324. ACM Press, Berlin, Germany (4–8 November 2013)
19. Pinkas, B., Reinman, T.: Oblivious RAM revisited. In: Rabin, T. (ed.) CRYPTO 2010. LNCS, vol. 6223, pp. 502–519. Springer, Heidelberg (2010)
20. Ren, L., Fletcher, C.W., Kwon, A., Stefanov, E., Shi, E., van Dijk, M., Devadas, S.: Constants count: practical improvements to oblivious RAM. In: Proceedings of the 24th USENIX Security Symposium (SECURITY 2015), pp. 415–430 (2015)
21. Ren, L., Yu, X., Fletcher, C.W., van Dijk, M., Devadas, S.: Design space exploration and optimization of path oblivious RAM in secure processors. In: The 40th Annual International Symposium on Computer Architecture, ISCA 2013, Tel-Aviv, Israel, June 23–27, 2013, pp. 571–582 (2013)

22. Serfozo, R.: Basics of Applied Stochastic Processes. Springer Science & Business Media, Berlin (2009). http://www.stat.yale.edu/~jtc5/251/readings/Basics%20of%20Applied%20Stochastic%20Processes_Serfozo.pdf

23. Shi, E., Chan, T.-H.H., Stefanov, E., Li, M.: Oblivious RAM with $O((\log N)^3)$ worst-case cost. In: Lee, D.H., Wang, X. (eds.) ASIACRYPT 2011. LNCS, vol. 7073, pp. 197–214. Springer, Heidelberg (2011)

24. Stefanov, E., Shi, E.: Oblivistore: high performance oblivious cloud storage. In: 2013 IEEE Symposium on Security and Privacy (SP), pp. 253–267. IEEE (2013)

25. Stefanov, E., Shi, E., Song, D.: Towards practical oblivious ram. In: NDSS (2012)

26. Stefanov E., van Dijk, M., Shi, E., Fletcher, C.W., Ren, L., Yu, X., Devadas, S.: Path ORAM: an extremely simple oblivious RAM protocol. In: Sadeghi, A.-R., Gligor, V.D., Yung, M. (eds.) ACM CCS 13, pp. 299–310. ACM Press, Berlin, Germany, (4–8 November 2013)

27. Wang, S., Ding, X., Deng, R.H., Bao, F.: Private information retrieval using trusted hardware. In: Gollmann, D., Meier, J., Sabelfeld, A. (eds.) ESORICS 2006. LNCS, vol. 4189, pp. 49–64. Springer, Heidelberg (2006)

28. Wang, X., Hubert Chan, T.-H., Shi, E.: Circuit ORAM: on tightness of the goldreich-ostrovsky lower bound. In: Ray, I., Li, N., Kruegel, C. (eds.) Proceedings of the 22nd ACM SIGSAC Conference on Computer and Communications Security. ACM, Denver, CO, USA, October 12–6, 2015, pp. 850–861 (2015)

29. Williams, P., Sion, R., Carbunar, B.: Building castles out of mud: practical access pattern privacy and correctness on untrusted storage. In: Ning, P., Syverson, P.F., Jha, S. (eds.) ACM CCS 2008, pp. 139–148. ACM Press, Alexandria, Virginia, USA (27–31 October 2008)

30. Williams, P., Sion, R., Tomescu, A.: PrivateFS: a parallel oblivious file system. In: Yu, T., Danezis, G., Gligor, V.D. (eds.) ACM CCS 2012, pp. 977–988. ACM Press, Raleigh, NC, USA (16–18 October 2012)

ABE and IBE

Déjà Q: Encore! Un Petit IBE

Hoeteck Wee$^{(\boxtimes)}$

ENS, Paris, France
wee@di.ens.fr

Abstract. We present an identity-based encryption (IBE) scheme in composite-order bilinear groups with essentially optimal parameters: the ciphertext overhead and the secret key are *one* group element each and decryption requires only *one* pairing. Our scheme achieves adaptive security and anonymity under standard decisional subgroup assumptions as used in Lewko and Waters (TCC '10). Our construction relies on a novel extension to the Déjà Q framework of Chase and Meiklejohn (Eurocrypt '14).

1 Introduction

In identity-based encryption (IBE) [5,27], ciphertexts and secret keys are associated with identities, and decryption is possible only when the identities match. IBE has been studied extensively over the last decade, with a major focus on obtaining constructions that simultanously achieve short parameters and full adaptive security under static assumptions in the standard model. This was first achieved in the works of Lewko and Waters [23,29], which also introduced the powerful dual system encryption methodology. The design of the Lewko-Waters IBE and the underlying proof techniques have since had a profound impact on both attribute-based encryption and pairing-based cryptography.

1.1 Our Contributions

In this work, we obtain the first efficiency improvement to the Lewko-Waters IBE in composite-order bilinear groups. We present an adaptively secure and anonymous identity-based encryption (IBE) scheme with essentially optimal parameters: the ciphertext overhead and the secret key are *one* group element each, and decryption only requires *one* pairing; this improves upon the Lewko-Waters IBE [23] in three ways: shorter parameters, faster decryption, and anonymity. Via Naor's transformation, we obtain a fully secure signature scheme where the signature is again only *one* group element. We stress that we achieve all of these improvements while relying on the same computational subgroup assumptions as in the Lewko-Waters IBE, notably in composite-order groups whose order is the product of three primes. We refer to Fig. 1 for a comparison with prior works.

CNRS (UMR 8548), INRIA and Columbia University. Supported in part by ERC Project aSCEND (639554) and NSF Award CNS-1445424.

E. Kushilevitz and T. Malkin (Eds.): TCC 2016-A, Part II, LNCS 9563, pp. 237–258, 2016.
DOI: 10.1007/978-3-662-49099-0_9

The Lewko-Waters IBE has played a foundational role in recent developments of IBE and more generally attribute-based encryption (ABE). Indeed, virtually all of the state-of-the-art prime-order IBE schemes in [2,22] —along with the subsequent extensions to ABE [1,12,24,30]— follow the basic design and proof strategy introduced in the Lewko-Waters IBE. For this reason, we are optimistic that our improvement to the Lewko-Waters IBE will lead to further advances in IBE and ABE. In fact, our improved composite-order IBE already hints at the potential of a more efficient prime-order IBE that subsumes all known schemes; we defer further discussion to Sect. 1.3.

We also present a selectively secure broadcast encryption scheme for n users where the ciphertext overhead is two group elements (independent of the number of recipients) and the user private key is a single group element.[1] To the best of our knowledge, this is the first broadcast encryption scheme to achieve constant-size ciphertext overhead, constant-size user private keys and linear-size public parameters under static assumptions; previously, such schemes were only known under q-type assumptions [6].

1.2 Our Techniques

The starting point of our constructions is the Déjà Q framework introduced by Chase and Meiklejohn [10]; this is an extension of Waters' dual system techniques to eliminate the use of q-type assumptions in settings beyond the reach of previous techniques. These settings include deterministic primitives such as pseudo-random functions (PRF) and —quite remarkably— schemes based on the inversion framework [4,8,26]. However, the Déjà Q framework is also limited in that it cannot be applied to advanced encryption systems such as identity-based and broadcast encryption, where certain secret exponents appear in both ciphertexts and secret keys on both sides of the pairing. We show how to overcome this limitation using several simple ideas.

IBE Overview. We describe our IBE scheme and the security proof next. We present a simplified variant of the constructions, suppressing many details pertaining to randomization and subgroups. Following the Lewko-Waters IBE [23], we rely on composite-order bilinear groups whose order N is the product of three primes p_1, p_2, p_3. We will use the subgroup G_{p_1} of order p_1 for functionality, and the subgroup G_{p_2} of order p_2 in the proof of security. The third subgroup corresponding to p_3 is used for additional randomization.

Recall that the Lewko-Waters IBE has the following form:

$$\mathsf{mpk} := (g, g^\beta, g^\gamma, e(g,u)), \quad \mathsf{ct}_{\mathsf{id}} := (g^s, g^{(\beta+\gamma\mathsf{id})s}, e(g,u)^s \cdot m), \quad \mathsf{sk}_{\mathsf{id}} := (u \cdot g^{(\beta+\gamma\mathsf{id})r}, g^r))$$

---

[1] Here, we ignore the additional overhead from specifying the set of recipients in the ciphertext, which requires n *bits*; decrypting also requires knowing some public parameters, which are not considered part of the user private keys.

Scheme	|mpk|	|sk|	|ct|	decryption	anonymous	no. of primes
LW10 [23]	$3|G_N|+|G_T|$	$2|G_N|$	$2|G_N|+|G_T|$	2 pairings	no	3
DIP10 [9]	$3|G_N|+|G_T|$	$2|G_N|$	$2|G_N|+|G_T|$	2 pairings	✓	4
YCZY14 [31]	$3|G_N|+|G_T|$	$2|G_N|$	$2|G_N|+|G_T|$	2 pairings	✓	4
this work (Fig 2)	$2|G_N|+|G_T|$	$|G_N|$	$|G_N|+|G_T|$	1 pairing	✓	3

Fig. 1. Comparison amongst adaptively secure IBEs in composite-order bilinear groups $e : G_N \times G_N \rightarrow G_T$.

Our IBE scheme has the following form:

$$\mathsf{mpk} := (g, g^\alpha, e(g,u)), \quad \mathsf{ct_{id}} := (g^{(\alpha+\mathsf{id})s}, e(g,u)^s \cdot m), \quad \mathsf{sk_{id}} := (u^{\frac{1}{\alpha+\mathsf{id}}})$$

Note that our scheme uses the "exponent inversion" framework [8], which has traditionally eluded a proof of security under static assumptions. In both schemes, g, u are random group elements of order p_1, and α, β, γ are random exponents over $\mathbb{Z}_N$. It is easy to see that decryption in our scheme only requires a single pairing to compute $e(g^{(\alpha+\mathsf{id})s}, u^{\frac{1}{\alpha+\mathsf{id}}}) = e(g,u)^s$.

IBE Security Proof. We rely on the same assumption as the Lewko-Waters IBE in [23], namely the $(p_1 \mapsto p_1 p_2)$-subgroup assumption, which asserts that random elements of order p_1 and those of order $p_1 p_2$ are computationally indistinguishable. In the proof of security, we rely on the assumption to introduce random G_{p_2}-components to the ciphertext and the secret keys.

We begin with the secret keys. We introduce a random G_{p_2}-component to the secret key $\mathsf{sk_{id}}$ following the Déjà Q framework [10] as follows:

$$\mathsf{sk_{id}} = u^{\frac{1}{\alpha+\mathsf{id}}} \overset{\text{subgroup}}{\longrightarrow} u^{\frac{1}{\alpha+\mathsf{id}}} g_2^{\frac{r_1}{\alpha+\mathsf{id}}} \overset{\text{CRT}}{\longrightarrow} u^{\frac{1}{\alpha+\mathsf{id}}} g_2^{\frac{r_1}{\alpha_1+\mathsf{id}}}, \tag{1}$$

where $\alpha_1 \leftarrow \mathbb{Z}_N$. In the first transition, we use the $(p_1 \mapsto p_1 p_2)$-subgroup assumption which says that $u \approx_c ug_2^{r_1}, r_1 \leftarrow_R \mathbb{Z}_N$, where g_2 is a generator of order p_2. In the second transition, we use the Chinese Reminder Theorem (CRT), which tell us $\alpha \mod p_1$ and $\alpha \mod p_2$ are independently random values, so we may replace $\alpha \mod p_2$ with $\alpha_1 \mod p_2$ for a fresh $\alpha_1 \leftarrow_R \mathbb{Z}_N$; this is fine as long as the challenge ciphertext and mpk reveal no information about $\alpha \mod p_2$, as is the case here. We may then repeat this transition q more times:

$$u^{\frac{1}{\alpha+\mathsf{id}}} \overset{\text{subgroup}}{\longrightarrow} u^{\frac{1}{\alpha+\mathsf{id}}} g_2^{\frac{r_1}{\alpha+\mathsf{id}}} \overset{\text{CRT}}{\longrightarrow} u^{\frac{1}{\alpha+\mathsf{id}}} g_2^{\frac{r_1}{\alpha_1+\mathsf{id}}}$$

$$\overset{\text{subgroup}}{\longrightarrow} u^{\frac{1}{\alpha+\mathsf{id}}} g_2^{\frac{r_2}{\alpha+\mathsf{id}}} g_2^{\frac{r_1}{\alpha_1+\mathsf{id}}} \overset{\text{CRT}}{\longrightarrow} u^{\frac{1}{\alpha+\mathsf{id}}} g_2^{\frac{r_2}{\alpha_2+\mathsf{id}} + \frac{r_1}{\alpha_1+\mathsf{id}}}$$

$$\overset{}{\longrightarrow} \quad \cdots \quad \overset{\text{CRT}}{\longrightarrow} u^{\frac{1}{\alpha+\mathsf{id}}} g_2^{\frac{r_{q+1}}{\alpha_{q+1}+\mathsf{id}} + \cdots + \frac{r_2}{\alpha_2+\mathsf{id}} + \frac{r_1}{\alpha_1+\mathsf{id}}}$$

where $r_1, \ldots, r_{q+1}, \alpha_1, \ldots, \alpha_{q+1} \leftarrow_R \mathbb{Z}_N$, and q is an upper bound on the number of key queries made by the adversary.[2]

---

[2] We use $q + 1$ values to account for the q key queries plus the challenge identity.

Next, we show that for distinct $x_1, \ldots, x_q$, the following matrix

$$
\begin{pmatrix}
\frac{1}{\alpha_1 + x_1} & \frac{1}{\alpha_1 + x_2} & \cdots & \frac{1}{\alpha_1 + x_q} \\
\vdots & \vdots & \ddots & \vdots \\
\frac{1}{\alpha_q + x_1} & \frac{1}{\alpha_q + x_2} & \cdots & \frac{1}{\alpha_q + x_q}
\end{pmatrix}
\tag{2}
$$

is invertible with overwhelming probability over $\alpha_1, \ldots, \alpha_q \leftarrow_R \mathbb{Z}_p$. We provide an explicit formula for the determinant of this matrix in Sect. 3.1; this is the only place in the proof where we crucially exploit the "exponent inversion" structure. We can then replace

$$
\mathsf{id} \mapsto \frac{r_{q+1}}{\alpha_{q+1} + \mathsf{id}} + \cdots + \frac{r_2}{\alpha_2 + \mathsf{id}} + \frac{r_1}{\alpha_1 + \mathsf{id}}
$$

by a truly random function $\mathsf{RF}(\cdot)$. Indeed, $\mathsf{sk_{id}}$ can now be written as $u^{\frac{1}{\alpha + \mathsf{id}}} g_2^{\mathsf{RF}(\mathsf{id})}$, which have independently random G_{p_2}-components.

So far, what we have done is the same as the use of Déjà Q framework for showing that $x \mapsto u^{\frac{1}{x+\alpha}}$ yields a PRF [10] (the explicit formula for the matrix determinant is new), and this is where the similarity ends. At this point, we still need to hide the message m in the ciphertext $(g^{(\alpha+\mathsf{id})s}, e(g, u)^s \cdot m)$. Towards this goal, we want to introduce a G_{p_2}-component into the ciphertext, which will then interact with newly random G_{p_2}-component in the keys to generate extra statistical entropy to hide m. At the same time, we need to ensure that the ciphertext still hides $\alpha \bmod p_2$ so that we may carry out the transition of the secret keys in (1). Indeed, naively applying the $(p_1 \mapsto p_1 p_2)$-subgroup assumption to g^s in the ciphertext would leak $\alpha \bmod p_2$.

To circumvent this difficulty, note that we can rewrite the ciphertext in terms of $\mathsf{sk_{id}}$ as

$$
\mathsf{ct_{id}} = (g^{(\alpha+\mathsf{id})s}, e(g^{(\alpha+\mathsf{id})s}, \mathsf{sk_{id}}) \cdot m)
$$

Moreover, as long as $\alpha + \mathsf{id} \neq 0$, we can replace $(\alpha + \mathsf{id})s$ with s without changing the distribution, which allows us to rewrite the challenge ciphertext as

$$
\mathsf{ct_{id}} = (g^s, e(g^s, \mathsf{sk_{id}}) \cdot m).
$$

This means that the challenge ciphertext leaks no information about α except through $\mathsf{sk_{id}}$. In addition, the challenge ciphertext also leaks no information about id, which allows us to prove anonymity. In contrast, the Lewko-Waters IBE is not anonymous, and anonymous variants there-of in [9,31] requires the use of 4 primes and additional assumptions.

We can now apply the $(p_1 \mapsto p_1 p_2)$-subgroup assumption to the ciphertext to replace g^s with $g^s g_2^{r'}$. Now, the ciphertext distribution is completely independent of α except what is leaked through $\mathsf{sk_{id}}$, so we can apply the secret key transitions as before, at the end of which the challenge ciphertext is given by:

$$
(g^s g_2^{r'}, e(g^s g_2^{r'}, u^{\frac{1}{\alpha+\mathsf{id}}} g_2^{\mathsf{RF}(\mathsf{id})}) \cdot m) = (g^s g_2^{r'}, e(g^s, u^{\frac{1}{\alpha+\mathsf{id}}}) \cdot \boxed{e(g_2^{r'}, g_2^{\mathsf{RF}(\mathsf{id})})} \cdot m)
$$

Recall that we only allow the adversary to request for secret keys corresponding to identities different from id, which means those keys leak no information about RF(id). We can then use the $\log p_2$ bits of entropy from RF(id) over G_{p_2} to hide m; this requires modifying the original scheme so that an encryption of m is given by $(g^{(\alpha + \mathrm{id})s}, \mathsf{H}(e(g, u)^s) \cdot m)$, where H denotes a strong randomness extractor whose seed is specified in mpk.

Broadcast Encryption. By rewriting the challenge ciphertext in terms of $\mathsf{sk}_{\mathsf{id}}$ in order to hide α, our technique for IBE seems inherently limited to IBE. We show how to extend our techniques to broadcast encryption in Sect. 4; however, we only achieve selective and not adaptive security. We briefly note that our broadcast encryption scheme is derived from Boneh-Gentry-Waters (BGW) scheme [6] based on the q-DBDHE assumption. This is the first scheme to asymptotically match the parameters of the BGW broadcast encryption scheme under static assumptions.

1.3   Discussion

Comparison with Déjà Q Framework [10]. The core of the Déjà Q framework is a beautiful technique which translates linear independence (and thus computational independence in the generic group model) amongst a set of monomials "in the exponent" into statistical independence, upon which security can be established using a purely information-theoretic argument. There are however three caveats to the prior instantiation in [10]: first, these monomials must appear on the same side of the pairing, which means the techniques cannot be applied to advanced encryption primitives where the same term often appears in the ciphertext and the secret key on both sides of the pairing; second, the statistical independence only holds within certain subgroups, and another subgroup assumption was used to spread this localized entropy over the entire group; third, the prior instantiation is limited to asymmetric composite-order groups. In this work, we showed how to overcome all of these three caveats.

In particular, we rely only on the $(p_1 \mapsto p_1 p_2)$-subgroup assumption and eliminated the additional use of the $(p_2 \mapsto p_1 p_2)$-subgroup assumption. This technique can also be applied to the PRF in [10]. We note that while simulating subgroup decisional assumptions in composite-order groups using the k-LIN assumption in prime-order groups, we can simulate the $(p_1 \mapsto p_1 p_2)$-subgroup assumption using $k + 1$ group elements whereas simulating both subgroup assumption requires $2k$ group elements.

Candidate Prime-Order IBE. As noted earlier, our composite-order IBE scheme constitutes the first evidence for an adaptively IBE based on SXDH with two group elements in the ciphertext and in the secret keys and constant-size public parameters, which would be a significant improvement over the state of the art, subsuming a long series of incomparable constructions, and giving us adaptive security at essentially the same cost as selective security! Moreover, such a IBE would in turn also yield a fully secure signature scheme based on SXDH with two group elements in the signature and constant-size public key. The optimism comes from combining our composite-order IBE scheme with the huge success we have had in converting composite-order schemes to prime-order ones [12,15,22,25]. In fact, we present a concrete candidate for a prime-order IBE in Sect. 3.3; we stress that we do not have a security proof for the scheme. We note that an improved SXDH-based signature scheme would likely yield further improvements to other related primitives, such as group signatures and structure-preserving signatures. These applications further motivate the open problem highlighted in [10] of finding prime-order analogues for the Déjà Q framework.

Perspective. We presented new constructions of "optimal" IBE and signatures and new IBE candidates that improve upon a long line of work; moreover, we achieve these via an extended Déjà Q framework which avoid the limitations of widely used techniques. We are optimistic and excited about challenges and possibilities that lie ahead.

2   Preliminaries

Notation. We denote by $s \leftarrow_R S$ the fact that s is picked uniformly at random from a finite set S. By PPT, we denote a probabilistic polynomial-time algorithm. Throughout, we use 1^λ as the security parameter.

2.1   Composite-Order Bilinear Groups and Cryptographic Assumptions

We instantiate our system in composite-order bilinear groups, which were introduced in [7] and used in [21,23,24]. A generator $\mathcal{G}$ takes as input a security parameter λ and outputs a description $\mathbb{G} := (N, G, G_T, e)$, where N is product of distinct primes of $\Theta(\lambda)$ bits, G and G_T are cyclic groups of order N, and $e : G \times G \to G_T$ is a non-degenerate bilinear map. We require that the group operations in G and G_T as well the bilinear map e are computable in deterministic polynomial time. We consider bilinear groups whose orders N are products of three distinct primes p_1, p_2, p_3 (that is, $N = p_1 p_2 p_3$). We can write $G = G_{p_1} G_{p_2} G_{p_3}$ where $G_{p_1}, G_{p_2}, G_{p_3}$ are subgroups of G of order p_1, p_2 and p_3 respectively. In addition, we use $G_{p_i}^*$ to denote $G_{p_i} \setminus \{1\}$. We will often write g_1, g_2, g_3 to denote random generators for the subgroups $G_{p_1}, G_{p_2}, G_{p_3}$.

Cryptographic Assumptions. Our construction relies on the following two decisional subgroup assumptions (also known as subgroup hiding assumptions).

We define the following two advantage functions:

$\mathsf{Adv}^{\mathrm{SD1}}_{\mathcal{G},\mathcal{A}}(\lambda) := |\Pr[\mathcal{A}(D, T_0) = 1] - \Pr[\mathcal{A}(D, T_1) = 1]|$

> where $\mathsf{G} \leftarrow \mathcal{G}, T_0 \leftarrow \boxed{G_{p_1}}, T_1 \leftarrow_{\mathrm{R}} \boxed{G_{p_1} G_{p_2}}$ $(p_1 \mapsto p_1 p_2)$
>
> and $D := (g_1, g_3, g_{\{1,2\}}), g_1 \leftarrow_{\mathrm{R}} G^*_{p_1}, g_3 \leftarrow_{\mathrm{R}} G^*_{p_3}, g_{\{1,2\}} \leftarrow_{\mathrm{R}} G_{p_1} G_{p_2}$

$\mathsf{Adv}^{\mathrm{SD2}}_{\mathcal{G},\mathcal{A}}(\lambda) := |\Pr[\mathcal{A}(D, T_0) = 1] - \Pr[\mathcal{A}(D, T_1) = 1]|$

> where $\mathsf{G} \leftarrow \mathcal{G}, T_0 \leftarrow \boxed{G_{p_1} G_{p_3}}, T_1 \leftarrow_{\mathrm{R}} \boxed{G_{p_1} G_{p_2} G_{p_3}}$ $(p_1 p_3 \mapsto N)$
>
> and $D := (g_1, g_3, g_{\{1,2\}}, g_{\{2,3\}}), g_1 \leftarrow_{\mathrm{R}} G^*_{p_1}, g_3 \leftarrow_{\mathrm{R}} G^*_{p_3}, g_{\{1,2\}} \leftarrow_{\mathrm{R}} G_{p_1} G_{p_2}, g_{\{2,3\}} \leftarrow_{\mathrm{R}} G_{p_2} G_{p_3}$

The decisional subgroup assumptions assert that that for all PPT adversaries $\mathcal{A}$, the advantages $\mathsf{Adv}^{\mathrm{SD1}}_{\mathcal{G},\mathcal{A}}(\lambda)$ and $\mathsf{Adv}^{\mathrm{SD2}}_{\mathcal{G},\mathcal{A}}(\lambda)$ are negligible functions in λ.

2.2 Anonymous Identity-Based Encryption

We define identity-based encryption (IBE) in the framework of key encapsulation. An identity-based encryption scheme consists of four algorithms (Setup, Enc, KeyGen, Dec):

Setup(1^λ) $\rightarrow$ (mpk, msk). The setup algorithm gets as input the security parameter λ and outputs the public parameter mpk, and the master key msk. All the other algorithms get mpk as part of its input.

Enc(mpk, id) $\rightarrow$ (ct, κ). The encryption algorithm gets as input mpk and an identity id $\in \{0,1\}^\lambda$. It outputs a ciphertext ct and a symmetric key $\kappa \in \{0,1\}^\lambda$.

KeyGen(msk, id) $\rightarrow$ sk$_{\mathrm{id}}$. The key generation algorithm gets as input msk and an identity id $\in \{0,1\}^\lambda$. It outputs a secret key sk$_{\mathrm{id}}$.

Dec(sk$_{\mathrm{id}}$, ct) $\rightarrow$ κ. The decryption algorithm gets as input sk$_{\mathrm{id}}$ and ct. It outputs a symmetric key κ.

Correctness. We require that for all id $\in \{0,1\}^\lambda$,

$$\Pr[(\mathsf{ct}, \kappa) \leftarrow \mathsf{Enc}(\mathsf{mpk}, \mathsf{id}); \; \mathsf{Dec}(\mathsf{sk}_{\mathrm{id}}, \mathsf{ct}) = \kappa)] = 1,$$

where the probability is taken over (mpk, msk) $\leftarrow$ Setup(1^λ) and the coins of Enc.

Security Definition. We require pseudorandom ciphertexts against adaptively chosen plaintext and identity attacks, which implies both anonymity and adaptive security. For a stateful adversary $\mathcal{A}$, we define the advantage function

$$\mathsf{Adv}^{\mathrm{A\text{-}IBE}}_{\mathcal{A}}(\lambda) := \Pr\left[b = b' : \begin{array}{l} (\mathsf{mpk}, \mathsf{msk}) \leftarrow \mathsf{Setup}(1^\lambda); \\ \mathsf{id}^* \leftarrow \mathcal{A}^{\mathsf{KeyGen}(\mathsf{msk}, \cdot)}(\mathsf{mpk}); \\ b \leftarrow_{\mathrm{R}} \{0,1\}; \mathsf{ct}_1 \leftarrow_{\mathrm{R}} \mathcal{C}; \kappa_1 \leftarrow_{\mathrm{R}} \{0,1\}^\lambda; \\ (\mathsf{ct}_0, \kappa_0) \leftarrow \mathsf{Enc}(\mathsf{mpk}, \mathsf{id}^*); \\ b' \leftarrow \mathcal{A}^{\mathsf{KeyGen}(\mathsf{msk}, \cdot)}(\mathsf{ct}_b, \kappa_b) \end{array} \right] - \frac{1}{2}$$

with the restriction that all queries id that $\mathcal{A}$ makes to KeyGen(msk, ·) satisfies id $\neq$ id*, and where $ct_1 \leftarrow_R \mathcal{C}$ denotes a random element from the ciphertext space.[3] An identity-based encryption (IBE) scheme is *adaptively secure and anonymous* if for all PPT adversaries $\mathcal{A}$, the advantage $\mathsf{Adv}_{\mathcal{A}}^{\text{A-IBE}}(\lambda)$ is a negligible function in λ.

2.3   Broadcast Encryption

A broadcast encryption scheme consists of three algorithms (Setup, Enc, Dec):

Setup$(1^\lambda, 1^n) \rightarrow (mpk, (sk_1, \ldots, sk_n))$. The setup algorithm gets as input the security parameter λ and 1^n specifying the number of users and outputs the public parameter mpk, and secret keys $sk_1, \ldots, sk_n$.

Enc$(mpk, \Gamma) \rightarrow (ct_\Gamma, \kappa)$. The encryption algorithm gets as input mpk and a subset $\Gamma \subseteq [n]$. It outputs a ciphertext ct_Γ and a symmetric key $\kappa \in \{0,1\}^\lambda$. Here, Γ is public given ct_Γ.

Dec$(mpk, sk_y, ct_\Gamma) \rightarrow \kappa$. The decryption algorithm gets as input mpk, sk_y and ct_Γ. It outputs a symmetric key κ.

Correctness. We require that for all $\Gamma \subseteq [n]$ and all $y \in [n]$ for which $y \in \Gamma$,

$$\Pr[(ct_\Gamma, \kappa) \leftarrow \mathsf{Enc}(mpk, \Gamma);\ \mathsf{Dec}(mpk, sk_y, ct_\Gamma) = \kappa] = 1,$$

where the probability is taken over $(mpk, (sk_1, \ldots, sk_n)) \leftarrow \mathsf{Setup}(1^\lambda, 1^n)$ and the coins of Enc.

Security Definition. For a stateful adversary $\mathcal{A}$, we define the advantage function

$$\mathsf{Adv}_{\mathcal{A}}^{\text{S-BCE}}(\lambda) := \Pr \left[b = b' : \begin{array}{l} \Gamma^* \leftarrow \mathcal{A}(1^\lambda); \\ (mpk, (sk_1, \ldots, sk_n)) \leftarrow \mathsf{Setup}(1^\lambda); \\ b \leftarrow_R \{0,1\}; \kappa_1 \leftarrow_R \{0,1\}^\lambda; \\ (ct_{\Gamma^*}, \kappa_0) \leftarrow \mathsf{Enc}(mpk, \Gamma^*); \\ b' \leftarrow \mathcal{A}(ct_{\Gamma^*}, \kappa_b, \{sk_y : y \notin \Gamma^*\}) \end{array} \right] - \frac{1}{2}$$

A broadcast encryption scheme is *selectively secure* if for all PPT adversaries $\mathcal{A}$, the advantage $\mathsf{Adv}_{\mathcal{A}}^{\text{S-BCE}}(\lambda)$ is a negligible function in λ.

3   Identity-Based Encryption

We present an adaptively secure and anonymous IBE scheme in Fig. 2, and a fully secure signature scheme in Fig. 3. The schemes here refer to symmetric composite-order bilinear groups; we present the variant for asymmetric composite-bilinear groups in Sect. A. The schemes and the proofs are the same as in the overview in the introduction (Sect. 1.2), except the secret keys in both the scheme and the proof have an extra random G_{p_3}-component and we will use the $(p_1 p_3 \mapsto N)$-subgroup assumption to switch the secret keys.

---

[3] This means that the distribution of ct_1 is independent of id*, which implies anonymity.

Comparison with Prior Schemes. We recall several IBE and signature schemes in the inversion framework which share a similar structure to our IBE and signature scheme. All of these schemes require an additional scalar in the key/signature, and both of the IBE schemes require an additional group element in the ciphertext.

BB$_2$ IBE [4]. The BB$_2$ IBE is selectively secure under the q-DBDHI assumption:

$$\mathsf{ct_{id}} := (g^{(\alpha+\mathsf{id})s}, g^{\beta s}, e(g,u)^s \cdot m), \ \mathsf{sk_{id}} := (u^{\frac{1}{\alpha+\mathsf{id}+\beta r}}, r)$$

Gentry's IBE [18]. Gentry's IBE is adaptively secure and anonymous under the q-ADBDHE assumption:

$$\mathsf{ct_{id}} := (g^{(\alpha+\mathsf{id})s}, e(g,g)^s, e(g,u)^s \cdot m), \ \mathsf{sk_{id}} := ((u \cdot g^{-r})^{\frac{1}{\alpha+\mathsf{id}}}, r)$$

Boneh-Boyen Signatures [3,10]. The Déjà Q analogue [10] of the Boneh-Boyen signatures is given by:

$$\mathsf{pk} := (g, g^\alpha, g^\beta, e(g,u)), \ \sigma := (u^{\frac{1}{\alpha+M+\beta r}}, r) \in \mathbb{G}_N \times \mathbb{Z}_N.$$

Our signature scheme in Fig. 3 is simpler and shorter, and the scheme can be also be instantiated in symmetric composite-order groups. In fact, our signature scheme may be viewed as applying the Déjà Q framework to the Boneh-Boyen weak signatures, which both "upgrades" the security from weak to full, and removes the use of q-type assumptions.

3.1 Core Lemma

The following lemma is implicit in the analysis of the PRF in [10, Theorem 4.2, Eq. 8].

Lemma 1. *Fix a prime p and define* $\mathsf{F}^q_{r_1,\ldots,r_q,\alpha_1,\ldots,\alpha_q} : \mathbb{Z}_p \to \mathbb{Z}_p$ *to be*

$$\mathsf{F}^q_{r_1,\ldots,r_q,\alpha_1,\ldots,\alpha_q}(x) := \sum_{i=1}^{q} \frac{r_i}{\alpha_i + x}$$

Then, for any (possibly unbounded) adversary $\mathcal{A}$ that makes at most q queries, we have

$$\left| \Pr_{r_1,\ldots,r_q,\alpha_1,\ldots,\alpha_q \leftarrow_R \mathbb{Z}_p} \left[\mathcal{A}^{\mathsf{F}^q_{r_1,\ldots,r_q,\alpha_1,\ldots,\alpha_q}(\cdot)}(1^q) = 1\right] - \Pr\left[\mathcal{A}^{\mathsf{RF}(\cdot)}(1^q) = 1\right] \right| \leq \frac{q^2}{p}$$

where $\mathsf{RF} : \mathbb{Z}_p \to \mathbb{Z}_p$ *is a truly random function.*

The proof in [10] directly rewrites the function $\mathsf{F}^q_{r_1,\ldots,r_q,\alpha_1,\ldots,\alpha_q}$ with a common denominator and then relates the numerator to the Lagrange interpolating polynomial for an appropriate choice of q points. We sketch an alternative proof which better explains the choice of the function $(\alpha, \mathsf{id}) \mapsto \frac{1}{\alpha+\mathsf{id}}$. We first consider the case where the queries $x_1, \ldots, x_q$ made by $\mathcal{A}$ are chosen non-adaptively. WLOG, we may assume that these queries are distinct. Then, it suffices to show that the following matrix

$$
\begin{pmatrix}
\frac{1}{\alpha_1 + x_1} & \frac{1}{\alpha_1 + x_2} & \cdots & \frac{1}{\alpha_1 + x_q} \\
\vdots & \vdots & \ddots & \vdots \\
\frac{1}{\alpha_q + x_1} & \frac{1}{\alpha_q + x_2} & \cdots & \frac{1}{\alpha_q + x_q}
\end{pmatrix}
$$

is invertible with overwhelming probability over $\alpha_1, \ldots, \alpha_q \leftarrow_R \mathbb{Z}_p$. (Such a statement follows from the proof in [10] but was not pointed out explicitly.) As it turns out, we can write the determinant of this matrix explicitly as:

$$
\frac{\Pi_{1 \leq i < j \leq q}(x_i - x_j)(\alpha_i - \alpha_j)}{\Pi_{1 \leq i,j \leq q}(\alpha_i + x_j)}
$$

which is non-zero as long as $\alpha_1, \ldots, \alpha_q$ are distinct, $x_1, \ldots, x_q$ are distinct, and the $\alpha_i + x_j$'s are all non-zero.

That is, we want to show that

$$
\Pi_{1 \leq i,j \leq q}(\alpha_i + x_j) \cdot \det
\begin{pmatrix}
\frac{1}{\alpha_1 + x_1} & \frac{1}{\alpha_1 + x_2} & \cdots & \frac{1}{\alpha_1 + x_q} \\
\vdots & \vdots & \ddots & \vdots \\
\frac{1}{\alpha_q + x_1} & \frac{1}{\alpha_q + x_2} & \cdots & \frac{1}{\alpha_q + x_q}
\end{pmatrix}
= \Pi_{1 \leq i < j \leq q}(x_i - x_j)(\alpha_i - \alpha_j)
$$

Using the standard formula for the determinant of the matrix, we can write the determinant above as a sum of inverses of homogenous polynomials of degree q in $x_1, \ldots, x_q, \alpha_1, \ldots, \alpha_q$. Upon multiplying by $\Pi_{1 \leq i,j \leq q}(\alpha_i + x_j)$, we would "clear the denominators" to obtain a homogeneous polynomial P in $x_1, \ldots, x_q, \alpha_1, \ldots, \alpha_q$ of degree $q^2 - q$. Moreover, the matrix has two equal rows (resp. columns) whenever we have $\alpha_i = \alpha_j$ (resp. $x_i = x_j$); when this happens, the matrix has determinant 0 and thus P vanishes. Therefore, the polynomial P must be a multiple of $\Pi_{1 \leq i < j \leq q}(x_i - x_j)(\alpha_i - \alpha_j)$, which also has degree $q^2 - q$. This means that P must be a constant multiple of $\Pi_{1 \leq i < j \leq q}(x_i - x_j)(\alpha_i - \alpha_j)$, and it is easy to check that the constant is 1.

To handle adaptive queries, observe that this corresponds to building the matrix one column at a time. As long as the partial selection of columns have full rank, the output of F is uniformly random, which then completely hides $\alpha_1, \ldots, \alpha_q$. Therefore, the probability that $\alpha_1, \ldots, \alpha_q$ are distinct, and that $\alpha_i + x_j$'s are all non-zero is at least $1 - q^2/p$, even for adaptive choices of distinct $x_1, \ldots, x_q$.

3.2   Our IBE Scheme

Theorem 1. *The scheme in Fig. 2 is an adaptively secure anonymous IBE under the decisional subgroup assumption in* $\mathbb{G}$.

Setup($\mathbb{G}$):
$\mathsf{msk} := (\alpha, u, g_3) \leftarrow_R \mathbb{Z}_N \times G_{p_1} \times G_{p_3}^*$;
$\mathsf{mpk} := (g_1, g_1^\alpha, e(g_1, u), \mathsf{H})$;
return $(\mathsf{mpk}, \mathsf{msk})$

KeyGen($\mathsf{msk}, \mathsf{id} \in \mathbb{Z}_N$):
pick $R_3 \leftarrow_R G_{p_3}$;
return $\mathsf{sk}_{\mathsf{id}} := u^{\frac{1}{\alpha + \mathsf{id}}} R_3$

Enc($\mathsf{mpk}, \mathsf{id} \in \mathbb{Z}_N$):
pick $s \leftarrow_R \mathbb{Z}_N$;
return $(\mathsf{ct}, \kappa) := (g_1^{(\alpha+\mathsf{id})s}, \mathsf{H}(e(g_1, u)^s))$

Dec($\mathsf{sk}_{\mathsf{id}}, \mathsf{ct}$):
return $\mathsf{H}(e(\mathsf{ct}, \mathsf{sk}_{\mathsf{id}}))$

Fig. 2. Adaptively secure anonymous IBE w.r.t. a composite-order bilinear group $\mathbb{G}$. Here, $\mathsf{H} : G_T \rightarrow \{0,1\}^\lambda$ is drawn from a family of pairwise-independent hash functions. In asymmetric groups, randomization with R_3 in KeyGen is not necessary (i.e., KeyGen is deterministic).

Setup($\mathbb{G}$):
$\mathsf{sk} := (\alpha, u, g_3) \leftarrow_R \mathbb{Z}_N \times G_{p_1} \times G_{p_3}^*$;
$\mathsf{pk} := (g_1, g_1^\alpha, e(g_1, u))$;
return $(\mathsf{pk}, \mathsf{sk})$

sign($\mathsf{sk}, M \in \mathbb{Z}_N$):
pick $R_3 \leftarrow_R G_{p_3}$;
return $\sigma := u^{\frac{1}{\alpha + M}} R_3$

verify(pk, M, σ):
check $e(g_1^M \cdot g_1^\alpha, \sigma) = e(g_1, u)$

Fig. 3. Fully secure signature scheme, obtained by applying Naor's transformation to the IBE scheme in Fig. 2. In asymmetric groups, randomization with R_3 in **sign** is not necessary (i.e., **sign** is deterministic).

Proof. Correctness follows readily from the equation

$$e(g_1^{(\alpha+\mathsf{id})s}, u^{\frac{1}{\alpha+\mathsf{id}}} R_3) = e(g_1, u)^s.$$

We show that for any adversary $\mathcal{A}$ that makes at most q queries against the IBE, there exist adversaries $\mathcal{A}_1, \mathcal{A}_2$ whose running times are essentially the same as that of $\mathcal{A}$, such that

$$\mathsf{Adv}_{\mathcal{A}}^{\text{A-IBE}}(\lambda) \leq \mathsf{Adv}_{\mathcal{G}, \mathcal{A}_1}^{\text{SD1}}(\lambda) + (q+1) \cdot \mathsf{Adv}_{\mathcal{G}, \mathcal{A}_2}^{\text{SD2}}(\lambda) + 2^{-\Omega(\lambda)}$$

We proceed via a series of games and we use Adv_i to denote the advantage of $\mathcal{A}$ in Game i.

Game 0. This is the real experiment as defined in Sect. 2.2. We will also make the following simplifying assumptions:

- We never encounter an identity id such that $\mathsf{id} = \alpha \mod p_1$; such an identity constitutes the discrete log of g_1^α and trivially breaks the subgroup assumption.
- The adversary's queries $\mathsf{id}_1, \ldots, \mathsf{id}_q \in \mathbb{Z}_N$ are distinct, since we can perfectly randomize the secret key $\mathsf{sk}_{\mathsf{id}} = u^{\frac{1}{\alpha+\mathsf{id}}} R_3$ given g_3 (we can add g_3 to mpk without affecting the security proof).

– $\mathsf{id}_1, \ldots, \mathsf{id}_q$ are distinct mod p_2; given $\mathsf{id}_i \neq \mathsf{id}_j \in \mathbb{Z}_N$ such that $\mathsf{id}_i = \mathsf{id}_j$ mod p_2, computing $\gcd(\mathsf{id}_i - \mathsf{id}_j, N)$ would allow us to factor N.

We can incorporate these simplifying assumptions by introducing an extra hybrid before Game 1 that aborts if the first or third condition is violated, and that uses randomization to handle repeated key queries.

Game 1. We change $(\mathsf{ct}_0, \kappa_0) \leftarrow_R \mathsf{Enc}(\mathsf{mpk}, \mathsf{id}^*)$ as follows: pick $C \leftarrow_R G_{p_1}$, output

$$(\mathsf{ct}_0, \kappa_0) := (C, H(e(C, \mathsf{sk}_{\mathsf{id}^*}))).$$

We claim that $\mathsf{Adv}_0 = \mathsf{Adv}_1$. This follows readily from the following two observations:

i. for all id, $e(g_1, u)^s = e(g_1^{(\alpha+\mathsf{id})s}, u^{\frac{1}{\alpha+\mathsf{id}}}) = e(g_1^{(\alpha+\mathsf{id})s}, \mathsf{sk}_{\mathsf{id}})$;

ii. if $\alpha + \mathsf{id} \neq 0$, $g_1^{(\alpha+\mathsf{id})s}$ and C are identically distributed.

Game 2. We change the distribution of C in $(\mathsf{ct}_0, \kappa_0)$ from $C \leftarrow_R G_{p_1}$ to $C \leftarrow_R G_{p_1} G_{p_2}$. We now construct $\mathcal{A}_1$ for which

$$\mathsf{Adv}_0 - \mathsf{Adv}_1 \leq \mathsf{Adv}^{\mathrm{SD1}}_{\mathcal{G}, \mathcal{A}_1}(\lambda).$$

$\mathcal{A}_1$ on input (g_1, g_3, C) where either $C \leftarrow_R G_{p_1}$ or $C \leftarrow_R G_{p_1} G_{p_2}$, simulates the experiment in Game 1 with the adversary $\mathcal{A}$ as follows: runs $\mathsf{Setup}(\mathbb{G})$ honestly to obtain (α, u), then uses (α, u) to answer all key queries honestly and to compute (ct, κ_0) as $(C, H(e(C, \mathsf{sk}_{\mathsf{id}^*})))$.

Game 3. We change the distribution of $\mathsf{sk}_{\mathsf{id}}$ from $u^{\frac{1}{\alpha+\mathsf{id}}} R_3$ to $u^{\frac{1}{\alpha+\mathsf{id}}} g_2^{\sum_{i=1}^{q+1} \frac{r_i}{\alpha_i+\mathsf{id}}} R_3$, where $r_1, \ldots, r_{q+1}, \alpha_1, \ldots, \alpha_{q+1} \leftarrow_R \mathbb{Z}_N$, as outlined in Sect. 1.1. We proceed via a series of sub-games $3.j.0$ and $3.j.1$ for $j = 1, 2, \ldots, q+1$, where

– In Sub-Game $3.j.0$, $\mathsf{sk}_{\mathsf{id}}$ is given by $u^{\frac{1}{\alpha+\mathsf{id}}} g_2^{\frac{r_j}{\alpha+\mathsf{id}} + \sum_{i=1}^{j-1} \frac{r_i}{\alpha_i+\mathsf{id}}} R_3$;

– In Sub-Game $3.j.1$, $\mathsf{sk}_{\mathsf{id}}$ is given by $u^{\frac{1}{\alpha+\mathsf{id}}} g_2^{\sum_{i=1}^{j} \frac{r_i}{\alpha_i+\mathsf{id}}} R_3$. Game 2 corresponds to Sub-Game $3.0.1$, and Game 3 corresponds to Sub-Game $3.q+1.1$.

First, observe that $\mathsf{Adv}_{3.j.0} = \mathsf{Adv}_{3.j.1}$. This follows readily from the fact that $\alpha \mod p_2$ is completely hidden given mpk and the challenge ciphertext, and therefore we may replace $\alpha \mod p_2$ with $\alpha_j \mod p_2$. Next, for $j = 1, \ldots, q+1$, we construct $\mathcal{A}_2$ for which

$$\mathsf{Adv}_{3.(j-1).1} - \mathsf{Adv}_{3.j.0} \leq \mathsf{Adv}^{\mathrm{SD2}}_{\mathcal{G}, \mathcal{A}_2}(\lambda).$$

$\mathcal{A}_2$ on input $(\mathbb{G}, g_1, g_{\{2,3\}}, g_3, C, T)$ where $C \leftarrow_R G_{p_1} G_{p_2}$ and either $T = uR_3 \leftarrow_R G_{p_1} G_{p_3}$ or $T = ug_2^{r_j} R_3 \leftarrow_R G_{p_1} G_{p_2} G_{p_3}$, simulates the experiment in Game 3 with the adversary $\mathcal{A}$ as follows:

– picks $\alpha \leftarrow_R \mathbb{Z}_N$ and publishes $\mathsf{mpk} := (g_1, g_1^\alpha, e(g_1, T), H)$, where $e(g_1, T) = e(g_1, u)$;

– picks $\alpha_1, \ldots, \alpha_{j-1}, r_1, \ldots, r_{j-1} \leftarrow_R \mathbb{Z}_N$;

– simulates KeyGen on input id by choosing $R_3' \leftarrow_R G_{p_3}$ and outputting

$$T^{\frac{1}{\alpha+\mathsf{id}}} g_{2,3}^{\sum_{i=1}^{j-1} \frac{r_i}{\alpha_i+\mathsf{id}}} R_3'$$

– uses C to compute $(\mathsf{ct}_0, \kappa_0)$;

Observe that if $T = uR_3$, then this is exactly Game $3.j - 1.1$, and if $T = ug_2^{r_j}R_3$, then this is exactly Game $3.j.0$. It follows readily that

$$\mathsf{Adv}_2 - \mathsf{Adv}_3 \leq (q+1) \cdot \mathsf{Adv}_{\mathcal{G},\mathcal{A}_2}^{\mathrm{SD2}}(\lambda).$$

Game 4. We replace $\sum_{i=1}^{q+1} \frac{r_i}{\alpha_i + \mathsf{id}}$ in sk_id with $\mathsf{RF}(\mathsf{id})$ where $\mathsf{RF} : \mathbb{Z}_N \to \mathbb{Z}_{p_2}$ is a truly random function; that is, sk_id is now given by $u^{\frac{1}{\alpha + \mathsf{id}}} g_2^{\mathsf{RF}(\mathsf{id})} R_3$. It follows readily from Lemma 1 that

$$\mathsf{Adv}_3 - \mathsf{Adv}_4 \leq O(q^2/p_2).$$

Game 5. We replace $\kappa_0 = \mathsf{H}(e(C, \mathsf{sk}_{\mathsf{id}^*}))$ with $\kappa_0 \leftarrow_\mathrm{R} \{0,1\}^\lambda$. Observe that the quantity (from which κ_0 is derived)

$$e(C, \mathsf{sk}_{\mathsf{id}^*}) = e(C, u^{\frac{1}{\alpha + \mathsf{id}^*}} g_2^{\mathsf{RF}(\mathsf{id}^*)}) = e(C, u^{\frac{1}{\alpha + \mathsf{id}^*}}) \cdot \boxed{e(C, g_2^{\mathsf{RF}(\mathsf{id}^*)})}$$

has $\log p_2 = \Theta(\lambda)$ bits of min-entropy coming from $\mathsf{RF}(\mathsf{id}^*)$, since $\mathsf{id}^* \notin \{\mathsf{id}_1, \ldots, \mathsf{id}_q\}$; this holds as long as the G_{p_2}-component of C is not 1, which happens with probability $1 - 1/p_2$. Then, by the left-over hash lemma, $\kappa_0 = \mathsf{H}(e(C, \mathsf{sk}_{\mathsf{id}^*}))$ is $2^{-\Omega(\lambda)}$-close to the uniform distribution over $\{0,1\}^\lambda$, even given $\mathsf{ct}_0 = C$.

In Game 5, the joint distribution of $(\kappa_0, \mathsf{ct}_0)$ is uniformly random over $\{0,1\}^\lambda \times \mathcal{C}$, where $\mathcal{C} := G_{p_1}G_{p_2}$. Therefore, the view of the adversary $\mathcal{A}$ is statistically independent of the challenge bit b. Hence, $\mathsf{Adv}_5 = 0$. This completes the proof. $\square$

Setup(G):	Enc(mpk, id $\in \mathbb{Z}_p$):
$\mathsf{msk} := (\mathbf{W}, \mathbf{u}) \leftarrow_\mathrm{R} \mathbb{Z}_p^{(k+1)\times(k+1)} \times \mathbb{Z}_p^k$;	pick $\mathbf{s} \leftarrow_\mathrm{R} \mathbb{Z}_p^k$;
$\mathsf{mpk} := ([\mathbf{A}]_1, [\mathbf{A}^\top\mathbf{W}]_1, [\mathbf{A}^\top\mathbf{Bu}]_T)$;	return (ct, κ) $:=$ $([\mathbf{s}^\top\mathbf{A}^\top(\mathbf{W} + $
return (mpk, msk)	$\mathsf{id}\mathbf{I}_{k+1})]_1, [\mathbf{s}^\top\mathbf{A}^\top\mathbf{Bu}]_T)$;
KeyGen(msk, id $\in \mathbb{Z}_p$):	Dec(sk$_\mathsf{id}$, ct):
return $\mathsf{sk}_\mathsf{id} := [(\mathbf{W} + \mathsf{id}\mathbf{I}_{k+1})^{-1}\mathbf{Bu}]_2$	return $e(\mathsf{ct}, \mathsf{sk}_\mathsf{id})$

Fig. 4. Candidate IBE in prime-order bilinear groups under the k-LIN assumption, following the Diffie-Hellman framework and notation in [13]. Here, $\mathbf{A}, \mathbf{B} \in \mathbb{Z}_p^{k\times(k+1)}$ denote the matrices for the k-LIN assumptions in G_1 and G_2 respectively. Both the keys and the ciphertext contain $k + 1$ group elements, i.e. 2 elements under SXDH = 1-LIN.

3.3   A Candidate Prime-Order Scheme

In Fig. 4, we present a *candidate* prime-order scheme obtained by applying the transformation in [12] to our composite-order IBE scheme; concretely, the transformation was used to obtain prime-order dual-system ABE schemes starting from composite-order ones based on the same decisional subgroup assumptions as used in this work. The ciphertext and secret keys in the candidate scheme contain $k+1$ group elements, which is a substantial improvement over the state-of-the-art, c.f. Fig. 5. Applying Naor's transformation then yields a signature scheme with signature size $k+1$ group elements. In contrast, a scheme that uses both the $(p_1 \mapsto p_1 p_2)$-subgroup and $(p_2 \mapsto p_1 p_2)$-subgroup assumptions as in [10] would likely require at least $2k$ group elements, which is another reason to eliminate the use of the $(p_2 \mapsto p_1 p_2)$-subgroup assumption.

We stress that we do not have a proof of security for this scheme. The main technical difficulties arise from having to understand the matrix inverse $(\mathbf{W} + \mathrm{id}\mathbf{I}_{k+1})^{-1}$ for general matrices $\mathbf{W}$. For this specific scheme, it appears that we can completely recover $\mathbf{W} \in \mathbb{Z}_p^{k \times k}$ given $(\mathbf{W} + \mathrm{id}\mathbf{I}_{k+1})^{-1}\mathbf{B} \in \mathbb{Z}_p^k$ for many choices of id, which ruins parameter-hiding in the secret key space. On the other hand, in the composite-order scheme, given $\frac{1}{\alpha + \mathrm{id}}$ mod p_1 for an unbounded number of id still completely hides α mod p_2. Nonetheless, we conjecture that a more judicious choice of a matrix distribution for $\mathbf{W}$ would yield a variant of this scheme which is adaptively secure under the k-linear assumption. We quickly point out here that diagonal matrices don't work.

Scheme	|mpk|	|sk|	|ct|	anon.	assumption
W05 [28]	$(4+\lambda)\|G_1\|$	$2\|G_2\|$	$2\|G_1\|$	–	DBDH
G06 [18]	$2\|G_1\|+2\|G_T\|$	$\|G_2\|+\|\mathbb{Z}_p\|$	$\|G_1\|+\|G_T\|$	✓	q-ABDHE
L12, W09 [22, 29]	$24\|G_1\|+\|G_T\|$	$6\|G_2\|$	$6\|G_1\|$	–	DLIN
CLLWW12 [11]	$8\|G_1\|+\|G_T\|$	$4\|G_2\|$	$4\|G_1\|$	✓	SXDH
JR13 [20]	$6\|G_1\|+\|G_T\|$	$5\|G_2\|$	$3\|G_1\|+\|\mathbb{Z}_p\|$	✓	SXDH
Fig 4 *	$4\|G_1\|+\|G_T\|$	$2\|G_2\|$	$2\|G_1\|$	?	*SXDH?*

Fig. 5. Comparison amongst adaptively secure IBEs from standard assumptions in prime-order bilinear groups. We refer to both groups of prime order p with pairing $e : G_1 \times G_2 \to G_T$. We included the candidate scheme in Fig. 4 for comparison, and we stress that we do not have a proof of security for the scheme.

4   Broadcast Encryption

In broadcast encryption [14], a sender broadcasts encrypted content in such a way that only a specified set of authorized receivers may decrypt the message. In this section, we present a selectively secure broadcast encryption scheme for n users, where the ciphertext overhead and the secret keys are a constant number of group elements, and security is based on the decisional subgroup assumption in composite-order groups. Previous dual-system broadcast encryption schemes [16,19,29] achieve adaptive security under static assumptions, but never better than a $(t, n/t)$-type trade-off between ciphertext overhead and key size [17].

4.1 Overview

We begin with an informal description of the scheme, ignoring randomization in the G_{p_3}-subgroup. The scheme is derived from the Boneh-Gentry-Waters (BGW) broadcast encryption scheme [6], which is also selectively secure under the q-DBDHE assumption. The public parameters in our scheme are given by

$$\mathsf{mpk} := (g_1^\gamma, g_1^\alpha, g_1^{\alpha^2}, \ldots, g_1^{\alpha^n}, u^\alpha, u^{\alpha^2}, \ldots, u^{\alpha^n}, u^{\alpha^{n+2}}, \ldots, u^{\alpha^{2n}})$$

The ciphertext for a subset $\Gamma \subseteq [n]$ and the key for a user $y \in [n]$ are given by

$$\mathsf{ct}_\Gamma := (g_1^s, g_1^{(\gamma + \sum_{k \in \Gamma} \alpha^k)s}, e(g_1, u^{\alpha^{n+1}})^s \cdot m), \quad \mathsf{sk}_y := u^{\alpha^{n-y+1}\gamma}$$

Decryption proceeds analogously to the BGW scheme, and requires a judicious choice of pairing-product equation to recover $e(g_1, u^{\alpha^{n+1}})^s$. We note that $u^{\alpha^{n+1}}$ is omitted from mpk. Indeed, given g_1^s and mpk, it is easy to compute $e(g_1, u^{\alpha^k})^s$ for any $k \neq n + 1$. We also note that the BGW scheme uses $u = g_1$.

To establish security, we will introduce random G_{p_2}-components to the $2n$ terms $u^\alpha, u^{\alpha^2}, \ldots, u^{\alpha^{2n}}$ (including $u^{\alpha^{n+1}}$), and the extra entropy from $u^{\alpha^{n+1}}$ will be used to hide the message m. That is, we apply the Déjà Q framework to the set of $2n$ linearly independent monomials $\{\alpha, \alpha^2, \ldots, \alpha^{2n}\}$, as encoded "in the exponent of u" in the secret keys. To achieve this, we proceed as follows:

$$u^{\alpha^k} \xrightarrow{\text{subgroup}} u^{\alpha^k} g_2^{r_1 \alpha^k} \xrightarrow{\text{CRT}} u^{\alpha^k} g_2^{r_1 \alpha_1^k}$$
$$\xrightarrow{\text{subgroup}} u^{\alpha^k} g_2^{r_2 \alpha^k} g_2^{r_1 \alpha_1^k} \xrightarrow{\text{CRT}} u^{\alpha^k} g_2^{r_2 \alpha_2^k + r_1 \alpha_1^k}$$
$$\longrightarrow \quad \cdots \quad \xrightarrow{\text{CRT}} u^{\alpha^k} g_2^{r_{2n} \alpha_{2n}^k + \cdots + r_2 \alpha_2^k + r_1 \alpha_1^k}$$

where $r_1, \ldots, r_{2n}, \alpha_1, \ldots, \alpha_{2n} \leftarrow_{\mathrm{R}} \mathbb{Z}_N$. We can then replace

$$k \mapsto r_{2n} \alpha_{2n}^k + \cdots + r_2 \alpha_2^k + r_1 \alpha_1^k$$

by a truly random function $\mathsf{RF}(\cdot)$. As with the IBE scheme, we need to avoid leaking $\alpha \bmod p_2$ in the ciphertext in order to carry out the transformation to the secret keys above. That is, we need to eliminate all occurrences of α in the polynomial $\gamma + \sum_{k \in \Gamma} \alpha^k$ which shows up in the ciphertext. Unfortunately, we do not know a transformation to the ciphertext distribution analogous to that for the IBE. Instead, we will need to settle for selective security where the adversary announces the subset Γ at the very beginning, so that we can use γ as a one-time pad. We will then select $\tilde{\gamma}$ at random (which is treated as a known scalar) and program γ so that $\tilde{\gamma} = \gamma + \sum_{k \in \Gamma} \alpha^k$. We can then rewrite the ciphertext and key as

$$\mathsf{ct}_\Gamma := (g^s, g^{\tilde{\gamma}s}, e(g, u^{\alpha^{n+1}})^s \cdot m), \quad \mathsf{sk}_y := (u^{\alpha^{n-y+1}\tilde{\gamma} - \sum_{k \in \Gamma} \alpha^{n+1-y+k}})$$

Now, the monomials in α only show up on the same side of the pairing in both the ciphertext and the secret keys in the exponents of u. As in the security proof for the BGW scheme, we will later use the fact that the monomial α^{n+1} does not show up in any sk_y for which $y \notin \Gamma$. We note that in the proof of security, the distribution of mpk changes, which is quite unusual for a proof based on the dual system methodology.

Setup($\mathbb{G}, 1^n$):	**Enc(mpk, $\Gamma \subseteq [n]$):**
$(\alpha, \gamma, u) \leftarrow_R \mathbb{Z}_N^2 \times G_{p_1}$;	pick $s \leftarrow_R \mathbb{Z}_N$;
pick $R'_{3,k} \leftarrow_R G_{p_3}$;	$\mathsf{ct}_\Gamma := (g_1^s, g_1^{(\gamma + \sum_{k \in \Gamma} \alpha^k)s})$;
$u'_k := u^{\alpha^k} R_{3,k}$, for $k = 1, \ldots, 2n$;	$\kappa := \mathsf{H}(e(g_1, u^{\alpha^{n+1}})^s)$;
pick $R_{3,y} \leftarrow_R G_{p_3}$;	return $(\mathsf{ct}_\Gamma, \kappa)$;
$\mathsf{sk}_y := u^{\alpha^{n-y+1}\gamma} R_{3,y}$, for $y = 1, \ldots, n$;	
$\mathsf{mpk} := (g_1, g_1^\gamma, e(g_1, u'_{n+1}), \mathsf{H},$	**Dec(mpk, sk_y, $\mathsf{ct}_\Gamma = (c_0, c_1)$):**
$g_1^\alpha, \ldots, g_1^{\alpha^n},$	$\kappa' := e(c_1, u'_{n-y+1}) \cdot$
$u'_1, u'_2, \ldots, u'_n, u'_{n+2}, \ldots, u'_{2n})$;	$e(c_0, \mathsf{sk}_y \prod_{k \in \Gamma, k \neq y} u'_{n+1+(k-y)})$;
return $(\mathsf{mpk}, (\mathsf{sk}_1, \ldots, \mathsf{sk}_n))$	return $\mathsf{H}(\kappa')$

Fig. 6. Broadcast encryption w.r.t. a composite-order bilinear group $\mathbb{G}$. Here, $\mathsf{H} : G_T \to \{0,1\}^\lambda$ is drawn from a family of pairwise-independent hash functions.

4.2 Our Broadcast Encryption Scheme

Theorem 2. *The scheme in Fig. 6 is a selectively secure broadcast encryption scheme under the decisional subgroup assumption in $\mathbb{G}$.*

Proof. Correctness follows readily from the fact that for all $y \in \Gamma$,

$$e\left(g_1^{(\gamma + \sum_{k \in \Gamma} \alpha^k)s}, u^{\alpha^{n-y+1}}\right) \cdot e\left(g_1^s, u^{\alpha^{n-y+1}\gamma} \prod_{k \in \Gamma, k \neq y} u^{\alpha^{n+1+(k-y)}}\right) = e(g_1, u^{\alpha^{n+1}})^s.$$

Note that for all $k \neq y$, $n + 1 + (k - y) \in \{2, \ldots, n, n+2, \ldots, 2n\}$, which means we can compute $u^{\alpha^{n+1+(k-y)}}$ given mpk. Next, we show that for any adversary $\mathcal{A}$ against the broadcast encryption scheme, there exist adversaries $\mathcal{A}_1, \mathcal{A}_2$ whose running times are essentially the same as that of $\mathcal{A}$, such that

$$\mathsf{Adv}_{\mathcal{A}}^{\text{s-BCE}}(\lambda) \leq \mathsf{Adv}_{\mathbb{G}, \mathcal{A}_1}^{\text{SD1}}(\lambda) + 2n \cdot \mathsf{Adv}_{\mathbb{G}, \mathcal{A}_2}^{\text{SD2}}(\lambda) + 2^{-\Omega(\lambda)}$$

We proceed via a series of games and we use Adv_i to denote the advantage of $\mathcal{A}$ in Game i.

Game 0. This is the real experiment as defined in Sect. 2.3.

Game 1. Pick $(\alpha, \tilde{\gamma}, u) \leftarrow_R \mathbb{Z}_N^2 \times G_{p_1}$ and set $\gamma := \tilde{\gamma} - \sum_{k \in \Gamma^*} \alpha^k$, where Γ^* is the selective challenge output by $\mathcal{A}$. Then,

- compute $u'_1, \ldots, u'_{2n}$ as in the honest Setup;
- compute mpk as in the honest Setup.
- compute $\mathsf{ct}_{\Gamma^*} = (g_1^s, (g_1^s)^{\tilde{\gamma}})$ and $\kappa_0 = \mathsf{H}(e(g_1^s, u'_{n+1}))$;
- simulate $\{\mathsf{sk}_y : y \notin \Gamma^*\}$ using $\tilde{\gamma}$ and $(u'_1, \ldots, u'_n, u'_{n+2}, \ldots, u'_{2n})$, by computing

$$\mathsf{sk}_y = (u'_{n-y+1})^{\tilde{\gamma}} \cdot \Big(\prod_{k \in \Gamma^*, k \neq y} u'_{n+1+(k-y)} \Big)^{-1} \cdot R_{3,y}$$

Clearly, Game 0 and 1 are identically distributed, so $\mathsf{Adv}_0 = \mathsf{Adv}_1$.

Game 2. We change the distribution of $(\mathsf{ct}_{\Gamma^*}, \kappa_0)$ by replacing g_1^s with $C \leftarrow_R G_{p_1} G_{p_2}$, that is

$$(\mathsf{ct}_{\Gamma^*}, \kappa_0) := ((C, C^{\tilde{\gamma}}), \mathsf{H}(e(C, u'_{n+1}))$$

It is straight-forward to construct $\mathcal{A}_1$ (following the proof for Theorem 1) for which

$$\mathsf{Adv}_1 - \mathsf{Adv}_2 \le \mathsf{Adv}_{\mathcal{G}, \mathcal{A}_1}^{\mathrm{SD1}}(\lambda).$$

Game 3. We change the distribution of $u'_1, \ldots, u'_{2n}$ from $u^{\alpha^k} R'_{3,k}$ to $u^{\alpha^k} g_2^{\sum_{i=1}^{2n} r_i \alpha_i^k} R'_{3,k}$, where $r_1, \ldots, r_{2n}, \alpha_1, \ldots, \alpha_{2n} \leftarrow_R \mathbb{Z}_N$, as outlined in Sect. 4.1; this in turn affects the distribution of mpk, κ_0 and $\{\mathsf{sk}_y : y \notin \Gamma^*\}$. We proceed via a series of sub-games 3.j.0 and 3.j.1 for $j = 1, 2, \ldots, 2n$, where

- In Sub-Game 3.j.0, u'_k is given by $u^{\alpha^k} g_2^{r_j \alpha^k + \sum_{i=1}^{j-1} r_i \alpha_i^k} R'_{3,k}$ for $k = 1, \ldots, 2n$;
- In Sub-Game 3.j.1, u'_k is given by $u^{\alpha^k} g_2^{\sum_{i=1}^{j} r_i \alpha_i^k} R'_{3,k}$ for $k = 1, \ldots, 2n$.

  Game 2 corresponds to Sub-Game 3.0.1, and Game 3 corresponds to Sub-Game 3.$2n$.1.

First, observe that $\mathsf{Adv}_{3.j.0} = \mathsf{Adv}_{3.j.1}$ as before. Next, for $j = 1, \ldots, 2n$, we construct $\mathcal{A}_2$ for which

$$\mathsf{Adv}_{3.(j-1).1} - \mathsf{Adv}_{3.j.0} \le \mathsf{Adv}_{\mathcal{G}, \mathcal{A}_2}^{\mathrm{SD2}}(\lambda).$$

$\mathcal{A}_2$ on input $(g_1, g_{\{2,3\}}, g_3, C, T)$ where $C \leftarrow_R G_{p_1} G_{p_2}$ and either $T = u R'_{3,k} \leftarrow_R G_{p_1} G_{p_3}$ or $T = u g_2^{r_j} R'_{3,k} \leftarrow_R G_{p_1} G_{p_2} G_{p_3}$, simulates the experiment in Game 2 with the adversary $\mathcal{A}$ as follows:

- picks $\alpha, \alpha_1, \ldots, \alpha_{j-1}, r_1, \ldots, r_{j-1} \leftarrow_R \mathbb{Z}_N$;
- for $k = 1, \ldots, 2n$, computes u'_k by choosing $R'_{3,k} \leftarrow_R G_{p_3}$ and outputting

$$T^{\alpha^k} g_{2,3}^{\sum_{i=1}^{j-1} r_i \alpha_i^k} R'_{3,k}$$

- proceed as in Game 2 using $\alpha, u'_1, \ldots, u'_{2n}$ as computed above to compute mpk and $\{\mathsf{sk}_y : y \notin \Gamma^*\}$, and using C as provided and u'_{n_1} as computed above to compute $(\mathsf{ct}_{\Gamma^*}, \kappa_0)$.

Observe that if $T = u R'_{3,k}$, then this is exactly Game 3.$j - 1.1$, and if $T = u g_2^{r_j} R'_{3,k}$, then this is exactly Game 3.j.0. It follows readily that

$$\mathsf{Adv}_2 - \mathsf{Adv}_3 \le 2n \cdot \mathsf{Adv}_{\mathcal{G}, \mathcal{A}_2}^{\mathrm{SD2}}(\lambda).$$

Game 4. We replace $\sum_{i=1}^{2n} r_i \alpha_i^k$ in u_k' with $\mathsf{RF}(k)$ where $\mathsf{RF} : [2n] \to \mathbb{Z}_{p_2}$ is a truly random function; that is, u_k' is now given by $u^{\alpha^k} g_2^{\mathsf{RF}(k)} R_{3,k}'$, for $k = 1, \ldots, 2n$. Now, we exploit the fact that the Vandermonde matrix

$$\begin{pmatrix} \alpha_1 & \alpha_2 & \cdots & \alpha_{2n} \\ \vdots & \vdots & \ddots & \vdots \\ \alpha_1^{2n} & \alpha_2^{2n} & \cdots & \alpha_{2n}^{2n} \end{pmatrix}$$

is invertible as long as $\alpha_1, \ldots, \alpha_{2n} \bmod p_2$ are distinct, which happens with overwhelming probability over $\alpha_1, \ldots, \alpha_{2n} \leftarrow_{\mathsf{R}} \mathbb{Z}_N$. It follows readily that

$$\mathsf{Adv}_3 - \mathsf{Adv}_4 \leq O(n^2/p_2).$$

Game 5. We replace $\kappa_0 = \mathsf{H}(e(C, u_{n+1}'))$ with $\kappa_0 \leftarrow_{\mathsf{R}} \{0,1\}^\lambda$. First, recall from Game 1 that $\{\mathsf{sk}_y : y \notin \Gamma^*\}$ only depend on $u_1', \ldots, u_n', u_{n+2}', \ldots, u_{2n}'$; therefore, they only depend on $\mathsf{RF}(1), \ldots, \mathsf{RF}(n), \mathsf{RF}(n+2), \ldots, \mathsf{RF}(2n)$ and do not reveal any information about $\mathsf{RF}(n+1)$. Then, the quantity (from which κ_0 is derived)

$$e(C, u_{n+1}') = e(C, u^{\alpha^{n+1}} g_2^{\mathsf{RF}(n+1)}) = e(C, u^{\alpha^{n+1}}) \cdot \boxed{e(C, g_2^{\mathsf{RF}(n+1)})}$$

has $\log p_2 = \Theta(\lambda)$ bits of min-entropy coming from $\mathsf{RF}(n+1)$; this holds as long as the G_{p_2}-component of C is not 1, which happens with probability $1 - 1/p_2$. Then, by the left-over hash lemma, $\kappa_0 = \mathsf{H}(e(C, u_{n+1}'))$ is $2^{-\Omega(\lambda)}$-close to the uniform distribution over $\{0,1\}^\lambda$.

In Game 5, both κ_0, κ_1 are uniformly random over $\{0,1\}^\lambda$. Therefore, the view of the adversary $\mathcal{A}$ is statistically independent of the challenge bit b. Hence, $\mathsf{Adv}_5 = 0$. This completes the proof. $\qquad\square$

Acknowledgments. I would like to thank Allison Bishop, Dan Boneh, Melissa Chase, Jie Chen, Sarah Meiklejohn and Alain Passelègue for helpful discussions.

A   Asymmetric Composite-Order Bilinear Groups

In this section, we outline the extension of our result to asymmetric composite-order bilinear groups. Here, we can work with groups whose group order is the product of two primes, and we obtain IBE and signature schemes (shown in Figs. 7 and 8) where the key generation and signing algorithms are deterministic. We state the underlying decisional subgroup assumptions, and the proofs are exactly analogous to the ones from before.

Asymmetric Composite-Order Bilinear Groups. The generator $\mathcal{G}$ takes as input a security parameter λ and outputs a description $\mathbb{G} := (N, G, H, G_T, e)$, where N is product of distinct primes of $\Theta(\lambda)$ bits, G, H and G_T are cyclic groups of order N, and $e : G \times H \to G_T$ is a non-degenerate bilinear map. We consider bilinear groups where N is the product of two distinct primes p_1, p_2 (that is, $N = p_1 p_2$). We can write $G = G_{p_1} G_{p_2}$ where G_{p_1}, G_{p_2} are subgroups of G of order p_1 and p_2 respectively. In addition, we use $G_{p_i}^*$ to denote $G_{p_i} \setminus \{1\}$. We will often write g_1, g_2 to denote random generators for the subgroups G_{p_1}, G_{p_2}. We can also write $H = H_{p_1} H_{p_2}$, where $H_{p_1}, H_{p_2}, h_1, h_2$ are defined analogously.

Cryptographic Assumptions. Our construction relies on the following two subgroup decisional assumptions. We define the following two advantage functions:

$$\mathsf{Adv}^{\mathrm{SD1}}_{\mathcal{G},\mathcal{A}}(\lambda) := \left| \Pr[\mathcal{A}(D, T_0) = 1] - \Pr[\mathcal{A}(D, T_1) = 1] \right|$$

$$\text{where } \mathbb{G} \leftarrow \mathcal{G}, T_0 \leftarrow \boxed{G_{p_1}}, T_1 \leftarrow_{\mathrm{R}} \boxed{G_{p_1} G_{p_2}}$$

$$\text{and } D := (g_1, g_{\{1,2\}}, h_1, h_{\{1,2\}}), g_1 \leftarrow_{\mathrm{R}} G_{p_1}^*, g_{\{1,2\}} \leftarrow_{\mathrm{R}} G_{p_1} G_{p_2},$$

$$h_1 \leftarrow_{\mathrm{R}} H_{p_1}^*, h_{\{1,2\}} \leftarrow_{\mathrm{R}} H_{p_1} H_{p_2}$$

$$\mathsf{Adv}^{\mathrm{SD2}}_{\mathcal{G},\mathcal{A}}(\lambda) := \left| \Pr[\mathcal{A}(D, T_0) = 1] - \Pr[\mathcal{A}(D, T_1) = 1] \right|$$

$$\text{where } \mathbb{G} \leftarrow \mathcal{G}, T_0 \leftarrow \boxed{H_{p_1}}, T_1 \leftarrow_{\mathrm{R}} \boxed{H_{p_1} H_{p_2}}$$

$$\text{and } D := (h_1, h_2, h_{\{1,2\}}, g_1, g_{\{1,2\}}), h_1 \leftarrow_{\mathrm{R}} H_{p_1}^*, h_2 \leftarrow_{\mathrm{R}} H_{p_2}^*,$$

$$h_{\{1,2\}} \leftarrow_{\mathrm{R}} H_{p_1} H_{p_2}, g_1 \leftarrow_{\mathrm{R}} G_{p_1}^*, g_{\{1,2\}} \leftarrow_{\mathrm{R}} G_{p_1} G_{p_2}$$

The decisional subgroup assumptions assert that that for all PPT adversaries $\mathcal{A}$, the advantages $\mathsf{Adv}^{\mathrm{SD1}}_{\mathcal{G},\mathcal{A}}(\lambda)$ and $\mathsf{Adv}^{\mathrm{SD2}}_{\mathcal{G},\mathcal{A}}(\lambda)$ are negligible functions in λ.

$\underline{\mathsf{Setup}(\mathbb{G}):}$	$\underline{\mathsf{Enc}(\mathsf{mpk}, \mathsf{id} \in \mathbb{Z}_N):}$
$\mathsf{msk} := (\alpha, u) \leftarrow_{\mathrm{R}} \mathbb{Z}_N \times H_{p_1};$	pick $s \leftarrow_{\mathrm{R}} \mathbb{Z}_N;$
$\mathsf{mpk} := (g_1, g_1^{\alpha}, e(g_1, u), \mathsf{H});$	return $(\mathsf{ct}, \kappa) := (g_1^{(\alpha + \mathsf{id})s}, \mathsf{H}(e(g_1, u)^s))$
return $(\mathsf{mpk}, \mathsf{msk})$	
	$\underline{\mathsf{Dec}(\mathsf{sk}_{\mathsf{id}}, \mathsf{ct}):}$
$\underline{\mathsf{KeyGen}(\mathsf{msk}, \mathsf{id} \in \mathbb{Z}_N):}$	return $\mathsf{H}(e(\mathsf{ct}, \mathsf{sk}_{\mathsf{id}}))$
return $\mathsf{sk}_{\mathsf{id}} := u^{\frac{1}{\alpha + \mathsf{id}}}$	

Fig. 7. Adaptively secure anonymous IBE w.r.t. an asymmetric composite-order bilinear group $\mathbb{G}$. Here, $\mathsf{H} : G_T \to \{0,1\}^{\lambda}$ is drawn from a family of pairwise-independent hash functions.

Setup(𝔾):	sign(sk, $M \in \mathbb{Z}_N$):
sk := $(\alpha, u) \leftarrow_R \mathbb{Z}_N \times H_{p_1}$; pk := $(g_1, g_1^\alpha, e(g_1, u))$; return (pk, sk)	return $\sigma := u^{\frac{1}{\alpha+M}}$ verify(pk, M, σ): check $e(g_1^M \cdot g_1^\alpha, \sigma) = e(g_1, u)$

Fig. 8. Fully secure signature scheme w.r.t. an asymmetric composite-order bilinear group 𝔾.

Remark 1. Note that Assumption 2 is false if the pairing is symmetric (i.e., there exists an efficiently computable isomorphism between G and H) since we can pair with h_2 to distinguish between T_0 and T_1. The term h_2 will play the role of $g_{2,3}$ in the transitions from Game $3.(j-1).1$ to $3.j.0$ in the proofs of Theorems 1 and 2.

References

1. Attrapadung, N.: Dual system encryption via doubly selective security: framework, fully secure functional encryption for regular languages, and more. In: Nguyen, P.Q., Oswald, E. (eds.) EUROCRYPT 2014. LNCS, vol. 8441, pp. 557–577. Springer, Heidelberg (2014)
2. Blazy, O., Kiltz, E., Pan, J.: (Hierarchical) identity-based encryption from affine message authentication. In: Garay, J.A., Gennaro, R. (eds.) CRYPTO 2014, Part I. LNCS, vol. 8616, pp. 408–425. Springer, Heidelberg (2014)
3. Boneh, D., Boyen, X.: Short signatures without random oracles. In: Cachin, C., Camenisch, J.L. (eds.) EUROCRYPT 2004. LNCS, vol. 3027, pp. 56–73. Springer, Heidelberg (2004)
4. Boneh, D., Boyen, X.: Efficient selective-ID secure identity-based encryption without random oracles. In: Cachin, C., Camenisch, J.L. (eds.) EUROCRYPT 2004. LNCS, vol. 3027, pp. 223–238. Springer, Heidelberg (2004)
5. Boneh, D., Franklin, M.K.: Identity-based encryption from the Weil pairing. SIAM J. Comput. **32**(3), 586–615 (2003)
6. Boneh, D., Gentry, C., Waters, B.: Collusion resistant broadcast encryption with short ciphertexts and private keys. In: Shoup, V. (ed.) CRYPTO 2005. LNCS, vol. 3621, pp. 258–275. Springer, Heidelberg (2005)
7. Boneh, D., Goh, E.-J., Nissim, K.: Evaluating 2-DNF formulas on ciphertexts. In: Kilian, J. (ed.) TCC 2005. LNCS, vol. 3378, pp. 325–341. Springer, Heidelberg (2005)
8. Boyen, X.: General *Ad Hoc* encryption from exponent inversion IBE. In: Naor, M. (ed.) EUROCRYPT 2007. LNCS, vol. 4515, pp. 394–411. Springer, Heidelberg (2007)
9. De Caro, A., Iovino, V., Persiano, G.: Fully secure anonymous HIBE and Secret-key anonymous IBE with short ciphertexts. In: Joye, M., Miyaji, A., Otsuka, A. (eds.) Pairing 2010. LNCS, vol. 6487, pp. 347–366. Springer, Heidelberg (2010)

10. Chase, M., Meiklejohn, S.: Déjà Q: using dual systems to revisit q-type assumptions. In: Nguyen, P.Q., Oswald, E. (eds.) EUROCRYPT 2014. LNCS, vol. 8441, pp. 622–639. Springer, Heidelberg (2014). (Cryptology ePrint Archive, Report 2014/570.)

11. Chen, J., Lim, H.W., Ling, S., Wang, H., Wee, H.: Shorter IBE and signatures via asymmetric pairings. In: Abdalla, M., Lange, T. (eds.) Pairing 2012. LNCS, vol. 7708, pp. 122–140. Springer, Heidelberg (2013)

12. Chen, J., Gay, R., Wee, H.: Improved dual system ABE in prime-order groups via predicate encodings. In: Oswald, E., Fischlin, M. (eds.) EUROCRYPT 2015. LNCS, vol. 9057, pp. 595–624. Springer, Heidelberg (2015)

13. Escala, A., Herold, G., Kiltz, E., Ràfols, C., Villar, J.: An algebraic framework for diffie-hellman assumptions. In: Canetti, R., Garay, J.A. (eds.) CRYPTO 2013, Part II. LNCS, vol. 8043, pp. 129–147. Springer, Heidelberg (2013)

14. Fiat, A., Naor, M.: Broadcast encryption. In: Stinson, D.R. (ed.) CRYPTO 1993. LNCS, vol. 773, pp. 480–491. Springer, Heidelberg (1994)

15. Freeman, D.M.: Converting pairing-based cryptosystems from composite-order groups to prime-order groups. In: Gilbert, H. (ed.) EUROCRYPT 2010. LNCS, vol. 6110, pp. 44–61. Springer, Heidelberg (2010)

16. Garg, S., Kumarasubramanian, A., Sahai, A., Waters, B.: Building efficient fully collusion-resilient traitor tracing and revocation schemes. In: ACM Conference on Computer and Communications Security, pp. 121–130 (2010)

17. Gay, R., Kerenidis, I., Wee, H.: Communication complexity of conditional disclosure of secrets and attribute-based encryption. In: Gennaro, R., Robshaw, M. (eds.) CRYPTO 2015. LNCS, vol. 9216, pp. 485–502. Springer, Heidelberg (2015)

18. Gentry, C.: Practical identity-based encryption without random oracles. In: Vaudenay, S. (ed.) EUROCRYPT 2006. LNCS, vol. 4004, pp. 445–464. Springer, Heidelberg (2006)

19. Gentry, C., Waters, B.: Adaptive security in broadcast encryption systems (with short ciphertexts). In: Joux, A. (ed.) EUROCRYPT 2009. LNCS, vol. 5479, pp. 171–188. Springer, Heidelberg (2009)

20. Jutla, C.S., Roy, A.: Shorter quasi-adaptive NIZK proofs for linear subspaces. In: Sako, K., Sarkar, P. (eds.) ASIACRYPT 2013, Part I. LNCS, vol. 8269, pp. 1–20. Springer, Heidelberg (2013)

21. Katz, J., Sahai, A., Waters, B.: Predicate encryption supporting disjunctions, polynomial equations, and inner products. In: Smart, N.P. (ed.) EUROCRYPT 2008. LNCS, vol. 4965, pp. 146–162. Springer, Heidelberg (2008)

22. Lewko, A.: Tools for simulating features of composite order bilinear groups in the prime order setting. In: Pointcheval, D., Johansson, T. (eds.) EUROCRYPT 2012. LNCS, vol. 7237, pp. 318–335. Springer, Heidelberg (2012)

23. Lewko, A., Waters, B.: New techniques for dual system encryption and fully secure HIBE with short ciphertexts. In: Micciancio, D. (ed.) TCC 2010. LNCS, vol. 5978, pp. 455–479. Springer, Heidelberg (2010)

24. Lewko, A., Okamoto, T., Sahai, A., Takashima, K., Waters, B.: Fully secure functional encryption: attribute-based encryption and (hierarchical) inner product encryption. In: Gilbert, H. (ed.) EUROCRYPT 2010. LNCS, vol. 6110, pp. 62–91. Springer, Heidelberg (2010)

25. Okamoto, T., Takashima, K.: Hierarchical predicate encryption for inner-products. In: Matsui, M. (ed.) ASIACRYPT 2009. LNCS, vol. 5912, pp. 214–231. Springer, Heidelberg (2009)

26. Sakai, R., Kasahara, M.: ID based cryptosystems with pairing on elliptic curve. Cryptology ePrint Archive, Report 2003/054 (2003)

27. Shamir, A.: Identity-based cryptosystems and signature schemes. In: Blakely, G.R., Chaum, D. (eds.) CRYPTO 1984. LNCS, vol. 196, pp. 47–53. Springer, Heidelberg (1985)

28. Waters, B.: Efficient identity-based encryption without random oracles. In: Cramer, R. (ed.) EUROCRYPT 2005. LNCS, vol. 3494, pp. 114–127. Springer, Heidelberg (2005)

29. Waters, B.: Dual system encryption: realizing fully secure IBE and HIBE under simple assumptions. In: Halevi, S. (ed.) CRYPTO 2009. LNCS, vol. 5677, pp. 619–636. Springer, Heidelberg (2009)

30. Wee, H.: Dual system encryption via predicate encodings. In: Lindell, Y. (ed.) TCC 2014. LNCS, vol. 8349, pp. 616–637. Springer, Heidelberg (2014)

31. Yuen, T.H., Chow, S.S., Zhang, C., Yiu, S.M.: Exponent-inversion signatures and IBE under static assumptions. Cryptology ePrint Archive, Report 2014/311 (2014)

A Study of Pair Encodings: Predicate Encryption in Prime Order Groups

Shashank Agrawal[1](✉) and Melissa Chase[2]

[1] University of Illinois Urbana-Champaign, Champaign, USA
sagrawl2@illinois.edu
[2] Microsoft Research, Redmond, USA
melissac@microsoft.com

Abstract. Pair encodings and predicate encodings, recently introduced by Attrapadung [2] and Wee [36] respectively, greatly simplify the process of designing and analyzing predicate and attribute-based encryption schemes. However, they are still somewhat limited in that they are restricted to composite order groups, and the information theoretic properties are not sufficient to argue about many of the schemes. Here we focus on pair encodings, as the more general of the two. We first study the structure of these objects, then propose a new relaxed but still information theoretic security property. Next we show a generic construction for predicate encryption in prime order groups from our new property; it results in either semi-adaptive or full security depending on the encoding, and gives security under SXDH or DLIN. Finally, we demonstrate the range of our new property by using it to design the first semi-adaptively secure CP-ABE scheme with constant size ciphertexts.

Keywords: Predicate encryption · Attribute-based encryption · Pair encoding schemes · Dual system technique · Short ciphertexts

1 Introduction

In traditional public key encryption systems, a message is encrypted under a particular public key, with the guarantee that it can only be decrypted by the party holding the corresponding secret key. Attribute based encryption (ABE), introduced in [30], instead allows us to use attributes to determine who has the power to decrypt. In these systems, there is a single entity which publishes system parameters and distributes the appropriate decryption keys to various parties. In key-policy ABE (KP-ABE) [18], a message is encrypted under a set of attributes describing that message, and each decryption key is associated with a policy describing which ciphertexts it can decrypt. Conversely, in ciphertext-policy ABE (CP-ABE) [8] each user is given a decryption key that depends on his attributes, and ciphertexts are encrypted with policies describing which users can decrypt them. ABE has been proposed for a variety of applications, from social

---

S. Agrawal—Part of this work was done when the author was at Microsoft Research.

E. Kushilevitz and T. Malkin (Eds.): TCC 2016-A, Part II, LNCS 9563, pp. 259–288, 2016.
DOI: 10.1007/978-3-662-49099-0_10

network privacy to pay-per-view broadcasting to health record access-control to cloud security (see e.g. [1, 6, 28, 31, 34]).

Recently there has been a lot of progress in terms of both security and functionality. Using the dual system framework introduced by Waters [35], several works [23, 25] have designed ABE schemes that satisfy the natural security definition, avoiding the restrictions of selective security[1]. Other works consider extra features like short ciphertexts whose length is independent of the size of the associated attribute set and policy [5, 37], or "unbounded" schemes that place no bounds on the space of possible attributes or the number of attributes that can be tied to a ciphertext or key [24, 27, 29]. Predicate encryption [10] generalizes the concept to require only that the ciphertext and key are associated with values x, y, and decryption succeeds iff some predicate $P(x, y)$ holds. Note that in this work we assume that x and y are revealed by the ciphertext and key respectively; we do not consider attribute-hiding [11, 21] or predicate-hiding [9, 32].

As these schemes have progressed, however, constructions and proofs have become increasingly complex. Many of the proposed schemes require composite order pairings, in which the order of the pairing groups is a product of two or more primes; since these schemes require that factoring the group order is hard, this in practice means that these groups must be at least an order of magnitude larger than prime order groups of comparable security level, and according to [19] composite order pairing computations are at least 2 orders of magnitude slower. This has prompted efforts to design schemes in prime order groups [17, 20, 22, 26, 27], but many of these schemes still have fairly high cost as compared to their selectively secure counterparts, and designing and analyzing security of such schemes can be quite challenging.

Two very recent works, by Wee [36] and Attrapadung [2] make significant progress in simplifying the design and analysis of new constructions. These works introduce simple new objects, called predicate encodings and pair encodings respectively in the two works, which can be used to construct ABE and other predicate encryption schemes. Essentially, they consider one decryption key and one ciphertext, and focus on what happens in the exponent space. Both formalisms introduce simple information theoretic properties on these objects and show that if these properties are met, they can be extended into fully secure ABE/predicate encryption schemes. The major advantage of this approach is that instead of having to design and prove security of a complex scheme, now all one has to do is design and analyze an appropriate encoding, which is a much simpler task. This vastly simplifies the design of new schemes, and in fact, both works resulted in new constructions and more efficient variants of previously known schemes.

Currently these works have two primary limitations. First, they both result in ABE schemes that rely on composite order pairings, which as explained above is

---

[1] The original construction of Sahai and Waters [30], and much of the following work, considers what is referred to as the selective security model, in which the adversary must commit to the attributes/policy used in the challenge ciphertext before requesting any decryption keys.

very undesirable from an efficiency standpoint. The second drawback is that the strict information theoretic properties they require from the underlying objects mean that there are many constructions that they cannot capture in their model. Attrapadung [2] addresses this by introducing a computational security notion, which allows several more interesting constructions to be captured in the framework. However, this security notion is much harder to analyze - it involves not only the encodings in the exponent space, but also elements in the composite order group in which it is embedded, and the proofs that the encodings satisfy this notion are not only computational (rather than information theoretic) but are based on much stronger assumptions.

Still these encodings seem extremely promising as a way to simplify the design and analysis of predicate encryption schemes. In our work we further study these objects, with the aim of understanding them better and beginning to address these limitations. In particular we focus on the pair encodings from [2], as they seem to be able to capture more constructions.

Our Contributions. First, we study the *structure of pair encodings*. Attrapadung's pair encodings have only limited structural requirements. This means that he is able to capture many existing constructions in his framework, although as mentioned above, in many cases the information theoretic security property he defines does not hold for these schemes. A better understanding of the natural structure of these schemes may help to design new schemes, by providing better intuition for what is important and simply by limiting the search space.

Here we consider two structural properties. First we assume a simple property that describes where the public parameters appear in the key and ciphertext. This seems to reflect some basic structure, as all the pair encodings in [2] have this property. Looking ahead, this property allows us to instantiate these schemes efficiently in prime order groups. We then show that this implies a second, seemingly unrelated property involving the use of random variables in the key and ciphertexts. We can use this second property to simplify our security definitions and analyses.

Using this understanding, we propose a *relaxation of the information theoretic security property* proposed in [2]. This property essentially allows us to consider the scheme at smaller granularity than an entire key or ciphertext. It is still information theoretic, and it does not depend on the group in which it will be used; this means it is still easy to analyze whether a given encoding satisfies this property. We consider two flavors of this property and show that the stronger of the two is implied by the security properties in [2]. However, we will see that our new property is indeed a relaxation in that it allows us to consider encodings that did not satisfy the original property. Thus, we make a first step towards addressing the limitations of the strict information theoretic property of previous work.

Next we present a *generic construction of predicate encryption* from pair encodings. Here we make use of the dual system groups introduced by [13]; although we must modify their properties slightly, we show that their instantiations are still sufficient[2]. We show that pair encodings which satisfy the stronger flavor of our new property result in fully secure predicate encryption schemes, while pair encodings which satisfy the weaker flavor result in schemes which can still be shown to be semi-adaptively secure[3]. While full security is preferable, we will see that this second result allows us to design schemes in areas in which even selectively secure constructions are hard to construct.

This approach has two advantages. First, this means that we can transform any pair encoding scheme which satisfies the information theoretic security properties in [2] into a fully secure ABE or predicate encryption scheme *in a prime order group* based only on the SXDH or DLIN assumption. This results in schemes which are of practical efficiency, with strong security guarantees based on mild assumptions. Moreover, the advantage of this approach is that while proof of our generic construction is fairly involved, analyzing a given pair encoding scheme to verify the necessary property is still quite straightforward.

Finally, to demonstrate how our relaxed security property allows us to consider additional functionalities, we present a new pair encoding for *CP-ABE with constant-size ciphertext*. When used in our generic construction, this results in a CP-ABE with constant size-ciphertext which is semi-adaptively secure and can be instantiated under either SXDH or DLIN. To the best of our knowledge, prior to our work there were no known schemes for constant-size CP-ABE, even considering only selectively security and allowing for very strong assumptions.[4] This shows then that our new techniques allow us to consider a strictly greater range of schemes; we hope that they will continue to prove useful and lead to other interesting constructions.

Other Related Work. As mentioned above, the original works of [2,36] gave constructions only in composite order groups. In a recent work, however, Chen, Gay, and Wee [12] proposed a transformation to go from pair encodings to prime order predicate encryption schemes, requiring the same strong information theoretic property on the underlying pair encoding as in [36]. However, they also require strict restrictions on the structure of pair encodings, which are not satisfied by most of the encodings which had previously been proposed; essentially this requires that there be only one unit of randomness in each ciphertext or key. They show that the previous encodings which satisfy the information theoretic property from [2] (the basic KP- and CP-ABE schemes) have counterparts which satisfy these stricter requirements. This results in the most efficient known constructions for a number of problems. As mentioned above, our generic

---

[2] Since we use these groups in a black box way, any improvement in the underlying instantiation will translate directly into an improvement in our generic construction. In particular we believe that the simplified new dual system groups proposed in [12] satisfy our modified definitions as well, so they could be used to simplify our construction.

[3] Unlike selective security, in semi-adaptive security an adversary is not forced to commit to the challenge before seeing the public parameters.

[4] Here we discount threshold access policies because when only threshold policies are considered, CP-ABE and KP-ABE are equivalent.

construction can be applied directly to the original pair encodings [2]; this will yield similar constructions, with slightly different tradeoffs (generally smaller public parameters but slower decryption). Interestingly, our relaxed perfect security property is designed to leverage exactly the kind of structure they prohibit, so perhaps it suggests another way forward for predicates that cannot be addressed under their model.

In concurrent work, Attrapadung [3] proposed a generic construction that compiles any secure (computational or information-theoretic) pair encoding scheme for a predicate R to a fully secure FE scheme for the same predicate in prime-order groups under Matrix Diffie-Hellman assumption [16] (of which DLIN is a special case) with an additional q-type assumption in the case of pair encodings that only satisfy the computational security definition from [2]. This then also gives prime order group constructions for any predicate encoding scheme satisfying the strong information theoretic property under DLIN, and for KP-ABE with short ciphertext (as well as unbounded KP-ABE and ABE for regular languages) under a q-type assumption. However, as compared to this work, our results have the following advantages: First, we use dual system groups in a black box way, which simplifies the transformation, unifies prime and composite order group constructions, and means that any new construction of dual system groups directly gives new constructions for ABE. Moreover, our relaxed perfect security property allows us to show semi-adaptive security for the short ciphertext schemes based only on SXDH or DLIN, without any q-type assumptions; in addition to giving us the new results on CP-ABE, we can also give a much simpler proof of semi-adaptive security for Attrapadung's KP-ABE with short ciphertexts, and this proof does not require q-type assumptions. (See the full version of the paper.)

Finally, we mention the concurrent work of Attrapadung, Hanaoka, and Yamada [4]. This work presents various conversions among pair encoding schemes. Among other things, they show that if one starts with the KP-ABE scheme with constant-size ciphertexts recently proposed by Takashima [33], then by applying the conversion one gets a CP-ABE scheme with constant-size ciphertexts, which is *selectively* secure under the DLIN assumption. On the other hand, we get a *semi-adaptive* scheme secure under any assumption which can be used to construct dual system groups (which includes SXDH, DLIN, etc.). Moreover, since Takashima's construction does not use any abstractions, our construction is significantly more modular, easier to analyze and easier to extend. As we view the CP-ABE more as a test-case for the utility of our new definition and transformation, having an approach that can extend easily to other types of ABE schemes seems particularly valuable.

2   Preliminaries

We use $\cong, \equiv$ and $\approx$ to denote statistical, perfect and computational indistinguishability respectively. Security parameter is denoted by λ, and $\mathsf{negl}(\lambda)$ denotes a negligible function in λ.

We normally use lower case letters in bold to denote vectors; but if a vector's elements are themselves vectors, we use upper case. For two vectors $\mathbf{u} = (u_1, \ldots, u_n)$ and $\mathbf{v} = (v_1, \ldots, v_n)$, we use $\mathbf{u} \cdot \mathbf{v}$ to denote the entry-wise product, i.e., $(u_1 v_1, \ldots, u_n v_n)$, and $\langle u, v \rangle$ to denote the inner-product, i.e., $\sum_{i=1}^{n} u_i v_i$. The $\cdot$ operator naturally extends to vectors of vectors (or matrices): if $\mathbf{U} = (\mathbf{u}_1, \ldots, \mathbf{u}_m)$ and $\mathbf{V} = (\mathbf{v}_1, \ldots, \mathbf{v}_m)$, then $\mathbf{U} \cdot \mathbf{V} = (\mathbf{u}_1 \cdot \mathbf{v}_1, \ldots, \mathbf{u}_m \cdot \mathbf{u}_m)$. $g^{\mathbf{u}}$ should be interpreted as the vector $(g^{u_1}, \ldots, g^{u_n})$. $g^{\mathbf{A}}$, where $\mathbf{A}$ is a matrix, should be interpreted in an analogous way.

We use $\mathbf{u}_1, \ldots, \mathbf{u}_m \leftarrow \mathsf{SampAlg}(\cdot)$ to denote that the algorithm $\mathsf{SampAlg}$ is run m times with independent coin tosses to generate samples $\mathbf{u}_1, \ldots, \mathbf{u}_m$. Since the output of this algorithm is a vector, we also use $(u_1, \ldots, u_n) \leftarrow \mathsf{SampAlg}(\cdot)$ to denote that a single sample with co-ordinates $u_1, \ldots, u_n$ is drawn from $\mathsf{SampAlg}$ (this should not be confused with the previous notation). Finally, $a \leftarrow_R S$ denotes drawing an element a uniformly at random from the set S.

Bilinear Pairings: Let $\mathbb{G}, \mathbb{H}$ and $\mathbb{G}_T$ be three multiplicative groups. A pairing $e : \mathbb{G} \times \mathbb{H} \to \mathbb{G}_T$ is bilinear if for all $g \in \mathbb{G}, h \in \mathbb{H}$ and $a, b \in \mathbb{Z}$, $e(g^a, h^b) = e(g, h)^{ab}$. This pairing is non-degenerate if whenever $e(g, h) = 1_{\mathbb{G}_T}$, then either $g = 1_{\mathbb{G}}$ or $h = 1_{\mathbb{H}}$ (where $1_{\mathbb{G}}$, for instance, denotes the identity element of $\mathbb{G}$.) We will only be interested in bilinear pairings that are efficiently computable.

The order of an element g of a group G is the smallest positive integer a such that $g^a = 1_G$. The exponent of a group is defined as the least common multiple of the orders of all elements of the group. One can show that if a non-degenerate bilinear pairing $e : \mathbb{G} \times \mathbb{H} \to \mathbb{G}_T$ can be defined over three groups $\mathbb{G}, \mathbb{H}$ and $\mathbb{G}_T$, then they all have the same exponent. We use $\exp(G)$ to denote the exponent of a group G.

Homomorphism: A homomorphism from a group $\langle G, \cdot \rangle$ to a group $\langle H, \oplus \rangle$ is a function $\psi : G \to H$ such that for all $g_1, g_2 \in G$, $\psi(g_1 \cdot g_2) = \psi(g_1) \oplus \psi(g_2)$. We define two sets with respect to a homomorphism: $\mathsf{Image}(\psi) = \{\psi(g) \mid g \in G\}$ and $\mathsf{Kernel}(\psi) = \{g \in G \mid \psi(g) = 1_H\}$.

2.1   Predicate Encryption (PE)

An encryption scheme for a predicate family $P = \{P_\kappa\}_{\kappa \in \mathbb{N}^c}$ over a message space $\mathcal{M} = \{\mathcal{M}_\lambda\}_{\lambda \in \mathbb{N}}$ consists of four PPT algorithms which satisfy a correctness condition defined below.

- $\mathsf{Setup}(1^\lambda, \mathsf{par}) \to (\mathsf{MPK}, \mathsf{MSK})$. The Setup algorithm takes as input the unary representation of the security parameter λ and some additional parameters par. It outputs a master public key MPK and a master secret key MSK. The output of Setup defines a number $N \in \mathbb{N}$ (perhaps implicitly), and κ is set to (N, par).
- $\mathsf{Encrypt}(\mathsf{MPK}, x, m) \to \mathsf{CT}$. The encryption algorithm takes public parameters MPK, an $x \in \mathcal{X}_\kappa$ and an $m \in \mathcal{M}_\lambda$ as inputs, and outputs a ciphertext CT.
- $\mathsf{KeyGen}(\mathsf{MPK}, \mathsf{MSK}, y) \to \mathsf{SK}$. The key generation algorithm takes as input the public parameters MPK, the master secret key MSK and a $y \in \mathcal{Y}_\kappa$, and outputs a secret key SK.

– Decrypt(MPK, SK, CT) → m'. The decryption algorithm takes as input the public parameters MPK, a secret key SK and a ciphertext CT, and outputs a message $m' \in \mathcal{M}_\lambda$.

Correctness: For all λ and par, MPK and MSK output by Setup(1^λ, par), $m \in \mathcal{M}_\lambda$, $x \in \mathcal{X}_\kappa$ and $y \in \mathcal{Y}_\kappa$ such that $P_\kappa(x, y) = 1$, if

$$\text{CT} \leftarrow \text{Encrypt}(\text{MPK}, x, m) \quad \text{SK} \leftarrow \text{KeyGen}(\text{MPK}, \text{MSK}, y),$$

then

$$\Pr[\text{Decrypt}(\text{MPK}, \text{CT}, \text{SK}) \neq m] \leq \text{negl}(\lambda),$$

where the probability is over the random coin tosses of Encrypt, KeyGen and Decrypt.

Security: Let Π be an encryption scheme for a predicate family $P = \{P_\kappa\}_{\kappa \in \mathbb{N}^c}$ over a message space $\mathcal{M} = \{\mathcal{M}_\lambda\}_{\lambda \in \mathbb{N}}$. Consider the following experiment $\text{Expt}^{(b)}_{\mathcal{A}, \Pi}(\lambda, \text{par})$ between an adversary $\mathcal{A}$ and a challenger Chl for $b \in \{0, 1\}$ when both are given input 1^λ and par:

1. Setup: Chl runs Setup(1^λ, par) to obtain MPK and MSK. It gives MPK to $\mathcal{A}$.
2. Query: $\mathcal{A}$ issues a key query by sending $y \in \mathcal{Y}_\kappa$ to Chl, and obtains SK ← KeyGen(MPK, MSK, y) in response. This step can be repeated any number of times $\mathcal{A}$ desires.
3. Challenge: $\mathcal{A}$ sends two messages $m_0, m_1 \in \mathcal{M}_\lambda$ and an $x \in \mathcal{X}_\kappa$ to Chl, and gets CT ← Encrypt(MPK, x, m_b) as the challenge ciphertext.
4. Query: This step is identical to step 2.

At the end of the experiment, $\mathcal{A}$ outputs a bit which is defined to be the output of the experiment. We call an adversary admissible if for every $y \in \mathcal{Y}_\kappa$ queried in steps 2 and 4, $P_\kappa(x, y) = 0$. This prevents $\mathcal{A}$ from succeeding in the experiment simply by decrypting CT.

Definition 1. *An encryption scheme Π is adaptively or fully secure for a predicate family $P = \{P_\kappa\}_{\kappa \in \mathbb{N}^c}$ if for every* PPT *admissible adversary $\mathcal{A}$ and every* par,

$$|\Pr[\text{Expt}^{(0)}_{\mathcal{A}, \Pi}(\lambda, \text{par}) = 1] - \Pr[\text{Expt}^{(1)}_{\mathcal{A}, \Pi}(\lambda, \text{par}) = 1]| \leq \text{negl}(\lambda),$$

where the probabilities are taken over the coin tosses of $\mathcal{A}$ and Chl. *On the other hand, Π is semi-adaptively secure if the above condition is satisfied w.r.t. to a modified experiment where $\mathcal{A}$ provides $x \in \mathcal{X}_\kappa$ to* Chl *right after the setup phase (instead of the challenge phase), i.e., before it starts querying* [15].

3   Pair Encoding Schemes

The notion of pair encoding schemes (PES) was introduced by Attrapadung [2]. Our definition of this scheme is slightly different from the one given by [2] in that we place a restriction on the structure. Though the latter definition is more general, we believe that our formulation mirrors the concrete design of such schemes more closely. In particular, all the constructions of pair encoding schemes given in [2] fit into our framework without any changes.

We first present the definition given by Attrapadung and discuss the restrictions we impose afterwards. A pair encoding scheme for a predicate family $P_\kappa : \mathcal{X}_\kappa \times \mathcal{Y}_\kappa \to \{0,1\}$ indexed by $\kappa = (N, \mathsf{par})$ consists of four polynomial-time *deterministic* algorithms which satisfy a correctness condition as defined below.

- $\mathsf{Param}(\mathsf{par}) \to n$. The Param algorithm takes the parameters par as input, and outputs a positive integer $n \in \mathbb{N}$ which specifies the number of common variables shared by the following two algorithms. Let $\mathbf{b} := (b_1, b_2, \ldots, b_n)$ denote the common variables.
- $\mathsf{EncC}(x, N) \to (\mathbf{c} := (c_1, c_2, \ldots, c_{w_1}); w_2)$. The EncC algorithm takes an $N \in \mathbb{N}$ and an $x \in \mathcal{X}_{(N,\mathsf{par})}$ as inputs, and outputs a sequence of w_1 polynomials $c_1, c_2, \ldots, c_{w_1}$ with coefficients in $\mathbb{Z}_N$ and a $w_2 \in \mathbb{N}$. Every polynomial c_ℓ is a linear combination of monomials of the form $s, s_i, sb_j, s_i b_j$ in variables $s, s_1, s_2, \ldots, s_{w_2}$ and $b_1, \ldots, b_n$. More formally, for $\ell \in [1, w_1]$,

$$c_\ell := \zeta_\ell s + \sum_{i \in [1, w_2]} \eta_{\ell, i} s_i + \sum_{j \in [1, n]} \theta_{\ell, j} sb_j + \sum_{i \in [1, w_2], j \in [1, n]} \vartheta_{\ell, i, j} s_i b_j,$$

where $\zeta_\ell, \eta_{\ell, i}, \theta_{\ell, j}, \vartheta_{\ell, i, j} \in \mathbb{Z}_N$ are constants which define c_ℓ.
- $\mathsf{EncK}(y, N) \to (\mathbf{k} := (k_1, k_2, \ldots, k_{m_1}); m_2)$. The EncK algorithm takes an $N \in \mathbb{N}$ and a $y \in \mathcal{Y}_{(N,\mathsf{par})}$ as inputs, and outputs a sequence of m_1 polynomials $k_1, k_2, \ldots, k_{m_1}$ with coefficients in $\mathbb{Z}_N$ and an $m_2 \in \mathbb{N}$. Every polynomial k_t is a linear combination of monomials of the form $\alpha, r_{i'}, r_{i'} b_j$ in variables $\alpha, r_1, r_2, \ldots, r_{m_2}$ and $b_1, \ldots, b_n$. More formally, for $t \in [1, m_1]$,

$$k_t := \tau_t \alpha + \sum_{i' \in [1, m_2]} \upsilon_{t, i'} r_{i'} + \sum_{i' \in [1, m_2], j \in [1, n]} \phi_{t, i', j} r_{i'} b_j,$$

where $\tau_t, \upsilon_{t, i'}, \phi_{t, i', j} \in \mathbb{Z}_N$ are constants which define k_t.
- $\mathsf{Pair}(x, y, N) \to \mathbf{E}$. The EncC algorithm takes an $N \in \mathbb{N}$, an $x \in \mathcal{X}_{(N,\mathsf{par})}$ and a $y \in \mathcal{Y}_{(N,\mathsf{par})}$ as inputs, and outputs a matrix $\mathbf{E} \in \mathbb{Z}_N^{m_1 \times w_1}$.

Correctness: A pair encoding scheme is correct if for every $\kappa = (N, \mathsf{par})$, $x \in \mathcal{X}_\kappa$ and $y \in \mathcal{Y}_\kappa$ such that $P_\kappa(x, y) = 1$, the following holds symbolically

$$\mathbf{k} \mathbf{E} \mathbf{c}^T = \sum_{\substack{t \in [1, m_1], \\ \ell \in [1, w_1]}} E_{t, \ell} k_t c_\ell = \alpha s.$$

Structural Restrictions. We impose an additional restriction on the form of $\mathbf{E}$. Essentially this says that if k_t has a monomial of the form $r_{i'} b_{j'}$ and a c_ℓ has a monomial of the form sb_j or $s_i b_j$ then $E_{t, \ell}$ must be 0. One can easily verify that *every* pair encoding scheme given in [2] (as well as the new one we propose) satisfies this. We also assume that the variable s is explicitly given out in the encoding of x, i.e., $s \in \mathbf{c}$.

Moreover, we can show that given the constraint on $\mathbf{E}$, we can assume w.l.o.g. that the set of polynomials output by EncC and EncK have a fairly restricted structure. In simple words, if a polynomial contains the monomial sb_j (or $s_i b_j$, $r_{i'} b_j$), then there must exist a polynomial which only contains the monomial s (resp. s_i, $r_{i'}$). More precisely, we show that for any pair encoding which satisfies the restriction on $\mathbf{E}$, there is a corresponding one in which EncC and EncK have this structure, and this correspondence preserves all of the security properties defined in [2].

For formal statements see the full version. For the rest of this work, we will assume that all pair encodings satisfy the properties listed above.

3.1   Security

Attrapadung provided two security notions for pair encoding schemes: perfect and computational. As discussed in Sect. 1, in this paper we focus on perfect security, which is the information theoretic property, for which we propose a relaxation. First, we restate here the original security definition given by Attrapadung (which is referred to as *perfectly master-key hiding* in his paper).

Definition 2 (Perfect Security [2]). *A pair encoding scheme* (Param, EncC, EncK, Pair) *for a predicate family* P_κ *is perfectly secure if for every* $\kappa = (N, \mathsf{par})$, $x \in \mathcal{X}_\kappa$ *and* $y \in \mathcal{Y}_\kappa$ *such that* $P_\kappa(x, y) = 0$,

$$(\mathbf{c}(\mathbf{s}, \mathbf{b}), \mathbf{k}(0, \mathbf{r}, \mathbf{b})) \equiv (\mathbf{c}(\mathbf{s}, \mathbf{b}), \mathbf{k}(\alpha, \mathbf{r}, \mathbf{b})), \tag{1}$$

where $\mathbf{s} \leftarrow_R \mathbb{Z}_N^{w_2+1}$, $\mathbf{b} \leftarrow_R \mathbb{Z}_N^n$, $\mathbf{r} \leftarrow_R \mathbb{Z}_N^{m_2}$ *and* $\alpha \leftarrow_R \mathbb{Z}_N$.

We propose a new relaxed notion of perfect security that allows more flexibility in the design of pair encoding schemes. Very roughly, this property will allow us to add noise gradually to the parameters used in the key, as long as this noise is not detectable given the relevant part of the key and the ciphertext. The goal is to eventually add sufficient noise to completely hide the master secret. Towards this, we define a new *randomized* polynomial-time sampling algorithm for pair encoding schemes. While the algorithms above are used in the generic construction, the Samp algorithm described below will be used in the security proof.

– Samp$(d, x, y, N) \to (\mathbf{b}_d := (b_{d,1}, b_{d,2}, \ldots, b_{d,n}))$. This algorithm takes a $d \in [1, m_2]$, an $N \in \mathbb{N}$, an $x \in \mathcal{X}_{(N,\mathsf{par})}$, and a $y \in \mathcal{Y}_{(N,\mathsf{par})}$ as inputs, and outputs a sequence of n numbers in $\mathbb{Z}_N$. We require that the probability of this algorithm producing $(u \cdot b_{d,1}, u \cdot b_{d,2}, \ldots, u \cdot b_{d,n})$ as output is equal to the probability that it produces $(b_{d,1}, b_{d,2}, \ldots, b_{d,n})$ as output, for any $u \in \mathbb{Z}_N^*$.

Jumping ahead, the dependence of Samp on its inputs will play a crucial role in the proof of security of our generic construction. We will see that if Samp doesn't depend on x, then we can prove our construction to be *fully* secure. But in case it does, we can only prove *semi-adaptive* security.

Recall that EncK on input y and N produces a sequence of polynomials $\mathbf{k}(\alpha, \mathbf{r}, \mathbf{b})$ with coefficients in $\mathbb{Z}_N$, where every polynomial is a linear combination of monomials of the form $\alpha, r_{i'}, r_{i'}b_j$ in variables $\alpha, r_1, r_2, \ldots, r_{m_2}$ and $b_1, \ldots, b_n$. In the following we use $\mathbf{k}_d(\alpha, r_d, \mathbf{b})$, for $d \in [1, m_2]$, to denote the polynomials in $\mathbf{k}$ obtained by setting all the variables in $\{r_1, r_2, \ldots, r_{m_2}\}$ except r_d to 0. We are now ready to define our new notion of perfect security.

Definition 3 (Relaxed Perfect Security). *A pair encoding scheme* $\Gamma = $ (Param, EncC, EncK, Pair) *for a predicate family* P_κ *is relaxed perfectly secure if there exists a* PPT *algorithm* Samp *(as defined above) such that for every* par, $x \in \mathcal{X}_\kappa$ *and* $y \in \mathcal{Y}_\kappa$ *such that* $P_\kappa(x, y) = 0$, *and every* $d \in [1, m_2]$:

$$\{\mathbf{c}(\mathbf{s}, \mathbf{b}), \mathbf{k}_d(0, r_d, \mathbf{b})\}_{N \in \mathbb{N}} \quad \cong \quad \{\mathbf{c}(\mathbf{s}, \mathbf{b}), \mathbf{k}_d(0, r_d, \mathbf{b} + \mathbf{b}_d)\}_{N \in \mathbb{N}}, \quad (2)$$

where $\mathbf{s} \leftarrow_R \mathbb{Z}_N^{w_2+1}$, $\mathbf{b} \leftarrow_R \mathbb{Z}_N^n$, $r_d \leftarrow_R \mathbb{Z}_N, \mathbf{b}_d \leftarrow$ Samp(d, x, y, N). *Furthermore,*

$$\left\{\mathbf{c}(\mathbf{s}, \mathbf{b}), \sum_{d \in [1, m_2]} \mathbf{k}_d(0, r_d, \mathbf{b} + \mathbf{b}_d)\right\}_{N \in \mathbb{N}} \cong \left\{\mathbf{c}(\mathbf{s}, \mathbf{b}), \sum_{d \in [1, m_2]} \mathbf{k}_d(\alpha, r_d, \mathbf{b} + \mathbf{b}_d)\right\}_{N \in \mathbb{N}}, \quad (3)$$

where $\mathbf{s} \leftarrow_R \mathbb{Z}_N^{w_2+1}$, $\mathbf{b} \leftarrow_R \mathbb{Z}_N^n$, $r_1, r_2, \ldots, r_{m_2} \leftarrow_R \mathbb{Z}_N$, $\alpha \leftarrow_R \mathbb{Z}_N$, $\mathbf{b}_d \leftarrow$ Samp(d, x, y, N) *for* $d \in [1, m_2]$, *and* $\cong$ *denotes statistical indistinguishability. We say* Γ *satisfies* strong *relaxed perfect security if* Samp *does not depend on* x.

Note that in Eqs. (2) and (3), we have distribution ensembles indexed by N, unlike the definition of perfect security where we are dealing with only one distribution. We require that the ensembles are statistically indistinguishable from each other, which means that for large enough values of N, the statistical distance between the distributions is negligible.

We now show that any pair encoding scheme that is perfectly secure under the original definition is also secure under the stronger flavor of the relaxed definition.

Lemma 1. *Let* $\Gamma = $ (Param, EncC, EncK, Pair) *be a pair encoding scheme. If* Γ *is prefectly secure (Definition 2), then* Γ *is also relaxed perfectly secure (Definition 3). Moreover, we can define a* Samp *algorithm for* Γ *that does not depend on the input* x.

Proof. For any pair encoding scheme Γ, define Samp to output a vector of zeroes on any input. With this definition, (2) is trivially satisfied for every $d \in [1, m_2]$, and the two distributions in (3) reduce to

$$\left\{\mathbf{c}(\mathbf{s}, \mathbf{b}), \sum_{d \in [1, m_2]} \mathbf{k}_d(0, r_d, \mathbf{b})\right\} \quad \text{and} \quad \left\{\mathbf{c}(\mathbf{s}, \mathbf{b}), \sum_{d \in [1, m_2]} \mathbf{k}_d(\alpha, r_d, \mathbf{b})\right\}. \quad (4)$$

Since Γ is perfectly secure, we know that if $\mathbf{s} \leftarrow_R \mathbb{Z}_N^{w_2+1}$, $\mathbf{b} \leftarrow_R \mathbb{Z}_N^n$, $\mathbf{r} \leftarrow_R \mathbb{Z}_N^{m_2}$ and $\alpha \leftarrow_R \mathbb{Z}_N$, then

$$\{\mathbf{c}(\mathbf{s}, \mathbf{b}), \mathbf{k}(0, \mathbf{r}, \mathbf{b})\} \equiv \{\mathbf{c}(\mathbf{s}, \mathbf{b}), \mathbf{k}(\alpha, \mathbf{r}, \mathbf{b})\}.$$

We can replace $\mathbf{k}(\alpha, \mathbf{r}, \mathbf{b})$ with $\mathbf{k}(m_2\alpha, \mathbf{r}, \mathbf{b})$ in the above without changing the joint distribution. Now, observe that $\mathbf{k}(0, \mathbf{r}, \mathbf{b}) = \sum_{d \in [1, m_2]} \mathbf{k}_d(0, r_d, \mathbf{b})$ and $\mathbf{k}(m_2\alpha, \mathbf{r}, \mathbf{b}) = \sum_{d \in [1, m_2]} \mathbf{k}_d(\alpha, r_d, \mathbf{b})$ symbolically. Therefore, the two distributions in (4) are identical.   □

4   Dual System Groups

Our construction of predicate encryption schemes from pair encodings is based on dual system groups (DSG), introduced by Chen and Wee [14] in a recent work. Our formulation of DSG, given below, can be seen as a generalization of theirs. However, as we will show, both their instantiations satisfy the new properties without making any changes.

A dual system group is parameterized by a security parameter λ and a number n. It consists of six PPT algorithms as described below.

4.1   Syntax

- SampP(1^λ, 1^n): On input 1^λ and 1^n, SampP outputs public parameters PP and secret parameters SP, which have the following properties:
  - PP contains a triple of groups $(\mathbb{G}, \mathbb{H}, \mathbb{G}_T)$ and a non-degenerate bilinear map $e : \mathbb{G} \times \mathbb{H} \to \mathbb{G}_T$, a homomorphism μ from $\mathbb{H}$ to $\mathbb{G}_T$, along with some additional parameters used by SampG, SampH. Given PP, we know the exponent of group $\mathbb{H}$ and how to sample uniformly from it. Let $N = \exp(\mathbb{H})$ (see Sect. 2). We require that N is a product of distinct primes of $\Theta(\lambda)$ bits.
  - SP contains $\tilde{h} \in \mathbb{H}$ (where $\tilde{h} \neq 1_{\mathbb{H}}$) along with additional parameters used by $\overline{\mathsf{SampG}}$ and $\overline{\mathsf{SampH}}$.
- SampGT takes an element in the image of μ and outputs another element from $\mathbb{G}_T$.
- SampG and SampH take PP as input and output a vector of $n + 1$ elements from $\mathbb{G}$ and $\mathbb{H}$ respectively.
- $\overline{\mathsf{SampG}}$ and $\overline{\mathsf{SampH}}$ take both PP and SP as inputs and output a vector of $n+1$ elements from $\mathbb{G}$ and $\mathbb{H}$ respectively.

4.2   Properties

We require that all the properties below hold for every PP and SP output by SampP. Let SampG_0 be the algorithm that outputs only the first element of SampG. Analogously, SampH_0, $\overline{\mathsf{SampG}}_0$ and $\overline{\mathsf{SampH}}_0$ can be defined. A dual system group is *correct* if it satisfies the following two properties[5]:

---

[5] Note that we have omitted the $\mathbb{H}$-subgroup property. It is required to construct encryption schemes with key delegation like HIBE. We do not use this property in our constructions.

Projective: For all $h \in \mathbb{H}$ and coin tosses σ, $\mathsf{SampGT}(\mu(h); \sigma) = e(\mathsf{SampG_0}$ $(\mathrm{PP}; \sigma), h)$.

Associative: If $(g_0, g_1, \ldots, g_n)$ and $(h_0, h_1, \ldots, h_n)$ are samples from $\mathsf{SampG}(\mathrm{PP})$ and $\mathsf{SampH}(\mathrm{PP})$ respectively, then for all $i \in [1, n]$, $e(g_0, h_i) = e(g_i, h_0)$.

For *security* we require the following three properties to hold:

Orthogonality: $\tilde{h} \in \mathsf{Kernel}(\mu)$, i.e., $\mu(\tilde{h}) = 1_{\mathbb{G}_T}$.

Non-degeneracy:

1. $\overline{\mathsf{SampH_0}}(\mathrm{PP}, \mathrm{SP}) \cong \tilde{h}^\delta$, where $\delta \leftarrow_R \mathbb{Z}_N$.
2. $\exists \, \tilde{g} \in \mathbb{G}$ s.t. $\overline{\mathsf{SampG_0}}(\mathrm{PP}, \mathrm{SP}) \cong \tilde{g}^\alpha$, where $\alpha \leftarrow_R \mathbb{Z}_N$.
3. For all $\hat{g}_0 \leftarrow \overline{\mathsf{SampG_0}}(\mathrm{PP}, \mathrm{SP})$, $e(\hat{g}_0, \tilde{h})^\beta$ is uniformly distributed over $\mathbb{G}_T$, where $\beta \leftarrow_R \mathbb{Z}_N$.

(Here $\cong$ denotes statistical indistinguishability.)

Remark 1. In [14], the non-degeneracy property is defined in a slightly different way. First, they require that for all $\hat{h}_0 \leftarrow \overline{\mathsf{SampH_0}}(\mathrm{PP}, \mathrm{SP})$, $\tilde{h}$ lies in the group generated by $\hat{h}_0$, instead of the first point above. And secondly, they do not have any constraint on the output of $\overline{\mathsf{SampG_0}}(\mathrm{PP}, \mathrm{SP})$ like in the second point above. The third property, though, is also present in their definition[6].

Indistinguishability. For two (positive) polynomials $\mathsf{poly}_1(\cdot)$ and $\mathsf{poly}_2(\cdot)$, define $\mathbf{G}, \mathbf{H}, \hat{\mathbf{G}}, \hat{\mathbf{H}}, \hat{\mathbf{G}}', \hat{\mathbf{H}}'$ as follows:

$$(\mathrm{PP}, \mathrm{SP}) \leftarrow \mathsf{SampP}(1^\lambda, 1^n); \quad \gamma_1, \gamma_2, \ldots, \gamma_n \leftarrow_R \mathbb{Z}_N;$$

$$\mathbf{g}_1, \mathbf{g}_2, \ldots, \mathbf{g}_{\mathsf{poly}_1(\lambda)} \leftarrow \mathsf{SampG}(\mathrm{PP}); \mathbf{G} := (\mathbf{g}_1, \mathbf{g}_2, \ldots, \mathbf{g}_{\mathsf{poly}_1(\lambda)});$$

$$\mathbf{h}_1, \mathbf{h}_2, \ldots, \mathbf{h}_{\mathsf{poly}_2(\lambda)} \leftarrow \mathsf{SampH}(\mathrm{PP}); \mathbf{H} := (\mathbf{h}_1, \mathbf{h}_2, \ldots, \mathbf{h}_{\mathsf{poly}_2(\lambda)});$$

$$\forall i \in [1, \mathsf{poly}_1(\lambda)], \quad \hat{\mathbf{g}}_i := (\hat{g}_{i,0}, \ldots) \leftarrow \overline{\mathsf{SampG}}(\mathrm{PP}, \mathrm{SP}); \quad \hat{\mathbf{g}}_i' := (1, \hat{g}_{i,0}^{\gamma_1}, \hat{g}_{i,0}^{\gamma_2}, \ldots, \hat{g}_{i,0}^{\gamma_n})$$

$$\forall j \in [1, \mathsf{poly}_2(\lambda)], \quad \hat{\mathbf{h}}_j := (\hat{h}_{j,0}, \ldots) \leftarrow \overline{\mathsf{SampH}}(\mathrm{PP}, \mathrm{SP}); \quad \hat{\mathbf{h}}_j' := (1, \hat{h}_{j,0}^{\gamma_1}, \hat{h}_{j,0}^{\gamma_2}, \ldots, \hat{h}_{j,0}^{\gamma_n})$$

$$\hat{\mathbf{G}} := (\hat{\mathbf{g}}_1, \hat{\mathbf{g}}_2, \ldots, \hat{\mathbf{g}}_{\mathsf{poly}_1(\lambda)}); \hat{\mathbf{H}} := (\hat{\mathbf{h}}_1, \hat{\mathbf{h}}_2, \ldots, \hat{\mathbf{h}}_{\mathsf{poly}_2(\lambda)});$$

$$\hat{\mathbf{G}}' := (\hat{\mathbf{g}}_1', \hat{\mathbf{g}}_2', \ldots, \hat{\mathbf{g}}_{\mathsf{poly}_1(\lambda)}'); \hat{\mathbf{H}}' := (\hat{\mathbf{h}}_1', \hat{\mathbf{h}}_2', \ldots, \hat{\mathbf{h}}_{\mathsf{poly}_2(\lambda)}').$$

We call a dual system group *Left Subgroup Indistinguishable* (LSI), *Right Subgroup Indistinguishable* (RSI) and *Parameter hiding* (PH) if for all polynomials $\mathsf{poly}_1(\cdot)$ and $\mathsf{poly}_2(\cdot)$,

$$\{\mathrm{PP}, \mathbf{G}\} \approx \{\mathrm{PP}, \mathbf{G} \cdot \hat{\mathbf{G}}\}, \tag{5}$$

$$\{\mathrm{PP}, \tilde{h}, \mathbf{G} \cdot \hat{\mathbf{G}}, \mathbf{H}\} \approx \{\mathrm{PP}, \tilde{h}, \mathbf{G} \cdot \hat{\mathbf{G}}, \mathbf{H} \cdot \hat{\mathbf{H}}\}, \text{and} \tag{6}$$

$$\{\mathrm{PP}, \tilde{h}, \hat{\mathbf{G}}, \hat{\mathbf{H}}\} \equiv \{\mathrm{PP}, \tilde{h}, \hat{\mathbf{G}} \cdot \hat{\mathbf{G}}', \hat{\mathbf{H}} \cdot \hat{\mathbf{H}}'\} \tag{7}$$

---

[6] In the composite-order instantiation of [14], this property holds only in a computational sense.

hold respectively. Observe that the two distributions in (5) and (6) are computationally indistinguishable, while the two distributions in (7) are identical.

Instantiations of DSG. The three indistinguishability properties defined above are generalizations of the corresponding ones in Chen and Wee [14]. In the full version we show that the two instantiations of DSG – in composite-order groups under the subgroup decision assumption and in prime-order groups under the decisional linear assumption (d-LIN) – given by [14] satisfy our generalized indistinguishability properties as well as our new definition of non-degeneracy.

Remark 2. In the prime-order instantiation of dual system groups under the d-LIN assumption given by [14], an element from groups $\mathbb{G}$ or $\mathbb{H}$ is represented by $d+1$ elements from a source prime-order group (an element from $\mathbb{G}_T$ is mapped to just one element of a target prime-order group). Now, suppose we have an encryption scheme in dual system groups where the ciphertext/key consists of elements from $\mathbb{G}$ or $\mathbb{H}$ (and possibly an element from $\mathbb{G}_T$). Then, a concrete instantiation in prime-order groups would only double the size of ciphertext/key, if we make the SXDH assumption (special case of d-LIN with $d=1$), and only triple it if we make the DLIN assumption (special case of d-LIN with $d=2$).

5 Predicate Encryption from Pair Encodings

In this section, we show how to construct a predicate encryption scheme $\Pi_P =$ (Setup, Encrypt, KeyGen, Decrypt) for any predicate family $P = \{P_\kappa\}_{\kappa \in \mathbb{N}^c}$ for which we have a pair encoding scheme $\Gamma_P = $ (Param, EncC, EncK, Pair), using dual system groups. The message space for Π_P would be the target group in DSG. Recall that κ specifies a number $N \in \mathbb{N}$ and some additional parameters par.

- Setup(1^λ, par): First run Param(par) to obtain n, then run SampP($1^\lambda, 1^n$) to obtain PP and SP. Recall that given PP, we know the exponent of group $\mathbb{H}$ and can sample uniformly from it. Output

$$\text{MSK} \leftarrow_R \mathbb{H} \qquad \text{MPK} := (\text{PP}, \mu(\text{MSK})).$$

Set $N = \exp(\mathbb{H})$ and $\kappa = (N, \text{par})$.
- Encrypt(MPK, x, m): On input an $x \in \mathcal{X}_\kappa$ and an $m \in \mathbb{G}_T$, run EncC(x, N) to obtain a sequence of w_1 polynomials $(c_1, c_2, \ldots, c_{w_1})$ and a $w_2 \in \mathbb{N}$. Draw $w_2 + 1$ samples from SampG:

$$(g_{0,0}, \ldots, g_{0,n}) \leftarrow \text{SampG}(\text{PP}; \sigma)$$

$$(g_{1,0}, \ldots, g_{1,n}) \leftarrow \text{SampG}(\text{PP}), \ldots, (g_{w_2,0}, \ldots, g_{w_2,n}) \leftarrow \text{SampG}(\text{PP}),$$

where σ denotes the coin tosses used in drawing the first sample from SampG. Recall that the polynomial c_ℓ is given by

$$\zeta_\ell s \quad + \quad \sum_{i \in [1,w_2]} \eta_{\ell,i} s_i \quad + \quad \sum_{j \in [1,n]} \theta_{\ell,j} s b_j \quad + \quad \sum_{i \in [1,w_2], j \in [1,n]} \vartheta_{\ell,i,j} s_i b_j,$$

where $\zeta_\ell, \eta_{\ell,i}, \theta_{\ell,j}, \vartheta_{\ell,i,j} \in \mathbb{Z}_N$ are constants. Output $\text{CT} := (\text{CT}_1, \dots, \text{CT}_{w_1}, \text{CT}_{w_1+1})$ as the encryption of m under x where

$$\text{CT}_\ell \;:=\; g_{0,0}^{\zeta_\ell} \;\cdot\; \prod_{i\in[1,w_2]} g_{i,0}^{\eta_{\ell,i}} \;\cdot\; \prod_{j\in[1,n]} g_{0,j}^{\theta_{\ell,j}} \;\cdot\; \prod_{i\in[1,w_2],j\in[1,n]} g_{i,j}^{\vartheta_{\ell,i,j}}$$

for $\ell \in [1, w_1]$ and $\text{CT}_{w_1+1} := m \cdot \mathsf{SampGT}(\mu(\text{MSK}); \sigma)$. Notice that the monomials s, s_i, sb_j, and $s_i b_j$ are *mapped* to group elements $g_{0,0}$, $g_{i,0}$, $g_{0,j}$, and $g_{i,j}$, respectively.

– KeyGen(MPK, MSK, y): On input a $y \in \mathcal{Y}_\kappa$, run $\mathsf{EncK}(y, N)$ to obtain a sequence of m_1 polynomials $(k_1, k_2, \dots, k_{m_1})$ and an $m_2 \in \mathbb{N}$. Draw m_2 samples from SampH:

$$(h_{1,0}, \dots, h_{1,n}) \leftarrow \mathsf{SampH}(\text{PP}), \dots, (h_{m_2,0}, \dots, h_{m_2,n}) \leftarrow \mathsf{SampH}(\text{PP}).$$

Output the key as $\text{SK} := (\text{SK}_1, \text{SK}_2, \dots, \text{SK}_{m_1})$ where for $t \in [1, m_1]$

$$\text{SK}_t \;:=\; \text{MSK}^{\tau_t} \;\cdot\; \prod_{i'\in[1,m_2]} h_{i',0}^{\upsilon_{t,i'}} \;\cdot\; \prod_{i'\in[1,m_2],j\in[1,n]} h_{i',j}^{\phi_{t,i',j}}.$$

In this case, the variables α, $r_{i'}$, and $r_{i'} b_j$ are mapped to MSK, $h_{i',0}$, and $h_{i',j}$, respectively.

– Decrypt(MPK, SK_y, CT_x): On input $\text{SK}_y := (\text{SK}_1, \text{SK}_2, \dots, \text{SK}_{m_1})$ and $\text{CT}_x := (\text{CT}_1, \dots, \text{CT}_{w_1+1})$, run $\mathsf{Pair}(x, y, N)$ to obtain an $m_1 \times w_1$ matrix $\mathbf{E}$. Output

$$\text{CT}_{w_1+1} \cdot \left[\prod_{t\in[1,m_1],\ell\in[1,w_1]} e(\text{CT}_\ell, \text{SK}_t^{E_{t,\ell}}) \right]^{-1}.$$

Correctness (Sketch). We know that if $P_\kappa(x, y) = 1$, then $\sum_{t\in[1,m_1],\ell\in[1,w_1]} E_{t,\ell} k_t c_\ell = \alpha s$. Consider two polynomials k_t and c_ℓ. When these polynomials are multiplied together, no two monomials – one from k_t and one from c_ℓ – combine to give the same monomial in the product polynomial $k_t c_\ell$, except when

– s is multiplied with $r_{i'} b_j$ and sb_j is multiplied with $r_{i'}$, or
– s_i is multiplied with $r_{i'} b_j$ and $s_i b_j$ is multiplied with $r_{i'}$,

because of the restriction on the form of $\mathbf{E}$. Now, s is mapped to $g_{0,0}$, $r_{i'} b_j$ is mapped to $h_{i',j}$, sb_j is mapped to $g_{0,j}$ and $r_{i'}$ is mapped to $h_{i',0}$. By the associativity property of dual system groups, we know that $e(g_{0,0}, h_{i',j}) = e(g_{0,j}, h_{i',0})$. Further, we mapped s_i to $g_{i,0}$ and $s_i b_j$ to $g_{i,j}$, and associativity guarantees that $e(g_{i,0}, h_{i',j}) = e(g_{i,j}, h_{i',0})$. Therefore, from the observations above, it follows that

$$\prod_{t\in[1,m_1],\ell\in[1,w_1]} e(\text{CT}_\ell, \text{SK}_t^{E_{t,\ell}}) = e(g_{0,0}, \text{MSK}).$$

Finally, by projective property we know that $e(g_{0,0}, \text{MSK}) = \mathsf{SampGT}(\mu(\text{MSK}); \sigma)$.

Remark 3 (Preserving Size). Observe that the output of Encrypt consists of $w_1 + 1$ elements, w_1 from $\mathbb{G}$ and 1 from $\mathbb{G}_T$, where w_1 is the number of polynomials output by EncC. Further, any key has the same number of elements from $\mathbb{H}$ as the number of polynomials output by EncK. Hence, in particular, if w_1 (resp. m_1) is a constant then ciphertexts (resp. keys) are also of constant size, in terms of dual system group elements. Further, if we instantiate dual system groups in prime-order groups under SXDH or DLIN assumption, then the ciphertexts (resp. keys) would still be of constant size (see Remark 2.)

6   Proof of Security

In this section, we show that the encryption scheme Π_P constructed for a predicate family $P = \{P_\kappa\}_{\kappa \in \mathbb{N}^c}$ in the previous section is secure using the properties of dual system groups and relaxed perfect security of pair encoding schemes. More formally, we prove the following theorem.

Theorem 1. *For any predicate family $P = \{P_\kappa\}_{\kappa \in \mathbb{N}^c}$, if $\Gamma_P = $ (Param, EncC, EncK, Pair) is a **relaxed perfectly secure** pair encoding scheme, then the encryption scheme $\Pi_P = $ (Setup, Encrypt, KeyGen, Decrypt) constructed in Sect. 5 (using Γ_P) is **semi-adaptively** secure. Furthermore, if the algorithm Samp does not depend on input x, then Π_P is **fully** secure (see Definition 1).*

Using Lemma 1, a corollary of the above theorem is that:

Corollary 1. *For any predicate family $P = \{P_\kappa\}_{\kappa \in \mathbb{N}^c}$, if $\Gamma_P = $ (Param, EncC, EncK, Pair, Samp) is a **perfectly secure** pair encoding scheme, then the encryption scheme $\Pi_P = $ (Setup, Encrypt, KeyGen, Decrypt) constructed in Sect. 5 (using Γ_P) is **fully** secure.*

Recall that dual system groups can be instantiated in prime-order groups under the d-LIN assumption. Together with the above corollary, this gives a useful and interesting result:

Corollary 2. *Every perfectly secure pair encoding scheme proposed by Attrapadung [2] has a fully secure predicate encryption scheme in prime order groups under the d-LIN assumption.*

The rest of this section is devoted to the proof of Theorem 1. We first define auxiliary algorithms for encryption and key generation.

- $\overline{\text{Encrypt}}$(PP, x, m; $(\mathbf{g}'_0, \mathbf{g}'_1, \ldots, \mathbf{g}'_{w_2})$, MSK): This algorithm is the same as Encrypt except that it uses the input $\mathbf{g}'_i \in \mathbb{G}^{n+1}$ instead of choosing samples $\mathbf{g}_i$ from SampG for $i \in [0, w_2]$, and sets $\text{CT}_{w_1+1} := m \cdot e(g'_{0,0}, \text{MSK})$, where $g'_{0,0}$ if the first element of the vector $\mathbf{g}'_0$.
- $\overline{\text{KeyGen}}$(PP, MSK, y; $(\mathbf{h}'_1, \ldots, \mathbf{h}'_{m_2})$): This algorithm is the same as KeyGen except that it uses $\mathbf{h}'_i$ instead of the samples $\mathbf{h}_i$ from SampH for $i \in [1, m_2]$.

Using these algorithms, we define alternate forms for the ciphertext and master secret key:

- *Semi-functional master secret key* is defined to be $\overline{\text{MSK}} := \text{MSK} \cdot \tilde{h}^\beta$ where $\beta \leftarrow_R \mathbb{Z}_N$.
- *Semi-functional ciphertext* is given by $\overline{\text{Encrypt}}(\text{PP}, x, m; \mathbf{G} \cdot \hat{\mathbf{G}}, \text{MSK})$ where $\mathbf{g}_1, \mathbf{g}_2, \ldots, \mathbf{g}_{w_2} \leftarrow \text{SampG}(\text{PP})$, $\hat{\mathbf{g}}_1, \hat{\mathbf{g}}_2, \ldots, \hat{\mathbf{g}}_{w_2} \leftarrow \overline{\text{SampG}}(\text{PP}, \text{SP})$, $\mathbf{G} := (\mathbf{g}_1, \mathbf{g}_2, \ldots, \mathbf{g}_{w_2})$, and $\hat{\mathbf{G}} := (\hat{\mathbf{g}}_1, \hat{\mathbf{g}}_2, \ldots, \hat{\mathbf{g}}_{w_2})$. Observe that $\overline{\text{Encrypt}}(\text{PP}, x, m; \mathbf{G}, \text{MSK})$ is identically distributed to $\text{Encrypt}(\text{MPK}, x, m)$ – the normal ciphertext – by the projective property of dual system groups.

Table 1 defines various forms of keys for $\rho \in [1, m_2]$ and the inputs that need to be passed to $\overline{\text{KeyGen}}$ (besides PP and y) in order to generate them. Intermediate-3 and SF-intermediate-3 keys are also defined for $\rho = 0$ (SF stands for semi-functional). In the table, $\mathbf{h}_1, \ldots, \mathbf{h}_{m_2} \leftarrow \text{SampH}(\text{PP})$, $\hat{\mathbf{h}}_1, \ldots, \hat{\mathbf{h}}_{m_2} \leftarrow \overline{\text{SampH}}(\text{PP}, \text{SP})$, and $\mathbf{z}_d := (1, z_{d,1}, \ldots, z_{d,n})$, where $(z_{d,1}, \ldots, z_{d,n}) \leftarrow \text{Samp}(d, x, y, N)$ for all $d \in [1, m_2]$. For convenience in the following, we define a slightly modified form of Samp, called $\overline{\text{Samp}}$, which just prepends 1 to the output of Samp. Note that 0-Intermediate-3 is distributed identically to a normal key and 0-SF-intermediate-3 is distributed identically to a SF noisy key. Since we have many forms of keys, (where appropriate) we use a box to highlight the part of a key which is different from the previous key.

Table 1. Various types of keys

Type of key	Inputs to $\overline{\text{KeyGen}}$ (besides PP and y)
Normal	$\text{MSK}; (\mathbf{h}_1, \ldots, \mathbf{h}_{m_2})$
ρ-Intermediate-1	$\text{MSK}; (\mathbf{h}_1 \cdot \tilde{h}^{\mathbf{z}_1}, \ldots, \mathbf{h}_{\rho-1} \cdot \tilde{h}^{\mathbf{z}_{\rho-1}}, \mathbf{h}_\rho \cdot \hat{\mathbf{h}}_\rho, \mathbf{h}_{\rho+1}, \ldots, \mathbf{h}_{m_2})$
ρ-Intermediate-2	$\text{MSK}; (\mathbf{h}_1 \cdot \tilde{h}^{\mathbf{z}_1}, \ldots, \mathbf{h}_{\rho-1} \cdot \tilde{h}^{\mathbf{z}_{\rho-1}}, \boxed{\mathbf{h}_\rho \cdot \hat{\mathbf{h}}_\rho \cdot \tilde{h}^{\mathbf{z}_\rho}}, \mathbf{h}_{\rho+1}, \ldots, \mathbf{h}_{m_2})$
ρ-Intermediate-3	$\text{MSK}; (\mathbf{h}_1 \cdot \tilde{h}^{\mathbf{z}_1}, \ldots, \mathbf{h}_{\rho-1} \cdot \tilde{h}^{\mathbf{z}_{\rho-1}}, \boxed{\mathbf{h}_\rho \cdot \tilde{h}^{\mathbf{z}_\rho}}, \mathbf{h}_{\rho+1}, \ldots, \mathbf{h}_{m_2})$
Pseudo-normal noisy	$\text{MSK}; (\mathbf{h}_1 \cdot \hat{\mathbf{h}}_1 \cdot \tilde{h}^{\mathbf{z}_1}, \ldots, \mathbf{h}_{m_2} \cdot \hat{\mathbf{h}}_{m_2} \cdot \tilde{h}^{\mathbf{z}_{m_2}})$
Pseudo-SF noisy	$\boxed{\overline{\text{MSK}}}; (\mathbf{h}_1 \cdot \hat{\mathbf{h}}_1 \cdot \tilde{h}^{\mathbf{z}_1}, \ldots, \mathbf{h}_{m_2} \cdot \hat{\mathbf{h}}_{m_2} \cdot \tilde{h}^{\mathbf{z}_{m_2}})$
SF noisy	$\overline{\text{MSK}}; (\mathbf{h}_1 \cdot \tilde{h}^{\mathbf{z}_1}, \ldots, \mathbf{h}_{m_2} \cdot \tilde{h}^{\mathbf{z}_{m_2}})$
ρ-SF-intermediate-1	$\overline{\text{MSK}}; (\mathbf{h}_1, \ldots, \mathbf{h}_{\rho-1}, \mathbf{h}_\rho \cdot \hat{\mathbf{h}}_\rho \cdot \tilde{h}^{\mathbf{z}_\rho}, \mathbf{h}_{\rho+1} \cdot \tilde{h}^{\mathbf{z}_{\rho+1}}, \ldots, \mathbf{h}_{m_2} \cdot \tilde{h}^{\mathbf{z}_{m_2}})$
ρ-SF-intermediate-2	$\overline{\text{MSK}}; (\mathbf{h}_1, \ldots, \mathbf{h}_{\rho-1}, \boxed{\mathbf{h}_\rho \cdot \hat{\mathbf{h}}_\rho}, \mathbf{h}_{\rho+1} \cdot \tilde{h}^{\mathbf{z}_{\rho+1}}, \ldots, \mathbf{h}_{m_2} \cdot \tilde{h}^{\mathbf{z}_{m_2}})$
ρ-SF-intermediate-3	$\overline{\text{MSK}}; (\mathbf{h}_1, \ldots, \mathbf{h}_{\rho-1}, \boxed{\mathbf{h}_\rho}, \mathbf{h}_{\rho+1} \cdot \tilde{h}^{\mathbf{z}_{\rho+1}}, \ldots, \mathbf{h}_{m_2} \cdot \tilde{h}^{\mathbf{z}_{m_2}})$
SF	$\overline{\text{MSK}}; (\mathbf{h}_1, \ldots, \mathbf{h}_{m_2})$

Proof Structure: The novelty in our proof is that instead of working at the level of a key, we work at the level of samples that form the key. Let ξ denote the number of queries made by the adversary, and let y_φ denote the φth query for $\varphi \in [1, \xi]$. Further, let $m_{2,\varphi}$ be the second output of $\text{EncK}(y_\varphi, N)$. We define the following hybrids for $\varphi \in [1, \xi]$ and $\rho \in [1, m_{2,\varphi}]$ (fix any $b \in \{0, 1\}$).

- Hyb_0: This is the real security game $\mathsf{Expt}_{\mathcal{A},\Pi_P}^{(b)}(\lambda, \mathsf{par})$ described in Sect. 2.1.
- Hyb_1: This game is same as the above except that the ciphertext is semi-functional.
- $\mathsf{Hyb}_{2,\varphi,i,\rho}$ for $i \in \{1, 2, 3\}$: This game is same as the above except that the first $\varphi - 1$ keys are semi-functional, φth key is of the form ρ-intermediate-i, and rest of the keys are normal.
- $\mathsf{Hyb}_{2,\varphi,4}$: This game is same as the above except that the φth key is Pseudo-normal noisy.
- $\mathsf{Hyb}_{2,\varphi,5}$: This game is same as the above except that the φth key is Pseudo-SF noisy.
- $\mathsf{Hyb}_{2,\varphi,6}$: This game is same as the above except that the φth key is SF noisy.
- $\mathsf{Hyb}_{2,\varphi,i,\rho}$ for $i \in \{7, 8, 9\}$: This game is same as the above except that the φth key is of the form ρ-SF-intermediate-$(i - 6)$.
- Hyb_3: This game is same as $\mathsf{Hyb}_{2,\xi,9,m_{2,\xi}}$ except that the ciphertext is a semi-functional encryption of a random message in $\mathbb{G}_T$.

Table 2. An outline of the proof structure.

Indistinguishability	Properties needed
$\mathsf{Hyb}_0 \approx \mathsf{Hyb}_1$	left subgroup indistinguishability
$\mathsf{Hyb}_{2,\varphi,3,\rho-1} \approx \mathsf{Hyb}_{2,\varphi,1,\rho}$	right subgroup indistinguishability
$\mathsf{Hyb}_{2,\varphi,1,\rho} \cong \mathsf{Hyb}_{2,\varphi,2,\rho}$	non-degeneracy, parameter-hiding, relaxed perfect security (2)
$\mathsf{Hyb}_{2,\varphi,2,\rho} \approx \mathsf{Hyb}_{2,\varphi,3,\rho}$	right subgroup indistinguishability
$\mathsf{Hyb}_{2,\varphi,3,m_2,\varphi} \approx \mathsf{Hyb}_{2,\varphi,4}$	right subgroup indistinguishability
$\mathsf{Hyb}_{2,\varphi,4} \cong \mathsf{Hyb}_{2,\varphi,5}$	non-degeneracy, parameter-hiding, relaxed perfect security (3)
$\mathsf{Hyb}_{2,\varphi,5} \approx \mathsf{Hyb}_{2,\varphi,6}$	right subgroup indistinguishability
$\mathsf{Hyb}_{2,\varphi,9,\rho-1} \approx \mathsf{Hyb}_{2,\varphi,7,\rho}$	right subgroup indistinguishability
$\mathsf{Hyb}_{2,\varphi,7,\rho} \cong \mathsf{Hyb}_{2,\varphi,8,\rho}$	non-degeneracy, parameter-hiding, relaxed perfect security (2)
$\mathsf{Hyb}_{2,\varphi,8,\rho} \approx \mathsf{Hyb}_{2,\varphi,9,\rho}$	right subgroup indistinguishability
$\mathsf{Hyb}_{2,\xi,9,m_{2,\xi}} \cong \mathsf{Hyb}_3$	projective, orthogonality, non-degeneracy

Our goal is to show that Hyb_0 and Hyb_3 are computationally indistinguishable from each other, for both values of the bit b used by Chl in the security game $\mathsf{Expt}_{\mathcal{A},\Pi_P}^{(b)}(\lambda, \mathsf{par})$. Since Chl encrypts a random message in Hyb_3, there would be no way for a PPT adversary to tell whether m_0 or m_1 was encrypted. This would imply that Π_P is a secure encryption scheme.

Our proof proceeds as follows. We first show that Hyb_0 and Hyb_1 are computationally indistinguishable due to the left subgroup indistinguishability (LSI) property of dual system groups; this takes the ciphertext from normal to semi-functional space (the form of the ciphertext doesn't change after this step).

After that, we take the keys one by one from normal to semi-functional space by going through a series of hybrids. We show that $\mathsf{Hyb}_{2,1,3,0}$ (or, equivalently, Hyb_1) is computationally indistinguishable from $\mathsf{Hyb}_{2,1,9,m_{2,1}}$ by following the steps shown in Table 2 for $\varphi = 1$; this makes the first key semi-functional while keeping the rest of the keys unchanged. Then, we show that $\mathsf{Hyb}_{2,2,3,0}$ (or, equivalently, $\mathsf{Hyb}_{2,1,9,m_{2,1}}$) is computationally indistinguishable from $\mathsf{Hyb}_{2,2,9,m_{2,2}}$ by once again following the steps shown in Table 2, but now for $\varphi = 2$; as a result, the second key also moves into the semi-functional space. We continue in the same fashion till all the keys are in the semi-functional space, i.e., we are in the hybrid $\mathsf{Hyb}_{2,\xi,9,m_{2,\xi}}$. The last step of the proof is to show that $\mathsf{Hyb}_{2,\xi,9,m_{2,\xi}}$ and Hyb_3 are statistically close to each other.

We formally prove the indistinguishability of hybrids that require relaxed perfect security, our new information-theoretic notion of security, in Lemmas 2 and 3 below, but defer the other proofs to the full version because they follow directly from the properties of dual system groups in a manner similar to Chen and Wee's security proof for HIBE [14].

Remark 4 (Full vs. Semi-adaptive Security.). In transitioning from $\mathsf{Hyb}_{2,\varphi,1,\rho}$ to $\mathsf{Hyb}_{2,\varphi,2,\rho}$ in Lemma 2, we add randomness using the algorithm $\overline{\mathsf{Samp}}$ to the ρ-th sample of the φ-th key. Observe that if $\overline{\mathsf{Samp}}$ depends on input x, then this transition can only take place if x is known *before* any key queries are issued. Therefore, in this case, we can prove semi-adaptive security. On the other hand, if $\overline{\mathsf{Samp}}$ does not depend on x, then we get full security (and as shown in Lemma 1, this is the case for all of the perfectly secure pair encoding schemes of [2]).

Remark 5 (Perfectly Secure Encodings). Recall from the proof of Lemma 1 that for any perfectly secure pair encoding scheme, we can define a dummy sampling algorithm that always outputs a vector of 0s. When this is the case, the security proof can be considerably simplified: we could directly go from Hyb_1 to $\mathsf{Hyb}_{2,\varphi,4}$ and also from $\mathsf{Hyb}_{2,\varphi,5}$ to $\mathsf{Hyb}_{2,\varphi,9,m_{2,\varphi}}$ using right subgroup indistinguishability.

Remark 6 (Cost of Our Reduction). There are many complex predicates for which we do not know any perfectly secure pair encoding schemes. But if one can design a scheme that is relaxed perfectly secure, then we show that an encryption scheme can be derived from it, which is secure under standard assumptions. The reduction cost of our security proof, however, is higher than usual: if an adversary makes ξ queries and m_2 is the maximum number of samples used in any key, then the cost is $O(\xi \cdot m_2)$. For instance, this cost only depends on the number of pre-challenge queries in the case of Attrapadung's computationally secure encodings (Theorem 1 in [2]). Note, however, that computational security of the encoding itself is proved under q-type assumptions.

Lemma 2. *For every $\varphi \in [1, \xi]$ and $\rho \in [1, m_{2,\varphi}]$, $\mathsf{Hyb}_{2,\varphi,1,\rho} \cong \mathsf{Hyb}_{2,\varphi,2,\rho}$.*

Proof. Given PP, MSK and $\tilde{h}$, one can generate MPK and every key except the φth (because in order to generate this key and the ciphertext, we need to be able to sample from $\overline{\mathsf{SampH}}$ and $\overline{\mathsf{SampG}}$, for which secret parameters SP are required).

Hence, it suffices to show that the following two distributions are statistically close (for clarity, we omit φ in the following):

$$\{\text{PP}, \text{MSK}, \tilde{h}, \overline{\text{Encrypt}}(\text{PP}, x, m; \mathbf{G} \cdot \hat{\mathbf{G}}, \text{MSK}),$$

$$\overline{\text{KeyGen}}(\text{PP}, \text{MSK}, y; (\mathbf{h}_1 \cdot \tilde{h}^{\mathbf{z}_1}, \ldots, \mathbf{h}_{\rho-1} \cdot \tilde{h}^{\mathbf{z}_{\rho-1}}, \mathbf{h}_\rho \cdot \hat{\mathbf{h}}_\rho, \mathbf{h}_{\rho+1}, \ldots, \mathbf{h}_{m_2}))\},$$

$$\{\text{PP}, \text{MSK}, \tilde{h}, \overline{\text{Encrypt}}(\text{PP}, x, m; \mathbf{G} \cdot \hat{\mathbf{G}}, \text{MSK}),$$

$$\overline{\text{KeyGen}}(\text{PP}, \text{MSK}, y; (\mathbf{h}_1 \cdot \tilde{h}^{\mathbf{z}_1}, \ldots, \mathbf{h}_{\rho-1} \cdot \tilde{h}^{\mathbf{z}_{\rho-1}}, \mathbf{h}_\rho \cdot \hat{\mathbf{h}}_\rho \cdot \tilde{h}^{\mathbf{z}_\rho}, \mathbf{h}_{\rho+1}, \ldots, \mathbf{h}_{m_2}))\}.$$

But observe that:

$$\overline{\text{Encrypt}}(\text{PP}, x, m; \mathbf{G} \cdot \hat{\mathbf{G}}, \text{MSK}) = \overline{\text{Encrypt}}(\text{PP}, x, m; \mathbf{G}, \text{MSK}) \cdot \overline{\text{Encrypt}}(\text{PP}, x, 1; \hat{\mathbf{G}}, \text{MSK}),$$

$$\overline{\text{KeyGen}}(\text{PP}, \text{MSK}, y; (\mathbf{h}_1 \cdot \tilde{h}^{\mathbf{z}_1}, \ldots, \mathbf{h}_{\rho-1} \cdot \tilde{h}^{\mathbf{z}_{\rho-1}}, \mathbf{h}_\rho \cdot \hat{\mathbf{h}}_\rho, \mathbf{h}_{\rho+1}, \ldots, \mathbf{h}_{m_2}))$$
$$= \overline{\text{KeyGen}}(\text{PP}, \text{MSK}, y; (\mathbf{h}_1 \cdot \tilde{h}^{\mathbf{z}_1}, \ldots, \mathbf{h}_{\rho-1} \cdot \tilde{h}^{\mathbf{z}_{\rho-1}}, \mathbf{h}_\rho, \mathbf{h}_{\rho+1}, \ldots, \mathbf{h}_{m_2})) \cdot$$
$$\overline{\text{KeyGen}}(\text{PP}, 1, y; (1, \ldots, 1, \hat{\mathbf{h}}_\rho, 1, \ldots, 1)),$$

$$\overline{\text{KeyGen}}(\text{PP}, \text{MSK}, y; (\mathbf{h}_1 \cdot \tilde{h}^{\mathbf{z}_1}, \ldots, \mathbf{h}_{\rho-1} \cdot \tilde{h}^{\mathbf{z}_{\rho-1}}, \mathbf{h}_\rho \cdot \hat{\mathbf{h}}_\rho \cdot \tilde{h}^{\mathbf{z}_\rho}, \mathbf{h}_{\rho+1}, \ldots, \mathbf{h}_{m_2}))$$
$$= \overline{\text{KeyGen}}(\text{PP}, \text{MSK}, y; (\mathbf{h}_1 \cdot \tilde{h}^{\mathbf{z}_1}, \ldots, \mathbf{h}_{\rho-1} \cdot \tilde{h}^{\mathbf{z}_{\rho-1}}, \mathbf{h}_\rho, \mathbf{h}_{\rho+1}, \ldots, \mathbf{h}_{m_2})) \cdot$$
$$\overline{\text{KeyGen}}(\text{PP}, 1, y; (1, \ldots, 1, \hat{\mathbf{h}}_\rho \cdot \tilde{h}^{\mathbf{z}_\rho}, 1, \ldots, 1)),$$

because of the way Encrypt and KeyGen are defined and bilinearity of e (see the construction in Sect. 5). The first component on the right hand side of each of the above equations can be generated given PP, MSK and $\tilde{h}$. Hence, we only need to focus on the second components, i.e., it is enough to show that the following two distributions are statistically close:

$$\{\text{PP}, \text{MSK}, \tilde{h}, \overline{\text{Encrypt}}(\text{PP}, x, 1; \hat{\mathbf{G}}, \text{MSK}), \overline{\text{KeyGen}}(\text{PP}, 1, y; (1, \ldots, 1, \hat{\mathbf{h}}_\rho, 1, \ldots, 1))\},$$
$$\tag{8}$$
$$\{\text{PP}, \text{MSK}, \tilde{h}, \overline{\text{Encrypt}}(\text{PP}, x, 1; \hat{\mathbf{G}}, \text{MSK}), \overline{\text{KeyGen}}(\text{PP}, 1, y; (1, \ldots, 1, \hat{\mathbf{h}}_\rho \cdot \tilde{h}^{\mathbf{z}_\rho}, 1, \ldots, 1))\}.$$
$$\tag{9}$$

Let us focus on the first distribution between the two above. By the parameter-hiding property of dual system groups we know that $\{\text{PP}, \tilde{h}, \hat{\mathbf{G}}, \hat{\mathbf{h}}_\rho\}$ and $\{\text{PP}, \tilde{h}, \hat{\mathbf{G}} \cdot \hat{\mathbf{G}}', \hat{\mathbf{h}}_\rho \cdot \hat{\mathbf{h}}'_\rho\}$ are identically distributed. Hence (8) is identically distributed to

$$\{\text{PP}, \text{MSK}, \tilde{h}, \overline{\text{Encrypt}}(\text{PP}, x, 1; \hat{\mathbf{G}} \cdot \hat{\mathbf{G}}', \text{MSK}), \overline{\text{KeyGen}}(\text{PP}, 1, y; (1, \ldots, 1, \hat{\mathbf{h}}_\rho \cdot \hat{\mathbf{h}}'_\rho, 1, \ldots, 1))\}.$$
$$\tag{10}$$

Let $\hat{\text{CT}} := (\hat{\text{CT}}_1, \ldots, \hat{\text{CT}}_{w_1+1})$ and $\hat{\text{SK}} := (\hat{\text{SK}}_1, \ldots, \hat{\text{SK}}_{m_1})$ denote the output of $\overline{\text{Encrypt}}$ and $\overline{\text{KeyGen}}$ respectively. We know that for $\ell \in [1, w_1]$,

$$\hat{\text{CT}}_\ell = \hat{g}_{0,0}^{\zeta_\ell} \cdot \prod_{i \in [1, w_2]} \hat{g}_{i,0}^{\eta_{\ell,i}} \cdot \prod_{j \in [1,n]} (\hat{g}_{0,j} \cdot \hat{g}_{0,0}^{\gamma_j})^{\theta_{\ell,j}} \cdot \prod_{i \in [1, w_2], j \in [1,n]} (\hat{g}_{i,j} \cdot \hat{g}_{i,0}^{\gamma_j})^{\vartheta_{\ell,i,j}},$$

where $(\hat{g}_{i,0}, \ldots, \hat{g}_{i,n}) \leftarrow \overline{\mathsf{SampG}}(\mathrm{PP}, \mathrm{SP})$ for $i \in [0, w_2]$ and $\gamma_1, \ldots, \gamma_n \leftarrow_R \mathbb{Z}_N$. Also, $\hat{\mathrm{CT}}_{w_1+1} = e(\hat{g}_{0,0}, \mathrm{MSK})$. Using the non-degeneracy property of dual system groups, we can write $\hat{g}_{0,0}$ and $\hat{g}_{i,0}$ as $\tilde{g}^\delta$ and $\tilde{g}^{\delta_i}$ respectively, for $i \in [1, w_2]$, where $\delta, \delta_1, \ldots, \delta_{w_2} \leftarrow_R \mathbb{Z}_N$. Then we consider $\hat{g}_{0,j}$ (and $\hat{g}_{i,j}$) for $j = 1, \ldots, n$ to be values sampled from $\overline{\mathsf{SampG}}$ conditioned on the value of $\hat{g}_{0,0}$ (resp. $\hat{g}_{i,0}$). (These values may not be efficiently sampleable.) Therefore, we have

$$\hat{\mathrm{CT}}_\ell = \tilde{g}^{\zeta_\ell \delta + \sum_{i \in [1,w_2]} \eta_{\ell,i} \delta_i + \sum_{j \in [1,n]} \theta_{\ell,j} \delta \gamma_j + \sum_{i \in [1,w_2], j \in [1,n]} \vartheta_{\ell,i,j} \delta_i \gamma_j}. \tag{11}$$
$$\prod_{j \in [1,n]} \hat{g}_{0,j}^{\theta_{\ell,j}} \cdot \prod_{i \in [1,w_2], j \in [1,n]} \hat{g}_{i,j}^{\vartheta_{\ell,i,j}}$$

Shifting our focus to the key, we know that its tth component is given by

$$\hat{\mathrm{SK}}_t = \hat{h}_{\rho,0}^{\upsilon_{t,\rho}} \cdot \prod_{j \in [1,n]} (\hat{h}_{\rho,j} \cdot \hat{h}_{\rho,0}^{\gamma_j})^{\phi_{t,\rho,j}},$$

for $t \in [1, m_1]$, where $(\hat{h}_{\rho,0}, \ldots, \hat{h}_{\rho,n}) \leftarrow \overline{\mathsf{SampH}}(\mathrm{PP}, \mathrm{SP})$. Using non-degeneracy once again, we can write $\hat{h}_{\rho,0}$ as $\tilde{h}^\omega$ for an $\omega \leftarrow_R \mathbb{Z}_N$, and consider $\hat{h}_{\rho,j}$ for $j = 1, \ldots, n$ to be sampled from $\overline{\mathsf{SampH}}$ conditioned on the value of $\hat{h}_{\rho,0}$. Hence,

$$\hat{\mathrm{SK}}_t = \tilde{h}^{\upsilon_{t,\rho}\omega + \sum_{j \in [1,n]} \phi_{t,\rho,j}\omega\gamma_j} \cdot \prod_{j \in [1,n]} \hat{h}_{\rho,j}^{\phi_{t,\rho,j}}. \tag{12}$$

Now, observe the superscripts of $\tilde{g}$ and $\tilde{h}$ in (11) and (12) respectively (over $\ell \in [1, w_1]$ and $t \in [1, m_1]$). We know that $\delta, \delta_1, \ldots, \delta_{w_2}, \gamma_1, \ldots, \gamma_n$ and ω are randomly chosen from $\mathbb{Z}_N$. Hence, we can use the first property (2) of relaxed perfect security to add noise to the ρ-th sample used in the key. But the problem is that in any sample drawn from $\overline{\mathsf{SampG}}$ and $\overline{\mathsf{SampH}}$, elements of the sample may depend on each other. In particular $\hat{g}_{0,j}$ may reveal some information about δ, and similarly for $\hat{g}_{i,j}$ and for $\hat{h}_{\rho,j}$, so we must ensure that (2) applies even given this information. Recall the discussion on *structural restrictions* after the definition of pair encoding schemes. We know that if $\vartheta_{\ell,i,j} \neq 0$ for any $\ell \in [1, w_1]$ and $j \in [1, n]$ (otherwise, we don't need to worry about $\hat{g}_{i,j}$), then δ_i is an *explicit* part of the encoding output by EncC. Similarly, if $\phi_{t,\rho,j} \neq 0$ for any $t \in [1, m_1]$ and $j \in [1, n]$, then ω is an explicit part of the encoding output by EncK. Further, δ is always explicit. Therefore, given a sample from either of the distributions in (2), one can compute the first element of the samples from $\overline{\mathsf{SampG}}$ and $\overline{\mathsf{SampH}}$, and then draw rest of the elements conditioned on the first ones.

In a nutshell, we can apply (2) to conclude that the distribution

$$\{\mathrm{PP}, \mathrm{MSK}, \tilde{h}, (\hat{\mathrm{CT}}_1, \ldots, \hat{\mathrm{CT}}_{w_1+1}), (\hat{\mathrm{SK}}_1, \ldots, \hat{\mathrm{SK}}_{m_1})\}$$

is statistically close to

$$\{\mathrm{PP}, \mathrm{MSK}, \tilde{h}, (\hat{\mathrm{CT}}_1, \ldots, \hat{\mathrm{CT}}_{w_1+1}), (\tilde{\mathrm{SK}}_1, \ldots, \tilde{\mathrm{SK}}_{m_1})\},$$

where

$$
\tilde{\mathrm{sK}}_t \quad := \quad \tilde{h}^{\upsilon_{t,\rho}\omega + \sum_{j\in[1,n]}\phi_{t,\rho,j}\omega(\gamma_j+z_j)} \quad \cdot \quad \prod_{j\in[1,n]} \hat{h}_{\rho,j}^{\phi_{t,\rho,j}}
$$

$$
= \quad \tilde{h}^{\upsilon_{t,\rho}\omega + \sum_{j\in[1,n]}\phi_{t,\rho,j}\omega\gamma_j} \quad \cdot \quad \prod_{j\in[1,n]} (\hat{h}_{\rho,j}\cdot\tilde{h}^{\omega z_j})^{\phi_{t,\rho,j}},
$$

for $t \in [1, m_1]$, and $\mathbf{z}_\rho = (z_1, \dots, z_n) \leftarrow \mathsf{Samp}(\rho, x, y, N)$. We use the fact that δ is always explicit to generate the $w_1 + 1$th component of the ciphertext.

Observe that the only difference between $\hat{\mathrm{sk}}_t$ and $\tilde{\mathrm{sk}}_t$ is that an extra $\tilde{h}^{\omega z_j}$ is multiplied with $\hat{h}_{\rho,j}$ in the latter case. Hence, the key $(\tilde{\mathrm{sk}}_1, \dots, \tilde{\mathrm{sk}}_{m_1})$ can be generated by giving $\hat{\mathbf{h}}_\rho \cdot \hat{\mathbf{h}}'_\rho \cdot \tilde{h}^{\mathbf{z}_\rho}$ as the ρ-th sample to $\overline{\mathsf{KeyGen}}$ ($\mathbf{z}_\rho$ has the same distribution as $\omega \cdot \mathbf{z}_\rho$ since $\omega \in \mathbb{Z}_N^*$ with high probability). Therefore, (10) is statistically close to

$$
\{\mathrm{PP}, \mathrm{MSK}, \tilde{h}, \overline{\mathsf{Encrypt}}(\mathrm{PP}, x, 1; \hat{\mathbf{G}} \cdot \hat{\mathbf{G}}', \mathrm{MSK}), \overline{\mathsf{KeyGen}}(\mathrm{PP}, 1, y; (1, \dots, 1, \hat{\mathbf{h}}_\rho \cdot \hat{\mathbf{h}}'_\rho \cdot \tilde{h}^{\mathbf{z}_\rho},
$$
$$
1, \dots, 1)).
$$

Using parameter-hiding once again, we can show that the above distribution is identical to

$$
\{\mathrm{PP}, \mathrm{MSK}, \tilde{h}, \overline{\mathsf{Encrypt}}(\mathrm{PP}, x, 1; \hat{\mathbf{G}}, \mathrm{MSK}), \overline{\mathsf{KeyGen}}(\mathrm{PP}, 1, y; (1, \dots, 1, \hat{\mathbf{h}}_\rho \cdot \tilde{h}^{\mathbf{z}_\rho}, 1, \dots, 1)),
$$

which completes the proof. $\square$

The above proof can be easily adapted to show that $\mathsf{Hyb}_{2,\varphi,7,\rho} \cong \mathsf{Hyb}_{2,\varphi,8,\rho}$. In this case, we want that the two distributions

$$
\{\mathrm{PP}, \mathrm{MSK}, \tilde{h}, \overline{\mathsf{Encrypt}}(\mathrm{PP}, x, m; \mathbf{G} \cdot \hat{\mathbf{G}}, \mathrm{MSK}),
$$
$$
\overline{\mathsf{KeyGen}}(\mathrm{PP}, \overline{\mathrm{MSK}}, y; (\mathbf{h}_1, \dots, \mathbf{h}_{\rho-1}, \mathbf{h}_\rho \cdot \hat{\mathbf{h}}_\rho \cdot \tilde{h}^{\mathbf{z}_\rho}, \mathbf{h}_{\rho+1} \cdot \tilde{h}^{\mathbf{z}_{\rho+1}}, \dots, \mathbf{h}_{m_2} \cdot \tilde{h}^{\mathbf{z}_{m_2}}))\},
$$

$$
\{\mathrm{PP}, \mathrm{MSK}, \tilde{h}, \overline{\mathsf{Encrypt}}(\mathrm{PP}, x, m; \mathbf{G} \cdot \hat{\mathbf{G}}, \mathrm{MSK}),
$$
$$
\overline{\mathsf{KeyGen}}(\mathrm{PP}, \overline{\mathrm{MSK}}, y; (\mathbf{h}_1, \dots, \mathbf{h}_{\rho-1}, \mathbf{h}_\rho \cdot \hat{\mathbf{h}}_\rho, \mathbf{h}_{\rho+1} \cdot \tilde{h}^{\mathbf{z}_{\rho+1}}, \dots, \mathbf{h}_{m_2} \cdot \tilde{h}^{\mathbf{z}_{m_2}}))\}.
$$

are indistinguishable from each other. Observe that the only difference now is that we have $\overline{\mathrm{MSK}}$ instead of MSK, and noise is present in the samples $\rho+1, \dots, n$ instead of $1, \dots, \rho-1$. So, we can split $\overline{\mathsf{Encrypt}}$ and $\overline{\mathsf{KeyGen}}$ in a way similar to the above proof, and once again it suffices to show that exactly the distributions in (8) and (9) are indistinguishable.

Lemma 3. *For every* $\varphi \in [1, \xi]$, $\mathsf{Hyb}_{2,\varphi,4} \cong \mathsf{Hyb}_{2,\varphi,5}$.

Proof. This proof proceeds in a manner similar to the proof of Lemma 2. To begin with, we observe as before that given $\mathrm{PP}, \mathrm{MSK}$ and $\tilde{h}$, one can generate

MPK and every key except the φth (for clarity, we omit φ below). Hence, it suffices to show that the distribution

$$\{\text{PP}, \text{MSK}, \tilde{h}, \overline{\text{Encrypt}}(\text{PP}, x, m; \mathbf{G} \cdot \hat{\mathbf{G}}, \text{MSK}), \overline{\text{KeyGen}}(\text{PP}, \text{MSK}, y; (\mathbf{h}_1 \cdot \hat{\mathbf{h}}_1 \cdot \tilde{h}^{\mathbf{z}_1}, \ldots,$$
$$\mathbf{h}_{m_2} \cdot \hat{\mathbf{h}}_{m_2} \cdot \tilde{h}^{\mathbf{z}_{m_2}}))\},$$

is statistically close to a distribution where MSK is replaced by $\overline{\text{MSK}}$, the semi-functional master secret key. Further,

$$\overline{\text{Encrypt}}(\text{PP}, x, m; \mathbf{G} \cdot \hat{\mathbf{G}}, \text{MSK}) = \overline{\text{Encrypt}}(\text{PP}, x, m; \mathbf{G}, \text{MSK}) \cdot \overline{\text{Encrypt}}(\text{PP}, x, 1; \hat{\mathbf{G}}, \text{MSK}),$$

$$\overline{\text{KeyGen}}(\text{PP}, \text{MSK}, y; (\mathbf{h}_1 \cdot \hat{\mathbf{h}}_1 \cdot \tilde{h}^{\mathbf{z}_1}, \ldots, \mathbf{h}_{m_2} \cdot \hat{\mathbf{h}}_{m_2} \cdot \tilde{h}^{\mathbf{z}_{m_2}}))$$
$$= \overline{\text{KeyGen}}(\text{PP}, \text{MSK}, y; (\mathbf{h}_1, \ldots, \mathbf{h}_{m_2})) \cdot \overline{\text{KeyGen}}(\text{PP}, 1, y; (\hat{\mathbf{h}}_1 \cdot \tilde{h}^{\mathbf{z}_1}, \ldots, \hat{\mathbf{h}}_{m_2} \cdot \tilde{h}^{\mathbf{z}_{m_2}})),$$

$$\overline{\text{KeyGen}}(\text{PP}, \overline{\text{MSK}}, y; (\mathbf{h}_1 \cdot \hat{\mathbf{h}}_1 \cdot \tilde{h}^{\mathbf{z}_1}, \ldots, \mathbf{h}_{m_2} \cdot \hat{\mathbf{h}}_{m_2} \cdot \tilde{h}^{\mathbf{z}_{m_2}}))$$
$$= \overline{\text{KeyGen}}(\text{PP}, \text{MSK}, y; (\mathbf{h}_1, \ldots, \mathbf{h}_{m_2})) \cdot \overline{\text{KeyGen}}(\text{PP}, \tilde{h}^{\beta}, y; (\hat{\mathbf{h}}_1 \cdot \tilde{h}^{\mathbf{z}_1}, \ldots, \hat{\mathbf{h}}_{m_2} \cdot \tilde{h}^{\mathbf{z}_{m_2}})),$$

where $\beta \leftarrow_R \mathbb{Z}_N$. The first component on the right hand side of each of the above equations can be generated given PP, MSK and $\tilde{h}$. Hence, it is enough to show that the following two distributions are statistically close:

$$\{\text{PP}, \text{MSK}, \tilde{h}, \overline{\text{Encrypt}}(\text{PP}, x, 1; \hat{\mathbf{G}}, \text{MSK}), \overline{\text{KeyGen}}(\text{PP}, 1, y; (\hat{\mathbf{h}}_1 \cdot \tilde{h}^{\mathbf{z}_1}, \ldots, \hat{\mathbf{h}}_{m_2} \cdot \tilde{h}^{\mathbf{z}_{m_2}}))\}, \tag{13}$$
$$\{\text{PP}, \text{MSK}, \tilde{h}, \overline{\text{Encrypt}}(\text{PP}, x, 1; \hat{\mathbf{G}}, \text{MSK}), \overline{\text{KeyGen}}(\text{PP}, \tilde{h}^{\beta}, y; (\hat{\mathbf{h}}_1 \cdot \tilde{h}^{\mathbf{z}_1}, \ldots, \hat{\mathbf{h}}_{m_2} \cdot \tilde{h}^{\mathbf{z}_{m_2}}))\}. \tag{14}$$

Let us focus on the first distribution between the two above. By the parameter-hiding property of dual system groups, it is identically distributed to

$$\{\text{PP}, \text{MSK}, \tilde{h}, \overline{\text{Encrypt}}(\text{PP}, x, 1; \hat{\mathbf{G}} \cdot \hat{\mathbf{G}}', \text{MSK}), \overline{\text{KeyGen}}(\text{PP}, 1, y; (\hat{\mathbf{h}}_1 \cdot \hat{\mathbf{h}}_1' \cdot \tilde{h}^{\mathbf{z}_1}, \ldots, \tag{15}$$
$$\hat{\mathbf{h}}_{m_2} \cdot \hat{\mathbf{h}}_{m_2}' \cdot \tilde{h}^{\mathbf{z}_{m_2}}))\}.$$

Let $\hat{\text{CT}} := (\hat{\text{CT}}_1, \ldots, \hat{\text{CT}}_{w_1+1})$ and $\hat{\text{SK}} := (\hat{\text{SK}}_1, \ldots, \hat{\text{SK}}_{m_1})$ denote the output of $\overline{\text{Encrypt}}$ and $\overline{\text{KeyGen}}$ respectively. We know that for $\ell \in [1, w_1]$,

$$\hat{\text{CT}}_\ell = \hat{g}_{0,0}^{\zeta_\ell} \cdot \prod_{i \in [1, w_2]} \hat{g}_{i,0}^{\eta_{\ell,i}} \cdot \prod_{j \in [1,n]} (\hat{g}_{0,j} \cdot \hat{g}_{0,0}^{\gamma_j})^{\theta_{\ell,j}} \cdot \prod_{i \in [1,w_2], j \in [1,n]} (\hat{g}_{i,j} \cdot \hat{g}_{i,0}^{\gamma_j})^{\vartheta_{\ell,i,j}},$$

where $(\hat{g}_{i,0}, \ldots, \hat{g}_{i,n}) \leftarrow \overline{\text{SampG}}(\text{PP}, \text{SP})$ for $i \in [0, w_2]$ and $\gamma_1, \ldots, \gamma_n \leftarrow_R \mathbb{Z}_N$. Using non-degeneracy property of dual system groups, we can write $\hat{g}_{0,0}$ and $\hat{g}_{i,0}$ as $\tilde{g}^{\delta}$ and $\tilde{g}^{\delta_i}$ respectively, for $i \in [1, w_2]$, where $\delta, \delta_1, \ldots, \delta_{w_2} \leftarrow_R \mathbb{Z}_N$. Therefore, we have

$$\hat{\text{CT}}_\ell = \tilde{g}^{\zeta_\ell \delta + \sum_{i \in [1,w_2]} \eta_{\ell,i} \delta_i + \sum_{j \in [1,n]} \theta_{\ell,j} \delta \gamma_j + \sum_{i \in [1,w_2], j \in [1,n]} \vartheta_{\ell,i,j} \delta_i \gamma_j}. \tag{16}$$
$$\prod_{j \in [1,n]} \hat{g}_{0,j}^{\theta_{\ell,j}} \cdot \prod_{i \in [1,w_2], j \in [1,n]} \hat{g}_{i,j}^{\vartheta_{\ell,i,j}}$$

Shifting our focus to the key, we know that its tth component is given by

$$\hat{\text{SK}}_t = \prod_{i'\in[1,m_2]} \hat{h}_{i',0}^{v_{t,i'}} \cdot \prod_{i'\in[1,m_2],j\in[1,n]} (\hat{h}_{i',j} \cdot \hat{h}_{i',0}^{\gamma_j} \cdot \tilde{h}^{z_{i',j}})^{\phi_{t,i',j}},$$

for $t \in [1, m_1]$, where $(\hat{h}_{i',0}, \ldots, \hat{h}_{i',n}) \leftarrow \overline{\text{SampH}}(\text{PP}, \text{SP})$ and $(z_{i',1}, \ldots, z_{i',n}) \leftarrow$ Samp(i', x, y, N) for $i' \in [1, m_2]$. Using non-degeneracy once again, we can write $\hat{h}_{i',0}$ as $\tilde{h}^{\omega_{i'}}$ for an $\omega_{i'} \leftarrow_R \mathbb{Z}_N$. Hence,

$$\hat{\text{SK}}_t = \tilde{h}^{\sum_{i'\in[1,m_2]}[v_{t,i'}\omega_{i'} + \sum_{j\in[1,n]}(\phi_{t,i',j}\omega_{i'}\gamma_j + \phi_{t,i',j}z_{i',j})]} \cdot \prod_{i'\in[1,m_2],j\in[1,n]} \hat{h}_{i',j}^{\phi_{t,i',j}}$$

$$= \tilde{h}^{\sum_{i'\in[1,m_2]}[v_{t,i'}\omega_{i'} + \sum_{j\in[1,n]}(\phi_{t,i',j}\omega_{i'}(\gamma_j + z_{i',j}))]} \cdot \prod_{i'\in[1,m_2],j\in[1,n]} \hat{h}_{i',j}^{\phi_{t,i',j}},$$

(17)

since the distribution of $(z_{i',1}, \ldots, z_{i',n})$ is statistically close to $(\omega_{i'} z_{i',1}, \ldots, \omega_{i'} z_{i',n})$ (with high probability $\omega_{i'} \in \mathbb{Z}_N^*$) for all $i' \in [1, m_2]$.

Now, observe the superscripts of $\tilde{g}$ and $\tilde{h}$ in (16) and (17) respectively (over $\ell \in [1, w_1]$ and $t \in [1, m_1]$). We know that $\delta, \delta_1, \ldots, \delta_{w_2}, \gamma_1, \ldots, \gamma_n$ and $\omega_1, \ldots, \omega_{m_2}$ are randomly chosen from $\mathbb{Z}_N$. Hence, we can use the second property (3) of relaxed perfect security to add noise to the master secret key. (The dependencies between the elements of the samples drawn from $\overline{\text{SampG}}$ and $\overline{\text{SampH}}$ can be handled as in the previous proof.) Therefore, we have that the distribution

$$\{\text{PP}, \text{MSK}, \tilde{h}, (\hat{\text{CT}}_1, \ldots, \hat{\text{CT}}_{w_1+1}), (\hat{\text{SK}}_1, \ldots, \hat{\text{SK}}_{m_1})\}$$

is statistically close to

$$\{\text{PP}, \text{MSK}, \tilde{h}, (\hat{\text{CT}}_1, \ldots, \hat{\text{CT}}_{w_1+1}), (\tilde{\text{SK}}_1, \ldots, \tilde{\text{SK}}_{m_1})\},$$

where

$$\tilde{\text{SK}}_t := \tilde{h}^{\tau_t\beta + \sum_{i'\in[1,m_2]}[v_{t,i'}\omega_{i'} + \sum_{j\in[1,n]}(\phi_{t,i',j}\omega_{i'}(\gamma_j + z_{i',j}))]} \cdot \prod_{i'\in[1,m_2],j\in[1,n]} \hat{h}_{i',j}^{\phi_{t,i',j}},$$

for $t \in [1, m_1]$, and $\beta \leftarrow_R \mathbb{Z}_N$. Observe that the only difference between $\hat{\text{SK}}_t$ and $\tilde{\text{SK}}_t$ is that an extra $\tau_t\beta$ is begin added to the exponent of $\tilde{h}$ in the latter case. Hence, the key $(\tilde{\text{SK}}_1, \ldots, \tilde{\text{SK}}_{m_1})$ can be generated by providing $\tilde{h}^\beta$ as master secret key to $\overline{\text{KeyGen}}$. Therefore, (15) is statistically close to

$$\{\text{PP}, \text{MSK}, \tilde{h}, \overline{\text{Encrypt}}(\text{PP}, x, 1; \hat{\mathbf{G}} \cdot \hat{\mathbf{G}}', \text{MSK}), \overline{\text{KeyGen}}(\text{PP}, \tilde{h}^\beta, y; (\hat{\mathbf{h}}_1 \cdot \hat{\mathbf{h}}_1' \cdot \tilde{h}^{\mathbf{z}_1}, \ldots,$$
$$\hat{\mathbf{h}}_{m_2} \cdot \hat{\mathbf{h}}_{m_2}' \cdot \tilde{h}^{\mathbf{z}_{m_2}}))\}.$$

Using parameter-hiding once again, we can show that the above distribution is identical to

$$\{\text{PP}, \text{MSK}, \tilde{h}, \overline{\text{Encrypt}}(\text{PP}, x, 1; \hat{\mathbf{G}}, \text{MSK}), \overline{\text{KeyGen}}(\text{PP}, \tilde{h}^\beta, y; (\hat{\mathbf{h}}_1 \cdot \tilde{h}^{\mathbf{z}_1}, \ldots, \hat{\mathbf{h}}_{m_2} \cdot \tilde{h}^{\mathbf{z}_{m_2}}))\},$$

which completes the proof. $\qquad\square$

7   Ciphertext-Policy ABE

In this section, we design a relaxed perfectly secure pair encoding scheme for Ciphertext-Policy Attribute Based Encryption (CP-ABE). The access policy is represented by a linear secret sharing (LSS) scheme $(\mathbf{A}, \pi)$, where $\mathbf{A}$ is a matrix of size $n_1 \times n_2$ with entries in $\mathbb{Z}_N$ and π is a mapping from $[1, n_1]$ to a universe of attributes $\mathcal{U}$. Let $\mathbf{a_i}$ denote the ith row of $\mathbf{A}$ for $i \in [1, n_1]$. Let $S \subseteq \mathcal{U}$ be a set of attributes and $\Upsilon = \{i \mid i \in [1, n_1], \pi(i) \in S\}$ be the indices of rows in $\mathbf{A}$ associated with S.

We say that the LSS scheme $(\mathbf{A}, \pi)$ accepts S if $\mathbf{e} = (1, 0, \ldots, 0)$ lies in the span of rows associated with S (otherwise the scheme rejects S). In other words, if S is acceptable, there exists constants $\{\varepsilon_i\}_{i \in \Upsilon}$ such that $\sum_{i \in \Upsilon} \varepsilon_i \mathbf{a_i} = \mathbf{e}$. (This set of constants can be easily computed given S.) An interesting property of LSS schemes that will be useful to us later in the proofs is that if (A, π) rejects S, then there must exist a vector $\mathbf{w} = (w_1, \ldots, w_{n_2})$ such that $\langle \mathbf{w}, \mathbf{a_i} \rangle = 0$ for all $i \in \Upsilon$ but $\langle \mathbf{w}, \mathbf{e} \rangle = 1$. This, in particular, implies that $w_1 = 1$. (See [7], Claim 2, for a proof of this and other properties below about secret sharing schemes.)

In order to share a secret $s \in \mathbb{Z}_N$, one picks $v_2, v_3, \ldots, v_{n_1} \leftarrow_R \mathbb{Z}_N$, and outputs $\langle \mathbf{a_i}, \mathbf{v} \rangle$ as the ith share for $i \in [1, n_1]$, where $\mathbf{v} = (s, v_2, v_3, \ldots, v_{n_1})$. This way of sharing a secret leads to two useful properties:

- **Correctness:** For every S accepted by $(\mathbf{A}, \pi)$, every secret $s \in \mathbb{Z}_N$ and any $v_2, v_3, \ldots, v_{n_1} \in \mathbb{Z}_N$, $\sum_{i \in \Upsilon} \varepsilon_i \langle \mathbf{a_i}, \mathbf{v} \rangle = \langle \mathbf{v}, \sum_{i \in \Upsilon} \varepsilon_i \mathbf{a_i} \rangle = s$.
- **Privacy:** For every S rejected by $(\mathbf{A}, \pi)$, the distribution of $\{\langle \mathbf{a_i}, \mathbf{v} \rangle\}_{i \in \Upsilon}$ is independent of the secret s being shared.

The predicate family for CP-ABE is indexed by $\kappa = (N, n_1, n_2, \mathcal{U}, T)$. $\mathcal{X}_\kappa$ is the set of all LSS schemes where the matrix is of size $n_1 \times n_2$ with entries in $\mathbb{Z}_N$ and the mapping is from $[1, n_1]$ to $\mathcal{U}$. $\mathcal{Y}_\kappa$ is given by the set $\{S \mid S \subseteq \mathcal{U}, |S| \leq T\}$. For all $x \in \mathcal{X}_\kappa$ and $y \in \mathcal{Y}_\kappa$, $P_\kappa(x, y) = 1$ if and only if x accepts y. It is clear from our definition of predicate family that there is a bound on the size of matrices and the number of attributes associated with a key. But there are no other restrictions: the size of attribute universe $\mathcal{U}$ could be arbitrary and π need not be injective. Without loss of generality, we assume $\mathcal{U}$ to be $\mathbb{Z}_N$.

We are now ready to design a relaxed perfectly secure pair encoding scheme $\Phi_{\mathsf{cp\text{-}abe}} = (\mathsf{Param}, \mathsf{EncC}, \mathsf{EncK}, \mathsf{Pair})$ for the CP-ABE predicate family.

7.1   Pair Encoding Scheme

- $\mathsf{Param}(\mathsf{par}) \rightarrow n_1(n_2 + T + 1)$. Let $\mathbf{b} = (\{b_{i,j}\}_{i \in [1,n_1], j \in [1,n_2]}, \{b'_{i,t}\}_{i \in [1,n_1], t \in [0,T]})$.
- $\mathsf{EncC}((A, \pi), N) \rightarrow \mathbf{c}(\mathbf{s}, \mathbf{b}) := (c_1, c_2)$ where

$$
c_1 = s \qquad c_2 = s \left(\sum_{\substack{i \in [1, n_1] \\ j \in [1, n_2]}} a_{i,j} b_{i,j} + \sum_{\substack{i \in [1, n_1] \\ t \in [0, T]}} \pi(i)^t b'_{i,t} \right),
$$

and $\mathbf{s} = (s)$, and $a_{i,j}$ denotes the entry in the ith row and jth column of A.

- $\mathsf{EncK}(S, N)$ $\longrightarrow$ $\mathbf{k}(\alpha, \mathbf{r}, \mathbf{b})$ $:=$ $(\{k_{1,i}, k_{2,i,j}$ $k_{3,i,\ell,j}, k_{4,i,y}$ $k_{5,i,\ell,t}\}$
  $i,\ell \in [1,n_1], i \neq \ell, j \in [1,n_2], y \in S, t \in [0,T])$ where

$$k_{1,i} = r_i \quad k_{2,i,j} = r_i b_{i,j} - v_j \quad k_{3,i,\ell,j} = r_i b_{\ell,j}$$

$$k_{4,i,y} = r_i \sum_{t \in [0,T]} y^t b'_{i,t} \quad k_{5,i,\ell,t} = r_i b'_{\ell,t}$$

and $\mathbf{r} = (r_1, r_2, \ldots, r_{n_1}, v_2, \ldots, v_{n_2})$ and $v_1 = \alpha$.

We informally discuss how to recover αs by combining the polynomials generated by EncC and EncK, with an intent to provide some intuition about the scheme, and defer a formal proof to the full version. We can think of $v_2, v_3, \ldots, v_{n_1}$ as the randomness picked in order to share $v_1 = \alpha$ according to the scheme (A, π). Hence, if we find $\langle \mathbf{a_i}, \mathbf{v} \rangle$ for all $i \in \Upsilon$, we can recover α (ignore s for now). One could start out by multiplying $a_{i,j}$ by $k_{2,i,j}$ and summing over j, for an $i \in \Upsilon$. This does give $\sum_j a_{i,j} v_j$ but also produces an extra term $r_i \sum_j a_{i,j} b_{i,j}$ (ignore r_i for now). We could try to get rid of this term by using c_2 but the product $a_{i,j} b_{i,j}$ there is also summed over i (since we want EncC to produce a constant number of polynomials, we are forced to pack as much into one polynomial as possible). Fortunately, we have the polynomials $k_{3,i,\ell,j}$ for $\ell \neq i$. We can multiply these by $a_{\ell,j}$ and remove the unwanted $a_{i,j} b_{i,j}$ terms. But we are not done yet: we must also remove the term $\sum_{i,t} \pi(i)^t b'_{i,t}$ left in the mix because we used c_2. If $\pi(i) \in S$, then this is easy: use $k_{4,i,\pi(i)}$ to remove $\sum_t \pi(i)^t b'_{i,t}$, and $k_{5,i,\ell,t} \cdot \pi(\ell)^t$ to remove the rest. However, if $\pi(i) \notin S$, there is no way to do this.

7.2 Relaxed Perfect Security

We now prove that the pair encoding scheme $\Phi_{\mathsf{cp-abe}}$ designed above is relaxed perfectly secure (Definition 3). Towards this, we first define a sampling algorithm Samp as follows. On input an $i \in [1, n_1]$, $(A, \pi) \in \mathcal{X}_\kappa$, $S \in \mathcal{Y}_\kappa$ and N, Samp checks whether $\pi(i) \notin S$. If yes, it picks elements $\hat{b}_{i,1}, \hat{b}_{i,2}, \ldots, \hat{b}_{i,n_2}$ independently and uniformly from $\mathbb{Z}_N$; otherwise it picks them uniformly but with the constraint that $\sum_{j \in [1,n_2]} a_{i,j} \hat{b}_{i,j} = 0$. Samp outputs

$$\hat{\mathbf{b}}_i := (\underbrace{0, \ldots, \ldots, \ldots, 0}_{(i-1)n_2}, \hat{b}_{i,1}, \hat{b}_{i,2}, \ldots, \hat{b}_{i,n_2}, \underbrace{0, \ldots, \ldots, \ldots, \ldots, 0}_{(n_1-i)n_2 + n_1(T+1)}). \quad (18)$$

Observe that the output of Samp depends on (A, π), the input to EncC. Hence, this sampling algorithm would lead to a semi-adaptively secure scheme.

We consider only those $N \in \mathbb{N}$ which are a product of distinct primes of $\Theta(\lambda)$ bits. This is sufficient for our purposes because the Setup algorithm of the generic construction in Sect. 5 defines N of exactly this form. We first show that for all $i \in [1, n_1]$ and $N \in \mathbb{N}$,

$$\big(\mathbf{c}(\mathbf{s}, \mathbf{b}), \mathbf{k}_i(0, r_i, \mathbf{b})\big) \quad \equiv \quad \big(\mathbf{c}(\mathbf{s}, \mathbf{b}), \mathbf{k}_i(0, r_i, \mathbf{b} + \hat{\mathbf{b}}_i)\big), \quad (19)$$

where $\mathbf{s} \leftarrow_R \mathbb{Z}_N^1$, $\mathbf{b} \leftarrow_R \mathbb{Z}_N^n$, $r_i \leftarrow_R \mathbb{Z}_N$, $\widehat{\mathbf{b}}_i \leftarrow \mathsf{Samp}(i, (A, \pi), S, N)$. Recall that $\mathbf{k}_i$ denotes the polynomials in $\mathbf{k}$ obtained by setting all the variables in $\mathbf{r} = (r_1, r_2, \ldots, r_{n_1}, v_2, \ldots, v_{n_2})$ except the ith to 0. For $i \in [n_1 + 1, n_1 + n_2 - 1]$, the only polynomial in $\mathbf{k}_i$ is $-v_{i-n_1+1}$, or, more importantly, there is no monomial with any b. Hence, the equation above trivially holds for i in this range irrespective of what Samp outputs. (That is why we don't care about defining Samp's behavior on such inputs.)

Let us refer to the left and right distributions in Eq. (19) as Δ_L and Δ_R respectively. Fix an arbitrary $i^* \in [1, n_1]$. By the definition of $\mathbf{k}_{i^*}$, we know that in these two distributions only those components of the key survive which have subscript i^*. Further, in the components $k_{2,i^*,1}, \ldots, k_{2,i^*,n_2}$, the variables $v_1, \ldots, v_{n_2}$ are all set to 0. Now, focus on the distribution Δ_R. It is clear from Eq. (18) that the added randomness $\widehat{\mathbf{b}}_{i^*}$ affects only $k_{2,i^*,1}, \ldots, k_{2,i^*,n_2}$ components. For $i \in [1, n_1]$ and $j \in [1, n_2]$, let $\delta_{i,j} := b_{i,j}$ if $i \neq i^*$ and $\delta_{i^*,j} := b_{i^*,j} + \widehat{b}_{i^*,j}$. Since $b_{i,j}$ are uniformly and independently distributed, so are $\delta_{i,j}$. The second component of ciphertext encoding, c_2, can now be rewritten as

$$
s \left(\sum_{\substack{i \in [1,n_1] \\ j \in [1,n_2]}} a_{i,j} \delta_{i,j} - \sum_{j \in [1,n_2]} a_{i^*,j} \widehat{b}_{i^*,j} + \sum_{t \in [0,T]} \pi(i^*)^t b'_{i^*,t} + \sum_{\substack{i \in [1,n_1], i \neq i^* \\ t \in [0,T]}} \pi(i)^t b'_{i,t} \right).
$$

Observe that the only difference between Δ_L and Δ_R is that in the latter case there is an additional term $\mathsf{rand} := \sum_{j \in [1,n_2]} a_{i^*,j} \widehat{b}_{i^*,j}$ in c_2. If $\pi(i^*) \in S$, then this term is 0 by our choice of Samp. On the other hand when $\pi(i^*) \notin S$, we show that $\sum_{t \in [0,T]} \pi(i^*)^t b'_{i^*,t}$ is an independent uniform random variable over $\mathbb{Z}_N$, and therefore, the additional term rand does not matter. Towards this, consider the polynomial $f(x) = b'_{i^*,T} \cdot x^T + b'_{i^*,T-1} \cdot x^{T-1} + \ldots + b'_{i^*,0}$. Since $b'_{i^*,T}, \ldots, b'_{i^*,0}$ are chosen at random, any $T + 1$ distinct points on $f(x)$ are uniformly distributed over $\mathbb{Z}_N^{T+1}$. The only components of the key which depend on $b'_{i^*,T}, \ldots, b'_{i^*,0}$ are $\{k_{4,i^*,y}\}_{y \in S}$, which could also be rewritten as $\{r_i \cdot f(y)\}_{y \in S}$. There could be at most T such components because $|S| \leq T$. Therefore, $\sum_{t \in [0,T]} \pi(i^*)^t b'_{i^*,t} = f(\pi(i^*))$ is independently and uniformly distributed.

The second and last step in proving relaxed perfect security is to show that when (A, π) does not accept S, Eq. (3) holds, i.e., for large enough values of N, the statistical distance between the distributions,

$$
\left(\mathbf{c}(\mathbf{s}, \mathbf{b}), \sum_{i \in [1,n_1+n_2-1]} \mathbf{k}_i(0, r_i, \mathbf{b} + \widehat{\mathbf{b}}_i) \right) \text{ and } \left(\mathbf{c}(\mathbf{s}, \mathbf{b}), \sum_{i \in [1,n_1+n_2-1]} \mathbf{k}_i(\alpha, r_i, \mathbf{b} + \widehat{\mathbf{b}}_i) \right),
$$
(20)

is negligible, where $\mathbf{s} \leftarrow_R \mathbb{Z}_N^1$, $\mathbf{b} \leftarrow_R \mathbb{Z}_N^n$, $\mathbf{r} \leftarrow_R \mathbb{Z}_N^{n_1+n_2-1}$, $\alpha \leftarrow_R \mathbb{Z}_N$, and $\widehat{\mathbf{b}}_i \leftarrow \mathsf{Samp}(i, (A, \pi), S, N)$ for $i \in [1, n_1 + n_2 - 1]$. Let us denote the left and right distributions in Eq. (20) above by Γ_L and Γ_R respectively. The second component of the key in these two distributions is given by

$$
k_{2,i,j} = r_i b_{i,j} + r_i \widehat{b}_{i,j} - v_j
$$

for $i \in [1, n_1]$ and $j \in [1, n_2]$. The only difference between the distributions is in the components $k_{2,1,1}, \ldots, k_{2,n_1,1}$. In the case of Γ_L, $v_1 = (n_1 + n_2 - 1)\alpha = 0$, while in the case of Γ_R, it is chosen independently and uniformly from $\mathbb{Z}_N$.

Let us focus on the distribution Γ_L. Recall that there exists a vector $\mathbf{w} = (w_1, \ldots, w_{n_2})$ orthogonal to all the rows associated with S such that $w_1 = 1$. We claim that if we replace the variables $\hat{b}_{i,j}$ by $\hat{b}_{i,j} - r_i^{-1} w_j \alpha$, where $\alpha \leftarrow_R \mathbb{Z}_N$, then Γ_L is not affected. (With high probability $r_i \in \mathbb{Z}_N^*$, so r_i^{-1} exists.) If $\pi(i) \notin S$, we know that $\hat{b}_{i,1}, \hat{b}_{i,2}, \ldots, \hat{b}_{i,n_2}$ are independently and uniformly distributed. Hence adding $-r_i^{-1} w_j \alpha$ has no effect on their joint distribution. On the other hand when $\pi(i) \in S$, $\hat{b}_{i,1}, \hat{b}_{i,2}, \ldots, \hat{b}_{i,n_2}$ are uniformly chosen with the constraint that $\sum_{j \in [1,n_2]} a_{i,j} \hat{b}_{i,j} = 0$. Now, when $-r_i^{-1} w_j \alpha$ is added,

$$\sum_{j \in [1,n_2]} a_{i,j}(\hat{b}_{i,j} - r_i^{-1} w_j \alpha) \quad = \quad \sum_{j \in [1,n_2]} a_{i,j} \hat{b}_{i,j} \quad - \quad r_i^{-1} \alpha \sum_{j \in [1,n_2]} a_{i,j} w_j \quad = \quad 0$$

because $\mathbf{w}$ is orthogonal to every $\mathbf{a}_i$ such that $\pi(i) \in S$. Hence, the variables $\hat{b}_{i,1}, \hat{b}_{i,2}, \ldots, \hat{b}_{i,n_2}$ still satisfy the constraint they did before.

After replacing $\hat{b}_{i,j}$ by $\hat{b}_{i,j} - r_i^{-1} w_j \alpha$, we have that $k_{2,i,j} = r_i b_{i,j} + r_i \hat{b}_{i,j} - w_j \alpha - v_j$ (where $v_1 = 0$). The final step in the proof is to replace the variables $w_1 \alpha, w_2 \alpha + v_2, \ldots, w_{n_2} \alpha + v_{n_2}$ by $\alpha, v_2, \ldots, v_{n_2}$. This does not affect Γ_L because $v_2, \ldots, v_{n_2}$ are picked independently and uniformly from $\mathbb{Z}_N$ (and $w_1 = 1$). But now Γ_L is exactly the distribution Γ_R.

7.3   Instantiation: Constant-Size Ciphertext

We briefly comment about instantiating the pair encoding scheme $\Phi_{\text{cp-abe}} = (\text{Param}, \text{EncC}, \text{EncK}, \text{Pair})$. Using the generic method in Sect. 5, one can construct a predicate encryption scheme $\Pi_{\text{cp-abe}} = (\text{Setup}, \text{Encrypt}, \text{KeyGen}, \text{Decrypt})$ for CP-ABE using $\Phi_{\text{cp-abe}}$. According to Theorem 1, $\Pi_{\text{cp-abe}}$ is semi-adaptively secure because the Samp algorithm we defined in the previous sub-section depends on the access structure. However, since EncC outputs only two polynomials, Encrypt outputs only two elements from $\mathbb{G}$ (and one element from $\mathbb{G}_T$). Now, from Remark 2, it follows that one can design a concrete scheme for CP-ABE in prime-order groups where the ciphertext contains *only 4 group elements under the* SXDH *assumption*, and only 6 elements under the DLIN assumption (plus an additional element from the target group). Furthermore, only a constant number of pairing operations would be required to decrypt a ciphertext.

References

1. Akinyele, J.A., Pagano, M.W., Green, M.D., Lehmann, C.U., Peterson, Z.N.J., Rubin, A.D.: Securing electronic medical records using attribute-based encryption on mobile devices. In: SPSM 2011, ACM Workshop Security and Privacy in Smartphones and Mobile Devices 2011, pp. 75–86 (2011)

2. Attrapadung, N.: Dual system encryption via doubly selective security: framework, fully secure functional encryption for regular languages, and more. In: Nguyen, P.Q., Oswald, E. (eds.) EUROCRYPT 2014. LNCS, vol. 8441, pp. 557–577. Springer, Heidelberg (2014)
3. Attrapadung, N.: Dual system encryption framework in prime-order groups. Cryptology ePrint Archive, Report 2015/390 (2015). http://eprint.iacr.org/2015/390
4. Attrapadung, N., Hanaoka, G., Yamada, S.: Conversions among several classes of predicate encryption and their applications. Cryptology ePrint Archive, Report 2015/431 (2015). http://eprint.iacr.org/2015/431
5. Attrapadung, N., Libert, B., de Panafieu, E.: Expressive key-policy attribute-based encryption with constant-size ciphertexts. In: Catalano, D., Fazio, N., Gennaro, R., Nicolosi, A. (eds.) PKC 2011. LNCS, vol. 6571, pp. 90–108. Springer, Heidelberg (2011)
6. Baden, R., Bender, A., Spring, N., Bhattacharjee, B., Starin, D.: Persona: an online social network with user-defined privacy. In: ACM SIGCOMM 2009 Conference on Applications, Technologies, Architectures, and Protocols for Computer Communications, pp. 135–146 (2009)
7. Beimel, A.: Secret-sharing schemes: a survey. In: Chee, Y.M., Guo, Z., Ling, S., Shao, F., Tang, Y., Wang, H., Xing, C. (eds.) IWCC 2011. LNCS, vol. 6639, pp. 11–46. Springer, Heidelberg (2011)
8. Bethencourt, J., Sahai, A., Waters, B.: Ciphertext-policy attribute-based encryption. In: 2007 IEEE Symposium on Security and Privacy, pp. 321–334. IEEE Computer Society Press (2007)
9. Boneh, D., Raghunathan, A., Segev, G.: Function-private identity-based encryption: hiding the function in functional encryption. In: Canetti, R., Garay, J.A. (eds.) CRYPTO 2013, Part II. LNCS, vol. 8043, pp. 461–478. Springer, Heidelberg (2013)
10. Boneh, D., Sahai, A., Waters, B.: Functional encryption: definitions and challenges. In: Ishai, Y. (ed.) TCC 2011. LNCS, vol. 6597, pp. 253–273. Springer, Heidelberg (2011)
11. Boneh, D., Waters, B.: Conjunctive, subset, and range queries on encrypted data. In: Vadhan, S.P. (ed.) TCC 2007. LNCS, vol. 4392, pp. 535–554. Springer, Heidelberg (2007)
12. Chen, J., Gay, R., Wee, H.: Improved dual system ABE in prime-order groups via predicate encodings. In: Oswald, E., Fischlin, M. (eds.) EUROCRYPT 2015. LNCS, vol. 9057, pp. 595–624. Springer, Heidelberg (2015)
13. Chen, J., Wee, H.: Fully, (almost) tightly secure IBE and dual system groups. In: Canetti, R., Garay, J.A. (eds.) CRYPTO 2013, Part II. LNCS, vol. 8043, pp. 435–460. Springer, Heidelberg (2013)
14. Chen, J., Wee, H.: Dual system groups and its applications – compact HIBE and more. Cryptology ePrint Archive, Report 2014/265 (2014). http://eprint.iacr.org/2014/265
15. Chen, J., Wee, H.: Semi-adaptive attribute-based encryption and improved delegation for Boolean formula. In: Abdalla, M., De Prisco, R. (eds.) SCN 2014. LNCS, vol. 8642, pp. 277–297. Springer, Heidelberg (2014)
16. Escala, A., Herold, G., Kiltz, E., Ràfols, C., Villar, J.: An algebraic framework for Diffie-Hellman assumptions. In: Canetti, R., Garay, J.A. (eds.) CRYPTO 2013, Part II. LNCS, vol. 8043, pp. 129–147. Springer, Heidelberg (2013)
17. Freeman, D.M.: Converting pairing-based cryptosystems from composite-order groups to prime-order groups. In: Gilbert, H. (ed.) EUROCRYPT 2010. LNCS, vol. 6110, pp. 44–61. Springer, Heidelberg (2010)

18. Goyal, V., Pandey, O., Sahai, A., Waters, B.: Attribute-based encryption for fine-grained access control of encrypted data. In: Juels, A., Wright, R.N., Vimercati, S. (eds.) ACM CCS 2006, pp. 89–98. ACM Press (2006). Available as Cryptology ePrint Archive Report 2006/309

19. Guillevic, A.: Comparing the pairing efficiency over composite-order and prime-order elliptic curves. In: Jacobson, M., Locasto, M., Mohassel, P., Safavi-Naini, R. (eds.) ACNS 2013. LNCS, vol. 7954, pp. 357–372. Springer, Heidelberg (2013)

20. Herold, G., Hesse, J., Hofheinz, D., Ràfols, C., Rupp, A.: Polynomial spaces: a new framework for composite-to-prime-order transformations. In: Garay, J.A., Gennaro, R. (eds.) CRYPTO 2014, Part I. LNCS, vol. 8616, pp. 261–279. Springer, Heidelberg (2014)

21. Katz, J., Sahai, A., Waters, B.: Predicate encryption supporting disjunctions, polynomial equations, and inner products. In: Smart, N.P. (ed.) EUROCRYPT 2008. LNCS, vol. 4965, pp. 146–162. Springer, Heidelberg (2008)

22. Lewko, A.: Tools for simulating features of composite order bilinear groups in the prime order setting. In: Pointcheval, D., Johansson, T. (eds.) EUROCRYPT 2012. LNCS, vol. 7237, pp. 318–335. Springer, Heidelberg (2012)

23. Lewko, A., Okamoto, T., Sahai, A., Takashima, K., Waters, B.: Fully secure functional encryption: attribute-based encryption and (hierarchical) inner product encryption. In: Gilbert, H. (ed.) EUROCRYPT 2010. LNCS, vol. 6110, pp. 62–91. Springer, Heidelberg (2010)

24. Lewko, A., Waters, B.: Unbounded HIBE and attribute-based encryption. In: Paterson, K.G. (ed.) EUROCRYPT 2011. LNCS, vol. 6632, pp. 547–567. Springer, Heidelberg (2011)

25. Lewko, A., Waters, B.: New proof methods for attribute-based encryption: achieving full security through selective techniques. In: Safavi-Naini, R., Canetti, R. (eds.) CRYPTO 2012. LNCS, vol. 7417, pp. 180–198. Springer, Heidelberg (2012)

26. Okamoto, T., Takashima, K.: Fully secure functional encryption with general relations from the decisional linear assumption. In: Rabin, T. (ed.) CRYPTO 2010. LNCS, vol. 6223, pp. 191–208. Springer, Heidelberg (2010)

27. Okamoto, T., Takashima, K.: Fully secure unbounded inner-product and attribute-based encryption. In: Wang, X., Sako, K. (eds.) ASIACRYPT 2012. LNCS, vol. 7658, pp. 349–366. Springer, Heidelberg (2012)

28. Pirretti, M., Traynor, P., McDaniel, P., Waters, B.: Secure attribute-based systems. In: Juels, A., Wright, R.N., Vimercati, S. (eds.) ACM CCS 2006, pp. 99–112. ACM Press (2006)

29. Rouselakis, Y., Waters, B.: Practical constructions and new proof methods for large universe attribute-based encryption. In: Sadeghi, A.R., Gligor, V.D., Yung, M. (eds.) ACM CCS 2013, pp. 463–474. ACM Press (2013)

30. Sahai, A., Waters, B.: Fuzzy identity-based encryption. In: Cramer, R. (ed.) EUROCRYPT 2005. LNCS, vol. 3494, pp. 457–473. Springer, Heidelberg (2005)

31. Santos, N., Rodrigues, R., Gummadi, K.P., Saroiu, S.: Policy-sealed data: a new abstraction for building trusted cloud services. In: USENIX Security Symposium 2012, pp. 175–188 (2012)

32. Shen, E., Shi, E., Waters, B.: Predicate privacy in encryption systems. In: Reingold, O. (ed.) TCC 2009. LNCS, vol. 5444, pp. 457–473. Springer, Heidelberg (2009)

33. Takashima, K.: Expressive attribute-based encryption with constant-size ciphertexts from the decisional linear assumption. In: Abdalla, M., De Prisco, R. (eds.) SCN 2014. LNCS, vol. 8642, pp. 298–317. Springer, Heidelberg (2014)

34. Traynor, P., Butler, K.R.B., Enck, W., McDaniel, P.: Realizing massive-scale conditional access systems through attribute-based cryptosystems. In: NDSS 2008. The Internet Society (2008)

35. Waters, B.: Dual system encryption: realizing fully secure IBE and HIBE under simple assumptions. In: Halevi, S. (ed.) CRYPTO 2009. LNCS, vol. 5677, pp. 619–636. Springer, Heidelberg (2009)

36. Wee, H.: Dual system encryption via predicate encodings. In: Lindell, Y. (ed.) TCC 2014. LNCS, vol. 8349, pp. 616–637. Springer, Heidelberg (2014)

37. Yamada, S., Attrapadung, N., Hanaoka, G., Kunihiro, N.: A framework and compact constructions for non-monotonic attribute-based encryption. In: Krawczyk, H. (ed.) PKC 2014. LNCS, vol. 8383, pp. 275–292. Springer, Heidelberg (2014)

Codes and Interactive Proofs

Optimal Amplification of Noisy Leakages

Stefan Dziembowski[1]([✉]), Sebastian Faust[2], and Maciej Skórski[1]

[1] University of Warsaw, Warsaw, Poland
s.dziembowski@crypto.edu.pl, maciej.skorski@gmail.com
[2] Ruhr University Bochum, Bochum, Germany
sebastian.faust@gmail.com

Abstract. During the last 15 years there have been intensive research efforts in constructing cryptographic algorithms resilient to the side-channel leakage. The most fundamental part of every such construction are the *leakage-resilient encoding schemes*. Usually the cryptographic secrets encoded by them are assumed to belong to some finite group $(\mathbb{G}, +)$. The most common encoding scheme is the n-out-of-n additive secret sharing: a secret X is encoded as $(X_1, \ldots, X_n)$ such that $X_1 + \cdots X_n = X$. Intuitively, if an adversary receives only small partial independent information about each X_i then his information about X should be even smaller, and should decrease (i.e. the noise should *amplify*) when n grows. However, of course, the concrete parameters (the amount of leakage that can be tolerated, and the number of shares needed to achieve a given level of security) depend on the exact model that is used.

One of the most prominent models used in this area is the so-called *noisy-leakage model* (Chari et al., CRYPTO'99, and Prouff and Rivain, EUROCRYPT'13), which is believed to correspond well to the real-life engineering experience, where the information that the adversary receives is always "noisy". In the Prouff and Rivain model the amount of information that the noise provides is measured using a parameter δ that is equal to 0 when the noise is "full", and equal to 1 when there is no noise. It is natural to ask how small δ needs to be to achieve the amplification of noise (in the additive encoding scheme described above). Until now it was known that such amplification can be achieved for $\delta < 1/16$. In this paper we show that:

- in the prime order groups $\mathbb{G}$ it suffices that $\delta < 1 - 1/|\mathbb{G}|$,
- in general it suffices that $\delta < 1/2$.

We also prove that these bounds are optimal. We then analyze the number n of shares needed to achieve security ϵ of the encoded value X (where ϵ is also defined in terms of "noisy information" that the adversary obtains about X). We give close lower and upper bounds on this value (that differ only in factor polylogarithmic in $|\mathbb{G}|$). We achieve our results using techniques from the additive combinatorics, the harmonic analysis, and the convex optimization.

The first and the last author were partly supported by the WELCOME/2010-4/2 grant founded within the framework of the EU Innovative Economy Operational Programme.

E. Kushilevitz and T. Malkin (Eds.): TCC 2016-A, Part II, LNCS 9563, pp. 291–318, 2016.
DOI: 10.1007/978-3-662-49099-0_11

1   Introduction

Leakage-resilient cryptography [1–3,5,6,9,11,12,14,15,19,20,24] aims at constructing cryptographic schemes that are secure against side-channel leakages of secret information Leakage-resilient encoding schemes are important building blocks for constructing such algorithms. They allow to encode a secret message X with a randomized encoding function $\mathsf{Enc}(X) = (X_1, \ldots, X_n)$ such that leakage from the codeword does not help the adversary to recover the secret message X. The simplest encoding function $\mathsf{Enc}(.)$ uses an additive secret sharing scheme, where the shares X_i are chosen uniformly at random from a group $\mathbb{G}$ such that $X := X_1 + \ldots + X_n$. Of course, the leakage from $\mathsf{Enc}(X)$ cannot be arbitrary as otherwise no security is possible. A common assumption that has been studied both in theoretical and practical works [4,8,12,22,24] is to assume that the leakage from the encoding is a "noisy" function. It has been shown in the work of Chari et al. [4] that the encoding scheme described above amplifies security when the leakage is "sufficiently" noisy. While several recent works improve the quantitative bounds on the amount of noise needed [10,22], the current best bounds require the leakage to be very noisy – in particular far away from the level of noise that is typically available in physical measurements [8]. The main contribution of our work is to give an optimal characterization of the noisy leakage model. We develop optimal bounds for the amount of noise that is needed to amplify security, and give matching upper bounds by showing that above this threshold amplification is impossible. Our results are particular important for the security analysis of masking schemes, which we further describe below.

Leakage resilient encodings for masking schemes. Leakage resilient encodings are prominently used to build masking schemes [4,15]. Masking schemes are widely used in practice to protect cryptographic implementations against side-channel leakage – in particular against leakage from the power consumption [16]. The basic idea of a masking scheme is simple: instead of computing directly on the sensitive information (e.g., the secret key), a cryptographic algorithm protected with the masking countermeasure computes on encoded values thereby concealing sensitive information, which makes it harder to extract relevant information from the leakage. The simplest and most widely used masking scheme is the Boolean masking scheme. The Boolean masking scheme introduced in the important work of Ishai et al. [15] uses the simple encoding function from above when $\mathbb{G} = \mathsf{GF}(2)$, but can easily be extended to work in larger groups. For instance, to protect an implementation of the AES algorithm we typically use $\mathbb{G} = \mathsf{GF}(2^8)$ as the AES algorithm can be implemented in a particular efficient way using operations in the Galois field. Another example is a protected implementation of discrete-log based crypto schemes that work in prime-order fields.

The noisy leakage model. While there are several variants of the "noisy leakage model" [12,20,22] most works that consider the security of masking schemes use the model of Chari et al. [4] and its generalization by Prouff and Rivain [22]. Informally, a noisy leakage function $f : \mathbb{G} \to \mathcal{Y}$ is called δ-noisy if the statistical distance between the uniform distribution X over $\mathbb{G}$ and the conditional distribution of X *given* $f(X)$ is bounded by a parameter $\delta \in [0, 1 - 1/|\mathbb{G}|]$. Here, $\mathcal{Y}$ is the domain of the leakage, which in general can be an infinite set. To better understand the Prouff-Rivain noise model, let us consider the two extreme cases. First, when δ is close to 0 then f is assumed to be very noisy, and hence the noisy leakage reveals little about the underlying sensitive information. On the other extreme, when δ is close to $1 - 1/|\mathbb{G}|$ then there is almost no noise in the leakage (i.e., the function f is "almost" deterministic). The Prouff-Rivain noise model is believed to model well real-world physical side-channel leakages. Moreover, as shown in [7,10] it is also a robust noise measure as it is equivalent to various other ways of describing the noise present in a leakage function.

Masking schemes in the noisy leakage model. A masking function is said to be secure in the noisy leakage model if for any $X, Y \in \mathbb{G}$, we have that noisy leakage from an encoding of $(X_1, \ldots X_n) \leftarrow \mathsf{Enc}(X)$ is statistically close to noisy leakage from $(Y_1, \ldots, Y_n) \leftarrow \mathsf{Enc}(Y)$. As discussed above recent works have significantly improved the Prouff-Rivain δ-bias for which security of the encoding function can be shown. While initial works [7,22] require that $\delta = O(1/|\mathbb{G}|)$, the recent work of Dziembowski et al. [10] show that noise amplifies security already when $\delta < 1/16$ (and, hence in particular independent of the size of the underlying group $\mathbb{G}$). In this work, we can show that for prime-order groups masking amplifies the noise for $\delta \le 1 - 1/p$. Notice that in case when p is super-polynomial in the security parameter (as in discrete-log based cryptosystems), then we achieve security under the optimal assumption of $1 - negl(n)$. For general groups (in particular, groups with small factorization) we show that amplification is possible when $\delta < 1/2$. We also show that both our bounds are optimal as for values above the threshold amplification is not possible. We provide further details on our contributions and techniques in the next two sections.

1.1 Our Contributions

We analyze the amplification of noisy leakage for the simple additive encoding function Enc. The quality of the amplification is measured by the ϵ-security of the encoding. We say that an encoding is ϵ-secure if the statistical distance between the δ-noisy leakage of $\mathsf{Enc}(x)$ and $\mathsf{Enc}(x')$ for two elements $x, x' \in \mathbb{G}$ is upper bounded by ϵ. We characterize how many shares n we need in order to amplify the noise available in the leakage from the shares. To this end we derive a value $\delta_{\max}$ which is the maximal value for δ until which we can still amplify the noise for sufficiently large n. Of course, as we show the number of share n needs to increase the closer we set δ to $\delta_{\max}$. One interesting observation arising from our analysis is that the value of $\delta_{\max}$ depends on the structure of the underlying group $\mathbb{G}$. We summarize our results regarding the upper bound until which amplification is possible in the following informal theorem.

Theorem (Noise amplification, informal version of Theorem 1). *Let $\mathbb{G}$ be a group (either prime order, or arbitrary) and let the adversary obtain δ-noisy* *leakage from the shares of the encoding. Define* the maximal noise parameter $\delta_{\max}$ *as*

$$\delta_{\max} = \begin{cases} 1 - \frac{1}{p}, & \text{when } \mathbb{G} \text{ is of prime order} \\ \frac{1}{2}, & \text{when } \mathbb{G} \text{ is arbitrary.} \end{cases} \tag{1}$$

Then for any $\delta < \delta_{\max}$ and any $\epsilon > 0$ we have that the encoding Enc *is ϵ-secure for*

$$n = \text{poly}\left(\log |\mathbb{G}|, \log(\epsilon^{-1}), (\delta_{\max} - \delta)^{-1}\right) \tag{2}$$

(where poly *is some polynomial).*

Let us explain the interplay of the parameters used in the above theorem. The number of shares grows polynomially with two parameters: (a) the logarithm of ϵ^{-1}, i.e., the target security we aim for, and (b) the gap $\theta = \delta_{\max} - \delta$ between the maximal possible noise value and the actual chosen noise level δ. The dependency on (a) is as expected: if we aim for better security meaning a smaller value of ϵ we require a larger security parameter n. The reason for the dependency stated in (b) is more technical, and essentially comes from a bias convergence in the harmonic analysis (when $\mathbb{G}$ has prime order), or in the XOR Lemma (when $\mathbb{G}$ has non-prime order), see Sect. 3 for details.[1]

Dependence on the group order. As already noted, our Theorem 1 distinguishes between two cases. In the first case the group $\mathbb{G}$ is of prime order. Interestingly, in this case it turns out that arbitrarily small noise (i.e. δ close to 1) can be amplified. Informally, this is thanks to the fact that prime-order groups have no non-trivial sub-groups. On the other hand, if a group has a non-prime order, i.e., it contains non-trivial subgroups, then we require a higher noise (more precisely: $\delta < 1/2$). On a practical level this means that in some sense the prime order groups "provide a better leakage resilience" than the general groups. This may be useful for discrete-log based cryptosystems that typically work in such groups.

Lower bounds on the necessary noisy level via homomorphic attacks. We show that the maximal noise parameter $\delta_{\max}$, as defined in Eq. (1), is optimal in the following sense. We show in Proposition 1 that δ has to be less than $1 - \frac{|\mathbb{H}|}{|\mathbb{G}|}$, where $\mathbb{H}$ is the largest proper subgroup of $\mathbb{G}$. An intuitive explanation for why the group structure is relevant here, is the existence of "homomorphic attacks", when given $X_1, \ldots, X_n$ being shares of X and their evaluations $\phi(X_1), \ldots, \phi(X_n)$ under a homomorphism $\phi : \mathbb{G} \to \mathbb{H}$, we can compute $\phi(X) = \phi(X_1) + \ldots + \phi(X_n)$.

The implications of Proposition 1 are two-fold. Firstly, as long as the general groups are considered (and hence no assumptions on the order can be made), then we prove that one needs to assume that the δ is less than $1/2$. This is because, since the largest proper subgroup $\mathbb{H}$ of $\mathbb{G}$ can be of size $|\mathbb{G}|/2$, thus

---

[1] We show also that the dependency on θ is necessary as otherwise we can provide attacks against the encoding.

$1 - |\mathbb{H}|/|\mathbb{G}|$ can be as small as $1/2$. Secondly, is $\mathbb{G}$ has prime order then $|\mathbb{H}| = 1$ and therefore $1 - |\mathbb{H}|/|\mathbb{G}| = 1 - 1/p$. Hence in this case the lower bound matches the upper bound from Theorem 1.

It is natural to ask if our results can be more fine-grained, and fully characterize the noise requirements in terms of $|\mathbb{H}|/|\mathbb{G}|$. We conjecture that in fact the upper bound $1 - |\mathbb{H}|/|\mathbb{G}|$ can also be always achieved (not only in the cases when $|\mathbb{H}| = |\mathbb{G}|/2$ and $\mathbb{H} = |1|$). We leave proving it as an open problem. We summarize the upper bounds and the relation to the number of shares in Table 1 below.

Table 1. Matching bounds for the necessary noise amount and the necessary number of shares

Group	Necessary noise	The gap	Necessary number of shares
$\lvert\mathbb{G}\rvert$ is even	$\delta < \delta_{\max} = \frac{1}{2}$	$\theta = \frac{1}{2} - \delta$	$n = \mathrm{poly}\left(\log(\epsilon^{-1}), \theta^{-1}\right)$
$\mathbb{G} = \mathbb{Z}_p$	$\delta < \delta_{\max} = 1 - \frac{1}{p}$	$\theta = 1 - \frac{1}{p} - \delta$	

Applications of our techniques outside of masking schemes. We show that our techniques also have applications outside of the domain of leakage resilient cryptography. In particular, using our results we can extend the following product theorem, due to Maurer et al. [17].

Lemma 1 (Product Theorem [17]). *Let $\mathbb{G}$ be a group, $d(\cdot)$ denote the distance from uniform and X_1, X_2 be arbitrary independent random variables on $\mathbb{G}$. Then we have*

$$d(X_1 + X_2) \leqslant 2 \cdot d(X_1) \cdot d(X_2)$$

We give a different proof and calculate optimal constants for any group.

Theorem (Product Theorem with optimal constants, informal version of Theorem 2). *Let $\mathbb{G}$ be a group, $d(\cdot)$ denote the distance from uniform and X_1, X_2 be arbitrary independent random variables on $\mathbb{G}$. Then we have*

$$d(X_1 + X_2) \leqslant c(\mathbb{G}) \cdot d(X_1) \cdot d(X_2)$$

where $c(\mathbb{G}) \leq 2$ is a constant depending on the structure of the underlying group $\mathbb{G}$.

For the exact value of $c(\mathbb{G})$ we refer the reader to Appendix A.6.

Comparing our results to previous works. Our work improves the previous state-of-the-art in the following aspects:

(a) For general groups the best upper bound was given in [10], where it was shown that $\delta < 1/16$. We improve this bound to $\delta < 1/2$, which is optimal for groups with even order. Moreover, for groups of prime-order p we give a novel bound of $1 - \frac{1}{p}$. Notice that for primes that are super-polynomial in the security parameter we achieve security for δ arbitrary close to 1. We notice that in contrast to earlier works our proof techniques also have the advantage of being more modular.

(b) We provide matching lower bounds showing that the noise threshold as well as the growth rate of the number of shares are optimal, which was not known before.

(c) Our proof techniques may be of independent interest and have not been used previously for analyzing the security of noisy leakages. In particular, our analysis uses techniques from convex optimization, additive combinatorics and harmonic analysis.

In Table 2 below we compare our results with related works.

Table 2. The initial amount of noise needed for the security of Enc.

	Proof techniques	Sufficient noise	Minimal Noise	Sufficient n	Minimal n
[22]	direct information theoretic analysis	$O\left(|\mathbb{G}|^{-1}\right)$	not discussed	poly $\left(\log(|\mathbb{G}|),\ \log(\epsilon^{-1}),\ (\delta_{\max} - \delta)^{-1}\right)$	not discussed
[7]	reduction to random probing	$O\left(|\mathbb{G}|^{-1}\right)$			
[10]	reduction to random walks, amplifying indistinguishability	$\frac{1}{16}$			
here	optimal reduction to random walks, harmonic analysis, additive combinatorics, convex optimization	$\frac{1}{2}$ for any G $1 - \frac{1}{p}$ for $\mathbb{Z}_p$	$\frac{1}{2}$ if $|\mathbb{G}|$ even $1 - \frac{1}{p}$ for $\mathbb{Z}_p$	poly $\left(\log(|\mathbb{G}|),\ \log(\epsilon^{-1}),\ (\delta_{\max} - \delta)^{-1}\right)$	poly $\left(\log(\epsilon^{-1}),\ (\delta_{\max} - \delta)^{-1}\right)$

Comparison with the binomial noise model and the XOR lemma. We stress that the noise model of Prouff and Rivain [22] we consider is significantly more general than the binomial noise model considered by Faust et al. in [12] (even in the binary case) and considers many other types of noisy functions. In particular, our noisy function f maps elements from the group $\mathbb{G}$ to a possibly infinite set Y.

If we restrict in Theorem 1 the noise model to the special case of binomial noise (i.e., the leakage function f is the binomial noise function), then we obtain comparable parameters with e.g., in [12] (cf. Lemma 4 in [12]).[2]

1.2 A High-Level Proof Outline

We now give an overview of our proof techniques (the details appear in Sect. 3). We prove our result in five steps illustrated on Fig. 1. We first show that it is

---

[2] The main restriction when comparing our result with a direct application of the XOR Lemma, is that we require the probability p of flipping the shares to be $> 1/4$ (in contrast to [12] where $p > 0$ is sufficient). The later restriction stems from the fact that leakages in the binomial noise model with parameter p are transformed in our noise model to a requirement of $delta = 1 - 2p$ noisy-function. We emphasize that for the general type of "noisy leakage" we consider, our bounds are optimal as shown by our lower bounds.

enough to consider uniform secrets X. The proof appears in Sect. 3.1.1. The cryptographic interpretation of this claim is that it suffices to consider only *random-plaintext attacks*, instead of *chosen-plaintext attacks*.

Then, in Sect. 3.1.2, we consider the distance between a uniform secret X given δ-noisy leakages $f_1(X_1), \ldots, f_n(X_n)$ from its encoding and a uniformly and independently chosen X'. We show that bounding this distance is equivalent to bounding the distance of a random sum of the form $Z = \sum_{i=1}^{n} Z_i$ with independent summands Z_i, conditioned on noisy information $\{f_i(Z_i)\}_{i=1}^{n}$, from the uniform distribution U. Here we use the fact that X is uniform (guaranteed by Step 1). The fact that leakage functions are δ-noisy guarantees that Z_i is δ-close to uniform given $f_i(Z_i)$, for $i = 1, \ldots, n$. Intuitively it is clear that if independent random variables on a group are close to the uniform distribution, then their sum is even closer to uniform (cf. XOR-lemmas [13], see also Appendix A.8). The main issue here is that our summands are conditional distributions, which means that Z_i is close to U only in average conditioned on concrete values of the leakage $f_i(Z_i) = y_i$.

Next in Sect. 3.1.3 we get rid of the conditional part $\{f_1(Z_1), \ldots, f_n(Z_n)\}$. This step is accomplished by considering concrete leakage values $f_1(Z_i) = y_i$ for $i = 1, \ldots, n$ and noticing that for most of them the distance from uniform is not much bigger than δ, which we conclude by the Chernoff Bound. This step results into an error term, which is exponential in $n\theta^2$ where $\theta = \delta_{\max} - \delta$ is the gap-parameter defined above. Informally speaking, the gap θ is what allows to further reduce the problem to study only the distance of sums of the form $Z = Z_1 + \ldots + Z_n$ from the uniform distribution.

Later, in Sect. 3.1.4, we characterize the distributions Z_i for which the distance between $Z = Z_1 + \ldots + Z_n$ and U is *maximal*. It turns out that they have a *simple shape*: they are a combination of a mass-point and a distribution uniform over a set of size $(1 - \delta)|\mathbb{G}| - 1$. A description of how these "worst-case" distributions look like, enables us to come up with concrete estimates for the statistical distance in the next step.

Finally, in Sect. 3.1.5 we prove concrete facts about the convergence speed of cumulative sums of random variables that are sufficiently close to the uniform distribution. Depending on the technique used and the assumption on the group structure imposed, we obtain different bounds in Theorem 1.

We note that for the case when we make no assumptions on the structure of $\mathbb{G}$, steps from Sects. 3.1.4 and 3.1.5 (but not from Sects. 3.1.1—3.1.3) could be replaced by a product theorem due to Maurer et al. [18]. Our technique allow us to extend this theorem, taking the group structure into account. In the last step we split the proof depending on the technique and the assumption about $\mathbb{G}$. The quantitative comparison of different bounds we get is given in Table 3. Note that the number n of shares is asymptotically larger when the additive combinatorics is used (second row of Table 3) than when the harmonic analysis is used (the third row). Nevertheless we think it is instructive the present both results since the proof techniques are different, and both can be of independent interest.

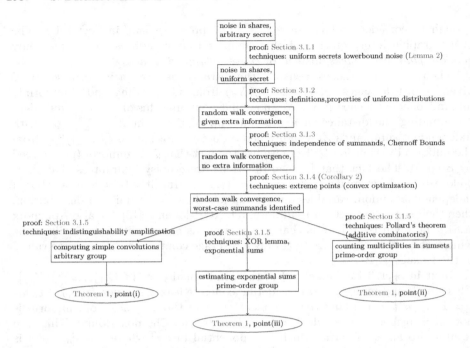

Fig. 1. An overview of the proof of Theorem 1 and applied techniques.

Table 3. The amount of shares needed to mask the secret state below the advantage ϵ, depending on the assumed group structure and proof technique. In the last column, θ is an arbitrarily small positive number.

Domain	Proof technique	Number n of shares	Assumption
$\mathbb{G}$ arbitrary	amplifying indistinguishability	$O\left(\log(|\mathbb{G}|/\epsilon)/\theta^2\right)$	$\delta \leqslant \frac{1}{2} - \theta$
$\mathbb{G} = \mathbb{Z}_p$	additive combinatorics	$O\left(\log(|\mathbb{G}|/\epsilon) \cdot 2^{12 \log(1/\theta)/\theta^2}\right)$	$\delta \leqslant 1 - \frac{1}{p} - \theta$
$\mathbb{G} = \mathbb{Z}_p$	harmonic analysis	$O\left(\theta^{-4} \log(|\mathbb{G}|/\epsilon) \log(1/\theta)\right)$	$\delta \leqslant 1 - \frac{1}{p} - \theta$

1.3 Our Techniques

In this section we summarize our main techniques used in the proof of Theorem 1.

Convex analysis. We use the convex analysis in Sect. 3.1.4 to deal with the problem of determining how fast the sum of independent components $Z_1 + \ldots + Z_n$ on a group $\mathbb{G}$ converges towards the uniform distribution. Our assumption on the noise guarantees that Z_i are at most δ-close to the uniform distribution with some parameter δ. Since we can think of distributions over $\mathbb{G}$ as vectors in $\mathbb{R}^{|\mathbb{G}|}$, we observe that the mapping

$$(Z_1, \ldots, Z_n) \longrightarrow \mathsf{SD}\,(Z_1 + \ldots + Z_n; U)$$

is a convex mapping, with respect to the distribution of any Z_i when the remaining components are fixed. Since the restrictions on the distance of Z_i's from uniform are also convex, we conclude that the maximum is achieved for one of the *extreme points* from the set of feasible Z_i. As a consequence we observe that they have a surprisingly *simple shape* (see Lemma 5). That simple structure will play an important role in the very last step of the proof. Also, it allows us to derive a general product theorem for groups, with an explicit expression with tight constants.

Additive combinatorics. In Sect. 3.1.5, when studying the convergence of random sums over a general group $\mathbb{G}$, we can find a proper subgroup $A \lhd \mathbb{G}$ which is by definition an *additive set*, that is

$$A + A = A.$$

Such a set may constitute a *trap* for our random walk $Z_1 + \ldots + Z_n$. If $Z_i \in A$ for all i, then $Z_1 + \ldots + Z_n \in A$. When $\mathbb{G} = \mathbb{Z}_p$ such a trap does not exist, so intuitively the sum takes all the elements with similar probability when n is large (because even one non-zero point generates the group when added multiple times). This is where we use some basic facts from additive combinatorics. The first result of this sort is the Cauchy-Davenport theorem which states that the sumset $A + B$, where A, B are arbitrary subsets of $\mathbb{Z}_p$ must be substantially bigger than A and B alone. More precisely

$$|A + B| \geqslant |A| + |B| - 1.$$

This result does not help us much because it gives no estimate on *repetitions* in the sumset $A + B$, that is how many of the expressions $a + b$ where $a \in A, b \in B$ hit the same place. To get more information about the distribution of repetitions in the sumset we use a more refined result due to Pollard. Combining it with the explicit form of Z_i (developed in the previous steps) we obtain a non-trivial upper bound on $\mathsf{SD}(Z_1 + Z_2; U)$ in terms of $\mathsf{SD}(Z_1; U), \mathsf{SD}(Z_2; U)$, which is then extended to the sum of n elements.

Harmonic analysis. Also, in Sect. 3.1.5, having reduced our problem to the convergence of a random walk with independent increments, we can use techniques from Fourier analysis. Recall that a character is a complex-valued function ϕ which is additive on $\mathbb{G}$, that is $\phi(x + y) = \phi(x) \cdot \phi(y)$. The expectations of characters on independent sums are especially easy to evaluate, because

$$\mathbb{E}\left[\phi\,(Z_1 + \ldots + Z_n)\right] = \prod_{i=1}^{n} \mathbb{E}\left[\phi(Z_i)\right].$$

For $\mathbb{Z}_p$ expectations are easier to calculate, because any character ϕ is of the simple form $\phi(x) = \exp(2\pi k i/p)$ for some k. Since we know the shape of the worst Z_i, we can obtain a concrete estimate for a nontrivial character ϕ, namely,

$$|\mathbb{E}\left[\phi\left(Z_i\right)\right]| < c \ll 1.$$

Intuitively, this comes from the fact that Z_i "contains" a large uniform component which doesn't allow the mass to concentrate at one point. Using a bound on geometric sums over unity roots, we conclude that $\mathbb{E}\left[\phi\left(Z_1 + \ldots + Z_n\right)\right] < c$. Finally we apply the XOR lemma which states that characters are "representative" distinguishers: if two distributions have close expectations under every character, they indeed are statistically close. In our case we apply this claim to $Z = Z_1 + \ldots + Z_n$ and U and the result follows since we have shown that $\mathbb{E}[\phi(Z)]$ is small and trivially we have $\mathbb{E}[\phi(U)] = 0$ (for non-trivial ϕ).

2 Preliminaries

If X and Y are random variables over the same set $\mathcal{X}$ then the statistical distance between X and Y is denoted as $\mathsf{SD}(X;Y)$, and defined as $\mathsf{SD}(X;Y) = \frac{1}{2}\sum_{x \in \mathcal{X}} |\Pr[X = x] - \Pr[Y = x]|$. If Z is a random variable then by $\mathsf{SD}(X;Y|Z)$ we mean $\mathsf{SD}((X,Z);(Y,Z))$, i.e., the statistical distance of the two joint distributions. If two distributions X and Y are equivalent, then we write $X \overset{d}{=} Y$. Below we formally define the encoding and decoding of a secret $X \in \mathbb{G}$.

Definition 1 (Encoding and Decoding). *Let* $(\mathbb{G}, +)$ *be a fixed group and* $n > 1$ *be a fixed natural number. For any* $X \in \mathbb{G}$, *we define the encoding function* Enc *by*

$$\mathsf{Enc}_{\mathbb{G}}^n(X) = (X_1, \ldots, X_{n-1}, X - (X_1 + \ldots + X_{n-1}))$$

where $X_1, \ldots, X_{n-1}$ *are independent and uniform over* $\mathbb{G}$, *and the decoding function* Dec *by*

$$\mathsf{Dec}_{\mathbb{G}}^n(X_1, \ldots, X_n) = X_1 + \ldots + X_n.$$

We will typically omit n *and* $\mathbb{G}$ *and simply write* $(\mathsf{Enc}, \mathsf{Dec})$ *when clear from the context.*

Noisy leakages. The noise in the observed version Y of a real distribution X, denoted by $\beta(X|Y)$, is measured by comparing how close is the product distribution $\mathbf{P}_X \cdot \mathbf{P}_Y$ to the joint distribution $\mathbf{P}_{(X,Y)}$. More formally, we have the following definition, which comes from [22] (see also [7]), where it was argued that it models physical noise in a realistic way.

Definition 2 (Noisy observations and noisy functions). *A random variable* $Y \in \mathcal{X}$ *is called a* δ-*noisy observation of* X *if*

$$\beta(X|Y) \overset{def}{=} \mathsf{SD}(X'; X|Y) \leqslant \delta$$

where X' *is an independent copy of* X. *A function* f *is called* δ-*noisy if* $f(U)$ *is a* δ-*noisy version of* U, *where* U *is uniform over* U.

Notice that [22] defined the noisy function as the $\sum_y \Pr[Y = y] \cdot \mathsf{SD}(X'; (X|Y = y))$. This is equivalent to the above when X and X' are independent and Y is the leakage from X. Note that this definition may seem a bit counterintuitive as more noise means a bias value closer to 0, while a value closer to 1 means that less noise is present in the leakage observations.

3 Our Main Result

We are now ready to present our main result that was already informally described in Sect. 1.1.

Theorem 1. *Let X be a random variable on a group $\mathbb{G}$ and let $\mathsf{Enc}(X) = (X_1, \ldots, X_n)$ be its encoding. Suppose that f_i for $i = 1, \ldots, n$ are all δ-noisy functions, i.e. $\beta(X_i | f_i(X_i)) \leqslant \delta$. Then we have the following bounds*

(i) *For arbitrary $\mathbb{G}$, if $\delta \leqslant \frac{1}{2} - \theta$, then $\beta(X | f_i(X_1), \ldots, f_i(X_n)) \leqslant \epsilon$ provided that*
$$n > 8\theta^{-2} \log(5|\mathbb{G}|\epsilon^{-1})$$

(ii) *If $\mathbb{G} = \mathbb{Z}_p$ and $\delta \leqslant 1 - \frac{1}{p} - \theta$ then $\beta(X | f_i(X_1), \ldots, f_i(X_n)) \leqslant \epsilon$, provided that*
$$n > \log(3|\mathbb{G}|\epsilon^{-1}) \cdot 2^{12 \log(1/\theta)/12\theta}$$

(iii) *If $\mathbb{G} = \mathbb{Z}_p$ and $\delta \leqslant 1 - \frac{1}{p} - \theta$ then $\beta(X | f_i(X_1), \ldots, f_i(X_n)) \leqslant \epsilon$, provided that*
$$n > 2\theta^{-4} \log(|\mathbb{G}|\epsilon^{-1})$$

3.1 Proof of Theorem 1

The proof of Theorem 1 was already outlined in Sect. 1.2. In the next sections we describe it in more detail. The final proof appears in Sect. 3.1.5.

3.1.1 Reducing to Uniform Secrets
Below we show that it sufficient to consider only uniform secrets X.

Lemma 2. *Suppose that X is uniform over $\mathbb{G}$ with the encoding $\mathsf{Enc}(X) = (X_1, \ldots, X_n)$. Let X' be an arbitrary distribution over $\mathbb{G}$ with the encoding $\mathsf{Enc}(X') = (X'_1, \ldots, X'_n)$. Then for arbitrary functions $f_1, \ldots, f_n$ we have*
$$\beta(X' | f_1(X'_1), \ldots, f_n(X'_n)) \leqslant 3|\mathbb{G}| \cdot \beta(X | f_1(X_1), \ldots, f_n(X_n)) \tag{3}$$

The proof appears in Appendix A.3. Note that we lose a factor of $|\mathbb{G}|$ in this transformation. However this does not actually affect the bound we want to prove, because we show that the main part $\beta(X | f_1(X_1), \ldots, f_n(X_n))$ converges to 0 exponentially fast with n.

3.1.2   Reducing to Random Walks Conditioned on Noisy Information.

We now show that the noise in a uniform secret X given δ-noisy leakages $f_1(X_1), \ldots, f_n(X_n)$ is equal to the distance of a random sum of the form $Z = \sum_{i=1}^{n} Z_i$ with independent summands Z_i, conditioned on noisy information $\{f_i(Z_i)\}_{i=1}^{n}$, from the uniform distribution U.

Lemma 3. *For X uniform on a set $\mathcal{X}$ and any functions f_i the following equality holds*

$$\beta(X|(f_i(X_i))_{i=1}^{n}) = \mathsf{SD}\left(\sum_{i=1}^{n} Z_i;\ U\ \middle|\ (f_i(Z_i))_{i=1}^{n}\right) \qquad (4)$$

where U and Z_i for $i = 1, \ldots, n$ are uniform and independent over $\mathcal{X}$.

The proof appears in Appendix A.4. Justifying the title, we note that we can think of the sum $\sum_{i=1}^{n} Z_i$ as a random walk which starts at 0, with increments Z_i.

3.1.3   Reducing to Unconditional Random Walks

The following lemma is an easy consequence of the Chernoff Bound.

Lemma 4. *Let $(Z_i, Y_i)_i$ for $i = 1, \ldots, n$ be independent random variables such that $\Delta(Z_i; U|Y_i) \leqslant \delta$. Then, for any $\gamma > 0$, $\delta' = \delta + 2\gamma$ and $n' = \gamma n$*

$$\mathsf{SD}\left(\sum_{i=1}^{n} Z_i; U\ \middle|\ (Y_i)_i\right) \leqslant \max_{(Z'_i)_i:\ \mathsf{SD}(Z'_i;U)\leqslant\delta'} \mathsf{SD}\left(\sum_{i=1}^{n'} Z'_i; U\right) + e^{-2n\gamma^2} \qquad (5)$$

where the maximum is taken over all independent random variables Z'_i.

The proof appears in Appendix A.5. The immediate corollary below shows that we can get rid of the conditinal part in the right-hand side of Eq. (4) in Lemma 3.

Corollary 1. *For $n' = \frac{\theta}{4} \cdot n$ we have*

$$\beta(X|(f_i(X_i))_{i=1}^{n}) \leqslant \max_{(Z'_i)_i:\ \mathsf{SD}(Z'_i;U)\leqslant\delta+\frac{\theta}{2}} \mathsf{SD}\left(\sum_{i=1}^{n'} Z'_i; U\right) + e^{-\frac{1}{8}n\theta^2}$$

where the maximum is taken over all independent random variables Z'_i.

Note that by combining Step 1, Step 2 and Step 3 with Lemma 1 we can already conclude part (i) of Theorem 1 (see Sect. 3.1.5).

3.1.4   Worst-Case Summands

We prove the following geometrical fact:

Lemma 5 (Shape of extreme points of distributions close to uniform).
Let $\mathcal{X}$ be a finite set and U be uniform over $\mathcal{X}$. Any distribution $X \in \mathcal{X}$ such that $\mathsf{SD}(X; U) \leqslant \delta$ can be written as a convex combination of "extreme" distributions X' of the following form: with probability $p = \delta + |\mathcal{X}|^{-1}$ the distribution X' takes a value a and with probability $q = 1 - p$ the distribution X' is uniform over a set A, where $a \notin A$, of size $|A| = (1 - \delta)|\mathcal{X}| - 1^3$. Equivalently, each of these distributions X' is of the following form:

$$\mu_{X'} = \mu_U + \delta\mu_b - \delta\mu_B \tag{6}$$

for some B such that $|B| = \delta|\mathbb{G}|$ and $b \notin B$.

Note that we always have $(1 - \delta)|\mathcal{X}| - 1 \geqslant 0$, as the range of the noise parameter is $0 \leqslant \delta \leqslant 1 - \frac{1}{\mathcal{X}}$, when we consider secrets over $\mathcal{X}$.

Corollary 2. *Let Z_i, for $i = 1, \ldots, n$ be independent random variables such that $\mathsf{SD}(Z_i; U) \leqslant \delta$ for every i. Then $\mathsf{SD}\left(\sum_i Z_i; U\right)$ is maximized for Z_i as in Lemma 5*

Proof (of Corollary 2). Note that the distribution of $\sum_i Z_i$ is a convolution of individual distributions $\mathbf{P}_{Z_i}$, and therefore it is multilinear in $\mathbf{P}_{Z_i}$. It follows that $\mathsf{SD}\left(\sum_i Z_i; U\right) = \frac{1}{2}\left\|\mathbf{P}_{\sum_i Z_i} - \mathbf{P}_U\right\|_1$ is convex in $\mathbf{P}_{Z_i}$ and the claim follows by the extreme point principle.

Using Lemma 5 we derive the following generalization of Lemma 1

Theorem 2. *Let Z_1, Z_2 be independent random variables on a group $\mathbb{G}$. Then we have*

$$\mathsf{SD}(Z_1 + Z_2; U) \leqslant c_{\max}(\mathbb{G}) \cdot \mathsf{SD}(Z_1; U) \cdot \mathsf{SD}(Z_2; U) \tag{7}$$

where the constant is given by

$$c_{\max}(\mathbb{G}) = \frac{1}{2} \max_{A,B: |A| = \delta_1|\mathbb{G}|, |B| = \delta_2|\mathbb{G}|} \|\mu_B + \mu_A - \mu_A * \mu_B - \mu_0\|_{\ell_1(\mathbb{G})} \tag{8}$$

where μ_0 is the point mass at 0, $\delta_i = \mathsf{SD}(Z_i; U)$, and μ_A, μ_B are uniform over the sets A and B. Moreover, the sharp constant is achieved for the following random variables: Z_i is constant with probability $\delta_i + \frac{1}{|\mathbb{G}|}$ and with probability $1 - \delta_i - \frac{1}{|\mathbb{G}|}$ is uniform on some set of size $(1 - \delta_i)|\mathbb{G}| - 1$.

Lemma 5 is a corollary from Theorem 2, whose proof appears in Appendix A.6. Note that Lemma 1 follows from Theorem 2, since $c_{\max}(\mathbb{G}) \leqslant 2$ trivially, since

$$\|\mu_B + \mu_A - \mu_A * \mu_B - \mu_0\|_1 \leqslant \|\mu_B\|_1 + \|\mu_B\|_1 + \|\mu_A * \mu_B\|_1 + \|\mu_0\|_1 = 4$$

by the triangle inequality and the fact that the total mass of a probability measure is 1.

---

[3] If $\delta|\mathcal{X}|$ is not an integer, then instead of a uniform distribution we consider the distribution flat over a set A such that $|A| = \lceil (1 - \delta)|\mathcal{X}| - 1 \rceil$, which assigns the mass of $\frac{1}{(1-\delta)|\mathcal{X}|-1}$ to all but one points, and the mass of $\frac{\lceil (1-\delta)|\mathcal{X}|-1\rceil - ((1-\delta)|\mathcal{X}|-1)}{(1-\delta)|\mathcal{X}|-1}$ to the ramining point.

3.1.5   Concrete Bounds

In view of Corollary 1 it remains to give an upper bound on the distance between sums of independent random variables which are not too far from uniform and the uniform distribution, i.e., on:

$$\mathrm{SD}\left(\sum_{i=1}^{n} Z_i; U\right).$$

To this end, we split our analysis depending on the structure of $\mathbb{G}$ and chosen technique.

Case: $\mathbb{G}$ is arbitrary. From Corollary 1 and Lemma 1 applied $(n-1)$ times it follows that

$$\beta(X|(f_i(X_i))_{i=1}^n) \leqslant \frac{1}{2}(2\delta + \theta)^{\frac{\theta}{4}n} + e^{-\frac{1}{8}n\theta^2}.$$

From the assumption $\delta < \frac{1}{2} - \theta$ and the elementary inequality $1 - u \leqslant e^{-u}$ we obtain $(2\delta + \theta)^{\frac{\theta}{4}n} \leqslant e^{-\frac{1}{4}\theta^2 n}$, which gives us

$$\beta(X|(f_i(X_i))_{i=1}^n) < \frac{3}{2} \cdot e^{-\frac{1}{8}n\theta^2}$$

for uniform X. Taking into account Step 1, we finally obtain

$$\beta(X|(f_i(X_i))_{i=1}^n) < 5|\mathbb{G}| \cdot e^{-\frac{1}{8}n\theta^2}.$$

for any X, which is equivalent to part (i) of Theorem 1.

Case: $\mathbb{G} = \mathbb{Z}_p$, for p prime (by additive combinatorics). When $\mathbb{G} = \mathbb{Z}_p$, we improve Lemma 1 in the following way, using Corollary 2 and some tools from additive combinatorics (see Appendix A.7 for a proof).

Lemma 6. *Let Z_1, Z_2 be independent random variables on $\mathbb{Z}_p$ such that $\mathrm{SD}(Z_i; U) \leqslant \delta_i$. Then*

$$\mathrm{SD}(Z_1 + Z_2, U_G) \leqslant h(\delta_1, \delta_2) \tag{9}$$

where

$$h(\delta_1, \delta_2) = \begin{cases} 2\delta_1\delta_2, & \phi(\delta_1, \delta_2) \leqslant 0 \\ 2\delta_1\delta_2 - \frac{1}{4}\phi(\delta_1, \delta_2)^2 + \frac{1}{4p^2}, & \phi(\delta_1, \delta_2) > 0 \end{cases} \tag{10}$$

and $\phi(\delta_1, \delta_2) \overset{def}{=} \delta_1 + \delta_2 + \min(\max(\delta_1, \delta_2), 1 - |\delta_1 - \delta_2|) - 1$.

We will use only the following consequence of Lemma 6.

Corollary 3. *Let Z_1, Z_2 be independent random variables on $\mathbb{Z}_p$ such that $\mathrm{SD}(Z_i; U) \leqslant \delta_i \leqslant \delta$. Suppose that $\delta > \frac{1}{3}$. Then we have*

$$\mathrm{SD}(Z_1 + Z_2; U_G) \leqslant 2\delta^2 - \frac{(3\delta - 1)^2}{4} + \frac{1}{4p^2} \tag{11}$$

Using recursively Corollary 3 and applying Corollary 1 we obtain the following bound for uniform X

$$\beta\left(X|(f_i(X_i))_{i=1}^n\right) \leqslant 2^{-n/(2^{16/\theta}\cdot 12\theta)} + e^{-\frac{1}{8}n\theta^2}$$

which, by Step 1, implies part (ii) of Theorem 1. For a detailed derivation, see Lemma 8 in Appendix A.7.

Case $\mathbb{G} = \mathbb{Z}_p$, *for a prime number p (by harmonic analysis).* We start by obtaining the following auxiliary estimate on trigonometric sums, valid for any $A \subset \mathbb{Z}_p$ such that $|A| = \theta p$ and $k \in \{1, 2, \ldots, p-1\}$:

$$\left|\sum_{x \in A} \exp\left(\frac{2k\pi i x}{p}\right)\right| \leqslant \frac{\sin \pi \theta}{p \sin \frac{\pi}{p}}$$

The proof uses a geometrical argument and some trigonometric identities and is given inside the proof of Lemma 10 in Appendix A.8. Based on Corollary 2 and the XOR lemma (see Lemma 9 in Appendix A.8), we prove that

$$\beta\left(X|(f_i(X_i))_{i=1}^n\right) \leqslant 3|\mathbb{G}|^{\frac{3}{2}} \cdot e^{-\frac{1}{8}\theta^3} + e^{-\frac{1}{8}n\theta^2}$$

for uniform X, which by Step 1 implies part (iii) of Theorem 1; the details are given in Lemma 10 in Appendix A.8.

4   Lower Bounds

Proposition 1 (The noise threshold (1) is optimal). *For any group $\mathbb{G}$, there exist a δ-noisy function f where*

$$\delta = 1 - \frac{|\mathbb{H}|}{|\mathbb{G}|},$$

$\mathbb{H}$ *being the biggest proper subgroup of $\mathbb{G}$, such that for every n we have*

$$\beta\left(X|f(X_1), \ldots, f(X_n)\right) \geqslant \delta.$$

The proof appears in Appendix A.1.

Proposition 2 (The growth rate in (2) is optimal). *For $\mathbb{G} = \mathbb{Z}_2$, X being uniform on $\mathbb{G}$, and any $\theta < \frac{1}{2}$ there exists a $\left(\frac{1}{2} - \theta\right)$-noisy leakage function f such that for every n satisfying*

$$n < \log\left((2\epsilon)^{-1}\right) / \log\left((1 - 2\theta)^{-1}\right) \tag{12}$$

we have

$$\beta\left(f(X)|f(X_1), \ldots, f(X_n)\right) \geqslant \epsilon.$$

In turn, for $\mathbb{G} = \mathbb{Z}_p$ *where* p *is prime,* X *being uniform on* $\mathbb{G}$*, and any* $\theta <$ $1 - \frac{1}{p}$ *there exists a* $\left(1 - \frac{1}{p} - \theta\right)$*-noisy leakage function* f *such that for every* n *satisfying*

$$n < \log(2\epsilon^{-1})/\log(1 - \theta) \tag{13}$$

we have

$$\beta\left(f(X)|f(X_1), \ldots, f(X_n)\right) \geqslant \epsilon.$$

The proof appears in Appendix A.2. Note that for $\theta < \frac{1}{4}$ we have $\log((1 - 2\theta)^{-1}) < \frac{1}{4}\theta^{-1}$, and then Eq. (12) can be replaced by

$$n < 4\theta^{-1}\log\left((2\epsilon)^{-1}\right)$$

Similarly, for $\theta < \frac{1}{2}$ we have $\log((1 - \theta)^{-1}) < \frac{1}{2}\theta^{-1}$, and then Eq. (13) can be replaced by

$$n < 2\log(2\epsilon^{-1}) \cdot \theta^{-1} \tag{14}$$

A   Proofs

A.1   Proof of Proposition 1

Proof. Let X be uniform over the set $\mathbb{G}$ and let ϕ be the canonical quotient homomorphism, that is $\phi(g) = g + \mathbb{H}$. For n shares $X_1, \ldots, X_n$ of X we have that $X_i|\phi(X_i)$ and X are $\left(1 - \frac{|\mathbb{H}|}{|\mathbb{G}|}\right)$-far, because $X_i|\phi(X_i) = y_i$ is uniform over a set of $|\mathbb{G}|/|\mathbb{H}|$ elements, for every choice of y_i. Similarly, $X = \sum_i X_i$ is $\left(1 - \frac{|\mathbb{H}|}{|\mathbb{G}|}\right)$-far from uniform given $\phi(X_i)$ for all i, because $\phi(X) = \sum_{i=1} \phi(X_i) = \sum_i y_i$. To see this, note that for independent uniform U we have

$$\mathsf{SD}\left(X, \phi(X_1), \ldots, \phi(X_n); U, \phi(X_1), \ldots, \phi(X_n)\right) \geqslant \mathsf{SD}\left(X, \sum_i \phi(X_i); U, \sum_i \phi(X_i)\right)$$

$$= \mathsf{SD}\left(X, \phi(X); U, \phi(X)\right)$$

$$= \mathsf{SD}\left(X; U|\phi(X)\right)$$

where the first line follows from the fact that applying a function to two random variables only decreases the statistical distance, and the second line uses the homomorphic property of ϕ. The last expression is at least $\left(1 - \frac{|\mathbb{H}|}{|\mathbb{G}|}\right)$ as already observed for uniform X.

A.2   Proof of Proposition 2

Proof. Fix $\mathbb{G} = \mathbb{Z}_2$ and consider a uniform secret X, its shares $X_1, \ldots, X_n$, and leakage functions $f_i = f$ for $i = 1, \ldots, n$ where $f(x)$ flips the bit x with probability $\theta < \frac{1}{2}$. It is easy to see that these functions are $\left(\frac{1}{2} - \theta\right)$-noisy, that is

$\mathsf{SD}\left(X_i; U \mid f(X_i)\right) = \frac{1}{2} - \theta$ where U is an independent uniform random variable. Note that for uniform X (and any functions f_i) we have the equality of distributions

$$(f_1(X_1), \ldots, f_n(X_n), X) \overset{d}{=} (f_1(Z_1), \ldots, f_n(Z_n), Z_1 + \ldots + Z_n).$$

where $\{Z_i\}_{i=1}^n$ are independent and uniform on $\mathbb{G}$ (see Sect. 3.1.2). As a consequence we get

$$\mathsf{SD}\left(X; U \mid f(X_1), \ldots, f(X_n)\right) = \mathsf{SD}\left(Z_1 + \ldots + Z_n; U \mid f_1(Z_1), \ldots, f_n(Z_n)\right)$$

One can check that every X_i' has bias $\delta = \frac{1}{2} - \theta$ (it outputs a bit with probability $\frac{1}{2} \pm \delta$) conditioned on $f(X_i') = y$ for every $y \in \{0, 1\}$. Since the xor-sum $Y_1 + Y_2 + \ldots + Y_n$ of δ-biased independent bits $Y_1, Y_2, \ldots, Y_n$ has bias exactly $2^{n-1}\delta^n$, we conclude that $\mathsf{SD}\left(X_1' + \ldots + X_n'; U \mid f_1(X_1'), \ldots, f_n(X_n')\right) = 2^{n-1}\delta^n$. To go below ϵ, we need $(1 - 2\theta)^n < 2\epsilon$ or

$$n > \ln\left((2\epsilon)^{-1}\right) / \ln\left((1 - 2\theta)^{-1}\right).$$

Finally, consider the case $\mathbb{G} = \mathbb{Z}_p$. We proceed as in the previous case, achieving

$$\mathsf{SD}\left(X; U \mid f(X_1), \ldots, f(X_n)\right) = \mathsf{SD}\left(Z_1 + \ldots + Z_n; U \mid f_1(Z_1), \ldots, f_n(Z_n)\right)$$

for arbitrary functions. We take the functions f_i so that the distribution of Z_i given $f(Z_i) = y_i$ for every i has the following form:

$$\mu_{Z_i \mid f(Z_i) = y_i} = \mu_{\mathbb{G}} + \delta\mu_a - \delta\mu_A$$

where a is a point and A is a set such that $a \notin A$, $|A| = \delta|\mathbb{G}|$. As it follows from the proof of Lemma 10 in Appendix A.8, we can choose A so that $|\mathbb{E}\phi(V_i)| \geqslant 1 - \theta$ for some character ϕ, where V_i is the distribution of Z_i conditioned on $f(Z_i)$. This means that the Fourier transform $\hat{V}_i$ of V_i is at least $1 - \theta$ in the supremum norm, that is $\|\hat{V}_i\|_\infty \geqslant 1 - \theta$. Since the Fourier transform is multiplicative under convolution (summing independent variables) we see that we can prepare functions f_i so that $\|\hat{V}\|_\infty \geqslant (1 - \theta)^n$, where $V = V_1 + \ldots + V_n$. The Parseval identity gives us $\|\hat{V}\|_2 = \|\mu_Z - \mu_U\|_2$. Since $\|\hat{Z}\|_\infty \leqslant \|\hat{V}\|_2$ and $\|\mu_V - \mu_U\|_2 \leqslant \|\mu_V - \mu_U\|_1$ we finally obtain

$$(1 - \theta)^n \leqslant \|\mu_V - \mu_U\|_1 = \mathsf{SD}(Z_1 + \ldots + Z_n; U \mid f(Z_1) = y_1, \ldots, f(Z_n) = y_n)$$

The claim follows now by averaging over different values of $y_1, \ldots, y_n$, exactly as in the previous case.

A.3   Proof of Lemma 2

We prove the following version, from which we conclude Lemma 2.

Suppose that X is uniform and X_i be the encoding of X. Let g be a probabilistic function, $(G_i)_i$ be the encoding of $G = g(X)$ and let f_i be noisy leakage functions. Then we have

$$\beta(X|(f_i(G_i))_i) \leqslant 3|\mathbb{G}| \cdot \beta(X|(f_i(X_i))_i) \tag{15}$$

Proof. Let V be uniform and $(V_i)_i$ be the encoding of V and let X', V' be independent copies of X, V. Note that $X, (f_i(G_i))_i$ is identically distributed as $X, (f_i(V_i))_i|V = g(X)$. Therefore

$$
\begin{aligned}
\Pr[X = x|(f_i(G_i))_i = (y_i)_i] &= \Pr[X = x|(f_i(V_i))_i = (y_i)_i, V = g(X)] \\
&= \frac{\Pr[X = x, (f_i(V_i))_i = (y_i)_i, V = g(x)]}{\Pr[(f_i(V_i))_i = (y_i)_i, V = g(X)]} \\
&= \frac{\Pr[X = x]\Pr[(f_i(V_i))_i = (y_i)_i, V = g(x)]}{\sum_{x'}\Pr[X = x']\Pr[(f_i(V_i))_i = (y_i)_i, V = g(x')]} \\
&= \frac{\Pr[V = g(x)|(f_i(V_i))_i = (y_i)_i]}{\sum_{x'}\Pr[V = g(x')|(f_i(V_i))_i = (y_i)_i]}
\end{aligned}
\tag{16}
$$

Let $\epsilon(x) = \Pr[V = x|(f_i(V_i))_i = (y_i)_i] - \frac{1}{|G|}$. Suppose first, that g is deterministic. We have

$$
\begin{aligned}
\Pr[X = x|(f_i(G_i))_i = (y_i)_i] - \frac{1}{|G|} &= \frac{\frac{1}{|G|} + \epsilon(g(x))}{1 + \sum_{x'}\epsilon(g(x'))} - \frac{1}{|G|} \\
&= \frac{|G|\epsilon(g(x)) - \sum_{x'}\epsilon(x')}{|G|(1 + \sum_{x'}\epsilon(g(x')))}
\end{aligned}
\tag{17}
$$

and

$$
\sum_x\left|\Pr[X = x|(f_i(G_i))_i = (y_i)_i] - \frac{1}{|G|}\right| = \frac{\frac{1}{|G|}\sum_x\left|\epsilon(g(x)) - \frac{1}{|G|}\sum_{x'}\epsilon(g(x'))\right|}{\frac{1}{|G|} + \frac{1}{|G|}\sum_{x'}\epsilon(g(x'))}
\tag{18}
$$

Note that $\left|\epsilon(g(x)) - \frac{1}{|G|}\sum_{x'}\epsilon(g(x'))\right| \leqslant \sum_{x'}|\epsilon(x')|$ and $\frac{1}{|G|}\sum_{x'}\epsilon(g(x')) \leqslant \sum_{x'}\epsilon(x')$. If $\sum_{x'}\epsilon(x') \leqslant \frac{1}{\frac{3}{2}|G|}$ then we obtain

$$
\sum_x\left|\Pr[X = x|(f_i(G_i))_i = (y_i)_i] - \frac{1}{|G|}\right| \leqslant \frac{\sum_{x'}|\epsilon(x')|}{\frac{1}{|G|} - \frac{1}{\frac{3}{2}|G|}} = 3|G|\sum_{x'}\epsilon(x')
\tag{19}
$$

otherwise

$$\sum_x \left| \Pr[X = x | (f_i(G_i))_i = (y_i)_i] - \frac{1}{|G|} \right| \leqslant 2 \leqslant 3|G| \sum_{x'} \epsilon(x') \qquad (20)$$

This way, we have shown

$$\Delta \left(X; X' | (f_i(G_i))_i = (y_i)_i \right) \leqslant 3|G| \Delta \left(V; V' | (f_i(V_i))_i = (y_i)_i \right) \qquad (21)$$

and by taking the average the result follows. If g is randomized, the proof is the same but $\epsilon(g(x))$ is replaced by $\mathbf{E}_g \epsilon(g(x))$ (note that we have $\beta(X|(f_i(G_i))_i) \leqslant \beta(g(X)|(f_i(G_i))_i)$).

A.4 Proof of Lemma 3

We start with the following observation: suppose that X_i for $i = 1, \ldots, n$ are shares of the *uniform* secret X. Let X_i' for $i = 1, \ldots, n$ be all uniform and independent. Then we have the following equality of distributions

$$(X, (X_1, \ldots, X_n)) \overset{d}{=} \left(\sum_{i=1}^n X_i', \ (X_1', \ldots, X_n') \right) \qquad (22)$$

Therefore,

$$(X, (f_i(X_i))_{i=1}^n) \overset{d}{=} \left(\sum_{i=1}^n X_i', \ (f_i(X_i'))_{i=1}^n \right). \qquad (23)$$

As a consequence we obtain the following equality

$$\beta(X|(f_i(X_i))_{i=1}^n) = \Delta \left(\sum_{i=1}^n X_i'; \ U \middle| (f_i(X_i'))_{i=1}^n \right) \qquad (24)$$

Thus, our problem reduces to investigate the random walk on $\mathbb{G}$ defined as $\sum_{i=1}^n X_i' | f_i(X_i')$. We need to show that it (under some restrictions) eventually approaches the uniform distribution as n increases, and estimate the convergence speed.

A.5 Proof of Lemma 4

Proof. We can assume that $\delta + 2\gamma < 1$. We start with the following observation:
Claim. Suppose that $\delta_1, \ldots, \delta_n$ are independent random variables with expected value at most $\delta < 1$. Then with probability $1 - \exp(-2n\gamma^2)$, at least $n' = \gamma n$ of them are smaller than $\delta + 2\gamma$.

Proof (Proof of Claim). With probability $1 - \exp(-2n\gamma^2)$ we have $\frac{1}{n} \sum_i \delta_i < \delta + \gamma$. Let n' be the number of i's for which $\delta_i < \delta + 2\gamma$. Since we have $\sum_i \delta_i > (n - n')(\delta + 2\gamma)$, with probability $1 - \exp(-2n\gamma^2)$ it holds that $n(\delta + \theta) > (n - n')(\delta + 2\gamma)$ or $n' > \frac{\gamma}{\delta + 2\gamma} \cdot n > \gamma n$.

By applying the claim we see that with probability $1 - \exp(-2n\theta^2)$ over $(y_i) \leftarrow (Y_i)_i$, there always exists a set $I \subset \{1, \ldots, n\}$ such that $|I| \geq n'$ (possibly depending on $(y_i)_i$) such that $\mathsf{SD}\,(Z_i; U \,|\, Y_i = y_i) \leq \delta + 2\theta$ for $i \in I'$. Since the distributions $(Z_i, Y_i)_i$ are independent for different i's and since $U + Z \overset{d}{=} U$ for any independent random variable Z, from the elementary properties of the statistical distance we obtain

$$
\mathsf{SD}\left(\sum_{i=1}^{n} Z_i; \; U \,\middle|\, (Y_i)_i = (y_i)_i\right) = \mathsf{SD}\left(\sum_{i\in I} Z_i + \sum_{i\notin I} Z_i; \; U + \sum_{i\notin I} Z_i \,\middle|\, (Y_i)_i = (y_i)_i\right)
$$

$$
= \mathsf{SD}\left(\sum_{i\in I} Z_i; \; U \,\middle|\, (Y_i)_i = (y_i)_i\right). \tag{25}
$$

The lemma now easily follows, as for every I as above we have

$$
\mathsf{SD}\left(\sum_{i\in I} Z_i; \; U \,\middle|\, (Y_i)_i = (y_i)_i\right) \leq \max_{(Z_i')_i:\ \mathsf{SD}(Z_i';U)\leq\delta'} \Delta\left(\sum_{i=1}^{n'} Z_i'; U\right). \tag{26}
$$

A.6   Proof of Theorem 2

Proof. Let μ_i be a distribution of Z_i for $i = 1, 2$ and let μ_U denotes the uniform measure. Let $\Delta(\mu_i, \mu_U) = \delta_i$. Note that we can decompose $\mu_i = \mu_U + \delta_i\mu_i^+ - \delta_i\mu_i^-$. Therefore

$$
\mu_1 * \mu_2 = \left(\mu_U + \delta_1\mu_1^+ - \delta_1\mu_1^-\right) * \left(\mu_U + \delta_2\mu_2^+ - \delta_2\mu_2^-\right)
$$
$$
= \mu_U + \delta_1\delta_2\left(\mu_1^+ * \mu_2^+ + \mu_1^- * \mu_2^- - \mu_1^+ * \mu_2^- - \mu_1^- * \mu_2^+\right) \tag{27}
$$

where we have made use of the fact that $\mu_U * \nu = \mu_U$ for any distribution ν. Now we have

$$
\mathsf{SD}(\mu_1 * \mu_2; \mu_U) = \frac{1}{2}\left\|\mu_1^+ * \mu_2^+ + \mu_1^- * \mu_2^- - \mu_1^+ * \mu_2^- - \mu_1^- * \mu_2^+\right\|_{\ell_1(G)} \tag{28}
$$

This is clearly at most 2. To identify the worst case choice of μ_i that maximizes this quantity, observe that we have to bound the last expression with respect to the constraints

$$
\left\|\mu_i^-\right\|_{\ell_\infty(G)} \leq \frac{1}{\delta_i|G|} \quad i = 1, 2 \tag{29}
$$

which come from the fact that μ_i, as decomposed, has to be positive. There is no restriction on μ_i^+. Note now that the form $\mu_1^+ * \mu_2^+ + \mu_1^- * \mu_2^- - \mu_1^+ * \mu_2^- - \mu_1^- * \mu_2^+$ is bilinear with respect to measures μ_i^+, μ_i^- and the real-valued function $\mu \rightarrow \|\mu\|_{\ell_1(G)}$ defined on *signed* measures is convex. It follows that $\left\|\mu_1^+ * \mu_2^+ + \mu_1^- * \mu_2^- - \mu_1^+ * \mu_2^- - \mu_1^- * \mu_2^+\right\|_{\ell_1(G)}$ attains its maximal value for measures that are extreme points of their domain. Looking at the restrictions in (29) we see that this is the case where μ_i^+ are a point mass and μ_i^- are uniform over the subset of cardinality $\delta_i |G|^4$. Thus we can assume that $\mu_1^+ = \mu_a$, $\mu_2^+ = \mu_b$ are point mass at a, b and $\mu_1^- = \mu_A$, $\mu_2^- = \mu_B$ are uniform over A, B where $|A| = \delta_1 |G|$ and $|B| = \delta_2 |G|$. This way our quantity simplifies to

$$
\|\mu_a * \mu_b - \mu_a * \mu_B - \mu_b * \mu_A + \mu_A * \mu_B\|_{\ell_1(G)} = \|\mu_{a+b} - \mu_{B+a} - \mu_{b+A} + \mu_A * \mu_B\|_{\ell_1(G)}
$$
$$
= \|\mu_0 - \mu_{B-b} - \mu_{A-a} + \mu_{A-a} * \mu_{B-b}\|_{\ell_1(G)}
$$
(30)

where we have used the fact that the norm $\ell_1(G)$ is shift invariant and that a point mass act as shifts under the convolution.

From this we easily derive the following result

Lemma 7 (Mixing time for a sum of random variables on a group).
Let $\{Z_i\}_{i=1,\dots,n}$ be independent random variables on an abelian group $\mathbb{G}$, such that $\Delta(Z_i; U) = \delta_i$ where $\delta_i \leqslant \frac{1}{2} - \theta$ and $\theta > 0$. Then for $n \geqslant \log(1/\epsilon)/(2\theta)$ it holds that

$$
\mathsf{SD}\left(\sum_{i=1}^n Z_i; U\right) \leqslant \epsilon
$$
(31)

A.7   Proof of Lemma 6

We will show that the constant given by (8) could be much better estimated when $\mathbb{G} = \mathbb{Z}_p$. The trivial estimate is 2, however this is possible only if $A + B$ is disjoint with A and B. Here we remind the following result due to Cauchy and Davenport

Theorem (Cauchy-Davenport Theorem). *For any $A, B \subset \mathbb{Z}_p$, where p is prime, we have $|A + B| \geqslant \min(|A| + |B| - 1, p)$.*

In view of this result, a better estimate is impossible if only $\delta_1 + \delta_2 + \max(\delta_1, \delta_2) > 1 + 1/p$. From this we know that the estimate (7) is not sharp for $\delta_1 + \delta_2 \geqslant \frac{2}{3} + \frac{2}{3p}$. Therefore we expect to improve the estimate for *sufficiently big* values of $\delta_1 + \delta_2$ whereas for the smaller we can still use the general result. To this end, we will need a result stronger than the Cauchy-Davenport Theorem

Theorem (Pollard's Theorem [21]). *For any $A, B \subset \mathbb{Z}_p$, where p is prime, we have*

$$
\sum_{x \in \mathbb{Z}_p} r_{A,B}(x) \mathbf{1}_{\{r_{A,B}(x) > t\}}(x) \leqslant |A||B| - t(|A| + |B| - t)
$$
(32)

where $r_{A,B}(x)$ counts in how many different ways can we represent x as a sum $a + b$ with $a \in A, b \in B$.

---

[4] Otherwise we could decompose either the positive part μ^+ into a combination of two distributions (when μ^+ is supported on more than one point) or the negative part μ^- (when the constraint Equation (29) is not binding at some point in the support).

Intuitively, Pollard's theorem says that the distribution of $r_{A,B}(x)$ cannot be too "heavy tailed".

Proof (of Lemma 6). In fact, we will show that $\mu_A * \mu_B$ *always* puts some large mass on every sufficiently big set C, essentially on A or B. Observe first that

$$\mu_A * \mu_B(x) = \frac{r_{A,B}(x)}{|A||B|} \tag{33}$$

where $r_{A,B}(x)$ counts for how many different ways can we represent x as a sum $a + b$ with $a \in A, b \in B$. By trivial estimates $r_{A,B}(x) \leqslant \min(|A|, |B|)$ we see that

$$\mu_A * \mu_B(x) \leqslant \min(\mu_A(x), \mu_B(x)), \quad x \in A \cup B \tag{34}$$

Using this we can estimate the expression in (8) as follows

$$\begin{aligned}
\frac{1}{2} \|\mu_A + \mu_B - \mu_A * \mu_B - \mu_0\|_{\ell_1(G)} &= \max_{S \subset G} (\mu_A(S) + \mu_B(S) - \mu_A * \mu_B(S) - \mu_0(S)) \\
&\leqslant \max_{S \subset G} (\mu_A(S) + \mu_B(S) - \mu_A * \mu_B(S)) \\
&= (\mu_A(A \cup B) + \mu_B(A \cup B) - \mu_A * \mu_B(A \cup B)) \\
&= 2 - \frac{1}{|A||B|} \sum_{x \in A \cup B} r_{A,B}(x) \tag{35}
\end{aligned}$$

From Pollard's theorem, for every set C we obtain

$$\sum_x r_{A,B}(x) 1_C(x) \geqslant \sum_x r_{A,B}(x) 1_{r_{A,B}(x) \leqslant t}(x) - t(|G| - |C|) \tag{36}$$

$$\geqslant t(|A| + |B| - t) - t(p - |C|) = t(|A| + |B| + |C| - p - t) \tag{37}$$

the maximum is for $t_{\max} = \frac{|A| + |B| + |C| - p}{2}$ provided that $|A| + |B| + |C| - p \geqslant 0$. We check that the required inequality $|A| + |B| - p \leqslant t_{\max} \leqslant \min(|A|, |B|)$ is true if only the set C satisfies

$$|A| + |B| - p \leqslant |C| \leqslant p - ||B| - |A||. \tag{38}$$

Note that if $t_{\max} \notin \mathbb{Z}$ then the conditions above are still sufficient provided that we replace $t_{\max}$ with $\lceil t_{\max} \rceil$ or $\lfloor t_{\max} \rfloor$. Considering the function $f(t) = t(|A| + |B| + |C| - p - t)$ by the mean-value theorem we see that

$$\begin{aligned}
|f(\lceil t_{\max} \rceil) - f(\lfloor t_{\max} \rfloor)| &\leqslant \max_{\xi \in [\lfloor t_{\max} \rfloor, \lceil t_{\max} \rceil]} f'(\xi) \\
&= \max_{\xi \in [\lfloor t_{\max} \rfloor, \lceil t_{\max} \rceil]} (-2\xi + |A| + |B| + |C| - p) \\
&\leqslant -2\left(t_{\max} + \frac{1}{2}\right) + |A| + |B| + |C| - p = 1. \tag{39}
\end{aligned}$$

Therefore, we obtain

$$\sum_x r_{A,B}(x)\mathbf{1}_C(x) \geqslant \left\lfloor \frac{(|A|+|B|+|C|-p)^2}{4} \right\rfloor \tag{40}$$

Setting $C \subset A \cup B$ such that $|C| = \min\left(\max(|A|,|B|), p - ||A| - |B||\right)$ we see that the condition $|C| \geqslant |A|+|B|-p$ is satisfied. Provided that $|A|+|B|+|C|-p \geqslant 0$ we obtain

$$2 - \frac{1}{|A||B|}\sum_{x \in A \cup B} r_{A,B}(x) \leqslant 2 - \frac{(\delta_1 + \delta_2 + \min(\max(\delta_1,\delta_2), 1 - |\delta_1 - \delta_2|) - 1)^2 + p^{-2}}{4\delta_1\delta_2} \tag{41}$$

and the result follows by (8).

From this result we obtain the following result, from which we conclude the part (ii) of Theorem 1 by replacing θ by $\frac{\theta}{4}$ and combining with Corollary 1 in the same way as in the derivation of part (ii).

Lemma 8 (Mixing time for a sum of random variables on $\mathbb{Z}_p$). *Let $\{Z_i\}_{i=1,\ldots,n}$ be independent random variables on $\mathbb{G} = \mathbb{Z}_p$, such that $\mathsf{SD}(Z_i; U) \leqslant \delta_i$ where $\delta_i \leqslant 1 - p^{-1} - \theta$ and $\theta > 0$. Then for $n \geqslant 3 \cdot 2^{4/\theta} \log(1/\epsilon)/\theta$ it holds that*

$$\mathsf{SD}\left(\sum_{i=1}^n Z_i; U\right) \leqslant \epsilon \tag{42}$$

Proof. First, using Corollary 3, we show that every sufficiently long sum has distance at most $\frac{1}{3}$. Once we have that, it is enough to split the entire sum into sufficiently many blocks and then apply Theorem 2. Consider $n_0 = 2^m$. By applying Lemma 6 several times we see that

$$\mathsf{SD}\left(\sum_{i=1}^{n_0} Z_i; \, U\right) \leqslant B_m \tag{43}$$

where B_i are numbers defined by the following recursion

$$B_0 = 1 - p^{-1} - \theta, \quad B_i = h(B_{i-1}, B_{i-1}) \text{ for } i \geqslant 1 \tag{44}$$

We will prove that $1 - p^{-1}$ is the *repelling point*: if we start from any B_0 satisfying $\frac{1}{3} \leqslant B_0 < 1 - p^{-1}$ then B_i decreases below $\frac{1}{3}$. Let $C_i = 1 - B_i$. If $B_{i-1} \geqslant \frac{1}{3}$, then by Corollary 3 we get

$$\begin{aligned} C_i &= 1 - B_i = 1 - h(B_{i-1}, B_{i-1}) \\ &= 2B_{i-1}^2 - \frac{(3B_{i-1}-1)^2}{4} - \frac{1}{4p^2} \\ &= C_{i-1} + \frac{C_{i-1}^2}{4} - \frac{1}{4p^2} \\ &= C_{i-1}\left(1 + \frac{C_{i-1}}{4}\left(1 - \frac{1}{C_{i-1}^2 p^2}\right)\right) \end{aligned} \tag{45}$$

From this we conclude that if $\frac{1}{3} \leqslant B_{i-1} < 1 - p^{-1}$ then $C_{i-1} > p^{-1}$ and hence $C_i > C_{i-1}$ or equivalently $B_i < B_{i-1}$. Moreover, if $C_{i-1} \geqslant p^{-1} + \theta$, we get

$$C_i \geqslant C_{i-1} \left(1 + \theta \cdot \frac{2 + p\theta}{4 + 4p\theta} \right)$$

$$\geqslant C_{i-1} \left(1 + \frac{\theta}{4} \right) \tag{46}$$

Since $C_i \leqslant 1$ and $C_0 \geqslant p^{-1} + \theta > \theta$, for some $j \leqslant \frac{4}{\theta} \log\left(\frac{1}{\theta}\right)$ we must have $B_j < \frac{1}{3}$. Thus for $m = \lceil \frac{4}{\theta} \log\left(\frac{1}{\theta}\right) \rceil$ we have

$$\mathsf{SD} \left(\sum_{i=1}^{2^m} Z_i; U \right) \leqslant \frac{1}{3} \tag{47}$$

Consider $\ell = \log(1/\epsilon)$ blocks of random variables $\{(Z_{2^m j + 1}, \ldots, (Z_{2^m j + 2^m}))\}_j$ for $j = 0, \ldots, N - 1$. For every such a 2^m-element block from the last observation it follows that

$$\mathsf{SD} \left(\sum_{i=1}^{2^m} Z_{2^m j + i}; U \right) \leqslant \frac{1}{3} \tag{48}$$

Applying ℓ times Lemma 2 yields the estimate

$$\mathsf{SD} \left(\sum_{i=1}^{\ell 2^m} Z_i; U \right) \leqslant \left(\frac{2}{3} \right)^{\ell} \tag{49}$$

which finishes the proof.

A.8   Harmonic Analysis

We need the following lemma, being a generalization of Vazirani's XOR lemma.

Lemma 9 (XOR lemma for abelian groups, [23]**).** *Let Z be a distribution over a finite abelian group $\mathbb{G}$, such that $|\mathbb{E}\phi(Z)| \leqslant \epsilon$ for every non-trivial character ϕ on $\mathbb{G}$. Then X is $\epsilon\sqrt{|\mathbb{G}|}$-close to uniform.*

Lemma 10 (Mixing times of random sums over $\mathbb{Z}_p$). *Let $\{Z_i\}_{i=1,\ldots,n}$ be independent random variables on $\mathbb{G} = \mathbb{Z}_p$, such that $\mathsf{SD}(Z_i; U) \leqslant 1 - p^{-1} - \theta$ and $\theta > 0$. Then for $n \geqslant 8 \cdot \log(|\mathbb{G}|/\epsilon)/\theta^3$ it holds that*

$$\mathsf{SD} \left(\sum_{i=1}^{n} Z_i; U \right) \leqslant \epsilon \tag{50}$$

Proof. We apply some facts from harmonic analysis. Let Z_i be the worst-case distributions that maximize $\mathsf{SD}(\sum_{i=1}^{n} X_i, U)$ under the constraints $\mathsf{SD}(Z_i, U) \leqslant 1 - p^{-1} - \theta$. By Eq. (6) that

$$\mu_{Z_i} = \left(1 - \frac{|A|}{p} \right) \cdot \mu_0 + \frac{|A|}{p} \cdot \frac{1}{|A|} \mu_A, \quad |A| = p\theta \tag{51}$$

Consider a non-trivial character $\phi(x) = \exp(2k\pi i/p)$ on $\mathbb{Z}_p$. Since $A \neq \emptyset$ we have $\theta \geqslant \frac{1}{p}$. We will show an upper bound on $\mathbf{E}\phi(X_i)$. First, observe that

$$|\mathbf{E}\phi(X_i)| = \left| 1 - \frac{|A|}{p} + \frac{|A|}{p} \cdot \frac{1}{|A|} \sum_{x \in A} \exp\left(\frac{2k\pi i x}{p}\right) \right| \tag{52}$$

is maximized exactly when $kA = \left\{ -\frac{|A|-1}{2}, \ldots, 0, \ldots, \frac{|A|-1}{2} \right\}$. Indeed, we have

Claim. For any subset A of $\mathbb{Z}_p$ and any non-trivial character ϕ over $\mathbb{Z}_p$ we have the following estimate

$$|\mathbf{E}\phi(X_i)| = \left| 1 - \frac{|A|}{p} + \frac{|A|}{p} \cdot \frac{1}{|A|} \sum_{x \in A} \exp(2k\pi i x/p) \right| \leqslant 1 - \theta + \frac{\sin \pi \theta}{p \sin \frac{\pi}{p}}$$

Proof. Note that every non-trivial character is of the form $\phi(x) = \exp(2k\pi i x/p)$ where $k \in \{1, 2, \ldots, p-1\}$. Next, we can assume that $k = 1$, by replacing A with $A' = k \cdot A$, which doesn't change the set size. Now, by the triangle inequality we have

$$\left| 1 - \frac{|A|}{p} + \frac{1}{p} \sum_{x \in A} \phi(x) \right| \leqslant 1 - \frac{|A|}{p} + \frac{|A|}{p} \cdot \left| \frac{1}{|A|} \sum_{x \in A} \phi(x) \right|$$

It remains to estimate $|m_A|$ where

$$m_A = \frac{1}{|A|} \sum_{x \in A} \phi(x)$$

is the mass center of the set $\phi(A) = \{\phi(x) : x \in A\}$. Note that $\phi(A)$ may be any arbitrary $|A|$-element subset of the set of all p-th roots of unity (because ϕ is a bijection), see Fig. 2 for an illustration. Our task is therefore to maximize the length of m_A which happens when A is the set of subsequent unity roots. In particular

$$|\mathbf{E}\phi(X_i)| \leqslant \left| 1 - \frac{|A|}{p} \right| + \frac{1}{p} \left| \sum_{|x| \leqslant \frac{|A|-1}{2}} \exp(2\pi i x) \right| = 1 - \theta + \left| \frac{\sin \pi \theta}{p \sin \frac{\pi}{p}} \right|, \tag{53}$$

where the last equality follows by known trigonometric identities. Since $\theta < 1$ we can omit the absolute value here, and this finishes the proof.

We will prove the following inequality

Claim. For any $\theta < 1$ and any $c \leqslant \frac{4}{3} - \frac{\pi^2}{18}$, we have $1 - \theta + \frac{\sin \pi \theta}{p \sin \frac{\pi}{p}} \leqslant 1 - c\theta^3$

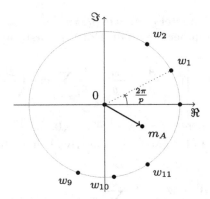

Fig. 2. The mass center of the set $\phi(A)$ should be as close to the circle as possible.

Proof. We want to prove that $f(\theta) = c\theta^3 - \theta + \frac{\sin \pi\theta}{p \sin \frac{\pi}{p}} \leqslant 0$. We have $f(0) = 0$ and $\frac{\partial f(\theta)}{\partial \theta} = -1 + 3c\theta^2 + \pi \frac{\cos \pi\theta}{p \sin \frac{\pi}{p}}$ Since for $t \in \left[0, \frac{\pi}{2}\right]$ it holds that $\cos t \leqslant 1 - \frac{4t^2}{\pi^2}$ and $\sin t \geqslant t - \frac{t^3}{6}$, we obtain

$$\frac{\partial f(\theta)}{\partial \theta} \leqslant -1 + 3c\theta^2 + \frac{1 - 4\theta^2}{1 - \frac{\pi^2}{6p^2}} = \frac{\frac{\pi^2}{6p^2} - \theta^2 \left(4 - 3c + \frac{3c\pi^2}{6p^2}\right)}{1 - \frac{\pi^2}{6p^2}} \tag{54}$$

and since $\theta \geqslant \frac{1}{p}$, the result follows.

From the last claim it follows that we can put $c = \frac{1}{2}$ and thus

$$|\mathbf{E}\phi(X)| = \left|\mathbf{E}\phi\left(\sum_{i=1}^{n} X_i\right)\right|$$

$$= \prod_{i=1}^{n} |\mathbf{E}\phi(X_i)|$$

$$\leqslant \left(1 - \theta^3/2\right)^n. \tag{55}$$

Now the result follows by Lemma 9.

References

1. Akavia, A., Goldwasser, S., Vaikuntanathan, V.: Simultaneous hardcore bits and cryptography against memory attacks. In: Reingold, O. (ed.) TCC 2009. LNCS, vol. 5444, pp. 474–495. Springer, Heidelberg (2009)
2. Alwen, J., Dodis, Y., Wichs, D.: Leakage-resilient public-key cryptography in the bounded-retrieval model. In: Halevi, S. (ed.) CRYPTO 2009. LNCS, vol. 5677, pp. 36–54. Springer, Heidelberg (2009)

3. Brakerski, Z., Kalai, Y. T., Katz, J., Vaikuntanathan, V.: Overcoming the hole in the bucket: public-key cryptography resilient to continual memory leakage. In: 51st FOCS, Las Vegas, Nevada, USA, pp. 501–510. IEEE Computer Society Press, October 23–26, 2010

4. Chari, S., Jutla, C.S., Rao, J.R., Rohatgi, P.: Towards sound approaches to counteract power-analysis attacks. In: Wiener, M. (ed.) CRYPTO 1999. LNCS, vol. 1666, pp. 398–412. Springer, Heidelberg (1999)

5. Davì, F., Dziembowski, S., Venturi, D.: Leakage-resilient storage. In: Garay, J.A., De Prisco, R. (eds.) SCN 2010. LNCS, vol. 6280, pp. 121–137. Springer, Heidelberg (2010)

6. Dodis, Y., Haralambiev, K., López-Alt, A., Wichs, D.: Cryptography against continuous memory attacks. In: 51st FOCS, Las Vegas, Nevada, USA, pp. 511–520. IEEE Computer Society Press, 23–26 October, 2010

7. Duc, A., Dziembowski, S., Faust, S.: Unifying leakage models: from probing attacks to noisy leakage. In: Nguyen, P.Q., Oswald, E. (eds.) EUROCRYPT 2014. LNCS, vol. 8441, pp. 423–440. Springer, Heidelberg (2014)

8. Duc, A., Faust, S., Standaert, F.-X.: Making masking security proofs concrete. In: Oswald, E., Fischlin, M. (eds.) EUROCRYPT 2015. LNCS, vol. 9056, pp. 401–429. Springer, Heidelberg (2015)

9. Dziembowski, S., Faust, S.: Leakage-resilient circuits without computational assumptions. In: Cramer, R. (ed.) TCC 2012. LNCS, vol. 7194, pp. 230–247. Springer, Heidelberg (2012)

10. Dziembowski, S., Faust, S., Skorski, M.: Noisy leakage revisited. In: Oswald, E., Fischlin, M. (eds.) EUROCRYPT 2015. LNCS, vol. 9057, pp. 159–188. Springer, Heidelberg (2015)

11. Dziembowski S., Pietrzak, K.: Leakage-resilient cryptography. In: 49th FOCS, Philadelphia, Pennsylvania, USA, pp. 293–302. IEEE Computer Society Press, 25–28 October, 2008

12. Faust, S., Rabin, T., Reyzin, L., Tromer, E., Vaikuntanathan, V.: Protecting circuits from leakage: the computationally-bounded and noisy cases. In: Gilbert, H. (ed.) EUROCRYPT 2010. LNCS, vol. 6110, pp. 135–156. Springer, Heidelberg (2010)

13. Goldreich, O.: Three XOR-lemmas — an exposition. In: Goldreich, O. (ed.) Studies in Complexity and Cryptography. LNCS, vol. 6650, pp. 248–272. Springer, Heidelberg (2011)

14. Goldwasser S., Rothblum, G.N.: How to compute in the presence of leakage. In: 53rd FOCS, New Brunswick, NJ, USA, pp. 31–40. IEEE Computer Society Press, 20–23 October, 2012

15. Ishai, Y., Sahai, A., Wagner, D.: Private circuits: securing hardware against probing attacks. In: Boneh, D. (ed.) CRYPTO 2003. LNCS, vol. 2729, pp. 463–481. Springer, Heidelberg (2003)

16. Kocher, P.C., Jaffe, J., Jun, B.: Differential power analysis. In: Wiener, M. (ed.) CRYPTO 1999. LNCS, vol. 1666, pp. 388–397. Springer, Heidelberg (1999)

17. Maurer, U.M., Pietrzak, K., Renner, R.S.: Indistinguishability amplification. IACR Cryptology ePrint Archive, 2006:456 (2006)

18. Maurer, U.M., Pietrzak, K., Renner, R.S.: Indistinguishability amplification. In: Menezes, A. (ed.) CRYPTO 2007. LNCS, vol. 4622, pp. 130–149. Springer, Heidelberg (2007)

19. Micali, S., Reyzin, L.: Physically observable cryptography. In: Naor, M. (ed.) TCC 2004. LNCS, vol. 2951, pp. 278–296. Springer, Heidelberg (2004)

20. Naor, M., Segev, G.: Public-key cryptosystems resilient to key leakage. In: Halevi, S. (ed.) CRYPTO 2009. LNCS, vol. 5677, pp. 18–35. Springer, Heidelberg (2009)
21. Pollard, J.M.: A generalisation of the theorem of cauchy and davenport. J. Lond. Math. Soc. **s2–8**(3), 460–462 (1974)
22. Prouff, E., Rivain, M.: Masking against side-channel attacks: a formal security proof. In: Johansson, T., Nguyen, P.Q. (eds.) EUROCRYPT 2013. LNCS, vol. 7881, pp. 142–159. Springer, Heidelberg (2013)
23. Rao, A.: An exposition of bourgain's 2-source extractor. In: Electronic Colloquium on Computational Complexity (ECCC), vol. 14, p. 034 (2007)
24. Standaert, F.-X., Malkin, T.G., Yung, M.: A unified framework for the analysis of side-channel key recovery attacks. In: Joux, A. (ed.) EUROCRYPT 2009. LNCS, vol. 5479, pp. 443–461. Springer, Heidelberg (2009)

Rational Sumchecks

Siyao Guo[1]([✉]), Pavel Hubáček[2], Alon Rosen[3], and Margarita Vald[4]

[1] Chinese University of Hong Kong, Hong Kong, China
syguo@cse.cuhk.edu.hk
[2] Weizmann Institute of Science, Rehovot, Israel
pavel.hubacek@weizmann.ac.il
[3] IDC Herzliya, Herzliya, Israel
alon.rosen@idc.ac.il
[4] Tel Aviv University, Tel Aviv, Israel
margarita.vald@cs.tau.ac.il

Abstract. Rational proofs, introduced by Azar and Micali (STOC 2012) are a variant of interactive proofs in which the prover is neither honest nor malicious, but rather rational. The advantage of rational proofs over their classical counterparts is that they allow for extremely low communication and verification time. In recent work, Guo *et al.* (ITCS 2014) demonstrated their relevance to delegation of computation by showing that, if the rational prover is additionally restricted to being computationally bounded, then every language in NC1 admits a single-round delegation scheme that can be verified in sublinear time.

We extend the Guo *et al.* result by constructing a single-round delegation scheme with sublinear verification for all languages in P. Our main contribution is the introduction of *rational sumcheck protocols*, which are a relaxation of classical sumchecks, a crucial building block for interactive proofs. Unlike their classical counterparts, rational sumchecks retain their (rational) soundness properties, *even if the polynomial being verified is of high degree* (in particular, they do not rely on the Schwartz-Zippel lemma). This enables us to bypass the main efficiency bottleneck in classical delegation schemes, which is a result of sumcheck protocols being inapplicable to the verification of the computation's input level.

Part of this work done while authors were visiting IDC Herzliya, supported by the European Research Council under the European Union's Seventh Framework Programme (FP 2007-2013), ERC Grant Agreement No. 307952.
S. Guo—Work partially supported by RGC GRF grants CUHK410112 and CUHK410113.
P. Hubáček—Supported by the I-CORE Program of the Planning and Budgeting Committee and The Israel Science Foundation (Grant No. 4/11).
A. Rosen—Supported by ISF Grant No. 1255/12 and by the ERC under the EU's Seventh Framework Programme (FP/2007-2013) ERC Grant Agreement No. 307952. Work in part done while the author was visiting the Simons Institute for the Theory of Computing, supported by the Simons Foundation and by the DIMACS/Simons Collaboration in Cryptography through NSF Grant #CNS-1523467.
M. Vald—Work supported by the Check Point Institute for Information Security and by ISF Grant No. 1255/12.

© International Association for Cryptologic Research 2016
E. Kushilevitz and T. Malkin (Eds.): TCC 2016-A, Part II, LNCS 9563, pp. 319–351, 2016.
DOI: 10.1007/978-3-662-49099-0_12

As an additional contribution we study the possibility of using rational proofs as efficient blocks within classical interactive proofs. Specifically, we show a composition theorem for substituting oracle calls in an interactive proof by a rational protocol.

1   Introduction

The availability of on-demand computational power and the ubiquitous connectivity of small devices are some of the main driving forces behind the move to the model of cloud computing. In this model a client faces a computationally demanding task and relies on the assistance of an external server with sufficient computational power, e.g. a cluster of machines. When the weak client asks the powerful server to perform a computation on its behalf it would like to have some guarantees on the correctness of the provided result. This scenario is addressed by the model of *verifiable delegation of computation*. In this setting, the server provides the client with the result of the computation together with a proof of its correctness. Since the client must be able to verify the proof despite its limited computational resources, the verification should be much easier than running the computation itself, or else there is no point in outsourcing it.

Interactive Proofs and Arguments. A setting where an all-powerful entity aims to convince a computationally bounded one of the correctness of a computational statement was studied in the context of *interactive proof systems*. In this model interaction and randomization enable the prover to efficiently convince the verifier. The IP = PSPACE theorem [22,28] showed that it is possible for the prover to convince the verifier about large classes of languages, in particular any language computable in polynomial time. However, this result is not efficient enough to be practically applicable to the problem of verifiable delegation. In this context, one aims to minimize multiple complexity measures at once, such as communication complexity (both in the number and size of exchanged messages), running time of the verifier and prover efficiency.

For higher complexity classes, the round-complexity/prover-efficiency of interactive proofs is a limiting factor to their use in practice. The notion of *interactive arguments* considers a setting where the prover is computationally bounded, allowing to circumvent these efficiency shortcomings. The work of Kilian [20] gave four round interactive arguments for all languages in NP. Micali [23], relying on random oracles proposed a non-interactive version of this protocol. More recently, there has been significant effort to obtain more efficient non-interactive arguments for NP (see e.g. [5,10] and the references therein). One limitation of all such known constructions is that they are based on non-standard assumptions (cf. [24]). The problem of constructing efficient non-interactive arguments for NP under standard assumptions is still open, though there is some evidence that non-standard assumptions are unavoidable [13].

Unlike in the case of arguments for non-deterministic computation, the situation for tractable languages (which actually correspond to problems common in real-life delegation scenarios) is significantly better. The first evidence that one can attain delegation schemes for restricted complexity classes is the work of Goldwasser *et al.* [14], who gave a single-round argument that allows to verifiably delegate any bounded depth computation with quasi-linear verification time. Recently, the work of Kalai *et al.* [18] achieved a single-round argument (under standard assumptions) with quasi-linear verification time for any language in P.

In some scenarios quasi-linear verification time may not be good enough. For instance, if the input $x \in \{0, 1\}^n$ is a large database and the output $f(x)$ of the outsourced computation is a concise aggregation of its statistics, then it is desirable if the verifier does not need to read the whole database to verify correctness. In such cases one would prefer to have a delegation scheme with verification time *sublinear in the input size* n, preferably even as low as polylog(n). As was pointed out in the literature, delegation schemes with sublinear verification are in general not achievable with respect to the standard notion of soundness (cf. Rothblum *et al.* [27]), which led to introduction of alternative relaxed models that would enable sublinear verification time.

Rational Proofs and Arguments. One recent notion that opens the door for sublinear verification is that of *rational arguments* [15]. This model follows the paradigm of rational proofs introduced by Azar and Micali [2], who relax the prover in interactive proof systems to be rational. In rational proofs the verifier pays the prover according to the quality of the provided answer, and the reward is set up so that it is irrational for the prover to report an incorrect result of the computation. Azar and Micali [2] illustrated the power of rational proofs by giving a single-round rational proof for any problem in #P and in general a constant round rational proof for any level of the counting hierarchy. In subsequent work, Azar and Micali [3] gave a "scaled-down" version of their #P-protocol that leads to constant round rational interactive proof with sublinear ($O(\log n)$ time) verification for the class of log-time uniform TC^0, i.e., the class of constant-depth, log-time uniform polynomial-size circuits with threshold gates. They also argue that such efficient rational proofs capture precisely the class of log-time uniform TC^0.

More recently, Guo *et al.* [15] put forward the notion of rational arguments, by further restricting the rational prover to be computationally bounded. They then showed how to construct single-round rational arguments with sublinear (polylog(n) time) verification for the class NC^1, of search problems computable by log-time uniform Boolean circuits of $O(\log n)$-depth.

1.1   Our Results

We extend the results of Guo *et al.* [15] and give a single-round rational argument with sublinear (polylog(n) time) verification for any language in P. Our initial observation is that both the non-interactive arguments for NC of Goldwasser *et al.* [14] and the non-interactive arguments for P of Kalai *et al.* [18] have for the most part sublinear verification time, with the exception of a single heavy verification step that ultimately induces quasi-linear running time for the verifier. If we could substitute this step by a more efficient procedure that does not dominate the rest of the protocol then we would achieve sublinear verification time.

Our proposal is to use a rational proof with sublinear verification for the heavy step and get a rational version of the original protocol which enjoys sublinear verification time. There are two main issues that we will need to address: (1) construct sublinear rational proofs for the heavy step; (2) argue how the rationality can be preserved under composition.

Our main contribution is the introduction of *rational sumcheck protocols*, which are a relaxation of classical sumchecks, a crucial building block for interactive proofs. To show that our approach yields the desired result, we pin down sufficient conditions for our transformation to work and prove that the protocol of Goldwasser *et al.* [14] (respectively Kalai *et al.* [18]), with the rational sumcheck replacing the heavy step, yields the sought after rational argument for NC (respectively for P).

It should be noted that our main efficiency gains are not due to the fact that rational sumcheck protocols are more efficient than their classical counterparts (though we do gain some efficiency by making sumcheck protocols noninteractive). Indeed, one of the key observations behind the works on efficient delegation [14,18] is that one could verify correctness of computation via very efficient sumcheck protocols. The one place where rational sumcheck protocols turn out to be more useful than classical ones is at the input layer, where usage of the latter would entail a total break-down of soundness.

We show that a rational version of sumcheck protocols is in fact sufficient to carry out verification, *even without reading the entire input*. This is something that was not possible to achieve using classical sumcheck protocols, since the input layer does not satisfy the structural properties (low-degree) that would guarantee soundness when verifying via classical sumchecks. Our (equally efficient) rational sumcheck protocols, on the other hand, give a meaningful soundness guarantee even when such structural properties are absent.

Sumcheck Protocols. At a high level, a classical sumcheck protocol allows the verifier to check a sum of evaluations of a given low-degree polynomial $h : \mathbb{F}_q^m \to \mathbb{F}_q$ on a certain subset $S \subset \mathbb{F}_q^m$ of its domain (e.g. $S = \{0,1\}^m$). The source of the protocol's power is that it makes it sufficient for the verifier to evaluate h on a *single* randomly chosen point $p \in \mathbb{F}_q^m$, rather than on the entire subset S. This results in significant efficiency gains, since instead of requiring the evaluation of h on $|S|$ points it reduces the problem of verification to the evaluation on a single point (at the cost of $m = \log(|S|)$ rounds of communication).

Previous works on delegation [14,18] make extensive use of sumcheck protocols in order to efficiently verify the low degree extensions $\tilde{W}$ of intermediate levels of computation.[1] Specifically, it is possible to write $\tilde{W}(z) = \sum_{p \in S} \beta_z(p) W(p)$, where $\beta_z(p)$ is a low-degree function and $W(p)$ is an appropriate encoding of the corresponding level. This reduces the task of verifying the correctness of evaluating $\tilde{W}$ on z to the problem of performing a sumcheck on individual inner summands $\beta_z(p) W(p)$. In intermediate levels of the computation, we are guaranteed that W is of low degree, and hence so is $\beta_z(p) W(p)$. However, at the input level the function $W(p) = W_x(p)$ corresponds to a straightforward bit-wise representation of the input $x \in \{0, 1\}^n$. The problem is that this representation might result in a high-degree polynomial. Not being of low degree, $\beta_z(p) W_x(p)$ cannot be verified by a classical sumcheck protocol. This means that the input x needs to be read in its entirety, or else the protocol is not sound.

Rational Sumcheck Protocols. To circumvent the above issue, we leverage the power of rational proofs, in which soundness relies on rationality of the prover. We give a rational sumcheck protocol that allows to efficiently verify summation of any function over a fixed set, as long as evaluating the function on a single point can be performed efficiently (see Sect. 3 for details). Not only that our rational sumcheck protocol preserves the efficiency of classical sumchecks, but it can also be performed without any communication overhead (it is in fact non-interactive). The main feature of rational sumchecks, however, is that they give a meaningful (rational) soundness guarantee *even if the degree of the polynomial is high*, which implies that unlike their classical counterparts they are also applicable at the input layer.

Technically speaking, the reason for which the new rational protocols work regardless of the polynomial's degree is because the soundness analysis does not necessitate invoking the Schwartz-Zippel lemma. Instead, we rely on a specially-tailored reward function that is designed to translate sums of finite field elements to numerical values that are used to determine the reward. The challenge in designing the reward function originates from the fact that modular sums lose information about the summands, whereas the reward is required to reflect this information in its entirety.

Composition of Classical and Rational Interactive Proofs. To make the above fit into a general purpose protocol, we need to carefully show how to plug a rational subprotocol into a larger one while retaining rational soundness. To this end, we show a composition theorem for substituting oracle calls in an interactive proof by a rational protocol. This allows us to use the classical interactive proofs almost as a black-box. This approach may turn out to be useful elsewhere.

Putting the Pieces Together. At a high level, the structure of our construction of single-round rational arguments for P follows the delegation scheme of

---

[1] In Goldwasser *et al.* [14] this is performed layer by layer over the circuit computing the function, whereas in Kalai *et al.* [18] the reduction to sumchecks is done via a global encoding of the transcript of the computation.

Kalai *et al.* [18]. In particular, we define and construct δ-no-signaling rational multi-prover proofs (RMIPs) by using our composition theorem and relying on rational sumchecks as a subprotocol. We then show a general efficient transformation that uses any sub-exponentially secure *Fully Homomorphic Encryption* (FHE) scheme to transform no-signaling RMIPs into single-round rational arguments (in a manner similar to Kalai *et al.* [18]). Crucial to our transformation is the *reward gap* of the underlying rational protocol, which roughly captures the utility loss of the prover as a result of misreporting the function's value. Unlike early rational proofs of Azar and Micali [2] and akin to Guo *et al.* [15], both our sumchecks and the overall composed protocol enjoy noticeable reward gap. This is sufficient for the overall transformation to go through (enabling a reduction from the security of the FHE scheme), and results in the sought-after single-round rational argument for P with sublinear verification time.

Beyond being of importance in the transformation from rational proofs to non-interactive rational arguments, noticeable reward gap is also crucial for incentivizing the prover to report the correct value of the computation, as otherwise he might be tempted to avoid performing the work while risking very little penalty (see Sect. 2 and Guo *et al.* [15] for an extended discussion of the subject).

1.2   Comparison to Alternative Delegation Schemes

The classical interactive proof for NC of Goldwasser *et al.* [14] has quasi-linear verification time. The running time of the verifier in their protocol appears to be optimal in the standard model, in the sense that achieving sublinear verification time with standard soundness guarantee seems unlikely without reading the whole input (even for a simple function such as parity). To circumvent this limitation Rothblum *et al.* [27] considered interactive proofs of proximity, a relaxation of interactive proofs motivated by property testing, and show that it is possible to achieve sublinear verification for NC in this new model (since the protocol does not need to provide soundness guarantee for all instances).

An alternative relaxation was studied by Azar and Micali [3] and Guo *et al.* [15]. These works considered delegation in the setting of rational proofs and proposed schemes whith both sublinear verification (as small as polylogarithmic) and (rational) soundness guarantees, which in contrast to proofs of proximity hold for all instances. Whereas their protocols work only for NC^1, our new rational proof, which is a combination of classical and rational proofs, works for the entirety of NC while preserving the desired properties of sublinear verification and rational soundness (see Table 1 for a detailed comparison). By composing classical and rational proofs, we obtain a rational multi-prover proof (secure against no-signaling provers) with sublinear verification for any deterministic computation akin to the classical proof of Kalai *et al.* [18] (see Table 2 for a detailed comparison). We remark it is possible to transform the above classical proofs and rational proofs into one-round classical and rational arguments.

Table 1. Efficiency comparison of results for NC

	Queries[a]	Rounds	Communication	Verification time	Depth
Goldwasser et al. [14] (interactive proofs)	n	$\tilde{O}(d)$	$\tilde{O}(d)$	$\tilde{O}(n)$	$d = \text{polylog}(n)$
Rothblum et al. [27] (proofs of D-proximity)	$(\frac{n}{D})^{1+o(1)}$	$\tilde{O}(d)$	$D(\frac{n}{D})^{o(1)} \cdot \tilde{O}(d)$	$(\frac{n}{D}+D)^{1+o(1)}\tilde{O}(d)$	$d = \text{polylog}(n)$
Azar and Micali [3] (rational proofs)	1	d	$\tilde{O}(d)$	$\tilde{O}(d)$	$d = O(\log n)$
Guo et al. [15] (rational proofs)	1	d	d	$\tilde{O}(d)$	$d = O(\log n)$
This work (rational proofs)	1	$\tilde{O}(d)$	$\tilde{O}(d)$	$\tilde{O}(d)$	$d = \text{polylog}(n)$

[a]By queries we denote the number of input bits read by the verifier.

Table 2. Efficiency comparison of results for P

	Queries	Number of provers	Communication	Verification time	Remarks
Kalai et al. [18] (MIP)	n	$\text{polylog}(t)$	$\text{polylog}(t)$	$n \cdot \text{polylog}(t)$	$\text{DTIME}(t)$
This work (rational MIP)	$\text{polylog}(t)$	$\text{polylog}(t)$	$\text{polylog}(t)$	$\text{polylog}(t)$	$\text{DTIME}(t)$

1.3 Other Related Work

To give a complete overview of works on verifiable delegation of computation is out of the scope of this paper, an interested reader can find many related results in the recent survey by Blumberg and Walfish [6].

An alternative approach for *interactive proofs with sublinear verification* was given in Rothblum et al. [27] who introduced *interactive proofs of proximity* and Gur and Rothblum [16] who considered their non-interactive analogues. Since both works studied a protocol analogue of property testing, their protocols provide guarantees only for instances that are either in the language or far from being in the language. Independently an in parallel to our work, Kalai and Rothblum [19] studied proofs of proximity with computationally bounded provers and introduced *arguments of proximity*.

Besides the mentioned works in the context of rational proofs, Zheng and Blanton [29] study the specific problem of delegating matrix multiplication and give also a rational argument for this task. The work of Chen et al. [9] introduces the model of rational interactive proofs with multiple provers.

Alternative approaches for incentivizing correct computation can be found in the work of Bentov and Kumaresan [21] who consider a model for incentivizing computation over Bitcoin. Alternatively, Belenkiy et al. [4] or Pham et al. [25] study a model where the verifier infrequently performs the whole computation to verify the correctness of prover's output.

The treatise of general composition of rational protocols in scientific literature is limited. The work of Garay et al. [12] provides some insights on composition of protocols secure in the presence of a single *central rational adversary*. The framework of Canetti and Vald [8] studies a notion sufficient for preserving rationality under composition by imposing strong restrictions on the information available to distinct adversarial entities.

2   Preliminaries

Throughout the rest of the paper we use the following notation and definitions. For $n \in \mathbb{N}$, let $[n]$ denote the set $\{1, \ldots, n\}$. A function $g : \mathbb{N} \to \mathbb{R}^+$ is *negligible* if it tends to 0 faster than any inverse polynomial, i.e., for all $c \in \mathbb{N}$ there exists $k_c \in \mathbb{N}$ such that for every $k > k_c$ it holds that $g(k) < k^{-c}$. We use $\mathrm{negl}(\cdot)$ to talk about negligible function if we do not need to specify its name.

Rational Proofs. In a rational proof, Arthur pays Merlin a randomized reward according to the transcript of the communication, and the communication constitutes a rational Merlin Arthur game if the correct evaluation $y = f(x)$ can be derived from a transcript that maximizes the expected reward.

For a pair of interactive Turing machines, P and V, we denote by $(P, V)(x)$ the random variable representing the transcript between P and V when interacting on common input x. Let $\mathrm{reward}(\cdot)$ denote a randomized function computed by V that given a transcript calculates a reward for P, and by $\mathrm{output}((P, V)(x))$ the output of V after interacting with P on common input x. In this setting, the goal of a *rational* P is to maximize the expected value of $\mathrm{reward}(\cdot)$, while the goal of V is to learn (and output) the true evaluation of the desired function f on x. We consider the setting where a rational prover first declares his answer to $f(x)$, and only then tries to prove the correctness of the reported value.

Definition 1. *[Functional Rational Merlin Arthur]. Let $C, T : \mathbb{N} \to \mathbb{R}$ be some functions. A function $f : \{0,1\}^* \to \{0,1\}^*$ is in* FRMA $[r, C, T]$ *if there exists an r-round public-coin protocol (P, V), referred as rational proof, and a randomized reward function* $\mathrm{reward} : \{0,1\}^* \to \mathbb{R}_{\geq 0}$ *such that for all inputs $x \in \{0,1\}^*$:*

(a) $\Pr[\mathrm{output}((P, V)(x)) = f(x)] = 1$.
(b) *For every round i and for any prover P^* that misreports $f(x)$ and behaves as P up to round i and differs on round i'th message it holds that:* $\mathrm{E}[\mathrm{reward}((P, V)(x))] > \mathrm{E}[\mathrm{reward}((P^*, V)(x))]$, *where the expectation is taken over the random coins of the verifier and the prover.*
(c) *The communication complexity of P is $C(|x|)$.*
(d) *The running time of V is $T(|x|)$.*

No-Signaling Provers. In this work we use the heuristic suggested by Aiello *et al.* [1] for transforming a multi-prover proof into a single round argument using an efficient Private Information Retrieval (PIR) scheme (or alternatively a Fully Homomorpic Encryption scheme), though in the rational setting. As pointed out in the work of Dwork *et al.* [11], the bottleneck when proving soundness of the resulting argument is the possibility for the prover to correlate the answers in an undetectable way. Such no-signaling strategies (introduced as "spooky interactions" in the work of Dwork *et al.* [11]) need to be accounted for in the proof of soundness, as shown in Kalai *et al.* [17].

Thus, we extend Definition 1 to the setting with multiple provers restricted to δ-no-signaling strategies. In contrast to the classical multi-prover setting, where each prover strategy is completely independent of other provers' queries, δ-no-signaling strategies can be correlated as long as for any subset of provers their answers do not contain information about the queries of provers outside the subset.

Definition 2 (Statistically No-Signaling Distributions). *Let D be a query alphabet and let Σ be an answer alphabet. For every $q = (q_1, \ldots, q_k) \in D^k$, let $\mathcal{A}_q$ be a distribution over Σ^k. We think of $\mathcal{A}_q$ as the distribution of the answers for queries q. We say that the family of distributions $\{\mathcal{A}_q\}_{q \in D^k}$ is δ-no-signaling if for every subset $S \subset [k]$ and every two sequences of queries $q, q' \in D^k$, such that $q_S = q'_S$, the following two random variables are δ-close: $\{a_S : a \leftarrow \mathcal{A}_q\}$ and $\{a'_S : a' \leftarrow \mathcal{A}_{q'}\}$.*

The rational no-signaling multi-prover proof consists of only one round. Given an input, the verifier generates queries, one for each prover, and sends them to the k provers. Each prover responds with an answer that might depend on all the queries, as long as the provers' strategies are no-signaling. Finally, the verifier computes the reward based on the received answers (as well as the input and the randomness used).

Definition 3 (One-Round Rational Multi-prover Interactive Proof). *Let $C, T : \mathbb{N} \to \mathbb{R}$ be some functions. A function $f : \{0,1\}^* \to \{0,1\}^*$ is in FRMIP$[k, \delta, C, T]$ if there exists a one-round public-coin protocol $(\overrightarrow{P}, V) = (P_1, \ldots, P_k, V)$, referred as multi-prover rational proof, and a randomized reward function* reward $: \{0,1\}^* \to \mathbb{R}_{\geq 0}$ *such that for all inputs $x \in \{0,1\}^*$:*

(a) Pr[output$((\overrightarrow{P}, V)(x)) = f(x)] = 1$.
(b) *For every set of provers $P_1^*, \ldots, P_k^*$ with δ-no-signaling distributions that misreport $f(x)$ it holds that:* E[reward$((P_1, \ldots, P_k, V)(x))] > $ E[reward$((P_1^*, \ldots, P_k^*, V)(x))]$, *where the expectation is taken over the random coins of the verifier and the provers.*
(c) *The communication complexity from any of the provers to V is at most is $C(|x|)$.*
(d) *The running time of V is $T(|x|)$.*

Reward Gap. We note that once computation incurs some cost to the prover the Definitions 1 and 3 of rational proofs do not rule out a "lazy behavior" of the prover corresponding to outputting a fixed default value. Having this in mind, Guo et al. [15] proposes the notion of *reward gap* that measures how big is the loss of a prover that always reports $f(x)$ incorrectly. A noticeable gap in expectation between such a prover and the prescribed behavior then assures that it is beneficial for the prover to perform the computation to significantly increase its expectation.

Definition 4 (Reward Gap). *Let $f \in \text{FRMA}[r, C, T]$ be some function and let (P, V) and* reward$(\cdot)$ *be the guaranteed protocol and reward function. The reward gap of* reward$(\cdot)$ *is a function $\Delta_{\text{reward}} : \mathbb{N} \to \mathbb{R}$, such that for every $n \in \mathbb{N}$,*

$$\Delta_{\text{reward}}(n) = \min_{x \in \{0,1\}^n} \min_{P^* \in S} \left(\text{E}[\text{reward}((P, V)(x))] - \text{E}[\text{reward}((P^*, V)(x))] \right),$$

where the expectation is taken over the random coins of the verifier and the prover, and S is the set of all P^* such that $\text{Pr}[\text{output}((P^*, V)(x)) \neq f(x)] = 1$.

We emphasize that scaling the reward does not imply a real improvement in the reward gap. In order to have a robust notion we always work with a *normalized reward gap*, i.e., reward gap divided by the maximal value of the reward function. An alternative approach (taken for example in Azar and Micali [3]) that prevents the use of scaling to improve the reward gap might be to assume that the verifier has a fixed budget. We use the natural extension of reward gap to rational multi-prover interactive proofs.

Rational Arguments. *Rational arguments* were defined by Guo *et al.* [15] to capture the behavior of a rational prover that is computationally bounded. The definition of rational arguments allows negligible gains over the reward guaranteed by the prescribed behavior (but not more), since the rational prover might not follow the prescribed strategy, and it would try to solve the underlying hard problems (see item (b) in Definition 5).

Another important issue needed to be addressed in the computational setting is the cost of computing $f(x)$. As in the unbounded setting, it must rule out a prover that always gives some default (possibly incorrect) output, without performing any computation, while getting just slightly less than the expectation of the prescribed behavior. To address this shortcoming the definition of rational arguments "pins down" the profitability of deviation explicitly by appropriately adapting the notion of reward gap to the computationally bounded setting (see item (c) in Definition 5).

Definition 5 (Rational Argument). *A function $f : \{0,1\}^* \to \{0,1\}^*$ admits a rational argument with security parameter $\kappa : \mathbb{N} \to \mathbb{N}$ if there exists a protocol (P, V) and a randomized reward function* reward $: \{0,1\}^* \to \mathbb{R}_{\geq 0}$ *such that for any input $x \in \{0,1\}^*$ and any prover P^* of size $\leq \text{poly}(2^{\kappa(|x|)})$ the following hold:*

(a) $\text{Pr}[\text{output}((P, V)(x)) = f(x)] = 1$.
(b) *There exists a negligible function $\epsilon(\cdot)$ such that $\text{E}[\text{reward}((P, V)(x))] + \epsilon(|x|) \geq \text{E}[\text{reward}((P^*, V)(x))]$.*
(c) *If there exists a polynomial $p(\cdot)$ such that $\text{Pr}[\text{output}((P^*, V)(x)) \neq f(x)] \geq p(|x|)^{-1}$ then there exists a polynomial $q(\cdot)$ such that $\text{E}[\text{reward}((P^*, V)(x))] + q(|x|)^{-1} \leq \text{E}[\text{reward}((P, V)(x))]$.*

The expectations and the probabilities are taken over the random coins of the respective prover and verifier. We say that the rational argument is efficient if the running time of V is $o(|x|)$ for every $x \in \{0,1\}^*$.

3   Rational Sumcheck Protocols

Sumcheck protocols are an important building block in many classical interactive proofs. In particular, they play a crucial role in the IP = PSPACE theorem [22, 28]. Informally, a sumcheck protocol allows a verifier to efficiently check that a summation of evaluations of a polynomial of low degree on a given set of points is equal to a certain value (e.g. zero). In this section we show how to construct a rational sumcheck protocol that is sound (against a rational prover) even when applied on a polynomial of high degree. An important property of rational proofs is the *reward gap*, that captures the minimal loss in reward of the prover that always misreports the value of the function (formal definitions of rational proofs and reward gap are provided in Sect. 2). All of our rational proofs achieve noticeable reward gap.

Before describing our rational sumchecks, we show how to solve a simpler related problem: the verifier is given a bound M and n integers $x_1, \ldots, x_n \in \{0, \ldots, M - 1\}$, the verifier's goal is to learn the sum of $x_1, \ldots, x_n$. In the even more restricted case when $x_1, \ldots, x_n$ are bits (i.e., $M = 2$), one could solve this binary counting problem using an analogue of the rational proof of Azar and Micali [2]. In particular, the verifier can use a strictly proper scoring rule (e.g. the Brier's score [7]) to reward the quality of the prover's answer $y = \sum_{i=1}^{n} x_i$ as a prediction of the binary random variable b defined by outputting a uniformly random x_i. The intuition behind such protocol is that the Boolean random variable b encodes the information about the number of ones within $x_1, \ldots, x_n$; specifically, the probability of $b = 1$ is exactly the number of ones divided by n. Since the reward is defined according to a strictly proper scoring rule, a rational prover will uniquely maximize its expected reward by reporting the correct $y = \sum_{i=1}^{n} x_i$ (it describes the true distribution of b) as long as it is possible to efficiently sample b.

When $M > 2$, the mean of the random variable defined by outputting a uniformly random x_i still encodes the sum of $x_1, \ldots, x_n$. However, b is not necessarily Boolean and, unlike in the case when $x_1 \ldots, x_n$ are bits, the problem can no longer be solved by the protocol of Azar and Micali [2]. In order to use the Brier's score, it is necessary to appropriately modify the procedure of sampling b. Our more general protocol is given in Fig. 1. The verifier picks a random i from $\{1, \ldots, n\}$, and sets $b = 1$ with probability x_i/M and otherwise sets $b = 0$. After this normalization the probability of $b = 1$ is $\sum_{i=1}^{n} x_i/(nM)$ which still encodes the sum of $x_1, \ldots, x_n$, and since b is a Boolean variable it is possible to use the same reward function to incentivize any rational prover to report correct description of b. Therefore, the protocol in Fig. 1 is a non-interactive rational proof for the simplified problem of summation of n bounded non-negative values.

Lemma 1 (Rational Proof for Summation). *For any integer $M \geq 2$, let $f(x_1, \ldots, x_n) = \sum_{i=1}^{n} x_i$ be the function that computes the sum of any n-tuple of integers $x_1, \ldots, x_n \in \{0, \ldots, M - 1\}$. Then $f \in \mathrm{FRMA}[1, \log(nM), O(\mathrm{polylog}(nM))]$ with reward gap at least $\frac{1}{(nM)^2}$.*

On common input $x_1, \ldots, x_n \in \{0, \ldots, M-1\}$:

1. The prover sends integer $y = \sum_{i=1}^{n} x_i$ to the verifier.
2. The verifier samples i from $\{1, \ldots, n\}$, it sets $b = 1$ with probability $\frac{x_i}{M}$ (otherwise it sets $b = 0$) and outputs the reward for the prover in the following way

$$
R(y) = \begin{cases} 2\left(\frac{y}{nM}\right) - \left(\frac{y}{nM}\right)^2 - \left(1 - \frac{y}{nM}\right)^2 + 1, & \text{if } b = 1, \\ 2\left(1 - \frac{y}{nM}\right) - \left(\frac{y}{nM}\right)^2 - \left(1 - \frac{y}{nM}\right)^2 + 1, & \text{if } b = 0. \end{cases}
$$

Fig. 1. Rational proof for summation of n non-negative integers.

Proof. Consider the protocol in Fig. 1. The expected reward when prover sends y is

$$
\mathrm{E}[R(y)] = -2\left(\frac{y}{nM} - \frac{\sum_{i=1}^{n} x_i}{nM}\right)^2 + 2\left(\frac{\sum_{i=1}^{n} x_i}{nM}\right)^2 - 2\left(\frac{\sum_{i=1}^{n} x_i}{nM}\right) + 2,
$$

therefore the expected reward of the prover is uniquely maximized when $y = \sum_{i=1}^{n} x_i$.

For any integer $y^* \neq \sum_{i=1}^{n} x_i$,

$$
\mathrm{E}\left[R\left(\sum_{i=1}^{n} x_i\right)\right] - \mathrm{E}[R(y^*)] = 2\left(\frac{y^*}{nM} - \frac{\sum_{i=1}^{n} x_i}{nM}\right)^2 \geq \frac{2}{(nM)^2},
$$

where the equality holds when $y^* = \sum_{i=1}^{n} x_i \pm 1$. The reward function has maximal value 2, hence the (normalized) reward gap is $\frac{1}{(nM)^2}$. Because $y = \sum_{i=1}^{n} x_i \leq nM$, y can be represented using $\log(nM)$ bits which upper bounds the total communication. The verifier only needs to access a single x_i where i is chosen uniformly and randomly from $\{1, \ldots, n\}$. After accessing to x_i, the computation of the reward can be done in $O(\text{polylog}(nM))$ time. $\square$

Note that for any polynomially bounded M, the protocol in Fig. 1 achieves sublinear verification (the verifier only needs to access a single value) and noticeable reward gap. Moreover, based on the protocol in Fig. 1, we can construct an efficient rational proof for any problem which can be reduced to summation of several bounded values. For example, we immediately obtain a rational proof for addition of n elements over a finite field $\mathbb{Z}_p$ of prime characteristic p. Given $x_1, \ldots, x_n \in \mathbb{Z}_p$:

1. The prover sends to the verifier the sum $s = \sum_{i=1}^{n} x_i$ over $\mathbb{Z}$ (i.e., without performing the modulo operation) together with $y = (s \mod p)$, where s serves as the proof of correctness of y.
2. If $y \neq (s \mod p)$ then the verifier pays reward 0, and otherwise the verifier computes the reward for s as in the rational proof for summation of $x_1, \ldots, x_n$ with $M = p$ (as described in Fig. 1).

To deal with general summation over a finite field $\mathbb{F}_q$ of prime power characteristic $q = p^m$, we leverage the fact that the additive group of $\mathbb{F}_{p^m}$ is isomorphic to $(\mathbb{Z}_p, +_{\bmod p})^m$, where $+_{\bmod p}$ denotes addition over $\mathbb{Z}_p$. Thus, we can work with the representation of elements in $\mathbb{F}_{p^m}$ as vectors over $\mathbb{Z}_p^m$, i.e., we represent any $x \in \mathbb{F}_{p^m}$ as $(x^1, \ldots, x^m) \in \mathbb{Z}_p^m$. This allows us to get a rational proof for the function $\sum_{i=1}^n x_i$ that computes the sum of any n-tuple of elements $x_1, \ldots, x_n \in \mathbb{F}_{p^m}$ over $\mathbb{F}_{p^m}$ simply by applying the rational protocol for summation over $\mathbb{Z}_p$ on a randomly chosen coordinate of the vector representation $(y^1, \ldots, y^m) \in \mathbb{Z}_p^m$ of the output $y \in \mathbb{F}_{p^m}$ declared by the prover. The protocol is given in Fig. 2.

On common input $x_1, \ldots, x_n \in \mathbb{F}_{p^m}$:

1. The prover sends to the verifier a vector $(y^1, \ldots, y^m) \in \mathbb{Z}_p^m$ corresponding to $y = \sum_{i=1}^n x_i \in \mathbb{F}_{p^m}$ and a vector $s = (s^1, \ldots, s^m)$ such that $s^j = \sum_{i=1}^n x_i^j$ for all $j \in \{1, \ldots, m\}$.
2. The verifier checks if $y^j = (s^j \bmod p)$ for all $j \in \{1, \ldots, m\}$. If not the verifier pays 0, and otherwise the verifier samples $j \in \{1, \ldots, m\}$ and $i \in \{1, \ldots, n\}$, it sets $b = 1$ with probability $\frac{x_i^j}{p}$ and computes the reward for the prover as

$$
R(s^j) = \begin{cases} 2\left(\frac{s^j}{np}\right) - \left(\frac{s^j}{np}\right)^2 - \left(1 - \frac{s^j}{np}\right)^2 + 1, & \text{if } b = 1, \\[2mm] 2\left(1 - \frac{s^j}{np}\right) - \left(\frac{s^j}{np}\right)^2 - \left(1 - \frac{s^j}{np}\right)^2 + 1, & \text{if } b = 0. \end{cases}
$$

Fig. 2. Rational proof for summation of n elements over a finite field.

Corollary 1 (Rational Proof for Addition over Finite Fields). *For any integer $m \geq 1$ and any prime $p \in \mathbb{N}$. Let $f(x_1, \ldots, x_n) = \sum_{i=1}^n x_i$ be the function that computes the sum of any n-tuple of elements $x_1, \ldots, x_n \in \mathbb{F}_{p^m}$ over the field $\mathbb{F}_{p^m}$. Then $f \in \mathrm{FRMA}\left[1, \log(np^m), O(m \cdot \mathrm{polylog}(np))\right]$ with reward gap at least $\frac{1}{m(np)^2}$.*

Proof. Consider the protocol in Fig. 2. Let y and s denote the vectors sent by the prover when he tells the truth. It is easy to check the expected reward of the prover is maximized at y, s. When prover answers $\tilde{y} \neq y$ and $\tilde{s}$, if $\tilde{y} \neq (\tilde{s} \bmod p^m)$ then the prover gets reward 0, otherwise s and $\tilde{s}$ must differ in at least one entry and the expected reward of the prover is

$$
\mathrm{E}_j[R(\tilde{s}^j)] = \mathrm{E}_j\left[-2\left(\frac{\tilde{s}^j}{np} - \frac{s^j}{np}\right)^2 + 2\left(\frac{s^j}{np}\right)^2 - 2\left(\frac{s^j}{np}\right) + 2\right]
$$

$$
\leq \mathrm{E}_j\left[2\left(\frac{s^j}{np}\right)^2 - 2\left(\frac{s^j}{np}\right) + 2\right] - \frac{2}{m(np)^2}.
$$

Note the reward function has maximal value 2 therefore the reward gap is at least $\frac{1}{m(np)^2}$. □

Note that a sumcheck protocol is used to verify a sum of evaluations of a polynomial on a given set of points. Corollary 1 immediatelly gives rise to a *non-interactive* rational sumcheck protocol, where the verifier needs to evaluate the polynomial on a single point from the subset.

Corollary 2 (Rational Sumcheck Protocol). *For any finite field $\mathbb{F}$ and integer $m \geq 1$. Let $S \subseteq \mathbb{F}^m$ be a non-empty subset of $\mathbb{F}^m$. Let $\sum_{z \in S} f(z)$ be the function that sums evaluations of a given polynomial $f : \mathbb{F}^m \to \mathbb{F}$ (of arbitrary degree) on S. Then $f \in \text{FRMA}[1, \log(|S||\mathbb{F}|), O(t + \text{polylog}(|S||\mathbb{F}|))]$, where t is the time it takes to evaluate f on any $z \in \mathbb{F}^m$. The rational proof has reward gap at least $1/(\log(|\mathbb{F}|) \cdot (|S||\mathbb{F}|)^2)$.*

Proof. Using the protocol in Fig. 2 with field $\mathbb{F}$ and setting $n = |S|$, we obtain a rational proof for $\sum_{z \in S} f(z)$ with reward gap $\frac{1}{\log(|\mathbb{F}|) \cdot (|S||\mathbb{F}|)^2}$, verification time $O(t + \text{polylog}(|S||\mathbb{F}|))$, and communication $\log(|S||\mathbb{F}|)$ bits. □

4 Composition of Classical and Rational Interactive Proofs

In this section we investigate on the possibility of composition of classical interactive proofs with rational interactive proofs. In particular, we show a composition theorem for replacing oracle calls in a certain type of classical interactive proofs by a rational proof implementing the oracle. The composition is presented for both interactive proofs and δ-no-signaling multi-prover interactive proofs (for formal definition see Definition 3 in Sect. 2) resulting in their respective rational counterparts. The obtained rational proof has minimal loss in the reward gap that is proportional to the soundness of the classical interactive proof.

4.1 Substituting Oracle by Rational Proof in Interactive Proof

Let $f : \{0,1\}^* \to \{0,1\}^*$ be a function implicitly defining language $L_f = \{(x,y)|y = f(x)\}$. Let $\pi^g = (P_\pi, V_\pi^g)$ be an interactive proof for L_f where the verifier has oracle access to function $g : \{0,1\}^* \to \{0,1\}^*$. Let $\varphi = (P_\varphi, V_\varphi)$ be a rational interactive proof for g with reward function reward_φ. We denote by $\pi^\varphi = (P, V)$ with a reward function R the protocol between the prover P and verifier V given in Fig. 3. We define the reward in the resulting protocol as the average of the rewards obtained for each rational proof implementing an oracle query, though we note that this is not crucial for our results. The new reward function can be defined in other natural ways depending on the application.

We concentrate on a class of *query independent interactive proofs* in which the queries to the oracle can depend only on the input and the randomness of the verifier. Aditionally, once a query is submitted to the oracle the prover also recives the query.

On common input x:

1. The prover P sends $y = f(x)$ to the verifier.
2. The prover and the verifier initiate the interactive protocol π^g with P_π proving that $(x,y) \in L_f$.
3. Whenever V_π^g would issue a query q to g, the verifier V sends q to the prover P and they run φ on input q. P answers the query of V_π^g using the output of P_φ and continues with simulating π^g.
4. After V_π^g terminates, V outputs $(y, R(\tau))$, where τ is the transcript of communication between P and V and R is computed as:

$$R(\tau) = \begin{cases} R(\tau_\phi) & \text{if } V_\pi^g \text{ on } \tau \text{ accepts} \\ 0 & \text{otherwise.} \end{cases}$$

where τ_ϕ is the transcript corresponding to executions of φ in π^φ and $R(\tau_\phi)$ is the average of rewards computed by applying reward$_\varphi$ to the transcript of each execution of φ separately.

Fig. 3. Rational proof $\pi^\varphi = (P, V)$ resulting from interactive proof $\pi^g = (P_\pi, V_\pi^g)$ with oracle calls to g substituted by a rational proof $\varphi = (P_\varphi, V_\varphi)$.

Definition 6 (Query Independent Interactive Proofs). *Let* $f : \{0,1\}^* \to \{0,1\}^*$ *be a function and let* $\pi^g = (P_\pi, V_\pi^g)$ *be an interactive proof for* $L_f = \{(x,y) | y = f(x)\}$ *with* V_π^g *having oracle access to some function* $g : \{0,1\}^* \to \{0,1\}^*$. *We say that* π^g *is a* query independent interactive proof *if for any input* x *the following holds:*

1. *Only one query is issued by* V_π^g *to* g *and it depends only on the input* x *and on the randomness of* V_π^g.
2. *The query issued by* V_π^g *is send to* P_π *in the next round.*

Theorem 1 (Oracle Substitution in IP). *Let* $f : \{0,1\}^* \to \{0,1\}^*$ *be a function and let* $\pi^g = (P_\pi, V_\pi^g)$ *be a query independent interactive proof for* $L_f = \{(x,y) | y = f(x)\}$ *with* V_π^g *having oracle access to some function* $g : \{0,1\}^* \to \{0,1\}^*$. *If* π^g *has perfect completeness and soundness* s *then for any rational interactive proof* $\varphi = (P_\varphi, V_\varphi)$ *for* g *with reward gap* Δ, *the composed protocol* $\pi^\varphi = (P, V)$ *is a rational proof for* f *with reward gap* $\Delta(1 - s)$.

Proof. The reward in the rational protocol π^φ (defined in Fig. 3) is equal to the reward in the rational proof φ for evaluating the oracle query if the verifier accepts and zero otherwise. In order to show that π^φ is a rational proof with the claimed reward gap, we show that for every x the expectation of any prover P^* that reports $y' \neq f(x)$ (i.e., (x, y') is not in L_f) can be bound. To simplify the notation, we define three events that might happen during the execution of the protocol π^φ:

- E_0 corresponds to the event when V_π^g (simulated by V) accepts and P^* supplies *a correct answer* to the oracle query q (i.e., $(P^*, V_\pi^g)(x) = 1 \land$ output$(P^*, V_\varphi)(q) = g(q)$).
- E_1 corresponds to the event when V_π^g (simulated by V) accepts and P^* supplies *an incorrect answer* to the oracle query q (i.e., $(P^*, V_\pi^g)(x) = 1 \land$ output$(P^*, V_\varphi)(q) \neq g(q)$).
- E_2 corresponds to the event when V_π^g (simulated by V) rejects.

We can express the expectation of P^* as

$$\mathrm{E}[\mathsf{reward}(P^*, V)(x)] = \Pr[E_0] \cdot \mathrm{E}[\mathsf{reward}(P^*, V)(x)|E_0]$$
$$+ \Pr[E_1] \cdot \mathrm{E}[\mathsf{reward}(P^*, V)(x)|E_1] + \Pr[E_2] \cdot \mathrm{E}[\mathsf{reward}(P^*, V)(x)|E_2].$$

Since the expected reward is zero in case of event E_2 (the verifier V_π^g rejects), the above is equal to

$$\Pr[E_0] \cdot \mathrm{E}[\mathsf{reward}(P^*, V)(x)|E_0] + \Pr[E_1] \cdot \mathrm{E}[\mathsf{reward}(P^*, V)(x)|E_1].$$

We can bound $\Pr[E_1]$ by $1 - \Pr[E_0]$, so

$$\mathrm{E}[\mathsf{reward}(P^*, V)(x)] \leq \Pr[E_0] \cdot \mathrm{E}[\mathsf{reward}(P^*, V)(x)|E_0]$$
$$+ (1 - \Pr[E_0]) \cdot \mathrm{E}[\mathsf{reward}(P^*, V)(x)|E_1].$$

We use the following two claims to conclude the proof.

Claim 1. $\Pr[E_0] \leq s$.

Proof (of Claim 1). The interactive protocol with oracle access (P_π, V_π^g) is query independent in the sense of Definition 6, hence the prover in the composed protocol π^φ does not gain any additional information from the verifier's query to the oracle for g. It follows that in the case when the prover P^* supplies a correct answer to the oracle query the verifier accepts at most with the same probability as in the interactive proof with an oracle access, and the claim follows from the soundness of the interactive proof (P_π, V_π^g). □

Claim 2. $\mathrm{E}[\mathsf{reward}(P^*, V)(x)|E_1] \leq \mathrm{E}_q[\mathsf{reward}(P_\varphi, V_\varphi)(q)] - \Delta$.

Proof (of Claim 2). Assume that the claim does not hold, then the prover P^* achieves for some q a higher reward than $\mathrm{E}[\mathsf{reward}(P_\varphi, V_\varphi)(q)] - \Delta$. P^* can be used in the rational proof (P_φ, V_φ) for evaluating the oracle in order to achieve a higher reward than what is guaranteed by the reward gap of (P_φ, V_φ), since the oracle query is completely independent of the transcript. □

We use Claim 2 to bound the expectation as:

$$\mathrm{E}[\mathsf{reward}(P^*, V)(x)] \leq \Pr[E_0] \cdot \mathrm{E}[\mathsf{reward}(P^*, V)(x)|E_0]$$
$$+ (1 - \Pr[E_0]) \cdot (\mathrm{E}_q[\mathsf{reward}(P_\varphi, V_\phi)(q)] - \Delta) .$$

Notice that the expectation when event E_0 materializes is equal to $\mathrm{E}_q[\mathrm{reward}(P_\varphi, V_\phi)(q)]$, and hence we can rewrite the right side of the above inequality:

$$\mathrm{E}[\mathrm{reward}(P^*, V)(x)] \leq \Pr[E_0] \cdot \mathrm{E}_q[\mathrm{reward}(P_\varphi, V_\phi)(q)]$$
$$+ (1 - \Pr[E_0]) \cdot (\mathrm{E}_q[\mathrm{reward}(P_\varphi, V_\phi)(q)] - \Delta).$$

The distribution of oracle queries q is independent of the communication between the prover and the verifier and we can merge the expressions on the right side of the inequality.

$$\mathrm{E}[\mathrm{reward}(P^*, V)(x)] \leq \mathrm{E}_q[\mathrm{reward}(P_\varphi, V_\varphi)(q)] - (1 - \Pr[E_0]) \cdot \Delta,$$

Finally, by Claim 1:

$$\mathrm{E}[\mathrm{reward}(P^*, V)(x)] \leq \mathrm{E}_q[\mathrm{reward}(P_\varphi, V_\varphi)(q)] - (1 - s) \cdot \Delta .$$

By observing that for all x it holds that $\mathrm{E}_q[\mathrm{reward}(P_\varphi, V_\varphi)(q)] = \mathrm{E}[\mathrm{reward}(P, V)(x)]$ (since the distribution of queries produced by V is independent of x), we get the sought after bound on the reward gap of the resulting rational proof.   □

4.2 Substituting Oracle by Rational Multi-prover Proof in Multi-prover Proof

The composition theorem holds for oracle substitution also in the setting of δ-no-signaling multi-prover proofs. Given a k-prover interactive proof $\pi^g = (\overrightarrow{P_\pi}, V_\pi^g)$ for function f with an oracle access to a function g and a "rational" k'-prover implementation $\varphi = (\overrightarrow{P_\varphi}, V_\varphi)$ of the function g, a new rational protocol $\pi^\varphi = (\overrightarrow{P}, V)$ with $(k + k')$ provers can be obtained by executing the rational protocol φ instead of the oracle call with a new set of k' provers, as defined in Fig. 4. We define the reward in the resulting protocol analogously to the single prover setting and take the average of the rewards.

Similarly to the previous setting we require the oracle queries to depend only on the input and the randomness of the verifier and the queries to the provers to be independent of the answers of the oracle. Definition 6 of query independent interactive proofs naturally extends to multi-prover interactive proofs and we refer to multi-prover interactive proofs with this analogous property as *query independent*. Note that item 1 in Definition 6 is no longer required since we only deal with no-signaling strategies. In order to enable submission of all queries at once in the composed protocol, we must require independence of the queries to the provers from the oracle answers.

Definition 7 (Query Independent Multi-prover Proofs). *Let $f: \{0,1\}^* \rightarrow \{0,1\}^*$ be a function and let $\pi^g = (\overrightarrow{P_\pi}, V_\pi^g)$ be a multi-prover proof for $L_f = \{(x,y)|y = f(x)\}$ with V_π^g having oracle access to some function $g : \{0,1\}^* \rightarrow \{0,1\}^*$. We say that π^g is a query independent multi-prover proof if for any input x the following holds:*

On common input x:

1. The prover P_1 sends $y = f(x)$ to the verifier V.
2. The verifier V simulates the verifier V_π^g on input (x, y) to obtain k queries $\mathbf{q}$ for $\overrightarrow{P_\pi}$ and sends them to k provers in $\overrightarrow{P}$ (one query to each prover). The k provers answer their queries $\mathbf{q}$ using the corresponding outputs of $\overrightarrow{P_\pi}$.
3. For every query q^* that V_π^g issues to g, the verifier V simulates the verifier V_φ on input q^* to obtain k' queries $\omega(q^*)$ for $\overrightarrow{P_\varphi}$ and sends them to a new set of k' provers in $\overrightarrow{P}$ (one query to each prover). The k' provers answer their queries $\omega(q^*)$ using the corresponding outputs of $\overrightarrow{P_\varphi}$.
4. After V receives answers to all the queries from $\overrightarrow{P}$, the verifier V outputs $(y, R(\tau))$, where τ is the transcript of communication between $\overrightarrow{P}$ and V and R is computed as:

$$R(\tau) = \begin{cases} R(\tau_\varphi) & \text{if } V_\pi^g \text{ on } \tau \text{ accepts} \\ 0 & \text{otherwise.} \end{cases}$$

where τ_φ is the transcript corresponding to executions of φ within π^φ and $R(\tau_\varphi)$ is the average of rewards computed by applying the reward function reward$_\varphi$ to the transcript of each execution of φ separately.

Fig. 4. Rational multi-prover proof $\pi^\varphi = (\overrightarrow{P}, V)$ resulting from multi-prover proof $\pi^g = (\overrightarrow{P_\pi}, V_\pi^g)$ with oracle calls to g substituted by a (multi-prover) rational proof $\varphi = (\overrightarrow{P_\varphi}, V_\varphi)$ for evaluating g.

1. *Only a single query q is issued by V_π^g to g and it depends only on the input x and on the randomness of V_π^g.*
2. *The queries of V_π^g to $\overrightarrow{P_\pi}$ are independent of the oracle answer to the query q.*

We show that the composition theorem holds also for oracle substitution in the setting of query independent multi-prover proofs. Note that our composition theorem shows that when dealing with δ-no-signaling strategies, a loss in the reward gap proportional to δ is incurred in the resulting composed protocol.

Theorem 2 (Oracle Substitution in MIP). *Let $f : \{0,1\}^* \to \{0,1\}^*$ be a function and let $\pi^g = (\overrightarrow{P_\pi}, V_\pi^g)$ be a query independent k-prover MIP for $L_f = \{(x,y) | y = f(x)\}$ with V_π^g having oracle access to some function $g : \{0,1\}^* \to \{0,1\}^*$. If π^g has perfect completeness and soundness s against δ-statistically no-signaling strategies then for any rational k'-prover RMIP $\varphi = (\overrightarrow{P_\varphi}, V_\varphi)$ for evaluating g with reward gap Δ in presence of δ'-statistically no-signaling strategies, the composed protocol $\pi^\varphi = (\overrightarrow{P}, V)$ is a $(k + k')$-prover RMIP for evaluating f with reward gap $\Delta(1 - s - \delta'')$ against δ''-no-signaling strategies, where $\delta'' = \min\{\delta, \delta'\}$.*

Proof. The reward in the rational protocol π^φ (defined in Fig. 4) is equal to the reward in the rational proof φ for evaluating the oracle query if the verifier accepts and zero otherwise. In order to show that π^φ is a multi-prover rational proof for evaluating f with the claimed reward gap, we show that for every x the expectation of any set of provers $\overrightarrow{P^*}$ that report $y' \neq f(x)$ (i.e., (x, y') is not in L_f) can be bounded. To simplify the notation, we define three events that might happen during the course of the protocol π^φ:

- E_0 corresponds to the event when V_π^g (simulated by V) accepts and $\overrightarrow{P^*}$ supply *a correct answer* to the oracle query q^* (i.e., $(\overrightarrow{P^*}, V_\pi^g)(x) = 1 \wedge \mathsf{output}(\overrightarrow{P^*}, V_\varphi)(q^*) = g(q^*))$.
- E_1 corresponds to the event when V_π^g (simulated by V) accepts and $\overrightarrow{P^*}$ supply *an incorrect answer* to the oracle query q^* (i.e., $(\overrightarrow{P^*}, V_\pi^g)(x) = 1 \wedge \mathsf{output}(\overrightarrow{P^*}, V_\varphi)(q^*) \neq g(q^*))$.
- E_2 corresponds to the event when V_π^g (simulated by V) rejects.

We can express the expectation of $\overrightarrow{P^*}$,

$$\mathrm{E}[\mathsf{reward}(\overrightarrow{P^*}, V)(x)] = \Pr[E_0] \cdot \mathrm{E}[\mathsf{reward}(\overrightarrow{P^*}, V)(x)|E_0]$$
$$+ \Pr[E_1] \cdot \mathrm{E}[\mathsf{reward}(\overrightarrow{P^*}, V)(x)|E_1] + \Pr[E_2] \cdot \mathrm{E}[\mathsf{reward}(\overrightarrow{P^*}, V)(x)|E_2].$$

Since the expected reward in case of event E_2 is zero, the above is equal to

$$\Pr[E_0] \cdot \mathrm{E}[\mathsf{reward}(\overrightarrow{P^*}, V)(x)|E_0] + \Pr[E_1] \cdot \mathrm{E}[\mathsf{reward}(\overrightarrow{P^*}, V)(x)|E_1].$$

We can bound the $\Pr[E_1]$ by $1 - \Pr[E_0]$, so

$$\mathrm{E}[\mathsf{reward}(\overrightarrow{P^*}, V)(x)] \leq \Pr[E_0] \cdot \mathrm{E}[\mathsf{reward}(\overrightarrow{P^*}, V)(x)|E_0]$$
$$+ (1 - \Pr[E_0]) \cdot \mathrm{E}[\mathsf{reward}(\overrightarrow{P^*}, V)(x)|E_1].$$

We use the two following claims to complete the proof.

Claim 3. $\Pr[E_0] \leq s + \delta''$.

Proof (of Claim 3). For any q^*, an oracle query of V_π^g, define $\omega(q^*)$ to be the queries to $\overrightarrow{P_\varphi}$ generated by V_φ on input q^*. Let $\mathsf{A} = \{\mathsf{A}_{\mathbf{q}, \omega(q^*)}\}$ denote the δ''-no-signaling family of distributions, where $\mathsf{A}_{\mathbf{q}, \omega(q^*)}$ is the distribution of answers of $\overrightarrow{P^*}$ given queries $(\mathbf{q}, \omega(q^*))$. We fix an arbitrary set of queries $\mathbf{w}$ of V_φ to $\overrightarrow{P_\varphi}$ and consider the family of distributions $\mathsf{B} = \{\mathsf{B}_{\mathbf{q}}\}$, where $\mathsf{B}_{\mathbf{q}}$ is defined by sampling uniformly and randomly $(\mathbf{a}, \mathbf{z}) \leftarrow \mathsf{A}_{\mathbf{q}, \mathbf{w}}$ and outputting $\mathbf{a}$.

First, we show that B is δ-no-signaling. Let S be an arbitrary subset of $[k]$ and $\mathbf{q}, \mathbf{q}'$ be two arbitrary queries such that $\mathbf{q}_S = \mathbf{q}'_S$. Since the projections of $\mathsf{A}_{\mathbf{q}, \mathbf{w}}$ and $\mathsf{A}_{\mathbf{q}', \mathbf{w}}$ on the coordinates in S are δ''-close (by the fact that A is δ''-no-signaling), the statistical distance between $\mathsf{B}_{\mathbf{q}}$ and $\mathsf{B}_{\mathbf{q}'}$ when projected on S is

$$\frac{1}{2} \sum_\beta \left| \Pr_{\mathbf{a} \leftarrow \mathsf{B}_{\mathbf{q}}} [\mathbf{a}_S = \beta] - \Pr_{\mathbf{a}' \leftarrow \mathsf{B}_{\mathbf{q}'}} [\mathbf{a}'_S = \beta] \right| = \frac{1}{2} \sum_\beta \left| \Pr_{\mathbf{a} \leftarrow \mathsf{A}_{\mathbf{q}, \mathbf{w}}} [\mathbf{a}_S = \beta] - \Pr_{\mathbf{a}' \leftarrow \mathsf{A}_{\mathbf{q}', \mathbf{w}}} [\mathbf{a}'_S = \beta] \right|$$
$$\leq \delta''$$
$$\leq \delta,$$

where the last inequality follows from δ'' being defined as $\min\{\delta, \delta'\}$. Hence, B is δ-no-signaling.

Let $\overrightarrow{P_\pi^*}$ be the set of provers in π^g that follow the δ-no-signaling strategies B. By the soundness of π^g in the presence of δ-no-signaling strategies, $\Pr[(\overrightarrow{P_\pi^*}, V_\pi^g)(x) = 1] \leq s$. Assume that the claim does not hold, then

$$\delta'' < \Pr[E_0] - \Pr[(\overrightarrow{P_\pi^*}, V_\pi^g)(x) = 1]$$
$$= \Pr_{(\mathbf{a}, \mathbf{z}) \leftarrow A_{\mathbf{q}, \omega(q^*)}} [V_\pi^g(x, \mathbf{a}, \mathbf{z_1}) = 1 \wedge \mathbf{z_1} = g(q^*)] - \Pr_{(\mathbf{a}, \mathbf{z}) \leftarrow A_{\mathbf{q}, \mathbf{w}}} [V_\pi^g(x, \mathbf{a}, g(q^*)) = 1]$$

A contradiction to A being δ''-no-signaling. $\square$

Claim 4. For all x it holds that $\mathrm{E}[\mathrm{reward}(\overrightarrow{P^*}, V)(x)|E_1] \leq \mathrm{E}_{q^*}[\mathrm{reward}(\overrightarrow{P_\varphi}, V_\varphi)(q^*)] - \Delta$.

Proof (of Claim 4). Assume that the claim does not hold. By an averaging argument over the randomness of the verifier V for generating queries to the provers $\overrightarrow{P}$, there exists an x and a fixed choice of randomness for generating the queries such that

$$\mathrm{E}[\mathrm{reward}(\overrightarrow{P^*}, V)(x)|E_1, (\mathbf{q}, \omega(q^*))] > \mathrm{E}[\mathrm{reward}(\overrightarrow{P_\varphi}, V_\varphi)(q^*)] - \Delta \ ,$$

where $\mathbf{q}$ and q^* are fixed. Let $A_{\mathbf{q}, \omega(q^*)}$ denote the δ''-no-signaling distribution of answers of $\overrightarrow{P^*}$ to the queries $(\mathbf{q}, \omega(q^*))$. Consider the family of distributions $B = \{B_{\omega(q^*)}\}$, where $B_{\omega(q^*)}$ is defined by sampling uniformly and randomly $(\mathbf{a}, \mathbf{z}) \leftarrow A_{\mathbf{q}, \omega(q^*)}$ and outputting $\mathbf{z}$.

First, we show that B is δ'-no-signaling. Let S be an arbitrary subset of $[k']$ and $\mathbf{w}, \mathbf{w}'$ be two sets of queries such that $\mathbf{w}_S = \mathbf{w}'_S$. Since the projections of $A_{\mathbf{q}, \mathbf{w}}$ and $A_{\mathbf{q}, \mathbf{w}'}$ on the coordinates $S' = \{k+i : i \in S\}$ are δ''-close, the statistical distance between $B_{\mathbf{w}}$ and $B_{\mathbf{w}'}$ is

$$\frac{1}{2} \sum_\beta \left| \Pr_{\mathbf{z} \leftarrow B_{\mathbf{w}}} [\mathbf{z}_S = \beta] - \Pr_{\mathbf{z}' \leftarrow B_{\mathbf{w}'}} [\mathbf{z}'_S = \beta] \right| = \frac{1}{2} \sum_\beta \left| \Pr_{\mathbf{z} \leftarrow A_{\mathbf{q}, \mathbf{w}}} [\mathbf{z}_S = \beta] - \Pr_{\mathbf{z}' \leftarrow A_{\mathbf{q}, \mathbf{w}'}} [\mathbf{z}'_S = \beta] \right|$$
$$\leq \delta''$$
$$\leq \delta' \ .$$

where the last inequality follows from δ'' being defined as $\min\{\delta, \delta'\}$, and hence B is δ'-no-signaling.

Let $\overrightarrow{P_\varphi^*}$ behave according to B, then on input q^*,

$$\mathrm{E}[\mathrm{reward}(\overrightarrow{P_\varphi^*}, V_\varphi)(q^*)] = \mathrm{E}[\mathrm{reward}(\overrightarrow{P^*}, V)(x)|E_1, (\mathbf{q}, \omega(q^*))]$$
$$> \mathrm{E}[\mathrm{reward}(\overrightarrow{P_\varphi}, V_\varphi)(q^*)] - \Delta \ .$$

Therefore, $\overrightarrow{P_\varphi^*}$ is a set of δ'-no-signaling provers that break the reward gap guarantee of φ, a contradiction. $\square$

We use Claim 4 to bound the expectation as:

$$\mathrm{E}[\mathrm{reward}(\overrightarrow{P^*}, V)(x)] \leq \Pr[E_0] \cdot \mathrm{E}[\mathrm{reward}(\overrightarrow{P^*}, V)(x)|E_0]$$
$$+ (1 - \Pr[E_0]) \cdot (\mathrm{E}_{q^*}[\mathrm{reward}(\overrightarrow{P_\varphi}, V_\varphi)(q^*)] - \Delta) .$$

Notice that due to the query independence of the protocol π^g the expectation when event E_0 materializes is equal to $\mathrm{E}_{q^*}[\mathrm{reward}(\overrightarrow{P_\varphi}, V_\varphi)(q^*)]$. Hence, we can rewrite the right side of the above inequality as

$$\Pr[E_0] \cdot \mathrm{E}_{q^*}[\mathrm{reward}(\overrightarrow{P_\varphi}, V_\varphi)(q^*)] + (1 - \Pr[E_0]) \cdot (\mathrm{E}_{q^*}[\mathrm{reward}(\overrightarrow{P_\varphi}, V_\varphi)(q^*)] - \Delta)$$
$$= \mathrm{E}_{q^*}[\mathrm{reward}(\overrightarrow{P_\varphi}, V_\varphi)(q^*)] - (1 - \Pr[E_0]) \cdot \Delta .$$

Finally, by Claim 3:

$$\mathrm{E}_{q^*}[\mathrm{reward}(\overrightarrow{P^*}, V)(q^*)] \leq \mathrm{E}_{q^*}[\mathrm{reward}(\overrightarrow{P_\varphi}, V_\varphi)(q^*)] - (1 - s - \delta'') \cdot \Delta .$$

Therefore, we get the sought after bound on the reward gap of the multi-prover rational proof π^φ resulting from the composition of π^g and φ.   □

5 Rational Delegation for NC

The work of Guo et al. [15] showed how to efficiently delegate computation performed by low-depth circuits in the rational setting, and in particular constructed a rational proof with noticeable reward gap for any language in NC^1. However, the reward gap in their construction is proportional to the depth of the evaluated circuit (the reward is scaled proportionally to the depth) and this prevents to use their rational proof with meaningful (noticeable) reward gap beyond the class NC^1. In this section we give a rational proof with sublinear verification time for any function computable by log-space uniform NC by composing the rational sumcheck protocol from Sect. 3 with the classical protocol of Goldwasser et al. [14].

5.1 The Protocol of Goldwasser, Kalai and Rothblum [15]

In their work Goldwasser et al. [14] gave a protocol that allows to delegate computation of any function computable by log-space uniform circuits via an interactive proof with a polynomial prover and a quasi-linear verifier. In particular, they showed the following theorem:

Theorem 3 (Theorem 1.1.1. in [26]). *Let L be a language computable by a family of $O\left(\log\left(S(n)\right)\right)$-space uniform boolean circuits of size $S(n)$ and depth $d(n)$. L has an interactive proof where:*

1. *The prover runs in time $\mathrm{poly}(S(n))$. The verifier runs in time $n \cdot \mathrm{poly}(d(n), \log(S(n)))$ and space $O(\log(S(n)))$. Moreover, if the verifier is given oracle access to the low degree extension of its input, then its running time is only $\mathrm{poly}(d(n), \log(S(n)))$.*

2. *The protocol has perfect completeness and soundness 1/2.*
3. *The protocol is public-coin, with communication complexity $d(n)$ · polylog($S(n)$).*
4. *Each message of the prover depends only on $O(\log(n))$ random bits sent by the verifier.*

Their interactive proof builds on arithmetization techniques and employs efficient sumcheck protocols in order to establish correctness of the output. The sumcheck is run on multivariate polynomials of low degree that encode the values of intermediate layers of computation to allow the verifier to efficiently check consistency of the prover's answers.

Let $w = (w_1, \ldots, w_k)$ be k bits. The vector w defines a function $W : \{1, \ldots, k\} \to \{0, 1\}$ such that $W(i) = w_i$ for all $i \in \{1, \ldots, n\}$. Let $\mathbb{H}$ be an extension field of $\mathbb{GF}[2]$, m be an integer such that $k \leq |\mathbb{H}|^m$, and let $\mathbb{F}$ be an extension field of $\mathbb{H}$. The *low degree extension* of w is the unique m-variate polynomial $\tilde{W} : \mathbb{F}^m \to \mathbb{F}$ of degree at most $|\mathbb{H}| - 1$ in each variable that agrees with W on $\mathbb{H}^m$. It is a useful fact that the low degree extension can be expressed as sum over $\mathbb{H}^m$, where each term is efficiently computable (for the details see Appendix A).

Here we provide a high-level overview of the protocol (for the full exposition see e.g. [26]):

1. The prover P evaluates the circuit C on input x received from the verifier V, and computes a low degree extension $\widetilde{W_i}$ for every layer i of the circuit C.
2. For $1 \leq i \leq d$, in each phase i the prover initiates an interactive sumcheck protocol to convince the verifier that $\widetilde{W_{i-1}}(z_{i-1}) = r_{i-1}$. In the first phase $z_0 = (0, \ldots, 0)$ and $r_0 = (C(x), 0, \ldots, 0)$. To complete the i-th sumcheck protocol the verifier would need to evaluate $\widetilde{W_i}$ on two random points ω_1, ω_2, but to avoid the related computational burden this task is reduced to another sumcheck performed in phase $i + 1$. In particular, the prover and the verifier run an interactive procedure using ω_1, ω_2, the verifier picks a random z_i and the prover reports a corresponding value $r_i = \widetilde{W_i}(z_i)$. The protocol proceeds to phase $i + 1$.
3. In phase $d + 1$ the verifier evaluates the low degree extension $\widetilde{W_d}$ (of the input x) on the random point z_d and checks that it is equal to r_d reported by the prover. This is the final phase and the only point at which the verifier evaluates a low degree extension.

The running time of the verifier in the first d phases is poly($d(n), \log(S(n))$), and it is the evaluation of the low degree extension of the input in the last step that induces the overall quasi-linear overhead of $n \cdot \text{poly}(d(n), \log(S(n)))$ for the verifier. Hence, given oracle access to the low degree extension of the input the verification can be performed in sublinear time. Moreover, the protocol of Goldwasser *et al.* [14] is query independent in the sense of Definition 6, i.e., after receiving the answer to its query the verifier can send the query (the random point z_d) to the prover and the soundness is preserved. This allows us to use our composition framework from Sect. 4 in order to substitute the oracle call with our rational sumcheck protocol.

Substituting the Low Degree Extension Oracle with a Rational Proof. First, we show that our rational sumcheck protocol from Sect. 3 can evaluate an arbitrary low degree extension.

Proposition 1 (Rational Protocol for Evaluating Low Degree Extension). *The low degree extension* $\tilde{W} : \mathbb{F}^m \to \mathbb{F}$ *of* $(w_1, \ldots, w_k) \in \{0,1\}^k$ *admits a rational proof with verification time* $\mathrm{poly}(|\mathbb{H}|, m)$, *assuming oracle access to* $(w_1, \ldots, w_k)$, *with reward gap* $1/4(\log|\mathbb{F}|)|\mathbb{H}^m|^2$.

Proof. By Proposition 2 (given in Appendix A), for any $z \in \mathbb{F}^m$, $\tilde{W}(z)$ is a summation of $|\mathbb{H}|^m$ terms of the form $\sum_{p \in \mathbb{H}^m} \tilde{\beta}(z, p) \cdot W(p)$, where the addition is over $\mathbb{F}$ and $\mathbb{F}$ is a extension field of $\mathrm{GF}[2]$. Moreover, for every (z, p), $\tilde{\beta}(z, p)$ can be computed in time $\mathrm{poly}(|\mathbb{H}|, m)$, therefore $\tilde{\beta}(z, p) \cdot W(p)$ can be computed in time $\mathrm{poly}(|\mathbb{H}|, m)$. By Corollary 2, $\tilde{W}(z)$ admits rational proof with reward gap $1/((\log|\mathbb{F}|)(2|\mathbb{H}^m|)^2) = 1/4(\log|\mathbb{F}|)|\mathbb{H}^m|^2$ and verification time $\mathrm{poly}(|\mathbb{H}|, m)$. □

Finally, we use the above efficient rational proof in the protocol of Goldwasser *et al.* [14] to allow the verifier to avoid reading the whole input when evaluating the low degree extension of the input.

Theorem 4 (Rational Interactive Proof for NC). *For any function* $f : \{0,1\}^* \to \{0,1\}$, *if* $L_f = \{(x,y)|y = f(x)\}$ *is computable by a family of* $O(\log(S(n)))$-*space uniform Boolean circuits of size* $S(n)$ *and depth* $d(n) = O(\mathrm{polylog}(n))$ *then* $f \in \mathrm{FRMA}[d(n) \cdot \mathrm{polylog}(n), d(n) \cdot \mathrm{polylog}(S(n)), \mathrm{poly}(d(n), \log(S(n)))]$ *with a public-coin rational interactive proof with a noticeable reward gap, where the prover runs in time* $\mathrm{poly}(S(n))$ *and the verifier runs in space* $O(\log(S(n)))$.

Proof. For $f \in NC$, we let $\pi^g = (P_\pi, V_\pi^g)$ be the interactive proof for $L_f = \{(x,y)|y = f(x)\}$ defined in Theorem 3 where g is the low degree extension of x with $|\mathbb{F}| = \mathrm{poly}(n, d)$ and $|\mathbb{H}^m| = \mathrm{poly}(n)$, the soundness is $1/2$ and the completeness is 1. Let $\varphi = (P_\varphi, V_\varphi)$ be the rational proof for g as defined in Proposition 1 with reward gap $\Delta = 1/(4\log|F|)(|\mathbb{H}|^{2m})$. Note that V_π^g only issues a single query and for all x the communication between P_π and V_π^g is independent of $(q, g(q))$. By Theorem 1, π^φ is a rational proof for f with regard gap $\Delta(1 - s) = \Delta/2 = 1/\mathrm{poly}(n)$.

The running time of the prover or verifier is at most the sum of the running time of P_π in Theorem 3 and the running time of P_φ. The total running time is $\mathrm{poly}(S(n))$. The verifier runs in at most V_π^g and the running time of V_φ. Therefore the running time of verifier is upper bounded by $\mathrm{poly}(d(n), \log(S(n)))$. The total communication is the communication of φ and the communication of π which is upper bounded by $d(n) \cdot \mathrm{poly}(S(n))$. □

5.2   Single-Round Rational Arguments for NC

Guo *et al.* [15] gave an efficient transformation from any rational proof with . noticeable reward gap to single-round rational argument. The transformation uses an efficient Private Information Retrieval (PIR) scheme (for formal definition see the full version) in order to submit all the round queries to the prover at once.

Theorem 5 (Theorem 6 in [15]). *Let* $f : \{0,1\}^n \to \{0,1\}$ *be a function in* FRMA $[r, C, T]$. *Assume the existence of a PIR scheme with communication complexity* $\text{poly}(\kappa)$ *and receiver work* $\text{poly}(\kappa)$, *where* $\kappa \geq \max\{C(n), \log n\}$ *is the security parameter. If* f *has an admissible rational proof with noticeable reward gap* Δ, *then* f *admits single-round rational argument which has the following properties:*

(a) The verifier runs in time $C(n) \cdot \text{poly}(\kappa) + O(T(n))$.
(b) The communication complexity is $r \cdot \text{poly}(\kappa, \lambda)$ *where* λ *is the longest message sent by the prover.*

By applying the above transformation of Guo *et al.* [15] on the rational interactive proofs in Theorem 4, we obtain single-round rational arguments for NC with sublinear verification.

Corollary 3 (Rational Argument for NC). *Let* $f : \{0,1\}^n \to \{0,1\}$ *be a function computable by log-space uniform* NC *of size* $S(n) = \text{poly}(n)$ *and depth* $d(n) = O(\text{polylog}(n))$. *Assume the existence of a PIR scheme with communication complexity* $\text{poly}(\kappa)$ *and receiver work* $\text{poly}(\kappa)$, *where* $\kappa \geq d(n) \cdot \text{polylog}(S(n))$ *is the security parameter. Then* f *admits single-round efficient rational argument which has the following properties:*

1. *The verifier runs in* $\text{poly}(\kappa, d(n), \log(S(n)))$ *and the prover runs in* $\text{poly}(\kappa, S(n))$.
2. *The length of the prover's message and the verifier's challenge is* $d(n) \cdot \text{poly}(\kappa, \log(S(n)))$. *The verifier's challenge depends only on his random coins and is independent of the input* x.

6   Rational Delegation for P

Recently, Kalai *et al.* [18] gave a single-round delegation scheme for every language computable in time $t(n)$, where the running time of the verifier is $n \cdot \text{polylog}(t(n))$. For languages in P where $t(n) = \text{poly}(n)$, the verification time is $O(n \cdot \text{polylog}(n))$. The efficiency bottleneck for achieving sublinear verification for P in Kalai *et al.* [18] (similarly to the protocol for NC of Goldwasser *et al.* [14]) is that the verifier needs to evaluate a low degree extension of the input which takes quasi-linear time. We show that it is possible to improve the verification time to be sublinear in the rational setting.

6.1   The Protocol of Kalai, Raz and Rothblum [19]

Recently, Kalai *et al.* [18] gave an MIP secure against no-signaling provers for any deterministic computation.

Theorem 6 (Theorem 4 in [18]). *Suppose that* $L \in \mathrm{DTIME}(t(n))$, *where* $t = t(n)$ *satisfies* $\mathrm{poly}(n) \leq t \leq \exp(n)$. *Then, for any integer* $(\log t)^c \leq k \leq \mathrm{poly}(n)$, *where* c *is some (sufficiently large) universal constant, there exists an MIP for* L *with* $k \cdot \mathrm{polylog}(t)$ *provers where:*

1. *The verifier runs in time* $n \cdot k^2 \cdot \mathrm{polylog}(t)$ *and the provers run in time* $\mathrm{poly}(t, k)$. *Moreover, if the verifier is given oracle access to the low degree extension of its input, then its running time is only* $t' \cdot k^2 \cdot \mathrm{polylog}(t)$, *where* t' *is the cost of the oracle access.*
2. *The protocol has perfect completeness and soundness* 2^{-k} *against* $2^{-k \cdot \mathrm{polylog}(t)}$-*no-signaling strategies.*
3. *Each query and answer is of length* $k \cdot \mathrm{polylog}(t)$.

Here we give a high level overview of the MIP construction of Kalai *et al.* [18]. It is obtained in three steps:

1. *No-signaling PCP with Oracle.* They first construct a Probabilisticaly Checkable Proof (PCP) with oracle access to a function which makes at most k queries and is secure against no-signaling provers. The construction of the PCP is the most technical part of their work and we refer to Kalai *et al.* [18] for the construction and analysis of this PCP. The total number of oracle queries is at most $k \cdot \mathrm{polylog}(t)$ and the running time of the verifier is $k \cdot \mathrm{polylog}(n)$.
2. *No-signaling MIP with Oracle.* Based on the PCP, they construct in a straightforward way an MIP with $k_{\max} \leq k \cdot \mathrm{polylog}(t)$ provers secure against no-signaling strategies given oracle access to the same function as for the PCP. In this MIP, the verifier simulates the PCP verifier, and the i-th prover prepares the PCP proof and answers the i-th query according to the PCP. The running time of the verifier is $O(k \cdot \mathrm{polylog}(t))$.
3. *No-signaling MIP without Oracle.* In order to remove the oracle, they employ an MIP for the oracle which is secure against no-signaling provers. They replace the number of queries to the oracle one by one, each time reducing one query to the oracle by letting the verifier run the MIP for the oracle with additional provers. At the end, they obtain an MIP without oracle access which is secure against no-signaling provers. To construct the MIP for the oracle, they observe that any interactive proof gives rise to an MIP secure against no-signaling provers by sending the first i messages to the i-th prover and letting the i-th prover answer the message in the i-th round. Observed that the oracle is computable by linear space, we have IP for this oracle so that we can obtain an MIP against no-signaling provers. The running time of the verifier is $k \cdot \mathrm{polylog}(t) + n \cdot k^2 \cdot \mathrm{polylog}(t)$. Moreover, if the verifier can compute the low degree extension of the input in time t', then the running time can be further improved into $k \cdot \mathrm{polylog}(n) + t' \cdot k^2 \cdot \mathrm{polylog}(t)$.

Note that for languages in P where $t = \text{poly}(n)$, we can let $k = \text{polylog}(t)$ so that the verifier runs in time $n \cdot \text{polylog}(n)$. Moreover, the running time can be improved to $t' \cdot \text{polylog}(n)$ when the verifier is given oracle access to evaluate the low degree extension and t' is the cost of the oracle access. Therefore, the task of constructing delegation scheme for P with sublinear verification can be reduced to constructing a delegation scheme for low degree extension with sublinear verification.

6.2   No-Signaling Rational Multi-prover Proofs for Deterministic Computations

In this section, we present our RMIPs for deterministic computations which are secure against no-signaling provers. Recall from the previous section that the efficiency bottleneck for achieving sublinear verification for P is that the evaluation of low degree extension runs in quasi-linear time. To overcome the efficiency bottleneck we combine the no-signaling MIP of Kalai et al. [18] with our sublinear rational proofs for evaluating the low degree extension of the input (Proposition 1). Unlike the oracle simulation mentioned in the third step of the work of Kalai et al. [18], we reduce all queries to the low degree extension oracle at once and only increase the number of provers by 1. To do this, we view the queries to the oracle as a single query consisting of many points to a larger oracle that evaluates the low degree extension of inputs on all the points and returns the answers at once.

For a function $g: \mathbb{F}^n \to \mathbb{F}$, we let $g^l: (\mathbb{F}^n)^l \to (\mathbb{F})^l$ be the function that on any l-tuple $(x_1, \ldots, x_l) \in (\mathbb{F}^n)^l$ outputs $(g(x_1), \ldots, g(x_l))$. For a rational proof $\varphi = (V_\varphi, P_\varphi)$ for g with input $x \in \mathbb{F}^n$, we define another rational proof $\varphi^l = (V_{\varphi^l}, P_{\varphi^l})$ for g^l with input $(x_1, \ldots, x_l) \in (\mathbb{F}^n)^l$, where the verifier V_{φ^l} simulates V_φ on x_i for all $i \in \{1, \ldots, l\}$ and pays the average reward outputted by V_ϕ on the l inputs and P_{φ^l} simulates P_φ on x_i for all $i \in \{1, \ldots, l\}$. It is easy to see that if g admits rational proof φ with reward gap Δ, then g^l admits a rational proof φ^l with reward gap Δ/l.

Theorem 7. *Suppose that $f : \{0,1\}^n \to \{0,1\}$ is a function computable by deterministic Turing machine in time $t(n)$, where $t = t(n)$ satisfies $\text{poly}(n) \leq t \leq \exp(n)$. Then, for any integer $(\log t)^c \leq k \leq \text{poly}(n)$, where c is some (sufficiently large) universal constant, there exists an RMIP for f with $k \cdot \text{polylog}(t) + 1$ provers where:*

1. *The provers run in time $\text{poly}(t, k)$ and the verifier runs in time $k^2 \cdot \text{polylog}(t)$.*
2. *The protocol has reward gap $1/k \cdot \text{poly}(\log(t), n)$ against $2^{-k \cdot \text{polylog}(t)}$-no-signaling strategies.*
3. *Each query and answer is of length $k \cdot \text{polylog}(t)$.*

Proof. Let $\pi^g = (\overrightarrow{P_\pi}, V_\pi^g)$ be the MIP for $L_f = \{(x, y) | y = f(x)\}$ from Theorem 6, which has soundness $s = 2^{-k}$ against $\delta = 2^{-k \cdot \text{polylog}(t)}$-no-signaling strategies and perfect completeness, where $g: \mathbb{F}^m \to \mathbb{F}$ is the low degree extension of inputs with parameters $\mathbb{F}, \mathbb{H}, m$ such that $|\mathbb{H}| \leq |\mathbb{F}| \leq \text{polylog}(t), |\mathbb{H}^m| = \text{poly}(n)$.

As noted in [18], the total number of the queries to g is $l \leq k \cdot \text{polylog}(t)$. We consider $\pi^{g^l} = (\overrightarrow{P_\pi}, V_\pi^{g^l})$ where $V_\pi^{g^l}$ behaves exactly as V^g except that V^{g^l} only makes a single query which consists all the queries of V^g to the oracle for g. Because the queries made by V_π^g are independent of each other, it is possible to query them at once and conclude that π^{g^l} is also an MIP for L_f with the same guarantee.

By Proposition 1, g admits a rational proof $\varphi = (P_\varphi, V_\varphi)$ with reward gap $\Delta = 1/(4 \log |\mathbb{F}|)(|\mathbb{H}^m|^2)$. Therefore g^l admits a rational proof φ^l with reward gap $\Delta' = \Delta/l$. Note that φ^l is also an RMIP with reward gap Δ in presence of $\delta' = 1$-no-signaling strategies.

Note that $V_\varphi^{g^l}$ only issues a single query and for all x the communication between $\overrightarrow{P_\pi}$ and $V_\pi^{g^l}$ is independent of $(q, g(q))$. By Theorem 2, π^{φ^l} is an RMIP for f with reward gap $\Delta'(1 - s - \min(\delta, \delta')) = \Omega(\Delta/l) = 1/k \cdot \text{poly}(\log(t), n)$, in presence of $\delta'' = \delta$-no-signaling strategies.

The running time of the prover is at most the sum of the running time of $\overrightarrow{P_\pi}$ in Theorem 6 and the running time of P_{φ^m} which is upper bounded by $\text{poly}(t, k)$. The verifier runs in at most $t' \cdot k^2 \cdot \text{polylog}(t)$ where the t' is the running time of V_φ upper bounded by $\text{poly}(\mathbb{H}, m) \leq \text{polylog}(t)$. Therefore the running time of verifier is $k^2 \cdot \text{polylog}(t)$. The maximal length of queries and answers in π^{g^l} is $k \cdot \text{polylog}(t)$ by Theorem 6, and the maximal length of queries and answers in π^{g^l} is $(m \log \mathbb{F}) \cdot l \leq k \cdot \text{polylog}(t)$. Therefore the maximal length of queries and answers in π^{g^l} is bounded by $k \cdot \text{polylog}(t)$.   □

6.3   Single-Round Rational Arguments for P

We show how to transform any RMIP secure against no-signaling provers into a single-round rational argument using a sub-exponentially secure Fully Homomorphic Encryption (see Definition 8), and as a result obtain a single-round rational argument with sublinear verification for any language in P. For that we extend the transformation of Guo et al. [15] to the multi-prover setting.

Theorem 8. *Let $f : \{0,1\}^n \rightarrow \{0,1\}$ be a function in $\text{FRMIP}[k, \delta, C, T]$. Assume f has a RMIP with noticeable reward gap Δ and negligible no-signaling parameter δ, and let λ denote the length of the longest message sent by the verifier. If there exists a secure FHE scheme, where $\kappa \geq \max\{\text{polylog}(n), \lambda, C\}$ is the security parameter, then f admits single-round rational argument which has the following properties:*

1. *The verifier runs in time $\text{poly}(\kappa) + O(T(n))$.*
2. *The prover runs in time $\text{poly}\left(\kappa, n, T_{\overrightarrow{P_{MIP}}}\right)$, where $T_{\overrightarrow{P_{MIP}}}$ is the sum of the running times of the provers in the RMIP.*
3. *The length of prover's message and the verifier's challenge is $\ell \cdot \text{poly}(\kappa)$.*

The proof of Theorem 8 follows by the following lemma (due to space restrictions, we provide the proof of Lemma 2 in the full version).

Lemma 2. *Let* $(\overrightarrow{P_{MIP}}, V_{MIP})$ *be a* δ-*no signaling* RMIP *protocol for a function* f *with* ℓ *provers. Let* λ *be the longest query size and* C *be the answer size. Let* reward$(\cdot)$ *and* Δ *be the reward function and the corresponding reward gap. Assume the existence of a* (Z, δ')-*secure FHE with correctness* $1 - \gamma$ *(where* γ *is some negligible function), and let* $\gamma_0 = \gamma \cdot \ell$. *If* $\delta' \leq \delta/\ell$ *and the security parameter* $\kappa = \kappa(n) \geq \max\{\text{poly}\log(n), \lambda, C\}$ *and* $Z = Z(\kappa) \geq \kappa$ *such that* $Z \geq \max\{n, 2^{\ell \cdot C}\}$, *then there exists a one-round protocol* (P_A, V_A) *with the following properties:*

(a) $\Pr[\text{output}((P_A, V_A)(x)) = f(x)] = 1$.
(b) $\text{E}[\text{reward}((P_A, V_A)(x))] \geq \text{E}[\text{reward}((\overrightarrow{P_{MIP}}, V_{MIP})(x))] \cdot (1 - \gamma_0))$.
(c) *The length of* P_A's *message and the* V_A's *challenge is* $\ell \cdot \text{poly}(\kappa)$.
(d) *The verifier* V_A *runs in time* $\text{poly}(\kappa) + O(T_{V_{MIP}})$, *where* $T_{V_{MIP}}$ *is the running time of* V_{MIP}.
(e) *The prover* P_A *runs in time* $\text{poly}\left(\kappa, n, T_{\overrightarrow{P_{MIP}}}\right)$, *where* $T_{\overrightarrow{P_{MIP}}}$ *is the sum of the running times of the provers in* $\overrightarrow{P_{MIP}}$.
(f) *For any prover* P^* *of size* $\leq \text{poly}(Z(\kappa))$ *that achieves*

$$\text{E}[\text{reward}((P^*, V_A)(x))] = \text{E}[\text{reward}((P_A, V_A)(x))] + \delta^* ,$$

let $\mu = \Pr[\text{output}((P^*, V_A)(x)) \neq f(x)]$. *It holds that*
(a) *(Utility gain)* $\delta^* \leq \gamma_0$, *and*
(b) *(Utility loss)* $(-\delta^*) \geq \mu\Delta - \gamma_0$.

From Interactive Rational Proofs to Rational Arguments. Let $(\overrightarrow{P_{MIP}}, V_{MIP})$ be a δ-no-signaling rational MIP with ℓ provers $P^1_{MIP}, \ldots, P^\ell_{MIP}$ for evaluating some function f, as in the statement of the Lemma 2. Recall that λ denotes length of the longest message sent by V_{MIP} in $(\overrightarrow{P_{MIP}}, V_{MIP})$. For simplicity of exposition (and without loss of generality) we assume that the first prover P^1_{MIP} sends $f(x)$, and all queries are of size exactly λ.

Fix any security parameter $\kappa \geq \max\{\text{polylog}(n), \lambda, C\}$ and let (Gen, Enc, Eval, Dec) be a (Z, δ')-secure FHE scheme, with respect to security parameter κ. The one-round rational argument (P_A, V_A) is constructed as follows:

1. On common input $x \in \{0, 1\}^n$, the verifier V_A proceeds as follows:
   (a) Emulate the verifier V_{MIP} and obtain queries $m_1, \ldots, m_\ell \in \{0, 1\}^\lambda$ to be sent by V_{MIP}.[2]
   (b) Compute key-pairs $(pk_i, sk_i) \leftarrow \text{Gen}(1^\kappa)$ and encryptions $q_i \leftarrow \text{Enc}(pk_i, m_i)$ for $1 \leq i \leq \ell$.
       Send $pk = (pk_1, \ldots, pk_\ell)$ and $q = (q_1, \ldots, q_\ell)$ to P_A.
2. Upon receiving keys $pk = (pk_1, \ldots, pk_\ell)$ and queries $q = (q_1, \ldots, q_\ell)$ from V_A, the prover P_A operates as follows:

---

[2] These queries can be computed in advance since in the protocol $(\overrightarrow{P_{MIP}}, V_{MIP})$ all the messages sent by V_{MIP} depend only on V_{MIP}'s random coin tosses.

(a) Emulate provers $\overrightarrow{P_{MIP}}$ to obtain $f(x)$.

(b) For each $1 \leq i \leq \ell$, compute $P_{x,i}$, a Boolean circuit that on input query m computes the function $P^i_{MIP}(x, m)$.

(c) For each $1 \leq i \leq \ell$, compute $a_i \leftarrow \mathsf{Eval}(pk_i, P_{x,i}, q_i)$ and send the message $(f(x), a_1, \ldots, a_\ell)$ to V_A.

3. Upon receiving the message $(f(x), a_1, \ldots, a_\ell)$ from P_A, the verifier V_A operates as follows:

(a) For every $1 \leq i \leq \ell$, compute $b'_i \leftarrow \mathsf{Dec}(sk_i, a_i)$.

(b) Emulate V_{MIP} on $(f(x), b'_1, \ldots, b'_\ell)$, as if each b'_i is P^i_{MIP}'s response.

(c) Output whatever V_{MIP} outputs (i.e., $f(x)$ and '1' with probability of the computed reward).

Proof (of Theorem 8). The running time of the verifier, the communication complexity, and property (a) of Definition 5 of rational arguments are all explicitly provided by Lemma 2. It remains to show property (b) and property (c) of definition of rational arguments.

The utility gain is $\delta^* \leq \gamma_0 \leq \kappa \cdot \mathsf{negl}(\kappa) = \mathsf{negl}(n)$. By the definition of δ^* we have, $\mathsf{negl}(n) + \mathrm{E}[\mathsf{reward}((P_A, V_A)(x))] \geq \delta^* + \mathrm{E}[\mathsf{reward}((P_A, V_A)(x))]$ which is equal to $\mathrm{E}[\mathsf{reward}((P^*, V_A)(x))]$. Hence, the property (a) of rational arguments holds.

To show property (c) of Definition 5, we assume that $\mu \geq p^{-1}(|x|)$ for some polynomial $p(\cdot)$. Due to the noticeable Δ, we know that $\mu\Delta \geq q_1^{-1}(|x|)$ for some polynomial $q_1(\cdot)$. From the utility loss bound we obtain that

$$(-\delta^*) \geq \mu\Delta - \gamma_0 = \mu\Delta - \mathsf{negl}(n) \geq q_1^{-1}(|x|) - \mathsf{negl}(n) \geq q_1^{-1}(|x|)/2 \ .$$

By defining polynomial $q(\cdot)$ to be $q(|x|) = 2q_1(|x|)$ we get

$$\mathrm{E}[\mathsf{reward}((P_A, V_A)(x))] = \mathrm{E}[\mathsf{reward}((P^*, V_A)(x))] - \delta^*$$
$$\geq \mathrm{E}[\mathsf{reward}((P^*, V_A)(x))] + q^{-1}(|x|) \ ,$$

as desired. $\square$

By applying the above transformation on the no-signaling RMIP protocol presented in Theorem 7, we obtain the following single-round rational arguments for P with sublinear verification.

Corollary 4 (Rational Argument for P). *Let $f : \{0,1\}^n \rightarrow \{0,1\}$ be a function computable by deterministic Turing machine in time $\mathsf{poly}(n) \leq T(n) \leq \exp(n)$ and let $k = \mathsf{polylog}(T(n))$. Let $\kappa \geq \mathsf{polylog}(T(n)) \cdot k$ be a security parameter and let $Z = Z(\kappa)$ be such that $2^{(\log T(n))^c} \leq Z \leq 2^\kappa$ for sufficiently large constant c. If there exists $(Z, 2^{-k^2 \cdot \mathsf{polylog}(T(n))})$-secure FHE scheme then f admits single-round efficient rational argument which has the following properties:*

1. The verifier runs in time $\mathsf{poly}(\kappa, \log(T(n)))$ and the prover runs in $\mathsf{poly}(\kappa, T(n))$.

2. *The length of prover's message and the verifier's challenge is* $k \cdot$
   $\mathrm{poly}(\kappa, \log(T(n)))$. *The verifier's challenge depends only on his random coins*
   and is independent of the input x.

Proof. Suppose that $f \in \mathrm{DTIME}(T)$, where $T = T(n)$ satisfies $\mathrm{poly}(n) \leq T \leq$
$\exp(n)$ and set $k = \mathrm{polylog}(T)$. Let $\kappa = \kappa(n)$ be a security parameter such that
$k \cdot \mathrm{polylog}(T) \leq \kappa$. Let $Z = Z(\kappa)$ such that $2^{(\log T)^c} \leq Z \leq 2^{\kappa}$ for sufficiently large
universal constant c satisfying $Z \geq \max\{n, 2^{k^2 \cdot \mathrm{polylog}(T)}\}$. Let $\delta' = 2^{-k^2 \mathrm{polylog}(T)}$.
By applying Theorem 7 (with respect to the parameter k) to the function f,
we obtain an RMIP for f with $k \cdot \mathrm{polylog}(T)$ provers and reward gap $1/k \cdot$
$\mathrm{poly}(\log(T), n)$ against $2^{-k \cdot \mathrm{polylog}(T)}$-no-signaling strategies. The verifier of the
RMIP runs in time $k^2 \cdot \mathrm{polylog}(T)$ and the provers run in time $\mathrm{poly}(T, k)$. Each
query and answer is of length $k \cdot \mathrm{polylog}(T)$. Assume that there exists an (Z, δ')-
secure FHE.

By Theorem 8, we obtain that f has a 1-round rational argument. The
running time of the verifier is $\mathrm{poly}(\kappa, \log(T)$ and the running time of the
prover is $\mathrm{poly}(\kappa, T)$. The message of the prover and the verifier is of length
$k \cdot \mathrm{poly}(\kappa, \log T)$. $\qquad\square$

We remark that Corollary 3 could be alternatively obtained using our new
transformation presented in Theorem 8. This is done by first transforming the
rational interactive proof for NC to RMIP (with only negligible loss in the reward
gap) and then applying Theorem 8 on the resulted RMIP.

A   Building Blocks

Here we provide only the main claim about efficiency of evaluation of a low
degree extension. For an in-depth exposition see e.g. Rothblum [26].

Proposition 2 ([26]). *There exists a Turing machine that takes as input an*
extension field $\mathbb{H}$ *of* $\mathrm{GF}[2]$, *an extension field* $\mathbb{F}$ *of* $\mathbb{H}$, *an integer* m, *and*
$w = (w_0, \ldots, w_{m-1}) \in \mathbb{H}^m$. *The machine runs in time* $\mathrm{poly}(|\mathbb{H}|, m)$ *and*
space $O(\log|\mathbb{H}| + \log m)$, *and it outputs the unique 2m-variate polynomial*
$\tilde{\beta}: \mathbb{F}^m \times \mathbb{F}^m \to \mathbb{F}$ *of degree at most* $|\mathbb{H}| - 1$ *in each variable (represented as*
an arithmetic circuit of degree at most $|\mathbb{H}| - 1$ *in each variable), such that for*
every $z \in \mathbb{F}^m$, *it holds for the unique low degree extension* $\tilde{W}: \mathbb{F}^m \to \mathbb{F}$ *of* w
that $\tilde{W}(z) = \sum_{p \in \mathbb{H}^m} \tilde{\beta}(z, p) \cdot W(p)$, *where* $W: \mathbb{H}^m \to \mathbb{F}$ *is the function corre-*
sponding to $(w_0, \ldots, w_{n-1})$ *defined using the lexicographic ordering* α *of* $\mathbb{H}^m$ *as*
$W(p) = w_{\alpha(p)}$ *if* $\alpha(p) \leq n - 1$ *and otherwise* 0. *Moreover,* $\tilde{\beta}$ *can be evaluated*
in time $\mathrm{poly}(|\mathbb{H}|, m)$ *and space* $O(\log|\mathbb{H}| + \log m)$. *Namely, there exists a Tur-*
ing machine with above time and space bounds, that takes an input parameters
$\mathbb{H}, \mathbb{F}, m$ *and a pair* $(z, p) \in \mathbb{F}^m \times \mathbb{F}^m$ *and outputs* $\tilde{\beta}(z, p)$.

Fully Homomorphic Encryption. A public-key fully homomorphic encryption scheme consists of four probabilistic polynomial-time algorithms (Gen, Enc,

Eval, Dec). The key generation algorithm Gen, when given as input a security parameter 1^κ, outputs a pair (pk, sk) of public and secret keys. The encryption algorithm, Enc, on input a public key pk and a message $m \in \{0,1\}^{poly(\kappa)}$, outputs a ciphertext q, The homomorphic evaluation algorithm, Eval, on input the public-key pk, a circuit $C : \{0,1\}^a \rightarrow \{0,1\}^b$, where $a, b \leq poly(\kappa)$, and a ciphertext q that is an encryption of a message $m \in \{0,1\}^a$ with respect to pk, outputs a ciphertext $\tilde{q}$ of length $poly(\kappa, a, b)$ that is an evaluation of C over q. The decryption algorithm, Dec, when given a ciphertext q and the secret key sk, outputs the original message m. We allow the decryption process to fail with negligible probability (over the randomness of all algorithms).

Definition 8 (Fully Homomorphic Encryption). *A public-key fully homomorphic encryption scheme* (Gen, Enc, Eval, Dec), *satisfies the following properties.*

Completeness. *For every security parameter κ, for every message $m \in \{0,1\}^{poly(\kappa)}$ and for every circuit C taking inputs of length $|m|$,*

$$\Pr_{(pk,sk)\leftarrow\text{Gen}(1^\kappa)} \left[C(m) = \text{Dec}(sk, \tilde{q}) \middle| \begin{array}{l} q \leftarrow \text{Enc}(pk, m) \\ \tilde{q} \leftarrow \text{Eval}(pk, C, q) \end{array} \right] = 1 - \text{negl}(\kappa).$$

Security. *For every polynomial $p(\cdot)$ and every polynomial size distinguisher $\mathcal{D}$, there exists a negligible function $\text{negl}(\cdot)$ such that for every sufficiently large security parameter κ and every pair of messages $m_0, m_1 \in \{0,1\}^{p(\kappa)}$*

$$\left| \Pr_{\substack{(pk,sk)\leftarrow\text{Gen}(1^\kappa) \\ b\leftarrow\{0,1\}}} [\mathcal{D}(pk, \text{Enc}(pk, m_b)) = b] - \frac{1}{2} \right| < \text{negl}(\kappa)$$

where the probability is also over the random coin tosses of Enc.

We say that the encryption scheme is (S, δ)-secure, for a function $S : \mathbb{N} \rightarrow \mathbb{N}$ and a negligible function $\delta : \mathbb{N} \rightarrow [0,1]$, if the security property holds for every adversary of size $poly(S(\kappa))$, with distinguishing gap at most $\delta(\kappa)$.

References

1. Aiello, W., Bhatt, S., Ostrovsky, R., Rajagopalan, S.R.: Fast verification of any remote procedure call: short witness-indistinguishable one-round proofs for NP. In: Welzl, E., Montanari, U., Rolim, J.D.P. (eds.) ICALP 2000. LNCS, vol. 1853, pp. 463–474. Springer, Heidelberg (2000)
2. Azar, P.D., Micali, S.: Rational proofs. In: Karloff, H.J., Pitassi, T. (eds.) STOC, pp. 1017–1028. ACM (2012)
3. Azar, P.D., Micali, S.: Super-efficient rational proofs. In: Kearns, M., Preston McAfee, R., Tardos, É. (eds.) ACM Conference on Electronic Commerce, pp. 29–30. ACM (2013)

4. Belenkiy, M., Chase, M., Erway, C.C., Jannotti, J., Küpçü, A., Lysyanskaya, A.: Incentivizing outsourced computation. In: Proceedings of the ACM SIGCOMM 2008 Workshop on Economics of Networked Systems, NetEcon 2008, Seattle, WA, USA, 22 August 2008, pp. 85–90 (2008)
5. Bitansky, N., Chiesa, A., Ishai, Y., Ostrovsky, R., Paneth, O.: Succinct non-interactive arguments via linear interactive proofs. In: Sahai, A. (ed.) TCC 2013. LNCS, vol. 7785, pp. 315–333. Springer, Heidelberg (2013)
6. Blumberg, A.J., Walfish, M.: Verifying computations without reexecuting them. Commun. ACM **58**(2), 74–84 (2015)
7. Brier, G.W.: Verification of forecasts expressed in terms of probability. Mon. Weather Rev. **78**(1), 1–3 (1950)
8. Canetti, R., Vald, M.: Universally composable security with local adversaries. In: Visconti, I., De Prisco, R. (eds.) SCN 2012. LNCS, vol. 7485, pp. 281–301. Springer, Heidelberg (2012)
9. Chen, J., McCauley, S., Singh, S.: Rational proofs with multiple provers. CoRR, abs/1504.08361 (2015)
10. Damgård, I., Faust, S., Hazay, C.: Secure two-party computation with low communication. In: Cramer, R. (ed.) TCC 2012. LNCS, vol. 7194, pp. 54–74. Springer, Heidelberg (2012)
11. Dwork, C., Langberg, M., Naor, M., Nissim, K., Reingold, O.: Succinct NP proofs and spooky interactions (2004). www.openu.ac.il/home/mikel/papers/spooky.ps
12. Garay, J.A., Katz, J., Maurer, U., Tackmann, B., Zikas, V.: Rational protocol design: cryptography against incentive-driven adversaries. In: 54th Annual IEEE Symposium on Foundations of Computer Science, FOCS 2013, 26–29 October 2013, Berkeley, CA, USA, pp. 648–657. IEEE Computer Society (2013)
13. Gentry, C., Wichs, D.: Separating succinct non-interactive arguments from all falsifiable assumptions. In: Fortnow, L., Vadhan, S.P. (eds.) STOC, pp. 99–108. ACM (2011)
14. Goldwasser, S., Kalai, Y.T., Rothblum, G.N.: Delegating computation: Interactive proofs for muggles. J. ACM **62**(4), 27 (2015)
15. Guo, S., Hubáček, P., Rosen, A., Vald, M.: Rational arguments: single round delegation with sublinear verification. In: Naor, M. (ed.) Innovations in Theoretical Computer Science, ITCS 2014, Princeton, NJ, USA, 12–14 January 2014, pp. 523–540. ACM (2014)
16. Gur, T., Rothblum, R.D.: Non-interactive proofs of proximity. In: Roughgarden, T. (ed.) Innovations in Theoretical Computer Science, ITCS 2015, Rehovot, Israel, 11–13 January 2015, pp. 133–142. ACM (2015)
17. Kalai, Y.T., Raz, R., Rothblum, R.D.: Delegation for bounded space. In: Boneh, D., Roughgarden, T., Feigenbaum, J. (eds.) Symposium on Theory of Computing Conference, STOC 2013, Palo Alto, CA, USA, 1–4 June 2013, pp. 565–574. ACM (2013)
18. Kalai, Y.T., Raz, R., Rothblum, R.D.: How to delegate computations: the power of no-signaling proofs. In: Shmoys, D.B. (ed.) Symposium on Theory of Computing, STOC 2014, New York, NY, USA, 31 May–03 June 2014, pp. 485–494. ACM (2014)
19. Kalai, Y.T., Rothblum, R.D.: Arguments of proximity. In: Gennaro, R., Robshaw, M. (eds.) CRYPTO 2015. LNCS, vol. 9216, pp. 422–442. Springer, Heidelberg (2015)
20. Kilian, J.: A note on efficient zero-knowledge proofs and arguments (extended abstract). In: Rao Kosaraju, S., Fellows, M., Wigderson, A., Ellis, J.A. (eds.) STOC, pp. 723–732. ACM (1992)

21. Kumaresan, R., Bentov, I.: How to use bitcoin to incentivize correct computations. In: Ahn, G.-J., Yung, M., Li, N. (eds.) Proceedings of the 2014 ACM SIGSAC Conference on Computer and Communications Security, Scottsdale, AZ, USA, 3–7 November 2014, pp. 30–41. ACM (2014)

22. Lund, C., Fortnow, L., Karloff, H.J., Nisan, N.: Algebraic methods for interactive proof systems. J. ACM **39**(4), 859–868 (1992)

23. Micali, S.: Computationally sound proofs. SIAM J. Comput. **30**(4), 1253–1298 (2000)

24. Naor, M.: On cryptographic assumptions and challenges. In: Boneh, D. (ed.) CRYPTO 2003. LNCS, vol. 2729, pp. 96–109. Springer, Heidelberg (2003)

25. Pham, V., Khouzani, M.H.R., Cid, C.: Optimal contracts for outsourced computation. In: Poovendran, R., Saad, W. (eds.) GameSec 2014. LNCS, vol. 8840, pp. 79–98. Springer, Heidelberg (2014)

26. Rothblum, G.N.: Delegating computation reliably: paradigms and constructions. Ph.D. thesis, Massachusetts Institute of Technology (2009)

27. Rothblum, G.N., Vadhan, S.P., Wigderson, A.: Interactive proofs of proximity: delegating computation in sublinear time. In: Boneh, D., Roughgarden, T., Feigenbaum, J. (eds.) Symposium on Theory of Computing Conference, STOC 2013, Palo Alto, CA, USA, 1–4 June 2013, pp. 793–802. ACM (2013)

28. Shamir, A.: IP = PSPACE. J. ACM **39**(4), 869–877 (1992)

29. Zhang, Y., Blanton, M.: Efficient secure and verifiable outsourcing of matrix multiplications. In: Chow, S.S.M., Camenisch, J., Hui, L.C.K., Yiu, S.M. (eds.) ISC 2014. LNCS, vol. 8783, pp. 158–178. Springer, Heidelberg (2014)

Interactive Coding for Interactive Proofs

Allison Bishop[1]($\boxtimes$) and Yevgeniy Dodis[2]

[1] Columbia University, New York, USA
allison@cs.columbia.edu
[2] New York University, New York, USA
dodis@cs.nyu.edu

Abstract. We consider interactive proof systems over adversarial communication channels. We show that the seminal result that **IP = PSPACE** still holds when the communication channel is malicious, allowing even a constant fraction of the communication to be arbitrarily corrupted.

1 Introduction

Interactive proofs are fundamental objects in both cryptography and complexity theory, and come with a rich history of exciting developments, such as the surprising characterization that **IP = PSPACE** [25]. This characterization assumes that a prover and a verifier communicate over a perfect communication channel, and crucially relies upon the fact that the number of rounds of the interaction can be polynomially long.

Recently, the study of interactive coding (pioneered by Schulman [22,23]) has emerged as a promising way to extend results involving lengthy interactions over perfect channels to analogous results over adversarial channels - even with a constant relative error rate. This high level of robustness cannot be achieved by simply applying an error correcting code to each message, a method which is limited to an error rate proportional to $\frac{1}{r}$, where r is the number of rounds. There has been much success in obtaining interactive coding protocols capable of performing any two party communication tasks over a noisy or adversarial channel [1,4–7,11,13,14,17,19,22,23]. However, all of these works assume that the task is described as a function of two inputs, and only correctness of the computation is required.

In the case of interactive proofs, it is not enough to ensure that an honest party "eventually" learns the real message the other party was attempting to send. Instead, we must ensure that the interference of the channel cannot prevent an honest prover from convincing a verifier of a true statement, and also cannot help a malicious prover convince a verifier of a false statement. This appears to be problematic if we consider the techniques employed by interactive coding

A. Bishop—Supported in part by NSF CNS 1413971 and NSF CCF 1423306.

Y. Dodis—Partially supported by gifts from VMware Labs and Google, and NSF Grants 1319051, 1314568, 1065288, 1017471.

E. Kushilevitz and T. Malkin (Eds.): TCC 2016-A, Part II, LNCS 9563, pp. 352–366, 2016.
DOI: 10.1007/978-3-662-49099-0_13

protocols, which enable parties to "replay" and "revise" their messages as the interactive coding mechanism runs. We must worry, then, that a malicious prover may use the excuse of potential channel errors to change its responses adaptively after peeking ahead at the verifier's future challenges. For this reason, it does not suffice to simply take an interactive proof system designed for an error-free channel and compile it blindly using an off-the-shelf interactive coding method.

An undaunted optimist might then ask for strong interactive coding mechanism, one that could provably compose with a wide variety of security properties, such as soundness for interactive proof systems or input privacy for multiparty computation. The most general version of this would achieve a notion ensuring that the participants in the error-resilient version of the protocol "do not learn" anything more than they would learn from executing the error-free protocol. A formalization of this called "knowledge-preserving interactive coding" was introduced and studied by Chung et al. [9], who showed that under this strong requirement, applying an error-correcting code in each round is essentially optimal. This means no error rate beyond $\frac{1}{r}$ is possible without making computational assumptions. Similarly, the work of Gelles et al. [12] proves an impossibility result for error-resilient secure multiparty computation.

In contrast, we show that *for interactive proofs a constant relative error rate can be achieved.* In other words, while positive results in traditional interactive coding only (successfully) handle *correctness*, and negative results rule out (strong forms of) *zero-knowledge/privacy*, we show that ensuring *soundness* is still feasible in the presence of adversarial communication noise. This is perhaps a bit counterintuitive, as the negative results proceed by proving that some amount of backtracking and replaying messages is inherent for this level of error-correction. Nonetheless, using amplification techniques, we can preprocess our interactive proofs to withstand a certain amount of backtracking, since the verifier is aware of the backtracking, and can sample fresh randomness to mitigate the potential advantage gained by a malicious prover as a consequence.

One additional challenge we face is that the verifier must remain efficient, meaning that the encoding and decoding for the interactive coding mechanism must be computable in polynomial time. Many of the interactive coding results are existential rather than efficient (e.g. [7,22,23]), but the recent work of Brakerski and Kalai [4] managed to obtain computationally efficient interactive coding protocols, even for constant rate adversarial errors. We employ a simplified version of their techniques, obtaining our simplifications due to the fact that the individual messages of our protocols can be taken to be not too short. This allows us to avoid the use of expensive "tree codes" that are needed in [4], and results in much easier to understand protocols.

Our Techniques. It is well-known how to design an interactive proof system for **PSPACE** that has both perfect completeness (an honest prover can convince a verifier of a true statement with probability 1) and very small soundness error (a malicious prover can only convince a verifier of a false statement with very small probability). Starting from such a system, we observe that even if a malicious prover could make a verifier re-sample a particular challenge polynomially many times, the chance of obtaining a value that would allow "cheating" remains reasonably small. We can thus hope to withstand a certain fraction of channel errors by allowing the protocol to "backtrack" when errors occur, while having the verifier resample its randomness to control the potential gain for a malicious prover. Our coding techniques will prevent a malicious prover from changing its answer to a previous challenge, instead requiring it to answer a new challenge if it uses the potential errors as an excuse to backtrack.

We organize our proof into two separate tasks: first obtaining an **IP** system that still works over perfect channels, but allows parties to arbitrarily signal that they want to back up one round of interaction at a time. We call such a system "backtracking-resilient." We then design a compilation procedure that takes any backtracking-resilient **IP** system and produces a new proof system that works over adversarial channels, allowing a constant error rate. This compilation relies upon hashing techniques that are reminiscent of the efficient compiler in [4]. In particular, we have both parties hash their current simulated transcripts with freshly chosen keys at each exchange, so that they can detect any disagreements whenever the channel does not introduce too many errors.

There are two noteworthy features of our compiler. First, since the backtracking-resilient **IP** system we use anyway has reasonably large individual messages, we can avoid the use of expensive tree codes that are needed in [4] to protect the "simulation units between the hash stages". Second, we give an explicit reduction between the soundness of our compiled protocol and the notion of backtracking resilience. This formalizes the intuition that the hashing techniques essentially limit a malicious prover impersonating an adversarial channel to choosing when to backtrack the protocol, and could be useful for future work.

2   Resilient Interactive Protocols

Interactive Protocols. We recall the notion of an r-round interactive protocol Π between a deterministic prover $\mathcal{P}$ and a probabilistic verifier $\mathcal{V}$ (who share some common input x which we omit, when clear). We view the verifier $\mathcal{V}$ as an algorithm that takes in a partial transcript and some fresh randomness and outputs a next message.[1] More formally, $\mathcal{V} : \mathcal{T} \times Rand \to \{0,1\}^*$, where $\mathcal{T}$ is the set of partial transcripts and $Rand$ is the set of random values. We view the prover $\mathcal{P}$ as a deterministic algorithm that takes in a partial transcript and outputs a next message: $\mathcal{P} : \mathcal{T} \to \{0,1\}^*$.

In the typical (error-free) setting, the protocol proceeds in some number of rounds, r, where in each round i the verifier sends a challenge C_i and the prover sends a response R_i. If we let τ_{i-1} denote the transcript after $i-1$ rounds, the ith round consists of the verifier sampling a fresh random value $c_i \in Rand$ and sending the challenge $C_i := \mathcal{V}(\tau_{i-1}, c_i)$. The prover then sends the response $R_i := \mathcal{P}(\tau_{i-1}||C_i)$. We then have $\tau_i = \tau_{i-1}||C_i||R_i$.

---

[1] This view is without loss of generality, since it is known that private-coin protocols can be simulated by public-coin ones [16], meaning that $\mathcal{V}$ never needs to keep any state beyond its partial transcript so far.

We assume each party locally stores its own copy of the partial transcript. We let τ_p denote the prover's (evolving) copy and τ_v denote the verifier's (evolving) copy of the transcript. (Note, in the error-free settings these values are always consistent; however, once we are allowing channel errors, these two partial transcripts may temporarily diverge.) At the end of round r, $\mathcal{V}$ either accepts or rejects the final transcript τ_v, and we denote the random variable (over the coins of $\mathcal{V}$) indicating this decision by $(\mathcal{P}, \mathcal{V})$.

Given a language L, we say that $\varPi$ is *(perfectly) complete* on L, if for any $x \in L$, we have $\Pr[(\mathcal{V}(x), \mathcal{P}(x)) \to accept] = 1$. Similarly, $\varPi$ is ε-sound on L, if for any $x \notin L$ and any potentially cheating prover $\tilde{\mathcal{P}}$, we have $\Pr[(\mathcal{V}(x), \tilde{\mathcal{P}}(x)) \to accept] \leq \varepsilon$.

Definition 1. *We say that L belongs to the class* **IP** *(Interactive Protocols), if there exist polynomial $r = r(n)$ and $t = t(n)$ and an r-round interactive protocol $(\mathcal{P}, \mathcal{V})$ where the running time of $\mathcal{V}$ on n-bit inputs x is at most $t(n)$ and: (a) $\varPi$ is (perfectly) complete; (b) $\varPi$ is $(1/2)$-sound.*

Of course, repeating $\varPi$ in parallel λ times, we can reduce the soundness error to $\varepsilon = 2^{-\lambda}$, for any polynomial $\lambda = \lambda(n)$. It is known [25] that **IP** = **PSPACE**, the class of languages decided with polynomial space.

Error-Resilient Protocols. To define error-resilient protocols, we must specify the power of an adversarial channel. We will model an adversarial channel as an algorithm $\mathcal{A}$ that intercepts messages as they are sent and may modify them arbitrarily in transit. We make no restrictions on the computational power of $\mathcal{A}$ (it may be unbounded) and also allow it to know the entire state of the prover, the verifier, and the partial transcript at any point during the execution. It does not know, however, the future randomness to be selected by the verifier.

Definition 2 (Completeness with Adversarial Channel Error). Given a language L, we say that a T-round protocol $\varPi$ is (α, δ)-error-complete on L, if the following condition holds for any $x \in L$. For any (unbounded) adversary $\mathcal{A}$ that can cause at most an α-fraction of errors throughout the entire communication, the honest prover $\mathcal{P}_\alpha$ will convince the honest verifier $\mathcal{V}_\alpha$ with probability $> \delta$.

For soundness in the presence of adversarial noise, we observe that we can always "merge" the adversarial prover $\tilde{\mathcal{P}}$ with our channel adversary $\mathcal{A}$, simply resulting in a different adversarial prover $\tilde{\mathcal{P}}'$. In other words, soundness with adversarial channel error is equivalent to traditional soundness!

Definition 3. *We say that L belongs to the class* **ERIP** *(Error-Resilient* **IP***), if there exists a constant $\alpha > 0$, polynomials $T = T(n)$ and $t = t(n)$, and a T-round interactive protocol $\varPi = (\mathcal{P}_\alpha, \mathcal{V}_\alpha)$ where the running time of $\mathcal{V}_\alpha$ on n-bit inputs x is at most $t(n)$ and: (a) $\varPi$ is $(\alpha, \frac{2}{3})$-error-complete; (b) $\varPi$ is $(1/3)$-sound.*

Just like for (imperfect completeness) **IP**, the completeness and soundness constants $2/3$ and $1/3$ of **ERIP** can be amplified to $(1 - 2^{-\Omega(p)})$ and $2^{-\Omega(p)}$, respectively, by doing p parallel repetitions $\Pi_1, \ldots, \Pi_p$, and taking the majority vote. A small subtlety in this (otherwise immediate) argument comes from analyzing completeness in the presence of errors (soundness is the same as for **IP**). This is because the attacker $\mathcal{A}$ can split his errors non-uniformly across the p repetitions Π_i, causing failures with high probability for repetitions where more than α-fraction of errors were introduced. Fortunately, by setting the robustness threshold α' of the new parallel protocol Π^* to be $\alpha' = \alpha/10$, we can apply Markov's inequality to conclude that $\mathcal{A}$ can cause more than α-fraction of errors on at most $p/10$ sub-protocols Π_i, meaning that $\mathcal{A}$ would still need to break $(\alpha, 2/3)$-error-completeness for at least $p/2 - p/10 = 2p/5$ protocols Π_i in order to break the error-completeness of Π^*. However, since in any of the p protocols $\mathcal{A}$'s success of doing so is at most $1/3$, the honest verifier acts independently across the p runs, and $2p/5 > p/3$, we can use the Chernoff bound to conclude that $\mathcal{A}$'s overall success probability will be $2^{-\Omega(p)}$.

Main Result. Our main result can then be stated as:

Theorem 1. (Main Result) ERIP = IP = PSPACE.

We will prove this over the course of the next three sections.

3   Backtracking-Resilient Protocols

As an intermediary step in achieving error-resilient proof systems, we will define proof systems that retain their completeness and soundness properties under a milder disruption we call "backtracking." In other words, we augment usual error-free protocols with an additional mechanism for backtracking. In addition to sending a challenge or response, either party may at any time transmit a special symbol B instead. Upon sending or receiving a B, the parties each remove the latest complete round from their partial transcripts. For example, suppose $\tau_p = \tau_v = \tau_i$ at the time a B is sent/received. Then both parties revert to τ_{i-1} and will start again with the verifier choosing *fresh randomness* to send a (potentially) new challenge to $\mathcal{P}$ for round i (it is as if the old version of round i never happened).

We let U be the maximal number of backtracking steps (i.e., the "budget") allowed by each party. Given any standard interactive protocol Π and any such budget U, we obtain U-backtracking extension of Π, Π_U, where the modified prover $\mathcal{P}_U$ (resp. verifier $\mathcal{V}_U$) is identical to the honest prover $\mathcal{P}$ (resp. verifier $\mathcal{V}$), except allowing allow up to U backtracking steps to the communicating partner, as described above. In particular, $\mathcal{V}_U$ will output the same decision as $\mathcal{V}$ when the transcript τ_v reaches the last round (for the first time), but also $\mathcal{V}_U$ will reject if more than U backtracking steps are attempted by the (possibly malicious) prover. Thus, if Π has r rounds, without loss of generality we can cap the number of rounds of Π_U by $T = r + 4U$, as each of the (at most $2U$) backtracking steps requires one extra "normal step" to get back, meaning that in at most $r + 4U$ rounds the transcript τ_v is guaranteed to reach the decision point for $\mathcal{V}$.

Of course, when playing against themselves, $\mathcal{P}_U$ and $\mathcal{V}_U$ will not use back backtracking steps, and the protocol will terminate in r rounds. However, we would like to extend completeness of soundness condition to hold even if (possibly malicious) backtracking is allowed. For the former, we will assume that $\mathcal{P}_U$ and $\mathcal{V}_U$ are honest, except for the adversarial backtracking steps (see below); for the latter, we will assume that the verifier $\mathcal{V}_U$ is honest, but the prover $\tilde{\mathcal{P}}_U$ is malicious, including (wlog, up to U) backtracking steps. This is formalized below.

Definition 4 (Perfect Completeness with U-Backtracking). *Given a language L, we say that an r-round protocol Π is (perfectly) U-backtracking-complete on L, if the following condition holds with probability 1, for any $x \in L$. Let $T = r + 4U$ and $C_p, C_v \in \{\perp, B\}^T$ be two strings of length T containing at most U occurrences of the symbol B. We let $C_p(i)$ denote the ith symbol of C_p, and same for $C_v(i)$. We require that if an honest prover $\mathcal{P}_U$ and verifier $\mathcal{V}_U$ run the protocol for a true statement, except with the prover sending B in round i whenever $C_p(i) = B$ and the verifier sending B in round i whenever $C_v(i) = B$, then the verifier accepts after at most T rounds with probability 1.*

Definition 5 (Soundness with U-Backtracking). *Given a language L, we say that an r-round protocol Π is (U, ε)-backtracking-sound on L, if for $x \notin L$, and any malicious prover $\tilde{\mathcal{P}}_U$ for the U-backtracking extension Π_U of Π, $\Pr[(\mathcal{V}_U(x), \tilde{\mathcal{P}}_U(x)) \to accept] \leq \varepsilon$.*

Lemma 1. *Assume Π is an r-round interactive protocol which is complete and ε-sound for some langueage L. Then, for any U, Π is U-backtracking-complete and (U, ε')-backtracking-sound, where*

$$\varepsilon' \leq \varepsilon \cdot 2^{r+4U}$$

Proof. Every final transcript produced with non-zero probability in Π_U must also occur with non-zero probability without backtracking, using the original algorithms. Thus, perfect completeness for the underlying algorithms implies perfect completeness with backtracking.

To prove soundness, we fix a false statement x and an arbitrary malicious prover $\tilde{\mathcal{P}}_U$ for Π_U. Let ε' be the probability that $\tilde{\mathcal{P}}_U(x)$ convinces $\mathcal{V}_U(x)$. Given any *fixed* sequence $C \in \{\perp, B\}^{r+4U}$ of possible backtracking steps of $\tilde{\mathcal{P}}_U$ containing at most U occurrences of the symbol B, we let ε'_C be the probability that $\tilde{\mathcal{P}}_U(x)$ succeeds and *precisely* respects the backtracking sequence C. Clearly, since all such events are disjoint, we have $\varepsilon' = \sum_C \varepsilon'_C$. To complete the proof, it suffices to show that $\varepsilon'_C \leq \varepsilon$, for any fixed C, as the number of C's is at most 2^{r+4U}.

To show that $\varepsilon'_C \leq \varepsilon$, we define a malicious prover $\tilde{\mathcal{P}}^C$ of the original (non-backtracking) protocol Π whose success probability is at precisely ε'_C, which implies that $\varepsilon'_C \leq \varepsilon$, by standard ε-soundness. Given C, $\tilde{\mathcal{P}}^C$ can pre-compute all the r rounds $1 \leq i_1, \ldots, i_r \leq r + 4U$ which will not be "erased" from the final transcript of $\mathcal{V}_U$ assuming that $\tilde{\mathcal{P}}_U$ follows C. Then $\tilde{\mathcal{P}}^C(x)$ emulates $\tilde{\mathcal{P}}_U(x)$ as follows: (a) the challenges for all the "non-erased" rounds $i_1, \ldots, i_r$ are obtained from the honest verifier $\mathcal{V}$; (b) the challenges from the remaining "erased" rounds are honestly generated by $\tilde{\mathcal{P}}^C$ himself; (c) if at any point $\tilde{\mathcal{P}}_U$ generates a backtracking step inconsistent with C, $\tilde{\mathcal{P}}^C$ aborts. Since the above emulation is identical to the real run of $\tilde{\mathcal{P}}_U$ when consistent with C, $\tilde{\mathcal{P}}^C$ succeeds with probability $\varepsilon'_C \leq \varepsilon$, as claimed.

Definition 6. *We say that L belongs to the class* **BRIP** *(Backtracking-Resilient* **IP***), if there exist polynomial $r = r(n)$ such that for any polynomial $U = U(n)$ there exists a polynomial $t = t(n)$ and an r-round interactive protocol $\Pi = (\mathcal{P}, \mathcal{V})$ where the running time of $\mathcal{V}$ on n-bit inputs x is at most $t(n)$ and: (a) Π is (perfectly) U-backtracking-complete; (b) Π is $(U, 1/2)$-backtracking-sound.*

Using Lemma 1, we observe that the class of interactive protocols is backtracking-resilient.

Corollary 1. BRIP = IP = PSPACE.

Proof. Take any $L \in$ **IP**. This means L has an r-round, $1/2$-sound interactive protocol Π for some polynomial r. Now take any polynomial U for the backtracking budget. By repeating Π in parallel $r + 4U + 1$ times, we get protocol Π' for L which still has r rounds, polynomial-time verifier, is complete and ε-sound, where $\varepsilon = 2^{-r-4U-1}$. By Lemma 1, Π' is U-backtracking-complete and $(U, 1/2)$-backtracking-sound, completing the proof.

Remark 1. We can easily reduce the soundness error $1/2$ to be exponentially small, either by directly adjusting the proof of Corollary 1, or by doing parallel repetition on any **BRIP** protocol.

4 Compiling Backtracking-Resilient Protocols Against Adversarial Channel Errors

We now present a method for taking a backtracking-resilient interactive proof system and compiling it into one that can resist a constant rate of adversarial channel errors. Intuitively, the prover and verifier will attempt to simulate the backtracking-resilient protocol over the adversarial channel. They will use hash functions with freshly chosen keys each time to check if they are in agreement on the partial transcript simulated so far. Every message and hash key will be encoded with an error correcting code, to ensure that the adversary must invest a high amount of errors to cause confusion between the parties. Of course, sometimes channel errors will still prevent the parties from detecting an inconsistency in the simulated transcript. But the adversary cannot afford to keep up this high error investment indefinitely, and eventually the parties will detect the problem and backtrack to fix it. This will result in a simulated transcript that mimics an execution of the backtracking-resilient protocol, and hence appropriate analogs of completeness and soundness for this compiled protocol can be reduced to backtracking-resilience of the underlying protocol.

We prove the following result, which, by Corollary 1, suffices to establish our main result in Theorem 1.

Theorem 2. ERIP = BRIP.

Since **ERIP** $\subseteq$ **IP** = **BRIP**, we only need to show that **BRIP** $\subseteq$ **ERIP**. Before proving this result, we need some standard tools from hashing and coding.

Hashing and Coding. We will use a family of hash functions indexed by keys $k \in \{0,1\}^\gamma$. More precisely, we invoke the following theorem also used in [4]:

Theorem 3 [2,20]. *There exists a constant $q > 0$ and an ensemble of hash families $\{H_N\}_{N \in \mathbb{N}}$ such that for every $N \in \mathbb{N}$ and for every $h \in H_N$, $h : \{0,1\}^{\leq 2^N} \to \{0,1\}^{qN}$ is poly-time computable, it is efficient to sample $h \leftarrow H_N$ using only qN random bits, and for all $y \neq z \in \{0,1\}^{\leq 2^N}$ it holds that*

$$\Pr_{h \to H_N}[h(y) = h(z)] \leq 2^{-N}.$$

We let $\gamma = qN = O(N)$ and write $h_k : \{0,1\}^{\leq 2^N} \to \{0,1\}^\gamma$ to denote the element of H_N sampled with the random string $k \in \{0,1\}^\gamma$. We also let *Encode* and *Decode* denote the encoding and decoding algorithms of an error-correcting code with a constant rate and a constant relative distance β.

Our Compiler. Take any $L \in$ **BRIP** which means that L has an r-round backtracking-resilient protocol Π, for any polynomially bounded budget U. (As we will see shortly, we will only use $U = O(r)$.) To show $L \in$ **ERIP**, we set $\alpha = \Omega(\beta)$ to be the constant error rate we will tolerate on the channel, and show how to build an error-resilient proof system for L tolerating an α-fraction of adversarial errors. Our new protocol $\tilde{\Pi}$ will run for $\tilde{T}$ rounds, where we define $\tilde{T}$ such that $\tilde{T}(1 - 18\alpha\beta^{-1}) = r$. In particular, when α is chosen to be a suitably small constant fraction of β (e.g., $\alpha = \beta/36$), this is possible with $\tilde{T} = O(r)$ (e.g., $\tilde{T} = 2r$).

We define the backtracking budget U of our original protocol $\Pi = (\mathcal{P}, \mathcal{V})$ by $U = 9\alpha\beta^{-1}\tilde{T} = O(r)$ (e.g., for $\alpha = \beta/36$, we have $U = r/2$), and assume that Π is $(U, \frac{1}{3})$-backtracking sound and perfectly U-backtracking complete (guaranteed possible by Lemma 1). We also denote by $T = r + 4U = O(r)$ (e.g., $T = 3r$ when $\alpha = \beta/36$) the maximal number of rounds of the U-backtracking extension Π_U of Π, by ℓ the length of the challenges to be sent by the verifier $\mathcal{V}$ in each round, and assume that the hashing parameters $\gamma, N = \Omega(\log r)$. Finally, we will use *Encode* and *Decode* for encoding/decoding messages of length $\ell + 2\gamma + \log(T)$.

We can now describe the new algorithms $\tilde{\mathcal{P}}$ and $\tilde{\mathcal{V}}$ for an interactive proof system that can resist adversarial channel errors at a constant rate α. These algorithms will run for $\tilde{T}$ message exchanges, where each exchange will still consist of a message sent by the verifier and then a response sent by the prover, and will require only black-box access to $\mathcal{P}$ and $\mathcal{V}$.

$\widetilde{\mathcal{P}}$ will maintain internal variables $\tilde{\tau}_p$ and $\tilde{i}_p$. These will function as the prover's internal views of the simulated transcript and round number respectively. $\widetilde{\mathcal{V}}$ will similarly maintain internal variables $\tilde{\tau}_v$, $\tilde{i}_v$, and $\tilde{C}$. These will function as the verifier's internal views of the simulated transcript, the round number, and the pending challenge. We initialize $\tilde{\tau}_p$, $\tilde{\tau}_v$ and $\tilde{C}$ to $\emptyset$, which denotes the empty string. We initialize $\tilde{i}_p$ and $\tilde{i}_v$ to 1.

The First Round: $\widetilde{\mathcal{V}}$ will start a run of $\mathcal{V}$ to obtain a first challenge C. It sets $\tilde{C} = C$. It also samples a uniformly random hash key $k_1 \in \{0,1\}^\gamma$. It will send to the prover: $Encode(C||k_1||h_{k_1}(\tilde{\tau}_v)||\tilde{i}_v)$.

The Prover's Algorithm: In any round, when the prover receives a message from the verifier, it decodes it as a challenge C, a key k, a hash value h, and a round index i. It then performs the following steps to update its internal variables and produce a response:

- If $h_k(\tilde{\tau}_p) \neq h$ and $i \leq \tilde{i}_p$, then decrement $\tilde{i}_p$, erase a round from $\tilde{\tau}_p$, and set R equal to the last prover response now reflected in $\tilde{\tau}_p$.
- If $h_k(\tilde{\tau}_p) \neq h$ and $i > \tilde{i}_p$, then keep $\tilde{i}_p, \tilde{\tau}_p$ the same, and set R equal to the last prover response reflected in $\tilde{\tau}_p$.
- If $h_k(\tilde{\tau}_p) = h$, then set $R = \mathcal{P}(\tilde{\tau}_p||C)$, concatenate $C||R$ onto $\tilde{\tau}_p$, and increment $\tilde{i}_p$.

The prover then chooses a new uniformly random key k' and sends

$$Encode(R||k'||h_{k'}(\tilde{\tau}_p)||\tilde{i}_p).$$

The Verifier's Algorithm: When the verifier receives a message from the prover, it decodes it as a response R, a key k, a hash value h, and a round index i. It then performs the following steps to update its internal variables and produce a response:

- If $h_k(\tilde{\tau}_v||\tilde{C}||R) \neq h$ and $i \leq \tilde{i}_v$, then decrement $\tilde{i}_v$, erase a round from $\tilde{\tau}_v$, and set $\tilde{C} = \mathcal{V}(\tilde{\tau}_v, rand)$, for a freshly chosen random value $rand$.
- If $h_k(\tilde{\tau}_v||\tilde{C}||R) \neq h$ and $i > \tilde{i}_v$, then keep $\tilde{i}_v, \tilde{\tau}_v, \tilde{C}$ the same.
- If $h_k(\tilde{\tau}_v||\tilde{C}||R) = h$, then concatenate $\tilde{C}||R$ onto $\tilde{\tau}_v$, increment $\tilde{i}_v$, and then set $\tilde{C} = \mathcal{V}(\tilde{\tau}_v, rand)$ for a freshly chosen random value $rand$.

The verifier then chooses a new uniformly random key k' and sends

$$Encode(\tilde{C}||k'||h_{k'}(\tilde{\tau}_v)||\tilde{i}_v).$$

At the end of the $\widetilde{T}$ message exchanges, the verifier outputs the decision of $\mathcal{V}(\tilde{\tau}_v)$.

Efficiency. Using any efficient hashing and coding scheme, our new prover and verifier algorithms are efficient given oracle access to the (next message function) of the original prover and verifier.

5   Analysis of the Compiled Algorithms

We now prove that the algorithms $\widetilde{P}, \widetilde{V}$ presented in the previous section satisfy completeness and soundness despite adversarial channel error.

5.1   Completeness

We first seek to prove completeness. For this, we assume an honest prover and an adversarial channel that can cause at most an α-fraction of errors throughout the entire communication. We prove:

Lemma 2. *If* $\mathcal{P}, \mathcal{V}$ *is perfectly* U-*backtracking complete, then* $\widetilde{\mathcal{P}}, \widetilde{\mathcal{V}}$ *is* $(\alpha, \frac{2}{3})$-*error-complete.*

Proof. We will define a measure of progress, M, that will potentially oscillate as the protocol runs. At the beginning of any particular exchange (just before the verifier sends its next message), we can determine the value of M as follows. First, we let m be the maximal number of rounds such that $\tilde{\tau}_p$ and $\tilde{\tau}_v$ agree on a prefix of m rounds. We then set

$$M := m - (\tilde{i}_p - m) - (\tilde{i}_v - m).$$

We define a *good exchange* as follows. First, we require that the verifier has correctly decoded the previous message sent by the prover. Additionally, we require that the two messages sent during the exchange (by the verifier and then by the prover) are also decoded correctly. We refer to any other exchange as a *bad exchange*.

Lemma 3. *A bad exchange decreases* M *by at most 3.*

Proof. In any exchange, the value of m can decrease by at most 1, since at most one round of the simulated transcripts is erased at a time. Since each of $\tilde{i}_p, \tilde{i}_v$ can be incremented by at most 1 in any exchange, we then have that the total decrement in M is bounded by 3.

Lemma 4. *Conditioned on the event that there are no hash collisions, a good exchange increases* M *by at least 1.*

Proof. Suppose at the beginning of a good exchange, the verifier has calculated that the current hash value agrees, and has just incremented $\tilde{i}_v$. Since we are assuming no hash collisions have occurred, this implies that $\tilde{\tau}_v = \tilde{\tau}_p$ and $\tilde{i}_v = \tilde{i}_p$ at this point. $\mathcal{V}$ will choose a new challenge $\tilde{C}$, send this to $\mathcal{P}$, who will form a response R, and both $\mathcal{V}, \mathcal{P}$ will concatenate $\tilde{C}||R$ onto their transcripts. In this case, m will increase by 1, and $\tilde{i}_p - m, \tilde{i}_v - m$ will both remain 0. Hence M will increase by 1.

Now suppose instead that at the beginning of a good exchange, the verifier has detected a disagreement in the hash values. In the case that $\tilde{i}_v \geq \tilde{i}_p$, the verifier will erase a round from $\tilde{\tau}_v$ and decrement $\tilde{i}_v$. This will lead to a decrease in $\tilde{i}_v - m$ but no decrease in m. When $\mathcal{P}$ correctly decodes the next message from $\mathcal{V}$, a decrease in m is impossible since an agreed upon round cannot be erased when $\tilde{i}_p = \tilde{i}_v$ (since the hash values will agree in this case) and when $\tilde{i}_p$ remains less than $\tilde{i}_v$, the prover will not erase a round from $\tilde{\tau}_p$. Also an increase in $\tilde{i}_p - m$ is impossible, since $\tilde{i}_p$ will only be incremented if a new agreed upon round is being added to the simulated transcript. Thus M will also increase in this case.

We next consider the case where the verifier has detected a disagreement in the hash values and $\tilde{i}_v < \tilde{i}_p$. The verifier will then leave $\tilde{\tau}_v$, $\tilde{i}_v$ unchanged, but the prover will decrement $\tilde{i}_p$ and erase a round from $\tilde{\tau}_p$, leading to a decrease in $\tilde{i}_p - m$, and hence an increase in M.

We now observe that a bad exchange can be extended to a *bad interval* containing three transmitted encodings - the previous message from the prover to the verifier, and the two messages in the bad exchange itself. At least one of these messages must have been corrupted beyond its capacity, resulting in a relative error rate within this interval of $> \frac{\beta}{3}$.

We note the following lemma stated in [24]:

Lemma 5 *(Lemma 7 in [24]). In any finite set of intervals on the real line whose union is of total length s, there is a subset of disjoint intervals whose union is of total length at least $\frac{s}{2}$.*

We suppose there are s bad exchanges during a run of $\tilde{T}$ total exchanges. Then there are at least $\frac{s}{2}$ bad exchanges whose corresponding bad intervals are disjoint. This results in a total relative error rate of $\frac{s\beta}{6T}$. We must have: $s \leq 6\alpha\beta^{-1}\tilde{T}$.

Thus, after $\tilde{T}$ exchanges if no hash collisions have occurred, the value of our progress measure M satisfies $M \geq \tilde{T} - 18\alpha\beta^{-1}\tilde{T}$, which we can rewrite as $M \geq \tilde{T}(1 - 18\alpha\beta^{-1})$. Recall that this is $\geq r$ by our choice of $\tilde{T}$. This implies that the simulated transcript will be a full r rounds of a transcript that occurs with non-zero probability in the backtracking resilient algorithms over clear channels. We also observe that if $\tilde{U}$ is the number of backtracks occurring during a particular execution, then $\tilde{T} - 2\tilde{U} \geq M$, so because we set $U := 9\alpha\beta^{-1}\tilde{T}$, we have ensured that at most U backtracks occur. Thus, completeness follows from the perfect completeness of the underlying backtracking resilient algorithms if we choose parameters that make the probability of a hash collision $< \frac{1}{3}$.

To bound the probability of hash collisions, we employ Theorem 3 and a union bound to conclude that the probability of a hash collision occurring at any time throughout the protocol simulation is $O(\tilde{T}2^{-N}) = O(r2^{-N})$. Thus it suffices to set N proportional to $\log(r)$ to achieve a bound $< \frac{1}{3}$.

Putting this all together, we have proven Lemma 2.

5.2   Soundness

Next, we show that soundness is preserved by our compiler, irrespective of the value U.

Lemma 6. *If $\mathcal{P}, \mathcal{V}$ is $(U, \frac{1}{3})$-backtracking-sound, then $\widetilde{\mathcal{P}}, \widetilde{\mathcal{V}}$ is $(\frac{1}{3})$-sound.*

Proof. We are considering a perfect channel and a malicious prover who seeks to convince the verifier of a false statement. The verifier, of course, does not know the errors are not coming from the channel. We fix a false statement and a malicious prover $\widetilde{\mathcal{P}}$ who manages to convince $\widetilde{\mathcal{V}}$ to accept with probability $> \frac{1}{3}$. From this, we will create a malicious prover $\mathcal{P}$ for the underlying back-tracking resilient algorithm that contradicts soundness with backtracking.

The Malicious $\mathcal{P}$: The malicious prover $\mathcal{P}$ for the backtracking-resilient proof system behaves as follows. It will run the malicious prover $\widetilde{\mathcal{P}}$ internally, simulating the messages from $\mathcal{V}$. It initializes $\widetilde{\mathcal{P}}$ with the same false statement to be proved, and initializes internal variables $\tilde{\tau}_v, \tilde{i}_v, \tilde{C}$.

When $\mathcal{V}$ submits a challenge C, $\mathcal{P}$ chooses a random hash key k and sends $Encode(C||k||H_k(\tilde{\tau}_v)||\tilde{i}_v)$. It updates $\tilde{C} = C$. Upon receiving a response from $\widetilde{\mathcal{P}}$, it decodes it and parses the result as a tuple (R, k, h, i). It then internally performs the algorithm of $\widetilde{\mathcal{V}}$. If the result is a decrement to $\tilde{i}_v$ and the erasure of round from $\tilde{\tau}_v$, then $\mathcal{P}$ sends B to $\mathcal{V}$. If the result is an increment to $\tilde{i}_v$, it sends R to $\mathcal{V}$. If the result is no change, it simulates the next message to $\widetilde{\mathcal{P}}$. It continues simulating $\widetilde{\mathcal{V}}$ in this way.

By construction, $\widetilde{V}$ will accept in this simulation only when $\mathcal{V}$ accepts. Hence this malicious prover $\mathcal{P}$ can falsely convince $\mathcal{V}$ with probability $> \frac{1}{3}$.

We observe that Lemmas 2 and 6 imply Theorem 2. Taken together, Theorem 2 and Corollary 1 imply Theorem 1.

6   Conclusions and Open Problems

We showed the feasibility of interactive coding for interactive protocols, tolerating a constant fraction of adversarial communication errors. Additionally, our compiled error-resilient protocol is within a constant factor from optimal in its round complexity, and has an honest prover/verifier which is efficient given oracle access to the original prover/verifier. We also believe that our result should "scale down" to the setting of "interactive proofs for muggles" considered by Goldwasser et al. [15], who showed how to achieve polynomial-time, communication-efficient interactive proofs with $O(n \cdot polylog(n))$ verification time for any language in **NC** (class of uniform, polynomial size and polylogarithmic depth circuits).

We now list several interesting open problems for future work.

Better Communication Complexity. Unlike its asymptotically optimal round complexity and error rate, our compiler incurs an $O(r)$ overhead in communication complexity, as compared to the error-free setting (where r is the number of rounds in the error-free setting). This is due to the $O(r)$ parallel repetition used to amplify the soundness of the original protocol Π in Corollary 1. We chose to consider communication complexity as a secondary constraint, as compared to achieving *constant* error-resiliency α. This is customary in the interactive coding literature, as, for example, Ghaffari et al. [14] show how to achieve optimal $\alpha = 2/7$ for traditional ("completeness-only") interactive coding, at the expense of quadratic blow-up in communication complexity. We could also use a more randomness efficient parallel repetition for public-coin **IP** due to Bellare et al. [3]. This would reduce the communication complexity from the verifier to the prover, but not from the prover to the verifier (hence only saving us a constant factor in communication complexity).

In our view, the seemingly large (but polynomial) communication complexity blow-up is largely a matter of a rather arbitrary historical tradition defining the class **IP** as having a *constant* soundness error. Traditionally, this was always justified by the parallel repetition, even though such repetition only reduces the soundness error at the expense of the communication complexity! To see this more clearly, imagine an alternative definition of **IP**, where the soundness error is $2^{-\Omega(r)}$ (where r is the round complexity). While quantitatively different, it clearly does not change the resulting class **IP** (due to parallel repetition). Yet, with this (qualitatively equivalent) definition we only need to run parallel repetition a *constant* number of times to gain the extra factor $2^{-\Omega(r)}$ needed to make our protocol backtracking-resilient. This means that our compiler would suddenly become "asymptotically optimal" even for communication complexity, even though nothing really changed from the conceptual point of view.

Hence, compared to the goal of achieving constant error-rate α, the question of achieving better communication complexity blow-up seems to be less well motivated and largely dependent on rather arbitrary definitional choices. Still, once constant α is achieved by our work, it is an interesting open problem if $O(r)$ communication overhead is inherent using the specific (constant soundness) variant of **IP** that we utilized following a historical tradition.

Error-Resilient Arguments. Another interesting direction is to add error-resilience to arguments, where the soundness condition only holds against a computationally sound prover. At first glance, this appears trivial, since our compiled protocol has honest prover/verifier which are efficient relative to the original prover/verifier. The subtlety comes from the fact our compiler must amplify the computational soundness of the original argument from $1/2$ to $2^{-\Omega(r)}$. For proofs, such amplification is trivial via parallel repetition. In contrast, hardness amplification for arguments must involve an *explicit reduction*. And although many such reductions exist [8,18,21] for public-coin arguments, all of them come at the expense of a horrible degradation in the running time of the malicious prover. In particular, for polynomially bounded provers these reductions can "only" amplify soundness to become negligible in the security parameter, which is not enough to absorb a factor $2^{O(r)}$ we need, when the number of rounds r is polynomial in the security parameter. Moreover, Dodis et al. [10] gave strong evidence that hardness amplification "beyond negligible" is false in general, suggesting that a radically new approach is required for adding error-resilience to arguments.

Other Security Properties? Finally, given the negative results of [9,12] for zero-knowledge/privacy in the presence of adversarial noise, coupled with our positive results for soundness, it is interesting to characterize which other security properties can withstand adversarial errors, and at what cost.

References

1. Agrawal, S., Gelles, R., Sahai, A.: Adaptive protocols for interactive communication, manuscript (2013). http://arxiv.org/abs/1312.4182
2. Alon, N., Goldreich, O., Håstad, J., Peralta, R.: Simple construction of almost k-wise independent random variables. Random Struct. Algorithms **3**(3), 289–304 (1992)
3. Bellare, M., Goldreich, O., Goldwasser, S.: Randomness in interactive proofs. Comput. Complex. **3**(4), 319–354 (1993)
4. Brakerski, Z., Kalai, Y.T.: Efficient interactive coding against adversarial noise. In: FOCS, pp. 160–166 (2012)
5. Braverman, M.: Towards deterministic tree code constructions. In: ITCS, pp. 161–167 (2012)
6. Braverman, M., Efremenko, K.: List and unique coding for interactive communication in the presence of adversarial noise. In: FOCS (2014)
7. Braverman, M., Rao, A.: Towards coding for maximum errors in interactive communication. In: STOC, pp. 159–166 (2011)
8. Chung, K.-M., Liu, F.-H.: Parallel repetition theorems for interactive arguments. In: Micciancio, D. (ed.) TCC 2010. LNCS, vol. 5978, pp. 19–36. Springer, Heidelberg (2010)
9. Chung, K.-M., Pass, R., Telang, S.: Knowledge-preserving interactive coding. In: FOCS (2013)
10. Dodis, Y., Jain, A., Moran, T., Wichs, D.: Parallel repetition theorems for interactive arguments. In: TCC, pp. 467–493 (2012)
11. Gelles, R., Moitra, A., Sahai, A.: Efficient and explicit coding for interactive communication. In: FOCS, pp. 768–777 (2011)
12. Gelles, R., Sahai, A., Wadia, A.: Private interactive communication across an adversarial channel. In: ITCS, pp. 135–144 (2014)
13. Ghaffari, M., Haeupler, B.: Optimal error rates for interactive coding ii: efficiency and list decoding. In: FOCS (2014)
14. Ghaffari, M., Haeupler, B., Sudan, M.: Optimal error rates for interactive coding: adaptivity and other settings. In: STOC (2014)
15. Goldwasser, S., Kalai, Y.T., Rothblum, G.N.: Delegating computation: interactive proofs for muggles. In: STOC, pp. 113–122 (2008)
16. Goldwasser, S., Sipser, M.: Private coins versus public coins in interactive proof systems. In: STOC, pp. 59–68 (1986)
17. Haeupler, B.: Interactive channel capacity revisited. In: FOCS (2014)
18. Håstad, J., Pass, R., Wikström, D., Pietrzak, K.: An efficient parallel repetition theorem. In: Micciancio, D. (ed.) TCC 2010. LNCS, vol. 5978, pp. 1–18. Springer, Heidelberg (2010)
19. Moore, C., Schulman, L.J.: Tree codes and a conjecture on exponential sums. CoRR, abs/1308.6007 (2013)
20. Naor, J., Naor, M.: Small-bias probability spaces: efficient constructions and applications. SIAM J. Comput. **22**(4), 838–856 (1993)

21. Pass, R., Venkitasubramaniam, M.: An efficient parallel repetition theorem for arthur-merlin games. In: STOC, pp. 420–429 (2007)
22. Schulman, L.J.: Communication on noisy channels: a coding theorem for computation. In: FOCS, pp. 724–733 (1992)
23. Schulman, L.J.: Deterministic coding for interactive communication. In: STOC, pp. 747–756 (1993)
24. Schulman, L.J.: Coding for interactive communication. IEEE Trans. Inf. Theory **42**(6), 1745–1756 (1996)
25. Shamir, A.: Ip = pspace. In: FOCS, pp. 11–15 (1990)

Information-Theoretic Local Non-malleable Codes and Their Applications

Nishanth Chandran[1]([✉]), Bhavana Kanukurthi[2], and Srinivasan Raghuraman[3]

[1] Microsoft Research, Bengaluru, India
nichandr@microsoft.com
[2] Department of Computer Science and Automation, Indian Institute of Science,
Bengaluru, India
bhavana@csa.iisc.ernet.in
[3] Massachusetts Institute of Technology, Cambridge, USA
srirag@mit.edu

Abstract. Error correcting codes, though powerful, are only applicable in scenarios where the adversarial channel does not introduce "too many" errors into the codewords. Yet, the question of having guarantees even in the face of many errors is well-motivated. Non-malleable codes, introduced by Dziembowski et al. (ICS 2010), address precisely this question. Such codes guarantee that even if an adversary completely over-writes the codeword, he cannot transform it into a codeword for a related message. Not only is this a creative solution to the problem mentioned above, it is also a very meaningful one. Indeed, non-malleable codes have inspired a rich body of theoretical constructions as well as applications to tamper-resilient cryptography, CCA2 encryption schemes and so on.

Another remarkable variant of error correcting codes were introduced by Katz and Trevisan (STOC 2000) when they explored the question of decoding "locally". Locally decodable codes are coding schemes which have an additional "local decode" procedure: in order to decode a bit of the message, this procedure accesses only a few bits of the codeword. These codes too have received tremendous attention from researchers and have applications to various primitives in cryptography such as private information retrieval. More recently, Chandran et al. (TCC 2014) explored the converse problem of making the "re-encoding" process local. Locally updatable codes have an additional "local update" procedure: in order to update a bit of the message, this procedure accesses/rewrites only a few bits of the codeword.

At TCC 2015, Dachman-Soled et al. initiated the study of locally decodable and updatable non-malleable codes, thereby combining all the important properties mentioned above into one tool. Achieving locality and non-malleability is non-trivial. Yet, Dachman-Soled et al. provide a

B. Kanukurthi—Research supported in part by a start-up grant from the Indian Institute of Science and in part by a grant from the Ministry of Communications and Information Technology, Government of India.

S. Raghuraman—Research done while this author was at Indian Institute of Science and Microsoft Research, India.

E. Kushilevitz and T. Malkin (Eds.): TCC 2016-A, Part II, LNCS 9563, pp. 367–392, 2016.
DOI: 10.1007/978-3-662-49099-0_14

meaningful definition of local non-malleability and provide a construction that satisfies it. Unfortunately, their construction is secure only in the computational setting.

In this work, we construct information-theoretic non-malleable codes which are locally updatable and decodable. Our codes are non-malleable against $\mathcal{F}_{\mathsf{half}}$, the class of tampering functions where each function is arbitrary but acts (independently) on two separate parts of the codeword. This is one of the strongest adversarial models for which explicit constructions of standard non-malleable codes (without locality) are known. Our codes have $\mathcal{O}(1)$ rate and locality $\mathcal{O}(\lambda)$, where λ is the security parameter. We also show a rate 1 code with locality $\omega(1)$ that is non-malleable against bit-wise tampering functions. Finally, similar to Dachman-Soled *et al.*, our work finds applications to information-theoretic secure RAM computation.

1   Introduction

Non-malleable Codes. The notion of error correcting codes allow a sender to encode a message $s \in \{0,1\}^k$ into a codeword $C \in \{0,1\}^n$ such that a receiver can then decode the original message s from a tampered codeword $\tilde{C} = f(C)$. Naturally, s cannot be recovered from arbitrarily tampered codewords, and hence traditional error correcting codes (for the Hamming distance metric) require that the tampering function f be such that $\tilde{C} = C + \Delta$, with $\Delta \in \{0,1\}^n$ and the Hamming weight of Δ is $\leq \delta n$ (for some constant $0 < \delta < 1$). While powerful, error correcting codes provide no guarantees for larger classes of tampering functions. In light of this, Dziembowski *et al.* [19], introduced the notion of *non-malleable codes*. Informally, non-malleable codes are codes such that for all messages $s \in \{0,1\}$, and for all f in the class of tampering functions $\mathcal{F}$, $\mathsf{Dec}(f(\mathsf{Enc}(s)))$ is either s or is *unrelated* to s. A little thought reveals that even in this case, $\mathcal{F}$ cannot be arbitrary – for example, if $\mathcal{F}$ includes the function $\mathsf{Enc}(\mathsf{Dec}(\cdot) + 1)$, then the output of $\mathsf{Dec}(f(\mathsf{Enc}(s)))$ would be $s + 1$ and clearly related to s. A rich line of work has explored the largest possible class of tampering functions $\mathcal{F}$ for which non-malleable codes can be constructed. Existential results [12,19,22], are known for large classes of tampering functions (essentially any function family whose size is less than $\mathcal{F}_{\mathsf{all}}$, the class of all functions). The works of [4,5,13] construct explicit non-malleable codes against the class of tampering functions $\mathcal{F}_{\mathsf{bit}}$ (i.e., functions that operate on every bit of the codeword separately) and $\mathcal{F}_{\mathsf{pertperm}}$ (i.e., functions that can perturb or permute bits of the codeword), while the works of [1,2,10,18] construct such codes against the class of tampering functions $\mathcal{F}_{\mathsf{half}}$ (i.e., functions that operate independently on two halves of the codeword). Non-malleable codes have found many applications in cryptography, such as in tamper resilient cryptography [3,26] and in constructing CCA secure encryption schemes [15].

Codes with Locality. Locally decodable codes (introduced formally by Katz and Trevisan [25]), are a class of error correcting codes, where every bit of the message can be decoded by reading only a few bits of the corrupted codeword. These codes have a wide range of applications and several constructions of such codes are known (see Yekhanin's survey [32] for further details). Locally updatable codes (introduced by Chandran *et al.* [8]) are error correcting codes with the property that in order to obtain a codeword of message s' from a codeword of message s (where s and s' differ only in one bit), one only needs to modify a few bits of the codeword.

Locally Updatable/Decodable Non-malleable Codes. A natural question to ask is whether we can construct non-malleable codes that can be locally decoded and updated. Indeed, Dachman-Soled *et al.* [16] consider the above question and show how to construct locally updatable/decodable non-malleable codes. Combining local decodability with non-malleability is challenging: indeed, local decodability gives us a way to read a bit of the message by only reading a few bits of the codeword. If these bits were precisely the ones which are tampered, then how can non-malleability be guaranteed? In particular, it is likely that these bits are not accessed while decoding some other bits of the message. At its core, the challenge is that the adversary could tamper the codeword in such a manner that decoding some of the bits of the message could return $\bot$, while the others may not. While this can be detected via a "global" decode, locally it will be undetected, thus resulting in a weak form of malleability. Dachman-Soled *et al.* capture these challenges by requiring that this weak form of malleability is all that the adversary will be able to accomplish. To be more specific, they show that their construction satisfies a (slightly) weaker form of non-malleability – in this, given a codeword $C = \mathsf{Enc}(s), s \in \{0,1\}^k$, an adversary may come up with a mauled codeword $\tilde{C}$ such that $\mathsf{Dec}(i, \tilde{C}) = s_i$ for $i \in [I]$, for some $[I] \subseteq [k]$ and $\mathsf{Dec}(i, \tilde{C}) = \bot$ for $i \notin [I]$. Otherwise, the standard definition of non-malleability holds.

Dachman-Soled *et al.* present a construction that is non-malleable in the split-state adversarial model *and* requires the adversary to be computationally bounded. Given the rich body of work in constructing information-theoretic non-malleable codes and local codes (individually), we believe the question of building local, non-malleable codes in the information-theoretic setting is very well motivated. This is the question which we investigate in this work.

1.1   Results

1. We construct a locally updatable and locally decodable non-malleable code that is non-malleable against the tampering class $\mathcal{F}_{\mathsf{half}}$, which denotes the class of tampering functions that operate independently on two different parts of the codeword, but can otherwise be arbitrary. Our code has constant rate and a decode/update locality of $\mathcal{O}(\lambda)$, where λ is the security parameter.
2. We can also obtain such non-malleable codes against the tampering class $\mathcal{F}_{\mathsf{bit}}$. In this case, our code has rate 1 and decode/update locality $\omega(1)$.
3. The work of Dachman-Soled *et al.* [16] showed how to use a local non-malleable code that is also leakage-resilient [26] to construct a protocol for

secure RAM computation that remains secure when the adversary can tamper and leak from memory. In a similar way, we show how to use a leakage-resilient version of our code to construct an *information-theoretic* protocol for secure RAM computation that remains secure when the adversary can tamper and leak from memory[1].

1.2   Techniques

Overview of [16]. Before we describe our techniques, we begin with a description of how Dachman-Soled *et al.* [16] construct their locally decodable/updatable non-malleable code. The idea is as follows: to encode a message $s \in \{0, 1\}^k$, pick a key key to a symmetric key encryption scheme and compute the codeword as $(\mathsf{Enc_{NM}}(\mathsf{key}), \mathsf{AEnc_{key}}(1, s_1), \cdots, \mathsf{AEnc_{key}}(1, s_1))$, where $\mathsf{Enc_{NM}}(\cdot)$ denotes a standard non-malleable code, $\mathsf{AEnc_{key}}(\cdot)$ denotes an authenticated encryption with key key, and s_i denotes the ith bit of s ($i \in [k]$). Now, suppose $\mathsf{Enc_{NM}}$ is a non-malleable code against a tampering function class $\mathcal{F_{NM}}$, then the claim is that the above construction is non-malleable against the tampering function class $\mathcal{F}$ of the form (f_1, f_2), where $f_1 \in \mathcal{F_{NM}}$ and f_2 is any polynomial-time computable function. To see why this is true, consider the following two cases: (a) the tampering function $f \in \mathcal{F}$ is such that f does not tamper with $\mathsf{Enc_{NM}}(\mathsf{key})$; (b) the tampering function $f \in \mathcal{F}$ is such that f tampers with $\mathsf{Enc_{NM}}(\mathsf{key})$. In the first case, note that the function f_2 does not have any information about the key key, and hence by the security of the authenticated encryption scheme, we have that any polynomial-time computable f_2 cannot tamper the authenticated encryptions of the s_i values to any related message[2]. In the second case, note that by the non-malleability of $\mathsf{Enc_{NM}}(\cdot)$, we have that f_1 can only compute an encoding of key' such that key' is unrelated to key. Since key' will be used to authenticate and decrypt the ciphertexts in the other part of the codeword, this essentially means that the output of the decode algorithm will be unrelated to s. Choosing $\mathsf{Enc_{NM}}$ to be the non-malleable code of Aggarwal *et al.* [1], gives a local non-malleable code that is secure against $\mathcal{F}^3_{\mathsf{split+poly}}$, which denotes the class of tampering functions that operate independently on three parts of the codeword, and additionally constrains the third function to be polynomial-time computable.

Challenges. A first attempt to convert the above code into an information-theoretically secure one is to use an information-theoretic authenticated encryption $\mathsf{ITAEnc_{itkey}}$ instead of $\mathsf{AEnc_{key}}$ above. We could follow a similar idea – encode itkey using a non-malleable encoding and encrypt+authenticate every bit of the

---

[1] Of course, in the case of single party RAM computation, our protocol is information-theoretic modulo the encryption that is used in the underlying oblivious RAM (ORAM) protocol; in the case of secure multi-party computation, we obtain a tamper and leakage resilient information-theoretic secure computation protocol.

[2] Of course, the adversary can always copy certain ciphertexts and have them decode to s_i and maul other indices to decode to $\perp$, but as noted earlier, this is allowed by their definition of non-malleability.

message. Unfortunately, this idea quickly runs into trouble – for the information-theoretic authenticated encryption to be secure, we require the size of itkey to be proportional to the message and hence |itkey| must be proportional to k^3. Now, if we encode itkey as a whole using a non-malleable code, we have lost all locality (since we would require locality of k to even decode the code and retrieve itkey). On the other hand, if we encode every part of itkey separately, then an adversary can always replace one of these parts with a (sub)key of his choice and appropriately replace the ciphertext to obtain a codeword that decodes to s_i in a few indices and decodes to (independent) $\tilde{s}_j$ in other indices (this violates the non-malleability definition from [16]). It seems that, in order to succeed, we must use an information-theoretic locally decodable code to encode itkey, thereby running into a circular problem!

Another approach that one might consider is to start with an information-theoretic non-malleable code and somehow make that code "local". Typical constructions of non-malleable codes make use of error-correcting codes with certain independence guarantees "across states". This independence is exploited to get non-malleability. Unfortunately, this approach doesn't yield any benefit as the locality of an error correcting code is orthogonal to its independence. Indeed, it is easy to see that a locally decodable code with *locality* r, necessarily has independence less than r.

Construction of Local Non-malleable Codes. To explain how we overcome these challenges, we explore the construction using (information-theoretic) authenticated encryption in more detail. The construction non-malleably encodes a itkey and uses it to authenticate encrypt the message block-wise. Non-malleability dictates that |itkey| $\geq k$ and this ruins locality. This tradeoff between non-malleability and locality is our main challenge. Our main observation is that this approach of using authenticated encryption is an overkill. In particular, we have existing constructions of non-malleable codes in the split-state model which we could use as a building block, except that it is unclear how to use them.

Consider this (insecure) construction: split the message s into k/t blocks each of size t, for some parameter t. Encode the message s as $(\text{Enc}_{\text{NM}}(s_1, \cdots, s_t), \cdots, \text{Enc}_{\text{NM}}(s_{k-t+1}, \cdots, s_k))$. To decode a bit s_i, decode $\text{Enc}_{\text{NM}}(s_{\lceil \frac{i}{t} \rceil}, \cdots, s_{\lceil \frac{i}{t} \rceil + t})$ and recover s_i appropriately. Let each block of the encoding be stored on separate states i.e., increase the number of states to $2k/t$. It is easy to see that this construction is not secure against $\mathcal{F}_{\text{split}}^{2k/t}$. Indeed, an adversary can always replace one block, say the first block, with an encoding of a known message, say all zeroes. Even though $\text{Enc}_{\text{NM}}(0^t)$ is independent of $s_1, \cdots, s_t$, the new message is related to the underlying message as a whole. The main problem is that an adversary is allowed to tamper certain parts of the encoding independently and still create a "globally related" codeword.

---

[3] One might think that we only require authentication and hence could use a shorter key; however non-malleable codes inherently imply that the underlying message be hidden, thus forcing us to use a key as long as the message.

This brings us to the following question: *how can we combine non-malleable encodings of different blocks of messages, so that the resulting construction is non-malleable?* The answer lies in preventing such isolated tampering or at least detecting it when it happens. To do this, we simply tie together all the encodings by using itkey to provide consistency across blocks. If an adversary changes one block independently, either it is detected or he needs to change all blocks to something independent. This use of itkey, as randomness that allows for consistency checks across blocks, and not as an encryption key, allows us to keep itkey short and achieve locality.

In retrospect, all our constructions are remarkably simple. We first give an overview of our non-malleable construction that is secure against $\mathcal{F}_{\mathsf{split}}^{2k/t+2}$ (i.e., the class of tampering functions that operate independently on $2k/t + 2$ parts of the codeword, for some chosen parameter t and message length k). We will discuss how to reduce the number of states later. The idea is as follows: to authenticate a part of the message, we will pick a random value r and encode it twice in 2 different states – once on its own and once with a message. In other words, to encode a message $s \in \{0,1\}^k$, split s into $\frac{k}{t}$ parts, each of length t, as before. Now, pick $\frac{k}{t}$ random r_i values (each r_i being of length λ). These r_i values correspond to the key itkey above.

Encode the message s as $(\mathsf{Enc}_{\mathsf{NM}}(r_1, \cdots, r_{\frac{k}{t}}), \mathsf{Enc}_{\mathsf{NM}}(r_1, s_1, \cdots, s_t), \cdots, \mathsf{Enc}_{\mathsf{NM}}(r_{\frac{k}{t}}, s_{k-t+1}, \cdots, s_k))$. To decode a bit s_i, decode $\mathsf{Enc}_{\mathsf{NM}}(r_1, \cdots, r_{\frac{k}{t}})$ to obtain $r_{\lceil \frac{i}{t} \rceil}$; then decode $\mathsf{Enc}_{\mathsf{NM}}(r_{\lceil \frac{i}{t} \rceil}, s_{\lceil \frac{i}{t} \rceil}, \cdots, s_{\lceil \frac{i}{t} \rceil + t})$. Now, check if the r values encoded in both these codewords match and if so, output s_i. The claim then is that if $\mathsf{Enc}_{\mathsf{NM}}(\cdot)$ is non-malleable against a tampering function class $\mathcal{F}_{\mathsf{NM}}$, then the above construction is non-malleable against a tampering function class $\mathcal{F}$ of the form $(f_0, f_1, \cdots, f_k)$, where $f_i \in \mathcal{F}_{\mathsf{NM}}, 0 \leq i \leq k$. At a very high level, to see why this is true, again consider two cases: (a) if the adversary does not maul the first component of the codeword, then if he mauls any other component of the codeword, the decode algorithm will output $\bot$ (except with probability $2^{-\lambda}$) as he must get "guess" an r_i value encoded in a different state; (b) if the adversary mauls the first component of the codeword, then he must maul all other components of the codeword (as otherwise the decode algorithm will output $\bot$) and by the non-malleability of the underlying code, the new codeword will be independent of the s_i values. We note that the r in our construction plays a role similar to the one played by the secret label L in the leakage and tamper-resilient RAM computation construction of Faust *et al.* [21]. While this indeed gives us a construction of a locally decodable/encodable non-malleable code, the locality of the code is $t + \frac{k}{t}$ (and is thus minimized with $t = \sqrt{k}$); also, using the $\mathcal{F}_{\mathsf{half}}$ code from Aggarwal *et al.* [1] this gives us a construction that is non-malleable against $\mathcal{F}_{\mathsf{split}}^{\sqrt{k}}$.

We now show how to reduce the number of states. Suppose $\mathsf{Enc}_{\mathsf{NM}}(\cdot)$ is non-malleable against the tampering class $\mathcal{F}_{\mathsf{half}}$, then $\mathsf{Enc}_{\mathsf{NM}}(\cdot)$ has the form (L, R) and hence our above construction has the form $(L_0, R_0, L_1, R_1, \cdots, L_t, R_t)$. In such a case, we show that the codeword can be written as $(C_1, C_2, C_3, C_4) = ([L_0], [R_0], [L_1, \cdots, L_t], [R_1, \cdots, R_t])$ and that this construction is non-malleable against $\mathcal{F}_{\mathsf{split}}^4$. While this code is secure against a larger class of tampering function, it still has locality $t + \frac{k}{t}$. However, we then show that a *single* r value can be reused across the encodings (instead of $\frac{k}{t}$ different r_i values) as long as we encode the s_i values with indices, and moreover that this r value does not even have to be encoded using a non-malleable code (as long as it is hidden). This can be accomplished by simply secret sharing r into r_L, r_R and storing them separately. In other words, our final construction has the form $([r_L, L_1, \cdots, L_k], [r_R, R_1, \cdots, R_k])$, where $(L_i, R_i) = \mathsf{Enc}_{\mathsf{NM}}(r, i, s_i)$ and $r_L \oplus r_R = r$. Instantiating the $\mathsf{Enc}_{\mathsf{NM}}(\cdot)$ with the code of [1] gives us our first result, while instantiating it with the code of [13] gives us our second result.

Tamper and Leakage-Resilient RAM Computation. In order to obtain a protocol for secure RAM computation that is tamper and leakage resilient, Dachman-Soled *et al.* [16] require the local non-malleable code to tolerate many-time leakage (i.e., the adversary can obtain an unbounded amount of leakage throughout the course of the protocol, but is bounded by the amount of leakage that can be obtained in between successive updates to the memory that will "refresh" the encoding). The challenge is to obtain such a construction even though the update algorithm is local and only updates a small part of the codeword. In their work, [16] do this by computing a Merkle hash of the ciphertexts and by encoding this Merkle hash along with the symmetric key key and by computing a fresh encoding of key together with the root of the Merkle Hash everytime. However, intuitively, obtaining such a guarantee seems a contradictory task for us – information theoretically, if we do not bound the total amount of leakage, and only refresh a part of the encoding, then the adversary over time can learn information about the various parts of the codeword (and hence the message itself, thereby defeating non-malleability). We show that by compromising on the leakage bound tolerated, and by using the information-theoretic leakage-resilient non-malleable codes of Aggarwal *et al.* [3], we can achieve both information-theoretic leakage/tamper-resilience along with locality, by periodically refreshing "different" parts of the codeword. We note here, that leakage and tamper-resilient RAM computation has also been studied by Faust *et al.* [21] in a model different from Dachman-Soled *et al.* [16] (and our work). In the model of Faust *et al.* [21], they allow an adversary to obtain and store past codewords and use that to tamper with the later encodings; on the other hand, they assume a tamper and leak-free component. Faust *et al.* [21] use continuous non-malleable codes [20], to obtain their construction. They show that if the underlying continuous non-malleable code is information-theoretic, then their final construction is also information-theoretic; however, no information-theoretic construction of continuous non-malleable codes are known. Furthermore, that construction would require a tamper/leak free component; in our case, as in [16], the memory of the RAM can be completely subjected to leakage and tampering.

1.3 Organization of the Paper

In Sect. 2, we present the formal definition of non-malleable coding schemes with locality. As a stepping stone towards our main construction, in Sect. 3, we present a construction of a non-malleable coding scheme with $\tilde{\mathcal{O}}(\sqrt{k})$ locality against $\mathcal{F}_{\mathsf{split}}^4$ adversaries. We present our main result namely, a constant rate non-malleable coding scheme with $\mathcal{O}(\lambda)$ locality against $\mathcal{F}_{\mathsf{half}}$, in Sect. 4. Section 5 contains our constructions which are also locally updatable and leakage-resilient. Finally, Sect. 6 presents the application of our non-malleable codes to secure RAM computation.

2 Preliminaries

2.1 Notation

We say that two probability distributions $\mathcal{X}$ and $\mathcal{Y}$ are ϵ-close if their statistical distance is $\leq \epsilon$ and this is denoted by $\mathcal{X} \approx_\epsilon \mathcal{Y}$. The formal definition is given below.

Definition 1. *Let $\mathcal{X}, \mathcal{Y}$ be two probability distributions over some set S. Their statistical distance is*

$$\mathbf{SD}\,(\mathcal{X}, \mathcal{Y}) \stackrel{def}{=} \max_{T \subseteq S}\{\Pr[\mathcal{X} \in T] - \Pr[\mathcal{Y} \in T]\} = \frac{1}{2}\sum_{s \in S}\left|\Pr_{\mathcal{X}}[s] - \Pr_{\mathcal{Y}}[s]\right|.$$

We say that $\mathcal{X}$ and $\mathcal{Y}$ are ϵ-close if $\mathbf{SD}\,(\mathcal{X}, \mathcal{Y}) \leq \epsilon$ and this is denoted by $\mathcal{X} \approx_\epsilon \mathcal{Y}$.

For a sequence $x = (x_1, \ldots, x_n)$ and set $S \subseteq [n]$, we use $x|_S$ to denote the subsequence of x_i values where $i \in S$. For any string y and $i \in [|y|]$, we use y_i to denote the ith bit of y. The security parameter is denoted by λ. We use $\tilde{\mathcal{O}}(\cdot)$ to denote asymptotic estimates that hide poly-logarithmic factors in the involved parameter.

2.2 Definitions

Definition 2 *(Coding schemes). A coding scheme consists of a pair of functions* $\mathsf{Enc} : \{0,1\}^k \to \{0,1\}^n$ *and* $\mathsf{Dec} : \{0,1\}^n \to \{0,1\}^k \cup \{\bot\}$ *where k is the message length, n is the block length and $k < n$.*

1. *The encoder Enc takes as input a message $s \in \{0,1\}^k$ and outputs a codeword $c = \mathsf{Enc}(s)$.*
2. *The decoder Dec when given a correct (untampered) codeword as input, outputs the corresponding message. The correctness requirement is that for all $s \in \{0,1\}^k, \mathsf{Dec}(\mathsf{Enc}(s)) = s$, with probability 1.*

The rate of the coding scheme is the ratio k/n. A coding scheme is said to have relative distance δ (or minimum distance δn), for some $\delta \in [0, 1)$, if for every $s \in \{0,1\}^k$ the following holds. Let $X := \mathsf{Enc}\,(s)$. Then, for any $\Delta \in \{0,1\}^n$ of

Hamming weight at most δn, $\mathsf{Dec}(X + \Delta) = s$ with probability 1. Standard error correcting codes, as defined above, are only applicable in settings where the adversarial channel cannot make too many (i.e., more than δn) errors. Non-malleable codes, introduced by Dziembowski et al. [19], provide a meaningful guarantee in situations where the adversarial channel may completely overwrite the codeword. Informally, a coding scheme is said to be non-malleable if an adversary cannot transform the codeword of a message s into a codeword of a related message s'. Note that such codes do not focus on error-tolerance and, therefore, the parameter δ is set to 0.

Definition 3 *(Non-malleable codes [19]). A coding scheme* $(\mathsf{Enc}, \mathsf{Dec})$ *with message length k and block length n is said to be non-malleable with error ϵ (also called* exact *security) with respect to a family $\mathcal{F}$ of tampering functions acting on $\{0,1\}^n$ (i.e., each $f \in \mathcal{F}$ maps $\{0,1\}^n$ to $\{0,1\}^n$) if for every $f \in \mathcal{F}$ there is a simulator $\mathcal{S}$ such that for all $s \in \{0,1\}^k$, we have*

$$\mathbf{Tamper}_s^f \approx_\epsilon \mathbf{Ideal}_{\mathcal{S},s} \equiv \left\{ \begin{array}{l} \tilde{s} \leftarrow \mathcal{S}^{f(\cdot)}, \text{ where } \tilde{s} \in \{0,1\}^k \cup \{\bot, \underline{\mathsf{same}}\} \\ \text{Output } s \text{ if } \tilde{s} = \underline{\mathsf{same}}, \text{ and } \tilde{s} \text{ otherwise} \end{array} \right\}$$

where $\mathbf{Tamper}_s^f$ *is the output of the tampering experiment defined by*

$$\mathbf{Tamper}_s^f \equiv \left\{ \begin{array}{c} C \leftarrow \mathsf{Enc}(s) ; \tilde{C} \leftarrow f(C) ; \tilde{s} \leftarrow \mathsf{Dec}\left(\tilde{C}\right) \\ \text{Output } \tilde{s} \end{array} \right\}$$

In this work, we focus on information-theoretic non-malleable codes i.e., the $\approx_\epsilon$ is measured by statistical distance. Our goal is to design information-theoretic non-malleable codes which are also local. Locally decodable codes (LDCs), introduced by Katz and Trevisan [25] are a class of error correcting codes, where every bit of the message can be probabilistically decoded by reading only a few bits of the (possibly corrupted) codeword. We now state the formal definition.

Definition 4 *(Local Decodability [25]). A coding scheme* $(\mathsf{Enc}, \mathsf{Dec})$ *with message length k and block length n is said to be (r, δ, ϵ)-locally decodable if there exists a randomized decoding algorithm Dec such that the following properties hold.*

1. *For all $s \in \{0,1\}^k$, $i \in [k]$ and all vectors $y \in \{0,1\}^n$ such that the Hamming distance between $\mathsf{Enc}(s)$ and y is not more than δn,*

$$\Pr[\mathsf{Dec}(y, i) = s_i] \geq 1 - \epsilon,$$

   where the probability is taken over the random coin tosses of the algorithm Dec.
2. *Dec reads at most r coordinates of y.*

Dachman-Soled et al. [16] introduced and designed codes which combine non-malleability and locality. While their coding scheme is in the computational setting, their definition is applicable even for the information-theoretic setting by simply using the appropriate notion of "closeness".

Definition 5 *(Local Decodability and Non-malleability, LDNMC* [16]*). A coding scheme* (Enc, Dec) *with message length k and block length n is said to be a* $(r, \epsilon_1, \epsilon_2)$*-locally decodable non-malleable coding scheme with respect to a family* $\mathcal{F}$ *of tampering functions acting on* $\{0,1\}^n$ *if it is* $(r, 0, \epsilon_1)$*-locally decodable and if for every $f \in \mathcal{F}$ there is a simulator $\mathcal{S}$ such that for all $s \in \{0,1\}^k$, we have*

$$\textbf{Tamper}_s^f \approx_{\epsilon_2} \textbf{Ideal}_{\mathcal{S},s}$$

where $\textbf{Tamper}_s^f$ is the output of the tampering experiment defined by

$$\textbf{Tamper}_s^f \equiv \left\{ \begin{array}{c} C \leftarrow \textsf{Enc}(s)\,; \tilde{C} \leftarrow f(C)\,; \forall i, \tilde{s}_i \leftarrow \textsf{Dec}\left(\tilde{C}, i\right) \\ Output\ \tilde{s} = \tilde{s}_1, \cdots, \tilde{s}_k \end{array} \right\}$$

and $\textbf{Ideal}_{\mathcal{S},s}$ is defined as

1. $(\mathcal{I}, s^*) \leftarrow \mathcal{S}^{f(\cdot)}\left(1^\lambda\right)$, *where $\mathcal{I} \subseteq [k]$ and $s^* \in \{0, 1, \bot\}^k$.*
2. *If $\mathcal{I} = [k]$, then $\tilde{s} = s^*$. Otherwise, $\tilde{s}|_{\mathcal{I}} = \bot$ and $\tilde{s}|_{\overline{\mathcal{I}}} = s|_{\overline{\mathcal{I}}}$, where $\overline{\mathcal{I}}$ denotes the complement of the set $\mathcal{I}$.*
3. *Output $\tilde{s}$.*

Dachman-Soled *et al.* apply local NMCs to the problem of secure RAM computation. Towards this end, they require NMCs that are also locally updatable. Locally updatable and decodable error correcting codes were formalized in the work of Chandran *et al.* [8]. Informally, such codes allow for a bit of the underlying message to be updated by rewriting just a few bits of the codeword. In the context of non-malleable codes, which do not require error-tolerance, a weaker definition [16] of local updatability suffices, which we present next.

Definition 6 *(Local Decodability and Updatability* [8,16]*). A coding scheme* (Enc, Dec, Update) *with message length k and block length n is said to be* $(r_1, r_2, \delta, \epsilon)$*-locally decodable and updatable if it is (r_1, δ, ϵ)-locally decodable and there exists a randomized algorithm* Update *such that:*

1. *For all $s \in \{0,1\}^k$, $i \in [k]$, $s_i' \in \{0, 1, \bot\}$ and all vectors $y \in \{0,1\}^n$ such that the Hamming distance between $\textsf{Update}^C(i, s_i')$ and y is not more than δn, where $C = \textsf{Enc}(s)$,*

$$\Pr[\textsf{Dec}(y, i) = s_i'] \geq 1 - \epsilon,$$

   where the probability is taken over the random coin tosses of the algorithm Dec.
2. Update *reads and changes at most r_2 coordinates of y.*

Remarks. We note that the above definition can be extended in a straightforward manner to account for the decoding of a codeword which has been updated multiple times as opposed to once (as above). Additionally, although we focus on the case of *correcting* zero errors in the codeword, we can modify our construction to get a construction that tolerates errors and is also non-malleable, by simply encoding each "state" of our non-malleable codeword using an LDC. This would reduce the error-tolerance of the code (by a fraction equal to the number of states) and the rate of the obtained code would now depend on the rate of the LDC. It suffices however here to discuss the case of correcting zero errors in the codeword.

Similar to [16], we also construct locally decodable/updatable leakage resilient non-malleable codes and use them to construct information-theoretic tamper and leakage resilient RAM computation. We refer the reader to the full version of this paper for details on these primitives.

3 Non-malleable Codes with $\widetilde{\mathcal{O}}(\sqrt{k})$ Locality Against $\mathcal{F}_{\mathsf{split}}^4$

In this section, we describe a construction of a locally updatable/decodable non-malleable code that is non-malleable against the tampering function class $\mathcal{F}_{\mathsf{split}}^4$ (i.e., the tampering function class that operates independently on 4 parts of the codeword), with locality $\widetilde{\mathcal{O}}(\sqrt{k})$, where k is the length of the message being encoded. The motivation for presenting this construction is two-fold: first, it has ideas which will lead to our main construction described in Sect. 4; second, this construction will be used to achieve the application to secure RAM computation. We remark that the four parts of the codeword seen by a 4-state adversary from the class $\mathcal{F}_{\mathsf{split}}^4$ need not be of equal sizes (in fact, they are not in this construction). We specify how a codeword is broken into 4 parts in the proof of Theorem 2. If one is so particular on requiring all parts to be of equal length, we note that it is trivial to achieve this via padding, although this would affect the rate of the final coding scheme (by at most a constant factor).

Recall that λ denotes the security parameter; t denotes a parameter that will be set appropriately later on. Let $\mathsf{NMC} = (\mathsf{Enc}_{\mathsf{NM}}, \mathsf{Dec}_{\mathsf{NM}})$ be a non-malleable coding scheme on strings of length $\lambda k/t$ and $\mathsf{NMC}' = (\mathsf{Enc}_{\mathsf{NM}}', \mathsf{Dec}_{\mathsf{NM}}')$ be a non-malleable coding scheme on strings of length $\lambda + t$. We assume without loss of generality that t divides k. We define the following coding scheme:

1. $\mathsf{Enc}(s)$: On input $s \in \{0,1\}^k$, the algorithm splits s into k/t blocks, say $\mathsf{s}_1, \ldots, \mathsf{s}_{k/t}$ of size t each. Then, the algorithm chooses k/t random strings $r_1, \ldots, r_{k/t} \in \{0,1\}^\lambda$, and computes $c = \mathsf{Enc}_{\mathsf{NM}}(r_1\|\ldots\|r_{k/t})$ and $e_i = \mathsf{Enc}_{\mathsf{NM}}'(r_i\|\mathsf{s}_i)$ for $i \in [k/t]$. The algorithm finally outputs the codeword $C = (c, e_1, \ldots, e_{k/t})$.
2. $\mathsf{Dec}(C, i)$: On input $i \in [k]$, the algorithm reads the first and $(\lceil i/t \rceil + 1)$th block of C, retrieving $c, e_{\lceil i/t \rceil}$. Then it runs $r_1\|\ldots\|r_{k/t} := \mathsf{Dec}_{\mathsf{NM}}(c)$. If the decoding algorithm outputs $\bot$, the algorithm outputs $\bot$. Otherwise, it computes $r_{\lceil i/t \rceil}^*\|\mathsf{s}_{\lceil i/t \rceil} = \mathsf{Dec}_{\mathsf{NM}}'(e_{\lceil i/t \rceil})$. If the decoding algorithm outputs $\bot$, the algorithm outputs $\bot$. If $r_{\lceil i/t \rceil}^* \neq r_{\lceil i/t \rceil}$, the algorithm outputs $\bot$. Otherwise, the algorithm outputs s_i from $\mathsf{s}_{\lceil i/t \rceil}$.

We instantiate this construction by instantiating the non-malleable codes NMC and NMC'. A natural and strong class of functions which we may assume

378   N. Chandran et al.

the schemes are non-malleable against is the class of split-state adversaries, $\mathcal{F}_{\mathsf{half}}$, that tamper two parts[4] of the codeword independently, that is, $f \in \mathcal{F}_{\mathsf{half}}$ iff $f : \{0,1\}^n$ can be written as $f(c_1, c_2) = (f_1(c_1), f_2(c_2))$ for $f_1, f_2 : \{0,1\}^{n/2} \to \{0,1\}^{n/2}$. The following result is known.

Theorem 1 [1]. *Let $\mathcal{F}_{\mathsf{half}}$ be the function family of split-state adversaries over $\{0,1\}^n$. Let $\epsilon > 0$ be an arbitrary value and $k, n > 0$ be integers such that $k/n \leq \gamma$, for some constant γ. Then there exists a non-malleable code with respect to $\mathcal{F}_{\mathsf{half}}$, with k-bit source-messages and n-bit codewords, and exact security ϵ.*

We now show the local-decodability and non-malleability of the above scheme instantiated using the non-malleable code in Lemma 1.

Theorem 2. *Assume that NMC, NMC' be non-malleable coding schemes of rate $1/\gamma$, and exact security $\epsilon > 0$, which is non-malleable against split-state adversaries. Then the above coding scheme is a $((\lambda(1 + k/t) + t)\gamma, 0, k(\epsilon + 2^{-\lambda})/t)$-locally decodable non-malleable coding scheme which is non-malleable against the tampering class $\mathcal{F}_{\mathsf{split}}^4$, for any $t \leq k$. The rate of the code is $1/\gamma'$, where $\gamma' = \lambda\gamma/t + (1 + \lambda/t)\gamma$.*

Proof. Clearly the decoding algorithm reads $(\lambda(1 + k/t) + t)\gamma$ positions of the codeword since $|c| = \lambda\gamma k/t$ and $|e_i| = (\lambda + t)\gamma$. Also, since the decoding algorithm is deterministic, the error probability in the local decoding procedure is 0. This justifies the first two parameters of the coding scheme.

The underlying non-malleable codes NMC and NMC' are non-malleable against split-state adversaries and let L_0, R_0 be the parts of c viewed by the two states corresponding to the split-state adversary for c, and let L_i, R_i be the parts of e_i viewed by the two states corresponding to the split-state adversary for e_i, for all $i \in [k/t]$. We define how a codeword is split into four parts – the four-state adversaries against which the above scheme is non-malleable consists of adversaries which are arbitrary functions over $L_0, R_0, L_1 \| \ldots \| L_{k/t}$ and $R_1 \| \ldots \| R_{k/t}$.

To show the theorem, for any suitable four-state adversary $f = (f_1, f_2, f_3, f_4)$ as described above, which we denote as $f_1(L_0)$, $f_2(R_0)$, $f_3(L_1, \ldots, L_{k/t})$ and $f_4(R_1, \ldots, R_{k/t})$, we need to construct a simulator $\mathcal{S}$. We describe the simulator with oracle access to f.

1. Let $\mathcal{S}'$ be the simulator for the non-malleable code NMC, and $\mathcal{S}''$ for NMC'. Now $\mathcal{S}^{f(\cdot)}$ simulates $\mathcal{S}'$ once and $\mathcal{S}''$ k/t times to obtain simulated codewords c and e_i for all $i \in [k/t]$. Note that the simulator described for the code instantiated from Lemma 1 does not need oracle access to the tampering function to produce simulated codewords assuming a super-polynomial message space, which is the case since the messages are of length $\lambda k/t$ and $\lambda + t$ respectively, where $0 \leq t \leq k$. Let $C = (c, e_1, \ldots, e_{k/t})$.
2. $\mathcal{S}^{f(\cdot)}$ then computes $\tilde{C} = f(C)$, where $\tilde{C} = (\tilde{c}, \tilde{e}_1, \ldots, \tilde{e}_{k/t})$.

---

[4] While we define these two parts to be of equal length, as remarked earlier, there is no such requirement.

3. Let L_i, R_i be the parts of $\tilde{e}_i$ viewed by the two states corresponding to the split-state adversary for e_i, for all $i \in [k/t]$. Let $f_i' = f_3(L_1, \ldots, L_{i-1}, \cdot, L_{i+1}, \ldots, L_{k/t})$ and $f_i'' = f_4(R_1, \ldots, R_{i-1}, \cdot, R_{i+1}, \ldots, R_{k/t})$ for all $i \in [k/t]$. Now $\mathcal{S}^{f(\cdot)}$ simulates $\mathcal{S}'^{f_1(\cdot), f_2(\cdot)}$ with $\tilde{c}$ and $\mathcal{S}''^{f_i'(\cdot), f_i''(\cdot)}$ with $\tilde{e}_i$ for each $i \in [k/t]$ internally. $\mathcal{S}'$ returns an output $r' = r_1' \| \ldots \| r_{k/t}'$, where $r' \in \{0,1\}^{\lambda k/t} \cup \{\perp, \underline{\text{same}}\}$ and $\mathcal{S}''$ returns an output $r_i'' \| s_i' \in \{0,1\}^{\lambda + t} \cup \{\perp, \underline{\text{same}}\}$ for each $i \in [k/t]$.
4. Set $\mathcal{I} = \emptyset$.
   (a) If $r' = \perp$, then set $\mathcal{I} = [k]$ and $s_i^* = \perp$ for all $i \in [k]$.
   (b) If $r' = \underline{\text{same}}$, then, for each $i \in [k]$, check if $r_{\lceil i/t \rceil}'' \| s_{\lceil i/t \rceil}' \neq \underline{\text{same}}$. If so, set $\mathcal{I} = \mathcal{I} \cup \{j : (i-1)t + 1 \leq j \leq it\}$ and $s_j^* = \perp$ for all j such that $(i-1)t + 1 \leq j \leq it$.
   (c) Otherwise, if $r' \notin \{\perp, \underline{\text{same}}\}$, set $\mathcal{I} = [k]$. Let $s' = s_1' \| \ldots \| s_{k/t}'$. For each $i \in [k]$,
      i. If $r_{\lceil i/t \rceil}'' \| s_{\lceil i/t \rceil}' = \perp$ or $r_{\lceil i/t \rceil}'' \| s_{\lceil i/t \rceil}' = \underline{\text{same}}$, then set $s_i^* = \perp$.
      ii. Otherwise, check if $r_{\lceil i/t \rceil}' = r_{\lceil i/t \rceil}''$. If so, set $s_i^* = s_i'$, otherwise set $s_i^* = \perp$.
5. Output $(\mathcal{I}, s^*)$.

The above simulator now defines $\mathbf{Ideal}_{\mathcal{S}, s}$. We must now show that $\mathbf{Tamper}_s^f \approx_{\epsilon'} \mathbf{Ideal}_{\mathcal{S}, s}$ for some negligible ϵ'. We proceed through a series of hybrids of the form $\mathbf{Ideal}_{\mathcal{S}_j, s}$ for $j \in [k/t]$, which is the same as $\mathbf{Ideal}_{\mathcal{S}_{j-1}, s}$ except that it randomly chooses an $r_i \in \{0,1\}^{\lambda}$, and generates $e_i = \mathrm{Enc}_{\mathsf{NM}}'(r_i \| s_i)$ for $i = j$ and obtains $r_i'' \| s_i' = \mathrm{Dec}_{\mathsf{NM}}'(g_i(e_i))$ for $i = j$, where $g_i = (f_i', f_i'')$; if $r_i'' \| s_i' = r_i \| s_i$, it outputs $\underline{\text{same}}$. This is to say that it obtains codewords and performs decoding as in the real experiment for index j (as well). Note that $\mathbf{Ideal}_{\mathcal{S}, s} \equiv \mathbf{Ideal}_{\mathcal{S}_0, s}$.

Lemma 1. *For all $j \in [k/t]$, $\mathbf{Ideal}_{\mathcal{S}_{j-1}, s} \approx_{\epsilon} \mathbf{Ideal}_{\mathcal{S}_j, s}$.*

Proof. Let $\mathcal{A} = (\mathcal{A}_1, \mathcal{A}_2, \mathcal{A}_3, \mathcal{A}_4)$ be a four-state adversary that can distinguish between the outputs of the experiments $\mathbf{Ideal}_{\mathcal{S}_{j-1}, s}$ and $\mathbf{Ideal}_{\mathcal{S}_j, s}$ for some $j \in [k/t]$, with an advantage of α. We describe a split-state adversary $\mathcal{B} = (\mathcal{B}_1, \mathcal{B}_2)$ (where $\mathcal{B}_1$ and $\mathcal{B}_2$ operate independently on two halves of the underlying codeword) that can break the non-malleability of the scheme NMC' with the same advantage α. However, since NMC' is non-malleable against split-state adversaries with exact security ϵ, $\alpha \leq \epsilon$, which completes the proof.

Let $\mathcal{C}$ be the challenger for the scheme NMC'. $\mathcal{B}$, using $\mathcal{A}$, executes as follows. First, $\mathcal{A}$ chooses a message $s \in \{0,1\}^k$ on which he will distinguish between the outputs of the experiments $\mathbf{Ideal}_{\mathcal{S}_{j-1}, s}$ and $\mathbf{Ideal}_{\mathcal{S}_j, s}$, and sends it to $\mathcal{B}_1$, which then splits s into k/t blocks, say $s_1, \ldots, s_{k/t}$ of size t each. $\mathcal{B}_1$ randomly chooses $r_i \in \{0,1\}^{\lambda}$, where λ is the security parameter, for all $i \in [j]$. It then generates $e_i = \mathrm{Enc}_{\mathsf{NM}}'(r_i \| s_i)$ for $i \in [j-1]$. Let $\mathcal{S}'$ be the simulator for the non-malleable code NMC, and $\mathcal{S}''$ for NMC' (on all indices but j). Now, $\mathcal{B}_1$ simulates $\mathcal{S}'$ once and $\mathcal{S}''$ $k/t - j$ times to obtain simulated codewords c and e_i for all $i \in \{j+1, \ldots, k/t\}$. $\mathcal{B}_1$ then sends the message $r_j \| s_j$ to the challenger $\mathcal{C}$.

$\mathcal{C}$ then either computes $e_j = \mathsf{Enc}'_{\mathsf{NM}}(r_j \| \mathsf{s}_j)$, or uses the simulator $\mathcal{S}''$ to obtain a simulated codeword e_j. It then splits e_j into two parts L_j, R_j and sends L_j to $\mathcal{B}_1$ and R_j to $\mathcal{B}_2$ respectively. $\mathcal{B}_1$ splits c into two parts, L_0 and R_0, and e_i into two parts, L_i and R_i, for each $i \in [k/t] \backslash \{j\}$. $\mathcal{B}_1$ then sends across L_0 to $\mathcal{A}_1$, R_0 to $\mathcal{A}_2$, L_i, for all $i \in [k/t]$, to $\mathcal{A}_3$, and R_i, for all $i \in [k/t] \backslash \{j\}$, to $\mathcal{A}_4$, and $\mathcal{B}_2$ sends across R_j to $\mathcal{A}_4$. $\mathcal{A}$ then chooses its four-state tampering function $f = (f_1, f_2, f_3, f_4)$, and computes $\tilde{L}_0 = f_1(L_0)$, $\tilde{R}_0 = f_2(R_0)$, $\tilde{L} = f_3(L_1, \ldots, L_{k/t})$ and $\tilde{R} = f_4(R_1, \ldots, R_{k/t})$. It then parses $\tilde{L}$ as $\tilde{L} = \left(\tilde{L}_1, \ldots, \tilde{L}_{k/t}\right)$ and $\tilde{R}$ as $\tilde{R} = \left(\tilde{R}_1, \ldots, \tilde{R}_{k/t}\right)$. $\mathcal{A}_1$, $\mathcal{A}_2$, $\mathcal{A}_3$ and $\mathcal{A}_4$ also determine the descriptions of the functions f_1, f_2, $f'_i = f_3(L_1, \ldots, L_{i-1}, \cdot, L_{i+1}, \ldots, L_{k/t})$ and $f''_i = f_4(R_1, \ldots, R_{i-1}, \cdot, R_{i+1}, \ldots, R_{k/t})$, respectively, for all $i \in \{j+1, \ldots, k/t\}$. Then, $\mathcal{A}_1$ sends across $\tilde{L}_0$ and the description of the function f_1 to $\mathcal{B}_1$, $\mathcal{A}_2$ sends across $\tilde{R}_0$ and the description of the function f_2 to $\mathcal{B}_1$, $\mathcal{A}_3$ sends across $\tilde{L}_i$, for all $i \in [k/t]$, and the descriptions of the functions f'_i, for all $i \in \{j+1, \ldots, k/t\}$, to $\mathcal{B}_1$, and $\mathcal{A}_4$ sends across $\tilde{R}_i$, for all $i \in [k/t] \backslash \{j\}$, and the descriptions of the functions f''_i, for all $i \in \{j+1, \ldots, k/t\}$, to $\mathcal{B}_1$, and $\tilde{R}_j$ to $\mathcal{B}_2$.

$\mathcal{B}_1$ then computes $r''_i \| \mathsf{s}'_i = \mathsf{Dec}'_{\mathsf{NM}}\left(\tilde{L}_i, \tilde{R}_i\right)$ for $i \in [j-1]$; if $r''_i \| \mathsf{s}'_i = r_i \| \mathsf{s}_i$, it renames the output $r''_i \| \mathsf{s}'_i$ as <u>same</u>. $\mathcal{B}_1$ then simulates $\mathcal{S}'^{f_1(\cdot), f_2(\cdot)}$ with $\tilde{c} = \left(\tilde{L}_0, \tilde{R}_0\right)$ and $\mathcal{S}''^{f'_i(\cdot), f''_i(\cdot)}$ with $\tilde{e}_i = \left(\tilde{L}_i, \tilde{R}_i\right)$ for each $i \in \{j+1, \ldots, k/t\}$, to obtain $r' = r'_1 \| \ldots \| r'_{k/t}$, where $r' \in \{0,1\}^{\lambda k/t} \cup \{\bot, \underline{\mathsf{same}}\}$ and $r''_i \| \mathsf{s}'_i \in \{0,1\}^{\lambda+t} \cup \{\bot, \underline{\mathsf{same}}\}$ for each $i \in \{j+1, \ldots, k/t\}$. $\mathcal{B}_1$ and $\mathcal{B}_2$ then send across $\tilde{L}_j$ and $\tilde{R}_j$ respectively to $\mathcal{C}$. $\mathcal{C}$ then responds back with $r''_j \| \mathsf{s}'_j \in \{0,1\}^{\lambda+t} \cup \{\bot, \underline{\mathsf{same}}\}$ to $\mathcal{B}_1$, by either running the real decode algorithm or by simulation (in coherence with the way it generated the codeword to begin with).

$\mathcal{B}_1$ then defines variables $\mathcal{I}$ and s^*, and sets $\mathcal{I} = \emptyset$.

1. If $r' = \bot$, then it sets $\mathcal{I} = [k]$ and $s^*_i = \bot$ for all $i \in [k]$.
2. If $r' = \underline{\mathsf{same}}$, then, for each $i \in [k]$, it checks if $r''_{\lceil i/t \rceil} \| \mathsf{s}'_{\lceil i/t \rceil} \neq \underline{\mathsf{same}}$. If so, it sets $\mathcal{I} = \mathcal{I} \cup \{\beta : (i-1)t + 1 \le \beta \le it\}$ and $s^*_\beta = \bot$ for all β such that $(i-1)t + 1 \le \beta \le it$.
3. Otherwise, if $r' \notin \{\bot, \underline{\mathsf{same}}\}$, it sets $\mathcal{I} = [k]$. Let $s' = \mathsf{s}'_1 \| \ldots \| \mathsf{s}'_{k/t}$. For each $i \in [k]$,
   (a) If $r''_{\lceil i/t \rceil} \| \mathsf{s}'_{\lceil i/t \rceil} = \bot$ or $r''_{\lceil i/t \rceil} \| \mathsf{s}'_{\lceil i/t \rceil} = \underline{\mathsf{same}}$, then it sets $s^*_i = \bot$.
   (b) Otherwise, it checks if $r'_{\lceil i/t \rceil} = r''_{\lceil i/t \rceil}$. If so, it sets $s^*_i = s'_i$, otherwise it sets $s^*_i = \bot$.

Finally, $\mathcal{B}_1$ defines $\tilde{s}$ as follows. If $\mathcal{I} = [k]$, then it sets $\tilde{s} = s^*$. Otherwise, it sets $\tilde{s}|_{\mathcal{I}} = \bot$ and $\tilde{s}|_{\overline{\mathcal{I}}} = s|_{\overline{\mathcal{I}}}$, where $\overline{\mathcal{I}}$ denotes the complement of the set $\mathcal{I}$. Then, $\mathcal{B}_1$ sends across $\tilde{s}$ to $\mathcal{A}$. $\mathcal{A}$ then replies back with a bit b to $\mathcal{B}_1$, where $b = 0$ denotes that the experiment run was $\mathbf{Ideal}_{\mathcal{S}_{j,s}}$, and $b = 1$ denotes that the experiment run was $\mathbf{Ideal}_{\mathcal{S}_{j-1,s}}$, which $\mathcal{B}_1$ forwards to $\mathcal{C}$.

Note that if the challenger $\mathcal{C}$ sent across a simulated codeword for e_j, then the experiment is identical to $\mathbf{Ideal}_{\mathcal{S}_{j-1,s}}$, while if $\mathcal{C}$ sent across a real codeword for the message $r_j \| s_j$ for e_j, then the experiment is identical to $\mathbf{Ideal}_{S_j,s}$. Hence, since $\mathcal{A}$ is able to distinguish between the outputs of the two experiments with advantage α, so can $\mathcal{B}$ between the outputs of the experiments $\mathbf{Tamper}_{r_j\|s_j}^{(f'_j, f''_j)}$ and $\mathbf{Ideal}_{S'',r_j\|s_j}$ as defined in Definition 3, where $f'_j = f_3\left(L_1, \ldots, L_{j-1}, \cdot, L_{j+1}, \ldots, L_{k/t}\right)$ and $f''_j = f_4\left(R_1, \ldots, R_{j-1}, \cdot, R_{j+1}, \ldots, R_{k/t}\right)$. Since $\mathcal{B}$ is a valid split-state adversary for the scheme NMC', as mentioned before, $\alpha \le \epsilon$, which completes the proof. $\square$

We define $\mathbf{Ideal}_{S^\dagger,s}$, which is the same as $\mathbf{Ideal}_{S_{k/t},s}$ except that the first two components of the codeword are generated using an actual encoding (i.e., $\mathsf{Enc}_{\mathsf{NM}}\left(r_1 \| \ldots \| r_{k/t}\right)$) and the decoding is done using the real decoding algorithm; i.e., $r' = \mathsf{Dec}_{\mathsf{NM}}\left((f_1, f_2)\left(\mathsf{Enc}_{\mathsf{NM}}\left(r_1 \| \ldots \| r_{k/t}\right)\right)\right)$. If $r' = r_1 \| \ldots \| r_{k/t}$, it outputs <u>same</u>.

Lemma 2. $\mathbf{Ideal}_{S_{k/t},s} \approx_\epsilon \mathbf{Ideal}_{S^\dagger,s}$.

Proof. Let $\mathcal{A} = (\mathcal{A}_1, \mathcal{A}_2, \mathcal{A}_3, \mathcal{A}_4)$ be a four-state adversary who can distinguish between the outputs of the experiments $\mathbf{Ideal}_{S_{k/t},s}$ and $\mathbf{Ideal}_{S^\dagger,s}$ with an advantage of α. We describe a split-state adversary $\mathcal{B} = (\mathcal{B}_1, \mathcal{B}_2)$ (where $\mathcal{B}_1$ and $\mathcal{B}_2$ do not communicate with each other) who can break the non-malleability of the the scheme NMC with the same advantage α. However, since NMC is non-malleable against split-state adversaries with exact security ϵ, $\alpha \le \epsilon$, which completes the proof.

Let $\mathcal{C}$ be the challenger for the scheme NMC. $\mathcal{B}$, using $\mathcal{A}$, executes as follows. First, $\mathcal{A}$ chooses a message $s \in \{0,1\}^k$ on which he will distinguish between the outputs of the experiments $\mathbf{Ideal}_{S_{k/t},s}$ and $\mathbf{Ideal}_{S^\dagger,s}$, and sends it to $\mathcal{B}_1$, which then splits s into k/t blocks, say $s_1, \ldots, s_{k/t}$ of size t each. $\mathcal{B}_1$ randomly chooses $r_i \in \{0,1\}^\lambda$, where λ is the security parameter, for all $i \in [k/t]$. It then generates $e_i = \mathsf{Enc}'_{\mathsf{NM}}\left(r_i \| s_i\right)$ for $i \in [k/t]$. $\mathcal{B}_1$ then sends the message $r_1 \| \ldots \| r_{k/t}$ to the challenger $\mathcal{C}$.

$\mathcal{C}$ then either computes $c = \mathsf{Enc}_{\mathsf{NM}}\left(r_1 \| \ldots \| r_{k/t}\right)$, or uses the simulator $\mathcal{S}'$ to obtain a simulated codeword c, where $\mathcal{S}'$ is the simulator for the non-malleable code NMC. It then splits c into two parts L_0, R_0 and sends L_0 to $\mathcal{B}_1$ and R_0 to $\mathcal{B}_2$ respectively. $\mathcal{B}_1$ splits e_i into two parts, L_i and R_i, for each $i \in [k/t]$. $\mathcal{B}_1$ then sends across L_0 to $\mathcal{A}_1$, L_i, for all $i \in [k/t]$, to $\mathcal{A}_3$, and R_i, for all $i \in [k/t]$, to $\mathcal{A}_4$, and $\mathcal{B}_2$ sends across R_0 to $\mathcal{A}_2$. $\mathcal{A}$ then chooses its four-state tampering function $f = (f_1, f_2, f_3, f_4)$, and computes $\tilde{L}_0 = f_1(L_0)$, $\tilde{R}_0 = f_2(R_0)$, $\tilde{L} = f_3\left(L_1, \ldots, L_{k/t}\right)$ and $\tilde{R} = f_4\left(R_1, \ldots, R_{k/t}\right)$. It then parses $\tilde{L}$ as $\tilde{L} = \left(\tilde{L}_1, \ldots, \tilde{L}_{k/t}\right)$ and $\tilde{R}$ as $\tilde{R} = \left(\tilde{R}_1, \ldots, \tilde{R}_{k/t}\right)$. Then, $\mathcal{A}_1$ sends across $\tilde{L}_0$ to $\mathcal{B}_1$, $\mathcal{A}_2$ sends across $\tilde{R}_0$ to $\mathcal{B}_2$, $\mathcal{A}_3$ sends across $\tilde{L}_i$, for all $i \in [k/t]$, to $\mathcal{B}_1$, and $\mathcal{A}_4$ sends across $\tilde{R}_i$, for all $i \in [k/t]$, to $\mathcal{B}_1$.

$\mathcal{B}_1$ then computes $r''_i \| s'_i = \mathsf{Dec}'_{\mathsf{NM}}\left(\tilde{L}_i, \tilde{R}_i\right)$ for $i \in [k/t]$; if $r''_i \| s'_i = r_i \| s_i$, it renames the output $r''_i \| s'_i$ as <u>same</u>. $\mathcal{B}_1$ and $\mathcal{B}_2$ then send across $\tilde{L}_0$ and $\tilde{R}_0$ respectively to $\mathcal{C}$. $\mathcal{C}$ then responds back with $r' = r'_1 \| \ldots \| r'_{k/t} \in \{0,1\}^{\lambda k/t} \cup \{\perp, \underline{\mathsf{same}}\}$

to $\mathcal{B}_1$, by either running the real decode algorithm or by simulation (in coherence with the way it generated the codeword to begin with).

$\mathcal{B}_1$ then defines variables $\mathcal{I}$ and s^*, and sets $\mathcal{I} = \emptyset$.

1. If $r' = \bot$, then it sets $\mathcal{I} = [k]$ and $s_i^* = \bot$ for all $i \in [k]$.
2. If $r' = \underline{\text{same}}$, then, for each $i \in [k]$, it checks if $r''_{\lceil i/t \rceil} \| s'_{\lceil i/t \rceil} \neq \underline{\text{same}}$. If so, it sets $\mathcal{I} = \mathcal{I} \cup \{\beta : (i-1)t + 1 \leq \beta \leq it\}$ and $s_\beta^* = \bot$ for all β such that $(i-1)t + 1 \leq \beta \leq it$.
3. Otherwise, if $r' \notin \{\bot, \underline{\text{same}}\}$, it sets $\mathcal{I} = [k]$. Let $s' = s'_1 \| \ldots \| s'_{k/t}$. For each $i \in [k]$,
    (a) If $r''_{\lceil i/t \rceil} \| s'_{\lceil i/t \rceil} = \bot$ or $r''_{\lceil i/t \rceil} \| s'_{\lceil i/t \rceil} = \underline{\text{same}}$, then it sets $s_i^* = \bot$.
    (b) Otherwise, it checks if $r'_{\lceil i/t \rceil} = r''_{\lceil i/t \rceil}$. If so, it sets $s_i^* = s_i'$, otherwise it sets $s_i^* = \bot$.

Finally, $\mathcal{B}_1$ defines $\tilde{s}$ as follows. If $\mathcal{I} = [k]$, then it sets $\tilde{s} = s^*$. Otherwise, it sets $\tilde{s}|_{\mathcal{I}} = \bot$ and $\tilde{s}|_{\overline{\mathcal{I}}} = s|_{\overline{\mathcal{I}}}$, where $\overline{\mathcal{I}}$ denotes the complement of the set $\mathcal{I}$. Then, $\mathcal{B}_1$ sends across $\tilde{s}$ to $\mathcal{A}$. $\mathcal{A}$ then replies back with a bit b to $\mathcal{B}_1$, where $b = 0$ denotes that the experiment run was $\mathbf{Ideal}_{S^\dagger, s}$, and $b = 1$ denotes that the experiment run was $\mathbf{Ideal}_{S_{k/t}, s}$, which $\mathcal{B}_1$ forwards to $\mathcal{C}$.

Note that if the challenger $\mathcal{C}$ sent across a simulated codeword for c, then the experiment is identical to $\mathbf{Ideal}_{S_{k/t}, s}$, while if $\mathcal{C}$ sent across a real codeword for the message $r_1 \| \ldots \| r_{k/t}$ for c, then the experiment is identical to $\mathbf{Ideal}_{S^\dagger, s}$. Hence, since $\mathcal{A}$ is able to distinguish between the outputs of the two experiments with advantage α, so can $\mathcal{B}$ between the outputs of the experiments $\mathbf{Tamper}_{r_1\|\ldots\|r_{k/t}}^{(f_1, f_2)}$ and $\mathbf{Ideal}_{S', r_1\|\ldots\|r_{k/t}}$ as defined in Definition 3. Since $\mathcal{B}$ is a valid split-state adversary for the scheme NMC, as mentioned before, $\alpha \leq \epsilon$, which completes the proof. $\square$

Lemma 3. $\mathbf{Ideal}_{S^\dagger, s} \approx_{k(\epsilon + 2^{-\lambda})/t} \mathbf{Tamper}_s^f$.

Proof. The only difference between the two experiments is step 4 of the simulator, which is the decoding step. In particular, differences only lie in steps 4(b) and 4(c)i where $r_i'' \| s_i' = \underline{\text{same}}$.

In step 4(b), $r' = \underline{\text{same}}$ while $r_i'' \| s_i' \neq \underline{\text{same}}$. By the non-malleability of NMC', $r_i'' \| s_i'$ is independent of $r_i \| s_i$, in particular, r_i'' is independent of r_i. Further, the split state adversaries see nothing else which has information about r_i (since the r_i's are all random). Hence, the probability that $r_i'' = r_i$ is atmost $2^{-\lambda}$, and with probability $1 - 2^{-\lambda}$, even the real decoding algorithm outputs $\bot$. Hence, for each $i \in [k/t]$, the output distributions of the two experiments differ only by $\epsilon + 2^{-\lambda}$.

In step 4(c)i. when $r_i'' \| s_i' = \underline{\text{same}}$, $r' \neq \underline{\text{same}}$. By the non-malleability of NMC, r_i' is independent of r_i and the split state adversaries see nothing else which has information about r_i (since the r's are all different and random). Hence, the probability that $r_i'' = r_i'$ is atmost $2^{-\lambda}$, and with probability $1 - 2^{-\lambda}$, even the real decoding algorithm outputs $\bot$. Hence, for each $i \in [k/t]$, the output distributions of the two experiments differ only by $\epsilon + 2^{-\lambda}$. $\square$

Combining all the hybrids, we see that $\mathbf{Tamper}_s^f \approx_{\epsilon'} \mathbf{Ideal}_{S,s}$ for $\epsilon' = \mathcal{O}\left(k\left(\epsilon + 2^{-\lambda}\right)/t\right)$. This completes the proof of non-malleability of the scheme. $\qquad\square$

Corollary 1. *For all* k, *there exists an explicit construction of a* $\left(\widetilde{\mathcal{O}}(\sqrt{k}), 0, \nu(\lambda)\right)$*-locally decodable non-malleable coding scheme over* k*-bit messages with constant rate (for some negligible function* $\nu(\cdot)$*) which is non-malleable against four-state adversaries.*

Proof. This follows by choosing $t = \sqrt{k}$ and using constant-rate non-malleable codes non-malleable against split state adversaries (from [1]) in Theorem 2. $\quad\square$

4   Non-malleable Codes with $\mathcal{O}(\lambda)$ Locality Against $\mathcal{F}_{\mathsf{half}}$

We now present our construction of LDNMC with $\mathcal{O}(\lambda)$ locality and against $\mathcal{F}_{\mathsf{half}}$. The key behind this improvement in locality is that we use just one random string r across all encodings instead of multiple r's as in the previous construction. Somewhat surprisingly, not only are we able to use this idea to build a non-malleable code, we are also able to secure it against a stronger adversarial model, i.e., $\mathcal{F}_{\mathsf{half}}$. Before we present this construction, for ease of exposition, we present a construction that is non-malleable against $\mathcal{F}_{\mathsf{split}}^3$ (and then show how to reduce the number of states to 2). Let $\mathsf{NMC} = (\mathsf{Enc}_{\mathsf{NM}}, \mathsf{Dec}_{\mathsf{NM}})$ be a non-malleable coding scheme on strings of length $\log k + \lambda + 1$, where λ is the security parameter. The construction works as follows:

1. $\mathsf{Enc}(s)$: On input $s \in \{0,1\}^k$, the algorithm chooses a random string $r \in \{0,1\}^\lambda$ and computes $e_i = \mathsf{Enc}_{\mathsf{NM}}(i, r\|s_i)$ for $i \in [k]$. The algorithm finally outputs the codeword $C = (r, e_1, \ldots, e_k)$.
2. $\mathsf{Dec}(C, i)$: On input $i \in [k]$, the algorithm reads the first and $(i + 1)$th block of C, retrieving r, e_i. Then it computes $i^*, r^*\|s_i = \mathsf{Dec}_{\mathsf{NM}}(e_i)$. If the decoding algorithm outputs $\bot$, the algorithm outputs $\bot$. If $r^* \neq r$ or $i^* \neq i$, the algorithm outputs $\bot$. Otherwise, the algorithm outputs s_i.

In order to prove the security of this construction, we digress and consider a modified construction which ignores r and merely encodes each bit of s along with its index i.e., $\mathsf{Enc}(s) = \{e_i = \mathsf{Enc}_{\mathsf{NM}}(i, s_i)\}_{i \in [k]}$. A quick inspection reveals that this does not satisfy our definition of non-malleability. Indeed, an adversary could replace e_1 with an encoding of a bit s_1' of his choosing and leave all other e_is the same. In other words, he can copy some bits of the encoding and replace the rest with encodings of bits chosen independently by him. While this construction is not non-malleable in the standard sense, we can show that the above mauling really is all that the adversary can do.

4.1   Quoted Non-malleability

To formalize this intuition, we introduce a new notion of non-malleability which we call "*Quoted Non-malleability*." This definition is similar in spirit to the definition "unquoted" CCA security (UCCA) defined in Myers and Shelat [27].

Definition 7 *(Quoted-non-malleability, QNMC). A coding scheme* $(\mathsf{Enc}, \mathsf{Dec})$
with message length k *and block length* n *is said to be* quoted-non-malleable *with*
error ϵ *with respect to a family* $\mathcal{F}$ *of tampering functions acting on* $\{0, 1\}^n$ *if for*
every $f \in \mathcal{F}$ *there is a simulator* S *such that for all* $s \in \{0, 1\}^k$, *we have*

$$QTamper_s^f \approx_\epsilon QIdeal_{S,s}$$

where $QTamper_s^f$ *is the output of the tampering experiment defined*[5] *by*

$$QTamper_s^f \equiv \left\{ \begin{array}{c} C \leftarrow \mathsf{Enc}\,(s)\,;\tilde{C} \leftarrow f\,(C)\,;\forall i,\tilde{s}_i \leftarrow \mathsf{Dec}\left(\tilde{C}, i\right) \\ Output\ \tilde{s} = \tilde{s}_1, \cdots, \tilde{s}_k \end{array} \right\}$$

and $QIdeal_{S,s}$ *is defined by*

$$QIdeal_{S,s} \equiv \left\{ \begin{array}{c} \overline{s} \leftarrow S^{f(\cdot)}\left(1^\lambda\right),\ where\ \overline{s} \in \left(\{0, 1\} \cup \{\bot, \underline{same}\}\right)^k \\ \forall i \in [k],\ if\ \overline{s}_i = \underline{same},\ set\ \tilde{s}_i = s_i,\ otherwise\ set\ \tilde{s}_i = \overline{s}_i \\ Output\ \tilde{s} = \tilde{s}_1, \cdots, \tilde{s}_k \end{array} \right\}$$

We now prove that the construction with the randomness r, i.e. $C = (r, \{e_i = \mathsf{Enc}_{\mathsf{NM}}\,(i, r\|s_i)\}_{i \in [k]})$ is quoted non-malleable.

Theorem 3. *Assume that* NMC *is a non-malleable coding scheme of exact secu-*
rity ϵ, *which is non-malleable against split-state adversaries. Then the above cod-*
ing scheme is a quoted-non-malleable coding scheme with exact security $k\epsilon$ *which*
is non-malleable against three-state adversaries.

Proof. The underlying non-malleable code NMC is non-malleable against split-
state adversaries and let L_i, R_i be the parts of e_i viewed by the two states
corresponding to the split-state adversary for e_i, for all $i \in [k]$. The three-
state adversaries against which the above scheme is non-malleable consists of
adversaries which are arbitrary functions over r, $L_1\|\ldots\|L_k$ and $R_1\|\ldots\|R_k$.

To show the theorem, for any function suitable three-state adversary $f = (f_1, f_2, f_3)$ as described above, which we denote as $f_1\,(r)$, $f_2\,(L_1, \ldots, L_k)$ and
$f_3\,(R_1, \ldots, R_k)$, we need to construct a simulator S. We describe the simulator
with oracle access to f.

1. $S^{f(\cdot)}$ first chooses a random string $r \in \{0, 1\}^\lambda$.
2. Let S' be the simulator for the non-malleable code NMC. Now $S'^{f(\cdot)}$ simulates S' k times to obtain simulated codewords e_i for all $i \in [k]$. Let
   $C = (r, e_1, \ldots, e_k)$.
3. Next $S^{f(\cdot)}$ obtains $r' = f_1\,(r)$, where $r' \in \{0, 1\}^\lambda$.

---

[5] Note that in this definition, we abuse notation mildly by allowing Dec to take the
index i as input, in addition to $\tilde{C}$. The output of $\mathsf{Dec}(\cdot, i)$ is in $\{0, 1\} \bigcup \bot$. Since the
definition of quoted non-malleability makes sense without locality, one can think of
$\mathsf{Dec}(\cdot, \cdot)$ as simply running the actual decode algorithm and simply outputting the
i^{th} bit (or $\bot$ if the decoding fails).

4. Let L_i, R_i be the parts of e_i viewed by the two states correspond-
   ing to the split-state adversary for e_i, for all $i \in [k]$. Let $f'_i = f_2(L_1, \ldots, L_{i-1}, \cdot, L_{i+1}, \ldots, L_k)$ and $f''_i = f_3(R_1, \ldots, R_{i-1}, \cdot, R_{i+1}, \ldots, R_k)$
   for all $i \in [k]$. Now $\mathsf{S}^{f(\cdot)}$ simulates $\mathcal{S}'^{f'_i(\cdot), f''_i(\cdot)}$ internally. At some point, $\mathcal{S}'$
   returns an output $(i', r''_i \| s'_i) \in \{0, 1\}^{\log k + \lambda + 1} \cup \{\bot, \underline{\mathsf{same}}\}$.
5. For each $i \in [k]$,
   (a) if $(i', r''_i \| s'_i) = \bot$, then set $\overline{s}_i = \bot$.
   (b) if $(i', r''_i \| s'_i) = \underline{\mathsf{same}}$,
       i. if $r' \neq r$, then set $\overline{s}_i = \bot$.
       ii. otherwise, set $\overline{s}_i = \underline{\mathsf{same}}$.
   (c) otherwise,
       i. if $r' = r''_i$ and $i' = i$, then set $\overline{s}_i = s'_i$.
       ii. otherwise, set $\overline{s}_i = \bot$.
6. Output $\overline{s}$.

The above simulator defines $\mathbf{QIdeal}_{\mathsf{S},s}$. We must now show that
$\mathbf{QTamper}_s^f \approx_{\epsilon'} \mathbf{QIdeal}_{\mathsf{S},s}$ for some ϵ'. We proceed through a series of hybrids of
the form $\mathbf{QIdeal}_{S_j,s}$ for $j \in [k]$, which is the same as $\mathbf{QIdeal}_{S_{j-1},s}$ except that it
generates $e_i = \mathsf{Enc}_{\mathsf{NM}}(i, r \| s_i)$ for $i = j$ and it obtains $(i', r''_i \| s'_i) = \mathsf{Dec}_{\mathsf{NM}}(g_i(e_i))$
for $i = j$, where $g_i = (f'_i, f''_i)$. If $(i', r''_i \| s'_i) = (i, r \| s_i)$, it outputs $\underline{\mathsf{same}}$. Note that
$\mathbf{QIdeal}_{\mathsf{S},s} \equiv \mathbf{QIdeal}_{S_0,s}$ and $\mathbf{QTamper}_s^f \equiv \mathbf{Ideals}_{S_k,s}$.

Lemma 4. *For all $j \in [k]$, $\mathbf{QIdeal}_{S_{j-1},s} \approx_{\epsilon} \mathbf{QIdeal}_{S_j,s}$.*

Proof. Let $\mathcal{A} = (\mathcal{A}_1, \mathcal{A}_2, \mathcal{A}_3)$ be a three-state adversary who can distinguish
between the outputs of the experiments $\mathbf{QIdeal}_{S_{j-1},s}$ and $\mathbf{QIdeal}_{S_j,s}$ for some
$j \in [k]$, with an advantage of α. We describe a split-state adversary $\mathcal{B} = (\mathcal{B}_1, \mathcal{B}_2)$
(where $\mathcal{B}_1$ and $\mathcal{B}_2$ do not communicate with each other) who can break the
non-malleability of the the scheme NMC with the same advantage α. However,
since NMC is non-malleable against split-state adversaries with exact security ϵ,
$\alpha \leq \epsilon$, which completes the proof.

Let $\mathcal{C}$ be the challenger for the scheme NMC. $\mathcal{B}$, using $\mathcal{A}$, executes as follows.
First, $\mathcal{A}$ chooses a message $s \in \{0, 1\}^k$ on which he will distinguish between
the outputs of the experiments $\mathbf{QIdeal}_{S_{j-1},s}$ and $\mathbf{QIdeal}_{S_j,s}$, and sends it to
$\mathcal{B}_1$. $\mathcal{B}_1$ randomly chooses $r \in \{0, 1\}^\lambda$, where λ is the security parameter. It then
generates $e_i = \mathsf{Enc}_{\mathsf{NM}}(i, r \| s_i)$ for $i \in [j - 1]$. Let $\mathcal{S}'$ be the simulator for the
non-malleable code NMC (on all indices but j). Now, $\mathcal{B}_1$ simulates $\mathcal{S}'$ $k - j$ times
to obtain simulated codewords e_i for all $i \in \{j + 1, \ldots, k\}$. $\mathcal{B}_1$ then sends the
message $(j, r \| s_j)$ to the challenger $\mathcal{C}$.
   $\mathcal{C}$ then either computes $e_j = \mathsf{Enc}_{\mathsf{NM}}(j, r \| s_j)$, or uses the simulator $\mathcal{S}'$ to
obtain a simulated codeword e_j. It then splits e_j into two parts L_j, R_j and
sends L_j to $\mathcal{B}_1$ and R_j to $\mathcal{B}_2$ respectively. $\mathcal{B}_1$ splits e_i into two parts, L_i and
R_i, for each $i \in [k] \backslash \{j\}$. $\mathcal{B}_1$ then sends across r to $\mathcal{A}_1$, L_i, for all $i \in [k]$,
to $\mathcal{A}_2$, and R_i, for all $i \in [k] \backslash \{j\}$, to $\mathcal{A}_3$, and $\mathcal{B}_2$ sends across R_j to $\mathcal{A}_3$. $\mathcal{A}$
then chooses its three-state tampering function $f = (f_1, f_2, f_3)$, and computes
$\tilde{r} = f_1(r)$, $\tilde{L} = f_2(L_1, \ldots, L_k)$ and $\tilde{R} = f_3(R_1, \ldots, R_k)$. It then parses $\tilde{L}$ as

$\tilde{L} = \left(\tilde{L}_1, \dots, \tilde{L}_k\right)$ and $\tilde{R}$ as $\tilde{R} = \left(\tilde{R}_1, \dots, \tilde{R}_k\right)$. $\mathcal{A}_2$ and $\mathcal{A}_3$ also determine the descriptions of the functions $f_i' = f_2(L_1, \dots, L_{i-1}, \cdot, L_{i+1}, \dots, L_k)$ and $f_i'' = f_3(R_1, \dots, R_{i-1}, \cdot, R_{i+1}, \dots, R_k)$, respectively, for all $i \in \{j+1, \dots, k\}$. Then, $\mathcal{A}_1$ sends across $\tilde{r}$ to $\mathcal{B}_1$, $\mathcal{A}_2$ sends across $\tilde{L}_i$, for all $i \in [k]$, and the descriptions of the functions f_i', for all $i \in \{j+1, \dots, k\}$, to $\mathcal{B}_1$, and $\mathcal{A}_3$ sends across $\tilde{R}_i$, for all $i \in [k]\backslash\{j\}$, and the descriptions of the functions f_i'', for all $i \in \{j+1, \dots, k\}$, to $\mathcal{B}_1$, and $\tilde{R}_j$ to $\mathcal{B}_2$.

$\mathcal{B}_1$ then computes $(i', r_i''\|s_i') = \mathsf{Dec}_{\mathsf{NM}}\left(\tilde{L}_i, \tilde{R}_i\right)$ for $i \in [j-1]$; if $(i', r_i''\|s_i') = (i, r\|s_i)$, it renames the output $(i', r_i''\|s_i')$ as <u>same</u>. $\mathcal{B}_1$ then simulates $\mathcal{S}'^{f_i'(\cdot), f_i''(\cdot)}$ with $\tilde{e}_i = \left(\tilde{L}_i, \tilde{R}_i\right)$ for each $i \in \{j+1, \dots, k\}$, to obtain $(i', r_i''\|s_i') \in \{0,1\}^{\lambda+t} \cup \{\perp, \underline{\mathsf{same}}\}$ for each $i \in \{j+1, \dots, k\}$. $\mathcal{B}_1$ and $\mathcal{B}_2$ then send across $\tilde{L}_j$ and $\tilde{R}_j$ respectively to $\mathcal{C}$. $\mathcal{C}$ then responds back with $(j', r_j''\|s_j') \in \{0,1\}^{\lambda+t} \cup \{\perp, \underline{\mathsf{same}}\}$ to $\mathcal{B}_1$, by either running the real decode algorithm or by simulation (in coherence with the way it generated the codeword to begin with).

$\mathcal{B}_1$ then defines the variable $\overline{s}$. For each $i \in [k]$,

1. if $(i', r_i''\|s_i') = \perp$, then it sets $\overline{s}_i = \perp$.
2. if $(i', r_i''\|s_i') = \underline{\mathsf{same}}$,
   (a) if $\tilde{r} \neq r$, then it sets $\overline{s}_i = \perp$.
   (b) otherwise, it sets $\overline{s}_i = \underline{\mathsf{same}}$.
3. otherwise,
   (a) if $\tilde{r} = r_i''$ and $i' = i$, then it sets $\overline{s}_i = s_i'$.
   (b) otherwise, it sets $\overline{s}_i = \perp$.

Finally, $\mathcal{B}_1$ defines $\tilde{s}$ as follows. For each $i \in [k]$, if $\overline{s}_i = \underline{\mathsf{same}}$, set $\tilde{s}_i = s_i$, otherwise set $\tilde{s}_i = \overline{s}_i$. Then, $\mathcal{B}_1$ sends across $\tilde{s}$ to $\mathcal{A}$. $\mathcal{A}$ then replies back with a bit b to $\mathcal{B}_1$, where $b = 0$ denotes that the experiment run was $\mathbf{QIdeal}_{\mathcal{S}_j, s}$, and $b = 1$ denotes that the experiment run was $\mathbf{QIdeal}_{\mathcal{S}_{j-1}, s}$, which $\mathcal{B}_1$ forwards to $\mathcal{C}$.

Note that if the challenger $\mathcal{C}$ sent across a simulated codeword for e_j, then the experiment is identical to $\mathbf{QIdeal}_{\mathcal{S}_{j-1}, s}$, while if $\mathcal{C}$ sent across a real codeword for the message $(j, r\|s_j)$ for e_j, then the experiment is identical to $\mathbf{QIdeal}_{\mathcal{S}_j, s}$. Hence, since $\mathcal{A}$ is able to distinguish between the outputs of the two experiments with advantage α, so can $\mathcal{B}$ between the outputs of the experiments $\mathbf{Tamper}_{(j, r\|s_j)}^{(f_j', f_j'')}$ and $\mathbf{Ideal}_{\mathcal{S}', (j, r\|s_j)}$ as defined in Definition 3, where $f_j' = f_2(L_1, \dots, L_{j-1}, \cdot, L_{j+1}, \dots, L_k)$ and $f_j'' = f_3(R_1, \dots, R_{j-1}, \cdot, R_{j+1}, \dots, R_k)$. Since $\mathcal{B}$ is a valid split-state adversary for the scheme NMC, as mentioned before, $\alpha \leq \epsilon$, which completes the proof. $\square$

Combining all the hybrids, we see that $\mathbf{QTamper}_s^f \approx_{\epsilon'} \mathbf{QIdeal}_{\mathsf{S}, s}$ for $\epsilon' = k\epsilon$. This completes the proof of quoted-non-malleability of the scheme. $\square$

4.2   Achieving Full Non-malleability

Recall that our ultimate goal is to construct a coding scheme which is non-malleable against split-state adversaries. As the theorem below states, we can show that the quoted non-malleable construction from the previous subsection is itself fully non-malleable. As a careful reader may have observed, the proof of quoted non-malleability does not use the randomness of r at all. Indeed, the construction, as we alluded to earlier, is quoted non-malleable even without using r in the encoding. Yet this randomness is precisely what makes the construction (fully) non-malleable. We first show how the construction from Sect. 4 is non-malleable against 3-state adversaries and then show how to modify the construction to achieve security against $\mathcal{F}_{\mathsf{half}}$.

Theorem 4. *Assume that* NMC *is a non-malleable coding scheme of rate* $1/\gamma$ *and exact security* ϵ, *which is non-malleable against split-state adversaries. Then the coding scheme from Sect. 4 is a* $\left(\lambda + (\lambda + \log k + 1)\gamma, 0, k\left(\epsilon + 2^{-\lambda}\right)\right)$-*locally decodable non-malleable coding scheme which is non-malleable against three-state adversaries. The rate of the code is* $1/\gamma'$, *where* $\gamma' = \lambda/k + (\lambda + \log k + 1)\gamma$.

Proof. Clearly the decoding algorithm reads $\lambda + (\lambda + \log k + 1)\gamma$ positions of the codeword since $|r| = \lambda$ and $|e_i| = (\lambda + \log k + 1)\gamma$. Also, since the decoding algorithm is deterministic, the error probability in the local decoding procedure is 0. This justifies the first two parameters of the coding scheme.

The underlying non-malleable code NMC is non-malleable against split-state adversaries and let L_i, R_i be the parts of e_i viewed by the two states corresponding to the split-state adversary for e_i, for all $i \in [k]$. The three-state adversaries against which the above scheme is non-malleable consists of adversaries which are arbitrary functions over r, $L_1 \| \ldots \| L_k$ and $R_1 \| \ldots \| R_k$.

To show the theorem, for any function suitable three-state adversary $f = (f_1, f_2, f_3)$ as described above, which we denote as $f_1(r)$, $f_2(L_1, \ldots, L_k)$ and $f_3(R_1, \ldots, R_k)$, we need to construct a simulator $\mathcal{S}$. We describe the simulator with oracle access to f.

1. $\mathsf{S}^{f(\cdot)}$ first chooses a random string $r \in \{0, 1\}^\lambda$.
2. Let $\mathcal{S}'$ be the simulator for the non-malleable code NMC. Now $\mathcal{S}^{f(\cdot)}$ simulates $\mathcal{S}'$ k times to obtain simulated codewords e_i for all $i \in [k]$. Note that the simulator described for the code instantiated from Lemma 1 does not need oracle access to the tampering function to produce simulated codewords assuming a super-polynomial message space, which is the case since the messages are of length $\log k + \lambda + 1$. Let $C = (r, e_1, \ldots, e_k)$.
3. Next $\mathsf{S}^{f(\cdot)}$ obtains $r' = f_1(r)$, where $r' \in \{0, 1\}^\lambda$.
4. Let L_i, R_i be the parts of e_i viewed by the two states corresponding to the split-state adversary for e_i, for all $i \in [k]$. Let $f'_i = f_2(L_1, \ldots, L_{i-1}, \cdot, L_{i+1}, \ldots, L_k)$ and $f''_i = f_3(R_1, \ldots, R_{i-1}, \cdot, R_{i+1}, \ldots, R_k)$ for all $i \in [k]$. Now $\mathsf{S}^{f(\cdot)}$ simulates $\mathcal{S}'^{f'_i(\cdot), f''_i(\cdot)}$ internally. At some point, $\mathcal{S}'$ returns an output $(i', r''_i \| s'_i) \in \{0, 1\}^{\log k + \lambda + 1} \cup \{\bot, \underline{\mathsf{same}}\}$.
5. Set $\mathcal{I} = \emptyset$.
   (a) If $r' = r$, then, for each $i \in [k]$,
      i. if $(i', r''_i \| s'_i) = \bot$, then set $\mathcal{I} = \mathcal{I} \cup \{i\}$ and $s^*_i = \bot$.
      ii. otherwise, if $(i', r''_i \| s'_i) \neq \underline{\mathsf{same}}$, then set $\mathcal{I} = \mathcal{I} \cup \{i\}$ and $s^*_i = \bot$.

(b) Otherwise, set $\mathcal{I} = [k]$. For each $i \in [k]$,

    i. if $(i', r_i''\|s_i') = \bot$ or $(i', r_i''\|s_i') = \underline{\text{same}}$, then set $s_i^* = \bot$.

    ii. otherwise,

        A. if $r' = r_i''$ and $i' = i$, then set $s_i^* = s_i'$.

        B. otherwise, set $s_i^* = \bot$.

6. Output $(\mathcal{I}, s^*)$.

We first note that for the construction in Sect. 4, $\mathbf{QTamper}_s^f \equiv \mathbf{Tamper}_s^f$, by definition. Hence, we only need to show the indistinguishability of $\mathbf{QIdeal}_{\mathcal{S},s}$ and $\mathbf{Ideal}_{\mathsf{S},s}$, where $\mathcal{S}$ is the simulator described above and S is the simulator described in the proof of Theorem 3.

Lemma 5. $\mathbf{QIdeal}_{\mathcal{S},s} \approx_{2^{-\lambda}k} \mathbf{Ideal}_{\mathsf{S},s}.$

Proof. The only difference between the two experiments is step 5, which is the decoding step. In particular, differences only lies in step 5(a)ii of the simulator $\mathcal{S}$.

In step 4(a)ii, $r_i''\|s_i' \neq \underline{\text{same}}$, and note that r_i'' is generated by the simulator $\mathcal{S}'$ without any knowledge of r. Hence, r_i'' is independent of r. Hence, the probability that $r_i'' = r$ is atmost $2^{-\lambda}$, and with probability $1 - 2^{-\lambda}$, even the decoding step in the simulator S in the proof of Theorem 3 outputs $\bot$. Hence, for each $i \in [k]$, the output distributions of the two experiments differ only by $2^{-\lambda}$. $\qquad\square$

Combining this hybrid with the proof of Theorem 3, we see that $\mathbf{Tamper}_s^f \approx_{\epsilon'} \mathbf{Ideal}_{\mathsf{S},s}$ for $\epsilon' = k\left(\epsilon + 2^{-\lambda}\right)$. This completes the proof of non-malleability of the scheme. $\qquad\square$

Reducing States to 2. The proof of Theorem 4 crucially relies on the *secrecy* of r (from the adversaries in states 2 and 3 above). This contributes to making the number of states to be 3. However, secrecy of r can also be preserved by simply secret sharing r into $r = r_L \oplus r_R$. The final encoding is as follows: $\mathsf{Enc}\,(s) = ([r_L, L_1, \cdots, L_k], [r_R, R_1, \cdots, R_k])$, where $r_L \oplus r_R = r$ for a random $r \in \{0,1\}^\lambda$ and $e_i = \mathsf{Enc}_{\mathsf{NM}}\,(i, r\|s_i) = (L_i, R_i)$ for $i \in [k]$. This gives us a construction that is non-malleable against $\mathcal{F}_{\mathsf{half}}$. Additionally, note that it is straight-forward to modify the construction to split s into blocks of size t as opposed to single bits (similar to the construction in Sect. 3) to obtain the following theorem, the proof of which is given in the full version of this paper.

Theorem 5. *Assume that* NMC *is a non-malleable coding scheme of rate* $1/\gamma$ *and exact security* ϵ, *which is non-malleable against split-state adversaries. Then there is an efficient* $\left(\lambda + (\lambda + \log(k/t) + t)\,\gamma, 0, k\left(\epsilon + 2^{-\lambda}\right)/t\right)$-*locally decodable non-malleable coding scheme which is non-malleable against* $\mathcal{F}_{\mathsf{half}}$. *The rate of the code is* $1/\gamma'$, *where* $\gamma' = \lambda/k + (1 + \lambda/t + \log(k/t)/t)\,\gamma$.

Corollary 2. *Assuming* $\lambda \geq \log k$, *there exists an explicit construction of a* $(\mathcal{O}\,(\lambda), 0, \nu\,(\lambda))$-*locally decodable non-malleable coding scheme over* k-*bit messages with constant rate (for some negligible function* $\nu(\cdot)$) *which is non-malleable against* $\mathcal{F}_{\mathsf{half}}$.

Proof. This follows by choosing $t = \lambda$ and using constant-rate non-malleable codes non-malleable against $\mathcal{F}_{\text{half}}$ (from the work of Aggarwal *et al.* [1]) in Theorem 5.   □

Corollary 3. *There exists an explicit construction of a $(\omega(1), 0, \nu(\lambda))$-locally decodable non-malleable coding scheme with rate 1 (for a negligible function $\nu(\cdot)$) which is non-malleable against the tampering function class $\mathcal{F}_{bit}$.*

Proof. The proof of this corollary follows by instantiating Enc_{NM} in Theorem 4 with the rate 1 non-malleable coding scheme from [13] that is non-malleable against $\mathcal{F}_{\text{bit}}$ and by splitting the k-bit input message into blocks of size $\omega(1)$ each and encoding these bits together.   □

5 Updatability and Security Against Continual Attacks

We now show how to modify the construction from Sect. 3 to get a code that is leakage and tamper-resilient against continual attacks. Note that if codewords are not periodically refreshed, then an adversary that obtains leakage that is unbounded, can, over time, leak one codeword completely and then tamper the codeword based on this codeword. At a high level, to prevent this, we must refresh codewords periodically (even if they are not updated). We do this, by cycling through the codewords that encode all s_i values one-by-one and "refresh" them. Of course, if the encoder and decoder maintain state, they can perform this refreshing in a cyclic manner. However, in order to perform this refresh in a stateless manner, we maintain a counter that is encoded along with all the r_i values. This ensures that we refresh all codewords periodically. Additionally, for technical reasons (that we describe later), we refresh codewords everytime we decode a particular index. By lowering the threshold of leakage tolerated in every "round", we ensure that our construction remains secure. We describe our construction (and the security) in more detail in the full version [9] of this paper.

6 Applications of Local Non-malleable Codes

Similar to the work of Dachman-Soled *et al.* [16], our locally updatable/decodable leakage-resilient non-malleable codes can be used in the construction of secure RAM computation protocols. At a very high level, if the memory and program code are encoded using a local leakage-resilient non-malleable code (that is resilient to tampering from the family $\mathcal{F}$ and leakage from the family $\mathcal{G}$) and the resulting codeword is then accessed through an oblivious RAM (ORAM) [23,24,28,29] protocol, one can show that the resulting protocol is a protocol for secure RAM computation that is secure against tampering of the memory from the same tampering family $\mathcal{F}$ and leakage from the same family $\mathcal{G}$. Now, if we instantiate the non-malleable code with our information-theoretic non-malleable code from Sect. 5, and instantiate the ORAM protocol with an ORAM that has information-theoretic guarantees [6,14,17,31], then one can show that the resulting RAM computation protocol has information-theoretic security. Of course, information-theoretic RAM protocols assume the existence of ideal encryption and our final compiler will make the same assumption. However, if the compiler is applied in the context of information-theoretic secure multi-party computation [7,11], then one can obtain an information-theoretic secure RAM computation protocol that is resilient to tampering from the class $\mathcal{F}$ and leakage from the class $\mathcal{G}$ (by replacing the ideal encryption with secret sharing [30]).

For further details of ORAM compilers, tamper/leakage resilient (information-theoretic) RAM computation, our construction and results, we refer the reader to the full version of this paper.

References

1. Aggarwal, D., Dodis, Y., Kazana, T., Obremski, M.: Non-malleable reductions and applications. In: Proceedings of the Forty-Seventh Annual ACM on Symposium on Theory of Computing, STOC 2015, Portland, OR, USA, 14–17 June 2015, pp. 459–468 (2015)
2. Aggarwal, D., Dodis, Y., Lovett, S.: Non-malleable codes from additive combinatorics. In: Symposium on Theory of Computing, STOC 2014, New York, NY, USA, 31 May–03 June 2014, pp. 774–783 (2014)
3. Aggarwal, D., Dziembowski, S., Kazana, T., Obremski, M.: Leakage-resilient non-malleable codes. In: Dodis, Y., Nielsen, J.B. (eds.) TCC 2015, Part I. LNCS, vol. 9014, pp. 398–426. Springer, Heidelberg (2015)
4. Agrawal, S., Gupta, D., Maji, H.K., Pandey, O., Prabhakaran, M.: Explicit non-malleable codes against bit-wise tampering and permutations. In: Gennaro, R., Robshaw, M. (eds.) CRYPTO 2015, Part I. LNCS, vol. 9215, pp. 538–557. Springer, Heidelberg (2015)
5. Agrawal, S., Gupta, D., Maji, H.K., Pandey, O., Prabhakaran, M.: A rate-optimizing compiler for non-malleable codes against bit-wise tampering and permutations. In: Dodis, Y., Nielsen, J.B. (eds.) TCC 2015, Part I. LNCS, vol. 9014, pp. 375–397. Springer, Heidelberg (2015)
6. Ajtai, M.: Oblivious RAMs without cryptographic assumptions. In: Proceedings of the 42nd ACM Symposium on Theory of Computing, STOC 2010, Cambridge, Massachusetts, USA, 5–8 June 2010, pp. 181–190 (2010)
7. Ben-Or, M., Goldwasser, S., Wigderson, A.: Completeness theorems for non-cryptographic fault-tolerant distributed computation (extended abstract). In: Proceedings of the 20th Annual ACM Symposium on Theory of Computing, Chicago, Illinois, USA, 2–4 May 1988, pp. 1–10 (1988)
8. Chandran, N., Kanukurthi, B., Ostrovsky, R.: Locally updatable and locally decodable codes. In: Lindell, Y. (ed.) TCC 2014. LNCS, vol. 8349, pp. 489–514. Springer, Heidelberg (2014)
9. Chandran, N., Kanukurthi, B., Raghuraman, S.: Information-theoretic local non-malleable codes and their applications. Cryptology ePrint Archive, Report 2015 (2015). http://eprint.iacr.org/
10. Chattopadhyay, E., Zuckerman, D.: Non-malleable codes against constant split-state tampering. In: 55th IEEE Annual Symposium on Foundations of Computer Science, FOCS 2014, Philadelphia, PA, USA, 18–21 October 2014, pp. 306–315 (2014)

11. Chaum, D., Crépeau, C., Damgård, I.: Multiparty unconditionally secure protocols (extended abstract). In: Proceedings of the 20th Annual ACM Symposium on Theory of Computing, Chicago, Illinois, USA, 2–4 May 1988, pp. 11–19 (1988)

12. Cheraghchi, M., Guruswami, V.: Capacity of non-malleable codes. In: Innovations in Theoretical Computer Science, ITCS 2014, Princeton, NJ, USA, 12–14 January 2014, pp. 155–168 (2014)

13. Cheraghchi, M., Guruswami, V.: Non-malleable coding against bit-wise and split-state tampering. In: Lindell, Y. (ed.) TCC 2014. LNCS, vol. 8349, pp. 440–464. Springer, Heidelberg (2014)

14. Chung, K.-M., Liu, Z., Pass, R.: Statistically-secure ORAM with $\tilde{O}(\log^2 n)$ overhead. In: Sarkar, P., Iwata, T. (eds.) ASIACRYPT 2014, Part II. LNCS, vol. 8874, pp. 62–81. Springer, Heidelberg (2014)

15. Coretti, S., Maurer, U., Tackmann, B., Venturi, D.: From single-bit to multi-bit public-key encryption via non-malleable codes. In: Dodis, Y., Nielsen, J.B. (eds.) TCC 2015, Part I. LNCS, vol. 9014, pp. 532–560. Springer, Heidelberg (2015)

16. Dachman-Soled, D., Liu, F.-H., Shi, E., Zhou, H.-S.: Locally decodable and updatable non-malleable codes and their applications. In: Dodis, Y., Nielsen, J.B. (eds.) TCC 2015, Part I. LNCS, vol. 9014, pp. 427–450. Springer, Heidelberg (2015)

17. Damgård, I., Meldgaard, S., Nielsen, J.B.: Perfectly secure oblivious RAM without random oracles. In: Ishai, Y. (ed.) TCC 2011. LNCS, vol. 6597, pp. 144–163. Springer, Heidelberg (2011)

18. Dziembowski, S., Kazana, T., Obremski, M.: Non-malleable codes from two-source extractors. In: Canetti, R., Garay, J.A. (eds.) CRYPTO 2013, Part II. LNCS, vol. 8043, pp. 239–257. Springer, Heidelberg (2013)

19. Dziembowski, S., Pietrzak, K., Wichs, D.: Non-malleable codes. In: Innovations in Computer Science, ICS 2010, Tsinghua University, Beijing, China, 5–7 January 2010, pp. 434–452 (2010)

20. Faust, S., Mukherjee, P., Nielsen, J.B., Venturi, D.: Continuous non-malleable codes. In: Lindell, Y. (ed.) TCC 2014. LNCS, vol. 8349, pp. 465–488. Springer, Heidelberg (2014)

21. Faust, S., Mukherjee, P., Nielsen, J.B., Venturi, D.: A tamper and leakage resilient von Neumann architecture. In: Katz, J. (ed.) PKC 2015. LNCS, vol. 9020, pp. 579–603. Springer, Heidelberg (2015)

22. Faust, S., Mukherjee, P., Venturi, D., Wichs, D.: Efficient non-malleable codes and key-derivation for poly-size tampering circuits. In: Nguyen, P.Q., Oswald, E. (eds.) EUROCRYPT 2014. LNCS, vol. 8441, pp. 111–128. Springer, Heidelberg (2014)

23. Goldreich, O.: Towards a theory of software protection and simulation by oblivious RAMs. In: Proceedings of the 19th Annual ACM Symposium on Theory of Computing, New York, USA, pp. 182–194 (1987)

24. Goldreich, O., Ostrovsky, R.: Software protection and simulation on oblivious RAMs. J. ACM **43**(3), 431–473 (1996)

25. Katz, J., Trevisan, L.: On the efficiency of local decoding procedures for error-correcting codes. In: Proceedings of the Thirty-Second Annual ACM Symposium on Theory of Computing, Portland, OR, USA, 21–23 May 2000, pp. 80–86 (2000)

26. Liu, F.-H., Lysyanskaya, A.: Tamper and leakage resilience in the split-state model. In: Safavi-Naini, R., Canetti, R. (eds.) CRYPTO 2012. LNCS, vol. 7417, pp. 517–532. Springer, Heidelberg (2012)

27. Myers, S., Shelat, A.: Bit encryption is complete. In: 50th Annual IEEE Symposium on Foundations of Computer Science, FOCS 2009, Atlanta, Georgia, USA, 25–27 October 2009, pp. 607–616 (2009)

28. Ostrovsky, R.: An efficient software protection scheme. In: Brassard, G. (ed.) CRYPTO 1989. LNCS, vol. 435, pp. 610–611. Springer, Heidelberg (1990)
29. Ostrovsky, R.: Efficient computation on oblivious RAMs. In: Proceedings of the 22nd Annual ACM Symposium on Theory of Computing, Baltimore, Maryland, USA, 13–17 May 1990, pp. 514–523 (1990)
30. Shamir, A.: How to share a secret. Commun. ACM **22**(11), 612–613 (1979)
31. Stefanov, E., van Dijk, M., Shi, E., Fletcher, C.W., Ren, L., Yu, X., Devadas, S.: Path ORAM: an extremely simple oblivious RAM protocol. In: 2013 ACM SIGSAC Conference on Computer and Communications Security, CCS 2013, Berlin, Germany, 4–8 November 2013, pp. 299–310 (2013)
32. Yekhanin, S.: Locally decodable codes. Found. Trends Theoret. Comput. Sci. **6**(3), 139–255 (2012)

Optimal Computational Split-state Non-malleable Codes

Divesh Aggarwal[1](✉), Shashank Agrawal[2], Divya Gupta[3], Hemanta K. Maji[4], Omkant Pandey[5], and Manoj Prabhakaran[2]

[1] EPFL, Lausanne, Switzerland
Divesh.Aggarwal@epfl.ch
[2] University of Illinois at Urbana-Champaign, Champaign, USA
{sagrawl2,mmp}@illinois.edu
[3] University of California at Los Angeles, Los Angeles, USA
divyag@cs.ucla.edu
[4] Purdue University, West Lafayette, USA
hmaji@purdue.edu
[5] University of California at Berkeley, Berkeley, USA
omkant@gmail.com

Abstract. Non-malleable codes are a generalization of classical error-correcting codes where the act of "corrupting" a codeword is replaced by a "tampering" adversary. Non-malleable codes guarantee that the message contained in the tampered codeword is either the original message m, or a completely unrelated one. In the common split-state model, the codeword consists of multiple *blocks* (or states) and each block is tampered with *independently*.

The central goal in the split-state model is to construct *high rate* non-malleable codes against all functions with only *two* states (which are necessary). Following a series of long and impressive line of work, *constant rate*, two-state, non-malleable codes against all functions were recently achieved by Aggarwal et al. [2]. Though constant, the rate of all known constructions in the split state model is very far from optimal (even with more than two states).

In this work, we consider the question of improving the rate of split-state non-malleable codes. In the "information theoretic" setting, it is not

S. Agrawal and M. Prabhakaran—Research supported in part by NSF grant 1228856.
D. Gupta and H.K. Maji—Research supported in part from a DARPA/ONR PRO-CEED award, NSF Frontier Award 1413955, NSF grants 1228984, 1136174, 1118096, and 1065276, a Xerox Faculty Research Award, a Google Faculty Research Award, an equipment grant from Intel, and an Okawa Foundation Research Grant. This material is based upon work supported by the Defense Advanced Research Projects Agency through the U.S. Office of Naval Research under Contract N00014-11- 1-0389. The views expressed are those of the author and do not reflect the official policy or position of the Department of Defense, the National Science Foundation, or the U.S. Government.
D. Gupta, O. Pandey and M. Prabhakaran—This work was done in part while the author was visiting the Simons Institute for the Theory of Computing, supported by the Simons Foundation and by the DIMACS/Simons Collaboration in Cryptography through NSF grant #CNS-1523467.

E. Kushilevitz and T. Malkin (Eds.): TCC 2016-A, Part II, LNCS 9563, pp. 393–417, 2016.
DOI: 10.1007/978-3-662-49099-0_15

possible to go beyond rate 1/2. We therefore focus on the standard computational setting. In this setting, each tampering function is required to be *efficiently* computable, and the message in the tampered codeword is required to be either the original message m or a "computationally" independent one.

In this setting, assuming only the existence of one-way functions, we present a compiler which converts any poor rate, two-state, (sufficiently strong) non-malleable code into a rate-1, two-state, computational non-malleable code. These parameters are asymptotically optimal. Furthermore, for the qualitative optimality of our result, we generalize the result of Cheraghchi and Guruswami [10] to show that the existence of one-way functions is necessary to achieve rate $> 1/2$ for such codes.

Our compiler requires a stronger form of non-malleability, called *augmented* non-malleability. This notion requires a stronger simulation guarantee for non-malleable codes and simplifies their modular usage in cryptographic settings where composition occurs. Unfortunately, this form of non-malleability is neither straightforward nor generally guaranteed by known results. Nevertheless, we prove this stronger form of non-malleability for the two-state construction of Aggarwal et al. [3]. This result is of independent interest.

Keywords: Non-malleable codes · Split-state · Explicit construction · Computational setting · One-way functions · Pseudorandom generators · Authenticated encryption schemes · Rate 1

1 Introduction

Non-Malleable Codes, introduced by Dziembowski, Pietrzak, and Wichs [18], are a generalization of the classical notion of error detection. Informally, a code is *non-malleable* if the message contained in a codeword that has been tampered with is either the original message, or a completely unrelated value. Non-Malleable Codes have emerged as a fundamental object at the intersection of coding theory and cryptography.

There are two main directions in this area: design *explicit codes* that can tolerate a large class of tampering functions, and achieve *high rate*[1] for such constructions.

Ideally, we would like to tolerate the class of all tampering functions that can be implemented in P/poly. However, this is impossible if the adversary has unrestricted access to the full codeword[2]. Therefore, one must either consider

---

[1] Rate refers to the asymptotic ratio of the length of a message to the length of its encoding (in bits), as the message length increases to infinity. The best rate possible is 1; if the length of the encoding is super-linear in the length of the message, the rate is 0.

[2] This is because a non-malleable code has efficient encoding and decoding procedures; an adversary can simply decode the message and encode a related value.

a (much weaker) class of tampering functions, or move to alternative models where the adversary has only restricted access to the codeword.

The most common model for tolerating arbitrary tampering functions is the *split state* model. In this model, the codeword is "split" into two or more states $c = (c_1, \ldots, c_k)$; a tampering function f is viewed as a list of k functions $(f_1, \ldots, f_k)$ fixed before c is sampled, where each function f_i tampers with the corresponding component c_i of the codeword independently, i.e., the tampered codeword is $c' = (f_1(c_1), \ldots, f_k(c_k))$. Ideally, we would like to achieve codewords with minimum number of states $k = 2$ while tolerating all possible tampering functions and achieving high-rate[3].

In a break-through result, Aggarwal et al. [3] presented an explicit non-malleable code for $k = 2$ states for messages of arbitrary length (significantly improving upon [17] which only encodes a single bit). However, their work only achieves rate $\Omega(n^{-6/7})$ (or rate 0, asymptotically) where n is the block length of the codeword. Chattopadhyay and Zuckerman [9] present an encoding which has constant rate by increasing the number of states to $k = 10$. Very recently, Aggarwal et al. [2] show that constant rate for such codes can in fact be achieved with only $k = 2$ states[4].

Though constant, the rate of codes in [2,9] is very far from optimal. A natural question is if we can achieve the best parameters, i.e.:

Can we construct explicit, 2-state, *non-malleable codes of rate 1 tolerating all tampering functions in* P/poly?

In the "information theoretic" setting it is impossible to go beyond rate 1/2. Recall that in the "information theoretic" setting, the tampering function is of the form (f_1, f_2) (restricting ourselves to 2 states), the component functions are not necessarily of polynomial size, and we require that the tampered codeword contains either the original message m or a message *statistically* independent of m.

We must therefore consider the "computational setting" which is a natural relaxation of the information theoretic setting. More specifically, we make two changes: first, we require that f_1, f_2 are both in P/poly; and second the tampered codeword either contains the original message m or a message that is only "computationally independent" of m[5].

In this work, we show that it is indeed possible to construct rate 1 non-malleable codes in the *computational setting* with only $k = 2$ states under the standard assumption that one-way functions exist. Our code is explicit and tolerates all tampering functions (in P/poly). Furthermore, we complement this

---

[3] We note that in this model, one can even tolerate tampering functions beyond P/poly. This is the so called "information theoretic" setting.

[4] Sometimes, the setting where $k = 2$ is commonly referred to as the split state setting; and when $k > 2$ it is explicitly mentioned and often called *multiple* split state setting.

[5] This is precisely defined by requiring a simulator whose output, in the case where the tampered message is not m, is computationally indistinguishable from a message in the (real) tampered codeword.

result by proving that the existence of non-malleable codes of rate better than 1/2 (in the information theoretic setting) implies one-way functions. Our motivation to rely on the computational setting to go beyond rate 1/2 comes from similar previous works [27,29] on classical error correcting codes where a computationally bounded channel is considered to correct more than 1/4th fraction of the errors.

Our approach actually yields a compiler which converts any 2-state, poor rate (potentially rate 0), non-malleable code into a rate-1 computational non-malleable code. However, this reduction requires the underlying code to have a stronger form of non-malleability: at a high level, it requires a stronger simulation guarantee where the simulator can not only simulate the distribution of the message in the tampered codeword, but also one of the states of the original codeword (say the second state). We formalize this stronger simulation requirement and call the resulting notion *augmented* non-malleability. Given this stronger form of non-malleability, we can prove the computational non-malleability of our 2-state construction. Augmented non-malleability simplifies the design of non-malleable codes by allowing us to compose them with other cryptographic constructs.

Unfortunately, augmented non-malleability is neither straightforward nor generally guaranteed to hold for known constructions [3,9]. Nevertheless, we prove this stronger form of non-malleability for the two-state construction of [3]. This gives an explicit code with the desired properties, i.e.:

Informal Theorem 1. *Assuming the existence of one-way functions, there exists a rate-1 split-state non-malleable code against computationally efficient tampering functions.*

We note that these parameters are asymptotically optimal in the computational setting. In addition, our extension of [3] to augmented non-malleability is of independent interest — it is particularly useful in settings where composition occurs. This is captured in the following theorem:

Informal Theorem 2. *For any k and ε, there exists an efficient (in k and $\log(\frac{1}{\varepsilon})$) information-theoretically secure ε-**augmented**-non-malleable code for encoding k-bit messages in the (two-partition) split-state model.*

We now present a technical overview of our approach.

1.1 Technical Overview

Improving Rate via Hybrid Encoding. The starting point of our work is to consider the standard "hybrid" approach where we first encode a short cryptographic key K using a low rate 2-state non-malleable code, and then use K along with an appropriate cryptographic object such as a "good rate" encryption scheme.

We note that this hybrid approach has been used in many different works to improve efficiency or the rate. For example, the most well-known example of this approach in cryptography is that of "hybrid encryption," which improves the efficiency of a (non-malleable) public-key encryption scheme by using it to encrypt a short key for a symmetric-key encryption scheme, and then using the latter to encrypt the actual message (e.g., see [16, 26]). In the context of error-correcting codes and non-malleable codes, this approach has been used to improve the rate in [11, 18, 22], and even by [5] (who obtain *information theoretic* non-malleability).

In our setting, let us start by considering the following construction: encode a fresh key K using a 2-state information-theoretic non-malleable code of low (potentially 0) rate to obtain the two states say (c_1, c_2), and then generate a third component c_3 which is an encryption of the message m to be encoded, under the key K, using a "high rate" symmetric authenticated encryption scheme. Such encryption schemes can be constructed from pseudorandom functions (implied by one-way functions).

At first, suppose that we can keep more than two states. Then, we can output $c = (c_1, c_2, c_3)$ as our *three*-state codeword. We argue that this is already a 3-state computational non-malleable code of rate 1. To see this, fix a tampering function $f = (f_1, f_2, f_3)$ and recall that each state of the codeword is tampered independently. Let $c' = (c_1', c_2', c_3')$ be the tampered codeword where $c_i' = f_i(c_i)$; let K' denote the "tampered" key in (c_1', c_2') and m' denote the tampered message defined by decryption of c_3' using K'. Then, intuitively, if $K' = K$ then m' must also be equal to m by the security of authenticated encryption. On the other hand, if $K' \neq K$ yet m' is not computationally independent of m, then it must be that K' is also not computationally independent of K. This will violate the non-malleability of the underlying 2-state code for K. This approach is reminiscent of the technique introduced by [18] for rate amplification for the restrictive class of bit-wise tampering functions.

To achieve a 2-state solution, we propose that c_2 and c_3 be kept in a single state and the resulting codeword is $(c_1, c_2 \| c_3)$. However, this creates a difficult situation: since $c_2 \| c_3$ are now available together, adversary might be able to generate $c_2' \| c_3'$ such that (c_1', c_2') encodes a key $K' \neq K$, K' stays independent of K by itself, yet decryption of c_3' yields a message that depends on m. Unlike the case of 3-states where c_3' was generated independently of c_2', in this setting, $(c_2' \| c_3')$ now depends on both c_2 and c_3. Therefore, we need a stronger guarantee from non-malleability where not only the distribution of K', but the distribution of K' *along with state* c_2 must be simulatable (and *computationally* independent of K).

We formalize this stronger simulation requirement and call the resulting notion *augmented* non-malleability. Given this stronger form of non-malleability, we can prove the computational non-malleability of our 2-state construction. We emphasize that the novelty of this augmented non-malleability is highlighted by the fact that our whole construction only uses one-way functions (in a fully black-box manner) while previous non-malleable code constructions by [19, 28] use CRS and extremely strong cryptographic primitives.

Achieving Augmented Non-malleability. As noted earlier, it is not clear if existing non-malleable codes also satisfy the augmented non-malleability property. In fact, we do not know if this is true in general for all non-malleable codes. We prove in Informal Theorem 2 that augmented non-malleability can be achieved from the 2-state code of [3]. We now describe how we achieve augmented non-malleability.

The main technical ingredient to prove Informal Theorem 2 is the following result.

Informal Theorem 3. *Assume* $\mathbb{F}_p$ *is a finite field of prime order, $n \geqslant$ poly$(\log p))$, L is uniformly random over $\mathbb{F}_p^n$, and $f, g : \mathbb{F}_p^n \to \mathbb{F}_p^n$ are two arbitrary functions. Then, for almost all $r \in \mathbb{F}_p^n$, the joint distribution $(\langle L, r \rangle, \langle f(L), g(r) \rangle)$ is "close" to a convex combination (that depends on r) of affine distributions $\{(U, aU + b) \mid a, b \in \mathbb{F}_p\}$, where U is uniformly random over $\mathbb{F}_p$.*

The formal statement appears in Theorem 3. A similar but weaker statement was shown in [3]. They showed that the above mentioned joint distribution is on average (over $r \in \mathbb{F}_p^n$) close to a convex combination of affine-distributions, while we show that this holds individually for almost all $r \in \mathbb{F}_p^n$.

The proof follows a similar structure as [3] where the ambient space $\mathbb{F}_p^n \times \mathbb{F}_p^n$ is partitioned into subsets depending on f, g, and then the joint distribution is analyzed over each of these subsets. One crucial difference from [3] is that several steps in their proof relied on the fact that the inner-product is a strong extractor, i.e., $\langle L, R \rangle$ is close to uniform conditioned on L. While this is sufficient to prove the result for R uniform in $\mathbb{F}_p^n$, we needed to be more careful since we needed to show the result for almost all $r \in \mathbb{F}_p^n$, and we cannot claim that $\langle L, R \rangle$ is close to uniform conditioned on both L, R. Fortunately, however, we could show (refer to Lemmas 3 and 4) that it is sufficient to show that $\langle L, R \rangle$ is close to uniform conditioned on R and $h(L)$ for some function $h : \mathbb{F}_p^n \mapsto \mathbb{F}_p$ and this holds since L has sufficient entropy conditioned on $h(L)$.

The proof of Informal Theorem 2 is relatively immediate from Informal Theorem 3 using affine-evasive sets [1,3].

Necessity of One-Way Functions. We sketch, at a very high level, how we extend the result of Cheraghchi and Guruswami [10] to show the existence of *distributional* one-way functions if 2-state (information theoretic) non-malleable codes of rate large than $1/2$ exist. See Sect. 5 for more details.

The following negative result is shown in [10]: Consider the set of tampering functions which depend only on the first αn bits of the code and tampers it arbitrarily. Then a non-malleable code which protects against this tampering class can have rate at most $1 - \alpha$.

In particular, k-split-state non-malleable code can have at most $1 - 1/k$ rate. Otherwise, one can use the same attack in [10] to tamper only the first state appropriately and violate the non-malleability condition.

The result in [10] uses the following idea. If the rate is higher than $1 - \alpha$ then there exists two messages s_0 and s_1, and a set $X \subseteq \{0,1\}^{\alpha n}$ such that the following condition holds: The first αn bits of an encoding of s_0 has higher probability to be in X than for an encoding of s_1. So, the tampering function just writes a dummy string w if the first αn bits belong in X; otherwise it keeps it intact. The decoding of the tampered code is, therefore, identical to the original message or it is an invalid string. Due to the property of X, the tampering function ensures that the decoding is $\perp$ with higher probability when the message is s_0.

Now consider the following function: $f(b, r) = Enc(s_b; r)|_{\alpha n}$, i.e. the function which outputs the first αn bits of the encoding of message s_b (using randomness r in the encoding procedure). Let y be any string in the domain of $f(\cdot, \cdot)$. Suppose B is an oracle which, when queried with y, provides a uniformly reverse sampled pre-image of y. Then we make t calls to B to create a set $S_y = \{(b_1, r_1), \ldots, (b_t, r_t)\}$. Counting the number of occurrences of $b = 0$ in S_y we can test whether $y \in X$ or not; when t is sufficiently large we have $y \in X$ implies $\mathsf{maj}\{b_1, \ldots, b_t\} = 0$ w.h.p. (by Chernoff bounds). Given access to the oracle B, we can emulate the tampering function which performs the tampering of [10] (except with $\mathsf{negl}(n)$ error).

Now, consider a setting where distributionally one-way functions do not exist. In this case, for $f(\cdot, \cdot)$ and suitably large $p(\cdot)$ (as a function of t), there exists an efficient inverter A which can simulate every call of B, except with error (at most) $1/p(n)$. Now, we can replace calls to algorithm B in the previous paragraph, with calls to A while incurring an error of at most $t(n)/p(n)$. By suitably choosing $t(n)$ and $p(n)$, we can construct an efficient tampering on the first αn bits of the encoding which emulates the tampering of [10] with error $t(n)/p(n)$.

1.2   Prior Work

Cramer et al. [14] introduced the notion of arithmetic manipulation detection (AMD) codes, which is a special case of non-malleable codes against tampering functions with a simple algebraic structure; explicit AMD codes with optimal (second order) parameters have been recently provided by [15]. Dziembowski et al. motivated and formalized the more general notion of non-malleable codes in [18]. They showed existence of a constant rate non-malleable code against the class of all bit-wise independent tampering functions (which are essentially multi-state codes with a large, non-constant, value of k).

The existence of rate-1 non-malleable codes against various classes of tampering functions is now known. For example, existence of such codes with rate $(1 - \alpha)$ was shown against any tampering function family of size $2^{2^{\alpha n}}$; but this scheme has inefficient encoding and decoding [10]. For tampering functions of size $2^{\mathsf{poly}(n)}$, rate-1 codes (with efficient encoding and decoding) exist, and can be obtained efficiently with overwhelming probability [20].

Very recently, an explicit rate-0 code against a more powerful class of tampering functions, which in addition to tampering with each bit of the codeword independently can also permute the bits of the resulting codeword after tampering, was achieved in [4]. This was further improved to rate 1 by [5].

In the "split state" setting where the codeword is partitioned into k separate blocks and each block can be tampered arbitrarily but independently, an encoding scheme was proposed in [12]. For the case of only two states, an explicit non-malleable code for encoding a *single bit* was proposed by [17]. Recently, in a break-through result, an explicit scheme (of rate 0) was proposed for arbitrary length messages by [3]. A constant rate construction for 10 states was provided in [9] (and later in [3]). Very recently, Aggarwal et al. [2] show that constant rate for such codes can in fact be achieved with only $k = 2$ states. We note that in this setting it is not possible to go beyond rate $1/2$ if one insists upon information theoretic non-malleability. Our present work shows that by relying on computational definition of non-malleability, we can achieve rate 1 with only 2 states (which are necessary). Asymptotically, these are the best possible parameters.

In the computational setting, there has been a sequence of works on improving the rate of error-correcting codes [8,22,23,27,29,30] as well as constructing non-malleable codes and its variants [19,28]. We also note that for the case of bit-wise tampering functions, a hybrid approach was suggested in [18] by relying on authenticated encryption. It is not clear if this approach works for a general class of functions. Chandran et al. [7] also rely on the computational setting in defining their new notion of *blockwise non-malleable codes*. Blockwise non-malleable codes are a generalization of the split-state model (and the recent lookahead model of [2]) where the adversary tampers with one state at a time.

Non-malleable codes have found interesting cryptographic applications like domain extension of self-destruct CCA-secure public-key encryption [13] and non-malleable commitments [4].

2   Preliminaries

Notation. We denote the security parameter by λ. Probability distributions are represented by capital letters. Given a distribution X, $x \sim X$ represents that x is sampled according to the distribution X. For a function $f(\cdot)$, the random variable $Y = f(X)$ represents the following distribution: sample $x \sim X$ and output $f(x)$.

For a randomized algorithm A, we write $A(z)$ to denote the distribution of the output of A on an input z. A function $f : \mathbb{N} \to \mathbb{R}^+$ is negligible if for every positive polynomial $\mathsf{poly}(\cdot)$ and all sufficiently large n, $f(n) \leqslant 1/\mathsf{poly}(n)$. We use $\mathsf{negl}(M)$ to denote an (unspecified) negligible function in M. Lastly, all logarithms in this paper are to the base 2.

For two variables X, X' their statistical distance is $\Delta(X; X') = \frac{1}{2} \sum_x | \Pr[X = x] - \Pr[X' = x] |$.

The min-entropy of a distribution is $\mathbf{H}_\infty(D) = \min_x \log(D[x]^{-1})$. For a finite set S, we denote by U_S the uniform distribution over S. Note that $\mathbf{H}_\infty(U_S) = \log |S|$. Moreover, if X is a distribution with min-entropy k then X is a convex combination of distributions uniform over sets of size 2^k.

Let E be an event. We denote by $X|E$ the conditional random variable, conditioned on E holding. For a set S we shorthand $X|_S = X|[X \in S]$.

2.1 Non-malleable Codes in the Split-State Model

In this section, we give a stronger definition of non-malleable codes in the split-state model (than what is considered in literature [2,3,18]). We call these augmented non-malleable codes, denoted my Aug-NMC. We define Aug-NMC both in the information theoretic setting as well as computational setting.

Let λ be the security parameter. Let $N_1(\lambda)$ and $N_2(\lambda)$ be some fixed polynomials in λ. These will denote the size of the states in the split state setting. We begin by defining the real tampering and ideal simulation experiments against any generic tampering class $\mathcal{F}$ in Fig. 1. We also define the advantage between the real and simulated experiments w.r.t. a class of distinguishers $\mathcal{D}$ in Fig. 1.

Let $\mathcal{D}_{\mathsf{all}}$ and $\mathcal{F}_{\mathsf{all}}$ denote the class of all distinguishers and all split-state tampering functions, respectively, as in Fig. 1. Similarly, let $\mathcal{D}_{\mathsf{eff}}$ and $\mathcal{F}_{\mathsf{eff}}$ denote the class of efficient distinguishers and efficient split-state tampering functions, respectively, as in Fig. 1. That is, there exists polynomials p, q such that for all $\lambda \in \mathbb{N}$, the running time of f_λ, g_λ is at most $p(\lambda)$ and running time of all $D \in \mathcal{D}_{\mathsf{eff},\lambda}$ is at most $q(\lambda)$. Next, we define Aug-NMC w.r.t. experiments defined in Fig. 1.

Definition 1 (Standard $[(N_1, N_2), M, \nu]$-Aug-NMC). *Suppose* Enc $: \{0,1\}^M \to \{0,1\}^{N_1} \times \{0,1\}^{N_2}$ *and* Dec $: \{0,1\}^{N_1} \times \{0,1\}^{N_2} \to \{0,1\}^M \cup \{\bot\}$ *are (possibly randomized) mappings. Then* (Enc, Dec) *is a (standard) $[(N_1, N_2), M, \nu]$-Aug-NMC if the following conditions hold:*

○ Correctness: $\forall s \in \{0,1\}^M$, $\Pr[\mathsf{Dec}(\mathsf{Enc}(s)) = s] = 1$.
○ Non-Malleability: $\mathsf{adv}_{\mathcal{F}_{\mathsf{all}}, \mathcal{D}_{\mathsf{all}}}^{\mathsf{Enc,Dec}} \leqslant \nu(\lambda)$. *(See Fig. 1 for description.)*

We say that the coding scheme is efficient *if (Enc,Dec) run in time bounded by a polynomial in M and λ.*

Definition 2 (Computational $[(N_1, N_2), M, \nu]$-Aug-NMC). *Suppose* Enc $: \{0,1\}^M \to \{0,1\}^{N_1} \times \{0,1\}^{N_2}$ *and* Dec $: \{0,1\}^{N_1} \times \{0,1\}^{N_2} \to \{0,1\}^M \cup \{\bot\}$ *are (possibly randomized) mappings. Then* (Enc, Dec) *is a computational $[(N_1, N_2), M, \nu]$-Aug-NMC if the following conditions hold:*

○ Correctness: $\forall s \in \{0,1\}^M$, $\Pr[\mathsf{Dec}(\mathsf{Enc}(s)) = s] = 1$.
○ Non-Malleability: $\mathsf{adv}_{\mathcal{F}_{\mathsf{eff}}, \mathcal{D}_{\mathsf{eff}}}^{\mathsf{Enc,Dec}} \leqslant \nu(\lambda)$. *(See Fig. 1 for description.)*

We say that the coding scheme is efficient *if (Enc,Dec) run in time bounded by a polynomial in M and λ.*

Remark 1. Note that the only difference between the two definitions is the class of tampering functions and class of distinguishers.

Remark 2. Note that the notion of non-malleable codes considered in literature is implied by our notion of Aug-NMC. In the original notion, the tampering and simulated experiments only output the result of decoding the tampered codeword (without outputting one of the original states).

Let λ be the security parameter. Let $N_1(\lambda)$ and $N_2(\lambda)$ be some fixed polynomials in λ. Consider two mappings $\mathsf{Enc} : \{0,1\}^M \to \{0,1\}^{N_1} \times \{0,1\}^{N_2}$ (possibly randomized) and $\mathsf{Dec} : \{0,1\}^{N_1} \times \{0,1\}^{N_2} \to \{0,1\}^M \cup \{\bot\}$. Let $\mathcal{F} = \{(f,g)\}$ be a set of functions of the form $f_\lambda : \{0,1\}^{N_1(\lambda)} \to \{0,1\}^{N_1(\lambda)}$ and $g_\lambda : \{0,1\}^{N_2(\lambda)} \to \{0,1\}^{N_2(\lambda)}$.

For $(f,g) \in \mathcal{F}$ and $s \in \{0,1\}^M$, define a random variable $\mathsf{Tamper}^+(f,g,s)$ over $\{0,1\}^{N_2} \times (\{0,1\}^M \cup \{\bot\})$ as:

$$\mathsf{Tamper}^+(f,g,s) = \left\{ \begin{array}{l} (L,R) \sim \mathsf{Enc}(s); \\ (\tilde{L},\tilde{R}) = (f(L), g(R)); \\ \tilde{m} = \mathsf{Dec}(\tilde{L},\tilde{R}) \\ Output : (R,\tilde{m}) \end{array} \right\}$$

Let Sim^+ be a map from $\mathcal{F}$ to distributions over the sample space $\{0,1\}^{N_2} \times (\{0,1\}^M \cup \{\mathsf{same}^*, \bot\})$. For $(f,g) \in \mathcal{F}$ and $s \in \{0,1\}^M$, define the random variable $\mathsf{Copy}^{(s)}_{\mathsf{Sim}^+(f,g)}$ as follows.

$$\mathsf{Copy}^{(s)}_{\mathsf{Sim}^+(f,g)} = \begin{cases} (R,s) & \text{if } \mathsf{Sim}^+(f,g) = (R, \mathsf{same}^*) \\ \mathsf{Sim}^+(f,g) & \text{otherwise.} \end{cases}$$

Let $\mathcal{D}$ be a class of distinguishers. The simulation error (or, advantage) $\mathsf{adv}^{\mathsf{Enc},\mathsf{Dec}}_{\mathcal{F},\mathcal{D}}$ w.r.t. $\mathcal{F}$ and $\mathcal{D}$ is defined to be

$$\inf_{\substack{\mathsf{Sim}^+}} \max_{\substack{s \in \{0,1\}^M \\ (f,g) \in \mathcal{F} \\ D \in \mathcal{D}}} \left| \Pr\left[D\left(\mathsf{Tamper}^+(f,g,s)\right) = 1\right] - \Pr\left[D\left(\mathsf{Copy}^{(s)}_{\mathsf{Sim}^+(f,g)}\right) = 1\right] \right|.$$

Fig. 1. Tampering and Simulation Experiments

2.2 Building Blocks

Our construction will build upon two ingredients. We describe these next.

Authenticated Encryption. We describe the notion of a secret key authenticated encryption scheme $(\mathsf{AEnc}, \mathsf{ADec})$. Later we will describe how such a scheme can be constructed using a secret key encryption scheme and a message authentication code, both of which can be based on one-way functions. Let $\mathcal{K}_\lambda$, $\mathcal{M}_\lambda$ and $\mathcal{C}_\lambda$ denote the key, message, and ciphertext space for the authenticated encryption scheme, respectively. The scheme should satisfy the following properties. In each of the following the probability is over the randomness of $\mathsf{AEnc}, \mathsf{ADec}$ and coins of the adversary.

1. Perfect Correctness: For every $k \in \mathcal{K}_\lambda$, $m \in \mathcal{M}_\lambda$, $\Pr[\mathsf{ADec}(k, (\mathsf{AEnc}(k,m)) = m] = 1$.
2. Semantic Security: For all PPT adversaries $\mathcal{A}$, for all messages $m, m' \in \mathcal{M}_\lambda$, over a random choice of $k \xleftarrow{\$} \mathcal{K}_\lambda$, $\{\mathsf{AEnc}(k,m)\} \approx_c \{\mathsf{AEnc}(k,m')\}$.

3. Unforgeability: For every PPT adversary $\mathcal{A} = (\mathcal{A}_1, \mathcal{A}_2)$,

$$\Pr\left[c' \neq c \wedge \mathsf{ADec}(k, c') \neq \perp \,\middle|\, \begin{array}{l} k \xleftarrow{\$} \mathcal{K}_\lambda; (m, \mathsf{st}) \leftarrow \mathcal{A}_1(1^\lambda); \\ c \sim \mathsf{AEnc}(k, m); c' \leftarrow \mathcal{A}_2(\mathsf{st}, c) \end{array} \right] \leqslant \mathsf{negl}(\lambda)$$

We call the above authenticated encryption scheme an $[M, K, C]$ scheme if $\mathcal{M} = \{0,1\}^M$, $\mathcal{K} \subseteq \{0,1\}^K$ and $\mathcal{C} \subseteq \{0,1\}^C$.

The scheme described above can be instantiated as follows: Let $(\mathsf{Encrypt}, \mathsf{Decrypt})$ be a semantically-secure secret key encryption scheme with perfect correctness. Let $\mathcal{K}_\lambda^{(1)}, \mathcal{M}_\lambda^{(1)}, \mathcal{C}_\lambda^{(1)}$ be the key, message and ciphertext space, respectively, for the encryption scheme. Let $(\mathsf{Tag}, \mathsf{Verify})$ be a message authentication scheme satisfying perfect correctness and unforgeability. Let $\mathcal{K}_\lambda^{(2)}$, $\mathcal{M}_\lambda^{(2)} = \mathcal{C}_\lambda^{(1)}$ and $\mathcal{T}_\lambda^{(2)}$ be the key, message and tag space, respectively. Then we can define an authenticated encryption naturally as follows: The key space will be $\mathcal{K}_\lambda = \mathcal{K}_\lambda^{(1)} \times \mathcal{K}_\lambda^{(2)}$, message space is $\mathcal{M}_\lambda = \mathcal{M}_\lambda^{(1)}$ and the ciphertext space is $\mathcal{C}_\lambda = \mathcal{C}_\lambda^{(1)} \times \mathcal{T}_\lambda^{(2)}$. For a key $k = (k_1, k_2) \xleftarrow{\$} \mathcal{K}_\lambda$, and $m \in \mathcal{M}_\lambda$, $\mathsf{AEnc}(k, m) = (c_1, c_2)$ such that $c_1 \sim \mathsf{Encrypt}(k_1, m)$ and $c_2 \sim \mathsf{Tag}(k_2, c_1)$.

It is easy to see that the described authenticated encryption scheme will satisfy the three desired properties. Moreover, such a scheme can be designed assuming only one-way functions. We describe one such construction in our proof of Corollary 1.

$[(N_1, N_2), M, \nu]$-**Aug-NMC with** $1/\text{poly}$ **rate** Based on [3] we prove the following theorem.

Theorem 1. *There exists a fixed polynomial p, such that for all $M \in \mathbb{N}$, there exists an efficient $[(N_1, N_2), M, \nu]$-Aug-NMC $(\mathsf{Enc}^+, \mathsf{Dec}^+)$ for the message space $\{0,1\}^M$ satisfying Definiion 1 such that $N_1 + N_2 \leqslant p(M, \lambda)$ and $\nu(\lambda) = \exp(-\lambda)$.*

For the proof of the above theorem, refer to Sect. 4.

3 Our Construction

In this section, we give a construction for rate-1 computational non-malleable codes in the split-state model and prove the following theorem.

Theorem 2. *Suppose there exists an $[M, K, C]$ authenticated encryption scheme and a standard or computational $[(N_1', N_2'), K, \nu']$-Aug-NMC satisfying Theorem 1. Then there exists a computational $[(N_1, N_2), M, \nu]$-Aug-NMC such that $N_1 + N_2 = N_1' + (N_2' + C)$ and $\nu = \mathsf{negl}(\lambda)$.*

Before we describe our construction, here is a corollary of the above theorem, which is our main result.

Corollary 1. *Assuming the existence of one-way functions, there exists a computational $[(N_1, N_2), M, \nu]$-Aug-NMC such that $N_1 + N_2 = M + \mathsf{poly}(\lambda)$.*

Proof. The corollary can be obtained from Theorem 2 by using an $[M, K, C]$ authenticated scheme where $K = 2\lambda$ and $C = M + \mathsf{poly}(\lambda)$. Consider $M = q(\lambda)$ for a fixed polynomial q. Consider a polynomial stretch PRG $G : \{0, 1\}^\lambda \rightarrow \{0, 1\}^M$ and a pseudorandom function $\mathsf{PRF} : \{0, 1\}^\lambda \times \{0, 1\}^M \rightarrow \{0, 1\}^\lambda$. Then the authenticated encryption scheme is as follows: $\mathcal{K} = \{0, 1\}^{2\lambda}$ and $\mathcal{C} = \{0, 1\}^{M+\lambda}$. For a key $(k_1, k_2) \in \{0, 1\}^{2\lambda}$, $\mathsf{AEnc}((k_1, k_2), m) = (c_1, c_2)$ such that $c_1 = G(k_1) \oplus m$ and $c_2 = \mathsf{PRF}(k_2, c_1)$. It can be seen that this is a valid authenticated encryption scheme.

Using this scheme in Theorem 2, we get that $N_1' + N_2' = p(2\lambda)$ and $N_1 + N_2 = p(2\lambda) + (M + \lambda) = M + r(\lambda)$, where r is some fixed polynomial in λ. The scheme is rate-1 if q is an asymptotically faster growing polynomial than r.

Construction. Let λ be the security parameter and $\mathcal{M}_\lambda = \{0, 1\}^M$ be the message space. Let $(\mathsf{AEnc}, \mathsf{ADec})$ be an authenticated encryption scheme for message space $\mathcal{M}_\lambda$ with key space $\mathcal{K}_\lambda \subseteq \{0, 1\}^K$ and ciphertext space $\mathcal{C}_\lambda \subseteq \{0, 1\}^C$. Let $(\mathsf{Enc}^+, \mathsf{Dec}^+)$ be a $[(N_1', N_2'), K, \nu']$ augmented non-malleable encoding scheme for message space $\{0, 1\}^K$ guaranteed by Theorem 1. Given these two ingredients, our scheme is as follows.

To encode a message $s \in \{0, 1\}^M$, sample a key k for authenticated encryption scheme and encode it using Enc^+ as (ℓ, r). Next, encrypt the message s using AEnc under key k, i.e. $c \sim \mathsf{AEnc}(k, s)$. Now, the encodings in two states are $L = \ell$ and $R = (r, c)$. The decoding function is natural, which first uses $\mathsf{Dec}^+(\ell, r)$ to obtain a key k, which is used to decrypt the ciphertext c using ADec.

A formal description of the scheme is provided in Fig. 2.

It is easy to see that the scheme is perfectly correct if the underlying authenticated encryption and augmented non-malleable codes are perfectly correct. In the next section, we prove its non-malleability.

3.1 Proof of Non-malleability

In this section, we prove that the construction in Fig. 2 is a $[(N_1, N_2), M, \nu]$ computational non-malleable code such that $\nu = \mathsf{negl}(\lambda)$ against the tampering functions $\mathcal{F}_{\mathsf{eff}}^{(N_1, N_2)}$ according to Definition 2.

We begin by describing our simulator Sim required by the definition and then argue via a sequence of hybrids that for any $s \in \{0, 1\}^M$ and any $(F, G) \in \mathcal{F}_{\mathsf{eff}}$, for any efficient distinguisher $D \in \mathcal{D}_{\mathsf{eff}}$,

$$\left| \Pr\left[D\left(\mathsf{Tamper}^+(F, G, s)\right) = 1\right] - \Pr\left[D\left(\mathsf{Copy}_{\mathsf{Sim}^+(F,G)}^{(s)}\right) = 1\right]\right| \leqslant \nu(\lambda).$$

The simulator Sim is defined formally in Fig. 3. At a high level, Sim does the following: It samples a key $k \xleftarrow{\$} \mathcal{K}_\lambda$ and generates a ciphertext for message 0^M, i.e., $c = \mathsf{AEnc}(k, 0^M)$. It defines a new tampering function g_c for the underlying augmented non-malleable code by hard-coding the value of c in tampering function G. Next, it runs the simulator $\mathsf{Sim}^+(F, g_c)$ to get (r, ans). Then, it computes

---

Ingredients:

1. (AEnc, ADec): An authenticated encryption scheme with key space $\mathcal{K}_\lambda \subseteq \{0,1\}^K$, message space $\mathcal{M}_\lambda = \{0,1\}^M$ and ciphertext space $\mathcal{C}_\lambda \subseteq \{0,1\}^C$.
2. (Enc$^+$, Dec$^+$): An $[(N'_1, N'_2), K, \nu']$-Aug-NMC satisfying Theorem 1.

Enc($s \in \{0,1\}^M$):

1. Sample $k \xleftarrow{\$} \mathcal{K}_\lambda$. Sample $(\ell, r) \sim$ Enc$^+(k)$.
2. Sample $c \sim$ AEnc(k, s).
3. Define $L = \ell$ and $R = (r, c)$. Output: (L, R).
   Note that this is a $[(N_1, N_2), M]$ code where $N_1 = N'_1$ and $N_2 = N'_2 + C$.

Dec($(L, R) \in \{0,1\}^{N_1} \times \{0,1\}^{N_2}$):

1. Parse $R = (r, c) \in \{0,1\}^{N'_2} \times \{0,1\}^C$. Let $\ell = L$.
2. Decode $k =$ Dec$^+(\ell, r)$.
3. If $k = \perp$, output $\perp$.
4. Else, output ADec(k, c).

Fig. 2. Construction for rate-1 non-malleable code in the split state model.

Sim(F, G) is defined as follows:

1. Sample $k \xleftarrow{\$} \mathcal{K}_\lambda$.
2. Define $c \sim$ AEnc($k, 0^M$).
3. Define a function $g_c : \{0,1\}^{N'_2} \to \{0,1\}^{N'_2}$ such that $g_c(x) = \tilde{x}$ if $G(x, c) = (\tilde{x}, \tilde{c})$.
4. Run Sim$^+(F, g_c)$ to obtain (r, ans).
5. Define $(\tilde{r}, \tilde{c}) = G(r, c)$.
6. We have the following cases for ans:
   o Case(a) ans $= \perp$, output $((r, c), \perp)$.
   o Case(b) ans $=$ same*: If $\tilde{c} = c$, output $((r, c), \text{same}^*)$. Else, output $((r, c), \perp)$.
   o Case(c) ans $= k^*$, output $((r, c), \text{ADec}(k^*, \tilde{c}))$.

Fig. 3. Description of Sim.

$\tilde{R} = G(r, c) = (\tilde{r}, \tilde{c})$. Finally, if ans $=$ same* and $\tilde{c} = c$, it outputs same*. Else, if ans $= k^*$, it outputs ADec($k^*, \tilde{c}$). Otherwise, it outputs $\perp$.

For ease of description of hybrids, below we first describe Hyb$_0$ which is same as Tamper$_{F,G}^{(s)}$.

Hyb$_0$: This is same as Tamper$_{F,G}^{(s)}$, where we also open up the description of Enc and Dec.

1. Sample $k \xleftarrow{\$} \mathcal{K}_\lambda$.
2. Sample $c \sim$ AEnc(k, s).
3. Sample $(\ell, r) \sim$ Enc$^+(k)$.

4. Define $L = \ell$ and $R = (r, c)$.
5. Define tampered codeword as: $\tilde{L} := F(L)$ and $\tilde{R} = (\tilde{r}, \tilde{c}) := G(R) = G(r, c)$.
6. Let $\tilde{k} = \mathsf{Dec}^+(\tilde{L}, \tilde{r})$.
7. If $\tilde{k} = \bot$, output $((r, c), \bot)$. Else, output $((r, c), \mathsf{ADec}(\tilde{k}, \tilde{c}))$.

Hyb_1: This hybrid is just a re-write of the previous experiment using $\mathsf{Tamper}^+(F, g_c, k)$. Hence, the outputs of the two experiments are identical.

1. Sample $k \xleftarrow{\$} \mathcal{K}_\lambda$.
2. Sample $c \sim \mathsf{AEnc}(k, s)$.
3. Define a function $g_c : \{0, 1\}^{N'_2} \to \{0, 1\}^{N'_2}$ such that $g_c(x) = \tilde{x}$ if $G(x, c) = (\tilde{x}, \tilde{c})$.
4. Define $(r, \tilde{k}) \sim \mathsf{Tamper}^+(F, g_c, k)$.
5. Define $(\tilde{r}, \tilde{c}) = G(r, c)$.
6. If $\tilde{k} = \bot$, output $((r, c), \bot)$. Else, output $((r, c), \mathsf{ADec}(\tilde{k}, \tilde{c}))$.

Hyb_2: In this hybrid, we use $\mathsf{Copy}^{(k)}_{\mathsf{Sim}^+(F, g_c)}$ instead of $\mathsf{Tamper}^{(s)}_{F, G}$. The two hybrids are statistically close by Theorem 1.

1. Sample $k \xleftarrow{\$} \mathcal{K}_\lambda$.
2. Sample $c \sim \mathsf{AEnc}(k, s)$.
3. Define a function $g_c : \{0, 1\}^{N'_2} \to \{0, 1\}^{N'_2}$ such that $g_c(x) = \tilde{x}$ if $G(x, c) = (\tilde{x}, \tilde{c})$.
4. Define $(r, \mathsf{ans}) \sim \mathsf{Sim}^+(F, g_c)$. Define $(r, \tilde{k}) = \mathsf{Copy}^{(k)}_{\mathsf{Sim}^+(F, g_c)}$.
5. Define $(\tilde{r}, \tilde{c}) = G(r, c)$.
6. If $\tilde{k} = \bot$, output $((r, c), \bot)$. Else, output $((r, c), \mathsf{ADec}(\tilde{k}, \tilde{c}))$.

Hyb_3: In this hybrid, we change last step of how we compute the output for the case when $\mathsf{ans} = \mathsf{same}^*$.

1. Sample $k \xleftarrow{\$} \mathcal{K}_\lambda$.
2. Sample $c \sim \mathsf{AEnc}(k, s)$.
3. Define a function $g_c : \{0, 1\}^{N'_2} \to \{0, 1\}^{N'_2}$ such that $g_c(x) = \tilde{x}$ if $G(x, c) = (\tilde{x}, \tilde{c})$.
4. Define $(r, \mathsf{ans}) \sim \mathsf{Sim}^+(F, g_c)$.
5. Define $(\tilde{r}, \tilde{c}) = G(r, c)$.
6. We have the following cases for ans:
   - Case(a) $\mathsf{ans} = \bot$, output $((r, c), \bot)$.
   - Case(b) $\mathsf{ans} = \mathsf{same}^*$: If $\tilde{c} = c$, output $((r, c), s)$. Else, output $((r, c), \bot)$.
   - Case(c) $\mathsf{ans} = k^*$, output $((r, c), \mathsf{ADec}(k^*, \tilde{c}))$.

First note that the cases (a) and (c) are identical in Hyb_2 and Hyb_3. By the unforgeability property of authenticated encryption scheme, case(b) in Hyb_3 is close to Hyb_2 for all efficient tampering functions.

Hyb_4: In this hybrid, we change how we compute the ciphertext c. Instead of computing an encryption of s, we start computing encryption of 0^M.

1. Sample $k \xleftarrow{\$} \mathcal{K}_\lambda$.
2. Sample $c \sim \mathsf{AEnc}(k, 0^M)$.
3. Define a function $g_c : \{0,1\}^{N_2'} \rightarrow \{0,1\}^{N_2'}$ such that $g_c(x) = \tilde{x}$ if $G(x, c) = (\tilde{x}, \tilde{c})$.
4. Define $(r, \mathsf{ans}) \sim \mathsf{Sim}^+(F, g_c)$.
5. Define $(\tilde{r}, \tilde{c}) = G(r, c)$.
6. We have the following cases for ans:
   ○ Case(a) $\mathsf{ans} = \bot$, output $((r, c), \bot)$.
   ○ Case(b) $\mathsf{ans} = \mathsf{same}^*$: If $\tilde{c} = c$, output $((r, c), s)$. Else, output $((r, c), \bot)$.
   ○ Case(c) $\mathsf{ans} = k^*$, output $((r, c), \mathsf{ADec}(k^*, \tilde{c}))$.

By semantic security of the authenticated encryption scheme, hybrid Hyb_3 is computationally close to Hyb_4.

Finally, note that Hyb_4 is identical to $\mathsf{Copy}^{(s)}_{\mathsf{Sim}^+(F,G)}$, where Sim is the simulator described in Fig. 3.

4   Proof of Theorem 1

For proving Theorem 1, we will need the following which is a stronger version of Theorem 3 from [3]. In particular, the proof structure of our result is similar.

Let $\mathbb{F}_p$ be a finite field of prime order. Let L be uniform in $\mathbb{F}_p^n$ and let $r \in \mathbb{F}_p^n$. Let $f, g : \mathbb{F}_p^n \rightarrow \mathbb{F}_p^n$ be a pair of functions. We consider the following family of distributions

$$\varphi_{f,g}(L, r) := (\langle L, r \rangle, \langle f(L), g(r) \rangle) \in \mathbb{F}_p^2$$

Theorem 3. *There exists absolute constants $c, c' > 0$ such that the following holds. For any finite field $\mathbb{F}_p$ of prime order, and any $n > c' \log^6 p$, let $L \in \mathbb{F}_p^n$ be uniform, and fix $f, g : \mathbb{F}_p^n \rightarrow \mathbb{F}_p^n$. Then there exists a set $\mathcal{R} \subset \mathbb{F}_p^n$ of cardinality at least $p^n \cdot (1 - 2^{-cn^{1/6}})$ such that for all $r \in \mathcal{R}$, there exist random variables $A, B \in \mathbb{F}_p$, and U uniform in $\mathbb{F}_p$ and independent of A, B such that*

$$\Delta(\varphi_{f,g}(L, r) ; (U, A \cdot U + B)) \leqslant 2^{-cn^{1/6}}.$$

To prove Theorem 3, we will need the following results from [3].

Claim. Let $X = (X_1, X_2) \in \mathbb{F}_p \times \mathbb{F}_p$ be a random variable. Assume that for all $a, b \in \mathbb{F}_p$ not both zero, $\Delta(aX_1 + bX_2 ; U_{\mathbb{F}_p}) \leqslant \varepsilon$. Then $\Delta((X_1, X_2) ; U_{\mathbb{F}_p^2}) \leqslant \varepsilon p^2$.

Claim. Let $X \in \mathbb{F}_p$ be a random variable. Assume that $\Delta(X ; U_{\mathbb{F}_p}) \geqslant \varepsilon$. Then if X' is an independent and i.i.d copy of X then

$$\Pr[X = X'] \geqslant \frac{1 + \varepsilon^2}{p}.$$

The following is a reformulation of the statement that the inner-product is a strong two-source extractor.

Lemma 1. *Let L be a random variable over $\mathbb{F}_p^n$, and let $\varepsilon > 0$. Then the number of $r \in \mathbb{F}_p^n$ such that $\Delta(\langle L, r\rangle \; ; \; U_{\mathbb{F}_p}) > \varepsilon$ is at most $\frac{p^{n+1}}{2\mathbf{H}_\infty(L) \cdot \varepsilon^2}$.*

We now prove Theorem 3. Let us fix functions $f, g : \mathbb{F}_p^n \to \mathbb{F}_p^n$ and shorthand $\varphi(L, r) = \varphi_{f,g}(L, r)$. We will use the following notation: for set $\mathcal{P} \subset \mathbb{F}_p^n$ let $\varphi(L, r)|_{\mathcal{P}}$ denote the conditional distribution of $\varphi(L, r)$ conditioned on $L \in \mathcal{P}$. Equivalently, it is the distribution of $\varphi(L, r)$ for uniformly chosen $L \in \mathcal{P}$.

The following is a reformulation of Lemma 5 from [3].

Lemma 2. *Let U be uniformly random in $\mathbb{F}_p$. Let $\mathcal{P} \subseteq \mathbb{F}_p^n$, and let $r \in \mathbb{F}_p^n$. Let $\mathcal{P}_1, \ldots, \mathcal{P}_k$ be a partition of $\mathcal{P}$. Assume that for all $1 \leqslant i \leqslant k$ there exist random variables $A_i, B_i \in \mathbb{F}_p$ independent of U such that,*

$$\Delta\left(\varphi(L, r)|_{L \in \mathcal{P}_i} \; ; \; (U, A_i \cdot U + B_i)\right) \leqslant \varepsilon_i.$$

Then there exist random variables $A, B \in \mathbb{F}_p$ independent of U such that

$$\Delta\left(\varphi(L, r)|_{L \in \mathcal{P}} \; ; \; (U, AU + B)\right) \leqslant \sum \varepsilon_i \frac{|\mathcal{P}_i|}{|\mathcal{P}|}.$$

Let $s = \lfloor \frac{n}{10} \rfloor$, and $t = \lfloor \frac{s^{1/6}}{c_1 \log p} \rfloor$, where c_1 is some constant that will be chosen later. Note that $s \gg t$. We choose the constant c' in the statement of Theorem 3 such that $t \geqslant 3$.

We call $r \in \mathbb{F}_p^n$ $(\mathcal{P}, \alpha)$-bad if for every pair of random variables $A, B \in \mathbb{F}_p$, and U uniform in $\mathbb{F}_p$ and independent of A, B

$$\Delta(\varphi_{f,g}(L, r)|_{\mathcal{P}} \; ; \; (U, A \cdot U + B)) > \alpha.$$

We consider a partition of $\mathbb{F}_p^n$ based on g to elements whose output is too popular; and the rest. For $y \in \mathbb{F}_p^n$ let $g^{-1}(y) = \{x \in \mathbb{F}_p^n : g(x) = y\}$ be the set of pre-images of y. Define

$$\mathcal{R}_0 := \{x \in \mathbb{F}_p^n : |g^{-1}(g(x))| \geqslant p^t\}.$$

and set $\mathcal{R}_1 := \mathbb{F}_p^n \setminus \mathcal{R}_0$.

g is close to a constant. We now bound the number of $r \in \mathcal{R}_0$ such that there $\varphi(L, r)$ is not close to affine.

Lemma 3. *The number of $r \in \mathcal{R}_0$ that are $(\mathbb{F}_p^n, p^{-t/4})$-bad is at most $p^{-t/3} \cdot |\mathcal{R}_0|$.*

Proof. Let $Y = \{y \in \mathbb{F}_p^n : |g^{-1}(y)| \geqslant p^t\}$. We can decompose $\mathcal{R}_0$ as the disjoint union over $y \in Y$ of $g^{-1}(y)$. Fix such a $y \in Y$ and let $\mathcal{R}^\star = \{r \in \mathbb{F}_p^n : g(r) = y\}$. Since the min-entropy of L conditioned on $\langle f(L), y\rangle$ is at least $(n-1)\log p$, using Lemma 1, we have that for all but at most $p^{t/2+2}$ different $r \in \mathcal{R}^\star$

$$\Delta(\varphi_{f,g}(L, r) \; ; \; (U, \langle f(L), y\rangle)) > p^{-t/4}.$$

Thus the total number of $p^{-t/4}$-bad $r \in \mathcal{R}_0$ is at most $p^{t/2+2} \cdot |Y|$ which is upper bounded by $p^{t/2+2} \cdot |\mathcal{R}_0| \cdot p^{-t} \leqslant p^{-t/3} \cdot |\mathcal{R}_0|$.

f is close to linear. We now define a partition $\mathcal{L}_1, \ldots, \mathcal{L}_a$ of $\mathbb{F}_p^n$ based on f. Intuitively, $\mathcal{L}_i$ for $1 \leqslant i < a$ will correspond to inputs on which f agrees with a popular linear function; and $\mathcal{L}_a$ will be the remaining elements.

We define $\mathcal{L}_1, \ldots, \mathcal{L}_a$ iteratively. For $i \geqslant 1$, given $\mathcal{L}_1, \ldots, \mathcal{L}_{i-1}$, if there exists a linear map $A_i : \mathbb{F}_p^n \to \mathbb{F}_p^n$ for which

$$\left| \{ x \in \mathbb{F}_p^n : f(x) = A_i x \} \setminus (\mathcal{L}_1 \cup \ldots \cup \mathcal{L}_{i-1}) \right| \geqslant p^{n-s},$$

then set $\mathcal{L}_i$ to be $\{ x \in \mathbb{F}_p^n : f(x) = A_i x \} \setminus (\mathcal{L}_1 \cup \ldots \cup \mathcal{L}_{i-1})$. If no such linear map exists, set $a := i$, $\mathcal{L}_a := \mathbb{F}_p^n \setminus (\mathcal{L}_0 \cup \ldots \cup \mathcal{L}_{a-1})$ and complete the process. Note we obtained a partition $\mathcal{L}_1, \ldots, \mathcal{L}_a$ of $\mathbb{F}_p^n$ with $a \leqslant p^s + 1$.

Lemma 4. *Fix $1 \leqslant i < a$. The number of $r \in \mathcal{R}_1$ that are $(\mathcal{L}_i, p^{-s})$-bad is at most p^{7s}.*

Proof. Let $\mathcal{R}^\star$ be the set of all $r \in \mathcal{R}_1$ such that $(\langle L', r \rangle, \langle f(L'), g(r) \rangle)$ is p^{-s}-close to $U_{\mathbb{F}_p^2}$. Clearly, no $r \in \mathcal{R}^\star$ is $(\mathcal{L}_i, p^{-s})$-bad.

Let L' be uniform in $\mathcal{L}_i$. Note that for any $r \in \mathcal{R}_1 \setminus \mathcal{R}^\star$,

$$\langle f(L'), g(r) \rangle = \langle AL', g(r) \rangle = \langle L', A^T g(r) \rangle.$$

If $(\langle L', r \rangle, \langle f(L'), g(r) \rangle)$ is not p^{-s}-close to $U_{\mathbb{F}_p^2}$ then by Claim 4 there exist $a, b \in \mathbb{F}_p$, not both zero, such that

$$\Delta(\langle L', ar + bA^T g(r) \rangle \; ; \; U_{\mathbb{F}_p}) > p^{-2-s}.$$

Now, by assumption, L' is uniform over a set of size at least p^{n-s}. By Lemma 1, this implies that $ar + bA^T g(r)$ can take at most p^{3s+4} different values. Let $Y_{a,b} \in \mathbb{F}_p^n$ be the set of distinct values taken by $ar + bA^T g(r)$.

Fix a, b and $y \in Y_{a,b}$ and let $\mathcal{R}' \subset \mathcal{R}_1 \setminus \mathcal{R}^\star$ be such that

$$ar + bA^T g(r) = y \qquad \forall r \in \mathcal{R}'.$$

We will upper bound the number of $r \in \mathcal{R}'$ that are $(\mathcal{L}_i, p^{-s})$-bad. If $b = 0$, then clearly $|\mathcal{R}'| = 1$. If $b \neq 0$, we can rewrite (and rename the constants for convenience) as

$$A^T g(r) = a_1 r + y_1 \qquad \forall r \in \mathcal{R}'.$$

We know that for any $r \in \mathcal{R}'$, $\langle f(L'), g(r) \rangle = \langle L', A^T g(r) \rangle = a_1 \langle L', r \rangle + \langle L', y_1 \rangle$.

We know that the min-entropy of L' given $\langle L', y_1 \rangle$ is at least $(n - s - 1) \log p$. Thus, by Lemma 1, the number of $r \in \mathcal{R}'$ that are $(\mathcal{L}_i, p^{-s})$-bad is at most p^{3s+2}.

Enumerating over various possible values of a, b, y, we get that the number of $r \in \mathcal{R}_1$ that are $(\mathcal{L}_i, p^{-s})$-bad is at most $p^{3s+2} \cdot p^{3s+4} \cdot p^2 \leqslant p^{7s}$.

f is far from linear and g is far from constant. The last partition we need to analyze is $\mathcal{L}_a \times \mathcal{R}_1$, corresponding to the case where f is far from linear and g is far from constant. For this, we need the following result that can be seen as a generalization of the linearity test from [31] that was proved in [3] using results from [6,21,32].

Theorem 4. *Let p be a prime, and $n \in \mathbb{N}$. For any $\varepsilon = \varepsilon(n, p) > 0$, $\gamma_1 = \gamma_1(n, p) \leqslant 1$, $\gamma_2 = \gamma_2(n, p) \geqslant 1$, the following is true. For any function $f : \mathbb{F}_p^n \mapsto \mathbb{F}_p^n$, let $\mathcal{A} \subseteq \{(x, f(x)) : x \in \mathbb{F}_p^n\} \subseteq \mathbb{F}_p^{2n}$. If $|\mathcal{A}| \geqslant \gamma_1 \cdot |\mathbb{F}_p^n|$ and there exists some set $\mathcal{B}$ such that $|\mathcal{B}| \leqslant \gamma_2 \cdot p^n$, and*

$$\Pr_{a, a' \in \mathcal{A}}[a - a' \in \mathcal{B}] \geqslant \varepsilon,$$

then there exists a linear map $M : \mathbb{F}_p^n \to \mathbb{F}_p^n$ such that

$$\Pr_{(x, f(x)) \in \mathcal{A}}[f(x) = Mx] \geqslant p^{-O(\log^6(\frac{\gamma_2}{\gamma_1 \varepsilon}))}.$$

We will now show that, $\varphi(L, r)|_{\mathcal{L}_a}$ is close to uniform over $\mathbb{F}_p \times \mathbb{F}_p$ for most $r \in \mathcal{R}_1$.

Lemma 5. *If $|\mathcal{L}_a| \geqslant p^{n-t}$ then the number of $r \in \mathcal{R}_1$ that are $(\mathcal{L}_a, p^{-t})$-bad is at most p^{n-t}.*

Proof. Let $L' \in \mathcal{L}_a$ be uniform. Let $\mathcal{R}'$ be the set of $r \in \mathcal{R}_1$ such that $\varphi(L', r)$ is not p^{-t}-close to $U_{\mathbb{F}_p \times \mathbb{F}_p}$. Assume that the cardinality of $\mathcal{R}'$ is more than p^{n-t}, and we will show a contradiction.

For any $r \in \mathcal{R}'$, by Claim 4 there exist $a, b \in \mathbb{F}_p$, not both zero, so that $\Delta(a\langle L', r\rangle + b\langle f(L'), g(r)\rangle \; ; \; U_{\mathbb{F}_p}) \geqslant p^{-t-2}$. Define functions $F, G : \mathbb{F}_p^n \to \mathbb{F}_p^{2n}$ as follows

$$F(x) = (x, f(x)), \quad G(y) = (ay, bg(y)).$$

We have that $\Delta(\langle F(L'), G(r)\rangle \; ; \; U_{\mathbb{F}_p}) \geqslant p^{-t-2}$. Applying Claim 4, we get that for L'' i.i.d to L' we have

$$\Pr[\langle F(L'), G(r)\rangle = \langle F(L''), G(r)\rangle] \geqslant \frac{1}{p} + \frac{1}{p^{2t+5}}.$$

This implies that for all $r \in \mathcal{R}'$

$$\Pr[\langle F(L') - F(L''), G(r)\rangle = 0] \geqslant \frac{1}{p} + \frac{1}{p^{2t+5}}.$$

Let R' be uniform in $\mathcal{R}'$ and define

$$\mathcal{B} := \left\{\alpha \in \mathbb{F}_p^{2n} : \Pr[\langle \alpha, G(R')\rangle = 0] \geqslant \frac{1}{p} + \frac{1}{p^{2t+6}}\right\}.$$

Let $B \in \mathcal{B}$ be uniform. Then $\Delta(\langle B, G(R')\rangle, U_{\mathbb{F}_p}) \geqslant \frac{1}{p^{2t+6}}$. Also, since $g(y)$ has at most p^t preimages for any $y \in \mathbb{F}_p^n$, $G(R')$ has min-entropy at least $\log(|\mathcal{R}'|p^{-t}) \geqslant (n - 2t) \log p$. Hence, by Lemma 1, we have $\mathbf{H}_\infty(B) \leqslant (n + 6t + 13) \cdot \log p$, which implies $|\mathcal{B}| \leqslant p^{n+6t+13}$. Furthermore, we have that

$$\Pr[\langle F(L') - F(L''), G(R')\rangle = 0] \leqslant \Pr[F(L') - F(L'') \in \mathcal{B}] + \frac{1}{p} + \frac{1}{p^{2t+6}}.$$

So we must have that

$$\Pr[F(L') - F(L'') \in \mathcal{B}] \geqslant \frac{1}{p^{2t+5}} - \frac{1}{p^{2t+6}} \geqslant \frac{1}{p^{2t+6}}.$$

Thus, using Theorem 4, we get that there exists a linear map $M : \mathbb{F}_p^n \to \mathbb{F}_p^n$ for which

$$\Pr_{x \in \mathbb{F}_p^n}[Mx = f(x)] \geqslant p^{-O(t^6 \log^6 p)}.$$

This violates the definition of $\mathcal{L}_a$ whenever $s \geqslant C(t^6 \log^6 p)$ for a big enough constant C.[6]

To conclude the proof of Theorem 3, note that from Lemmas 3, 4 and 5, and applying Lemma 2, we have that apart from $p^{n-t/3} + p^{7s} \cdot p^s + p^{n-t} \leqslant p^{n-t/4}$ different elements in $\mathbb{F}_p^n$, for every other $r \in \mathbb{F}_p^n$, there exist random variables $A, B \in \mathbb{F}_p$, and U uniform in $\mathbb{F}_p$ and independent of A, B, such that the statistical distance of $\varphi(L, r)$ and $(U, AU + B)$ is at most

$$\max\left(p^{-t/4}, \sum_{i=1}^{a-1} p^{-s} \cdot \frac{|\mathcal{L}_i|}{p^n} + \frac{p^{n-t}}{p^n} \cdot 1\right) \leqslant p^{-t/4}.$$

To complete the proof of Theorem 1, we will need the notion of an affine-evasive set modulo p and the following result from [1].

Definition 3. *A surjective function* $h : \mathbb{F}_p \mapsto \mathcal{M} \cup \{\perp\}$ *is called* (γ, δ)-*affine-evasive if for any* $a, b \in \mathbb{F}_p$ *such that* $a \neq 0$, *and* $(a, b) \neq (1, 0)$, *and for any* $m \in \mathcal{M}$,

1. $\Pr_{U \leftarrow \mathbb{F}_p}(h(aU + b) \neq \perp) \leqslant \gamma$
2. $\Pr_{U \leftarrow \mathbb{F}_p}(h(aU + b) \neq \perp \mid h(U) = m) \leqslant \delta$
3. *A uniformly random* X *such that* $h(X) = m$ *is efficiently samplable.*

Lemma 6 ([1, Lemma 2]). *There exists an efficiently computable* $(p^{-3/4}, \Theta(K \log p \cdot p^{-1/4}))$-*affine-evasive function* $h : \mathbb{F}_p \mapsto \mathcal{M} \cup \{\perp\}$.

Additionally, we will need the following from [3].

Claim. Let $X_1, X_2, Y_1, Y_2 \in \mathcal{A}$ be random variables such that $\Delta((X_1, X_2); (Y_1, Y_2)) \leqslant \varepsilon$. Then, for any non-empty set $\mathcal{A}_1 \subseteq \mathcal{A}$, we have

$$\Delta(X_2 \mid X_1 \in \mathcal{A}_1 \ ; \ Y_2 \mid Y_1 \in \mathcal{A}_1) \leqslant \frac{2\varepsilon}{\Pr(X_1 \in \mathcal{A}_1)}.$$

Proof (Proof of Theorem 1). We construct a ν-augmented-non-malleable encoding scheme from $\mathcal{M} = \{1, \ldots, K\}$ to $\mathbb{F}_p^n \times \mathbb{F}_p^n$, where $\mathbb{F}_p$ is a finite field of

---

[6] The constant C here determines the choice of the constant c_1 used while defining the parameter t.

prime order p such that $p \geqslant (\frac{2K}{\nu})^8$, and n chosen as $\left(\lceil \frac{2\log p}{c} \rceil\right)^6$ (i.e., such that $2^{cn^{1/6}} \geqslant p^2$), where c is the constant from Theorem 3.

The decoding function $\mathsf{Dec}^+ : \mathbb{F}_p^n \times \mathbb{F}_p^n \mapsto \mathcal{M} \cup \{\bot\}$ is defined using the affine-evasive function h from Lemma 6 as:

$$\mathsf{Dec}^+(L, R) := h(\langle L, R \rangle) .$$

The encoding function is defined as $\mathsf{Enc}^+(m) := (L, R)$ where L, R are chosen uniformly at random from $\mathbb{F}_p^n \times \mathbb{F}_p^n$ conditioned on the fact that $h(\langle L, R \rangle) = m$.

We will show that our scheme is ν-non-malleable with respect to the family of all functions $(f, g) : \mathbb{F}_p^n \times \mathbb{F}_p^n \mapsto \mathbb{F}_p^n \times \mathbb{F}_p^n$, where f and g are functions from $\mathbb{F}_p^n \mapsto \mathbb{F}_p^n$, and $(f, g)(x, y) = (f(x), g(y))$, for all $x, y \in \mathbb{F}_p^n$.

Simulator. For any functions $f, g : \mathbb{F}_p^n \mapsto \mathbb{F}_p^n$, we define the distribution $D_{f,g}$ over $\mathcal{M} \cup \{\bot, \mathsf{same}^*\}$ as the output of the following sampling procedure:

1. Choose $L, R \leftarrow \mathbb{F}_p^n$.
2. If $\langle f(L), g(R) \rangle = \langle L, R \rangle$, then output (R, same^*), else output $(R, h(\langle f(L), g(R) \rangle))$.

Note that this distribution is efficiently samplable given oracle access to f and g. The distribution $D_{f,g}$ can also be expressed as:

$$D_{f,g} = \begin{cases} (r, \mathsf{same}^*) & \text{with prob. } \frac{1}{p^n} \cdot \Pr_{L \leftarrow \mathbb{F}_p^n}(\langle f(L), g(r) \rangle = \langle L, r \rangle) \\ (r, m') & \text{with prob. } \Pr_{L \leftarrow \mathbb{F}_p^n}(h(\langle f(L), g(r) \rangle) = m', \text{ and} \\ & \qquad \langle f(L), g(r) \rangle \neq \langle L, r \rangle), \end{cases}$$

where $m' \in \mathcal{M} \cup \{\bot\}$.

Security Proof. The random variable corresponding to the tampering experiment $\mathsf{Tamper}^+(f, g, m)$ has the following distribution for all $m' \in \mathcal{M} \cup \{\bot\}$.

$$\Pr(\mathsf{Tamper}^+(f, g, m) = (r, m')) = \frac{1}{p^n} \cdot \Pr\left(h(\langle f(L), g(r) \rangle) = m' \mid h(\langle L, r \rangle) = m\right) . \quad (1)$$

The random variable corresponding to the simulator $\mathsf{Copy}^{(m)}_{\mathsf{Sim}^+(f,g)}$ has the following distribution for all $m' \in \mathcal{M} \cup \{\bot\}$.

$$\Pr(\mathsf{Copy}^{(m)}_{\mathsf{Sim}^+(f,g)} = (r, m')) = \begin{cases} \frac{1}{p^n} \cdot \Pr\left(h(\langle f(L), g(r) \rangle) = m' \wedge \overline{E}\right) & \text{if } m' \neq m \\ \frac{1}{p^n} \cdot \Pr\left(E \vee \left(h(\langle f(L), g(r) \rangle) = m \wedge \overline{E}\right)\right) & \text{if } m' = m \end{cases}, \quad (2)$$

where E is the event $\langle f(L), g(r) \rangle = \langle L, r \rangle$

From Theorem 3, we get that for all but at most p^{n-2} different $r \in \mathbb{F}_p^n$ (call these $\mathcal{R}_{\mathsf{bad}}$), there exists random variables $A, B \in \mathbb{F}_p$ and U uniform in $\mathbb{F}_p$ and independent of A, B such that

$$\Delta\left(\langle L, r \rangle, \langle f(L), g(r) \rangle ; U, AU + B\right) \leqslant \frac{1}{p^2}.$$

At the cost of an additional error of at most $\frac{1}{p^2}$, we assume that $r \notin \mathcal{R}_{\mathsf{bad}}$ for the remainder of the proof.

Using Claim 4 and that $\Delta\left(\langle L, r \rangle, \langle f(L), g(r) \rangle \; ; \; U, aU + b\right) \leqslant \frac{1}{p^2}$, we get that

$$\Delta(\mathsf{Tamper}^+(f, g, m) \; ; \; T) \leqslant \frac{2}{p} \quad \text{and} \quad \Delta(\mathsf{Copy}^{(m)}_{\mathsf{Sim}^+(f,g)} \; ; \; S) \leqslant \frac{1}{p^2} \, ,$$

where S and T are defined as follows for all $m' \in \mathcal{M} \cup \{\bot\}$:

$$\Pr(T = (r, m')) = \frac{1}{p^n} \cdot \Pr\left(h(AU + B) = m' \mid h(U) = m\right)$$

$$\Pr(S = (r, m')) = \begin{cases} \frac{1}{p^n} \cdot \Pr\left(h(AU + B) = m' \wedge AU + B \neq U\right) & \text{if } m' \neq m \\ \frac{1}{p^n} \cdot \Pr\left(AU + B = U \vee (h(AU + B) = m \wedge U \neq AU + B)\right) & \text{if } m' = m \end{cases} .$$

Note that if $(A, B) = (1, 0)$, then for all $m' \in \mathcal{M}$, $\Pr(T = (r, m')) = \Pr(S = (r, m'))$. Thus, we have that

$$\Delta(S, T) = \sum_{m' \in \mathcal{M}, r \in \mathbb{F}_p^n} |\Pr(T = (r, m')) - \Pr(S = (r, m')|$$

$$\leqslant \frac{1}{p} + p^{-3/4} + \Theta(K \log p \cdot p^{-1/4})$$

$$\leqslant \nu/4 \, ,$$

where the first inequality uses Lemma 6.

Therefore, using the triangle inequality, and including the error $\frac{1}{p^2}$ that occurs due to $r \in \mathcal{R}_{\mathsf{bad}}$, we have that

$$\Delta\left(\mathsf{Tamper}^+(f, g, m) \; ; \; \mathsf{Copy}^{(m)}_{\mathsf{Sim}^+(f,g)}\right) \leqslant \Delta\left(\mathsf{Tamper}^+(f, g, m) \; ; \; T\right) + \Delta\left(T; S\right)$$

$$+ \Delta\left(S \; ; \; \mathsf{Copy}^{(m)}_{\mathsf{Sim}^+(f,g)}\right)$$

$$\leqslant \frac{\nu}{4} + \frac{1}{p^2} + \frac{2}{p} + \frac{1}{p^2} \leqslant \nu \, ,$$

thus completing the proof.

5   Necessity of One-Way Functions

We start by recalling the definition of distributional one-way functions.

Definition 4 (Distributionally One-way Functions [24,25]). *A function* $f \colon \{0,1\}^* \to \{0,1\}^*$ *is a* distributionally one-way function *if there exists a positive polynomial* $p(\cdot)$ *such that for every probabilistic polynomial-time algorithm* A *and all sufficiently large* n's *we have:*

$$\mathsf{SD}\left((U_n, f(U_n)), (A(1^n, f(U_n)), f(U_n))\right) \geqslant \frac{1}{p(n)}.$$

Intuitively, it says that if f is a distributionally one-way function, then there exists an associated (fixed) polynomial $p(\cdot)$ such that no algorithm can uniformly reverse-sample from the pre-image set on average with at most $1/p(n)$ error. It was shown that if one-way functions do not exist, then distributionally one-way functions also do not exist [25]. If distributionally one-way functions do not exist, then for every function f and polynomial $p(\cdot)$, there exists an algorithm A such that (for large enough n) it can ensure: $\mathsf{SD}\left((U_n, f(U_n)), (A(1^n, f(U_n)), f(U_n))\right) < \frac{1}{p(n)}$.

We briefly recall the overview of our result (already presented in Sect. 1.1).

Cheraghchi and Guruswami [10] show the following negative result. Consider the set of tampering functions which depend only on the first αn bits of the code and tampers it arbitrarily. Then a non-malleable code which protects against this tampering class can have rate at most $1 - \alpha$. In particular, k-split-state non-malleable code can have at most $1 - 1/k$ rate. Otherwise, one can use the attack of [10] to show that one can tamper only the first state appropriately to violate the non-malleability condition.

The result in [10] uses the following idea. If the rate is higher than $1 - \alpha$ then there exists two messages s_0 and s_1, and a set $X \subseteq \{0,1\}^{\alpha n}$ such that the following condition holds: The first αn bits of encoding of s_0 has higher probability to be in X than for an encoding of s_1. So, the tampering function just writes a dummy string w if the first αn bits belong in X; otherwise it keeps it intact. The decoding of the tampered code is, therefore, identical to the original message or it is an invalid string. Due to the property of X, the tampering function ensures that the decoding is $\perp$ with higher probability when the message is s_0.

Now consider the following function: $f(b, r) = Enc(s_b; r)|_{\alpha n}$, i.e. the function which outputs the first αn bits of the encoding of message s_b (using randomness r in the encoding procedure). Let y be any string in the domain of $f(\cdot, \cdot)$. Suppose B is an oracle which, when queried with y, provides a uniformly reverse sampled pre-image of y. Then we make t calls to B to create a set $S_y = \{(b_1, r_1), \ldots, (b_t, r_t)\}$. Counting the number of occurrences of $b = 0$ in S_y we can test whether $y \in X$ or not; when t is sufficiently large we have $y \in X$ implies $\mathsf{maj}\{b_1, \ldots, b_t\} = 0$ w.h.p. (by Chernoff bounds). Given access to the oracle B, we can emulate the tampering function which performs the tampering of [10] (except with $\mathsf{negl}(n)$ error).

Now, consider a setting where distributionally one-way functions do not exist. In this case, for $f(\cdot, \cdot)$ and suitably large $p(\cdot)$ (as a function of t), there exists an efficient inverter A which can simulate every call of B, except with error (at most) $1/p(n)$. Now, we can replace calls to algorithm B in the previous paragraph with calls to A while incurring an error of at most $t(n)/p(n)$. By suitably choosing $t(n)$ and $p(n)$, we can construct an efficient tampering on the first αn bits of the encoding which emulates the tampering of [10] with error $t(n)/p(n)$.

Formally, this proves the following theorem:

Theorem 5. *Let $k \in \mathbb{N}$ and suppose there exists a k-split-state non-malleable code with rate $\geqslant 1 - (1/k) + \delta(n)$ and simulation error $\varepsilon(n)$. Then there exists*

$\delta_0(n) = \Theta(\log n / n)$ *such that if* $\delta(n) \in [\delta_0(n), 1/k]$ *and* $\varepsilon(n) < k\delta/96 - n^{-c}$ *(for some* $c \geqslant 1$*) then one-way functions exist.*

Proof. Suppose one-way functions do not exist, $\delta(n) \in [\delta_0(n), 1/k]$ and $\varepsilon(n) < k\delta/96 - n^{-c}$. Set $\eta = k\delta/4$ and $f(b,r) = Enc(s_b; r)|_{\alpha n}$. Cheraghchi and Guruswami [10] proved that there exists a $w \in \{0,1\}^{\alpha n}$ such that there is no valid codeword which is consistent with w. Let y be the αn bits in the encoding. Given y in the image of $f(\cdot, \cdot)$, the tampering functions does the following: Consider $t(n) = n^c$ uniformly sampled pre-images such that their image under $f(\cdot, \cdot)$ is y. To reverse sample, set $p(n) = t(n)^2$ to obtain a corresponding efficient reverse sampler A. Let the obtained samples be $S_y = \{(b_1, r_1), \ldots, (b_{t(n)}, r_{t(n)})\}$. Let n_0 and n_1 be the, respective, number of samples with $b_i = 0$ and $b_i = 1$. If $n_0 / n_1 \geqslant 3/2 - n^{2/3}$, then write w otherwise leave it untampered.

Cheraghchi and Guruswami [10] show that there exists a set $X_\eta \subseteq \{0,1\}^{\alpha n}$ and inputs s_0 and s_1 such that

1. $\Pr[f(0, U) \in X_\eta] \geqslant \eta$,
2. $\Pr[f(1, U) \in X_\eta] \leqslant \eta/2$,

and therefore, there exists a set $Y_\eta \subseteq X_\eta$ (by pigeon hole principle) such that

1. $\Pr[f(0, U) \in Y_\eta] \geqslant (3/2) \cdot \Pr[f(1, U) \in Y_\eta]$, and
2. $\Pr[f(0, U) \in Y_\eta] \geqslant \eta/4$.

Note that $\Pr[f(0, U) \in Y_\eta] - \Pr[f(1, U) \in Y_\eta] \geqslant \eta/12$. Instead of X_η, using Y_η in the argument of [10] we get a contradiction because $\varepsilon(n) \geqslant k\delta/(16 \cdot (12/2)) - n^{-c}$. Hence, we get the theorem.

References

1. Aggarwal, D.: Affine-evasive sets modulo a prime. Inf. Process. Lett. **115**(2), 382–385 (2015). http://dx.doi.org/10.1016/j.ipl.2014.10.015
2. Aggarwal, D., Dodis, Y., Kazana, T., Obremski, M.: Non-malleable reductions and applications. In: Servedio, R.A., Rubinfeld, R. (eds.) Proceedings of the Forty-Seventh Annual ACM on Symposium on Theory of Computing, STOC 2015, Portland, OR, USA, 14–17 June 2015, pp. 459–468. ACM (2015). http://doi.acm.org/10.1145/2746539.2746544
3. Aggarwal, D., Dodis, Y., Lovett, S.: Non-malleable codes from additive combinatorics. In: STOC, pp. 774–783 (2014)
4. Agrawal, S., Gupta, D., Maji, H.K., Pandey, O., Prabhakaran, M.: Explicit non-malleable codes against bit-wise tampering and permutations. In: Gennaro, R., Robshaw, M. (eds.) CRYPTO 2015, Part I. LNCS, vol. 9215, pp. 538–557. Springer, Heidelberg (2015)
5. Agrawal, S., Gupta, D., Maji, H.K., Pandey, O., Prabhakaran, M.: A rate-optimizing compiler for non-malleable codes against bit-wise tampering and permutations. In: Dodis, Y., Nielsen, J.B. (eds.) TCC 2015, Part I. LNCS, vol. 9014, pp. 375–397. Springer, Heidelberg (2015). http://dx.doi.org/10.1007/978-3-662-46494-6_16

6. Balog, A., Szemeredi, E.: A statistical theorem for set addition. Combinatorica **14**(3), 263–268 (1994)
7. Chandran, N., Goyal, V., Mukherjee, P., Pandey, O., Upadhyay, J.: Block-wise non-malleable codes. Cryptology ePrint Archive, Report 2015/129 (2015). http://eprint.iacr.org
8. Chandran, N., Kanukurthi, B., Ostrovsky, R.: Locally updatable and locally decodable codes. In: Lindell, Y. (ed.) TCC 2014. LNCS, vol. 8349, pp. 489–514. Springer, Heidelberg (2014)
9. Chattopadhyay, E., Zuckerman, D.: Non-malleable codes against constant split-state tampering. In: 55th IEEE Annual Symposium on Foundations of Computer Science, FOCS 2014, Philadelphia, PA, USA, 18–21 October 2014, pp. 306–315. IEEE Computer Society (2014). http://dx.doi.org/10.1109/FOCS.2014.40
10. Cheraghchi, M., Guruswami, V.: Capacity of non-malleable codes. In: Naor, M. (ed.) ITCS, pp. 155–168. ACM (2014)
11. Cheraghchi, M., Guruswami, V.: Non-malleable coding against bit-wise and split-state tampering. In: Lindell, Y. (ed.) TCC 2014. LNCS, vol. 8349, pp. 440–464. Springer, Heidelberg (2014)
12. Choi, S.G., Kiayias, A., Malkin, T.: BiTR: built-in tamper resilience. In: Lee, D.H., Wang, X. (eds.) ASIACRYPT 2011. LNCS, vol. 7073, pp. 740–758. Springer, Heidelberg (2011)
13. Coretti, S., Maurer, U., Tackmann, B., Venturi, D.: From single-bit to multi-bit public-key encryption via non-malleable codes. In: Dodis, Y., Nielsen, J.B. (eds.) TCC 2015, Part I. LNCS, vol. 9014, pp. 532–560. Springer, Heidelberg (2015). http://dx.doi.org/10.1007/978-3-662-46494-6_22
14. Cramer, R., Dodis, Y., Fehr, S., Padró, C., Wichs, D.: Detection of algebraic manipulation with applications to robust secret sharing and fuzzy extractors. In: Smart, N.P. (ed.) EUROCRYPT 2008. LNCS, vol. 4965, pp. 471–488. Springer, Heidelberg (2008)
15. Cramer, R., Padró, C., Xing, C.: Optimal algebraic manipulation detection codes (2014). http://eprint.iacr.org/2014/116
16. Cramer, R., Shoup, V.: Design and analysis of practical public-key encryption schemes secure against adaptive chosen ciphertext attack. SIAM J. Comput. **33**(1), 167–226 (2003). http://dx.doi.org/10.1137/S0097539702403773
17. Dziembowski, S., Kazana, T., Obremski, M.: Non-malleable codes from two-source extractors. In: Canetti, R., Garay, J.A. (eds.) CRYPTO 2013, Part II. LNCS, vol. 8043, pp. 239–257. Springer, Heidelberg (2013)
18. Dziembowski, S., Pietrzak, K., Wichs, D.: Non-malleable codes. In: Yao, A.C.C. (ed.) ICS, pp. 434–452. Tsinghua University Press (2010)
19. Faust, S., Mukherjee, P., Nielsen, J.B., Venturi, D.: Continuous non-malleable codes. In: Lindell, Y. (ed.) TCC 2014. LNCS, vol. 8349, pp. 465–488. Springer, Heidelberg (2014)
20. Faust, S., Mukherjee, P., Venturi, D., Wichs, D.: Efficient non-malleable codes and key-derivation for poly-size tampering circuits. In: Nguyen, P.Q., Oswald, E. (eds.) EUROCRYPT 2014. LNCS, vol. 8441, pp. 111–128. Springer, Heidelberg (2014)
21. Gowers, T.: A new proof of Szemeredi's theorem for arithmetic progression of length four. Geom. Func. Anal. **8**(3), 529–551 (1998)
22. Guruswami, V., Smith, A.: Codes for computationally simple channels: explicit constructions with optimal rate. In: FOCS, pp. 723–732. IEEE Computer Society (2010)
23. Hemenway, B., Ostrovsky, R.: Public-key locally-decodable codes. In: Wagner, D. (ed.) CRYPTO 2008. LNCS, vol. 5157, pp. 126–143. Springer, Heidelberg (2008)

24. Impagliazzo, R.: Pseudo-random generators for cryptography and for randomized algorithms. Ph.D. thesis, University of California at Berkeley (1989)
25. Impagliazzo, R., Levin, L.A., Luby, M.: Pseudo-random generation from one-way functions (extended abstracts). In: Johnson, D.S. (ed.) STOC, pp. 12–24. ACM (1989)
26. Kurosawa, K.: Hybrid encryption. In: Encyclopedia of Cryptography and Security, 2nd edn., pp. 570–572 (2011). http://dx.doi.org/10.1007/978-1-4419-5906-5_321
27. Lipton, R.J.: A new approach to information theory. In: STACS, pp. 699–708 (1994)
28. Liu, F.-H., Lysyanskaya, A.: Tamper and leakage resilience in the split-state model. In: Safavi-Naini, R., Canetti, R. (eds.) CRYPTO 2012. LNCS, vol. 7417, pp. 517–532. Springer, Heidelberg (2012)
29. Micali, S., Peikert, C., Sudan, M., Wilson, D.A.: Optimal error correction against computationally bounded noise. In: Kilian, J. (ed.) TCC 2005. LNCS, vol. 3378, pp. 1–16. Springer, Heidelberg (2005)
30. Ostrovsky, R., Pandey, O., Sahai, A.: Private locally decodable codes. In: Arge, L., Cachin, C., Jurdziński, T., Tarlecki, A. (eds.) ICALP 2007. LNCS, vol. 4596, pp. 387–398. Springer, Heidelberg (2007)
31. Samorodnitsky, A.: Low-degree tests at large distances. In: ACM Symposium on Theory of Computing, pp. 506–515. ACM (2007)
32. Sanders, T.: On the Bogolyubov-Ruzsa lemma. Anal. PDE 5, 627–655 (2012)

Limitations of Obfuscation and Obfuscation-Avoiding Constructions

How to Avoid Obfuscation Using Witness PRFs

Mark Zhandry[✉]

Massachusetts Institute of Technology, Cambridge, MA, USA
mzhandry@gmail.com

Abstract. We propose a new cryptographic primitive called *witness pseudorandom functions* (witness PRFs). Witness PRFs are related to witness encryption, but appear strictly stronger: we show that witness PRFs can be used for applications such as multi-party key exchange without trusted setup, polynomially-many hardcore bits for any one-way function, and several others that were previously only possible using obfuscation. Thus we improve the minimal assumptions required for these applications. Moreover, current candidate obfuscators are far from practical and typically rely on unnatural hardness assumptions about multilinear maps. We give a construction of witness PRFs from multilinear maps that is simpler and much more efficient than current obfuscation candidates, thus bringing several applications of obfuscation closer to practice. Our construction relies on new but very natural hardness assumptions about the underlying maps that appear to be resistant to a recent line of attacks.

Keywords: Witness PRFs · Multilinear maps · Multiparty key exchange

1 Introduction

Program obfuscation is the act of "scrambling" a program such that the functionality is preserved, but the inner workings of the program are completely hidden even given the scrambled code. Recently, Garg et al. [GGH+13b] proposed the first construction of a general purpose program obfuscator [GGH+13b], which has sparked significant advances in cryptographic capabilities. Obfuscation has been used to construct a plethora of surprising and powerful cryptographic applications, including functional encryption [GGH+13b], deniable encryption [SW14], multiparty non-interactive key agreement [BZ14], multiparty computation in very few rounds [GGHR14], and much more. Thus, obfuscation is a "heavy hammer" by which, it seems, most of cryptography can be built. This leads to a natural question:

> *To what extent is obfuscation actually needed for various applications?*

M. Zhandry—Work done while the author was a graduate student at Stanford University. Supported by the DARPA PROCEED program.

E. Kushilevitz and T. Malkin (Eds.): TCC 2016-A, Part II, LNCS 9563, pp. 421–448, 2016.
DOI: 10.1007/978-3-662-49099-0_16

This is a very important question, as using obfuscation for applications has some major drawbacks. For one, current candidate obfuscators [GGH+13b, BR14, BGK+14, PST14, AGIS14, GLSW14, SZ14, Zim15, AB15] are incredibly inefficient, to the point that they are utterly *unimplementable* for all except the simplest of functionalities. This is even despite significant improvements in efficiency obtained by several recent works. Second, we do not know how to base the security of obfuscation on any traditional assumptions, but must instead make strong new assumptions on multilinear maps.

Therefore, obfuscation is likely too general of a tool for practical protocols with reasonable underlying security assumptions. Instead, obfuscation serves as a proof of concept, showing that a particular application is plausible. Then, more application-specific tools and techniques are required to actually obtain a usable protocol.

This Work. In this work, we make progress toward answering the above question by showing that obfuscation is *not* necessary for several applications. We do this by introducing a new technical tool called *witness pseudorandom functions (witness PRFs)* that abstracts an obfuscation technique used by several recent applications. We show that witness PRFs maintains enough of the power of obfuscation that it can still be used for these tasks, which were only previously possible using obfuscation. We also give a very simple construction of witness PRFs using multilinear maps that is significantly more efficient that current obfuscators. For security, our construction relies on new assumptions on the underlying maps. Our assumptions are very simple, and we argue that they are in some ways "better" than the assumptions on which obfuscation is based.

While applications of our witness PRFs remain impractical and security is still based on relatively untested multilinear map assumptions, our work provides a significant step towards improving the efficiency of some applications of obfuscation, and potentially towards basing applications on better assumptions. Moreover, our work provides a more refined view of the cryptographic landscape by showing that weaker primitives suffice for some applications.

1.1 Motivating Example: Non-interactive Key Exchange Without Setup

We motivate the following discussion using a specific application of obfuscation: multiparty non-interactive key exchange (NIKE) without trusted setup. In such a protocol protocol, n users each generate a secret and public value and simultaneously publishes their public values to a public bulletin board. All of the users then read off the values from the bulletin board and are each able to derive the same shared key with no further interaction. Non-interaction is crucial to obtaining a *re-usable* protocol: $N \gg n$ users can each publish their public value, and then at a later point *any* subset of n of them can establish a shared secret key *without any additional interaction*. In contrast, in an interactive scheme, the protocol needs to be carried out once *for every* subset of users that wishes to derive a key.

The first key exchange protocol for $n = 2$ users is the celebrated Diffie-Hellman protocol. Joux [Jou04] shows how to use pairings to extend this to $n = 3$ users. Boneh and Silverberg [BS02] generalize Joux's work to obtain multiparty NIKE for arbitrary n from (symmetric) multilinear maps as follows. Recall that a symmetric n-linear map consists of a source group $\mathbb{G}$ with generator g of order p, a target group $\mathbb{G}_T$ with generator g_T of order p, and a multilinear "pairing" operation $e : \mathbb{G}^n \to \mathbb{G}_T$ with the property that $e(g^{a_1}, g^{a_2}, \ldots, g^{a_n}) = g_T^{a_1 a_2 \cdots a_n}$. We call n the multilinearity of the map. Ideally any operation except the group and pairing operations should be computationally infeasible. Using an n-linear map, the Boneh-Silverberg protocol for $n + 1$ users is as follows: user i chooses a random $a_i \in \mathbb{Z}_p$, and publishes $h_i = g^{a_i}$. The shared secret is $K = g_T^{a_1 a_2 \cdots a_{n+1}}$. User i can compute K as $e(h_1, h_2, \ldots, h_{i-1}, h_{i+1}, \ldots, h_{n+1})^{a_i}$ by pairing the other n public values, and then exponentiating by her secret value. However, an eavesdropper that only sees the h_i would have to pair all $n + 1$ of the public values to obtain K, but the pairing operation only supports pairing n elements together. Security can be proved based on the *multilinear DDH* assumption, a natural generalization of the DDH assumption to the multilinear setting, and one of the most basic assumptions made on multilinear maps.

Garg, Gentry, and Halevi [GGH13a] give the first candidate multilinear map construction, thus giving the first multiparty NIKE protocol. However, in their construction, generating g and e requires secrets, knowledge of which completely breaks any security of the maps. The protocol is therefore only non-interactive in a *trusted* setup model, where setup must be performed by a central authority, and the authority will also be able to learn the shared key. Moreover, since g is needed for users to compute their public value, the setup must take place *before* the protocol is carried out. The need for a trusted central authority is a serious limitation of the protocol, and also for all protocols for $n > 3$ users prior to the obfuscation-based protocol we explain next.

Multiparty NIKE Without Setup. Boneh and Zhandry [BZ14] show how to use obfuscation to remove the setup phase entirely. In their protocol, each party generates a seed s_i of length λ for a pseudorandom generator G with output size 2λ, and publishes the corresponding output x_i. In addition, a designated master party (say, party 1) chooses a random key fk for a PRF F, and builds the following program P:

- On input $x_1, \ldots, x_n, s, i$, check that $\mathsf{G}(s) = x_i$.
- If the check fails, output $\perp$. Otherwise, output $\mathsf{F}(\mathsf{fk}, x_1, \ldots, x_n)$.

The master party then publishes an obfuscation of P along with their public x_i. Each party i can now compute $K = \mathsf{F}(\mathsf{fk}, x_1, \ldots, x_n)$ by feeding $x_1, \ldots, x_n, s_i, i$ into the obfuscation of P. Thus, all parties establish the same shared key K. An eavesdropper meanwhile only gets to see the obfuscation of P and the x_i, and tries to determine K. He can do so in one of two ways: either run the obfuscation of P on inputs of his choice, hoping that one of the outputs is K, or inspect the obfuscated code of P to try to learn K.

The one-wayness of G means the first approach is not viable. Boneh and Zhandry show that when using an "indistinguishability" obfuscator and "puncturable" PRF, the value of K is still hidden, even if the adversary inspects the obfuscated code for P. The proof works roughly as follows: first, all of the public values x_i are replaced with truly random strings. The security of G shows that this change is undetectable. Then, since G is expanding, with high probability, none of the x_i have pre-images under G. This means there is no input to the program P that causes it to pass the check and output $K = F(fk, x_1, \ldots, x_n)$. Then, using indistinguishability obfuscation and the puncturing property of F, it is possible to show the adversary learns no information about K.

Implementing the Boneh Zhandry Protocol. There are two ways to instantiate the obfuscator in the protocol above using multilinear maps:

- Directly on a "core obfuscator" for shallow circuits. The multilinearity required for the underlying map will be approximately 2^d for input circuit of depth d. This presents a serious implementation barrier, as parameters in current multilinear maps grow polynomially with the multilinearity. In an asymptotic sense, using circuits of logarithmic depth will result in polynomial-sized programs. In the case of the Boneh-Zhandry protocol, the bottleneck is clearly the PRF. While there exist puncturable PRFs that are computable in log-depth (for example, it is folklore that the Naor-Reignold PRF is puncturable), the constant term is moderate. Thus, if the depth of the PRF is, say $c \log(2n\lambda)$ ($2n\lambda$ being roughly the input size to the PRF), the resulting program requires multilinearity at least $(2n\lambda)^c$, a polynomial. However, for even moderate c, this polynomial becomes extremely large.
- By boosting the "core obfuscator" to a general obfuscator for all circuits. Depending on the conversion used, this at best requires obfuscating a low-depth PRF [App13] with the core obfuscator anyway, and at worst obfuscating the decryption function of a fully homomorphic encryption scheme [GGH+13b]. Therefore, this approach seems unlikely to yield significant improvements.

In terms of security, current obfuscators can be separated into two categories:

- Schemes with heuristic security. This includes the first candidate scheme of Garg et al. [GGH+13b] as well as several subsequent constructions [BR14, BGK+14, PST14, AGIS14, SZ14, Zim15, AB15]. Some of these schemes can be proven secure in idealized models of computation [BR14, BGK+14, AGIS14, SZ14, Zim15, AB15], but such a proof does not translate into a standard model proof under any assumptions. Thus for these constructions, the security assumption is "tautological" and basically matches the scheme. While there has been significant progress towards simplifying obfuscation, these candidates still are complicated and require several techniques (straddling sets, Kilian randomization, etc.) that yield unnatural security assumptions.
- Schemes with security proved relative to a "nice" assumption. There are basically two examples. The first is a construction due to Pass, Seth, and

Telang [PST14], based on the "semantic security" assumption on multilinear maps. Unfortunately, this conjectured assumption is an "uber assumption" that is so general that it comes close assuming the scheme itself is secure. The second is a construction of Gentry et al. [GLSW14] based on a single assumption, the multilinear subgroup elimination (MSE) assumption. While this is a significant advance in terms of basing the security of obfuscation on better assumptions, there are some notable drawbacks. First, the assumption requires introducing subgroups, which complicates the scheme and makes it less efficient. Second, the MSE assumption is a "source group" assumption on multilinear maps, which has proven very problematic on current map candidates. In particular, the MSE assumption is broken on all other multilinear maps due to a recent line of attacks [CHL+14, GHMS14, BWZ14b, CLT14][1]. Finally, the proof uses complexity leveraging, which seems inherent to basing obfuscation on simple assumptions [GGSW13]. This means that, for the proof to hold, the security parameter must be set quite large, compounding the efficiency issues above. Thus the most efficient obfuscators are likely to require complicated "tautological" security assumptions.

Thus, we pay a very steep price for eliminating the setup, both in terms of efficiency and in terms of assumptions. Using multilinearity as a proxy for efficiency, we see that the multilinearity for an n-user protocol increases from $n-1$ to $(2n\lambda)^c$ for a moderate constant c. Moreover, whereas the security of the basic multilinear map protocol is based on the very simple MDDH assumption, the setupless protocol requires somewhat more complicated assumptions. Outside of this work, all setupless key exchange protocols (even in subsequent work [HJK+14, Rao14]) require obfuscation, and therefore suffer from these weaknesses.

1.2 Our Contributions: Witness PRFs

Abstracting the Needed Functionality. We now ask, what features of obfuscation are needed for setupless key exchange? Observe that we do not need to hide the entire program P in the protocol: for example, the entire computation up until the PRF can be leaked. Thus, we do not necessarily need the full power of obfuscation. In fact, obfuscation is used in a very particular way:

- First, the input is separated into two parts. The first part, the "instance", consists of the $y_1, \ldots, y_n$. The second part, the "witness" or "token", consists of s, i.
- The program has the following structure: check some relation between the instance and witness and then apply a PRF to the instance (but not the witness) if the check passes.

---

[1] These assumptions are only broken on these maps if certain "re-randomization" terms are published, and it is possible to state the MSE assumption without these terms in which case the assumption may hold on *all* candidate multilinear maps. The obfuscator of [GLSW14] does not rely on such re-randomization parameters, but the security proof *does* need the parameters. Hence, the form of the assumption needed to prove security *is* broken.

– The security we desire is that if the instance has no witness, no information about the output of the PRF value at that input is revealed.

Thus, obfuscation is acting as an access control to the PRF, only allowing evaluation at a point if the user can supply a valid token.

Witness PRFs. We now define our new primitive called *witness pseudorandom functions* (witness PRFs) that captures the functionality and security properties needed above. Informally, a witness PRF for an NP language L is a PRF F such that anyone with a valid witness that $x \in L$ can compute $F(x)$ without the secret key, but for all $x \notin L$, $F(x)$ is computationally hidden without knowledge of the secret key. More precisely, a witness PRF consists of the following three algorithms:

– $\mathsf{Gen}(L, n)$ takes as input (a description of) an NP language L and instance length n (and implicitly a security parameter), and outputs a secret function key fk and public evaluation key ek.
– $\mathsf{F}(\mathsf{fk}, x)$ takes as input the function key fk, an instance $x \in \{0, 1\}^n$, and produces an output y.
– $\mathsf{Eval}(\mathsf{ek}, x, w)$ takes the evaluation key ek, and instance x, and a witness w that $x \in L$, and outputs $F(\mathsf{fk}, x)$ if w is a valid witness, $\bot$ otherwise.

For security, we require that for any $x \in \{0, 1\}^n \setminus L$, the value $F(\mathsf{fk}, x)$ is pseudorandom even given ek. In Sect. 3, we also consider many variants of this definition. For example, an interactive variant allows the adversary to make polynomially many PRF queries to $F(\mathsf{fk}, \cdot)$, and still requires that $F(\mathsf{fk}, x)$ is indistinguishable from random (conditioned, of course, on x not being one of the PRF queries). We also define an extractable variant that allows $x \in L$, but if the adversary can distinguish $F(\mathsf{fk}, x)$ from random, then the adversary must "know" a witness that $x \in L$.

Witness PRFs are closely related to the concept of smooth projective hash functions (a comparison is given in Sect. 1.5), and can be seen as a generalization of constrained PRFs [BW13, KPTZ13, BGI14] to arbitrary NP languages[2].

We first show how to replace obfuscation with witness PRFs for certain applications, including a no-setup multiparty key exchange protocol. We then show how to build witness PRFs from multilinear maps. Our witness PRFs are more efficient than current obfuscation candidates, and rely on very natural, though new, assumptions about the underlying maps. We stress that all of our applications can be instantiated using obfuscation, and the applications are therefore not "new." However, instantiating the applications with witness PRFs result in significant efficiency improvements compared to obfuscation. Our witness PRFs

---

[2] This is not strictly true, as constrained PRFs generate the secret function key independent of any language and multiple evaluation keys can be generated for multiple languages. Witness PRFs, on the other hand, only permit one evaluation key, and the language for the key must be known when the function key is generated. In the full version [Zha14b], we discuss how to obtain *multi-relation* witness PRFs which get around these issues.

rely on assumptions that appear to be weaker than those needed for obfuscation, and are qualitatively better in several ways. Our assumptions are very natural and simple, and while they essentially match the security of a component of our scheme, that component is much simpler than current obfuscation candidates. Our assumptions are also a very restricted case of the semantic security [PST14] assumption on multilinear maps, and do not seem general enough to imply obfuscation. Lastly, our assumption is a "target group" assumption, which appear to be more resilient to recent attacks on multilinear maps, whereas all assumptions required for obfuscation are "source group" assumptions (more details below).

Therefore, our work can be seen as (1) improving the minimal assumption under which several applications are possible and (2) providing significant efficiency improvements for those applications.

Our Results. Below, we list our main results:

- We show how to realize the following primitives from witness PRFs
  - **Multiparty Non-Interactive Key Exchange (NIKE) Without a Trusted Setup** (Sect. 5.2). We give a construction closely related to the Boneh-Zhandry [BZ14] protocol, where the obfuscator is replaced with a witness PRF, and prove that security still holds.
  - **Poly-Many Hardcore Bits.** Bellare, Stepanovs, and Tessaro [BST14] construct a hardcore function of arbitrary output size for any one-way function. They require differing inputs obfuscation [BGI+01, BCP14, ABG+13], which is a form of knowledge assumption for obfuscators. In the full version [Zha14b], we show how to replace the obfuscator with a witness PRF that satisfies our extractable notion of security.
  - **Reusable Witness Encryption.** In witness encryption, messages are encrypted to instances x of some NP language L, and any user that knows a witness that $x \in L$ can decrypt the ciphertext. Security says that if $x \notin L$, the ciphertext reveals no information about the plaintext. Garg, Gentry, Sahai, and Waters [GGSW13] define and build the first witness encryption scheme from multilinear maps. Later, Garg et al. [GGH+13b] show that indistinguishability obfuscation implies witness encryption. In the full version [Zha14b], we show that witness PRFs are actually sufficient, showing that witness PRFs are essentially a generalization of witness encryption.

    We also define a notion of re-usability for witness encryption, and give a construction from witness PRFs. Our re-usable witness encryption scheme has very short ciphertexts: namely proportional to the security parameter and independent of the size of the relation. Combining with the witness encryption-to-attribute-based encryption conversion of Garg et al. [GGSW13], this allows us to build attribute-based encryption (ABE) for circuits with similarly short ciphertexts (namely independent of the size of the access policy). No other ABE construction with such succinct ciphertexts is known without using obfuscation; it is not known how to construct such an ABE scheme from the (non-reusable) witness encryption scheme of [GGSW13].

- **Rudich Secret Sharing for** mNP. Rudich secret sharing is a generalization of secret sharing to the case where the sets of "qualified" users correspond to instances of a monotone NP (mNP) language L. In other words, n users are each given a share of a secret s. Any set $S \subseteq [n]$ of users corresponds to an instance $x \in \{0,1\}^n$, and if the users in S know a witness that $x \in L$, they can collectively reconstruct the secret using their shares. However, if $x \notin L$, the secret remains hidden. Monotonicity implies that adding users to a qualified set S does not affect the ability of S to compute the secret. Komargodski, Naor, and Yogev [KNY14] give the first construction for all of mNP using witness encryption[3]. In the full version [Zha14b], we give a related protocol using witness PRFs that is reusable, which results in much shorter shares than in [KNY14].
- **Fully Distributed Broadcast Encryption.** In broadcast encryption, n users each have a user-specific secret key, and anyone can encrypt a message to an arbitrary subset $S \subseteq [n]$ of users. Each user in S can decrypt using their individual secret, but users outside of S, even if they all collude, learn nothing about the message. The measures of interest for broadcast encryption are the sizes of the ciphertext, user secret keys, and public broadcast key as a function of the number of users n. Boneh and Zhandry [BZ14] observe that multiparty NIKE protocols with small messages give rise to broadcast encryption with constant-size ciphertexts and secret keys, but with large public keys. The resulting scheme has the novel property of being distributed, where users generate their own secret keys. In Boneh and Zhandry's notion of distributed broadcast encryption, the large public keys are inherent because there is a component of the public key corresponding to each user. In the full version [Zha14b], we put forward a new notion of *fully distributed* broadcast encryption which does not suffer from this issue, and give a construction from our extractable notion of witness PRFs where secret keys, public keys, and ciphertexts are all poly-logarithmic in n. Our scheme even obtains the strong notion of adaptive security[4]. We note that our construction could have been instantiated using (extractable) witness encryption, but witness PRFs give a protocol with better parameters.
- Next, we show how to build witness PRFs from multilinear maps. We first define an intermediate notion of a subset-sum encoding, and construct such encodings from multilinear maps. Our construction is very simple, and we argue security based on new assumptions on multilinear maps. While our assumptions basically match the security of the subset-sum encodings, the assumptions are very simple and natural due to the simplicity of our scheme. Our full construction is given in Sect. 4.

---

[3] Originally, [KNY14] used obfuscation, but in a later update showed that witness encryption was sufficient.

[4] Of course, obtaining adaptive security from an interactive assumption is not that interesting. However, our construction relies only on a *non*-interactive variant. Therefore, obtaining adaptive security is non-trivial.

In the full version [Zha14b], we then show how to build witness PRFs from subset-sum encodings. The resulting construction is much more efficient that what is currently possible with obfuscation. In particular, we can build witness PRFs for arbitrary relations directly without the costly boosting step required for obfuscation. The multilinearity required for the underlying multilinear maps is roughly equal to the size of the circuit defining the relation, rather than exponential in the depth, as in current obfuscators. While implementing our construction is still impractical for all except the most basic relations, future research in improving the efficiency of multilinear maps will bring our construction closer to practice.

– Finally, in the full version [Zha14b] we discuss how to obtain a multi-language variant of witness PRFs, where multiple evaluation keys ek_{L_i} corresponding to multiple language L_i can be produced. A witness for x relative to any of the L_i can be used to evaluate the PRF on x, and if $x \notin L_i$ for any i, then the value of the PRF on x is pseudorandom. We do not need such multi-language witness PRFs for any of our applications, but we believe they are an interesting object, and may be useful in other situations.

1.3 Techniques

Secure Subset-Sum Encodings. As a first step to building witness PRFs, we construct a primitive called a subset-sum encoding. Roughly, such an encoding corresponds to a (multi-)set S of n integers, and consists of a secret encoding function which maps integers t into encodings $\hat{t}$. Additionally, there is a public evaluation function which takes as input a subset $T \subseteq S$, and can compute the encoding $\hat{t}$ of the sum of the elements in T: $t = \sum_{i \in T} i$. For security, we ask that for any t that does not correspond to a subset-sum of elements of S, the encoding $\hat{t}$ is indistinguishable from a random element.

We provide a simple candidate subset-sum encoding from asymmetric cryptographic multilinear maps. We use asymmetric maps, though it is straightforward to adapt our protocol to the symmetric setting. Recall that in an asymmetric n-linear map, instead of a single source group $\mathbb{G}$, there are n source groups $\mathbb{G}_1, \ldots, \mathbb{G}_n$ with generators $g_1, \ldots, g_n$, and the pairing operation only allows for one element from each group. That is, $e : \mathbb{G}_1 \times \cdots \times \mathbb{G}_n \to \mathbb{G}_T$ where[5]

$$e(g_1^{a_1}, g_2^{a_2}, \ldots, g_n^{a_n}) = g_T^{a_1 a_2 \ldots a_n}.$$

To generate a subset-sum encoding for a collection $S = \{v_1, \ldots, v_n\}$ of n integers, choose a random $\alpha \xleftarrow{R} \mathbb{Z}_p$, and compute $V_i = g_i^{\alpha^{v_i}}$ for $i = 1, \ldots, n$. Publish each V_i, while α is kept secret.

---

[5] This is the asymmetric variant of the multilinear map notion proposed by Boneh and Silverberg [BS02]. Current multilinear map candidates actually support a much richer set of operations, but our construction does not require this additional structure.

The encoding of a target integer t is $\hat{t} = g_T^{\alpha^t}$. Given the secret α it is easy to compute $\hat{t}$[6]. Moreover, if $t = \sum_{i \in T} i$ for some subset $T \subseteq S$, then given the public values V_i, it is also easy to compute $\hat{t}$ using the multilinear operation: define $V_{i,1} = V_i$ and $V_{i,0} = g_i$ so that $V_{i,b} = g_i^{\alpha^{bv_i}}$. Then set b_i to be the indicator function for $i \in T$ (so that $t = \sum_{i \in [n]} b_i v_i$) and compute

$$\hat{t} = e(V_{1,b_1}, \dots, V_{n,b_n}) = e(g_1^{\alpha^{b_1 v_1}}, \dots, g_n^{\alpha^{b_n v_n}}) = g_T^{\left(\alpha^{\sum_{i \in [n]} b_i v_i}\right)} = g_T^{\alpha^t}$$

However, if t cannot be represented as a subset-sum of elements in S, then the multilinear map operations do not allow for computing $\hat{t}$: there is no way to pair or multiply the V_i and g_i together so that the result is $\hat{t}$. We conjecture that in this case, $\hat{t}$ is hard to compute. This gives rise to a new complexity assumption on multilinear maps: we say that the *multilinear subset-sum Diffie-Hellman assumption* holds for a multilinear map if, for any set of integers $S = \{v_1, \dots, v_n\}$ and any target t that cannot be represented as a subset-sum of elements in S, that $g_T^{\alpha^t}$ is indistinguishable from a random group element, even given the elements $\{g_i^{\alpha^{v_i}}\}_{i \in [n]}$[7]. In the full version [Zha14b], we show that this assumption holds in a generic model of multilinear maps, the same model that has been used to argue the security of current obfuscators [BR14, BGK+14]. We leave for future work the problem of proving security in the more refined generic model of Gentry et al. [GHMS14], which captures the recent line of "zero-izing" attacks. However, while we do not prove security in the zero-izing model, we stress that these attacks do not appear to apply to our assumptions.

Our assumption can be seen as an "uber-assumption", containing exponentially-many assumptions, one per SUBSETSUM instance (S, t). For example, setting S to be $\{1, 2, 3\}$ and t to be -1, our assumption states that $g_T^{\alpha^{-1}}$ is indistinguishable from random, given the elements $\{g_1^{\alpha^1}, g_2^{\alpha^2}, g_3^{\alpha^3}\}$. The assumptions in this family have the flavor of several existing assumptions on bilinear and multilinear maps, such as the Diffie-Hellman inversion and Diffie-Hellman Exponent assumptions.

Notice that the element that must be distinguished from random, namely $g_T^{\alpha^t}$, is in the target group $\mathbb{G}_T$. Therefore, our assumption is a *target-group* assumption, which appear more plausible on currently multilinear map candidates than *source-group* assumptions involving only elements in the groups $\mathbb{G}_1, \dots, \mathbb{G}_n$. Indeed, the focus of recent attacks [CHL+14, GHMS14, BWZ14b, CLT14] is usually the source-group assumptions. For all current obfuscators, the assumption

---

[6] Current multilinear map candidates do not allow all users to perform exponentiation by arbitrary elements of $\mathbb{Z}_p$, which makes computing V_i and $\hat{t}$ potentially problematic. However, whomever sets up the subset-sum encoding will also set up the multilinear map, and will thus have a trapdoor that *does* allow computing V_i and $\hat{t}$. Therefore, the secret key should also include this trapdoor along with α.

[7] We can also use an even stronger assumption that also allows the adversary to adaptively ask for values $g_T^{\alpha^{t'}}$ for $t' \neq t$. This will result in a stronger security guarantee for the subset-sum encodings and our derived witness PRFs.

that the scheme itself is secure is a source-group assumption, so while the recent line of attacks does not appear to break current obfuscators, the attacks do decrease our confidence in their security. Target-group assumptions, on the other hand, appear much more resistant to attack.

Application to Witness Encryption. Recall that in a witness encryption scheme as defined by Garg et al. [GGSW13], a message m is encrypted to an instance x, which may or may not be in some NP language L. Given a witness w that $x \in L$, it is possible to decrypt the ciphertext and recover m. However, if $x \notin L$, m should be computationally hidden.

Our subset-sum encodings immediately give us witness encryption for the language L of SUBSETSUM instances. Let (S, t) be a SUBSETSUM instance. To encrypt a message m to (S, t), generate a subset-sum encoding for set S. Then, using the secret encoding algorithm, compute $\hat{t}$. The ciphertext is the public evaluation function, together with $c = \hat{t} \oplus m$. To decrypt using a witness subset $T \subseteq S$, use the public evaluation procedure on T to obtain $\hat{t}$, and then XOR with c to obtain m. If $(S, T) \notin$ SUBSETSUM, then the security of our subset-sum encoding implies that $\hat{t}$, and hence m, is hidden from the adversary.

Since SUBSETSUM is NP-complete, we can use NP reductions to obtain witness encryption for any NP language L. Our scheme may be more efficient than [GGSW13] for languages L that have simpler reductions to SUBSETSUM than to the EXACTCOVER problem used by [GGSW13]. For example, the language L_{LWE} of learning-with-errors instances admits a very simple algebraic reductions to SUBSETSUM. Also, while our assumptions are new, they are no more or less plausible than the assumptions used in [GGSW13].

We can also obtain a special case of Rudich secret sharing. Given a SUBSETSUM instance (S, t), compute the elements $V_i, \hat{t}$ as above, and compute $c = \hat{t} \oplus s$ where s is the secret. Hand out share (V_i, c) to user i. Notice that a set U of users can learn s if they know a subset $T \subseteq U$ such that $\sum_{j \in T} j = t$. If no such subset exists, then our subset-sum Diffie-Hellman assumption implies that s is hidden from the group U of users.

Witness PRFs for NP. As defined above, witness PRFs are PRFs that can be evaluated on any input x for which the user knows a witness w that $x \in L$. For any $x \notin L$, the value of the PRF remains computationally hidden. Notice that subset-sum encodings *almost* give us witness PRFs for the SUBSETSUM problem. Indeed, the setup algorithm for a subset-sum encoding only depends on the subset S of integers, and not the target value t. Thus, a subset-sum encoding for a set S gives us a witness PRF for the language L_S of all integers t that are subset-sums of the integers in S.

To turn a subset-sum encoding into a witness PRF for an arbitrary language, we give a reduction from any NP language L to SUBSETSUM with the following property: the set S is independent of the instance x itself, but is instead determined entirely by the NP relation defining L (and the instance length). The instance x instead only affects the target t. Therefore, to build a witness PRF for any fixed NP relation R, run our reduction algorithm to obtain a set S_R, and then build a subset-sum encoding for S_R.

A notable feature of our resulting witness PRF is that its efficiency is comparable to that of existing witness encryption schemes for general relations R. In particular, the level of multilinearity required and the number of group elements in the evaluation key are equal to the size of the set S_R, which is roughly equal to the number of gates in R. The original witness encryption scheme of Garg et al. [GGSW13] required the level of multilinearity and the number of ciphertext group elements to roughly correspond to the EXACTCOVER instance size, which similarly grows linearly with R. Therefore, we get the added functionality of witness PRFs essentially "for free" in terms of efficiency.

Replacing Obfuscation with Witness PRFs. We return to attention to multiparty non-interactive key exchange without setup to demonstrate how witness PRFs can be used in place of obfuscation.

We now explain how witness PRFs actually suffice for this application. As in the Boneh-Zhandry protocol, each user chooses a random seed s_i for the PRG G, and publishes the output x_i. Simultaneously, we define an NP language L consisting of all tuples $(x_1, \ldots, x_n)$ where at least one of the x_i has a pre-image under G. Instead of obfuscating a program, the master party can simply produce a witness PRF F for the language L, and publishes the corresponding evaluation key ek. All users then set the shared key to be $F(fk, x_1, \ldots, x_n)$, which all the honest parties can compute using ek since they know a witness.

To argue security, as in the Boneh-Zhandry protocol we replace the x_i with random elements, and rely on the security of G to show that this change is undetectable. Then with overwhelming probability none of the x_i have preimages under G. This means that with overwhelming probability $(x_1, \ldots, x_n)$ is no longer in L. Therefore, the security of the witness PRF shows that the value $K = F(fk, x_1, \ldots, x_n)$ is computationally indistinguishable from a random string, as desired.

Notice that the master party does not know the instance $(x_1, \ldots, x_n)$ until *after* all parties have published their values; in particular, he does not know the instance when setting up the witness PRF. This is crucial to obtaining a noninteractive scheme. Witness encryption, on the other hand, requires knowing the instance when generating the ciphertext, and therefore appears insufficient for non-interactive key exchange.

Efficiency Comparison. Let $p(\lambda)$ be the circuit size for computing G. It is straightforward to implement a relation for L with circuits of size $8n\lambda + O(\lambda) + p(\lambda)$ (the bottleneck is the muxing operation to select one of the inputs to check). Using fast PRGs, we can take $p(\lambda) = O(\lambda)$. Thus, our witness PRF uses multilinear maps with linearity $8n\lambda + O(\lambda)$. While this is somewhat worse than the multilinearity $n - 1$ required for the direct protocol with trusted setup, it is many orders of magnitude better than the $(2n\lambda)^c$ multilinearity required for the obfuscation-based construction, and only about two orders of magnitude away from what is currently achievable [ACLL14]. We note that, using knowledge variants of obfuscation as in [ABG+13], it is possible to reduce the multilinearity required for the obfuscation construction to $(\lambda \log n)^{c'}$ for a larger constant c'. Using our knowledge variant of witness PRFs, we can similarly reduce the multilinearity of our protocol to $O(\lambda \log n)$. In either case (using knowledge assumptions or not), our witness PRFs are currently (by far) the most efficient multiparty key exchange protocols that do not require a trusted setup.

The reasons for the efficiency gains are two-fold:

- Our witness PRF construction grows polynomially with circuit size, rather than exponentially in the depth as in current obfuscators. Thus we will get immediate improvements for all except the shallowest circuits.
- For the applications discussed in this work, the original constructions required obfuscating a PRF. This translates to using the underlying multilinear map operations to simulate the evaluation of the PRF, which is quite costly. In contrast, our witness PRFs use the multilinear map elements themselves as the PRF outputs, eliminating the need for a separate PRF computation. Thus only the relation checking needs to be carried out with multilinear operations. For cases such as key exchange where the PRF evaluation is the bottleneck, this results in significant additional efficiency gains.

1.4 Directions for Future Work

Our work raises several intriguing open questions:

- We give several applications of witness PRFs that previously required the full power of obfuscation. For what other applications of obfuscation do witness PRFs suffice?
- Witness PRFs do not appear sufficient for many applications of obfuscation, including some that seem well-suited for witness PRFs on the surface. For example, obfuscation plays a similar role of gatekeeper to a PRF in the traitor tracing scheme of Boneh and Zhandry [BZ14]. However, in there scheme, the underlying relation must actually be kept secret for security to hold. In our notion of witness PRFs, the relation is not a secret, and our construction explicitly requires the relation to be public. A natural goal is to devise a stronger notion of witness PRFs that would suffice for these applications (say, by hiding some information about the relation) but yet has efficiency similar to that of witness PRFs and witness encryption.
- While our assumptions are natural, they are *instance dependent*, meaning that the assumption depends on the challenge instance. This means our scheme relies on an exponential number of assumptions, one per instance. An important goal is therefore to construct witness PRFs from simple instance independent assumptions. We note that the since witness PRFs imply witness encryption, the arguments of Garg et al. [GGSW13] indicate that such a construction would likely involve complexity leveraging.

Indistinguishability obfuscation (iO) can be used to build witness PRFs[8], and iO can in turn be based on simple assumptions following the work of Gentry et al. [GLSW14]. However, such an approach defeats the efficiency gains of building witness PRFs directly. A natural starting point to look for a construction would be the witness encryption scheme of Gentry et al. [GLW14], which is also based on instance independent assumptions.

– How do Witness PRFs relate to other advanced cryptographic primitives? For instance, are witness PRFs indeed weaker than obfuscation, and can witness encryption be used generically to build witness PRFs? In a subsequent work, Komargodski and Zhandry [KZ15] make progress in this direction by showing that witness PRFs are equivalent to a notion of secret sharing called *distributed secret sharing*. An interesting direction for future work would be to find more equivalences, or to give black box separations between witness PRFs and other primitives.

1.5   Other Related Work

Removing Obfuscation. Very recently, a few works have shown how to remove obfuscation from certain applications. Garg et al. [GGHZ14] build the first many-key functional encryption schemes that do not rely on obfuscation, though their construction is obfuscation-inspired. Boneh et al. [BLR+14] build a near-practical order revealing encryption scheme; the only other known construction requires obfuscation and is therefore far from practical.

Smooth Projective Hash Functions and Functional PRFs. Cramer and Shoup [CS02] define the notion of *smooth projective hash functions* (SPHFs), a concept similar to that of witness PRFs. Concurrently and independently of our work, Chen and Zhang [CZ14] define the notion of *publicly evaluable PRFs* (PEPRFs), which are again similar in concept to witness PRFs. The main differences between SPHFs and PEPRFs and our witness PRFs are that existing constructions of SPHFs and PEPRFs are only for certain classes of languages, such as certain group-theoretic languages. Witness PRFs on the other hand, can handle arbitrary NP languages, and such flexibility is required for the applications in this work. The trade-off is that witness PRFs are much less efficient and require much stronger assumptions. There are also minor differences in security notions.

Boyle et al. [BGI14] define *functional* PRFs, where the evaluation key corresponds to a function f, and given the evaluation key it is possible to compute $F(f(x))$, but $F(y)$ is pseudorandom for y not in the image of f. Functional PRFs

---

[8] To see this, start with any "puncturable" PRF, and obufscate the program that takes an input and a witness, checks the witness relation, and outputs the PRF evaluated on the input. The resulting obfuscated program is the evaluation key, and the PRF key is the secret key. Correctness is straightforward to verify, and the static security definition described above can be shown easily through the punctured programming technique of Sahai and Waters [SW14].

in their full generality equivalent to witness PRFs. In one direction, we can set $f(\ (x, w)\ ) = x$ if $R(x, w) = 1$ and $f(\ (x, w)\ ) = \perp$ otherwise. In the other, we can set $R(y; x) = 1$ if $f(x) = y$ and $R(y; x) = 0$ otherwise. We note, however, that [BGI14] only construct functional PRFs for very limited functions f related to prefix matching, which are insufficient for our applications. In particular, the functions f considered all correspond to languages that are in P (and so correspond exactly to constrained PRFs), where our construction supports general NP relations, as needed by our applications.

Witness Encryption. Garg et al. [GGSW13] define witness encryption and give the first candidate construction for the NP-Complete EXACTCOVER problem, whose security is based on the *multilinear no-exact-cover problem.* Goldwasser et al. [GKP+13] define a stronger notion, called extractable witness encryption, which stipulates that anyone who can distinguish the encryption of two messages relative to an instance x must actually be able to produce a witness for x. Our extractable notion for witness PRFs can be seen as a generalization of extractable witness encryption. Subsequently, Garg et al. [GGHW14] cast doubt on the plausibility of the most general forms of extractable witness encryption (and thus extractable witness PRFs), though their results do not apply to most potential applications of the primitives.

Hard-Core Bits. The Goldreich-Levin theorem [GL89] shows how to build a single hard-core bit for any one-way function. This result can be extended to logarithmically-many bits, and polynomially-many hard-core bits have been constructed for *specific* one-way functions [CGH01]. Bellare et al. [BST14] give poly-many hard-core bits for *any* one-way function using obfuscation, which is the only construction prior to this work.

Broadcast Encryption. There has been an enormous body of work on broadcast encryption, and we only mention a few specific works. Boneh et al. [BGW05] use bilinear maps to give a broadcast scheme with short ciphertexts and secret keys, though public broadcast keys grew linearly with the number of users. Some subsequent schemes based on bilinear maps were able to achieve adaptive security [GW09], but the public parameters always grew linearly with the number of recipients. Boneh and Zhandry [BZ14] give a broadcast scheme from indistinguishability obfuscation which achieves similarly short ciphertexts and secret keys. Their broadcast scheme has the novel property of being distributed, where every user chooses their own secret key. However, their public keys are obfuscated programs, and are quite large (namely, linear in the number of users), and security is proved in a weaker *static* model. Ananth et al. [ABG+13] show how to shrink the public key (while maintaining secret key and ciphertext size), though they lose the distributed property. Boneh et al. [BWZ14a] give several broadcast schemes whose concrete parameter sizes are much better directly from multilinear maps, and very recently Zhandry [Zha14a] gives a variant that is adaptively secure. However, these schemes are not distributed.

Secret Sharing. The first secret sharing schemes due to Blakely [Bla79] and Shamir [Sha79] are for the *threshold* access structure, where any set of users of size at least some threshold t can recover the secret, and no set of size less than t can learn anything about the secret. In an unpublished work, Yao shows how to perform (computational) secret sharing where the allowable sets are decided by a polynomial-sized monotone circuit. Komargodski, Naor and Yogev [KNY14] use witness encryption to build the first protocol for arbitrary NP access structures, answering a question of Rudich.

2   Preliminaries

2.1   Subset-Sum

Let $\mathbf{A} \in \mathbb{Z}^{m \times n}$ be an integer matrix, and $\mathbf{t} \in \mathbb{Z}^m$ be an integer vector. The *subset-sum* search problem is to find an $\mathbf{w} \in \{0,1\}^n$ such that $\mathbf{t} = \mathbf{A} \cdot \mathbf{w}$. The decision problem is to decide if such an $\mathbf{w}$ exists.

We define several quantities related to a subset-sum instance. Given a matrix $\mathbf{A} \in \mathbb{Z}^{m \times n}$, let $\mathsf{SubSums}(\mathbf{A})$ be the set of all subset-sums of columns of $\mathbf{A}$. That is, $\mathsf{SubSums}(\mathbf{A}) = \{\mathbf{A} \cdot \mathbf{w} : \mathbf{w} \in \{0,1\}^n\}$. Define $\mathsf{Span}(\mathbf{A})$ as the convex hull of $\mathsf{SubSums}(\mathbf{A})$. Equivalently, $\mathsf{Span}(\mathbf{A}) = \{\mathbf{A} \cdot \mathbf{w} : \mathbf{w} \in [0,1]^n\}$. We define the integer range of $\mathbf{A}$, or $\mathsf{IntRange}(\mathbf{A})$, as $\mathsf{Span}(\mathbf{A}) \bigcap \mathbb{Z}^m$. We note that given an instance $(\mathbf{A}, \mathbf{t})$ of the subset-sum problem, it is efficiently decidable whether $\mathbf{t} \in \mathsf{IntRange}(\mathbf{A})$. Moreover, $\mathbf{t} \notin \mathsf{IntRange}(\mathbf{A})$ implies that $(\mathbf{A}, \mathbf{t})$ is unsatisfiable. The only "interesting" instances of the subset sum problem therefore have $\mathbf{t} \in \mathsf{IntRange}(\mathbf{A})$. From this point forward, we only consider $(\mathbf{A}, \mathbf{t})$ a valid subset sum instance if $\mathbf{t} \in \mathsf{IntRange}(\mathbf{A})$.

2.2   Multilinear Maps

An asymmetric multilinear map [BS02] is defined by an algorithm Setup which takes as input a security parameter λ, a multilinearity n, and a minimum group order p_{min}[9]. It outputs (the description of) $n + 1$ groups $\mathbb{G}_1, \ldots, \mathbb{G}_n, \mathbb{G}_T$ of prime order $p \geq \max(2^\lambda, p_{min})$, corresponding generators $g_1, \ldots, g_n, g_T$, and a map $e : \mathbb{G}_1 \times \cdots \times \mathbb{G}_n \to \mathbb{G}_T$ satisfying

$$e(g_1^{a_1}, \ldots, g_n^{a_n}) = g_T^{a_1 \cdots a_n}$$

Cryptographic multilinear maps are multilinear maps where certain computations not expressly allowed by the map are computationally difficult. For example, it should at a minimum be computationally infeasible to compute $a \in \mathbb{Z}_p$ given g_i^a for a random a. An example of the type of computational assumption we make in this work is that the following problem is hard: given $g_i^{ab^i}$ for $i \in [n]$, distinguish g_T^a from a random element of $\mathbb{G}_T$.

Another requirement we make on multilinear maps is that a random element of $\mathbb{G}_T$ is statistically indistinguishable from a uniform random bit string.

---

[9] It is easy to adapt multilinear map constructions [GGH13a, CLT13] to allow setting a minimum group order.

Approximate Multilinear Maps. Current candidate multilinear maps [GGH13a, CLT13] are only *approximate* and do not satisfy the ideal model outlined above. In particular, the maps are noisy, resulting in several implications. First, representations of group elements are not unique. Current map candidates provide an extraction procedure that takes a representation of a group element in the the target group $\mathbb{G}_T$ and outputs a canonical representation. This allows multiple users with different representations of the same element to arrive at the same value. The extraction procedure satisfies the requirement that, when applied to a random element of the target group, the result is statistically close to a uniform random bit string.

A more significant limitation is that noise grows with the number of multiplications and pairing operations. If the noise term grows too large, then there will be errors in the sense that the extraction procedure above will fail to output the canonical representation. In our application, the number of multiplications is equal to the multilinearity, which current candidates natively support without needing to adjust the parameter settings[10].

Lastly, and most importantly for our use, current map candidates do not allow regular users to compute g_i^α for any $\alpha \in \mathbb{Z}_p$ of the user's choice. Instead, the user computes a "level-0 encoding" of a random (unknown) $\alpha \in \mathbb{Z}_p$, and then pairs the "level-0 encoding" with g_i, which amounts computing the exponentiation g_i^α. To compute terms like $g_i^{\alpha^k}$ would require repeating this operation k times, resulting in a large blowup in the error. Thus, for large k, computing terms like $g_i^{\alpha^k}$ is infeasible for regular users. However, whomever sets up the map knows secret parameters about the map and *can* compute g_i^α for any $\alpha \in \mathbb{Z}_p$ without blowing up the error. Thus, the user who sets up the map can pick α, compute α^k in $\mathbb{Z}_p$, and then compute $g_i^{\alpha^k}$ using the map secrets. This will be critical for our construction.

3   Witness PRFs

Informally, a witness PRF is a generalization of constrained PRFs [BW13, KPTZ13, BGI14] to arbitrary NP relations. That is, for an NP language L, a user can evaluate the function F at an instance x only if $x \in L$ *and* the user can provide a witness w that $x \in L$. More formally, a witness PRF is the following:

Definition 1. *A witness PRF is a triple of algorithms* (Gen, F, Eval) *such that:*

- Gen *is a randomized algorithm that takes as input a security parameter λ and a circuit $R : \mathcal{X} \times \mathcal{W} \to \{0,1\}$[11], and produces a secret function key* fk *and a public evaluation key* ek.

---

[10] In fact, the parameters can be set more aggressively since our application does not need to support re-randomization. Re-randomizing elements adds significant noise in current encodings, and the native parameter settings support this noise growth.

[11] By accepting relations as circuits, our notion of witness PRFs only handles instances of a fixed size. It is also possible to consider witness PRFs for instances of arbitrary size, in which case R would be a Turing machine.

- F *is a deterministic algorithm that takes as input the function key* fk *and an input* $x \in \mathcal{X}$, *and produces some output* $y \in \mathcal{Y}$ *for some set* $\mathcal{Y}$.
- Eval *is a deterministic algorithm that takes as input the evaluation key* ek *and input* $x \in \mathcal{X}$, *and a witness* $w \in \mathcal{W}$, *and produces an output* $y \in \mathcal{Y}$ *or* $\perp$.
- *For correctness, we require* $\mathsf{Eval}(\mathsf{ek}, x, w) = \begin{cases} \mathsf{F}(\mathsf{fk}, x) & \text{if } R(x, w) = 1 \\ \perp & \text{if } R(x, w) = 0 \end{cases}$ *for all* $x \in \mathcal{X}, w \in \mathcal{W}$.

We note one significant way in which our notion of witness PRFs is *weaker* than constrained PRFs: our notion only allows a single evaluation key ek for a relation R that must be chosen at setup time. In contrast, constrained PRFs allow arbitrarily-many ek for different circuits, and the circuits can be chosen after setup. This limitation will be inherent to our construction: the function defined by $\mathsf{F}(\mathsf{fk}, \cdot)$ will depend on the relation R. Nonetheless, this definition will be sufficient for our applications. In the full version [Zha14b], we define a multi-relation variant, discuss a possible approach to building such enhanced primitives.

3.1   Security

The simplest and most natural security notion we consider is a direct generalization of the security notion for constrained PRFs, which we call adaptive instance interactive security. Consider the following experiment $\mathsf{EXP}_{\mathcal{A}}^{R}(b, \lambda)$ between an adversary $\mathcal{A}$ and challenger, parameterized by a relation $R : \mathcal{X} \times \mathcal{W} \to \{0, 1\}$, a bit b and security parameter λ.

- Run $(\mathsf{fk}, \mathsf{ek}) \xleftarrow{R} \mathsf{Gen}(\lambda, R)$ and give ek to $\mathcal{A}$.
- $\mathcal{A}$ can adaptively make queries on instances $x_i \in \mathcal{X}$, to which the challenger response with $\mathsf{F}(\mathsf{fk}, x_i)$.
- $\mathcal{A}$ can make a single challenge query on an instance $x^* \in \mathcal{X}$. The challenger computes $y_0 \leftarrow \mathsf{F}(\mathsf{fk}, x^*)$ and $y_1 \xleftarrow{R} \mathcal{Y}$, and responds with y_b.
- After making additional F queries, $\mathcal{A}$ produces a bit b'. The challenger checks that $x^* \notin \{x_i\}$, and that there is no witness $w \in \mathcal{W}$ such that $R(x, w) = 1$ (in other words, $x \notin L$)[12]. If either check fails, the challenger outputs a random bit. Otherwise, it outputs b'.

Define W_b as the event the challenger outputs 1 in experiment b. Let

$$\mathsf{WPRF.Adv}_{\mathcal{A}}^{R}(\lambda) = |\Pr[W_0] - \Pr[W_1]|$$

Definition 2. WPRF $= (\mathsf{Gen}, \mathsf{F}, \mathsf{Eval})$ *is adaptive instance interactively secure for a relation* R *if, for all PPT adversaries* $\mathcal{A}$, *there is a negligible function* negl *such that.*

---

[12] This check in general cannot be implemented in polynomial time, meaning our challenger is not efficient.

We can also define a weaker notion of *static instance* security where $\mathcal{A}$ commits to x^* before seeing ek or making any F queries. Independently, we can also define *non-interactive* security where the adversary is not allowed any F queries. In the full version [Zha14b], we also consider more fine-grained security notions, similar to the obfuscation-based notions of [BST14]. In the full version, we also consider *extractability* notions of witness PRFs, where pseudorandomness holds even for x^* in the language, as long as the adversary does not "know" a witness for x.

4   An Abstraction: Subset-Sum Encoding

Now that we have seen many applications of witness PRFs, we begin our construction. In this section, we give an abstraction of functionality we need from multilinear maps. Our abstraction is called a *subset-sum* encoding. Roughly, a subset sum encoding is a way to encode vectors $\mathbf{t}$ such that (1) the encoding of $\mathbf{t} = \mathbf{A} \cdot \mathbf{w}$ for $\mathbf{w} \in \{0,1\}^n$ is efficiently computable given $\mathbf{w}$ and (2) the encoding of $\mathbf{t} \notin \mathsf{SubSums}(\mathbf{A})$ is indistinguishable from a random string. More formally, a subset-sum encoding is the following:

Definition 3. *A* subset-sum encoding *is a triple of efficient algorithms* (Gen, Encode, Eval) *where:*

- Gen *takes as input a security parameter* λ *and an integer matrix* $\mathbf{A} \in \mathbb{Z}^{m \times n}$, *and outputs an encoding key* sk *and an evaluation key* ek.
- Encode *takes as input the secret key* sk *and a vector* $\mathbf{t} \in \mathbb{Z}^m$, *and produces an encoding* $\hat{\mathbf{t}} \in \mathcal{Y}$. Encode *is deterministic.*
- Eval *takes as input the encoding key* ek *and a bit vector* $\mathbf{w} \in \{0,1\}^n$, *and outputs a value* $\hat{\mathbf{t}}$ *satisfying* $\hat{\mathbf{t}} = \mathsf{Encode}(\mathsf{sk}, \mathbf{t})$ *where* $\mathbf{t} = \mathbf{A} \cdot \mathbf{w}$.

Security Notions. The security notions we define for subset-sum encodings are very similar to those for witness PRFs. Consider the following experiment $\mathsf{EXP}_{\mathcal{A}}^{\mathbf{A}}(b, \lambda)$ between an adversar $\mathcal{A}$ and challenger, parameterized by a matrix $\mathbf{A} \in \mathbb{Z}^{m \times n}$, a bit b, and a security parameters λ:

- Run $(\mathsf{sk}, \mathsf{ek}) \xleftarrow{R} \mathsf{Gen}(\lambda, \mathbf{A})$, and give ek to $\mathcal{A}$
- $\mathcal{A}$ can adaptively make queries on targets $\mathbf{t}_i \in \{0,1\}^m$, to which the challenger responds with $\hat{\mathbf{t}_i} \leftarrow \mathsf{Encode}(\mathsf{sk}, \mathbf{t}_i) \in \mathcal{Y}$.
- $\mathcal{A}$ can make a single challenge query on a target $\mathbf{t}^*$. The challenger computes $y_0 = \hat{\mathbf{t}^*} \leftarrow \mathsf{Encode}(\mathsf{sk}, \mathbf{t}^*)$ and $y_1 \xleftarrow{R} \mathcal{Y}$, and responds with y_b.
- After making additional Encode queries, $\mathcal{A}$ produces a bit b'. The challenger checks that $\mathbf{t}^* \notin \{\mathbf{t}_i\}$ and $\mathbf{t}^* \notin \mathsf{SubSums}(\mathbf{A})$. If either check fails, the challenger outputs a random bit. Otherwise, it outputs b'.

Define W_b as the event the challenger outputs 1 in experiment b. Let

$$\mathsf{SS.Adv}_{\mathcal{A}}^{\mathbf{A}}(\lambda) = |\Pr[W_0] - \Pr[W_1]|$$

Definition 4. (Gen, Encode, Eval) *is adaptive target interactively secure for a matrix* $\mathbf{A}$ *if, for all adversaries* $\mathcal{A}$, *there is a negligible function* negl *such that* $\mathsf{SS.Adv}_{\mathcal{A}}^{\mathbf{A}}(\lambda) < \mathsf{negl}(\lambda)$.

We can also define a weaker notion of *static target* security where $\mathcal{A}$ commits to $\mathbf{t}^*$ before seeing ek or making any Encode queries. Independently, we can also define *non-interactive* security where the adversary is not allowed to make any Encode queries.

4.1   A Simple Instantiation from Multilinear Maps

We now construct subset-sum encodings from asymmetric multilinear maps.

Construction 1. *Let* Setup *be the generation algorithm for an asymmetric multilinear map. We build the following subset-sum encoding:*

- Gen$(\lambda, \mathbf{A})$: *on input a matrix* $\mathbf{A} \in \mathbb{Z}^{m \times n}$, *let* $B = \|\mathbf{A}\|_\infty$, *and* $p_{min} = 2nB + 1$. *Run* params$\xleftarrow{R}$Setup(λ, n, p_{min}) *to get the description of a multilinear map* $e : \mathbb{G}_1 \times \cdots \times \mathbb{G}_n \to \mathbb{G}_T$ *on groups of prime order* p, *together with generators* $g_1, \ldots, g_m, g_T$. *Choose random* $\boldsymbol{\alpha} \in (\mathbb{Z}_p^*)^m$. *Denote by* $\boldsymbol{\alpha}^{\mathbf{v}}$ *the product* $\prod_{i \in [m]} \alpha_i^{v_i}$ *(since each component of* $\boldsymbol{\alpha}$ *is non-zero, this operation is well-defined for all integer vectors* $\mathbf{v}_i$). *Let* $V_i = g_i^{\boldsymbol{\alpha}^{\mathbf{v}_i}}$ *where* $\mathbf{v}_i$ *are the columns of* $\mathbf{A}$. *Publish* ek $= (\mathsf{params}, \{V_i\}_{i \in [n]})$ *as the public parameters and* sk $= \boldsymbol{\alpha}$
- Encode$(\mathsf{sk}, \mathbf{t}) = g_T^{\boldsymbol{\alpha}^{\mathbf{t}}}$, *where* $\mathbf{t} \in \mathsf{IntRange}(\mathbf{A})$.
- Eval$(\mathsf{ek}, \mathbf{w})$: *define* $V_{i,1} = V_i$ *and* $V_{i,0} = g_i$. *Then output*

$$e(V_{1,w_1}, V_{2,w_2}, \ldots, V_{n,w_n})$$

For correctness, observe that $V_{i,w_i} = g_i^{\boldsymbol{\alpha}^{\mathbf{v}_i w_i}}$, and therefore

$$e(V_{1,w_1}, V_{2,w_2}, \ldots, V_{n,w_n}) = e(g_1^{\boldsymbol{\alpha}^{\mathbf{v}_1 w_1}}, \ldots, g_n^{\boldsymbol{\alpha}^{\mathbf{v}_n w_n}}) = g_T^{\boldsymbol{\alpha}^{\sum_{i \in [n]} \mathbf{v}_i w_i}} = g_T^{\boldsymbol{\alpha}^{\mathbf{A} \cdot \mathbf{w}}}$$
$$= \mathsf{Encode}(\mathsf{sk}, \mathbf{A} \cdot \mathbf{w})$$

Security. We assume the security of our subset-sum encodings, which translates to a new security assumption on multilinear maps, which we call the *(adaptive target interactive) multilinear subset-sum Diffie Hellman assumption*. For completeness, we formally define the assumption as follows. Let $\mathsf{EXP}_{\mathcal{A}}^{\mathbf{A}}(b, \lambda)$ be the following experiment between an adversary $\mathcal{A}$ and challenger, parameterized by a matrix $\mathbf{A} \in \mathbb{Z}^{m \times n}$, a bit b, and a security parameter λ:

- Let $B = \|\mathbf{A}\|_\infty$, and $p_{min} = 2nB + 1$. Run params$\xleftarrow{R}$Setup(λ, n, p_{min}).
- Choose a random $\boldsymbol{\alpha} \in \mathbb{Z}_p^m$, and let $V_i = g_i^{\boldsymbol{\alpha}^{\mathbf{v}_i}}$ where $\mathbf{v}_i$ are the columns of $\mathbf{A}$. Give $(\mathsf{params}, \{V_i\}_{i \in [n]})$ to $\mathcal{A}$.
- $\mathcal{A}$ can make oracle queries on targets $\mathbf{t}_i \in \mathsf{IntRange}(\mathbf{A})$, to which the challenger responds with $g_T^{\boldsymbol{\alpha}^{\mathbf{t}_i}}$.

– $\mathcal{A}$ can make a single challenge query on a target $\mathbf{t}^* \in \mathsf{IntRange}(\mathbf{A})$. The challenger computes $y_0 = g_T^{\alpha^{\mathbf{t}^*}}$ and $y_1 = g_T^r$ for a random $r \xleftarrow{R} \mathbb{Z}_p$, and responds with y_b.

– After making additional Encode queries, $\mathcal{A}$ produces a bit b'. The challenger checks that $\mathbf{t}^* \notin \{t_i\}$ and $t^* \notin \mathsf{SubSums}(\mathbf{A})$. If either check fails, the challenger outputs a random bit. Otherwise, it outputs b'.

Define W_b as the event that the challenger outputs 1 in experiment b. Let $\mathsf{SSDH.Adv}_{\mathcal{A}}^{\mathbf{A}}(\lambda) = |\Pr[W_0] - \Pr[W_1]|$.

Definition 5. *The adaptive target interactive multilinear subset-sum Diffie Hellman (SSDH) assumption holds relative to* Setup *if, for all adversaries $\mathcal{A}$, there is a negligible function* negl *such that* $\mathsf{SSDH.Adv}_{\mathcal{A}}^{\mathbf{A}}(\lambda) < \mathsf{negl}(\lambda)$.

Security of our subset-sum encodings immediately follows from the assumption:

Fact 2. *If the adaptive target interactive multilinear SSDH assumptions holds for* Setup*, the Construction 1 is an adaptive target interactively secure subset-sum encoding.*

Flattening the Encodings. We can convert any subset-sum encoding for $m = 1$ into a subsetsum encoding for any m. Let $\mathbf{A} \in \mathbb{Z}^{m \times n}$ and define $B = \|\mathbf{A}\|_\infty$. Then, for any $\mathbf{w} \in \{0, 1\}^n$, $\|\mathbf{A} \cdot \mathbf{w}\|_\infty \leq nB$. Therefore, we can let $\mathbf{A}' = (1, nB + 1, (nB + 1)^2, \ldots, (nB + 1)^{m-1}) \cdot \mathbf{A}$ be a single row, and run $\mathsf{Gen}(\lambda, \mathbf{A}')$ to get $(\mathsf{sk}, \mathsf{ek})$. To encode an element $\mathbf{t}$, compute $\mathbf{t}' = (1, nB, (nB)^2, \ldots, (nB)^{m-1}) \cdot \mathbf{t}$, and encode $\mathbf{t}'$. Finally, to evaluate on vector $\mathbf{w}$, simply run $\mathsf{Eval}(\mathsf{ek}, \mathbf{w})$.

Security translates since left-multiplying by $(1, nB, (nB)^2, \ldots, (nB)^{m-1})$ does not introduce any collisions. Therefore, we can always rely on subset-sum encodings, and thus the subset-sum Diffie-Hellman assumption, for $m = 1$. However, we recommend *not* using this conversion for two reasons:

– To prevent the exponent from "wrapping" mod $p - 1$, $p - 1$ needs to be larger than the maximum L_1-norm of the rows of $\mathbf{A}$. In this conversion, we are multiplying rows by exponential factors, meaning p needs to correspondingly be set much larger.

– In the full version [Zha14b], we prove the security of our encodings in the generic multilinear map model. Generic security is only guaranteed if $\|\mathbf{A}\|_\infty/p$ is negligible. This means for security, p will have to be substantially larger after applying the conversion.

4.2 Witness PRFs from Subset-Sum Encodings

We note that subset-sum encodings immediately give us witness PRFs for restricted classes. In particular, for a matrix $\mathbf{A}$, a subset-sum encoding is a witness PRF for the language $\mathsf{SubSums}(\mathbf{A})$. The various security notions for subset-sum encodings correspond exactly to the security notions for witness PRFs. In the full version [Zha14b], we show how to extend this to witness PRFs for any NP language, obtaining the following theorem:

Theorem 3. *If adaptive/static target interactively/non-interactively secure subset-sum encodings exist, then adaptive/static instance interactively/non-interactively secure witness PRFs exist.*

Roughly, we prove this Theorem 3 by providing a reduction from an instance x of any NP language L to subset-sum instance $(\mathbf{A}, \mathbf{t})$, where the matrix $\mathbf{A}$ is determined entirely by the language L, and is independent of x (except for its length). Thus, $\mathsf{SubSums}(\mathbf{A})$ corresponds exactly with L.

Our witness PRF for a language L is then a subset-sum encoding for the corresponding matrix $\mathbf{A}$. The value of the PRF on instance x is the encoding of the corresponding target $\mathbf{t}$. Given a witness w for x, the reduction gives a corresponding subset S of columns of $\mathbf{A}$ that sum to $\mathbf{t}$. This allows anyone with a witness to evaluate the PRF at x.

5   Applications

In this section, we show that for several applications of obfuscation, the obfuscator can be replaced with witness PRFs.

5.1   CCA-secure Public Key Encryption

We demonstrate that witness PRFs give a simple construction of CCA-secure public key encryption that is similar to the obfuscation-based construction of Sahai and Waters [SW14]. Given the similarities of witness PRFs to smooth projective hash functions (SPHFs) [CS02], and that the original motivation for SPHFs was CCA-secure public key encryption, this result is not surprising. Instead, we present the construction as a warm-up for the more interesting applications that follow.

Construction 4. *Let* $\mathsf{WPRF} = (\mathsf{WPRF.Gen}, \mathsf{F}, \mathsf{Eval})$ *be a witness PRF, and let* $\mathsf{G} : \mathcal{S} \to \mathcal{Z}$ *be a pseudorandom generator with* $|\mathcal{S}|/|\mathcal{Z}| < \mathsf{negl}$. *Build the following key encapsulation mechanism* $(\mathsf{Enc.Gen}, \mathsf{Enc}, \mathsf{Dec})$:

- $\mathsf{Enc.Gen}(\lambda)$: *Let* $R(z, s) = 1$ *if and only if* $\mathsf{G}(s) = z$. *In other words, R defines the language L of strings* $z \in \mathcal{Z}$ *that are images of G, and witnesses are the corresponding pre-images. Run* $(\mathsf{fk}, \mathsf{ek}) \xleftarrow{R} \mathsf{WPRF.Gen}(\lambda, R)$. *Set fk to be the secret key and ek to be the public key.*
- $\mathsf{Enc}(\mathsf{ek})$: *sample* $s \xleftarrow{R} \mathcal{S}$ *and set* $z \leftarrow \mathsf{G}(s)$. *Output z as the header and* $k \leftarrow \mathsf{Eval}(\mathsf{ek}, z, s) \in \mathcal{Y}$ *as the message encryption key.*
- $\mathsf{Dec}(\mathsf{fk}, z)$: *run* $k \leftarrow \mathsf{F}(\mathsf{fk}, z)$.

Correctness is immediate. For security, we have the following:

Theorem 5. *If* WPRF *is interactively secure, then Construction 4 is a CCA secure key encapsulation mechanism. If* WPRF *is static instance non-interactively secure, then Construction 4 is CPA secure.*

Proof. We prove the CCA case, the CPA case being almost identical. Let $\mathcal{B}$ be a CCA adversary with non-negligible advantage ϵ. Define **Game 0** as the standard CCA game, and define **Game 1** as the modification where the challenge header z^* is chosen uniformly at random in $\mathcal{Z}$. The security of G implies that $\mathcal{B}$ still has advantage negligibly-close to ϵ. Let **Game 2** be the game where z^* is chosen at random, but the game outputs a random bit and aborts if z^* is in the image space of G. Since $\mathcal{Z}$ is much larger than $\mathcal{S}$, the abort condition occurs with negligible probability. Thus $\mathcal{B}$ still has advantage negligibly close to ϵ in **Game 2**. Now we construct an adversary $\mathcal{A}$ for WPRF. $\mathcal{A}$ chooses a random z^*, and makes a challenge query on z^*, obtaining k. Then it simulates $\mathcal{B}$, answering decryption queries using its F oracle. When $\mathcal{B}$ makes a challenge query, and $\mathcal{A}$ responds with z^* as the header and k as the encapsulated key. When $\mathcal{B}$ outputs a bit b', $\mathcal{A}$ outputs the same bit. $\mathcal{A}$ has advantage equal to that of $\mathcal{B}$ in **Game 2**, which is non-negligible, thus contradicting the security of WPRF.

5.2 Non-interactive Multiparty Key Exchange

A multiparty key exchange protocol allows a group of g users to simultaneously post a message to a public bulletin board, retaining some user-dependent secret. After reading off the contents of the bulletin board, all the users establish the same shared secret key. Meanwhile, and adversary who sees the entire contents of the bulletin board should not be able to learn the group key. More precisely, a multiparty key exchange protocol consists of:

- Publish(λ, g) takes as input the security parameter and the group order, and outputs a user secret s and public value pv. pv is posted to the bulletin board.
- KeyGen$(\{pv_j\}_{j\in[g]}, s_i, i)$ takes as input g public values, plus the corresponding user secret s_i for the ith value. It outputs a group key $k \in \mathcal{Y}$.

For correctness, we require that all users generate the same key:

$$\mathsf{KeyGen}(\{\mathsf{pv}_j\}_{j\in[g]}, s_i, i) = \mathsf{KeyGen}(\{\mathsf{pv}_j\}_{j\in[g]}, s_{i'}, i')$$

for all $(s_j, \mathsf{pv}_j) \xleftarrow{R} \mathsf{Publish}(\lambda, g)$ and $i, i' \in [g]$. For security, we have the following:

Definition 6. *A non-interactive multiparty key exchange protocol is statically secure if the following distributions are indistinguishable:*

$$\{\mathsf{pv}_j\}_{j\in[g]}, k \ \text{where} \ (s_j, \mathsf{pv}_j) \xleftarrow{R} \mathsf{Publish}(\lambda, g) \forall j \in [g], k \xleftarrow{R} \mathcal{Y} \ \text{and}$$

$$\{\mathsf{pv}_j\}_{j\in[g]}, k \ \text{where} \ (s_j, \mathsf{pv}_j) \xleftarrow{R} \mathsf{Publish}(\lambda, g) \forall j \in [g], k \leftarrow \mathsf{KeyGen}(\{\mathsf{pv}_j\}_{j\in[g]}, s_1, 1)$$

Notice that our syntax does not allow a trusted setup, as constructions based on multilinear maps [BS02, GGH13a, CLT13] require. Boneh and Zhandry [BZ14] give the first multiparty key exchange protocol without trusted setup, based on obfuscation. We now give a very similar protocol using witness PRFs.

Construction 6. *Let* $G : \mathcal{S} \to \mathcal{Z}$ *be a pseudorandom generator with* $|\mathcal{S}|/|\mathcal{Z}| <$ *negl. Let* WPRF $=$ (Gen, F, Eval) *be a witness PRF. Let* $R_g : \mathcal{Z}^g \times (\mathcal{S} \times [g]) \to \{0,1\}$ *be a relation that outputs 1 on input* $((z_1, \ldots, z_g), (s, i))$ *if and only if* $z_i = G(s)$. *We build the following key exchange protocol:*

- Publish(λ, g): *compute* (fk, ek)$\xleftarrow{R}$Gen(λ, R_g). *Also pick a random seed* $s\xleftarrow{R}\mathcal{S}$ *and compute* $z \leftarrow G(s)$. *Keep* s *as the secret and publish* (z, ek).
- KeyGen($\{(z_i, \text{ek}_i)\}_{i \in [g]}, s$): *Each user sorts the pairs* (z_i, ek_i) *by* z_i, *and determines their index* i *in the ordering. Let* ek $=$ ek$_1$, *and compute* $k =$ Eval(ek, $(z_1, \ldots, z_g), (s, i)$)

Correctness is immediate. For security, we have the following:

Theorem 7. *If* WPRF *is static witness non-interactively secure, the Construction 6 is statically secure.*

Proof. Let $\mathcal{B}$ be an adversary for the key exchange protocol with non-negligible advantage. Then $\mathcal{B}$ sees $\{(z_i, \text{ek}_i)\}_{i \in [g]}$ where $z_i \leftarrow G(s_i)$ for a random $s_i \xleftarrow{R}\mathcal{S}$, as well as a key $k \in \mathcal{Y}$, and outputs a guess b' for whether $k = F(\text{ek}_1, \{(z_i)\}_{i \in [g]}$ or $k\xleftarrow{R}\mathcal{Y}$. Call this **Game 0**. Define **Game 1** as the modification where $z_i\xleftarrow{R}\mathcal{Z}$. The security of G implies that **Game 0** and **Game 1** are indistinguishable. Next define **Game 2** as identical to **Game 1**, except that the challenger outputs a random bit and aborts if any of the z_i are in the range of G. Since $|\mathcal{S}|/|\mathcal{Z}| <$ negl, this abort condition occurs with negligible probability, meaning $\mathcal{B}$ still has non-negligible advantage in **Game 2**. We construct an adversary $\mathcal{A}$ for WPRF as follows: $\mathcal{A}$ choses random $z_i \in \mathcal{Z}$ for $i \in [g]$, sorts the z_i, and makes a challenge query on $(z_1, \ldots, z_g)$, obtaining key k. Then after receiving ek, it sets ek$_1 =$ ek. For $i > 1$, $\mathcal{A}$ runs (fk$_i$, ek$_i$)$\xleftarrow{R}$Gen(λ, R_g). It then gives $\mathcal{A}$ $\{(z_i, \text{ek}_i)\}_{i \in [g]}, k$. Note that for key generation, ek$_1 =$ ek is chosen. Also, $(z_1, \ldots, z_g)$ is chosen at random in $\mathcal{Z}^g$, and $\mathcal{A}$'s challenger aborts if any of the z_g are in the range of G (that is, if $(z_1, \ldots, z_g)$ has a witness under R_g). Therefore, the view of $\mathcal{B}$ as a subroutine of $\mathcal{A}$ and the view of $\mathcal{B}$ in **Game 2** are identical. Therefore, the advantage of $\mathcal{A}$ is also non-negligible, a contradiction.

Adaptive Security. In semi-static or active security (defined by Boneh and Zhandry [BZ14]), the same published values pv$_j$ are used in many key exchanges, some involving the adversary. Obtaining semi-static or adaptive security from even the strongest forms of witness PRFs is not immediate. The issue, as noted by Boneh and Zhandry in the case of obfuscation, is that, even in the semi-static setting, the adversary may see the output of Eval on honest secrets, but using a malicious key ek. It may be possible for a malformed key to leak the honest secrets, thereby allowing the scheme to be broken. In more detail, consider an adversary $\mathcal{A}$ playing the role of user i, and suppose the maximum number of users in any group is 2. $\mathcal{A}$ generates and publishes params$_i$ in a potentially malicious way (and also generates and publishes some z_i). Meanwhile, an honest user j publishes an honest ek$_j$ and $z_j = G(s_j)$. Now, if $z_i < z_j$, user j computes the shared key for the group $\{i, j\}$ as $\mathsf{Eval}(\mathsf{ek}_i, (z_i, z_j), s_j, 2)$. While an honest ek_i would cause Eval to be independent of the witness, it may be possible for a dishonest ek_i to cause Eval to leak information about the witness.

Boneh and Zhandry circumvent this issue by using a special type of signature scheme, which they call a *puncturable* signature scheme, and only inputting signatures into Eval. Even if the entire signature leaks, it will not help the adversary produce the necessary signature to break the scheme. Such signature schemes can be built from witness indistinguishable proofs. It is straightforward to adapt Boneh and Zhandry's construction to use witness PRFs instead of obfuscation. We omit the details.

References

[AB15]   Applebaum, B., Brakerski, Z.: Obfuscating circuits via composite-order graded encoding. In: Dodis, Y., Nielsen, J.B. (eds.) TCC 2015, Part II. LNCS, vol. 9015, pp. 528–556. Springer, Heidelberg (2015)

[ABG+13] Ananth, P., Boneh, D., Garg, S., Sahai, A., Zhandry, M.: Differing-inputs obfuscation and applications. Cryptology ePrint Archive, Report 2013/689 (2013). http://eprint.iacr.org/2013/689

[ACLL14] Albrecht, M.R., Cocis, C., Laguillaumie, F., Langlois, A.: Implementing candidate graded encoding schemes from ideal lattices. Cryptology ePrint Archive, Report 2014/928 (2014). http://eprint.iacr.org/2014/928

[AGIS14] Ananth, P.V., Gupta, D., Ishai, Y., Sahai, A.: Optimizing obfuscation: avoiding Barrington's theorem. In Ahn, G.-J., Yung, M., Li, M. (eds.) ACM CCS 14: 21st Conference on Computer and Communications Security, Scottsdale, AZ, USA, 3–7 November 2014, pp. 646–658. ACM Press (2014)

[App13] Applebaum, B.: Bootstrapping obfuscators via fast pseudorandom functions. Cryptology ePrint Archive, Report 2013/699 (2013). http://eprint.iacr.org/2013/699

[BCP14] Boyle, E., Chung, K.-M., Pass, R.: On extractability obfuscation. In: Lindell, Y. (ed.) TCC 2014. LNCS, vol. 8349, pp. 52–73. Springer, Heidelberg (2014)

[BGI+01] Barak, B., Goldreich, O., Impagliazzo, R., Rudich, S., Sahai, A., Vadhan, S.P., Yang, K.: On the (Im)possibility of obfuscating programs. In: Kilian, J. (ed.) CRYPTO 2001. LNCS, vol. 2139, pp. 1–18. Springer, Heidelberg (2001)

[BGI14] Boyle, E., Goldwasser, S., Ivan, I.: Functional signatures and pseudorandom functions. In: Krawczyk, H. (ed.) PKC 2014. LNCS, vol. 8383, pp. 501–519. Springer, Heidelberg (2014)

[BGK+14] Barak, B., Garg, S., Kalai, Y.T., Paneth, O., Sahai, A.: Protecting obfuscation against algebraic attacks. In: Nguyen, P.Q., Oswald, E. (eds.) EUROCRYPT 2014. LNCS, vol. 8441, pp. 221–238. Springer, Heidelberg (2014)

[BGW05] Boneh, D., Gentry, C., Waters, B.: Collusion resistant broadcast encryption with short ciphertexts and private keys. In: Shoup, V. (ed.) CRYPTO 2005. LNCS, vol. 3621, pp. 258–275. Springer, Heidelberg (2005)

[Bla79] Blakley, G.R.: Safeguarding cryptographic keys. In: Proceedings of AFIPS 1979 National Computer Conference, vol. 48, pp. 313–317 (1979)

[BLR+14] Boneh, D., Lewi, K., Raykova, M., Sahai, A., Zhandry, M., Zimmerman, J.: Semantically secure order-revealing encryption: multi-input functional encryption without obfuscation. Cryptology ePrint Archive, Report 2014/834 (2014). http://eprint.iacr.org/2014/834

[BR14] Brakerski, Z., Rothblum, G.N.: Virtual black-box obfuscation for all circuits via generic graded encoding. In: Lindell, Y. (ed.) TCC 2014. LNCS, vol. 8349, pp. 1–25. Springer, Heidelberg (2014)

[BS02] Boneh, D., Silverberg, A.: Applications of multilinear forms to cryptography. Cryptology ePrint Archive, Report 2002/080 (2002). http://eprint.iacr.org/2002/080

[BST14] Bellare, M., Stepanovs, I., Tessaro, S.: Poly-many hardcore bits for any one-way function and a framework for differing-inputs obfuscation. In: Sarkar, P., Iwata, T. (eds.) ASIACRYPT 2014, Part II. LNCS, vol. 8874, pp. 102–121. Springer, Heidelberg (2014)

[BW13] Boneh, D., Waters, B.: Constrained pseudorandom functions and their applications. In: Sako, K., Sarkar, P. (eds.) ASIACRYPT 2013, Part II. LNCS, vol. 8270, pp. 280–300. Springer, Heidelberg (2013)

[BWZ14a] Boneh, D., Waters, B., Zhandry, M.: Low overhead broadcast encryption from multilinear maps. In: Garay, J.A., Gennaro, R. (eds.) CRYPTO 2014, Part I. LNCS, vol. 8616, pp. 206–223. Springer, Heidelberg (2014)

[BWZ14b] Boneh, D., Wu, D.J., Zimmerman, J.: Immunizing multilinear maps against zeroizing attacks. Cryptology ePrint Archive, Report 2014/930 (2014). http://eprint.iacr.org/2014/930

[BZ14] Boneh, D., Zhandry, M.: Multiparty key exchange, efficient traitor tracing, and more from indistinguishability obfuscation. In: Garay, J.A., Gennaro, R. (eds.) CRYPTO 2014, Part I. LNCS, vol. 8616, pp. 480–499. Springer, Heidelberg (2014)

[CGH01] Catalano, D., Gennaro, R., Howgrave-Graham, N.: The bit security of Paillier's encryption scheme and its applications. In: Pfitzmann, B. (ed.) EUROCRYPT 2001. LNCS, vol. 2045, pp. 229–243. Springer, Heidelberg (2001)

[CHL+14] Cheon, J.H., Han, K., Lee, C., Ryu, H., Stehlé, D.: Cryptanalysis of the multilinear map over the integers. Cryptology ePrint Archive, Report 2014/906 (2014). http://eprint.iacr.org/2014/906

[CLT13] Coron, J.-S., Lepoint, T., Tibouchi, M.: Practical multilinear maps over the integers. In: Canetti, R., Garay, J.A. (eds.) CRYPTO 2013, Part I. LNCS, vol. 8042, pp. 476–493. Springer, Heidelberg (2013)

[CLT14] Coron, J.-S., Lepoint, T., Tibouchi, M.: Cryptanalysis of two candidate fixes of multilinear maps over the integers. Cryptology ePrint Archive, Report 2014/975 (2014). http://eprint.iacr.org/2014/975

[CS02] Cramer, R., Shoup, V.: Universal hash proofs and a paradigm for adaptive chosen ciphertext secure public-key encryption. In: Knudsen, L.R. (ed.) EUROCRYPT 2002. LNCS, vol. 2332, pp. 45–64. Springer, Heidelberg (2002)

[CZ14] Chen, Y., Zhang, Z.: Publicly evaluable pseudorandom functions and their applications. In: Abdalla, M., De Prisco, R. (eds.) SCN 2014. LNCS, vol. 8642, pp. 115–134. Springer, Heidelberg (2014)

[GGH13a] Garg, S., Gentry, C., Halevi, S.: Candidate multilinear maps from ideal lattices. In: Johansson, T., Nguyen, P.Q. (eds.) EUROCRYPT 2013. LNCS, vol. 7881, pp. 1–17. Springer, Heidelberg (2013)

[GGH+13b] Garg, S., Gentry, C., Halevi, S., Raykova, M., Sahai, A., Waters, B.: Candidate indistinguishability obfuscation and functional encryption for all circuits. In: 54th Annual Symposium on Foundations of Computer Science, Berkeley, CA, USA, 26–29 October 2013, pp. 40–49. IEEE Computer Society Press (2013)

[GGHR14] Garg, S., Gentry, C., Halevi, S., Raykova, M.: Two-round secure MPC from indistinguishability obfuscation. In: Lindell, Y. (ed.) TCC 2014. LNCS, vol. 8349, pp. 74–94. Springer, Heidelberg (2014)

[GGHW14] Garg, S., Gentry, C., Halevi, S., Wichs, D.: On the implausibility of differing-inputs obfuscation and extractable witness encryption with auxiliary input. In: Garay, J.A., Gennaro, R. (eds.) CRYPTO 2014, Part I. LNCS, vol. 8616, pp. 518–535. Springer, Heidelberg (2014)

[GGHZ14] Garg, S., Gentry, C., Halevi, S., Zhandry, M.: Fully secure functional encryption without obfuscation. Cryptology ePrint Archive, Report 2014/666 (2014). http://eprint.iacr.org/2014/666

[GGSW13] Garg, S., Gentry, C., Sahai, A., Waters, B.: Witness encryption and its applications. In: Boneh, D., Roughgarden, T., Feigenbaum, J. (eds.) 45th Annual ACM Symposium on Theory of Computing, Palo Alto, CA, USA, 1–4 June 2013, pp. 467–476. ACM Press (2013)

[GHMS14] Gentry, C., Halevi, S., Maji, H.K., Sahai, A.: Zeroizing without zeroes: cryptanalyzing multilinear maps without encodings of zero. Cryptology ePrint Archive, Report 2014/929 (2014). http://eprint.iacr.org/2014/929

[GKP+13] Goldwasser, S., Kalai, Y.T., Popa, R.A., Vaikuntanathan, V., Zeldovich, N.: How to run turing machines on encrypted data. In: Canetti, R., Garay, J.A. (eds.) CRYPTO 2013, Part II. LNCS, vol. 8043, pp. 536–553. Springer, Heidelberg (2013)

[GL89] Goldreich, O., Levin, L.A.: A hard-core predicate for all one-way functions. In: 21st Annual ACM Symposium on Theory of Computing, Seattle, Washington, USA, 15–17 May 1989, pp. 25–32. ACM Press (1989)

[GLSW14] Gentry, C., Lewko, A., Sahai, A., Waters, B.: Indistinguishability obfuscation from the multilinear subgroup elimination assumption. Cryptology ePrint Archive, Report 2014/309 (2014). http://eprint.iacr.org/2014/309

[GLW14] Gentry, C., Lewko, A., Waters, B.: Witness encryption from instance independent assumptions. In: Garay, J.A., Gennaro, R. (eds.) CRYPTO 2014, Part I. LNCS, vol. 8616, pp. 426–443. Springer, Heidelberg (2014)

[GW09] Gentry, C., Waters, B.: Adaptive security in broadcast encryption systems (with short ciphertexts). In: Joux, A. (ed.) EUROCRYPT 2009. LNCS, vol. 5479, pp. 171–188. Springer, Heidelberg (2009)

[HJK+14] Hofheinz, D., Jager, T., Khurana, D., Sahai, A., Waters, B., Zhandry, M.: How to generate and use universal samplers. Cryptology ePrint Archive, Report 2014/507 (2014). http://eprint.iacr.org/2014/507

[Jou04] Joux, A.: A one round protocol for tripartite Diffie-Hellman. J. Cryptol. 17(4), 263–276 (2004)

[KNY14] Komargodski, I., Naor, M., Yogev, E.: Secret-sharing for NP. In: Sarkar, P., Iwata, T. (eds.) ASIACRYPT 2014, Part II. LNCS, vol. 8874, pp. 254–273. Springer, Heidelberg (2014)

[KPTZ13] Kiayias, A., Papadopoulos, S., Triandopoulos, N., Zacharias, T.: Delegatable pseudorandom functions and applications. In: Sadeghi, A.-Z., Gligor, V.D., Yung, M. (eds.) ACM CCS 2013: 20th Conference on Computer and Communications Security, Berlin, Germany, 4–8 November 2013 pp. 669–684. ACM Press (2013)

[KZ15] Komargodski, I., Zhandry, M.: Modern cryptography through the lens of secret sharing. Cryptology ePrint Archive, Report 2015/735 (2015). http:// eprint.iacr.org/2015/735

[PST14] Pass, R., Seth, K., Telang, S.: Indistinguishability obfuscation from semantically-secure multilinear encodings. In: Garay, J.A., Gennaro, R. (eds.) CRYPTO 2014, Part I. LNCS, vol. 8616, pp. 500–517. Springer, Heidelberg (2014)

[Rao14] Rao, V.: Adaptive multiparty non-interactive key exchange without setup in the standard model. Cryptology ePrint Archive, Report 2014/910 (2014). http://eprint.iacr.org/2014/910

[Sha79] Shamir, A.: How to share a secret. Commun. Assoc. Comput. Mach. **22**(11), 612–613 (1979)

[SW14] Sahai, A., Waters, B.: How to use indistinguishability obfuscation: deniable encryption, and more. In: Shmoys, D.B. (ed.) 46th Annual ACM Symposium on Theory of Computing, 31 May– 3 June 2014, pp. 475–484. ACM Press, New York (2014)

[SZ14] Sahai, A., Zhandry, M.: Obfuscating low-rank matrix branching programs. Cryptology ePrint Archive, Report 2014/773 (2014). http://eprint.iacr. org/2014/773

[Zha14a] Zhandry, M.: Adaptively secure broadcast encryption with small system parameters. Cryptology ePrint Archive, Report 2014/757 (2014). http:// eprint.iacr.org/2014/757

[Zha14b] Zhandry, M.: How to avoid obfuscation using witness PRFs. Cryptology ePrint Archive, Report 2014/301 (2014). http://eprint.iacr.org/2014/301

[Zim15] Zimmerman, J.: How to obfuscate programs directly. In: Oswald, E., Fischlin, M. (eds.) EUROCRYPT 2015. LNCS, vol. 9057, pp. 439–467. Springer, Heidelberg (2015)

Cutting-Edge Cryptography Through the Lens of Secret Sharing

Ilan Komargodski[1]([⊠]) and Mark Zhandry[2]

[1] Weizmann Institute of Science, Rehovot, Israel
ilan.komargodski@weizmann.ac.il
[2] MIT, Cambridge, USA
mzhandry@gmail.com

Abstract. Secret sharing is a mechanism by which a trusted dealer holding a secret "splits" the secret into many "shares" and distributes the shares to a collection of parties. Associated with the sharing is a monotone access structure, that specifies which parties are "qualified" and which are not: any qualified subset of parties can (efficiently) reconstruct the secret, but no unqualified subset can learn anything about the secret. In the most general form of secret sharing, the access structure can be any monotone NP language.

In this work, we consider two very natural extensions of secret sharing. In the first, which we call *distributed* secret sharing, there is no trusted dealer at all, and instead the role of the dealer is distributed amongst the parties themselves. Distributed secret sharing can be thought of as combining the features of multiparty non-interactive key exchange and standard secret sharing, and may be useful in settings where the secret is so sensitive that no one individual dealer can be trusted with the secret. Our second notion is called *functional* secret sharing, which incorporates some of the features of functional encryption into secret sharing by providing more fine-grained access to the secret. Qualified subsets of parties do not learn the secret, but instead learn some function applied to the secret, with each set of parties potentially learning a *different* function.

Our main result is that both of the extensions above are *equivalent* to several recent cutting-edge primitives. In particular, general-purpose distributed secret sharing is equivalent to witness PRFs, and general-purpose functional secret sharing is equivalent to indistinguishability obfuscation. Thus, our work shows that it is possible to view some of the recent developments in cryptography through a secret sharing lens, yielding new insights about both these cutting-edge primitives and secret sharing.

I. Komargodski—Supported in part by a grant from the Israel Science Foundation, the I-CORE Program of the Planning and Budgeting Committee, BSF and the Israeli Ministry of Science and Technology.
M. Zhandry—Supported by NSF.

E. Kushilevitz and T. Malkin (Eds.): TCC 2016-A, Part II, LNCS 9563, pp. 449–479, 2016.
DOI: 10.1007/978-3-662-49099-0_17

1   Introduction

Secret sharing is a mechanism by which a trusted dealer holding a secret "splits" the secret into many "shares" and distributes the shares to a collections of parties. Associated with the sharing is a monotone access structure, that specifies which parties are "qualified" and which are not: any qualified subset of parties can (efficiently) reconstruct the secret, but no unqualified subset can learn anything about the secret.[1] The first secret sharing schemes, due to Shamir [33] and Blakley [7], were for the threshold access structure, where the subsets that can reconstruct the secret are all the sets whose cardinality is at least a certain threshold. Such secret sharing schemes provide a digital analog of the "two-man rule", and are useful for splitting a sensitive key among several individuals so that no single individual knows the key. Secret sharing schemes, even for the simple threshold access structure, have found numerous applications in computer science (see [4] for a thorough survey).[2]

Since their introduction, it has been a major open problem to determine which access structures can secret sharing be realized for. Benaloh and Leichter [6] constructed a secret sharing scheme for any access structure that can be computed by a monotone formula. This result was generalized and improved by Karchmer and Wigderson [23] for access structures that can be computed by a monotone span program. In an unpublished work, Andrew Yao constructed a secret sharing scheme for any access structure that can be computed by a *monotone* circuit (see [4,28]), assuming any one-way function. Recently, Komargodski, Naor and Yogev [25] constructed secret sharing schemes for all of monotone NP (denoted mNP),[3] assuming one-way functions and a recent new primitive called *witness encryption* [18].[4] Monotone NP is essentially the largest class of access structures that we can hope for: if we cannot even efficiently identify a qualified set, we cannot hope to have qualified sets reconstruct the secret.

In this work we take secret sharing even further, by pursuing two very natural directions. First, we ask if the trusted dealer is required, or whether it is possible to *distribute* the role of the dealer amongst the parties themselves. Second, we ask if we can provide more fine-grained access mechanism to the shared secret, whereby qualified sets of parties only learn some function of the secret, each set of parties learning a possibly different function. Surprisingly, in both cases we

---

[1] In secret sharing, we always restrict our attention to *monotone* access structures, where a superset of a qualified set must be qualified. This is necessary because, if a set of parties contains a qualified subset, they can always "pretend" to be the smaller subset, discard the shares outside that subset, and reconstruct the secret.

[2] Most of the literature on secret sharing treats it as an information-theoretic primitive and insists on perfect security. In this work we consider the computational analog in which we only require security against computationally bounded adversaries. The survey of Beimel [4] discusses extensively both notions.

[3] For access structures in mNP, a qualified set of parties needs to know an NP witness that they are qualified.

[4] We note that the schemes of [6,23] are unconditionally secure, while the schemes of Yao and [25] are only secure against adversaries that run in polynomial-time.

show equivalences between these natural extensions of secret sharing and several cutting-edge cryptographic primitives that have recently been developed.

Distributed Secret Sharing. The usefulness of secret sharing schemes, as defined above, is limited to settings in which there exists a *trusted* dealer who knows the secret. What if we do not want any one individual to know the secret outright? What if our secret is so sensitive that we cannot afford anybody to know it? In this paper, we study the necessity of the trusted dealer in the setting of secret sharing and ask the question:

Is it possible to secret share a secret without anybody knowing it?

To address this question, we introduce the concept of *distributed* secret sharing schemes. Specifically, given an access structure, each party can generate for itself a public share (which is published) and a secret share (which is kept private). Then, there is a string S such that every qualified subset of parties can compute S (using their private shares and all public shares), whereas for every unqualified subset the secret S remains hidden.[5] Similarly to standard secret sharing schemes for mNP, for an access structure M in mNP, a qualified subset X should also provide a witness for the statements $X \in M$. Intuitively, one can view distributed secret sharing schemes as a hybrid of secret sharing schemes and non-interactive key-exchange: Indeed, non-interactive key-exchange is exactly the special case where M is set to be the threshold access structure with threshold $t = 1$.

In this paper we construct and explore distributed secret sharing schemes. Our main result is that distributed secret sharing schemes for access structures in mNP are *equivalent* to witness pseudorandom functions (witness PRFs) for NP. A witness PRF for a language $L \in$ NP is a function F such that anyone with a valid witness that $x \in L$ can compute $F(x)$ without the secret key, but for all $x \notin L$, $F(x)$ is computationally hidden to anybody that does not know the secret key. Witness PRFs were recently introduced by Zhandry [34] and shown to be very useful in constructing several important cryptographic primitives (including non-interactive multi-party key exchange without setup) that were previously only known to exist assuming seemingly much stronger assumptions.

In addition, we explore the possibility of distributed secret sharing for restricted classes of access structures based on weaker assumptions. To start, we consider the possibility of *information-theoretic* security for distributed secret sharing scheme (that is, security against unbounded adversaries). We show that such information-theoretic security is typically impossible: we prove that a distributed secret sharing scheme for any non-trivial access structure implies the existence of one-way functions.[6]

---

[5] We note that we do not assume secure point-to-point channels, a standard PKI or additional rounds of interaction (beyond publishing a public key) between the parties. With any of these assumptions the problem can be reduced to standard secret sharing.

[6] We call an access structure M *trivial* if M is empty or if there exists a subset of parties $X \in M$ which is contained in any qualified set. For trivial access structures, we show that there is a simple perfectly-secure distributed secret sharing scheme.

Next, we present a distributed secret sharing scheme for the threshold access structure, and prove its security based on the multilinear decisional Diffie-Hellman (MDDH) assumption. As an interesting application, we show that distributed secret sharing schemes for threshold access structures imply *constrained PRFs* that can be constrained to a Hamming ball around an arbitrary point and are secure for adversaries that obtain a single constrained key. Even though it is known that the MDDH assumption implies constrained PRFs for all circuits which are secure with respect to arbitrary collusions [10], our transformation is generic and applies to any threshold distributed secret sharing scheme, which perhaps can be based on simpler assumptions than multilinear maps.

Functional Secret Sharing. Traditional secret sharing schemes offer an all-or-nothing guarantee when reconstructing a shared secret — a qualified subset of parties can learn the entire secret, while unqualified subsets learn nothing about the secret. For many applications, especially in a distributed setting common to secret sharing, this notion is insufficient. Concretely, standard secret sharing schemes will not help in scenarios in which a dealer wants to share a secret such that every qualified subset of parties will learn a specific function of the secret (and nothing else). For example, a dealer holding a secret S, may want to distribute it such that any qualified subset X will be able to learn only the inner product of X and S, while making sure S remains computationally hidden for unqualified subsets.

A related issue has appeared in the context of encryption schemes, giving rise to the concept of *functional encryption* and a very fruitful line of work (see e.g., [8,30]). We study whether secret sharing schemes can be extended in an analogous way to support such functionalities: Given an efficiently computable two-input function F (that can be thought of as a family of functions indexed by the first input), we ask the question:

Is it possible to secret share a secret S such that any qualified subset of parties X can compute only $F(X,S)$, but for unqualified subsets, S will be computationally hidden?

To study this question, we introduce the concept of *functional* secret sharing schemes. Informally, such a scheme allows to secret share a secret S with respect to a function F and an access structure M, such that any qualified subset of parties X can pool their shares together and compute $F(X, S)$. Security is formalized by requiring that for any function F, any subset of parties X and any two secrets S_0 and S_1, as long as either $M(X) = 0$ or $F(X', S_0) = F(X', S_1)$ for any $X' \subseteq X$, secret shares corresponding to F, X and S_0 cannot be distinguished from secret shares corresponding to F, X and S_1. Notice that the condition that $F(X', S_0) = F(X', S_1)$ for any $X' \subseteq X$ in the case that $M(X) = 1$ is necessary, as otherwise, by evaluating $F(X', S_b)$ an adversary can distinguish between the case that $b = 0$ and $b = 1$.

Our main result is that functional secret sharing schemes for access structures in mNP and functions in P are *equivalent* to indistinguishability obfuscation (iO) for P.[7] An indistinguishability obfuscator [3,17] guarantees that if two circuits compute the same function, then their obfuscated version are computationally indistinguishable. This primitive was introduced by Barak et al. [3] and later proven to be extremely useful for construction of cryptographic primitives some of which were unknown before (see e.g., [11,17,31]). To complement this, several candidate constructions of indistinguishability obfuscators were recently proposed [1,2,13,17,19,29].

Note that when the function F is defined to be the identity function over its second input parameter (i.e., $F(\cdot, S) = S$) we get the standard definition of secret sharing for mNP of [25]. Moreover, when the access structure is the set of all subsets, the secret S is a description of a function and F is the universal circuit (i.e., $F(X, S) = S(X)$), we obtain a definition of a *function secret sharing scheme*. In such a scheme, the goal is to split a function (and not a secret) into shares that hide the function under some conditions. Our construction gives a way to split a function F into shares such that any subset of parties X can compute $F(X')$ for every $X' \subseteq X$ and "nothing" else. We note that other forms of function secret sharing have been studied in the literature (cf. [5,12,15,32]). However, our notion is quite different from (and incomparable to) these other notions. In particular, our notion is the first to allow for fine-grained access control to the secret by guaranteeing that any qualified set learns a *possibly different* function of the secret. Moreover, previous notions were mostly studied in the context of threshold access structures, only with very specific function classes or insisted on schemes with additional properties.[8]

Conclusions. Recent advances in cryptography, including the first constructions of multilinear maps [16] and obfuscation [17], have lead to the development of many incredible new cryptographic objects. Applications include functional encryption, witness encryption, witness PRFs, deniable encryption, multi-party computation in very few rounds, traitor-tracing schemes with very short messages, and many more. Our work can thus be seen as establishing a close connection between several of these advanced cryptographic capabilities and types of secret sharing, which at first appear totally unrelated. The known relationships, including our work, are depicted and summarized in Fig. 1. Our hope is that the connections we develop can help shed light on the relationships between advanced primitives, or between types of secret sharing: which are equivalent, why do some tasks appear difficult, and so on.

For example, our results indicate why witness PRFs, which are closely related to witness encryption, may be the "right" primitive for building non-interactive

---

[7] To show that iO implies functional secret sharing schemes, we also assume the existence of one-way functions. By a result of [25] we can actually only assume iO and NP $\not\subseteq$ io-BPP. Moreover, we note that in this paper we assume functions are represented as circuits, so we actually work with functions in P/poly (and not P).

[8] For example, the functional secret sharing notion of [12] is similar to ours but requires an additional homomorphic property for the reconstruction procedure. Our scheme does not have this extra property, however, our construction relies on iO while their construction relies on subexponentially-secure iO.

key exchange, and why witness encryption may be insufficient. Indeed, distributed secret sharing essentially combines the features of secret sharing for mNP (which is equivalent to witness encryption [25]) with non-interactive key exchange. If these non-interactive key exchange features could be obtained from witness encryption, then perhaps witness encryption could also imply witness PRFs. In addition, at first it may not be obvious what is the relationship between functional secret sharing and distributed secret sharing. Our results and the simple observation that indistinguishability obfuscation implies witness PRFs, show that functional secret sharing *implies* distributed secret sharing (assuming one-way functions).

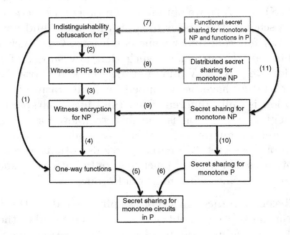

Fig. 1. The secret sharing zoo. (1) Holds assuming that NP $\not\subseteq$ io-BPP [24]. (2) Holds assuming one-way functions. (3) [34]. (4) Holds assuming the existence of a hard-on-average NP-problem [24]. (5) Yao's unpublished work. (6) By definition. (7) This work; the left-to-right arrow assumes one-way functions. (8) This work. (9) [25]; the left-to-right arrow assumes one-way functions. (10) By definition. (11) By definition.

1.1   Overview of Our Techniques

Distributed Secret Sharing and Witness PRFs. Here, we provide a high-level overview of our technique for transforming distributed secret sharing schemes into witness PRFs. At first, this seems like a difficult task. Indeed, distributed secret sharing only specifies a single secret: the shared secret for the groups of qualified parties. In contrast, in a witness PRF each instance corresponds to a secret, namely the output of the PRF on this instance. How can we obtain many secrets out of one?

Our main observation is that distributed secret sharing schemes are *reusable*. Suppose a set $\mathcal{P}_1$ of n parties runs the distributed secret sharing protocol, each party in $\mathcal{P}_1$ generating a secret/public share pair, and publishing the public share. Now, suppose a second set of n parties $\mathcal{P}_2$ wishes to run the distributed secret sharing protocol, and that there is some party i that is in both $\mathcal{P}_1$ and $\mathcal{P}_2$. Distributed secret sharing is reusable in the sense that party i does not need to generate a fresh secret/public share pair for the second invocation of the protocol, but can instead *reuse* the shares he already has. Thus, party i does not need to publish any additional material to take part in the second sharing. Taking this a step further, $N \gg n$ parties can each generate secret/public shares and publish the public shares. Then, various sets of n of them can engage in the distributed secret sharing protocol *without any additional setup or interaction*. This observation can be seen as a generalization of the fact that non-interactive key exchange (both in the two-party and multi-party setting) is reusable.

Since distributed secret sharing schemes are reusable, there are really many implicit secrets, one for every possible subset of the N parties of size n. This will be the source of our many secrets for our witness PRFs. To show how we use this idea of reusability, we sketch our approach for a simpler task: using *threshold* distributed secret sharing to build Hamming ball constrained PRFs.

Threshold Distributed Sharing Schemes to Hamming Ball Constrained PRFs. Recall that a constrained PRF (as defined by Boneh and Waters [10]) is a normal PRF with some additional requirements: First, given the secret key k, and a subset $T \subset \mathcal{X}$ where $\mathcal{X}$ is the domain of the PRF, it is possible to *constrain* the key k to the set T, producing a constrained key k_T. Next, given k_T and a point $x \in T$, it is possible to compute $\mathsf{PRF}_k(x)$. For security, we require that, even given k_T, for all $x \notin T$, $\mathsf{PRF}_k(x)$ is pseudorandom. For this exposition, we will consider Hamming ball constraints, where $\mathcal{X} = \{0,1\}^n$, and the possible sets T consist of all points withing Hamming distance r of some center point c.

Suppose that r is fixed a priori (this is assumed here for simplicity – our actual scheme handles the case in which r is not fixed a priori). Our Hamming ball constrained PRF is defined as follows. Let $N = 2n$ be the total number of parties, and label each party by a pair $(i, b) \in [n] \times \{0,1\}$. Generate secret/public shares $(\Pi_{i,b}, P_{i,b})$ for each of the N parties for the threshold distributed secret sharing scheme on n parties and threshold $n - r$. The secret key consists of all the public and secret shares. For every input $x \in \{0,1\}^n$, let $\mathcal{P}_x$ be the subset of n parties labeled by (i, x_i) for $i \in [n]$. $\mathsf{PRF}(x)$ is defined to be the shared secret S for the set of parties $\mathcal{P}_x$ defined by x. Since the secret key consists of $n \geq n - r$ of the secret shares for $\mathcal{P}_x$, the secret key allows for computing $\mathsf{PRF}(x)$.

The constrained key k_T for the Hamming ball T of radius r around center c consists of all of the public shares, as well as the secret shares for the set $\mathcal{P}_c$. For any input x with Hamming distance at most r from c, k_T contains at least $n - r$ of the secret shares for $\mathcal{P}_x$, and so $\mathsf{PRF}(x)$ can be computed. For x at distance more than r away, k_T contains fewer than $n - r$ secret shares for $\mathcal{P}_x$, so the security of the threshold distributed secret sharing scheme implies that $\mathsf{PRF}(x)$ is hidden.

For the general distributed secret sharing to witness PRF construction, we will make use of a similar strategy, defining the output of the PRF to be the shared secret S corresponding to a subset of parties. However, the construction becomes somewhat more complicated. For starters, the class of Hamming balls is very simple, and moreover has a lot of symmetry. In contrast, the general NP languages are much more complex and have no simple structural properties we can use. Additionally, we will need to allow the parties to be able to input a witness. We refer to Sect. 3.4 for the full details.

Functional Secret Sharing and iO. The fact that general-purpose functional secret sharing implies iO is rather straight-forward. Indeed, as we mentioned, function secret sharing is a special case of functional secret sharing, and thus, an obfuscation of a circuit is just the shares generated by the function secret sharing. Security of the obfuscator follows directly from the security of the function secret sharing scheme.

The other direction (namely, from iO to functional secret sharing) is more complicated. To this end, we rely on ideas developed by [25] in order to show that witness encryption implies (standard) secret sharing for mNP. Specifically, when sharing the secret S with respect to a function F and an access structure M, the share of party i will be an opening of a commitment and the iO of a circuit that given as input the secret openings of a subset of parties X verifies the openings, verifies the validity of the instance (together with a witness) with respect to M, and if all tests pass, it outputs the value $F(X, S)$. The security of this scheme relies on the perfect binding of the commitments and the indistinguishability guarantee of the obfuscator.

We note that multi-input functional encryption (MIFE) [20] provides another natural path to functional secret sharing. In an MIFE scheme, a secret key SK_G corresponds to an k-input function G, and message can be encrypted to any one of the k inputs to G. Denote the encryption of a message m to the ith input as $\mathsf{Enc}_i(m)$. With the secret key and ciphertexts $\mathsf{Enc}_i(m_i)$ for $i = 1, \ldots, k$, it is possible to compute $f(m_1, \ldots, m_k)$, but impossible to learn anything else the plaintexts. For simplicity, we will sketch the construction of functional secret sharing where both access structure M and function F are in P, the case of more general access structures being a straightforward extension. Let $G(x_1, \ldots, x_n, S) = M(x_1, \ldots, x_n) \wedge F(x_1, \ldots, x_n, S)$. The secret share for party $i \in [n]$ consists of $\mathsf{SK}_G, \mathsf{Enc}_1(0), \cdots, \mathsf{Enc}_n(0), \mathsf{Enc}_{n+1}(S), \mathsf{Enc}_i(1)$. Then, any subset X of parties can use SK_G together with ciphertexts $\{\mathsf{Enc}_i(X_i)\}_{i \in [n]}, \mathsf{Enc}_{n+1}(S)$ to compute $M(X) \wedge F(X, S)$. If X is qualified, this will give $F(X, S)$, whereas if X is unqualified, this will give 0. Since iO and MIFE are equivalent for general-purpose functionalities (assuming one-way functions), this construction gives an alternative way to build functional secret sharing from iO.[9]

2   Preliminaries

In this section we present the notation and basic definitions that are used in this work. For a distribution X we denote by $x \leftarrow X$ the process of sampling a value

---

[9] We thank a reviewer for pointing out this alternative solution.

x from the distribution X. Similarly, for a set $\mathcal{X}$ we denote by $x \leftarrow \mathcal{X}$ the process of sampling a value x from the uniform distribution over $\mathcal{X}$. For a randomized function f and an input $x \in \mathcal{X}$, we denote by $y \leftarrow f(x)$ the process of sampling a value y from the distribution $f(x)$. For an integer $n \in \mathbb{N}$ we denote by $[n]$ the set $\{1, \ldots, n\}$. A function $\mathsf{neg} : \mathbb{N} \to \mathbb{R}$ is *negligible* if for every constant $c > 0$ there exists an integer N_c such that $\mathsf{neg}(\lambda) < \lambda^{-c}$ for all $\lambda > N_c$. Throughout this paper we denote by λ the security parameter.

Two sequences of random variables $X = \{X_\lambda\}_{\lambda \in \mathbb{N}}$ and $Y = \{Y_\lambda\}_{\lambda \in \mathbb{N}}$ are *computationally indistinguishable* if for any probabilistic polynomial-time algorithm $\mathcal{A}$ there exists a negligible function $\mathsf{neg}(\cdot)$ such that for all $\lambda \in \mathbb{N}$ it holds that $\big|\Pr[\mathcal{A}(1^\lambda, X_\lambda) = 1] - \Pr[\mathcal{A}(1^\lambda, Y_\lambda) = 1]\big| \leq \mathsf{neg}(\lambda)$.

2.1 Monotone-NP and Access Structures

A function $f : 2^{[n]} \to \{0, 1\}$ is said to be monotone if for every $X \subseteq [n]$ such that $f(X) = 1$ it also holds that $\forall Y \subseteq [n]$ such that $X \subseteq Y$ it holds that $f(Y) = 1$. Given a potentially non-monotone function $f : 2^{[n]} \to \{0, 1\}$, we define the *monotone closure* of f, denoted $\overline{f}$, such that $\overline{f}(Y) = 1$ if and only if there is some $X \subset Y$ such that $f(X) = 1$.

A monotone Boolean circuits is a Boolean circuit with AND and OR gates (without negations). A non-deterministic circuit is a Boolean circuit whose inputs are divided into two parts: standard inputs and non-deterministic inputs. A non-deterministic circuit accepts a standard input if and only if there is some setting of the non-deterministic input that causes the circuit to evaluate to 1. A monotone non-deterministic circuit is a non-deterministic circuit where the monotonicity requirement applies only to the standard inputs, that is, every path from a standard input wire to the output wire does not have a negation gate.

Definition 1 ([21]). *A function L is in* mNP *if there exists a uniform family of polynomial-size monotone non-deterministic circuit that computes L.*

Lemma 1 ([21, Theorem 2.2]). mNP = NP $\cap$ mono, *where* mono *is the set of all monotone functions.*

A computational secret-sharing scheme involves a dealer who has a secret, a set of n parties, and a collection A of qualified subsets of parties called the access structure. A computational secret-sharing scheme for A is a method by which the dealer efficiently distributes shares to the parties such that (1) any subset in A can efficiently reconstruct the secret from its shares, and (2) any subset not in A cannot efficiently reveal any partial information on the secret. For more information on secret-sharing schemes we refer to [4] and references therein.

Throughout this paper we deal with secret-sharing schemes for access structures over n parties $\mathcal{P} = \mathcal{P}_n = \{\mathsf{p}_1, \ldots, \mathsf{p}_n\}$.

Definition 2 (Access structure). *An access structure M on $\mathcal{P}$ is a monotone set of subsets of $\mathcal{P}$. That is, for all $X \in M$ it holds that $X \subseteq \mathcal{P}$ and for all $X \in M$ and X' such that $X \subseteq X' \subseteq \mathcal{P}$ it holds that $X' \in M$.*

2.2 Commitment Schemes

In some of our constructions we need a non-interactive commitment scheme such that commitments of different strings has disjoint support. Jumping ahead, since the dealer in the setup phase of a secret-sharing scheme is not controlled by an adversary (i.e., it is honest), we can relax the foregoing requirement and use non-interactive commitment schemes that work in the CRS (common random string) model (for ease of notation, we usually ignore the CRS).

Definition 3 (Commitment scheme in the CRS model). *Let $\lambda \geq 0$ be a parameter. Let $\mathsf{Com}\colon \{0,1\} \times \{0,1\}^{\lambda} \times \{0,1\}^{\lambda} \to \{0,1\}^{q(\lambda)}$ be polynomial-time computable function. We say that Com is a (non-interactive perfectly binding) commitment scheme in the CRS model if the following two conditions hold:*

1. ***Computational Hiding:*** *Let $\mathsf{CRS} \leftarrow \{0,1\}^{\lambda}$ be chosen uniformly at random. The random variables $\mathsf{Com}(0, \mathbf{U}_{\lambda}, \mathsf{CRS})$ and $\mathsf{Com}(1, \mathbf{U}_{\lambda}, \mathsf{CRS})$ are computationally indistinguishable (given CRS).*
2. ***Perfect Binding:*** *With all but negligible fraction of the CRSs, the supports of the above random variables are disjoint.*

As usual, the above definition can be generalized to commitments of strings of polynomial size (rather than bits) by commiting to each bit separately.

Commitment schemes that satisfy the above definition, in the CRS model, can be constructed based on any pseudorandom generator [27] (which can be based on any one-way functions [22]). For simplicity, throghout the paper we ignore the CRS and simply write $\mathsf{Com}(\cdot, \cdot)$. We say that $\mathsf{Com}(x, r)$ is the commitment to the value x with the opening r.

2.3 Multilinear Maps

Definition 4 (Multilinear maps). *We say that a map $e\colon \mathbb{G}_1^n \to \mathbb{G}_2$ is an n-multilinear map if it is satisfies the following:*

1. *$\mathbb{G}_1$ and $\mathbb{G}_2$ are groups of the same prime order.*
2. *If $a_1, \ldots, \in \mathbb{Z}$ and $x_1, \ldots, x_n \in \mathbb{G}_1$, then*

$$e(x_1^{a_1}, \ldots, x_n^{a_n}) = e(x_1, \ldots, x_n)^{\prod_{i=1}^{n} a_i}.$$

3. *The map e is non-degenerate in the following sense: if $g \in \mathbb{G}_1$ is a generator of $\mathbb{G}_1$, then $e(g, \ldots, g)$ is a generator of $\mathbb{G}_2$.*

We say that e is an efficient n-multilinear map if it is effiently computable, namely, there exists a polynomial-time algorithm that computes $e(x_1^{a_1}, \ldots, x_n^{a_n})$ for any $a_1, \ldots, a_n \in \mathbb{Z}$ and $x_1, \ldots, x_n \in \mathbb{G}_1$.

An efficient mulilinear map generator $\mathsf{MMap.Gen}(1^\lambda, n)$ is a probabilistic polynomial-time algorithms that gets as input two inputs 1^λ and n, and outputs a tuple (Γ, g, ℓ), where Γ is the description of an efficient n-multlilinear map $e: \mathbb{G}_1^n \to \mathbb{G}_2$, g is a generator of $\mathbb{G}_1$, and ℓ is the order of the groups $\mathbb{G}_1$ and $\mathbb{G}_2$.

Next, we define the multilinear Diffie-Hellman assumption. Roughly, the assumption is that given $g, g^{a_1}, \ldots, g^{a_n}$, it is hard to compute $e(g, \ldots, g)^{\prod_{i=1}^n a_i}$, or even distinguish it from a random value.

Definition 5 (Multilinear decisional Diffie-Hellman assumption [9]). *We say that an efficient n-multilinear map generator* $\mathsf{MMap.Gen}$ *satisfies the* multilinear decisional Diffie-Hellman *(MDDH) assumption if for every polynomial time algorithm $\mathcal{A}$ there exists a negligible function* $\mathsf{neg}(\cdot)$ *such that for $\lambda \in \mathbb{N}$ it holds that*

$$\mathsf{Adv}^{\mathsf{mDH}}_{\mathsf{MMap.Gen}, \mathcal{A}, n, \lambda} = \left| \Pr\left[\mathcal{A}\left(g, g^{a_0}, \ldots, g^{a_n}, e(g, \ldots, g)^{\prod_{i=0}^n a_i} \right) = 1 \right] - \right.$$
$$\left. \Pr\left[\mathcal{A}\left(g, g^{a_0}, \ldots, g^{a_n}, K \right) = 1 \right] \right| \leq \mathsf{neg}(\lambda),$$

where the probability is over the execution of $(\Gamma, g, \ell) \leftarrow \mathsf{MMap.Gen}(1^\lambda, n)$, the choice of $a_0, \ldots, a_n \leftarrow (\mathbb{Z}/\ell\mathbb{Z})^{n+1}$, $K \leftarrow \mathbb{G}_2$, and the internal randomness of $\mathcal{A}$.

We note that we do not know of any "ideal" multilinear maps as described above that plausibly support the MDDH assumption. Instead, current candidates are "noisy" [14,16]. In particular, the group elements have some noise, and only a certain number of group operations are allowed before the multilinear identity fails. Moreover, each group element actually has many representations, and a special extraction procedure is required to obtain a unique "canonical" representation for a particular element. The extraction is only allowed in $\mathbb{G}_2$. Despite this departure from the ideal notion described above, it is usually straightforward (though often tedious) to use current candidate maps in place of the ideal maps. Therefore, for ease of exposition, we will describe our applications of multilinear maps in terms of the ideal abstraction, noting that the applications can be adapted to use the noisy candidate multilinear maps from the literature.

2.4   Witness Pseudorandom Functions

Witness pseudorandom functions (witness-PRFs) were recently introduced by Zhandry [34]. He showed that several important primitives, that were previously only known from iO (see Definition 7), follow from this seemingly weaker assumption. We note that witness-PRFs are related to witness encryption [18], but seem to be stronger.

Definition 6 (Witness-PRFs [34]). *A witness pseudorandom function is a tuple* (Gen, PRF, Eval) *where:*

1. $\mathsf{Gen}(1^\lambda, R)$ *is a polynomial-time randomized procedure that takes as input a security parameter and a relation* $R : \{0,1\}^n \times \{0,1\}^m \to \{0,1\}$ *represented as a circuit, and outputs a private function key* fk *and a public evaluation key* ek. *The relation* R *defines an* NP *language* L.
2. $\mathsf{PRF}(\mathsf{fk}, x)$ *is a polynomial-time deterministic procedure that takes as input the function key* fk *and an instance* $x \in \{0,1\}^n$.
3. $\mathsf{Eval}(\mathsf{ek}, x, w)$ *is a polynomial-time deterministic procedure that takes as input the evaluation key* ek, *an instance* $x \in \{0,1\}^n$, *and a witness* $w \in \{0,1\}^m$.
4. **Correctness:** *If* $x \in L$, *and moreover* w *is a valid witness for* x *(that is,* $R(x, w) = 1$), *then*

$$\Pr[\mathsf{Eval}(\mathsf{ek}, x, w) = \mathsf{PRF}(\mathsf{fk}, x)] = 1,$$

   where $(\mathsf{fk}, \mathsf{ek}) \leftarrow \mathsf{Gen}(1^\lambda, R)$ *and the probability is taken over the randomness* Gen.
5. **Security:** *For any relation* R *and any probabilistic polynomial-time algorithm* D, *there exists a negligible function* $\mathsf{neg}(\cdot)$ *such that for any* $\lambda \in \mathbb{N}$ *and any* $x \notin L$, *it holds that*

$$|\Pr[D(\mathsf{ek}, \mathsf{PRF}(\mathsf{fk}, x)) = 1] - \Pr[D(\mathsf{ek}, y) = 1]| < \mathsf{neg}(\lambda),$$

   where $(\mathsf{fk}, \mathsf{ek}) \leftarrow \mathsf{Gen}(1^\lambda, R)$, y *is chosen uniformly over the codomain of* PRF, *and the probabilities are taken over the randomness of* Gen, D, *and the choice of* y.

2.5   Indistinguishability Obfuscation

We say that two circuits C and C' are *equivalent* and denote it by $C \equiv C'$ if they compute the same function (i.e., $\forall x : C(x) = C'(x)$).

Definition 7 (Indistinguishability obfuscation [3]). *Let* $\mathcal{C} = \{\mathcal{C}_n\}_{n \in \mathbb{N}}$ *be a class of polynomial-size circuits, where* $\mathcal{C}_n$ *is a set of circuits operating on inputs of length* n. *A uniform polynomial-time algorithm* iO *is called an* indistinguishability obfuscator *for the class* $\mathcal{C}$ *if it takes as input a security parameter and a circuit in* $\mathcal{C}$ *and outputs a new circuit so that following properties are satisfied:*

1. **Preserving functionality:** *There exists a negligible function* α *such that for any input length* $n \in \mathbb{N}$, *any* λ *and any* $C \in \mathcal{C}_n$ *it holds that*

$$\Pr_{\mathsf{iO}}[C \equiv \mathsf{iO}(1^\lambda, C)] = 1,$$

   where the probability is over the internal randomness of iO.
2. **Polynomial slowdown:** *There exists a polynomial* $p(\cdot)$ *such that: For any input length* $n \in \mathbb{N}$, *any* λ *and any circuit* $C \in \mathcal{C}_n$ *it holds that* $|\mathsf{iO}(1^\lambda, C)| \leq p(|C|)$.

3. **Indistinguishable obfuscation:** *For any probabilistic polynomial-time algorithm D and any polynomial $p(\cdot)$, there exists a negligible function $\mathsf{neg}(\cdot)$, such that for any $\lambda, n \in \mathbb{N}$, any two equivalent circuits $C_1, C_2 \in \mathcal{C}_n$ of size $p(\lambda)$, it holds that*

$$\left| \Pr\left[D\left(\mathsf{iO}\left(1^\lambda, C_1 \right) \right) = 1 \right] - \Pr\left[D\left(\mathsf{iO}\left(1^\lambda, C_2 \right) \right) = 1 \right] \right| \leq \mathsf{neg}(\lambda),$$

where the probabilities are over the internal randomness of iO and D.

3 Distributed Secret Sharing

In this section we define the notion of distributed secret sharing schemes.

Definition 8 (Distributed secret sharing). *A distributed secret sharing (DSS) scheme consists of a probabilistic setup procedure SETUP, a probabilistic sharing procedure SHARE and a deterministic reconstruction procedure RECON that satisfy the following requirements:*

- $\mathsf{SETUP}(1^\lambda, 1^n, V_M)$ *takes as input a security parameter λ (in unary representation) the number n of parties (also in unary), the verification procedure V_M for an mNP access structure M on n parties. SETUP outputs a common reference string CRS.*
- $\mathsf{SHARE}(1^\lambda, 1^n, \mathsf{CRS}, V_M, i)$ *takes as input λ, n, the common reference string CRS, the verification procedure V_M for an mNP language M, and a party index $i \in [n]$. It outputs a public share $P(i)$ and a secret share $\Pi(i)$. For $X \subseteq \mathcal{P}_n$ we denote by $\Pi(X)$ the random variable that corresponds to the set of secret shares of parties in X. We denote by P the random variable that corresponds to the set of public shares of parties in $\mathcal{P}_n$.*
- $\mathsf{RECON}(1^\lambda, 1^n, \mathsf{CRS}, V_M, P, \Pi(X), w)$ *gets as input $\lambda, n, \mathsf{CRS}, V_M$, the public shares P of all n parties, the secret shares $\Pi(X)$ of a subset of parties $X \subseteq \mathcal{P}_n$, and a witness w, and outputs a shared secret. We will sometimes abuse notation, and also write $X \subseteq [n]$ to refer to the subset of the party indices appearing in X.*
- **Correctness:** *For every set of parties $\mathcal{P}_n$ with corresponding public shares P, there is a string S such that any set of qualified parties $X \subseteq \mathcal{P}_n$ with valid witness w (i.e., $V_M(X, w) = 1$) can recover S. That is,*

$$\Pr[\mathsf{RECON}(1^\lambda, 1^n, \mathsf{CRS}, V_M, P, \Pi(X), w) = S] = 1,$$

where the probability is taken over the generation of the shares — namely, over $(P(i), \Pi(i)) \leftarrow \mathsf{SHARE}(1^\lambda, 1^n, \mathsf{CRS}, V_M, i)$ for $i \in [n]$ — and the choice of S (which will typically be information-theoretically determined by P). We will sometimes refer to S as the shared secret.
- **Pseudorandomness of the secret:** *For any language $M \in \mathsf{mNP}$ and any probabilistic polynomial-time algorithm D, there exists a negligible function*

$\text{neg}(\cdot)$ *such that for any* $\lambda \in \mathbb{N}$ *and any unqualified set* $X \subseteq \mathcal{P}_n$ *(that is,* $X \notin M$*), it holds that*

$$|\Pr[D(P, \Pi(X), S) = 1] - \Pr[D(P, \Pi(X), K) = 1]| \leq \text{neg}(\lambda),$$

where the probability is taken over the generation of the shares, namely, over $(P(i), \Pi(i)) \leftarrow \text{SHARE}(1^\lambda, 1^n, \text{CRS}, V_M, i)$ *for* $i \in [n]$*,* K *is sampled uniformly at random, and* S *is the shared secret defined above.*

The Shared Secret S. Suppose M is non-empty, which is true for any interesting access structure M. In this case, by the monotonicity of M, $\mathcal{P}_n \in M$ and there exists a witness w attesting to this fact. Then, the shared secret S is well defined and information-theoretically determined, as we can use the correctness requirement for the set $\mathcal{P}_n$ as the definition of S: $S = \text{RECON}(1^\lambda, 1^n, \text{CRS}, V_M, P, \Pi, w)$.[10]

In the case where M is empty, correctness is trivially satisfied for *any* definition of S. We can therefore take S to be a uniformly random variable that is completely independent of the scheme, and *unconditional* security will be trivially satisfied as well. Interestingly, this means that, when analyzing schemes, it is only necessary to analyze correctness and security for non-empty access structures M, as *any* scheme will automatically be correct and secure for empty M.

In Sect. 3.2, we show how to obtain unconditional security for a slightly wider class of access structures, which we call *trivial* access structures.

Reusability. In this work, it will be useful to distinguish between *party* and *index*. A *party* is an entity that has run SHARE, and obtained a secret and public share. That party's *index* is the input i that was fed into SHARE. Multiple parties may share the same index. We will say a set X of parties is *complete* if, for every index i, there is exactly one party. Complete sets of parties are those for which RECON can be run, and therefore there is a shared secret S_X associated with every complete set of parties. In this sense, a DSS scheme is *reuseable*: an individual party with index i can take part in multiple sharings as part of different complete sets of parties, while only running SHARE once and publishing a single public share. This observation generalizes the fact that non-interactive key exchange (in the 2-party or multi-party setting) is reusable. This reusability property will be crucial for building witness PRFs from DSS.

Restricted Access Structures. The above definition requires that the DSS algorithms work for *any* access structure M recognized by a polynomial-sized verification circuit V_M. It is also possible to consider weaker versions where M is required to have a specific structure. For example, it is possible to consider M that are recognized by polynomial-size circuits (that is, $M \in \text{P}$). In Sect. 3.3, we consider an even more restricted setting where M is just a threshold function: $X \in M$ if and only if $|X| \geq t$ for some threshold t. We call these restrictions

---

[10] We note that to compute S we need to know w which may be computationally hard for some languages.

DSS for P or *DSS for threshold*, respectively. When distinguishing DSS for these limited classes from the standard definition above, we call the standard definition *DSS for* mNP. Finally, one can consider DSS for a specific, fixed access structure M, which we call *DSS for M*. For example, if M consists of all non-empty subsets (a special case of threshold where $t = 1$), then DSS for M is exactly multiparty non-interactive key exchange with trusted setup [9].

3.1 Alternative Definitions

We introduce several alternative definitions for distributed secret sharing. We first give a *strong* variant in which the sharing procedure is independent of the access structure V_M and of the party index i. Our second alternative is a *witnessless* version in which qualified sets are defined by an arbitrary circuit (possibly a non-monotone one). Our last variant is a definition of distributed secret sharing that has no setup (also known as no common reference string).

Definition 9 (Strong distributed secret sharing). *A strong distributed secret sharing scheme is a special case of a regular distributed secret sharing scheme (as in Definition 8) with the following differences:*

- SETUP$(1^\lambda, 1^n, V_M) = $ SETUP$(1^\lambda, 1^n, 1^{|V_M|})$. *That is,* SETUP *does not depend on V_M, except through the size of the circuit for V_M, but is otherwise independent of V_M or the language M.*
- SHARE$(1^\lambda, 1^n, CRS, V_M, i) = $ SHARE$(1^\lambda, 1^n, CRS, 1^{|V_M|})$. *That is,* SHARE *does not depend on V_M except for its size, and also does not depend on the party index i.*
- RECON$(1^\lambda, 1^n, CRS, V_M, P, \Pi(X), w)$ *now interprets P as a being ordered, and uses the order to determine the party index corresponding to each public share. From this information and $\Pi(X)$,* RECON *can determine the subset $X \subseteq [n]$ of indices for which secret shares are provided.*
- *For each verification circuit V_M and set of n parties $\mathcal{P}_n$, correctness is defined using an associated secret $S_{V_M, \mathcal{P}_n}$ that potentially varies for different V_M and $\mathcal{P}_n$ pairs. Notice that since a party is not assigned an index at sharing time, the only restriction we place on $\mathcal{P}_n$ is its size (i.e., $|\mathcal{P}_n| = n$), but we do not need $\mathcal{P}_n$ to be a complete set.*

The advantage of a strong DSS scheme is that the access structure does not need to be specified at sharing time. This allows parties to play multiple roles in different sharing executions without having to generate new shares, and allows a single sharing to be used for many different access structures. This will result in significant communication savings if many sharings with different access structures are being executed. When differentiating between the strong and regular variants, we will call the regular distributed secret sharing variant a *weak* scheme.

Definition 10 (Witnessless distributed secret sharing). *A witnessless distributed secret sharing is the following modification to (weak) distributed secret* *sharing, where the access structure M is set to be the monotone closure $\overline{C}$ of some (potentially non-monotone) function C.[11] In addition, we make the following modifications to the algorithms of the scheme:*

- SETUP$(1^\lambda, 1^n, C)$, *instead of taking as input the verification circuit V_M, now takes as input a circuit for the function C, which is potentially non-monotone. For the strong variant,* SETUP *takes as input $|C|$ instead of $|V_M|$.*
- SHARE$(1^\lambda, 1^n, \mathsf{CRS}, C, i)$ *also takes as input C instead of V_M. For the strong variant,* SHARE *takes as input $|C|$ instead of $|V_M|$, and does not take i as input.*
- RECON$(1^\lambda, 1^n, \mathsf{CRS}, C, P, \Pi(X))$ *similarly takes as input C instead of V_M. Also,* RECON *does not take as input a witnesses, hence the term* witnessless.
- *Correctness is modified so that $\Pr[\mathsf{RECON}(1^\lambda, 1^n, \mathsf{CRS}, C, P, \Pi(X)) = S] = 1$ for any $X \subseteq \mathcal{P}_n$ such that $C(X) = 1$.*

A set X of qualified parties in $M = \overline{C}$ cannot simply feed in all of the secret shares $\Pi(X)$ into RECON to obtain the secret, as $C(X)$ may not be 1. Instead, if they know a subset $X' \subseteq X$ such that $C(X') = 1$ (which must exist since X is qualified), they may simply feed the subset of their secret shares corresponding to X', namely $\Pi(X')$, into RECON, and correctness guarantees that they will learn the secret. Thus, even though the algorithms in a witnessless distributed secret sharing scheme do not take a witness as input, reconstructing the secret still requires knowing a witness, namely the subset X'.

We note that the access structure M is monotone, and is clearly in NP. Therefore, Lemma 1 shows that M is recognized by a monotone nondeterministic verification procedure V_M. Thus, the above formulation of distributed secret sharing is equivalent to regular DSS (with witnesses) where we restrict to access structures of this form. Therefore, this notion is no stronger than regular DSS.

We note that many NP languages naturally are represented using a circuit C, such as Hamiltonian Cycle (where C checks that the set of edges forms a Hamiltonian cycle) and Subset Sum (where C checks that the subset of integers sums to 0).

Definition 11 (Distributed secret sharing without setup). *In a distributed secret sharing scheme* without setup, *there is no* SETUP *algorithm, and* SHARE *and* RECON *do not take* CRS *as input. When distinguishing between schemes with and without setup, we call the standard notion (Definition 8)* distributed secret sharing with trusted setup.

Immediate Relations Between Definitions. All of the above variations are orthogonal, giving us 8 variants of distributed secret sharing. We make the following observation:

- Any of the 4 variants of *strong* distributed secret sharing imply the corresponding variant of *weak* distributed secret sharing. This is because being a strong scheme just imposes constraints on the form of the algorithms.

---

[11] Recall that the monotone closure $\overline{C}$ of a function C includes all sets X such that some subset $X' \subseteq X$ satisfies $C(X') = 1$ (see Sect. 2.1).

- Any of the 4 variants of distributed secret sharing *with witnesses* imply the corresponding *witnessless* variant, since the witnessless condition imposes a restriction on the languages allowed.
- Any of the 4 variants of distributed secret sharing *without trusted setup* imply the corresponding variant *with trusted setup*, where SETUP outputs an empty string.

In Sect. 3.4, we will show that all of the above notions are equivalent, and moreover that they are equivalent to witness PRFs.

3.2 Distributed Secret Sharing Implies One-Way Functions

Witness PRFs trivially imply one-way functions, and therefore by our equivalence in Sect. 3.4, information-theoretic distributed secret sharing is impossible for general access structures.

In this section, we consider DSS for specific access structures, and ask: for what access structures M is information-theoretic DSS possible? To answer this question we first define trivial access structures, and then in Theorem 1 we show that a DSS scheme for any non-trivial access structures implies one-way functions. DSS for trivial access structures, on the other hand, are shown to have a very simple information-theoretically secure construction.

Definition 12 (Trivial access structures). *We say that an access structure M for a set of parties $\mathcal{P}$ is trivial if either M is empty, or there exists a subset $X \subseteq \mathcal{P}$ such that $Y \in M$ if and only if $X \subseteq Y$.*

We call such access structures trivial due to the following reasons:

- Parties outside of X are irrelevant to the access structure, as they can be added or removed from a set of parties without changing the set's qualified status. Therefore, such a protocol is morally equivalent to the case where $X = \mathcal{P}$.
- When $X = \mathcal{P}$, all parties must get together to reconstruct the shared secret. In this case, there appears to be no reason to engage in the protocol in the first place, as the parties can just choose the group secret when they all coordinate at reconstruction time.

Note that trivial access structures are in P, so there is no distinction between standard DSS and witnessless DSS.

Theorem 1. *For an access structure M, the following hold:*

- *If M is trivial, then there exists a perfectly-secure DSS for M in the strongest possible sense (that is, strong DSS without setup)*
- *If M is non-trivial, then the existence of any DSS for M in the weakest possible sense (that is, weak DSS with trusted setup) implies the existence of one-way functions.*

Proof. Let M be trivial, with subset X such that $Y \in M$ if and only if $X \subseteq Y$. We then get the following strong DSS scheme for M without setup that has single-bit shared secrets:

- SHARE(): sample a random $\Pi(i) \leftarrow \{0,1\}$, and publish an empty string as the public share $P(i) = \emptyset$.
- RECON($P, \Pi(Y)$): if $X \subset Y$, simply XOR and output the shares for parties in X, namely $S \leftarrow \text{XOR}_{i \in X} \Pi(i)$. If X is not a subset of Y, abort.

The correctness of the protocol is trivial. For security, note that for any set Y that does not contain X, there is some party $i \in X \setminus Y$ such that the set of shares for Y does not contain the secret share $\Pi(i)$. Therefore, $\Pi(i)$ is independent of the shares $\Pi(Y)$. Thus, S is independent of $\Pi(Y)$. Perfect security follows.

The proof of the other case (in which M is a non-trivial access structure) can be found in the full version [26]. ∎

3.3   Distributed Secret Sharing for Threshold

In this section we present a distributed secret sharing scheme for the threshold access structure. The proof of security relies on the multilinear decisional Diffie-Hellman assumption (see Definition 5). This construction works in the trusted setup model (which is used for the setup of the multilinear map). Assume there are n parties and the threshold condition says that any t of them should be able to reconstruct the secret.

Lemma 2. *Assuming an $(n - t)$-multilinear map that satisfied the MDDH assumption, there is a t-out-of-n (weak) distributed secret sharing scheme (with trusted setup).*[12]

Proof. We start with the description of the scheme. The trusted setup will consists of an $n-t$ multilinear map. For the sharing, party p_i generates a random s_i and published $h_i = g^{s_i}$. The shared secret key is $S = e(g, \ldots, g)^{\prod_{i=1}^{n} s_i}$. With t of the s_i's one can easily compute S by pairing the other h_i's, and then raising the result by each of the s_i's. Security in the case of fewer than t shares follows from the security of the multilinear DH assumption.

More precisely, in the trusted setup we run MMap.Gen($1^\lambda, n$) to get (Γ, g, ℓ) which we set as the public parameters. The sharing procedure of party p_i samples a random $s_i \leftarrow \mathbb{Z}$ (which is kept secret) and outputs $h_i = g^{s_i}$. The shared secret key is $S = e(g, \ldots, g)^{\prod_{i=1}^{n} s_i}$. For correctness, we observe that given the secret shares of any subset of the t parties one can compute S. Indeed, given $h_{i_1}, \ldots, h_{i_{n-t}}$ one can compute

$$e(h_{i_1}, \ldots, h_{i_{n-t}}) = e(g, \ldots, g)^{\prod_{j=1}^{n-t} s_{i_j}}$$

---

[12] Since threshold is in P, there are no witnessness, so there is no distinction between the standard and witnessless notions of DSS.

and then, by raising the right-hand side to the powers $s_{i_{n-t+1}}, \ldots, s_{i_n}$, compute

$$\left(e(g, \ldots, g)^{\prod_{j=1}^{n-t} s_{i_j}} \right)^{\prod_{j=n-t+1}^{n} s_{i_j}} = S$$

The proof of security can be found in the full version [26].   ∎

Hamming Ball Constrained PRFs. We show that any distributed secret sharing scheme for threshold implies constrained PRFs that can be constrained to a Hamming ball around an arbitrary point. One limitation of our construction is that the PRF only allows a single collusion: an adversary that sees the PRF constrained to two Hamming balls can potentially recover the entire secret key.

Of course, our construction of DSS for threshold relies on the multilinear Diffie-Hellman assumption, which already implies constrained PRFs for all circuits with arbitrary collusions [10]. However, our conversion here is generic and applies to any threshold DSS scheme, which perhaps can be based on simpler assumptions than multilinear maps. Perhaps more importantly, the ideas presented here will be used in Sect. 3.4 to show the equivalence of general DSS and witness PRFs. Thus, this construction can be viewed as a warm-up to Theorem 3.

Definition 13 (One-time constrained PRFs for Hamming balls). *A constrained PRFs for Hamming balls is a tuple of algorithms* (Gen, PRF, Constrain, Eval) *where:*

- Gen($1^\lambda, 1^n$) *is a polynomial-time randomized procedure that takes as input a security parameter λ and a bit length n, and outputs a function key* fk.
- PRF(fk, x) *is a polynomial-time deterministic procedure that takes as input the function key* fk *and a bit string $x \in \{0,1\}^n$.*
- Constrain(fk, c, r) *is a polynomial-time (potentially randomized) procedure that takes as input the function key* fk*, a point $c \in \{0,1\}^n$, and a radius $r \in [0, n]$, and outputs the constrained evaluation key* ek *corresponding to the Hamming ball of radius r centered at c.*
- Eval(ek, x) *is a polynomial-time deterministic procedure that takes as input the evaluation key* ek *and a bit string $x \in \{0,1\}^n$.*
- **Correctness:** *If x and c differ on at most r bits, then*

$$\Pr[\mathsf{Eval}(\mathsf{ek}, x) = \mathsf{PRF}(\mathsf{fk}, x)] = 1,$$

where fk $\leftarrow$ Gen($1^\lambda, 1^n$), ek $\leftarrow$ Constrain(fk, c, r) *and the probability is taken over the randomness of* Gen, Constrain.
- **One-time security:** *For any probabilistic polynomial time algorithm D, there exists a negligible function* neg($\cdot$) *such that for any $\lambda \in \mathbb{N}$ and any $r \in [0, n]$, $x \in \{0,1\}^n$ and $c \in \{0,1\}^n$ such that x and c differ in strictly more than r points, it holds that*

$$|\Pr[D(\mathsf{ek}, \mathsf{PRF}(\mathsf{fk}, x)) = 1] - \Pr[D(\mathsf{ek}, y) = 1]| < \mathsf{neg}(\lambda),$$

where the probabilities are taken over the choice of fk $\leftarrow$ Gen($1^\lambda, 1^n$), ek $\leftarrow$ Constrain(fk, c, r), *and y which is chosen uniformly at random over the co-domain of* PRF.

Theorem 2. *If secure distributed secret sharing for threshold access structures exists, then secure one-time constrained PRFs for Hamming balls exists.*

At first glance, building a Hamming ball constrained PRFs from threshold DSS appears to be a difficult task. Indeed, the natural approach to constructing witness PRFs would be have the public evaluation key be the set of public shares P, and perhaps some subset of secret shares $\Pi(X)$ for $X \subseteq P$; the secret function key would naturally be the complete set of secret shares $\Pi(\mathcal{P})$. However, it is unclear how to define the PRF $\mathsf{PRF}(\cdot)$. One possibility is to try to set the outputs of the PRF to be the shared secret S. However, our threshold DSS only explicitly has a single S. Yet, we need many secret outputs, one for each possible input.

To get around these limitations, we make use of the fact that distributed secret sharing is *reusable*, as discussed in the beginning of Sect. 3. For example, suppose two distinct sets of parties $\mathcal{P}_0 \neq \mathcal{P}_1$ wish to carry out the protocol, and there is some party i that is a member of both sets. Then, party i could *reuse* his public share for both runs of the protocol. More generally, for a large collection $\mathcal{C}$ of parties with $|\mathcal{C}| \gg n$, all parties can run SHARE exactly once, and then any subset $\mathcal{P} \subseteq \mathcal{C}$ of n parties can then run the distributed secret sharing protocol without any interaction (assuming that $\mathcal{P}$ is complete, meaning every party index is present exactly once).

Our idea, then, is to have the PRF value be the shared secret for a subset of $\mathcal{C}$, and the input to the PRF selects which subset to use. We need to be careful, though, as we need to ensure that the subset is complete and contains every party index exactly once. We show that such valid subsets can still be used to construct witness PRFs.

Proof of Theorem 2. Let (SETUP, SHARE, RECON) be a distributed secret sharing scheme for threshold. We start with the construction of the constrained PRF.

– **Gen($1^\lambda, 1^n$):** First, run CRS $\leftarrow$ SETUP($1^\lambda, 1^{2n}$, thr $= n$). That is, initialize the setup procedure for the threshold DSS scheme with $2n$ parties and threshold n. Next, we will define a set $\mathcal{P} = \{(i, b)\}_{i \in [n]} \cup [n+1, 2n]$ of parties, where party (i, b) for $i \in [n]$ has index i, and party i for $i \in [n+1, 2n]$ has index i. Now run SHARE for each party. That is, run

$$(P_{i,b}, \Pi_{i,b}) \leftarrow \mathsf{SHARE}(1^\lambda, 1^{2n}, \mathsf{CRS}, \mathsf{thr} = n, i) \text{ for } i \in [n],$$
$$(P_i, \Pi_i) \leftarrow \mathsf{SHARE}(1^\lambda, 1^{2n}, \mathsf{CRS}, \mathsf{thr} = n, i) \text{ for } i \in [n+1, 2n].$$

Let $\Pi = \{\Pi_{i,b}\}_{i \in [n]} \cup \{\Pi_i\}_{i \in [n+1,2n]}$ be the set of secret shares, and P the corresponding set of public shares. Output the function key fk $= (\mathsf{CRS}, \Pi, P)$.
– **PRF(fk, x):** Define $\mathcal{P}_x$ to be the collection of parties (i, x_i) for $i \in [n]$, together with parties i for $i \in [n+1, 2n]$. Define

$$P(\mathcal{P}_x) = \{P_{i,x_i}\}_{i \in [n]} \cup \{P_i\}_{i \in [n+1,2n]} \text{ and } \Pi(\mathcal{P}_x) = \{\Pi_{i,x_i}\}_{i \in [n]} \cup \{\Pi_i\}_{i \in [n+1,2n]}.$$

Notice that $\mathcal{P}_x$ is complete, in that each party index is present. Now, use the secret shares to reconstruct the shared secret for $\mathcal{P}_x$:

$$S \leftarrow \mathsf{RECON}(1^\lambda, 1^{2n}, \mathsf{CRS}, \mathsf{thr} = n, P(\mathcal{P}_x), \Pi(\mathcal{P}_x))$$

and output S.

- **Constrain(fk, c, r)**: Let

$$\mathsf{ek} = (c, r, P, \{\Pi_{i,c_i}\}_{i \in [n]} \cup \{\Pi_i\}_{i \in [n+1, n+r]})$$

be the set of secret shares Π_{i,c_i} for parties $(i, c_i), i \in [n]$, as well as r of the secret shares Π_i for for parties $i \in [n+1, 2n]$. Output ek.

- **Eval(ek, x)**: Check that x and c differ in at most r points, and otherwise abort. Let $T \subseteq [n]$ be the set of indices where x and c agree. Then, the set of parties $X = \{(i, x_i)\}_{i \in T} \cup [n+1, n+r]$ forms a subset of $\mathcal{P}_x$. Moreover, X consists of $|T| + r \geq n = t$ parties (since x and c agree on at least $n-r$ points), and ek contains the secret shares $\Pi(X)$ for all of these parties. Therefore, run

$$K \leftarrow \mathsf{RECON}(1^\lambda, 1^{2n}, \mathsf{CRS}, \mathsf{thr} = n, P(\mathcal{P}_x), \Pi(X))$$

and output K.

An example of our construction for the case $n = 5$ is given in Fig. 2.

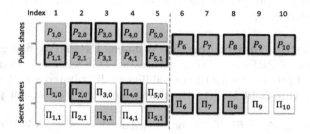

Fig. 2. Example instantiation for $n = 5$. The underlying threshold DSS scheme is instantiated with 10 indices and threshold $t = 5$. For indices 1 through 5, SHARE is run twice, returning two sets of secret/public pairs for each index 1 through 5. For indices 6 through 10, SHARE is run once. The secret key fk consists of *all* public shares and secret shares. The shares highlighted in green correspond to the evaluation key ek for the Hamming ball centered at $c = 00101$ with radius $r = 3$. The public shares outlined in bold purple indicate the public shares whose shared secret S is $\mathsf{PRF}(\mathsf{fk}, x = 10001)$. Notice that x and c have a Hamming distance $2 \leq r$, so S should be computable from ek. Indeed ek contains $6 \geq t$ of the corresponding secret shares (also outlined in bold purple), meaning that it is possible to construct $S = \mathsf{PRF}(\mathsf{fk}, x)$ from ek.

Correctness follows immediately from the observations above. Indeed, given x and r that differ on at most r coordinates, one can generate the secret shares for the set of parties X defined above. Now, the correctness of the distributed secret sharing scheme implies that K must be equal to S, where K and S are as defined in the scheme above. For security, we have the following claim whose proof can be found in the full version [26]:

Claim. If (SETUP, SHARE, RECON) is a secure distributed secret sharing scheme for threshold, then (Gen, PRF, Constrain, Eval) is a one-time secure constrained PRF for Hamming balls.

This completes the proof of the theorem.

3.4   Distributed Secret Sharing Is Equivalent to Witness PRFs

In this section, we prove that all variants of distributed secret sharing are actually equivalent to witness PRFs. Together with Zhandry's construction of witness PRFs [34], this gives a construction of distributed secret sharing from simple assumptions on multilinear maps.

Theorem 3. *The existence of the following are equivalent:*

- *Witness PRFs for* NP.
- *Any of the 8 variants of distributed secret sharing for* mNP.

Proof. To prove the theorem, it suffices to prove the following:

1. Weak distributed secret sharing without witnesses and with trusted setup implies witness PRFs.
2. Witness PRFs imply strong distributed secret sharing with witnesses and without trusted setup.

Distributed Secret Sharing Implies Witness PRFs. We first give the construction of witness PRFs from weak witnessless DSS with a trusted setup. Our construction and proof leverage the reusability of distributed secret sharing, and is based on the threshold DSS to Hamming ball PRF conversion presented in Sect. 3.3.

Let (SETUP, SHARE, RECON) be a witnessless weak distributed secret sharing scheme with trusted setup. We build the following witness PRF (Gen, PRF, Eval):

- **Gen(R):** Let n be the instance size and m the witness size. We will use a DSS scheme over a set of parties $\mathcal{P}$ with $2n + m$ party indices. We will generally think of the index set as containing $2n$ pairs $(i, b) \in [n] \times \{0, 1\}$, as well as m integers $j \in [m]$. The set of pairs $[n] \times \{0, 1\}$ we will call the "instance set", and the set of integers $[m]$ we will call the "witness set".
  Define a circuit $C \colon 2^{\mathcal{P}} \to \{0, 1\}$ that operates, given an input $S \subseteq \mathcal{P}$, as follows. If $S = \mathcal{P}$, output 1. For any i, if either both $(i, 0), (i, 1)$ from the instance set are in S or neither are in S, then C outputs 0. Otherwise if $(i, b) \in S$ (and therefore $(i, 1 - b) \notin S$), set $x_i = b$. Let x be the bit string $x_1 x_2 \ldots x_n$. Let w_j be 1 if $j \in S$ and let w be the bit string $w_1 w_2 \ldots w_n$. Then, C outputs $R(x, w)$. Recall that the monotone closure of C, $M = \overline{C}$, satisfies $X \in M$ if some subset $X' \subseteq X$ causes C to accept.
  First, we generate the CRS by running

$$\mathsf{CRS} \leftarrow \mathsf{SETUP}(1^{\lambda}, 1^{2n+m}, C).$$

Now, we define the set $\mathcal{P}$ to consist of the following parties: for each index (i, b) in the instance set, we will associate two parties $\{(i, b, c)\}_{c \in \{0, 1\}}$, and for each index $j \in [m]$ in the witness set, we will associate a party j. Next, we run SHARE for each party. That is, for each $i \in [n], b \in \{0, 1\}$ and $c \in \{0, 1\}$, run

$$(P_{i,b,c}, \Pi_{i,b,c}) \leftarrow \mathsf{SHARE}(1^\lambda, 1^{2n+m}, \mathsf{CRS}, C, (i, b))$$

and for each $j \in [m]$, run

$$(P_j, \Pi_j) \leftarrow \mathsf{SHARE}(1^\lambda, 1^{2n+m}, \mathsf{CRS}, C, j).$$

Let $P = \{P_{i,b,c}\}_{i \in [n], b, c \in \{0,1\}} \cup \{P_j\}_{j \in [m]}$ and $\Pi = \{\Pi_{i,b,c}\}_{i \in [n], b, c \in \{0,1\}} \cup \{\Pi_j\}_{j \in [m]}$ be the set of public and secret shares, respectively. Output the function key

$$\mathsf{fk} = (\mathsf{CRS}, P, \Pi)$$

and the evaluation key

$$\mathsf{ek} = (\mathsf{CRS}, P, \{\Pi_{i,b,b}\}_{i \in [n], b \in \{0,1\}}, \{\Pi_j\}_{j \in [m]}).$$

That is, the evaluation key consists of all of the public shares, all of the secret shares for indices in the witness set, and one of the secret shares for each index (i, b) in the instance set (recall that for each index in the instance set, we have two parties).
- **PRF(fk, x)**: Let

$$\mathcal{P}_x = \{(i, b, x_i)\}_{i \in [n], b \in \{0,1\}} \cup [m]$$

so that $P(\mathcal{P}_x) = \{P_{i,b,x_i}\}_{i \in [n], b \in \{0,1\}} \cup \{P_j\}_{j \in [m]}$ and $\Pi(\mathcal{P}_x) = \{\Pi_{i,b,x_i}\}_{i \in [n], b \in \{0,1\}} \cup \{\Pi_j\}_{j \in [m]}$.
Notice that $\mathcal{P}_x$ is complete, in the sense that each index is represented exactly once. Therefore, run

$$K \leftarrow \mathsf{RECON}(1^\lambda, 1^{2n+m}, \mathsf{CRS}, C, P(\mathcal{P}_x), \Pi(\mathcal{P}_x))$$

and output K.
That is, out of the entire collection of $4n + m$ parties, use the input x to select the appropriate set of parties $\mathcal{P}_x$ of size $2n + m$. Then, compute the shared key for that set of parties.
- **Eval(ek, x, w)**: Let $\mathcal{P}_x$ and $P(\mathcal{P}_x)$ be as above. Let $S_{x,w} = \{(i, x_i, x_i)\}_{i \in [n]} \cup \{j\}_{j:w_j=1}$ and $\Pi(S_{x,w}) = \{\Pi_{i,x_i,x_i}\}_{i \in [n]} \cup \{\Pi_j\}_{j:w_j=1}$. Run

$$K \leftarrow \mathsf{RECON}(1^\lambda, 1^{2n+m}, \mathsf{CRS}, C, P(\mathcal{P}_x), \Pi(S_{x,w}))$$

and output K.

To show correctness, we need to argue that $\mathsf{Eval}(\mathsf{ek}, x, w) = \mathsf{PRF}(\mathsf{fk}, x)$ for all w such that $R(x, w) = 1$. Indeed, $\mathsf{Eval}(\mathsf{ek}, x, w)$ attempts to compute the shared secret for the set of parties $\mathcal{P}_x$. Notice that the set $S_{x,w}$ is a subset of the set $\mathcal{P}_x$, and consists of the parties in $\mathcal{P}_x$ with indices in $T_{x,w} = \{(i, x_i)\}_{i \in [n]} \cup \{j\}_{j:w_j=1}$.

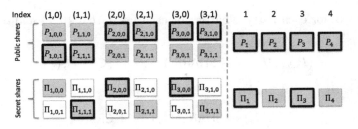

Fig. 3. Example instantiation for instance size $n = 3$ and witness size $m = 4$. The underlying threshold DSS scheme is instantiated with 10 indices, 6 for the instance set having the form (i, b), and 4 for the witness set having the form j. For each instance set index (i, b), SHARE is run twice, returning two sets of secret/public pairs for parties $(i, b, 0), (i, b, 1)$. For witness set indices, SHARE is run once. The secret key fk consists of *all* public keys and secret shares, and the evaluation key consists of the green highlighted shares. An example evaluation on $x = 100$ is given. The instance x selects the subset $\mathcal{P}_x$, whose public shares are bolded in purple. For these shares, there is a shared secret S, and the value of PRF on x is defined to be S. Suppose $w = 1010$ is a valid witness for x. Then, the secret shares for parties in $S_{x,w}$ are boxed in bold purple and represent the set of secret shares inside ek that can be fed into RECON to yield S. Notice that, among the instance set of indices, ek only contains secret shares for the parties in $\mathcal{P}_x$ that have indices (i, x_i).

Now notice that $C(T_{x,w})$ computes exactly $R(x, w) = 1$. Thus, the set of secret shares $\Pi(S_{x,w})$ is sufficient to reconstruct the shares secret S for $\mathcal{P}_x$. Notice that S is also the value outputted by PRF(fk, x). Therefore, Eval(ek, x, w) = PRF(fk, x) as desired. An example instantiation is given in Fig. 3.

It remains to prove that the scheme is secure. The proof of the following claim can be found in the full version [26].

Claim. If (SETUP, SHARE, RECON) is a secure weak distributed secret sharing scheme without witnesses and with trusted setup, then (Gen, PRF, Eval) is a secure witness PRF.

Witness PRFs Imply Distributed Secret Sharing. Given a Witness PRF (Gen, PRF, Eval), we can easily obtain a one-way function, and from this we can obtain a pseudorandom generator f [22]. We construct the following *strong* distributed secret sharing scheme (SHARE, RECON) without trusted setup.

- **SHARE($1^\lambda, 1^n, 1^k$):** Run (fk, ek) $\leftarrow$ Gen(R) where R is the following NP circuit. R takes as input an instance $(V_M, \{y_i\}_{i \in \mathcal{P}_n})$, where V_M is the description of an mNP circuit of size at most k, and witness $w' = (w, \{s_i\}_{i \in X})$ for some subset $X \subseteq \mathcal{P}_n$. It outputs 1 if (1) $V_M(X, w) = 1$ and (2) $y_i = f(s_i)$ for each $i \in X$. Otherwise, R outputs 0.
  Let $s \leftarrow \mathcal{S}$ where $\mathcal{S}$ is the domain of f, and $y = f(s)$. Output public share $P(i) = (\text{ek}, y)$ and secret share $\Pi(i) = s$.

- **RECON$(1^\lambda, 1^n, V_M, P, \Pi(X), w)$:** Write $\Pi(X) = \{s_i\}_{i \in X}$ and $P = \{(\text{ek}_i, y_i)\}_{i \in \mathcal{P}}$. Let x be the instance $(V_M, \{y_i\}_{i \in \mathcal{P}})$, and let $w' = (w, \{s_i\}_{i \in X})$ be a witness. For each i, compute

$$S_i = \text{Eval}(\text{ek}_i, x, w'),$$

and then compute $S = S_1 \oplus S_2 \oplus \cdots \oplus S_n$. Output S.

The correctness of the scheme follows immediately from the correctness of the underlying witness PRF. The security of the scheme follows from the following claim whose proof can be found in the full version [26].

Claim. If (Gen, PRF, Eval) is a secure witness PRF and f is a secure PRG, then (SHARE, RECON) is a secure *strong* distributed secret sharing scheme with witnesses and without trusted setup.

We have shown that the weakest variant of distributed secret sharing implies witness PRFs, which in turn imply the strongest variant of distributed secret sharing. Thus, all variants of DSS and witness PRFs are equivalent, completing the proof.

4   Functional Secret Sharing

We start this section with a definition of functional secret sharing. Later, in Theorem 4, we show that general-purpose functional secret sharing is equivalent to indistinguishability obfuscation for polynomial-size circuits.

Definition 14 (Functional secret sharing). *Let $\mathcal{F} = \{F \colon 2^{\mathcal{P}_n} \to \{0,1\}^*\}$ be a class of functions. Let $M : 2^{\mathcal{P}_n} \to \{0,1\}$ be an access structure corresponding to a language $L \in \text{mNP}$ and let V_M be a verifier for L. A functional secret sharing scheme for M and $\mathcal{F}$ consists of a setup procedure SETUP and a reconstruction procedure RECON that satisfy the following requirements:*

1. *SETUP$(1^\lambda, F, S)$ gets as input an efficiently computable function $F \colon 2^{\mathcal{P}_n} \times \{0,1\}^* \to \{0,1\}^*$ and a secret $S \in \{0,1\}^*$, and distributes a share for each party. For $i \in [n]$ denote by $\Pi(F, S, i)$ the random variable that corresponds to the share of party p_i. Furthermore, for $X \subseteq \mathcal{P}_n$ we denote by $\Pi(F, S, X)$ the random variable that corresponds to the set of shares of parties in X.*
2. *Completeness: If RECON$(1^\lambda, \Pi(F, S, X), w)$ gets as input the shares of a "qualified" subset of parties and a valid witness, and outputs the value of F on X and the shared secret. Namely, for $X \subseteq \mathcal{P}_n$ such that $M(X) = 1$ and any valid witness w such that $V_M(X, w) = 1$, it holds that:*

$$\Pr\left[\text{RECON}(1^\lambda, \Pi(F, S, X), w) = F(X, S)\right] = 1,$$

where the probability is over the internal randomness of the scheme and of RECON.

3. **Indistinguishability of the Secret:** *For every probabilistic polynomial-time algorithm D, every function $F \in \mathcal{F}$, every subset of parties $X \subseteq \mathcal{P}_n$ and every pair of secrets S_0, S_1, as long as either $M(X) = 0$ or $F(X', S_0) = F(X', S_1)$ for every $X' \subseteq X$, there exists a negligible function $\mathsf{neg}(\cdot)$ such that for $\lambda \in \mathbb{N}$ it holds that*

$$\left| \Pr\left[D(1^\lambda, \Pi(F, S_b, X)) = b \right] - \frac{1}{2} \right| \le \mathsf{neg}(\lambda),$$

where the probability is over the internal randomness of the scheme, the internal randomness of D and $b \leftarrow \{0, 1\}$ chosen uniformly at random.

A Remark on the Condition in the Security Definition. We note that in Definition 14, given a set of shares $\Pi(F, S, X)$, it is possible to derive for any $X' \subseteq X$ the set of shares $\Pi(F, S, X')$ simply by removing the shares for parties not in X'. Feeding $\Pi(F, S, X')$ into RECON then gives $F(X', S)$ for any $X' \subseteq X$. Thus, in the security definition above, the condition that $F(X', S_0) = F(X', S_1)$ for all $X' \subseteq X$ is required to have a satisfiable assumption. Our definition states that this is the only requirement.

Two Relaxations of Definition 14. We remark that when the function F is defined to be the identity function over its second input parameter (i.e., $F(\cdot, S) = S$) we get the definition of Rudich secret sharing for NP of [25].[13] Moreover, when $M = 2^{\mathcal{P}_n}$ (i.e., the access structure includes all subsets of parties), the secret S is a description of a function and F is the universal circuit (i.e., $F(X, S) = S(X)$), then Definition 14 boils down to the definition of *function secret sharing* which we formalize next.

Definition 15 (Function secret sharing). *Let $\mathcal{F} = \{F \colon 2^{\mathcal{P}_n} \to \{0, 1\}^*\}$ be a class of functions. A functional secret sharing scheme for $\mathcal{F}$ consists of a setup procedure SETUP and a reconstruction procedure RECON that satisfy the following requirements:*

1. *SETUP$(1^\lambda, F)$ gets as input a function $F \in \mathcal{F}$, and distributes a share for each party. For $i \in [n]$ denote by $\Pi(F, i)$ the random variable that corresponds to the share of party p_i. Furthermore, for $X \subseteq \mathcal{P}_n$, we denote by $\Pi(F, X)$ the random variable that corresponds to the set of shares of parties in X.*
2. *Completeness: RECON$(1^\lambda, \Pi(F, X))$ gets as input the shares of some subset X of parties, and outputs $F(X)$. More precisely,*

$$\Pr[\mathsf{RECON}(1^\lambda, \Pi(F, X)) = F(X)] = 1,$$

where the probability is over the internal randomness of the scheme and of RECON.

---

[13] [25] considered a uniform version of the above definition. We remark that our definitions from above can also be given in a uniform version and our results also apply to them (using ideas from [25]). For simplicity, we focus on the non-uniform versions.

3. **Indistinguishability of the function:** For every probabilistic polynomial-time algorithm D, every equal size $F_0, F_1 \in \mathcal{F}$ and $X \subseteq 2^{\mathcal{P}_n}$ such that $F_0(X') = F_1(X')$ for all $X' \subseteq X$, there exists a negligible function $\mathsf{neg}(\cdot)$ such that for $\lambda \in \mathbb{N}$ it holds that

$$\left| \Pr[D\left(1^\lambda, \Pi(F_b, X)\right) = b] - \frac{1}{2} \right| \le \mathsf{neg}(\lambda),$$

where the probability is over the internal randomness of the scheme, the internal randomness of D and $b \leftarrow \{0, 1\}$ chosen uniformly at random.

4.1 Functional Secret Sharing Is Equivalent to iO

In this section we state and prove our main result.

Theorem 4. *The following holds:*

1. *Function secret sharing (Definition 15) for polynomial-size circuits implies* iO *for polynomial-size circuits.*
2. iO *for polynomial-size circuits and one-way functions imply functional secret sharing (Definition 14) for access structures in* mNP *and functions computed by polynomial-size circuits.*

Recall that Definition 14 is a generalization of Definition 15. Thus, Theorem 4 implies that functional secret sharing is equivalent to function secret sharing and is equivalent to iO.[14]

Next, we provide a proof for each of the items in Theorem 4 separately.

Proof of Item 1 in Theorem 4. Given a circuit C with n inputs the indistinguishability obfuscator works as follows. We first run the SETUP($1^\lambda, C$) procedure with the circuit C as input and get back a list of n shares $\Pi(C, 1), \ldots, \Pi(C, n)$. The obfuscation consists of these n shares.

To evaluate an obfuscated circuit at a point $x \in \{0, 1\}^n$, we run RECON($1^\lambda, \Pi(C, x)$) and get a value y that we output. By the correctness of the functional secret sharing scheme, we have that $y = C(x)$, as required.

To prove security consider two equal size *functionally equivalent* circuits C_1 and C_2 and an adversary $\mathcal{A}$ that can distinguish their obfuscations with noticeable probability. Hence, $\mathcal{A}$ can distinguish secret shares corresponding to SETUP($1^\lambda, C_1$) from secret shares corresponding to SETUP($1^\lambda, C_2$). Since the circuits are equal size and functionally equivalent, this is a contradiction to the security guarantee of the function secret sharing scheme. ∎

Proof of Item 2 in Theorem 4. We start with the description of the functional secret sharing scheme. For every $i \in [n]$, the *share* of party p_i is composed of 2 components: (1) $r_i \in \{0, 1\}^\lambda$, an opening of a commitment to the value i,

---

[14] One of the directions requires one-way functions which can be relaxed to require a worst-case hardness assumption by [24].

and (2) an *obfuscated* circuit $iO(C)$. The circuit C to be obfuscated has the following hardwired: the function F, the secret S and the commitments of all parties (i.e., $c_i = \mathsf{Com}(i, r_i)$ for $i \in [n]$). We stress that the openings $r_1, \ldots, r_n$ of the commitments are *not* hardwired into the circuit. The input to the circuit C consists of alleged k openings $r'_{i_1}, \ldots, r'_{i_k}$ corresponding to a set of parties $X \in 2^{\mathcal{P}_n}$ denoted $p_{i_1}, \ldots, p_{i_k}$ where $k, i_1, \ldots, i_k \in [n]$ and an alleged witness w. The circuit C first checks that the openings are valid, i.e., verifies that for every $j \in [k] : c_{i_j} = \mathsf{Com}(i_j, r'_{i_j})$. Then, it verifies that the given w is a valid witness, i.e., that $V_M(X, w) = 1$. If all the tests pass, C outputs $F(X, S)$; otherwise, if any of the tests fail, the circuit C outputs NUL. The secret sharing scheme is formally described next.

Let iO be an efficient indistinguishability obfuscator (see Definition 7). Let $\mathsf{Com}: [2n] \times \{0, 1\}^\lambda \to \{0, 1\}^{q(\lambda)}$ be a string commitment scheme where $q(\cdot)$ is a polynomial (see Definition 3). Let $M \in \mathsf{NP}$ be an access structure.

The SETUP$(1^\lambda, F, S)$ procedure. Gets as input a function F represented as a polynomial-size circuits, a secret S and does the following:

1. For $i \in [n]$:
   (a) Sample uniformly at random an opening $r_i \in \{0, 1\}^\lambda$.
   (b) Compute the commitment $c_i = \mathsf{Com}(i, r_i)$.
2. Compute the circuit C from Fig. 4, where $C = C^{F,S,c_1,\ldots,c_n}$ has the function F, the secret S and the list of commitments $c_1, \ldots, c_n$ hardwired.
3. Set the share of party p_i to be $\Pi(S, i) = \langle r_i, iO(C) \rangle$.

The RECON(X, w) procedure. Gets as input a non-empty subset of parties $X \subseteq \mathcal{P}_n$ together with their shares and a witness w of X for M.

The Circuit $C^{F,S,c_1,\ldots,c_n}(r'_1, \ldots, r'_n, w)$

Hardwired: The function F, the secret S and the commitments of all parties $c_1, \ldots, c_n$.

Input: Secret shares corresponding to a subset of parties X and an alleged witness w. The secret shares are a sequence of n values $r'_1, \ldots, r'_n \in \{0, 1\}^\lambda \cup \mathsf{NUL}$ such that for any $i \in [n]$ if $p_i \in X$, then r'_i is the alleged opening of party p_i, and otherwise $r'_i = \mathsf{NUL}$.

Algorithm:

1. Execute the following tests:
   (a) For every $i \in [n]$ such that $r_i \neq \mathsf{NUL}$, verify that the opening r'_i is valid. That is, verify that $c_i = \mathsf{Com}(i, r'_i)$.
   (b) Verify that the given alleged witness w is a valid one. That is, verify that $V_M(X, w) = 1$.
2. If any of the above tests fails, output NUL; otherwise, output $F(X, S)$.

Fig. 4. The circuit to be obfuscated as part of the secret shares.

1. Let $iO(C)$ be the obfuscated circuit in the shares of X.
2. Evaluate the circuit $iO(C)$ with the shares of X and w and return its output.

Observe that if iO and Com are both probabilistic polynomial-time algorithms, then the scheme is efficient (i.e., SETUP and RECON are probabilistic polynomial-time algorithms). SETUP generates n commitments and an obfuscated circuit of polynomial-size. RECON only evaluates this polynomial-size obfuscated circuit once.

Security. Fix two secrets S_0, S_1, a subset of parties X, and a function F such that $F(X', S_0) = F(X', S_1)$ for every $X' \subseteq X$. The proof of security follows by a sequence of hybrid experiments that can be found in the full version [26]. ∎

Acknowledgments. We thank Moni Naor and the anonymous reviewers of TCC 2016A for helpful remarks.

References

1. Applebaum, B., Brakerski, Z.: Obfuscating circuits via composite-order graded encoding. In: Dodis, Y., Nielsen, J.B. (eds.) TCC 2015, Part II. LNCS, vol. 9015, pp. 528–556. Springer, Heidelberg (2015)
2. Barak, B., Garg, S., Kalai, Y.T., Paneth, O., Sahai, A.: Protecting obfuscation against algebraic attacks. In: Nguyen, P.Q., Oswald, E. (eds.) EUROCRYPT 2014. LNCS, vol. 8441, pp. 221–238. Springer, Heidelberg (2014)
3. Barak, B., Goldreich, O., Impagliazzo, R., Rudich, S., Sahai, A., Vadhan, S.P., Yang, K.: On the (im)possibility of obfuscating programs. In: Kilian, J. (ed.) CRYPTO 2001. LNCS, vol. 2139, pp. 1–18. Springer, Heidelberg (2001)
4. Beimel, A.: Secret-sharing schemes: a survey. In: Chee, Y.M., Guo, Z., Ling, S., Shao, F., Tang, Y., Wang, H., Xing, C. (eds.) IWCC 2011. LNCS, vol. 6639, pp. 11–46. Springer, Heidelberg (2011)
5. Beimel, A., Burmester, M., Desmedt, Y., Kushilevitz, E.: Computing functions of a shared secret. SIAM J. Discrete Math. **13**(3), 324–345 (2000)
6. Benaloh, J.C., Leichter, J.: Generalized secret sharing and monotone functions. In: Goldwasser, S. (ed.) CRYPTO 1988. LNCS, vol. 403, pp. 27–35. Springer, Heidelberg (1990)
7. Blakley, G.R.: Safeguarding cryptographic keys. Proc. AFIPS Natl. Comput. Conf. **48**, 313–317 (1979)
8. Boneh, D., Sahai, A., Waters, B.: Functional encryption: definitions and challenges. In: Ishai, Y. (ed.) TCC 2011. LNCS, vol. 6597, pp. 253–273. Springer, Heidelberg (2011)
9. Boneh, D., Silverberg, A.: Applications of multilinear forms to cryptography. IACR Cryptology ePrint Archive, p. 80 (2002)
10. Boneh, D., Waters, B.: Constrained pseudorandom functions and their applications. In: Sako, K., Sarkar, P. (eds.) ASIACRYPT 2013, Part II. LNCS, vol. 8270, pp. 280–300. Springer, Heidelberg (2013)
11. Boneh, D., Zhandry, M.: Multiparty key exchange, efficient traitor tracing, and more from indistinguishability obfuscation. In: Garay, J.A., Gennaro, R. (eds.) CRYPTO 2014, Part I. LNCS, vol. 8616, pp. 480–499. Springer, Heidelberg (2014)

12. Boyle, E., Gilboa, N., Ishai, Y.: Function secret sharing. In: Oswald, E., Fischlin, M. (eds.) EUROCRYPT 2015. LNCS, vol. 9057, pp. 337–367. Springer, Heidelberg (2015)

13. Brakerski, Z., Rothblum, G.N.: Virtual black-box obfuscation for all circuits via generic graded encoding. In: Lindell, Y. (ed.) TCC 2014. LNCS, vol. 8349, pp. 1–25. Springer, Heidelberg (2014)

14. Coron, J., Lepoint, T., Tibouchi, M.: New multilinear maps over the integers. In: Gennaro, R., Robshaw, M. (eds.) CRYPTO 2015. LNCS, vol. 9215, pp. 267–286. Springer, Heidelberg (2015)

15. Desmedt, Y.G., Frankel, Y.: Threshold cryptosystems. In: Brassard, G. (ed.) CRYPTO 1989. LNCS, vol. 435, pp. 307–315. Springer, Heidelberg (1990)

16. Garg, S., Gentry, C., Halevi, S.: Candidate multilinear maps from ideal lattices. In: Johansson, T., Nguyen, P.Q. (eds.) EUROCRYPT 2013. LNCS, vol. 7881, pp. 1–17. Springer, Heidelberg (2013)

17. Garg, S., Gentry, C., Halevi, S., Raykova, M., Sahai, A., Waters, B.: Candidate indistinguishability obfuscation and functional encryption for all circuits. In: 54th Annual IEEE Symposium on Foundations of Computer Science, FOCS, pp. 40–49 (2013)

18. Garg, S., Gentry, C., Sahai, A., Waters, B.: Witness encryption and its applications. In: Symposium on Theory of Computing Conference, STOC, pp. 467–476 (2013)

19. Gentry, C., Lewko, A.B., Sahai, A., Waters, B.: Indistinguishability obfuscation from the multilinear subgroup elimination assumption. IACR Cryptology ePrint Archive p. 309 (2014), to appear in FOCS 2015

20. Goldwasser, S., Gordon, S.D., Goyal, V., Jain, A., Katz, J., Liu, F.-H., Sahai, A., Shi, E., Zhou, H.-S.: Multi-input functional encryption. In: Nguyen, P.Q., Oswald, E. (eds.) EUROCRYPT 2014. LNCS, vol. 8441, pp. 578–602. Springer, Heidelberg (2014)

21. Grigni, M., Sipser, M.: Monotone complexity. In: Proceedings of LMS Workshop on Boolean Function Complexity. pp. 57–75 (1992)

22. Håstad, J., Impagliazzo, R., Levin, L.A., Luby, M.: A pseudorandom generator from any one-way function. SIAM J. Comput. 28(4), 1364–1396 (1999)

23. Karchmer, M., Wigderson, A.: On span programs. In: 8th Annual Structure in Complexity Theory Conference, pp. 102–111 (1993)

24. Komargodski, I., Moran, T., Naor, M., Pass, R., Rosen, A., Yogev, E.: One-way functions and (im)perfect obfuscation. In: 55th Annual IEEE Symposium on Foundations of Computer Science, FOCS, pp. 374–383 (2014)

25. Komargodski, I., Naor, M., Yogev, E.: Secret-sharing for NP. In: Sarkar, P., Iwata, T. (eds.) ASIACRYPT 2014, Part II. LNCS, vol. 8874, pp. 254–273. Springer, Heidelberg (2014)

26. Komargodski, I., Zhandry, M.: Cutting-edge cryptography through the lens of secret sharing. IACR Cryptology ePrint Archive p. 735 (2015)

27. Naor, M.: J. Cryptol. Bit commitment using pseudorandomness 4(2), 151–158 (1991)

28. Naor, M.: Secret sharing for access structures beyond P (2006), slides: http://www.wisdom.weizmann.ac.il/~naor/PAPERS/minicrypt.html

29. Pass, R., Seth, K., Telang, S.: Indistinguishability obfuscation from semantically-secure multilinear encodings. In: Garay, J.A., Gennaro, R. (eds.) CRYPTO 2014, Part I. LNCS, vol. 8616, pp. 500–517. Springer, Heidelberg (2014)

30. Sahai, A., Waters, B.: Slides on functional encryption (2008). http://www.cs.utexas.edu/~bwaters/presentations/files/functional.ppt

31. Sahai, A., Waters, B.: How to use indistinguishability obfuscation: deniable encryption, and more. In: Symposium on Theory of Computing, STOC, pp. 475–484 (2014)
32. Santis, A.D., Desmedt, Y., Frankel, Y., Yung, M.: How to share a function securely. In: Symposium on Theory of Computing, STOC, pp. 522–533 (1994)
33. Shamir, A.: How to share a secret. Commun. ACM **22**(11), 612–613 (1979)
34. Zhandry, M.: How to avoid obfuscation using witness PRFs. IACR Cryptology ePrint Archive p. 301 (2014). To appear. In: Kushilevitz, E., Malkin, T. (eds.) TCC 2016-A, Part II, LNCS 9563, pp. 421–448 (2016)

Functional Encryption Without Obfuscation

Sanjam Garg[1]([✉]), Craig Gentry[2], Shai Halevi[2], and Mark Zhandry[3]

[1] UC Berkeley, Berkeley, USA
sanjamg@berkeley.edu
[2] IBM Research, New York, USA
craigbgentry@gmail.com, shaih@alum.mit.edu
[3] MIT, Cambridge, USA
mzhandry@gmail.com

Abstract. Previously known functional encryption (FE) schemes for general circuits relied on indistinguishability obfuscation, which in turn either relies on an exponential number of assumptions (basically, one per circuit), or a polynomial set of assumptions, but with an exponential loss in the security reduction. Additionally most of these schemes are proved in the weaker selective security model, where the adversary is forced to specify its target before seeing the public parameters. For these constructions, full security can be obtained but at the cost of an exponential loss in the security reduction.

In this work, we overcome the above limitations and realize an adaptively secure functional encryption scheme without using indistinguishability obfuscation. Specifically the security of our scheme relies only on the polynomial hardness of simple assumptions on composite order multilinear maps. Though we do not currently have secure instantiations for these assumptions, we expect that multilinear maps supporting these assumptions will discovered in the future. Alternatively, follow up results may yield constructions which can be securely instantiated.

As a separate technical contribution of independent interest, we show how to add to existing graded encoding schemes a new *extension function*, that can be thought of as dynamically introducing new encoding levels.

1 Introduction

In traditional encryption schemes, decryption control is all or nothing: the sender encrypts its message under a particular key, and anyone with the corresponding secret key can recover the message. In contrast, functional encryption (FE) schemes [BSW11, O'N10] allow the sender to embed sophisticated functions into

---

S. Garg—Work supported in part from a DARPA/ARL SAFEWARE award, AFOSR Award FA9550-15-1-0274, and NSF CRII Award 1464397. The views expressed are those of the authors and do not reflect the official policy or position of the Department of Defense, the National Science Foundation, or the U.S. Government.
M. Zhandry—Work done while the author was a graduate student at Stanford University. Supported by the DARPA PROCEED program.

© International Association for Cryptologic Research 2016
E. Kushilevitz and T. Malkin (Eds.): TCC 2016-A, Part II, LNCS 9563, pp. 480–511, 2016.
DOI: 10.1007/978-3-662-49099-0_18

secret keys. More specifically, an FE scheme includes an authority, which holds a master secret key and publishes public system parameters. The sender uses the public parameters to encrypt its message m to obtain a ciphertext ct. A user may obtain a secret key sk_f for the function f from the authority (if the authority deems that the user is entitled). This key sk_f can be used to decrypt ct to recover $f(m)$; and nothing more. In a recent result, Garg et al. constructed the first FE scheme for general circuits using indistinguishability obfuscation ($i\mathcal{O}$) [GGH+13b].

While tremendous progress has been made on justifying the security of $i\mathcal{O}$ [BR14, BGK+14, PST14, GLW14, GLSW14], ultimately the security of the resulting constructions still either relies on an exponential number of assumptions [BR14, BGK+14, PST14] (basically, one per circuit), or a polynomial set of assumptions, but with an exponential loss in the security reduction [GLW14, GLSW14]. For example, the recent $i\mathcal{O}$ scheme based on the MSE assumption [GLSW14] crucially uses complexity leveraging in its proof — specifically, the number of hybrids in the proof is proportional to $2^{|x|}$ where x is the input, and each hybrid "examines" a particular input x and implicitly "verifies" that the circuits C_0, C_1 in question satisfy $C_0(x) = C_1(x)$. Garg et al. [GGSW13] provide an intuitive argument suggesting that either of these shortcoming might be inherent when realizing indistinguishability obfuscation,[1] though this argument is not applicable to FE schemes. In this work we ask the following fundamental question:

Can we construct a functional encryption scheme for general circuits assuming only polynomial hardness of simple computational assumptions?

Another limitation of the Garg et al. [GGH+13b] scheme is that it is only *selectively secure* – that is, they have been proved secure only in a weaker model in which the adversary is required to specify the message m for its challenge ciphertext before it sees the public parameters of the FE scheme. We would like FE for circuits that is *fully secure* — i.e., that allows the adversary to choose m^* adaptively after seeing the public parameters and even responses to some of its private key queries. In general, one can trivially reduce full security to selectively security via *complexity leveraging* – essentially the reduction tries to guess the adversary's chosen m, and succeeds with probability $2^{-|m|}$ – but complexity leveraging loses a $2^{|m|}$ factor in the reduction to the underlying hard problem that we would like to avoid.

Can we construct a fully secure functional encryption scheme for general circuits without an exponential loss in the security reduction?

Achieving full security without the lossiness of complexity leveraging is just as important for FE for circuits as it was for identity-based encryption (IBE) ten years ago [Wat05, Gen06, Wat09], for both efficiency and conceptual reasons.

---

[1] Garg et al. [GGSW13] only provide the intuition for witness encryption but it extends to $i\mathcal{O}$.

1.1   Our Results

In this work, we give positive answers to both questions above. Specifically we construct the first fully secure FE scheme for circuits without using indistinguishability obfuscation or any exponential loss in security reductions. Our scheme uses composite order multilinear maps in the asymmetric settings [BS02, GGH13a, CLT13, CLT15a] and security is based on polynomial hardness of fixed, relatively simple assumptions on a variant of the new CLT [CLT15a] maps.

We extend the existing graded encoding schemes [GGH13a, CLT13, CLT15a] with a new *extension* function that serves as a crucial ingredient in our construction. This extension function serves a role similar to that of the straddling set systems of [BGK+14], binding various encodings so that only certain subsets can be paired together. The important difference is that the extension function allows the binding to happen dynamically and publicly. This allows, for example, an encrypter to bind ciphertext encodings together so that encodings from different ciphertexts cannot be "mixed and matched." We believe that this new technique will be useful in other contexts as well. We provide details on this in the full version [GGHZ14b].

Theorem 1 (informal). *Assuming (1) simple polynomial assumptions on extendable composite order graded encodings and (2) the existence of PRFs that are both puncturable (in the sense of [BW13, BGI14, KPTZ13]) and can be evaluated in NC^1, then fully secure functional encryption for all polynomial-sized circuits exists.*

An immediate consequence of our scheme is a traitor tracing scheme where ciphertexts, secret keys, and public keys are short, namely logarithmic in the number of users. Previous such schemes [GGH+13b, BZ14] all relied on $i\mathcal{O}$. Our scheme is therefore the first traitor tracing scheme with small parameters whose security does not rely on $i\mathcal{O}$ or an exponential loss in the security reductions.

As an important intermediate step in our construction, we introduce the notion of *slotted* functional encryption, which allows for multiple independent execution paths, or slots, in functional encryption. We believe slotted FE may be of independent interest; in particular, several recent works [BS15, ABSV14] implicitly construct variations of slotted FE as an intermediate step.

1.2   Overview of Our Techniques

In this section we describe the high-level ideas behind our construction. We start by providing general intuition on how we avoid obfuscation. Subsequently, we will elaborate on our methodology and the intermediate abstraction of slotted FE that we use.

Though the final aim of this work is to avoid the use of obfuscation in realizing functional encryption, we build upon techniques that have previously been used to realize indistinguishability obfuscation. We start by recalling some of these tools. An indistinguishability obfuscator $i\mathcal{O}$ guarantees that given two functionally equivalent circuits C_1 and C_2, i.e. for every input x we require that $C_1(x) = C_2(x)$, the two distributions of obfuscations $i\mathcal{O}(C_1)$ and $i\mathcal{O}(C_2)$ are computationally indistinguishable. Known constructions of obfuscation build on the information theoretic argument of Kilian [Kil88] which provides security only when evaluation on a single input is allowed. In more detail, consider a circuit C that takes n bits as input. Kilian provides a mechanism for garbling C into garbled components $\{\tilde{C}_{i,b}\}_{i\in[n],b\in\{0,1\}}$, such that access to the components $\{\tilde{C}_{i,x_i}\}_{i\in[n]}$ allow computation of $C(x)$ while simultaneously preserving perfect secrecy of the circuit C. Note that here for each $i \in [n]$ only one of the two values $\tilde{C}_{i,0}$ and $\tilde{C}_{i,1}$ is disclosed. This is similar to Yao's [Yao82] garbled circuits construction except that Kilian's construction is limited to log depth circuits but achieves a stronger information theoretic security. However, obfuscation schemes need to enable secure evaluation on potentially any input and not just on one pre-specified input. All known constructions of obfuscation achieve this additional functionality as follows: the obfuscation of a circuit C consists of the terms $\{\hat{C}_{i,b}\}_{i\in[n],b\in\{0,1\}}$ where all these values are simultaneous disclosed. Just like Kilian, terms $\{\hat{C}_{i,x_i}\}_{i\in[n]}$ allow for evaluation of $C(x)$. This new garbling method, denoted by notation $\hat{C}$, has the additional property that it hides the circuit C in the sense of indistinguishability obfuscation.

Intuition behind previous constructions of Functional Encryption. Typical obfuscation based functional encryption schemes are constructed as follows. The setup procedure of the functional encryption scheme generates a public-secret key pair (pk, sk) of a public key encryption scheme and sets the public parameters for the functional encryption scheme to be pk. A message m is encrypted under the functional encryption scheme by just encrypting it to pk. Finally a private key for a function f is set to be the obfuscation of a circuit that outputs the evaluation of the function f on the message obtained by decrypting the ciphertext provided to it as input. The secret key sk is embedded inside this circuit for enabling decryption.

Our Starting Idea. Our starting idea in trying to avoid the use of obfuscation in realizing functional encryption is that even though a private key (which is an obfuscation) should work for arbitrary ciphertexts, the security requirement is much weaker — specifically, security is required only for the challenge ciphertext. We build on this observation; isolating the specific input for which security is desired and using the Kilian's information theoretic argument just for this input. Doing this isolation and enabling the Kilian's information theoretic argument is technically quiet challenging and requires us to build new techniques. We elaborate on this next.

As described earlier obfuscation of a circuit C consists of $\{\hat{C}_{i,b}\}_{i\in[n],b\in\{0,1\}}$ and knowledge of $\{\hat{C}_{i,x_i}\}_{i\in[n]}$ allow for evaluation of $C(x)$. The starting point for our new functional encryption scheme is to split these components of garbled C being generated as part of the obfuscation between the ciphertext and the private key. In other words the ciphertext and secret key provide parts of the obfuscation, that when put together allow for computation.

We interpret the input x to consist of two parts m and f and the circuit C to be universal circuit that evaluates and outputs $f(m)$. Here m is the message being encrypted and the encrypter is expected to provide the components corresponding to these parts. The components for the private key are provided by the trusted authority. More concretely, denoting $I_m = \{0, 1, \ldots, |m| - 1\}$ and $I_f = \{m, m+1, \ldots, |m| + |f| - 1\}$, the public key consists of $\{\hat{C}_{i,b}\}_{i \in I_m, b \in \{0,1\}}$. In order to encrypt a message m the encrypter chooses the components $\{\hat{C}_{i,m_i}\}_{i \in I_m}$ and further randomizes and *bundles* them (using an extension function that is explained later) to obtain the ciphertext $\{\overline{C}_{i,m_i}\}_{i \in I_m}$. The trusted authority generates the private keys analogously by randomizing and bundling together appropriate components, namely $\{\hat{C}_{i,f_i}\}_{i \in I_f}$ and obtaining $\{\overline{C}_{i,f_i}\}_{i \in I_f}$ as the secret key. Additional private keys can be generated in an analogous manner. Note that $\{\overline{C}_{i,m_i}\}_{i \in I_m}$ and $\{\overline{C}_{i,f_i}\}_{i \in I_f}$ together form a whole program that is executable on one input alone, bringing us closer to Kilian for arguing security.

Making this idea work involves a careful hybrid argument, isolating one secret key and a ciphertext at a time in order to apply Kilian's information theoretic argument. We specifically achieve this via a primitive that we call *slotted FE*:

Slotted FE. In a slotted FE scheme, ciphertexts and secret keys contain multiple slots, and each slot i can either be "active" (i.e., contain an actual message or function) or "inactive" (empty). Decryption is defined by taking all slots that are active in both the ciphertext and secret key, and computing $f_i(m_i)$ for those slots. If all slots agree on the result, that result is the output of decryption. If the slots do not agree, the output is unspecified. Ciphertexts and secret keys are generated by the following procedures:

- **Slotted encryption** is a procedure requiring the master secret, and it can produce an arbitrary ciphertext, containing any number of active slots with any messages in those slots.
- **Unslotted encryption** is a public procedure that can produce a ciphertext where a special slot 0 contains an arbitrary message, and the rest of the slots are inactive.
- **Slotted key generation** is a procedure requiring the master secret, and it can produce an arbitrary secret key containing any number of active slots with any functions in those slots.
- **Unslotted key generation** is a convenient shorthand for the special case of slotted key generation, producing a secret key with active slot 0 and the rest of the slots inactive.

Clearly, slotted FE is a strict generalization of standard FE, we can recover the standard notion by only using slot 0 and the unslotted procedures. However the new primitive lets us consider more refined security properties. Specifically, we define a small set of "local security properties" that can be mapped to simple assumptions on the underlying graded-encoding scheme, and prove that they imply our desired security notion for the induced FE scheme. Importantly, these properties should be strong enough to yield adaptive security, but not too strong so as to imply function-hiding (and thus obfuscation). This is somewhat similar on a high level to the approach from [GLW14, GLSW14] (e.g., the notion of "tribes schemes"), but the technical details are very different.

Our security properties for slotted FE are defined in Sects. 4.1 and 4.2. They all follow the standard indistinguishability game between the FE adversary and a challenger, but limit the types of queries that the adversary can use. For example, one such notion requires indistinguishability only when each key-pair-query that the adversary makes contains two identical sets of slots, the two challenge plaintexts only differ in a single pair of slots in which one plaintext has $(x^*, \bot)$ and the other has $(\bot, x^*)$, and moreover all the secret-key queries have the same function between these two slots. (We call this property "Ciphertext moving," see Sect. 4.1.)

Another advantage of using slotted FE is that it allows us to "bootstrap" the construction from NC^1 to all circuits. Our basic slotted FE scheme in Sect. 5 can only handle log-depth circuits (NC^1), and unfortunately it was previously unknown how to securely boost FE for NC^1 into FE for all circuits in a black-box way without requiring function hiding (and thus obfuscation)[2]. However, we show that the "local properties" of our slotted FE can be used for this "bootstrapping" transformation. In this sense, slotted FE seems to be "the right level of abstraction" for this construction.

Our Slotted FE for NC^1. Our slotted FE for NC^1 is related to current constructions of $i\mathcal{O}$ for NC^1 [GGH+13b, BR14, BGK+14, PST14, GLSW14]. Roughly, we choose a universal NC^1 circuit $U(f, m) = f(m)$, and convert U into a branching program BP. We then randomize BP using Kilian randomization, and place the resulting matrices "in the exponent" of an asymmetric graded encoding roughly as follows:

- In order to implement slots, we use a composite-order graded encoding, where each slot corresponds to a subgroup.
- The setup procedure generates the public parameters by taking the matrices corresponding to the m input, projecting them down into the first subgroup (corresponding to slot 0), and publishing encodings of these matrices in the appropriate levels.
- The key generation procedure takes as input a vector $(f_0, \ldots, f_{n-1})$, where some of the $f_i = \bot$. For all $f_i \neq \bot$, it selects the matrices corresponding to f_i, and projects them down to the ith subgroup, and encodes these matrices in the appropriate levels. Then it adds the encodings for different f_i together,

---

[2] We note that Gorbunov et al. [GVW12] show a general transformation from NC^1 to poly-size circuits, but the security proof relies on the underlying FE scheme being simulation secure. Such security is impossible in the setting where the number of secret key queries in unbounded [AGVW13], which is the setting studied in this work. Subsequent to our work, Ananth et al. [ABSV14] show that FE for NC^1 *can* be boosted to FE for all circuit.

and outputs the resulting encodings. By the Chinese Remainder Theorem, the
ith subgroup of the resulting encoding will contain the matrices for function
f_i. The result is that the secret key encodes function f_i in slot i.

- The slotted encryption procedure is analogous to the slotted key generation
  procedure, except that it operates on the matrices corresponding to the message input.
- The unslotted encryption procedure on input m takes the public parameters,
  selects the matrices corresponding to m, and re-randomizes and outputs those
  matrices.
- Finally, the decryption procedure multiplies the matrices for a secret key and
  ciphertext together, and then performs a zero test on one entry of the resulting
  matrix. Each of the subgroups act independently, and the result of multiplication will be a matrix where subgroup i contains the matrix corresponding
  to $f_i(m_i)$ (or the subgroup is empty if either ciphertext or secret key are inactive). If all of the $f_i(m_i) = 0$, the zero test gives 0. If all of the $f_i(m_i) = 1$,
  then the zero test gives 1.

Using subgroup-decision assumptions on multilinear graded encodings, we are
able to prove various security properties for our scheme, such as the "ciphertext
moving" property mentioned above. These properties allow us to move messages
and secret keys between slots. However, for the application to (un-slotted) functional encryption, we actually want the ability to change the values of messages.
To accomplish this, we first use the existing properties to isolate the ciphertext and one secret key in their own slot. At this point, we can invoke Kilian's
information-theoretic argument in the corresponding subgroup, since the matrices given out all correspond to a single input. We prove a new property called
"single-use hiding" which allows us to arbitrarily change the ciphertext and
secret key in this slot, provided decryption is unaffected. By carefully repeating
this process for each secret key, we are ultimately able to change the message
encrypted, thus proving the security of the derived un-slotted functional encryption scheme.

Extending graded encodings. A major issue with the above sketch is that matrices
from different ciphertexts can be "mixed and matched" (in particular, a target
matrix can be mixed with a ciphertext generated from the public parameters)
which may allow the adversary to learn more than he should. Different secret
keys can be mixed and matched as well. Similar problems arose in the obfuscation setting, and one way it was solved was by using so-called straddling set
systems [BGK+14].

In our setting, this would involve assigning a different set of levels to each
ciphertext, and requiring that the levels assigned to two different ciphertext are
incompatible. However, ciphertext generation is a public procedure, meaning the
public parameters must include enough information to encrypt into any possible
level that a ciphertext component will be in. But then the adversary can always
generate a ciphertext in levels matching the target ciphertext, which then allows
mixing the ciphertexts together. Roughly, the problem is that access control to

levels is all or nothing: either anyone can generate encodings in a level, or no one except the master party can.

We solve this problem by developing a new extension procedure on graded encodings, which lets any user extend the graded encoding by generating new levels. The user that ran the extension procedure will have to ability to map components from existing levels to the new level, but other users will not. If we apply the procedure to ciphertext components, the components will effectively be bound together in the new extended levels, since the adversary cannot move other ciphertexts into these levels.

In order to allow decryption, the new levels need to be mapped back to the original set of levels. However, the extension procedure publishes just enough information to map back to the original levels only after all the ciphertext components have been combined. Once the ciphertext components are all combined, it is impossible to mix the ciphertext with another ciphertext.

While the extension procedure falls outside of the traditional graded encoding abstraction, we point out that most graded encoding candidates [GGH13a, CLT13, CLT15b] support this procedure. We provide details in the full version [GGHZ14b].

Using our new notion of extendable graded encodings, we prove the following:

Lemma 1 (informal). *Assuming simple polynomial assumptions on extendable graded encodings, then fully secure slotted functional encryption exists for NC^1 circuits.*

Boosting to FE for all circuits. In order to boost to functional encryption for all circuits, we proceed in two steps.

- We first build functional encryption for NC^1 randomized functionalities from our slotted functional encryption scheme. This is accomplished by including a secret key k for a PRF in the ciphertext, and generating the randomness for the functionality by applying the PRF to a seed s contained in the secret key. In order to prove security, we will need to puncture the key k at s, so we need puncturable PRFs that can be evaluated in NC^1 [BLMR13][3]. The conversion is very similar to the bootstapping technique of Gorbunov et al. [GVW12], but we need the slotted property of our FE scheme in order to prove security in our setting.
- Next, we boost to FE for all circuits. Basically, a secret key for a function f will output not $f(m)$, but instead a randomized encoding [IK00] $\hat{f}(m)$, from which $f(m)$ can be computed, but m itself is hidden. Notably, $\hat{f}(m)$ can be computed in log-depth, so our randomized functional encryption for NC^1 suffices.

---

[3] This observation that [BLMR13] is puncturable appears in the full version of the paper: http://theory.stanford.edu/~klewi/papers/homprf-full.pdf. It is also folklore that the Naor-Reingold PRF is puncturable while maintaining NC^1 evaluation.

Lemma 2 (informal). *Assuming fully secure slotted functional encryption for NC^1 and PRFs that are both puncturable and can be evaluated in NC^1, then fully secure functional encryption for all polynomial-sized circuits exists.*

1.3   Instantiating Our Assumptions

Unfortunately, several recent attacks on multilinear maps [CHL+15, BWZ14, CGH+15] have broken many assumptions on known multilinear maps; the assumptions broken include our own, as well as all simple assumptions that have been used to build obfuscation. Nonetheless, constructing functional encryption from simple assumptions, without obfuscation, and without complexity leveraging remains an important problem. Fortunately, our assumptions are generic in the sense that they can be instantiated on any expressive-enough multilinear maps. It seems plausible that candidates satisfying these assumptions will be found in the future, either by modifying current candidates or by completely different means. Our work shows that any multilinear map supporting our assumptions and functionality requirements yields secure functional encryption, thereby motivating the search for and study of such maps.

1.4   Independent Work

In a very recent independent work, Waters [Wat14] constructs a fully secure functional encryption (FE) scheme using indistinguishability obfuscation ($i\mathcal{O}$) [GGH+13b] and one-way functions. Water's result has the advantage of being generic: any indistinguishability obfuscator or one-way function will suffice for his construction, whereas we require multilinear maps with specific properties. However, the focus of this work is to avoid indistinguishability obfuscation altogether and to build fully secure functional encryption using simpler, though less generic tools (multilinear maps and simple assumptions involving them).

One may try to combine Waters [Wat14] fully secure FE scheme with the indistinguishability obfuscator of Gentry et al. [GLSW14], whose security is based on simple assumptions on multilinear maps. The result would be a fully secure functional encryption scheme whose security is based on simple assumptions on multilinear maps. However, the reduction in [GLSW14] involves an exponential loss of security, meaning complexity leveraging is required and the assumptions on multilinear maps must be assumed secure against subexponential time adversaries. In this setting, static security and full adaptive security are equivalent, and so a fully secure scheme can be obtained by combining [GLSW14] with any selectively secure FE scheme, such as the original scheme of Garg et al. [GGH+13b].

In contrast, all reductions for our scheme are *polynomial*, meaning we only require polynomial hardness of the underlying multilinear map assumptions. Ours is the first scheme to obtain security in this setting, even among selectively secure schemes.

1.5   Subsequent Work

Subsequent to our work, there have been several developments regarding functional encryption. First, a few works [BV15, AJ15] show how to build obfuscation from sub-exponentially secure functional encryption, thus showing that in some sense obfuscation and functional encryption are equivalent. However, these results require complexity leveraging, and therefore only apply in the setting of sub-exponential hardness assumptions and exponential reductions. They do not apply to the polynomial security setting, which is the focus of this work. Moreover, their results require *compact* FE. Our construction is not compact, and it is currently still unknown how to obtain compact functional encryption without using obfuscation.

Second, Ananth et al. [ABSV14] show how to both obtain adaptive security from selective security for functional encryption, and also "bootstrap" functional encryption for NC^1 to functional encryption to all circuits. Their conversions need only regular functional encryption, whereas our bootstrapping requires the seemingly stronger notion of slotted functional encryption. While their techniques are quite different than ours, at a high level their proof can be seen as (1) implicitly showing how to add slots to regular (unslotted) functional encryption, and then (2) using slotted functional encryption for bootstrapping. This shows that our notion of slotted functional encryption serves as a useful abstraction in the context of functional encryption.

2   Preliminaries: Graded Encoding Schemes

In Sect. 3, we recall the basic definitions of functional encryption and branching programs. Here we describe the graded encoding scheme abstraction that will be needed in our context, mostly following [GGH13a, CLT13, GLW14]. To instantiate the abstraction, we can use Gentry et al.'s variant [GLW14] of the Coron-Lepoint-Tibouchi (CLT) graded encodings [CLT13]. This variant is designed to emulate multilinear groups of composite order, and to allow assumptions regarding subgroups of the multilinear groups. One key difference in our abstraction is a new *extension* function that we add to the GGH graded encoding abstraction. This new functionality will be crucial in our scheme. In the full version [GGHZ14b], we briefly recall the CLT graded encodings and show how they can be adapted to also support this extension functionality.[4]

Definition 1 (U-Graded Encoding System). *A U-Graded Encoding System consists of a ring $\mathfrak{R}$ and a system of sets $\mathcal{S} = \{S_T^{(\alpha)} \subset \{0,1\}^* : \alpha \in \mathfrak{R}, \ T \subseteq \mathbf{U}\}$, with the following properties:*

1. *For every fixed set T, the sets $\{S_T^{(\alpha)} : \alpha \in \mathfrak{R}\}$ are disjoint (hence they form a partition of $S_T \overset{\text{def}}{=} \bigcup_\alpha S_T^{(\alpha)}$).*

---

[4] We note that the GGH encodings can also be extended to deal with this functionality as well but here we provide this only for the CLT encodings.

2. *There is an associative binary operation '+' and a self-inverse unary operation '−' (on $\{0,1\}^*$) such that for every $\alpha_1, \alpha_2 \in \mathfrak{R}$, every set $T \subseteq \mathbb{U}$, and every $u_1 \in S_T^{(\alpha_1)}$ and $u_2 \in S_T^{(\alpha_2)}$, it holds that $u_1 + u_2 \in S_T^{(\alpha_1+\alpha_2)}$ and $-u_1 \in S_T^{(-\alpha_1)}$ where $\alpha_1 + \alpha_2$ and $-\alpha_1$ are addition and negation in $\mathfrak{R}$.*

3. *There is an associative binary operation '×' (on $\{0,1\}^*$) such that for every $\alpha_1, \alpha_2 \in \mathfrak{R}$, every T_1, T_2 with $T_1 \cup T_2 \subseteq \mathbb{U}$, and every $u_1 \in S_{T_1}^{(\alpha_1)}$ and $u_2 \in S_{T_2}^{(\alpha_2)}$, it holds that $u_1 \times u_2 \in S_{T_1 \cup T_2}^{(\alpha_1 \cdot \alpha_2)}$. Here $\alpha_1 \cdot \alpha_2$ is multiplication in $\mathfrak{R}$, and $T_1 \cup T_2$ is set union.*

CLT (and GGH) encodings do not quite meet the definition of graded encoding systems above, since the homomorphisms required in the definition eventually fail when the "noise" in the encodings becomes too large, analogously to how the homomorphisms may eventually fail in lattice-based homomorphic encryption. However, these noise issues are relatively straightforward (though tedious) to deal with.

Now, we define some procedures for graded encoding schemes. We start with the procedures standard in the graded encoding literature [GGH13a, CLT13].

Instance Generation. The randomized $\mathsf{InstGen}(1^\lambda, \mathbb{U}, r)$ takes as inputs the parameters $\lambda, \mathbb{U}, r$, and outputs params, where params is a description of a $\mathbb{U}$-Graded Encoding System as above for a ring $\mathfrak{R} = \mathfrak{R}_1 \times \ldots \times \mathfrak{R}_r$. We assume $\mathfrak{R}$ is chosen such that the density of zero divisors in each $\mathfrak{R}_i$ is negligible. Note that setting $r = 1$ corresponds to the prime order setting, while $r > 1$ corresponds to the composite-order setting.

Ring Sampler. The randomized $\mathsf{samp}(\text{params})$ outputs a "level-zero encoding" $a \in S_\phi^{(\alpha)}$ for a nearly uniform element $\alpha \in_R \mathfrak{R}$. (Note that we require that the "plaintext" $\alpha \in \mathfrak{R}$ is nearly uniform, but not that the encoding a is uniform in $S_\phi^{(\alpha)}$.)

Encoding. The (possibly randomized) $\mathsf{enc}(\text{params}, T, a)$ takes a "level-zero" encoding $a \in S_\phi^{(\alpha)}$ for some $\alpha \in \mathfrak{R}$ and index $T \subseteq \mathbb{U}$, and outputs the "level-T" encoding $u \in S_T^{(\alpha)}$ for the same α.

Re-Randomization. The randomized $\mathsf{reRand}(\text{params}, T, u)$ re-randomizes encodings relative to the same index. Specifically, for an index $T \subseteq \mathbb{U}$ and encoding $u \in S_T^{(\alpha)}$, it outputs another encoding $u' \in S_T^{(\alpha)}$. Moreover for any two $u_1, u_2 \in S_T^{(\alpha)}$, the output distributions of $\mathsf{reRand}(\text{params}, T, u_1)$ and $\mathsf{reRand}(\text{params}, T, u_2)$ are statistically indistinguishable.

Addition and negation. Given params and two encodings relative to the same index, $u_1 \in S_T^{(\alpha_1)}$ and $u_2 \in S_T^{(\alpha_2)}$, we have an addition function $\mathsf{add}(\text{params}, T, u_1, u_2) = u_1 + u_2 \in S_T^{(\alpha_1+\alpha_2)}$, and a negation function $\mathsf{neg}(\text{params}, T, u_1) = -u_1 \in S_T^{(-\alpha_1)}$.

Multiplication. For $u_1 \in S_{T_1}^{(\alpha_1)}$, $u_2 \in S_{T_2}^{(\alpha_2)}$ such that $T_1 \cup T_2 \subseteq \mathbb{U}$ and $T_1 \cap T_2 = \emptyset$, we have a multiplication function $\mathsf{mul}(\text{params}, T_1, u_1, T_2, u_2) = u_1 \times u_2 \in S_{T_1 \cup T_2}^{(\alpha_1 \cdot \alpha_2)}$.

Zero-test. The procedure isZero(params, u) outputs 1 if $u \in S_{\mathbb{U}}^{(0)}$ and 0 otherwise. Note that in conjunction with the subtraction procedure, this lets us test if $u_1, u_2 \in S_{\mathbb{U}}$ encode the same element $\alpha \in \mathfrak{R}$.

Next, we define two new *extension* procedures on graded encodings that we will use. Informally, these procedures allow the creation of new levels, using only the public parameters of the graded encoding. In particular, they take as input a subset of levels $\mathbb{V}$ of the universe $\mathbb{U}$, and create a new "clone" $\mathbb{V}'$ of the levels in $\mathbb{V}$ that is disjoint from $\mathbb{U}$. Since the levels lie outside $\mathbb{U}$, they cannot be zero-tested. Instead, the procedures output a function $f_{\mathbb{V}' \to \mathbb{V}}$ which maps the level $\mathbb{V}'$ back to $\mathbb{V}$, but does not allow mapping levels corresponding to any subsets of $\mathbb{V}'$. Thus, the entire set $\mathbb{V}'$ must be "filled out" before zero testing can happen. In particular, it is impossible to multiply an element encoded at a subset of $\mathbb{V}'$ with an element encoded at a subset of $\mathbb{V}$ and still be able to perform zero-testing. In effect, this binds the encodings in $\mathbb{V}'$ together, similar to how straddling sets [BGK+14] where used in obfuscation.

Extension. This procedure allows extending the graded encoding system by fresh asymmetric levels. Specifically, extend(params, $\mathbb{V}, \{e_i\}_i$) takes as input a set $\mathbb{V} \subseteq \mathbb{U}$ and a sequence of encodings e_i each at level $v_i \subseteq \mathbb{V}$ and outputs a new set $\mathbb{V}'$ where $\mathbb{V}' \cap \mathbb{U} = \emptyset$ and encodings e'_i each at level $v'_i \subseteq \mathbb{V}'$ along with a public transformation function $f_{\mathbb{V}' \to \mathbb{V}}$ such that:-
- Addition and multiplication procedures from above can be applied to encodings at these new levels as well. Thus, given $u_1 \in S_T^{(\alpha_1)}$ and $u_2 \in S_T^{(\alpha_2)}$ where $T \subseteq (\mathbb{U} \setminus \mathbb{V}) \cup \mathbb{V}'$, we have add(params, T, u_1, u_2) $= u_1 + u_2 \in S_T^{(\alpha_1 + \alpha_2)}$, and neg(params, T, u_1) $= -u_1 \in S_T^{(-\alpha_1)}$. Similarly, given $u_1 \in S_{T_1}^{(\alpha_1)}$ and $u_2 \in S_{T_2}^{(\alpha_2)}$ such that $T_1 \cup T_2 \subseteq (\mathbb{U} \setminus \mathbb{V}) \cup \mathbb{V}'$ and $T_1 \cap T_2 = \emptyset$, we have a multiplication function mul(params, T_1, u_1, T_2, u_2) $= u_1 \times u_2 \in S_{T_1 \cup T_2}^{(\alpha_1 \cdot \alpha_2)}$. Notice that we do not need to support adding or multiplying elements if the final level is some $\mathbb{W}$ such that both $\mathbb{W} \cap \mathbb{V} \neq \emptyset$ and $\mathbb{W} \cap \mathbb{V}' \neq \emptyset$.
- The new levels v'_i are obtained by mapping the old levels v_i into the clone $\mathbb{V}'$. Specifically, let $\mathbb{V} = \{j_1, \ldots j_t\}$ and $\mathbb{V}' = \{j'_1, \ldots j'_t\}$. For each i we have that if $v_i = \{j_{k_1}, \ldots j_{k_\ell}\}$ then $v'_i = \{j'_{k_1}, \ldots j'_{k_\ell}\}$
- $f_{\mathbb{V}' \to \mathbb{V}}(e', \mathbb{W}')$ takes as input a set $\mathbb{W}'$ such that $\mathbb{V}' \subseteq \mathbb{W}' \subseteq (\mathbb{U} \setminus \mathbb{V}) \cup \mathbb{V}'$ and an element $e' \in S_{\mathbb{W}'}^{(\alpha)}$. It outputs an encoding $e \in S_{\mathbb{V} \cup (\mathbb{W}' \setminus \mathbb{V}')}^{(\alpha)}$ obtained by mapping each element in $\mathbb{V}'$ back to $\mathbb{V}$. Specifically, if $\mathbb{W}' = \mathbb{X} \cup \{j'_{k_1}, \ldots j'_{k_\ell}\}$ where $j'_k \in \mathbb{V}'$ as above and $\mathbb{X} \subseteq \mathbb{U} \setminus \mathbb{V}$, then the output will be an element e encoded relative to set $\mathbb{W} = \mathbb{X} \cup \{j_{k_1}, \ldots, j_{k_\ell}\} \subseteq \mathbb{U}$, which will be in the original universe $\mathbb{U}$.

Extension†. This function extend† is the same as the previous function extend(params, $\mathbb{V}, \{e_i\}_i$) except that it also outputs additionally randomizers (encodings of 0) for each level it outputs an encoding at.

In the full version [GGHZ14b], we demonstrate how to obtain the above extension procedures from the new CLT encodings. We stress that, except for

the new extension procedures, all the procedures above are exactly the same
as an optimized variant in [CLT15b]. The extension functions are built *on top*
of the underlying graded encoding without any modifications to the existing
procedures — in particular, no extra terms are needed in the public parameters.
The extension functions can also be applied to any multilinear map that has a
similar form to the GGH or CLT maps. For that reason, while the complexity
assumptions we will be making currently do not hold on any multilinear map
candidate, it is very likely that future maps which may support our assumptions
will also support this extension procedure.

In order to simplify notation, we will denote encodings as $[\alpha]_T^i$ where T
denotes the level of the encoding, and i denotes that only the $\mathfrak{R}_i$ component of
α is preserved and the $\mathfrak{R}_j$ components for $j \neq i$ are zeroed out. Similarly, we
use $[\alpha]_T^{i_1,i_2,i_3}$ to denote that the $\mathfrak{R}_{i_1} \times \mathfrak{R}_{i_2} \times \mathfrak{R}_{i_3}$ component is preserved and all
other components are zeroed out. This notation is due to [GGHZ14a].

Our complexity assumptions. We now describe the complexity assumptions we
will be making in this work. Fix a universe $\mathbb{U}$, a dimension d, and a partition of
$\mathbb{U}$ into subsets $\mathbb{V}, \mathbb{W}$. For the assumptions below we will assume that randomizers
(encodings of zero) are provided for each index in $\mathbb{U}$.

For our first assumption, the adversary is given elements in every level and in
every subring except subring $\mathfrak{R}_0$. The adversary is additionally given challenge
elements in every level that either lie in the subring $\mathfrak{R}_1$, or lie in the subring
$\mathfrak{R}_0 \times \mathfrak{R}_1$, and is asked to distinguish the two cases. Using only multilinear operations, distinguishing those cases is impossible: pairing either challenge element
with anything in $\mathfrak{R}_1$ results in an element in $\mathfrak{R}_1$, while pairing either with anything in $\mathfrak{R}_i$ for $i > 1$ results in 0. Thus, only pairing with an element in $\mathfrak{R}_0$ will
allow for distinguishing the two cases, and such elements are not given to the
adversary.

Definition 2 (Assumption 1). *The following distributions are indistinguishable:*

$$\left(\left([s_{i,j}]_{\{i\}}^j\right)_{i\in\mathbb{U},j>0}, \left([t_i]_{\{i\}}^1\right)_{i\in\mathbb{U}} \right) \text{ and } \left(\left([s_{i,j}]_{\{i\}}^j\right)_{i\in\mathbb{U},j>0}, \left([t_i]_{\{i\}}^{0,1}\right)_{i\in\mathbb{U}} \right)$$

In our second assumption, the universe $\mathbb{U}$ is split into two disjoint sets: $\mathbb{V}$ and
$\mathbb{W}$. For levels in $\mathbb{V}$, the adversary is given elements encoded in each $\mathfrak{R}_i$ for $i > 1$,
as well as elements in $\mathfrak{R}_0 \times \mathfrak{R}_1$. No elements are provided in $\mathbb{V}$ that are encoded in
$\mathfrak{R}_0$ but not $\mathfrak{R}_1$, or vice versa. For levels in $\mathbb{W}$, the adversary is given elements in
all of the subrings. Additionally, a clone set of levels $\mathbb{W}'$ is created disjoint from
$\mathbb{U}$ using the extension function. The adversary is given the function $f_{\mathbb{W}'\to\mathbb{W}}$, also
outputted by the extension procedure, which allows him to translate elements
from the entire $\mathbb{W}'$ into $\mathbb{W}$. For each level in $\mathbb{W}'$, the adversary is given encodings
in $\mathfrak{R}_i$ for $i > 1$, as well as challenge encodings that are either all in $\mathfrak{R}_0$ or all in
$\mathfrak{R}_1$. The adversary is then asked to distinguish the two cases. To distinguish the
two cases, the adversary has to first "fill up" the set $\mathbb{W}'$ so that it can be mapped
back into the universe $\mathbb{U}$. If he pairs a challenge element with any non-challenge

element in $\mathbb{W}'$, the result will always be an encoding of zero since the challenge elements and non-challenge elements in $\mathbb{W}'$ lie in different subrings. Therefore, his only choice it to pair all of the challenge elements together and map back to $\mathbb{U}$, obtaining an element at level $\mathbb{W}$ encoded in either subring $\mathfrak{R}_0$ or $\mathfrak{R}_1$. At this point, he can only pair with elements in $\mathbb{V}$, and crucially, all the elements in $\mathbb{V}$ are either encoded in $\mathfrak{R}_0 \times \mathfrak{R}_1$, or are disjoint from $\mathfrak{R}_0$ and $\mathfrak{R}_1$. Therefore, there is no way to distinguish the two cases using only the multilinear operations.

In the following, let $[d]$ denote the set $\{0, 1, \ldots, d-1\}$.

Definition 3 (Assumption 2). *The following two distributions are indistinguishable:*

$$
\left(\left([s_{i,j}]^j_{\{i\}}\right)_{i\in\mathbb{V}, j>1}, \left([s_i]^j_{\{i\}}\right)_{i\in\mathbb{W}, j\in[d]}, \left([t_i]^{0,1}_{\{i\}}\right)_{i\in\mathbb{V}}, \right.
$$
$$
\left. \mathsf{extend}^\dagger \left(\mathsf{params}, \mathbb{W}, \left\{ \left([u_{i,j}]^j_{\{i\}}\right)_{i\in\mathbb{W}, j>1}, \left([v_i]^0_{\{i\}}\right)_{i\in\mathbb{W}} \right\} \right) \right) \quad and
$$

$$
\left(\left([s_{i,j}]^j_{\{i\}}\right)_{i\in\mathbb{V}, j>1}, \left([s_i]^j_{\{i\}}\right)_{i\in\mathbb{W}, j\in[d]}, \left([t_i]^{0,1}_{\{i\}}\right)_{i\in\mathbb{V}}, \right.
$$
$$
\left. \mathsf{extend}^\dagger \left(\mathsf{params}, \mathbb{W}, \left\{ \left([u_{i,j}]^j_{\{i\}}\right)_{i\in\mathbb{W}, j>1}, \left([v_i]^1_{\{i\}}\right)_{i\in\mathbb{W}} \right\} \right) \right)
$$

3   Additional Background

In this section, we start by providing the definition of adaptively secure FE for general circuits. Then we recall the notions of branching programs and develop notation that will be needed in our context.

3.1   Adaptively Secure FE

A functional encryption system consists of four algorithms: $Setup, KeyGen, Encrypt,$ and $Decrypt$.

- **Setup**(λ): The setup algorithm takes in the security parameter λ as input and outputs the public parameters MPK and a master secret key MSK.
- **KeyGen**(MSK, y): The key generation algorithm takes in the master secret key MSK, and an attribute string y as input. It outputs a private key SK_y for y. y is included as part of the secret key.
- **Encrypt**(MPK, x): The encryption algorithm takes the public parameters MPK and a message x as input. It outputs a ciphertext C.
- **Decrypt**(SK_y, C): The decryption algorithm takes a private key SK_y for attribute string y and a ciphertext C (encrypting say the message x) as input and outputs the value $\mathsf{C}(x, y)$, where C is a fixed universal circuit.

Correctness of the scheme requires that for correctly generated private keys for y and correctly generated ciphertexts encrypting x, decryption yields $C(x, y)$ except with negligible probability.

We will now give the security definition for *adaptive* FE. This is described by a security game between a challenger and an attacker that proceeds as follows.

- **Setup:** The challenger runs the Setup algorithm and gives the public parameters MPK to the attacker.
- **Query Phase I:** The attacker queries the challenger for private keys corresponding to attribute strings $y_1, \ldots, y_{q_1}$, which the challenger provides.
- **Challenge:** The attacker declares two messages x_0, x_1. We require that $\forall i \in [q_1]$ we have that $C(x_1, y_i) = C(x_0, y_i)$. The challenger flips a random coin $\beta \in \{0, 1\}$ and runs $C \leftarrow \textbf{Encrypt}(MPK, x_\beta)$. The challenger gives the ciphertext C to the adversary.
- **Query Phase II:** The attacker queries the challenger for private keys corresponding to the attribute strings $y_{q_1+1}, \ldots, y_q$, with the added restriction that $\forall i \in \{q_1, \ldots, q\}$ we have $C(x_1, y_i) = C(x_0, y_i)$.
- **Guess:** The attacker outputs a guess β' for β.

The advantage of an attacker in this game is defined to be $\Pr[\beta = \beta'] - \frac{1}{2}$.

3.2 Branching Programs

A branching program consists of a sequence of steps, where each step is defined by a pair of permutations. In each step the program examines one input bit, and depending on its value the program chooses one of the permutations. The program outputs 1 if and only if the multiplications of the permutations chosen in all steps is the identity permutation. In our setting, just like in previous work it will be easier to work with matrix branching programs that we define next.

Definition 4 (Matrix Branching Program). *A branching program of width w and length ℓ on n-bit inputs is given by two 0/1 permutation matrices $M_0, M_1 \in \{0, 1\}^{w \times w}$, $M_0 \neq M_1$ and by a sequence:*

$$BP = \big(\mathsf{inp}(i), B_{i,0}, B_{i,1}\big)_{i=1}^{\ell},$$

where each $B_{i,b}$ is a permutation matrix in $\{0, 1\}^{w \times w}$, and $\mathsf{inp}(i) \in [n]$ is the input bit position examined in step i. We require that, for all inputs $x \in \{0, 1\}^n$,

$$\prod_{i=1}^{\ell} B_{i, x_{\mathsf{inp}(i)}} \in \{M_0, M_1\}$$

Let (α, β) be a position where $M_1[\alpha, \beta] = 1$ and $M_0[\alpha, \beta] = 0$. Call (α, β) a distinguishing coordinate. The output of the branching program on input $x \in \{0, 1\}^n$ is as follows:

$$BP(x) = \left(\prod_{i=1}^{\ell} B_{i, x_{\mathsf{inp}(i)}}\right)[\alpha, \beta]$$

Theorem 2 [Bar86]. *For any depth-d fan-in-2 boolean circuit C, there exists an oblivious branching program of width 5 and length at most 4^d that computes the same function as the circuit C.*

Remark 1. In our functional encryption construction we do not require that the branching program is of constant width. In particular we can use any reductions that result in a polynomial size branching program.

For simplicity of notation, it will be convenient to consider two-input branching programs.[5] Here, the input $x \in \{0,1\}^{2n}$ is split into two inputs $(x[0], x[1])$. We then split inp into two functions:

- inp$'$: $[\ell] \to \{0,1\}$ where inp$'(i) = \lceil \text{inp}(i)/n \rceil - 1$. Basically, inp$'$ chooses which of the inputs $x[0]$ and $x[1]$ inp points to.
- bit : $[\ell] \to [n]$ where bit$(i) = \text{inp}(i) \mod n$. Basically, bit chooses which bit of $x[b]$ inp points to, where b is the bit chosen by inp$'$.

Then we can write the branching program evaluation as

$$BP(x) = \left(\prod_{i=1}^{\ell} B_{i,x[\text{inp}'(i)]_{\text{bit}(i)}} \right) [\alpha, \beta]$$

Remark 2. It is also straightforward to consider two-input branching programs where $x[0]$ and $x[1]$ have different sizes. We treat them as the same size for convenience.

Kilian Randomization of Branching Programs. Let BP be a branching program as above. Fix a ring $\mathfrak{R}$. Choose random invertible matrices $R_1, \ldots, R_{\ell-1}$, and define a new branching program BP' which is identical to BP, except that the matrices $B_{i,b}$ are replaced with $\tilde{B}_{i,b} = R_{i-1} \cdot B_{i,b} \cdot R_i^{-1}$, where we take $R_0 = R_\ell = I_w$. We observe that

$$\prod_{i=1}^{\ell} \tilde{B}_{i,x_{\text{inp}(i)}} = \prod_{i=1}^{\ell} B_{i,x_{\text{inp}(i)}}$$

so that for every x we have that $BP'(x) = BP(x)$. Moreover, we have the following:

Theorem 3 [Kil88]. *Fix any input $x \in \{0,1\}^\ell$, and let $b = BP(x) = BP'(x)$. Then the set of matrices multiplied together to evaluate $BP'(x)$, namely the set*

$$\left\{ \tilde{B}_{i,x_{\text{inp}(i)}} \right\}_{i \in [\ell]}$$

are distributed as uniform random $w \times w$ invertible matrices over $\mathfrak{R}$, conditioned on their product being M_b.

---

[5] Not to be confused with *dual-input* branching programs from [BGK+14].

4 Slotted Functional Encryption

In this section, we define the notion of *slotted functional encryption*. Later we will show how this scheme can be used to realize a functional encryption scheme for general circuits. A slotted functional encryption scheme, is roughly a functional encryption with multiple "slots," where each slot roughly serves as an independent copy of the functional encryption scheme. For any ciphertext or secret key, each slot is either active or inactive, and active slots will contain some bit string that potentially varies from slot to slot. Decryption is well-defined only if all slots that are active in both the ciphertext and the secret key agree on the output, in which case the result of decryption is the agreed-upon output. Otherwise, the output is undefined. Slot 0 is a special slot and where the public parameters rest. This is the slot that anyone can encrypt a message to; all the other slots require secret parameters.

- **Setup**(λ, d, C): The setup algorithm takes in the security parameter λ, a number d of slots, and a fixed universal circuit description C as inputs and outputs the public parameters MPK and a master secret key MSK.
- **KeyGen**$_S(MSK, \mathbf{y})$: The slotted key generation algorithm takes in the master secret key MSK, and a vector of attribute strings $\mathbf{y} \in \{\{0,1\}^n \cup \bot\}^d$ as input. It outputs a private key SK for $\mathbf{y}$.
- **KeyGen**(MSK, y): The unslotted version of the key generation is just a convenient shorthand, it runs **KeyGen**$(MSK, \mathbf{y})$ where $\mathbf{y} = (y, \bot, \dots)$.
- **Encrypt**$_S(MSK, \mathbf{x})$: A private slotted encryption algorithm takes in the secret parameters MSK, and a vector of messages $\mathbf{x} \in \{\{0,1\}^n \cup \bot\}^d$ as input. It outputs a ciphertext C.
- **Encrypt**(MPK, x): a public unslotted encryption algorithm takes in the public parameters MPK, and a single message $x \in \{0,1\}^n$ as input. It outputs an encryption of the message vector $(x, \bot, \bot, \dots)$
- **Decrypt**(SK, C): The decryption algorithm takes a private key SK for attribute string $\mathbf{y}$ and a ciphertext C (encrypting say the messages $\mathbf{x}$). Let $S \subseteq [d]$ be the set of *active* indices, namely those $i \in [d]$ where $x[j] \neq \bot$ and $y[j] \neq \bot$. If $\mathsf{C}(x[j], y[j]) = b$ for all active indices $i \in S$, it outputs b. Otherwise, the output is undefined.

We note that a slotted functional encryption scheme yields in particular a functional encryption using only the unslotted versions of the KeyGen and Encrypt procedures. Our goal will be to prove security of the derived (unslotted) functional encryption scheme, using various security properties of the full slotted scheme.

For security of slotted FE, consider the following general security game, parameterized by a predicate P (which encodes the security property that we want to capture).

- **Setup:** The challenger runs the Setup algorithm and gives the public parameters MPK to the attacker. The challenger also flips a random coin $\beta \in \{0,1\}$, which it keeps secret.

- **Query Phase I:** The attacker adaptively queries the challenger for private keys corresponding to attribute vectors pairs $\mathbf{y}_i^{(0)}, \mathbf{y}_i^{(1)} \in \{\{0,1\}^n \cup \perp\}^d$ for $i = 1, ..., q_1$. The challenger responds with the secret keys for $\mathbf{y}_i^{(\beta)}$.
- **Challenge:** The attacker declares two message s vector $\mathbf{x}^{(0)}, \mathbf{x}^{(1)} \in \{\{0,1\}^n \cup \perp\}^d$. The challenger responds with the ciphertext $C \leftarrow$ **Encrypt**$_S(MSK, \mathbf{x}^{(\beta)})$.
- **Query Phase II:** The attacker continues to adaptively queries the challenger for private keys corresponding to attribute vectors pairs $\mathbf{y}_i^{(0)}, \mathbf{y}_i^{(1)} \in \{\{0,1\}^n \cup \perp\}^d$ for $i = q_1 + 1, ..., q$. The challenger responds with the secret keys for $\mathbf{y}_i^{(\beta)}$.
- **Guess:** The attacker outputs a guess β' for β.
- **Check:** The challenger evaluates a predicate P on the secret-key and challenge queries: $c = P(\{\mathbf{y}_i^{(b)}\}_{i \in [q], b \in \{0,1\}}, \mathbf{x}^{(0)}, \mathbf{x}^{(1)})$. If the predicate holds ($c = 1$) then the challenger outputs $\beta'' = \beta'$. Otherwise the challenger outputs a random independent bit β''.

The advantage of an attacker in this game is defined to be $\Pr[\beta = \beta''] - \frac{1}{2}$ (and note that if $c = 0$ then the advantage is 0). The scheme is secure relative to the given predicate if feasible adversaries can only have a negligible advantage.

The predicate P. The security game varies depending on the predicate P, with more permissive predicates yielding stronger notions of security. At a minimum, we need P to exclude queries that let the adversary trivially distinguish the left and right sides by applying the decryption procedure on the secret keys and ciphertext received. Similarly, P must also exclude queries that let the adversary distinguish the left and right sides by generating its own ciphertexts.

However, it is not hard to see that using a permissive predicate P that only excludes these trivial attacks results in a security notion that is too strong: such permissive P would allow arbitrary secret-key queries (y, y') so long as $\mathsf{C}(x, y) = \mathsf{C}(x, y')$ for all $x \in \{0,1\}^n$, which means that we directly get indistinguishability obfuscation. Specifically, for a universal circuit U, we obfuscate a function $f(x) = U(f, x)$ by publishing the FE secret key SK_f. This lets anyone evaluate $f(x)$ for any x by encrypting x under the scheme, and then using SK_f to decrypt $f(x)$, and the security notion would say that any two functionally equivalent f and f' are indistinguishable.

Next, we therefore describe some simple predicates which are more restrictive, and hence they correspond to weaker notions of security (which are still strong enough for our purposes). Very roughly speaking, they all require that most of the time we have $\mathbf{y}_i^{(0)} = \mathbf{y}_i^{(1)}$ and/or $\mathbf{x}^{(0)} = \mathbf{x}^{(1)}$, and they differ only in a handful of slots and/or a handful of queries.

4.1   Core Predicates

We begin by describing some simple core predicates that our slotted FE scheme should satisfy. In the next section we show that the corresponding security properties imply also stronger properties, including adaptively security of the induced unslotted FE scheme.

0. **Slot Symmetry.** P checks that there are two distinct non-special slots $\alpha \neq \beta$, $\alpha, \beta \neq 0$ such that:
   - $\mathbf{x}^{(0)}, \mathbf{x}^{(1)}$ are equal in all the slots other than α, β, and they swap the content of these two slots. Namely $\mathbf{x}^{(0)}[j] = \mathbf{x}^{(1)}[j] := \mathbf{x}[j]$ for all $j \notin \{\alpha, \beta\}$, and $\mathbf{x}^{(b)}[\alpha] = \mathbf{x}^{(1-b)}[\beta] := x^{(b*)}$ for $b = 0, 1$.
   - Similarly for all i $\mathbf{y}_i^{(0)}, \mathbf{y}_i^{(1)}$ are equal in all the slots other than α, β, and they swap the content of these two slots. Namely $\mathbf{y}_i^{(0)}[j] = \mathbf{y}_i^{(1)}[j] := \mathbf{y}_i[j]$ for all $j \notin \{\alpha, \beta\}$, and $\mathbf{y}_i^{(b)}[\alpha] = \mathbf{y}_i^{(1-b)}[\beta] := y_i^{(b*)}$ for $b = 0, 1$.

$b = 0$	$\mathbf{x}^{(0)}[j]$	$\mathbf{y}_i^{(0)}[j]$
$j = \alpha$	$x^{(0*)}$	$y_i^{(0*)}$
$j = \beta$	$x^{(1*)}$	$y_i^{(1*)}$
$j \neq \alpha, \beta$	$\mathbf{x}[j]$	$\mathbf{y}_i[j]$

$b = 1$	$\mathbf{x}^{(1)}[j]$	$\mathbf{y}_i^{(1)}[j]$
$j = \alpha$	$x^{(1*)}$	$y_i^{(1*)}$
$j = \beta$	$x^{(0*)}$	$y_i^{(0*)}$
$j \neq \alpha, \beta$	$\mathbf{x}[j]$	$\mathbf{y}_i[j]$

Intuitively, this allows us to permute the contents of different slots without the adversary's notice.

1. **Single-Use Message and Function Hiding.** P checks that there is a non-special slot $\alpha \neq 0$ and a secret key query $\gamma \in [q]$ such that:
   - All key-queries other than γ contain two identical functions, $\mathbf{y}_i^{(0)} = \mathbf{y}_i^{(1)} := \mathbf{y}_i \ \forall i \neq \gamma$.
   - Key-query γ has two keys that differ only in slot α, $\mathbf{y}_\gamma^{(0)}[j] = \mathbf{y}_\gamma^{(1)}[j] := \mathbf{y}_\gamma[j] \ \forall j \neq \alpha$.
   - The challenge query has two plaintexts that differ only in slot α, $\mathbf{x}^{(0)}[j] = \mathbf{x}^{(1)}[j] := \mathbf{x}[j] \ \forall j \neq \alpha$.
   - We have either the same functionality $\mathsf{C}(\mathbf{x}^{(0)}[\alpha], \mathbf{y}_\gamma^{(0)}[\alpha]) = \mathsf{C}(\mathbf{x}^{(1)}[\alpha], \mathbf{y}^{(1)}[\alpha])$, or the two plaintext slots are inactive $\mathbf{x}^{(0)}[\alpha] = \mathbf{x}^{(1)}[\alpha] = \bot$, or the two key slots are inactive $\mathbf{y}_\gamma^{(0)}[\alpha] = \mathbf{y}_\gamma^{(1)}[\alpha] = \bot$.

$b = 0$	$\mathbf{x}^{(0)}[j]$	$\mathbf{y}_i^{(0)}[j]$	
		$i = \gamma$	$i \neq \gamma$
$j = \alpha$	$x^{(0*)}$	$y^{(0*)}$	$\mathbf{y}_i[\alpha]$
$j \neq \alpha$	$\mathbf{x}[j]$	$\mathbf{y}_i[j]$	

$b = 1$	$\mathbf{x}^{(1)}[j]$	$\mathbf{y}_i^{(1)}[j]$	
		$i = \gamma$	$i \neq \gamma$
$j = \alpha$	$x^{(1*)}$	$y^{(1*)}$	$\mathbf{y}_i[\alpha]$
$j \neq \alpha$	$\mathbf{x}[j]$	$\mathbf{y}_i[j]$	

Requirements:
$\mathsf{C}(x^{(0*)}, y^{(0*)}) = \mathsf{C}(x^{(1*)}, y^{(1*)})$ or
$x^{(0*)} = x^{(1*)} = \bot$ or
$y^{(0*)} = y^{(1*)} = \bot$

This allows us to argue both message and function hiding for one slot in one query, as long as that slot is not the special slot that the public parameters can encrypt to.

2. **Slot Duplication.** P checks that there are distinct slots $\alpha \neq \beta$ with $\beta \neq 0$ such that:

– All the slots other than β are the same between left and right, $\mathbf{x}^{(0)}[j] = \mathbf{x}^{(1)}[j] := \mathbf{x}[j]$ for all $j \neq \beta$, and $\mathbf{y}_i^{(0)}[j] = \mathbf{y}_i^{(1)}[j] := \mathbf{y}_i[j]$ for all i and all $j \neq \beta$.

– Slots β on the left are inactive, $\mathbf{x}^{(0)}[\beta] = \bot$ and $\mathbf{y}_i^{(0)}[\beta] = \bot$ for all i

– Slots β on the right are either inactive or equal to slots α, $\mathbf{x}^{(0)}[\beta] \in \{\mathbf{x}[\alpha], \bot\}$ and $\mathbf{y}_i^{(0)}[\beta] \in \{\mathbf{y}_i[\alpha], \bot\}$ for all i.

$b = 0$	$\mathbf{x}^{(0)}[j]$	$\mathbf{y}_i^{(0)}[j]$
$j = \alpha$	x^*	y_i^*
$j = \beta$	$\bot$	$\bot$
$j \neq \alpha, \beta$	$\mathbf{x}[j]$	$\mathbf{y}_i[j]$

$b = 1$	$\mathbf{x}^{(1)}[j]$	$\mathbf{y}_i^{(1)}[j]$
$j = \alpha$	x^*	y_i^*
$j = \beta$	x^* or $\bot$	y_i^* or $\bot$
$j \neq \alpha, \beta$	$\mathbf{x}[j]$	$\mathbf{y}_i[j]$

We stress that slot duplication can duplicate the slots of the ciphertext and secret keys *simultaneously*. We can choose to duplicate the slots of all keys and the ciphertext, or any subset of them.

3. **Ciphertext Moving.** P checks that there are two distinct slots $\alpha \neq \beta$ such that:

– For each secret key, all slots (including α and β) are the same on the left and right: $\mathbf{y}_i^{(0)}[j] = \mathbf{y}_i^{(1)}[j] := \mathbf{y}_i[j]$ for all i and j.

– For each secret key, slot α is identical to slot β on both the left and right: $\mathbf{y}_i[\alpha] = \mathbf{y}_i[\beta] := \mathbf{y}_i^*$ ($\mathbf{y}_i^*$ is potentially $\bot$).

– For the challenge ciphertext, all slots other than α, β are the same between left and right: $\mathbf{x}^{(0)}[j] = \mathbf{x}^{(1)}[j] := \mathbf{x}[j]$ for all $j \notin \{\alpha, \beta\}$.

– For the challenge ciphertext, slot β on the left and slot α on the right are inactive: $\mathbf{x}^{(0)}[\beta] = \mathbf{x}^{(1)}[\alpha] = \bot$.

– For the challenge ciphertext, slot α on the left is equal to slot β on the right: $\mathbf{x}^{(0)}[\alpha] = \mathbf{x}^{(1)}[\beta] = \mathbf{x}^*$.

$b = 0$	$\mathbf{x}^{(0)}[j]$	$\mathbf{y}_i^{(0)}[j]$
$j = \alpha$	x^*	y_i^*
$j = \beta$	$\bot$	y_i^*
$j \neq \alpha, \beta$	$\mathbf{x}[j]$	$y_i[j]$

$b = 1$	$\mathbf{x}^{(1)}[j]$	$\mathbf{y}_i^{(1)}[j]$
$j = \alpha$	$\bot$	y_i^*
$j = \beta$	x^*	y_i^*
$j \neq \alpha, \beta$	$\mathbf{x}[j]$	$y_i[j]$

This lets us rearrange the slots of the challenge ciphertext, as long as each secret keys is identical among the affected slots. We stress that ciphertext moving allows one of the slots being rearranged to be the special slot.

4. **Weak key moving.** P checks that there are two distinct non-special slots $\alpha \neq \beta$, $\alpha, \beta \neq 0$ and secret-key query γ such that:

– For the challenge ciphertext, all slots (including α and β) are the same between left and right: $\mathbf{x}^{(0)}[j] = \mathbf{x}^{(1)}[j] := \mathbf{x}[j]$ for all j.

– For the challenge ciphertext, slot α is identical to slot β on both the left and right: $\mathbf{x}[\alpha] = \mathbf{x}[\beta] := x^*$

– For each secret key query other than γ, all slots (including α and β) are the same on the left and right: $\mathbf{y}_i^{(0)}[j] = \mathbf{y}_i^{(1)}[j] := \mathbf{y}_i[j]$ for all $i \neq \gamma$ and all j.

- For secret key query γ, all slots *other than* α, β are the same on the left and right: $\mathbf{y}_\gamma^{(0)}[j] = \mathbf{y}_\gamma^{(1)}[j] := \mathbf{y}_\gamma[j]$ for all $j \notin \{\alpha, \beta\}$.
- For secret key query γ, slot β on the left and slot α on the right are inactive: $\mathbf{y}_\gamma^{(0)}[\beta] = \mathbf{y}_\gamma^{(1)}[\alpha] = \bot$.
- For secret key query γ, slot α on the left is identical to slot β on the right: $\mathbf{y}_\gamma^{(0)}[\alpha] = \mathbf{y}_\gamma^{(1)}[\beta] = \mathbf{y}_\gamma^* := y^*$.

		$b = 0$					$b = 1$	
	$\mathbf{x}^{(0)}[j]$	$\mathbf{y}_i^{(0)}[j]$ $i = \gamma$	$i \neq \gamma$			$\mathbf{x}^{(1)}[j]$	$\mathbf{y}_i^{(1)}[j]$ $i = \gamma$	$i \neq \gamma$
$j = \alpha$	x^*	y^*			$j = \alpha$	x^*	$\bot$	
$j = \beta$	x^*	$\bot$	$\mathbf{y}_i[j]$		$j = \beta$	x^*	y^*	$\mathbf{y}_i[j]$
$j \neq \alpha$	$\mathbf{x}[j]$	$\mathbf{y}_\gamma[j]$			$j \neq \alpha$	$\mathbf{x}[j]$	$\mathbf{y}_\gamma[j]$	

This is the secret key version of ciphertext moving, allowing us to rearrange the slots of a secret key, as long as the challenge ciphertext is identical among the affected slots. The main difference from ciphertext moving is that weak key moving does not allow us to modify the special slot 0.

We observe that the above properties, even in combination, will never allow the changing of a secret key in slot 0. Thus, we will not be able to obtain any form of function hiding for the derived unslotted functional encryption scheme just from the properties above. This serves as a sanity check that the above properties are not too strong, and might be obtainable from simple assumptions, and indeed we give a construction meeting these in Sect. 5.

In the following sections, we present several other more complex predicates, and show that security relative to the complex predicates is implied by the security relative only to the predicates above. The proofs "consume" some slots, so extra slots are needed to obtain security for the more complex predicates.

One of the predicates we prove security for corresponds exactly to regular functional encryption. The total number of slots consumed in the proof from the basic predicates is 3. Combining with our slotted FE construction in Sect. 5 for 4 slots, we obtain adaptively secure functional encryption for NC^1 functionalities.

In the full version [GGHZ14b], we show how to use our predicates, together with puncturable PRFs and randomized encodings (defined in Sect. 3) to obtain functional encryption for all circuits. The total number of slots consumed is 5, meaning we need a 6-slotted FE. In particular, the number of slots is constant, which translates to a constant number (namely 6) of subgroups in the underlying composite-order multilinear maps.

4.2   Additional Derivable Predicates

Now we describe several additional properties that can be derived from the core properties above, potentially "using up" several additional slots.

5. **New Slot**. P checks that there are distinct slots $\alpha \neq \beta$ with α not being the special 0 slot (but β may be), such that:

- For each secret key, all slots (including α and β) are the same on the left and right: $\mathbf{y}_i^{(0)}[j] = \mathbf{y}_i^{(1)}[j]$ for all i and j.
- For each secret key, slot α is inactive on both the left and the right: $\mathbf{y}_i^{(0)}[\alpha] = \mathbf{y}_i^{(1)}[\alpha] = \bot$ for all i
- For the challenge ciphertext, all slots other than slot α are the same on the left and right: $\mathbf{x}^{(0)}[j] = \mathbf{x}^{(1)}[j]$ for all $j \neq \alpha$.
- For the challenge ciphertext, slot β is active on both the left and the right: $\mathbf{x}^{(0)}[\beta] = \mathbf{x}^{(1)}[\beta] \neq \bot$.
- For the challenge ciphertext, slot α is inactive on the left: $\mathbf{x}^{(0)}[\alpha] = \bot$

<table>
<tr><td colspan="3">$b = 0$</td><td colspan="3">$b = 1$</td></tr>
<tr><td></td><td>$\mathbf{x}^{(0)}[j]$</td><td>$\mathbf{y}_i^{(0)}[j]$</td><td></td><td>$\mathbf{x}^{(1)}[j]$</td><td>$\mathbf{y}_i^{(1)}[j]$</td></tr>
<tr><td>$j = \alpha$</td><td>$\bot$</td><td>$\bot$</td><td>$j = \alpha$</td><td>x^*</td><td>$\bot$</td></tr>
<tr><td>$j = \beta$</td><td>$\mathbf{x}[\beta] \neq \bot$</td><td rowspan="2">$\mathbf{y}_i[j]$</td><td>$j = \beta$</td><td>$\mathbf{x}[\beta] \neq \bot$</td><td rowspan="2">$\mathbf{y}_i[j]$</td></tr>
<tr><td>$j \neq \alpha, \beta$</td><td>$\mathbf{x}[j]$</td><td>$j \neq \alpha, \beta$</td><td>$\mathbf{x}[j]$</td></tr>
</table>

Notice that there is no restriction to the value in slot α of the ciphertext on the right. Thus, the allows us to take a slot that is inactive for all secret keys and the challenge ciphertext, and place an arbitrary value in the slot for the ciphertext.

6. **Strong key moving.** P checks that there are distinct non-special slots $\alpha \neq \beta$, $\alpha, \beta \neq 0$, and secret key query γ such that:
  - For the challenge ciphertext, all slots (including α and β) are the same between left and right: $\mathbf{x}^{(0)}[j] = \mathbf{x}^{(1)}[j] := \mathbf{x}[j]$ for all j.
  - For each secret key query other than γ, all slots (including α and β) are the same on the left and right: $\mathbf{y}_i^{(0)}[j] = \mathbf{y}_i^{(1)}[j] := \mathbf{y}_i[j]$ for all $i \neq \gamma$ and all j.
  - For secret key query γ, all slots *other than* α, β are the same on the left and right: $\mathbf{y}_\gamma^{(0)}[j] = \mathbf{y}_\gamma^{(1)}[j] := \mathbf{y}_\gamma[j]$ for all $j \notin \{\alpha, \beta\}$.
  - For secret key query γ, slot β on the left and slot α on the right are inactive: $\mathbf{y}_\gamma^{(0)}[\beta] = \mathbf{y}_\gamma^{(1)}[\alpha] = \bot$.
  - For secret key query γ, slot α on the left is identical to slot β on the right: $\mathbf{y}_\gamma^{(0)}[\alpha] = \mathbf{y}_\gamma^{(1)}[\beta] := \mathbf{y}_\gamma^*$.
  - When decrypting the challenge with secret key γ, slot α on the left and slot β on the right give the same result. In other words, $\mathsf{C}(\mathbf{x}[\alpha], \mathbf{y}_\gamma^*) = \mathsf{C}(\mathbf{x}[\beta], \mathbf{y}_\gamma^*)$

<table>
<tr><td colspan="4">$b = 0$</td><td colspan="4">$b = 1$</td><td></td></tr>
<tr><td></td><td>$\mathbf{x}^{(0)}[j]$</td><td colspan="2">$\mathbf{y}_i^{(0)}[j]$</td><td></td><td>$\mathbf{x}^{(1)}[j]$</td><td colspan="2">$\mathbf{y}_i^{(1)}[j]$</td><td>Requirements:</td></tr>
<tr><td></td><td></td><td>$i = \gamma$</td><td>$i \neq \gamma$</td><td></td><td></td><td>$i = \gamma$</td><td>$i \neq \gamma$</td><td>$\mathsf{C}(x_0^*, y^*) =$</td></tr>
<tr><td>$j = \alpha$</td><td>x_0^*</td><td>y^*</td><td rowspan="2">$\mathbf{y}_i[j]$</td><td>$j = \alpha$</td><td>x_0^*</td><td>$\bot$</td><td rowspan="2">$\mathbf{y}_i[j]$</td><td>$\mathsf{C}(x_1^*, y^*)$</td></tr>
<tr><td>$j = \beta$</td><td>x_1^*</td><td>$\bot$</td><td>$j = \beta$</td><td>x_1^*</td><td>y^*</td><td></td></tr>
<tr><td>$j \neq \alpha$</td><td>$\mathbf{x}[j]$</td><td>$\mathbf{y}_\gamma[j]$</td><td></td><td>$j \neq \alpha$</td><td>$\mathbf{x}[j]$</td><td>$\mathbf{y}_\gamma[j]$</td><td></td><td></td></tr>
</table>

This is a stronger form of secret key moving where we can actually rearrange secret key slots even if the challenge ciphertext differs in those slots, as long as decryption is unaffected.

7. **Weak ciphertext indistinguishability.** P checks that there is a non-special slot $\alpha \neq 0$ such that:
   - For each secret key, all slots (including slot α) are the same on the left and right: $\mathbf{y}_i^{(0)}[j] = \mathbf{y}_i^{(1)}[j] := \mathbf{y}_i[j]$ for all i and j.
   - For the challenge ciphertext, all slots except slot α are the same on the left and right: $\mathbf{x}_i^{(0)}[j] = \mathbf{x}_i^{(1)}[j] := \mathbf{x}[j]$ for all $j \neq \alpha$.
   - For the challenge ciphertext, slot α decrypts to the same result for each secret key query: $\mathsf{C}(\mathbf{x}^{(0)}[\alpha], \mathbf{y}_i[\alpha]) = \mathsf{C}(\mathbf{x}^{(1)}[\alpha], \mathbf{y}_i[\alpha])$.

	$b = 0$			$b = 0$		Requirements:
	$\mathbf{x}^{(0)}[j]$	$\mathbf{y}_i^{(0)}[j]$		$\mathbf{x}^{(1)}[j]$	$\mathbf{y}_i^{(1)}[j]$	$\mathsf{C}(x_0^*, y_i^*) = \mathsf{C}(x_1^*, y_i^*) \forall i$
$j = \alpha$	x_0^*	y_i^*	$j = \alpha$	x_1^*	y_i^*	
$j \neq \alpha$	$\mathbf{x}[j]$	$\mathbf{y}_i[j]$	$j \neq \alpha$	$\mathbf{x}[j]$	$\mathbf{y}_i[j]$	

In other words, we can change the value of the ciphertext in any slot other than the special 0 slot as long as decryption is unaffected. This almost gives us functional encryption, except for the requirement that the slot is not the special slot.

8. **Strong ciphertext indistinguishability.** Same as above, except α can be 0.

4.3   Reductions

Now we describe several reductions showing that core properties described above are sufficient for obtaining the additional derivable properties also described above, at the cost of "using up" several additional slots. We note that in all of the reductions below, any existing property, whether core or derived, is preserved in the reduction.

Lemma 3. *(1) Single-use hiding and (2) slot duplication imply (5) new slot.*

Proof. Use slot duplication to duplicate contents of the β slot into the originally empty α slot of the ciphertext (don't duplicate the secret keys), and then use single-use message and function hiding to change the message to x^*, which is possible since there are no secret keys components in the α slot.

Lemma 4. *(1) Single-use hiding, (2) slot duplication, (3) and weak key moving for $d + 1$ slots implies (6) strong key moving for d slots (all existing properties being preserved).*

Proof. We prove for $\alpha = 1, \beta = 2$, the other cases being identical. We will move secret key $\gamma \in [q]$. Let slot $d+1$ be a "scratch" slot, that is unused by the normal scheme. We will use slot $d+1$ in the security proof. Below is the table of hybrids. For secret keys $i \in [q], i \neq \gamma$ not included in the table, slot $d + 1$ is inactive, and the rest of the slots remain the same throughout all hybrids. Similarly, slots $j \neq 1, 2, d + 1$ remain the same for the ciphertext and the γth secret key.

Hybrid	$x[j]$			$y_\gamma[j]$			comments
	$j=1$	$j=2$	$j=d+1$	$j=1$	$j=2$	$j=d+1$	
H_0	x_0^*	x_1^*	$\perp$	y^*	$\perp$	$\perp$	
H_1	x_0^*	x_1^*	x_0^*	y^*	$\perp$	$\perp$	Slot duplication
H_2	x_0^*	x_1^*	x_0^*	$\perp$	$\perp$	y^*	Weak secret key moving
H_3	x_0^*	x_1^*	x_1^*	$\perp$	$\perp$	y^*	Single-use message hiding
H_4	x_0^*	x_1^*	x_1^*	$\perp$	y^*	$\perp$	Weak secret key moving
H_5	x_0^*	x_1^*	$\perp$	$\perp$	y^*	$\perp$	Slot duplication

Lemma 5. *(0) Slot symmetry, (5) new slot, and (6) strong key moving for $d+1$ slots implies weak (7) weak ciphertext indistinguishability for d slots (all existing properties being preserved).*

Proof. We prove for $\alpha = 1$, the other cases being identical. The slot $d+1$ will be the "scratch" slot, that is unused by the normal scheme but used in the security proof. In the hybrids below we will use the strong key moving property. Note that the strong key moving only allows for changing one key at a time, while in the hybrids below we will need to change all the keys. This can be done by changing one key at a time.

Hybrid	$x[j]$		$\forall \gamma \in [q], y_\gamma[j]$		comments
	$j=1$	$j=d+1$	$j=1$	$j=d+1$	
H_0	x_0^*	$\perp$	y^*	$\perp$	
H_1	x_0^*	x_1^*	y^*	$\perp$	New slot
H_2	x_0^*	x_1^*	$\perp$	y^*	Strong key moving ($\times q$)
H_3	$\perp$	x_1^*	$\perp$	y^*	New slot
H_4	x_1^*	$\perp$	y^*	$\perp$	Slot Symmetry

Lemma 6. *(2) Slot duplication, (3) weak ciphertext moving, and (7) weak ciphertext indistinguishability for $d+1$ slots implies (8) strong ciphertext indistinguishability for d slots (all existing properties preserved).*

Proof. Only need to add the case for slot 0. Just as before, the slot $d+1$ will be the "scratch" slot, that is unused by the normal scheme but used in the security proof.

Hybrid	$x[j]$		$y_i[j]$		Comments
	$j=0$	$j=d+1$	$j=0$	$j=d+1$	
H_0	x_0^*	$\perp$	y_i^*	$\perp$	
H_1	x_0^*	$\perp$	y_i^*	y_i^*	Slot duplication
H_2	$\perp$	x_0^*	y_i^*	y_i^*	Weak ciphertext moving
H_3	$\perp$	x_1^*	y_i^*	y_i^*	Weak ciphertext indistinguishability
H_4	x_1^*	$\perp$	y_i^*	y_i^*	Weak ciphertext moving
H_5	x_1^*	$\perp$	y_i^*	$\perp$	Slot duplication

5   Slotted Functional Encryption for NC^1

We now give our slotted FE scheme for NC^1. We will describe our scheme in terms of matrix branching programs, using Barrington's Theorem (Theorem 2) to realize slotted FE for NC^1 circuits. We describe our scheme for single bit outputs — it can easily be extended to multi-bit outputs by running multiple instances of the scheme in parallel.

Setup(λ, BP, d): Given a universal 2-input matrix branching program

$$BP = \left(\mathsf{bit}, \mathsf{inp}, (B_{i,b})_{i\in[\ell], b\in\{0,1\}} \right)$$

run params $\leftarrow \mathsf{InstGen}(1^\lambda, \{1, \ldots, \ell\}, d)$. Then, choose random matrices $R_i \in \mathfrak{R}$ for $i \in [\ell-1]$, as well as random $\alpha_{i,b}$ for $i \in [\ell], b \in \{0,1\}$. Let $\tilde{B}_{i,b} = \alpha_{i,b} \cdot R_{i-1} \cdot B_{i,b} \cdot R_i^{-1}$ for $i \in [2, \ell-1]$, and $\tilde{B}_{1,b} = \alpha_{1,b} \cdot B_{1,b} \cdot R_1^{-1}$ and $\tilde{B}_{\ell,b} = \alpha_{\ell,b} \cdot R_{\ell-1} \cdot B_{\ell,b}$[6]. Compute $A_{i,b}^j = [\tilde{B}_{i,b}]_{\{i\}}^j$ for $j \in [d]$. (Here R_0 and R_ℓ are set to identity.)

Let $\mathbb{V}$ be the subset of $[\ell]$ that corresponds to the secret key: $\mathbb{V} = \{i \in [\ell] : \mathsf{inp}(i) = 0\}$, and $\mathbb{W}$ be the subset of $[\ell]$ that corresponds to the ciphertext: $\mathbb{W} = \{i \in [\ell] : \mathsf{inp}(i) = 1\}$. Then the universe $\mathbb{U} = \mathbb{V} \cup \mathbb{W}$.

The master public key is $MPK = (\mathsf{params}, (A_{i,b}^0)_{i\in\mathbb{W}, b\in\{0,1\}})$

The master secret key consists of the $A_{i,b}^j$ for $i \in \mathbb{V} \cup \mathbb{W}$.

KeyGen$_S(MSK, \mathbf{y})$: Given an attribute $y \in \{\{0,1\}^n \cup \bot\}^d$, choose random $\overline{\beta_i} \in \mathfrak{R}$ for $i \in \mathbb{V}, b \in \{0,1\}$, and output the secret key

$$SK_y = \mathsf{extend}\left(\mathsf{params}, \mathbb{V}, \left(\beta_i \cdot \left(\sum_{j : y[j] \neq \bot} A_{i, y[j]_{\mathsf{bit}(i)}}^j \right) \right)_{i \in \mathbb{V}} \right)$$

Encrypt$_S(MSK, \mathbf{x})$: Given an attribute $x \in \{\{0,1\}^n \cup \bot\}^d$, choose random $\overline{\beta_i} \in \mathfrak{R}$ for $i \in \mathbb{W}, b \in \{0,1\}$, and output the ciphertext

$$C = \mathsf{extend}\left(\mathsf{params}, \mathbb{W}, \left(\beta_i \cdot \left(\sum_{j : x[j] \neq \bot} A_{i, x[j]_{\mathsf{bit}(i)}}^j \right) \right)_{i \in \mathbb{W}} \right)$$

Encrypt(MPK, m): Given a message $m \in \{0,1\}^n$, choose random $\beta_i \in \mathfrak{R}$ for $i \in \mathbb{W}$, and output the ciphertext

$$C = \mathsf{extend}\left(\mathsf{params}, \mathbb{W}, \left(\beta_i \cdot A_{i, m_{\mathsf{bit}(i)}}^0 \right)_{i \in \mathbb{W}} \right)$$

---

[6] Using current graded encodings, it is not possible to *publicly* compute matrix inverses since users do not have direct access to the underlying ring. However, the setup procedure would know a trapdoor for the graded encodings that *does* allow computing the matrix inverse. Alternatively, we can replace R_i^{-1} with the *adjugate* matrix R_i^{adj}, encodings of which *can* be computed publicly. The adjugate and matrix inverse only differ by a scalar multiple (namely, the determinant), and since we multiply everything by a random scalar anyway, the distributions of encodings obtained are identical in both approaches.

Remark 3. Note that all the encodings given out in the ciphertext can be re-randomized (to noise σ') using the randomizer provided in the public parameters. We do not mention the re-randomization above explicitly, for the sake of simplicity of notation.

Decrypt(MPK, SK, C): Given a secret key $SK = f_{\mathbb{V}' \to \mathbb{V}}, (K_i)_{i \in \mathbb{V}'}$ and a ciphertext $C = f_{\mathbb{W}' \to \mathbb{W}}, (C_i)_{i \in \mathbb{W}'}$, let $D_i = \begin{cases} K_i & \text{if } i \in \mathbb{V}' \\ C_i & \text{if } i \in \mathbb{W}' \end{cases}$, and compute the product

$$D = f_{\mathbb{V}' \to \mathbb{V}}\left(f_{\mathbb{W}' \to \mathbb{W}}\left(\prod_{i \in \mathbb{U}} D_i \right)\right)$$

Then run the zero-test procedure on a distinguishing coordinate of D.

Correctness. Evaluation is carried out slot by slot. In slot j, if either K or C is inactive, then the corresponding ring will be empty. Therefore, the result of the computation is 0 in slot j. In a slot j where K and C are both active, then write $K_i[j] = [\beta_i \alpha_{i,y[j]_{\text{bit}(i)}} \tilde{B}_{i,y_{\text{bit}(i)}}]^j_{\{i'\}}$ and $C_i[j] = [\beta_i \alpha_{i,m_{\text{bit}(i)}} \tilde{B}_{i,m_{\text{bit}(i)}}]^j_{\{i'\}}$ for some index elements i' to be the components of K, C in the ring $\mathfrak{R}_j$. Let $d[j] = (y[j], m[j]) \in \{0,1\}^{2n}$. Then we can write $D_i[j] = [\beta_i \alpha_{i,d[j]_{\text{inp}(i)},\text{bit}(i)} \tilde{B}_{i,d[j]_{\text{inp}(i)},\text{bit}(i)}]^j_{\{i\}}$.

Therefore, the product $D'[j] = \prod_{i \in \mathbb{U}} D_i[j]$ is equal to

$$D'[j] = \left[\prod_{i \in \mathbb{U}} \left(\beta_i \alpha_{i,d[j]_{\text{inp}(i)},\text{bit}(i)} \right) \prod_{i \in \mathbb{U}} \tilde{B}_{i,d[j]_{\text{inp}(i)},\text{bit}(i)} \right]^j_{\mathbb{U}'}$$

$$= \left[\prod_{i \in \mathbb{U}} \left(\beta_i \alpha_{i,d[j]_{\text{inp}(i)},\text{bit}(i)} \right) \prod_{i \in \mathbb{U}} B_{i,d[j]_{\text{inp}(i)},\text{bit}(i)} \right]^j_{\mathbb{U}'}$$

where $\mathbb{U}' = \mathbb{V}' \cup \mathbb{W}'$. Applying $f_{\mathbb{W}' \to \mathbb{W}}$ to this encoding gives an encoding of the same product, but relative to the set $\mathbb{V}' \cup \mathbb{W}$, and then applying $f_{\mathbb{V}' \to \mathbb{V}}$ gives the encoding relative to $\mathbb{U}$. Therefore, $D = f_{\mathbb{V}' \to \mathbb{V}}(f_{\mathbb{W}' \to \mathbb{W}}(D'))$ satisfies

$$D[j] = \left[\prod_{i \in \mathbb{U}} \left(\beta_i \alpha_{i,d[j]_{\text{inp}(i)},\text{bit}(i)} \right) \prod_{i \in \mathbb{U}} B_{i,d[j]_{\text{inp}(i)},\text{bit}(i)} \right]^j_{\mathbb{U}}$$

$$= \left[\prod_{i \in \mathbb{U}} \left(\beta_i \alpha_{i,d[j]_{\text{inp}(i)},\text{bit}(i)} \right) M_{BP(d[j])} \right]^j_{\mathbb{U}}$$

We only care about ciphertexts and secret keys where the branching program evaluates the same in every slot, so $BP(d[j])$ is the same for all active slots j; call the result b. Define $\gamma[j] = \beta_i \alpha_{i,d[j]_{\text{inp}(i)},\text{bit}(i)}$ projected down to ring $\mathfrak{R}_j$, and $\gamma = \sum_{j \in S} \gamma[j]$ where S is the set of active slots. Note that we only care about secret keys and ciphertext where there is at least one active slot. Therefore with overwhelming probability $\gamma \neq 0$.

We can now write $D = [\gamma M_b]_{\mathbf{U}}$. Then when we zero test a distinguishing coordinate of D, with overwhelming probability, the result will match b.

5.1 Security Proof

Theorem 4. *Assuming Assumptions 1 and 2, the scheme described above satisfies the core properties of the slotted FE scheme.*

Slot Symmetry. Our scheme satisfies *perfect* slot symmetry, where the advantage of an even infinitely powerful adversary is 0. This follows from the fact that slots correspond to sub-rings in our scheme, and our subrings are generated in a totally symmetric manner.

Single-use Message and Function hiding. In our scheme, the matrices are just the matrices from Kilian-randomized branching programs, where the randomization in each sub-ring is independent. In the single slot j where changes are made, only the ciphertext and a single public key are active. Let $z = (x_0, y_0)$ be the ciphertext and secret key values active on the left side, and $z' = (x_1, y_1)$ be the values on the right side. Then on the left side, only the matrices $\tilde{B}_{i,z[\mathsf{inp}(i)]\mathsf{bit}(i)}$ are handed out in ring $\mathfrak{R}_j$, and by Theorem 3, these matrices are uniform random matrices subject to their product being $M_{\mathsf{C}(x_0, y_0)}$. Similarly, on the left size, the matrices handed out are uniform random matrices subject their product being $M_{\mathsf{C}(x_1, y_1)}$. Since $\mathsf{C}(x_0, y_0) = \mathsf{C}(x_1, y_1)$, these distributions are identical, so our scheme satisfies *perfect* single use hiding.

Slot duplication. We will prove slot duplication from Assumption 1. Let $\alpha \in [d]$ and $\beta \neq \alpha, 0$. Obtain the challenge for assumption 1, and re-order the rings so that the challenge has the form $\left(S_{i,j} = [s_{i,j}]^j_{\{i\}} \right)_{i \in \mathbf{U}, j \neq \beta}, (T_i)_{i \in \mathbf{U}}$ where $T_i = [t_i]^\alpha_{\{i\}}$ or $T_i = [t_i]^{\alpha, \beta}_{\{i\}}$. We now simulate the view of the adversary as follows. Given a 0/1 matrix B and an encoding e, let $e \cdot B$ be the matrix of encodings, where $e \cdot B$ has e in any position where B has a 1, and an encoding of 0 in any position where B has a 0 (note that we will be multiplying $e \cdot B$ by other matrices of encodings, so the encodings of 0 do not actually have to be computed, but merely serve as placeholders in the computation).

Choose random matrices $R_i \in \mathfrak{R}$ for $i \in [\ell - 1]$, as well as random $\alpha'_{i,b}$, and set $A^j_{i,b} = \alpha'_{i,b} \cdot R_{i-1} \cdot (S_{i,j} \cdot B_{i,b}) \cdot R_i^{-1}$ for $j \neq \beta$[7]. This formally sets $\alpha_{i,b} = \alpha'_{i,b} s_{i,j}$ in ring $\mathfrak{R}_j$, which leaves $\alpha_{i,b}$ in ring β undetermined. Define $D^j_{i,b} = \alpha'_{i,b} \cdot R_{i-1} \cdot (T_i \cdot B_{i,b}) \cdot R_i^{-1}$.

---

[7] We actually cannot compute the quantities R_i^{-1} since we do not have access to the trapdoor for the encodings. Therefore, we must actually compute R_i^{adj} instead of R_i^{-1}. However, since we multiply by a random scalar anyway, the distribution of encodings is exactly the same as if we had computed the matrix inverse.

Using the $A_{i,b}^j$, we can simulate the public paramters as in the scheme. To answer the challenge ciphertext query, there are two cases. If slot β is empty, then we can answer the challenge ciphertext query as in the slotted FE scheme with the $A_{i,b}^j$ (since β is empty, we do not need $A_{i,b}^\beta$). If slot β is not a copy of slot α on either side of the challenge, then we answer the challenge query by choosing a random $\beta_i' \in \mathfrak{R}$ for $i \in \mathbb{W}, b \in \{0,1\}$, and output the ciphertext

$$C = \text{extend}\left(\text{params}, \mathbb{W}, \left(\beta_i' \cdot \left(\sum_{j: x[j] \neq \bot, j \notin \{\alpha, \beta\}} A_{i,x[j]_{\text{bit}(i)}}^j + D_{i,x[\alpha]_{\text{bit}(i)}}^j \right) \right)_{i \in \mathbb{W}} \right)$$

If the T_i are only encodings in ring $\mathfrak{R}_\alpha$, then this correctly simulates the ciphertext when slot β empty, formally setting $\beta_i = \beta_i'$ in rings other that $\mathfrak{R}_\alpha, \mathfrak{R}_\beta$, and setting $\beta_i = \beta_i' t_i$ in rings $\mathfrak{R}_\alpha, \mathfrak{R}_\beta$ (the value in $\mathfrak{R}_\beta$ is irrelevant in this case). If the T_i are encodings in $\mathfrak{R}_\alpha \times \mathfrak{R}_\beta$, then this correctly simulates the ciphertext when slot β is a copy of slot α, with the same formal settings of variables as before.

We can perform a similar procedure to simulate the secret key queries. In the end, if T_i are only encodings in $\mathfrak{R}_\alpha$, then this correctly simulates the left side in slot duplication, where slot β is empty. If T_i are encodings in $\mathfrak{R}_\alpha \times \mathfrak{R}_\beta$, then this correctly simulates the right side of slot duplication, where slot β is sometimes a copy of slot α. Thus, if Assumption 1 holds, the two cases are indistinguishable.

Ciphertext moving. We will prove ciphertext moving from Assumption 2. Let $\alpha \neq \beta$, where α is the slot the ciphertext is in, and β is the slot we wish to move the ciphertext to. Obtain the challenge for assumption 2, and re-order the rings so that the challenge has the form

$$\left(S_{i,j} = [s_{i,j}]_{\{i\}}^j \right)_{i \in \mathbb{V}, j \notin \{\alpha, \beta\}}, \left(S_{i,j} = [s_{i,j}]_{\{i\}}^j \right)_{i \in \mathbb{W}, j \in [d]}, \left(T_i = [t_i]_{\{i\}}^{\alpha, \beta} \right)_{i \in \mathbb{V}},$$

$$E = \text{extend}^\dagger \left(\text{params}, \mathbb{W}, \left\{ \left(U_{i,j} = [u_{i,j}]_{\{i\}}^j \right)_{i \in \mathbb{W}, j > 1}, \left(V_i = [v_i]_{\{i\}}^\gamma \right)_{i \in \mathbb{W}} \right\} \right)$$

where $\gamma = \alpha$ or $\gamma = \beta$.

We now simulate the view of the adversary. Choose random matrices $R_i \in \mathfrak{R}$ for $i \in [\ell - 1]$, random $\alpha_{i,b}'$, and set $A_{i,b}^j = \alpha_{i,b}' \cdot R_{i-1} \cdot (S_{i,j} \cdot B_{i,b}) \cdot R_i^{-1}$ for $i \in \mathbb{V}, j \notin \{\alpha, \beta\}$, and all $i \in \mathbb{W}, j \in [d]$. This formally sets $\alpha_{i,b} = \alpha_{i,b}' s_{i,j}$ in ring $\mathfrak{R}_j$, which leaves $\alpha_{i,b}$ in rings α and β undetermined for $i \in \mathbb{V}$. Define $A_{i,b}^\alpha + A_{i,b}^\beta = \alpha_{i,b}' \cdot R_{i-1} \cdot (T_i \cdot B_{i,b}) \cdot R_i^{-1}$ for $i \in \mathbb{V}$, which formally sets $\alpha_{i,b} = \alpha_{i,b}' T_i$ in rings $\mathfrak{R}_\alpha$ and $\mathfrak{R}_\beta$.

Now using the $A_{i,b}^j$ values, we can simulate the public parameters (since we have all the values for $i \in \mathbb{W}, j = 0$), as well as all the secret key queries (since all the secret key queries are identical in slots α and β, meaning we will always have $A_{i,b}^\alpha + A_{i,b}^\beta$ together, neither being used separately). To generate the challenge ciphertext, we use the result E of extension. Let $U_{i,j}'$ be the components in E corresponding to the $U_{i,j}$, and V_i' the components corresponding to the V_i. Then the challenge ciphertext is set as

$$C = f_{\mathbb{W}' \to \mathbb{W}},$$

$$\left(\beta_i \cdot R_{i-1} \cdot \left((V_i' \cdot B_{i,x^*_{\mathsf{bit}(i)}}) + \sum_{j:x[j] \neq \perp, j \notin \{\alpha,\beta\}} (U_{i,j}' \cdot B_{i,x[j]_{\mathsf{bit}(i)}}) \right) \cdot R_i^{-1} \right)_{i \in \mathbb{W}}$$

Note that the randomization terms given in E must be used to randomize the components above.

Where x^* is the ciphertext term that is either in slot α or slot β. It is straightforward to show that if the V_i are encodings in $\mathfrak{R}_\alpha$, then this simulates the challenge ciphertext with x^* in slot α, and similarly if V_i are encodings in $\mathfrak{R}_\beta$, the challenge ciphertext has x^* in slot β. Therefore, the two cases are indistinguishable and ciphertext moving follows.

Weak key moving. This is basically the same as ciphertext moving, except that we swap the roles of $\mathbb{W}$ and $\mathbb{V}$. The main difference is that, because now the public parameters lie in $\mathbb{V}$, and we are not given terms in $\mathbb{V}$ containing α separate from β, we must have $\alpha, \beta \neq 0$ so that we can still generate the public parameters in $\mathfrak{R}_0$.

5.2 Adaptively Secure FE for NC^1

Our slotted FE scheme easily gives adaptively secure FE for $N\mathcal{C}^1$:

Theorem 5. *If assumptions 1 and 2 above hold, then adaptively secure FE for NC^1 exists.*

Proof. Set $d = 4$ in our slotted FE scheme. Then Lemmas $3, 4, 5,$ and 6 gives a slotted scheme with $d = 1$ that satisfies strong ciphertext indistinguishability, which implies adaptive FE security.

References

[ABSV14] Ananth, P., Brakerski, Z., Segev, G., Vaikuntanathan, V.: From selective to adaptive security in functional encryption. Cryptology ePrint Archive, Report 2014/917 (2014). http://eprint.iacr.org/2014/917

[AGVW13] Agrawal, S., Gorbunov, S., Vaikuntanathan, V., Wee, H.: Functional encryption: new perspectives and lower bounds. In: Canetti, R., Garay, J.A. (eds.) CRYPTO 2013, Part II. LNCS, vol. 8043, pp. 500–518. Springer, Heidelberg (2013)

[AJ15] Ananth, P., Jain, A.: Indistinguishability obfuscation from compact functional encryption. Cryptology ePrint Archive, Report 2015/173 (2015). http://eprint.iacr.org/2015/173

[Bar86] Barrington, D.A.: Bounded-width polynomial-size branching programs recognize exactly those languages in nc_1. In: STOC (1986)

[BGI14] Boyle, E., Goldwasser, S., Ivan, I.: Functional signatures and pseudorandom functions. In: Krawczyk, H. (ed.) PKC 2014. LNCS, vol. 8383, pp. 501–519. Springer, Heidelberg (2014)

[BGK+14] Barak, B., Garg, S., Kalai, Y.T., Paneth, O., Sahai, A.: Protecting obfuscation against algebraic attacks. In: Nguyen, P.Q., Oswald, E. (eds.) EURO-CRYPT 2014. LNCS, vol. 8441, pp. 221–238. Springer, Heidelberg (2014)

[BLMR13] Boneh, D., Lewi, K., Montgomery, H., Raghunathan, A.: Key homomorphic PRFs and their applications. In: Canetti, R., Garay, J.A. (eds.) CRYPTO 2013, Part I. LNCS, vol. 8042, pp. 410–428. Springer, Heidelberg (2013)

[BR14] Brakerski, Z., Rothblum, G.N.: Virtual black-box obfuscation for all circuits via generic graded encoding. In: Lindell, Y. (ed.) TCC 2014. LNCS, vol. 8349, pp. 1–25. Springer, Heidelberg (2014)

[BS02] Boneh, D., Silverberg, A.: Applications of multilinear forms to cryptography. Cryptology ePrint Archive, Report 2002/080 (2002). http://eprint. iacr.org/2002/080

[BS15] Brakerski, Z., Segev, G.: Function-private functional encryption in the private-key setting. In: Dodis, Y., Nielsen, J.B. (eds.) TCC 2015, Part II. LNCS, vol. 9015, pp. 306–324. Springer, Heidelberg (2015)

[BSW11] Boneh, D., Sahai, A., Waters, B.: Functional encryption: definitions and challenges. In: Ishai, Y. (ed.) TCC 2011. LNCS, vol. 6597, pp. 253–273. Springer, Heidelberg (2011)

[BV15] Bitansky, N., Vaikuntanathan, V.: Indistinguishability obfuscation from functional encryption. Cryptology ePrint Archive, Report 2015/163 (2015). http://eprint.iacr.org/2015/163

[BW13] Boneh, D., Waters, B.: Constrained pseudorandom functions and their applications. In: Sako, K., Sarkar, P. (eds.) ASIACRYPT 2013, Part II. LNCS, vol. 8270, pp. 280–300. Springer, Heidelberg (2013)

[BWZ14] Boneh, D., Wu, D.J., Zimmerman, J.: Immunizing multilinear maps against zeroizing attacks. Cryptology ePrint Archive, Report 2014/930 (2014). http://eprint.iacr.org/2014/930

[BZ14] Boneh, D., Zhandry, M.: Multiparty key exchange, efficient traitor tracing, and more from indistinguishability obfuscation. In: Garay, J.A., Gennaro, R. (eds.) CRYPTO 2014, Part I. LNCS, vol. 8616, pp. 480–499. Springer, Heidelberg (2014)

[CGH+15] Coron, J.-S., et al.: Zeroizing without low-level zeroes: New MMAP attacks and their limitations. In: Gennaro, R., Robshaw, M.J.B. (eds.) CRYPTO 2015, Part I. LNCS, vol. 9215, pp. 247–266. Springer, Heidelberg (2015)

[CHL+15] Cheon, J.H., Han, K., Lee, C., Ryu, H., Stehlé, D.: Cryptanalysis of the multilinear map over the integers. In: Oswald, E., Fischlin, M. (eds.) EUROCRYPT 2015. LNCS, vol. 9056, pp. 3–12. Springer, Heidelberg (2015)

[CLT13] Coron, J.-S., Lepoint, T., Tibouchi, M.: Practical multilinear maps over the integers. In: Canetti, R., Garay, J.A. (eds.) CRYPTO 2013, Part I. LNCS, vol. 8042, pp. 476–493. Springer, Heidelberg (2013)

[CLT15a] Coron, J.-S., Lepoint, T., Tibouchi, M.: New multilinear maps over the integers. Cryptology ePrint Archive, Report 2015/162 (2015). http:// eprint.iacr.org/2015/162

[CLT15b] Coron, J.-S., Lepoint, T., Tibouchi, M.: New multilinear maps over the integers. In: Gennaro, R., Robshaw, J.B. (eds.) CRYPTO 2015, Part I. LNCS, vol. 9215, pp. 267–286. Springer, Heidelberg (2015)

[Gen06] Gentry, C.: Practical identity-based encryption without random oracles. In: Vaudenay, S. (ed.) EUROCRYPT 2006. LNCS, vol. 4004, pp. 445–464. Springer, Heidelberg (2006)

[GGH13a] Garg, S., Gentry, C., Halevi, S.: Candidate multilinear maps from ideal lattices. In: Johansson, T., Nguyen, P.Q. (eds.) EUROCRYPT 2013. LNCS, vol. 7881, pp. 1–17. Springer, Heidelberg (2013)

[GGH+13b] Garg, S., Gentry, C., Halevi, S., Raykova, M., Sahai, A., Waters, B.: Candidate indistinguishability obfuscation and functional encryption for all circuits. In: 54th Annual Symposium on Foundations of Computer Science, pp. 40–49. IEEE Computer Society Press, Berkeley, CA, USA, 26–29 October 2013

[GGHZ14a] Garg, S., Gentry, C., Halevi, S., Zhandry, M.: Fully secure attribute based encryption from multilinear maps. Cryptology ePrint Archive, Report 2014/622 (2014). http://eprint.iacr.org/2014/622

[GGHZ14b] Garg, S., Gentry, C., Halevi, S., Zhandry, M.: Fully secure functional encryption without obfuscation. Cryptology ePrint Archive, Report 2014/666 (2014). http://eprint.iacr.org/2014/666

[GGSW13] Garg, S., Gentry, C., Sahai, A., Waters, B.: Witness encryption and its applications. In: Boneh, D., Roughgarden, T., Feigenbaum, J. (eds.) 45th Annual ACM Symposium on Theory of Computing, pp. 467–476. ACM Press, Palo Alto (2013)

[GLSW14] Gentry, C., Lewko, A., Sahai, A., Waters, B.: Indistinguishability obfuscation from the multilinear subgroup elimination assumption. Cryptology ePrint Archive, Report 2014/309 (2014). http://eprint.iacr.org/2014/309

[GLW14] Gentry, C., Lewko, A., Waters, B.: Witness encryption from instance independent assumptions. In: Garay, J.A., Gennaro, R. (eds.) CRYPTO 2014, Part I. LNCS, vol. 8616, pp. 426–443. Springer, Heidelberg (2014)

[GVW12] Gorbunov, S., Vaikuntanathan, V., Wee, H.: Functional encryption with bounded collusions via multi-party computation. In: Safavi-Naini, R., Canetti, R. (eds.) CRYPTO 2012. LNCS, vol. 7417, pp. 162–179. Springer, Heidelberg (2012)

[IK00] Ishai, Y., Kushilevitz, E.: Randomizing polynomials: a new representation with applications to round-efficient secure computation. In: 41st Annual Symposium on Foundations of Computer Science, pp. 294–304. IEEE Computer Society Press, Redondo Beach, California, USA, 12–14 November 2000

[Kil88] Kilian, J.: Founding cryptography on oblivious transfer. In: 20th Annual ACM Symposium on Theory of Computing, pp. 20–31. ACM Press, Chicago, Illinois, USA, 2–4 May 1988

[KPTZ13] Kiayias, A., Papadopoulos, S., Triandopoulos, N., Zacharias, T.: Delegatable pseudorandom functions and applications. In: Sadeghi, A.-R., Gligor, V.D., Yung, M. (eds.) ACM CCS 2013: 20th Conference on Computer and Communications Security, pp. 669–684. ACM Press, Berlin, Germany, 4–8 November 2013

[O'N10] O'Neill, A.: Definitional issues in functional encryption. Cryptology ePrint Archive, Report 2010/556 (2010). http://eprint.iacr.org/2010/556

[PST14] Pass, R., Seth, K., Telang, S.: Indistinguishability obfuscation from semantically-secure multilinear encodings. In: Garay, J.A., Gennaro, R. (eds.) CRYPTO 2014, Part I. LNCS, vol. 8616, pp. 500–517. Springer, Heidelberg (2014)

[Wat05] Waters, B.: Efficient identity-based encryption without random oracles. In: Cramer, R. (ed.) EUROCRYPT 2005. LNCS, vol. 3494, pp. 114–127. Springer, Heidelberg (2005)

[Wat09] Waters, B.: Dual system encryption: realizing fully secure IBE and HIBE under simple assumptions. In: Halevi, S. (ed.) CRYPTO 2009. LNCS, vol. 5677, pp. 619–636. Springer, Heidelberg (2009)

[Wat14] Waters, B.: A punctured programming approach to adaptively secure functional encryption. Cryptology ePrint Archive, Report 2014/588 (2014). http://eprint.iacr.org/

[Yao82] Yao, A.C.-C.: Protocols for secure computations (extended abstract). In: 23rd Annual Symposium on Foundations of Computer Science, pp. 160–164. IEEE Computer Society Press, Chicago, Illinois, 3–5 November 1982

On Constructing One-Way Permutations from Indistinguishability Obfuscation

Gilad Asharov[✉] and Gil Segev

Hebrew University of Jerusalem, 91904 Jerusalem, Israel
{asharov,segev}@cs.huji.ac.il

Abstract. We prove that there is no black-box construction of a one-way permutation family from a one-way function and an indistinguishability obfuscator for the class of all oracle-aided circuits, where the construction is "domain invariant" (i.e., where each permutation may have its own domain, but these domains are independent of the underlying building blocks).

Following the framework of Asharov and Segev (FOCS '15), by considering indistinguishability obfuscation for *oracle-aided* circuits we capture the common techniques that have been used so far in constructions based on indistinguishability obfuscation. These include, in particular, *non-black-box* techniques such as the punctured programming approach of Sahai and Waters (STOC '14) and its variants, as well as sub-exponential security assumptions. For example, we fully capture the construction of a trapdoor permutation family from a one-way function and an indistinguishability obfuscator due to Bitansky, Paneth and Wichs (TCC '16). Their construction is *not* domain invariant and our result shows that this, somewhat undesirable property, is unavoidable using the common techniques.

In fact, we observe that constructions which are not domain invariant circumvent all known negative results for constructing one-way permutations based on one-way functions, starting with Rudich's seminal work (PhD thesis '88). We revisit this classic and fundamental problem, and resolve this somewhat surprising gap by ruling out *all* such black-box constructions – even those that are not domain invariant.

1 Introduction

One-way permutations are among the most fundamental primitives in cryptography, enabling elegant constructions of a wide variety of central cryptographic primitives. Although various primitives, such as universal one-way hash functions and pseudorandom generators, can be constructed based on any one-way

This work was supported by the European Union's 7th Framework Program (FP7) via a Marie Curie Career Integration Grant, by the Israel Science Foundation (Grant No. 483/13), by the Israeli Centers of Research Excellence (I-CORE) Program (Center No. 4/11), by the US-Israel Binational Science Foundation (Grant No. 2014632), and by a Google Faculty Research Award.

© International Association for Cryptologic Research 2016
E. Kushilevitz and T. Malkin (Eds.): TCC 2016-A, Part II, LNCS 9563, pp. 512–541, 2016.
DOI: 10.1007/978-3-662-49099-0_19

function [40, 56], their constructions based on one-way permutations are much simpler and significantly more efficient [15, 51].

Despite the key role of one-way permutations in the foundations of cryptography, only very few candidates have been suggested over the years. Whereas one-way functions can be based on an extremely wide variety of assumptions, candidate one-way permutation families are significantly more scarce. Up until recently, one-way permutation families were known to exist only based on the hardness of problems related to discrete logarithms and factoring [53, 55]. Moreover, the seminal work by Rudich [57], within the framework of Impagliazzo and Rudich [43], initiated a line of research showing that a one-way permutation cannot be constructed in a black-box manner from a one-way function or from various other cryptographic primitives [24, 44, 49, 50].

Very recently, a one-way (trapdoor!) permutation family was constructed by Bitansky, Paneth and Wichs [13] based on indistinguishability obfuscation [6, 31] and one-way functions. Their breakthrough result provides the first trapdoor permutation family that is not based on the hardness of factoring, and motivates the task of studying the extent to which indistinguishability obfuscation can be used for constructing one-way permutations. Specifically, their work leaves completely unresolved the following question, representing to a large extent the "holy grail" of constructing one-way permutations:

*Is there a construction of a one-way permutation over $\{0,1\}^n$
based on indistinguishability obfuscation and one-way functions?*

While exploring this intriguing question, one immediately identifies two somewhat undesirable properties in the construction of Bitansky, Paneth and Wichs:

- Even when not aiming for trapdoor invertibility, their approach seems limited to providing a *family* of permutations instead of a *single* permutation[1].
- Their construction provides permutations that are defined over domains which both depend on the underlying building blocks and are extremely sparse[2].

From the theoretical perspective, one-way permutation families with these two properties are typically still useful for most constructions that are based on one-way permutations. However, such families lack the elegant structure that makes constructions based on one-way permutations more simple and significantly more efficient when compared to constructions based on one-way functions.

---

[1] Moreover, Bitansky et al. note that their permutations do not seem certifiable. That is, they were not able to provide an efficient method for certifying that a key is well-formed and describes a valid permutation. In contrast, a single permutation is certifiable by its nature.

[2] Each permutation in their construction is defined over a domain of elements of the form $(x, \mathsf{PRF}_K(x))$, where PRF is a pseudorandom function, and each permutation is associated with a different key K. This domain depends on the underlying building block, i.e., the pseudorandom function (equivalently, one-way function).

1.1   Our Contributions

Motivated by the recent construction of Bitansky et al. [13], we study the limitations of using indistinguishability obfuscation for constructing one-way permutations. Following the framework of Asharov and Segev [3], we consider indistinguishability obfuscation for *oracle-aided* circuits, and thus capture the common techniques that have been used so far in constructions based on indistinguishability obfuscation. These include, in particular, *non-black-box* techniques such as the punctured programming approach of Sahai and Waters [58] and its variants, as well as sub-exponential security assumptions. For example, we fully capture the construction of a trapdoor permutation family from a one-way function and an indistinguishability obfuscator due to Bitansky et al. [13]. We refer the reader to Sect. 1.3 for an overview of our framework and of the type of constructions that it captures.

Our work considers three progressively weaker one-way permutation primitives: (1) a *domain-invariant* one-way permutation, (2) a domain-invariant one-way permutation *family*, and (3) a one-way permutation family (which may or may not be domain invariant). Roughly speaking, we say that a construction of a one-way permutation (or a one-way permutation family) is domain invariant if the domain of the permutation is independent of the underlying building blocks (in the case of a permutation family we allow each permutation to have its own domain, but these domains have to be independent of the underlying building blocks).

Within our framework we prove the following two impossibility results, providing a tight characterization of the feasibility of constructing these three progressively weaker one-way permutation primitives based on one-way functions and indistinguishability obfuscation using the common techniques (we summarize this characterization in Fig. 1).

$i\mathcal{O}$+OWF $\not\Rightarrow$ Domain-Invariant OWP Family. Bitansky et al. [13] showed that any sub-exponentially-secure indistinguishability obfuscator and one-way function imply a one-way permutation family which is *not* domain invariant. We show that using the common techniques (as discussed above) one cannot construct the stronger primitive of a *domain-invariant* one-way permutation family (even when assuming sub-exponential security). In particular, we show that the above-described undesirable properties of their construction are unavoidable unless new non-black-box techniques are introduced.[3]

Theorem 1.1. *There is no fully black-box construction of a domain-invariant one-way permutation family from a one-way function f and an indistinguishability obfuscator for the class of all oracle-aided circuits C^f.*

OWF $\not\Rightarrow$ OWP Family. In fact, we observe that constructions which are not domain invariant circumvent the known negative results for constructing one-way

---

[3] In addition to the above-described undesirable properties, our impossibility result holds even for constructions of one-way permutation families that have a "pseudo" input-sampling procedure instead of an "exact" input-sampling procedure (as in [13]), as well as to constructions that are not necessarily certifiable (again, as in [13]).

permutations based on one-way functions, starting with Rudich's seminal work [44,50,52,57]. We revisit this classic and fundamental problem, and resolve this surprising gap by ruling out *all* black-box constructions of one-way permutation *families* from one-way functions – even those that are *not* domain invariant.

Theorem 1.2. *There is no fully black-box construction of a one-way permutation family (even a non-domain-invariant one) from a one-way function.*

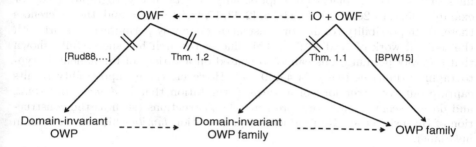

Fig. 1. A dashed arrow from a primitive A to a primitive B indicates that A implies B by definition. Bitansky et al. [13] showed that any sub-exponentially-secure indistinguishability obfuscator and one-way function imply a one-way permutation family (which is not domain invariant), and we show that one cannot construct the stronger primitive of a *domain-invariant* one-way permutation family unless new non-black-box techniques are introduced (even when assuming sub-exponential security). The line of research starting with Rudich [57] showed that one cannot construct a *domain-invariant* one-way permutation from a one-way function in a black-box manner. We improve this result, showing that one cannot construct the weaker primitive of a one-way permutation *family* (even one that is *not* domain invariant) from a one-way function in a black-box manner (again, even when assuming sub-exponential security).

1.2 Related Work

The recent line of research focusing on new constructions based on indistinguishability obfuscation has been extremely fruitful so far (e.g., [1,2,8–14,17, 19,21–23,26,27,30–32,38,41,45,58,60] and the references therein). However, the extent to which indistinguishability obfuscation can be used as a building block has been insufficiently explored. Our approach for proving meaningful impossibility results for constructions based on indistinguishability obfuscation is based on that of Asharov and Segev [3] (which, in turn, was inspired by that of Brakerski, Katz, Segev and Yerukhimovich [18]). They showed that the common techniques (including non-black-box ones) that are used in constructions based on indistinguishability obfuscation can be captured by considering the stronger notion of indistinguishability obfuscation for oracle-aided circuits (see Sect. 1.3 for an elaborate discussion). Generalizing the work of Simon [59] and

Haitner et al. [39], they showed that using these common techniques one cannot construct a collision-resistant hash function family from a general-purpose indistinguishability obfuscator (even when assuming sub-exponential security). In addition, generalizing the work of Impagliazzo and Rudich [43] and Brakerski et al. [18], they showed a similar result from constructing a perfectly-complete key-agreement protocol from a private-key functional encryption scheme (again, even when assuming sub-exponential security).

It is far beyond the scope of this paper to provide an overview of the lines of research on black-box impossibility results in cryptography (see, for example, [5,7,16,20,25,28,29,33–35,42,43,47,48,54,59,61] and the references therein). Impossibility results for constructing one-way permutations start with the seminal work of Rudich [57]. This line of research has successfully shown that one-way permutations cannot be based on a variety of fundamental cryptographic primitives (e.g., [24,44,49,50]). However, these impossibility results capture only constructions of a *single* permutation that is *domain invariant*, and do not seem to capture more general constructions (such as the construction of Bitansky et al. [13] producing a permutation *family* which is *not* domain invariant).

The notion of "domain invariance" that we consider in this work for black-box constructions is somewhat related to that of "function obliviousness" that was introduced by Dachman-Soled, Mahmoody and Malkin [29] for coin-flipping protocols. They proved an impossibility result for constructing an optimally-fair coin-flipping protocol based on any one-way function, as long as the outcome of the protocol is completely independent of the specific one-way function that is used.

1.3   Overview of Our Results

In this section we provide a high-level overview of our two results. First, we describe the framework that enables us to prove a meaningful impossibility result for constructions that are based on indistinguishability obfuscation. Next, we describe Rudich's attack for inverting any domain-invariant permutation relative to a random oracle. Extending Rudich's approach, we then discuss the main technical ideas underlying our results: We present an attack on any domain-invariant permutation family relative to our, significantly more structured, oracle, and we generalize Rudich's attack to non-domain-invariant permutation families in the random-oracle model.

Capturing Non-Black-Box Constructions via $i\mathcal{O}$ for Oracle-Aided Circuits. The fact that constructions that are based on indistinguishability obfuscation are almost always *non-black-box* makes it extremely challenging to prove any impossibility results. For example, a typical such construction would apply the obfuscator to a function that uses the evaluation circuit of a pseudorandom generator or a pseudorandom function, and this requires *specific implementations* of its underlying building blocks.

However, as observed by Asharov and Segev [3], most of the non-black-box techniques that are used on such constructions have essentially the same flavor: The obfuscator is applied to functions that can be constructed in a fully black-box manner from a low-level primitive, such as a one-way function. In particular, the vast majority of constructions rely on the obfuscator itself in a black-box manner. By considering the stronger primitive of an indistinguishability obfuscator for *oracle-aided* circuits (see Definition 2.4), Asharov and Segev showed that such non-black-box techniques in fact directly translate into black-box ones. These include, in particular, non-black-box techniques such as the punctured programming approach of Sahai and Waters [58] and its variants (as well as sub-exponential security assumptions – which are already captured by most frameworks for black-box impossibility results).

Example: The Sahai-Waters Approach. Consider, for example, the construction of a public-key encryption scheme from a one-way function and a general-purpose indistinguishability obfuscator by Sahai and Waters [58]. Their construction relies on the underlying one-way function in a non-black-box manner. However, relative to an oracle that allows the existence of a one-way function f and indistinguishability obfuscation $i\mathcal{O}$ for *oracle-aided circuits*, it is in fact a fully black-box construction. Specifically, Sahai and Waters use the underlying indistinguishability obfuscator for obfuscating a circuit that invokes a puncturable pseudorandom function and a pseudorandom generator as sub-routines. Given that puncturable pseudorandom functions and pseudorandom generators can be based on any one-way function in a fully black-box manner, from our perspective such a circuit is a polynomial-size oracle-aided circuit C^f – which can be obfuscated using $i\mathcal{O}$ (we refer to reader to [3, Sec. 4.6] for an in-depth technical treatment).

This reasoning extends to various variants of the punctured programming approach by Sahai and Waters [58], and in particular fully captures the construction of a trapdoor permutation family from a one-way function and an indistinguishability obfuscator due to Bitansky, Paneth and Wichs [13]. As noted in [3], this approach does not capture constructions that rely on the obfuscator itself in a non-black-box manner (e.g., [11])[4], or constructions that rely on zero-knowledge techniques and require using NP reductions[5].

The Oracle. Our first result is obtained by presenting an oracle Γ relative to which the following two properties hold: (1) there is no domain-invariant one-way permutation family, and (2) there exist an *exponentially-secure* one-way function f and an *exponentially-secure* indistinguishability obfuscator $i\mathcal{O}$ for the

---

[4] With the exception of obfuscating a function that may invoke an indistinguishability obfuscator in a black-box manner. This is captured by our approach – see [3, Sec. 3.1].

[5] Such techniques are captured by the work of Brakerski et al. [18], and we leave it as an intriguing open problem to see whether the two approaches for capturing non-black-box techniques can be unified.

class of all polynomial-size oracle-aided circuits C^f. Our oracle is quite intuitive and consists of three functions: (1) a random function f that will serve as the one-way function, (2) a random injective length-increasing function $\mathcal{O}$ that will serve as the obfuscator (an obfuscation of an oracle-aided circuit C is a "handle" $\mathcal{O}(C, r)$ for a uniformly-chosen string r), and (3) a function Eval that enables evaluations of obfuscated circuits (Eval has access to both f and $\mathcal{O}$): Given a handle $\mathcal{O}(C, r)$ and an input x, it "finds" C and returns $C^f(x)$. We refer the reader to Sect. 3.2 for more details.

The vast majority of our effort is in showing that relative to Γ there is no domain-invariant one-way permutation family. Specifically, as for the second part, our oracle Γ is somewhat similar to the oracle introduced by [3], relative to which they proved the existence of an exponentially-secure one-way function and an exponentially-secure indistinguishability obfuscator (see Sect. 3.2 for the differences between the oracles).

In the remainder of this section we first provide a high-level overview of Rudich's attack on any single domain-invariant permutation in the random-oracle model. Inspired by this attack, we explain the main challenges in extending Rudich's attack to domain invariant constructions relative to our oracle, and to non-domain invariant constructions in the random-oracle model. We again refer the reader to Fig. 1 which summarizes our characterization of the feasible constructions.

Warm-up: Rudich's Attack in the Random-Oracle Model. Following [44, 50, 57] we show that for any oracle-aided polynomial-time algorithm P, if P^f implements a permutation over the same domain $\mathcal{D}$ for all functions f (i.e., P is domain invariant), then there exists an oracle-aided algorithm $\mathcal{A}$ that for any function f inverts P^f with probability 1 by querying f for only a polynomial number of times. The algorithm $\mathcal{A}$ is given some string $y^* \in \mathcal{D}$ and oracle access to f, and is required to find the unique $x^* \in \mathcal{D}$ such that $P^f(x^*) = y^*$. It first initializes a set of queries/answers Q, which will contain the actual queries made by $\mathcal{A}$ to the true oracle f. It repeats the following steps polynomially many times:

1. **Simulation:** $\mathcal{A}$ finds an input $x' \in \mathcal{D}$ and a set of oracle queries/answers f' that is consistent with Q (i.e., $f'(w) = f(w)$ for every $w \in Q$) such that $P^{f'}(x') = y^*$.
2. **Evaluation:** $\mathcal{A}$ evaluates $P^f(x')$ (i.e., evaluation with respect to the true oracle f). If the output is y^*, it terminates and outputs x'.
3. **Update:** $\mathcal{A}$ asks f for all queries in f' that are not in Q, and updates the set Q.

The proof relies on the following observation: In each iteration, either (1) $\mathcal{A}$ finds the pre-image x^* such that $P^f(x^*) = y^*$ or (2) in the update phase, $\mathcal{A}$ queries f with at least one new query that is also made by P during the computation of $P^f(x^*) = y^*$.

Intuitively, if neither of the above holds, then we can construct a "hybrid" oracle $\tilde{f}$ that behaves like f in the evaluation of $P^f(x^*) = y^*$ and behaves like

f' in the evaluation of $P^{f'}(x') = y^*$. This hybrid oracle can be constructed since the two evaluations $P^{f'}(x')$ and $P^f(x^*)$ have no further intersection queries rather than the queries which are already in Q. According to this hybrid oracle $\widetilde{f}$ it holds that $P^{\widetilde{f}}(x') = P^{\widetilde{f}}(x^*) = y^*$ but yet $x^* \neq x'$, and thus relative to $\widetilde{f}$ the value y^* has two pre-images, in contradiction to the fact that P always implements a permutation. Using this claim, since there are only polynomially many f-queries in the evaluation of $P^f(x^*) = y^*$, the algorithm $\mathcal{A}$ must output x^* after a polynomial number of iterations (more specifically, after at most $q + 1$ iterations, where q is the number of oracle gates in the circuit P).

Attacking Domain-Invariant Permutation Families Relative to Our Oracle. We extend the attack described above in two different aspects. First, we rule out constructions of domain-invariant permutation *families* and not just a single permutation. Second, we extend the attack to work relative to our oracle, which is a significantly more structured oracle than a random oracle and therefore raises new technical challenges. Indeed, by the discussion in Sect. 1.3, relative to our oracle *there exists a non-domain-invariant construction of one-way permutation family* [13]. This mere fact represents the subtleties we have to deal with in our setting. In the following overview we focus our attention on the challenges that arise due to the structure of our oracle, as these are the most important and technically challenging ones.

Recall that our oracle Γ consists of three oracles: A length-preserving function f, an *injective* length-increasing function $\mathcal{O}$, and an "evaluation" oracle Eval that depends on both f and $\mathcal{O}$. We now sketch the challenges that these oracles introduce. The first challenge is that the evaluation oracle Eval is not just a "simple" function. This oracle performs (by definition) exponential time computations (e.g., an exponential number of queries to f and $\mathcal{O}$) which may give immense power to the construction P. Specifically, unlike in Rudich's case, here it is no longer true that the computation $P^\Gamma(x^*)$ performs a polynomial number of oracle queries (although P itself is of polynomial size). The second challenge is that since the oracle Eval depends on both f and $\mathcal{O}$, each query to Eval determines many other queries to f and $\mathcal{O}$ implicitly, which we need to make sure that they are considered in the attack. Specifically, given the structured dependencies between f, $\mathcal{O}$ and Eval, in some cases it may not be possible to construct a hybrid oracle even if there are no more intersection queries (in Rudich's case a hybrid oracle always exists).

Finally, the third challenge is the fact that $\mathcal{O}$ is *injective*, which causes the following problem (somewhat similar to [50]). In our case, we are forced to assume that P^Γ is a permutation only when $\mathcal{O}$ is an *injective length-increasing* function and not just any arbitrary function as in Rudich's case (as otherwise our obfuscator may not preserve functionality). Therefore, when constructing the hybrid oracle $\widetilde{\mathcal{O}}$, we must ensure that it is also *injective* in order to reach a contradiction. However, the hybrid oracle $\widetilde{\mathcal{O}}$ might be non-injective when there is some overlap between the images of the true oracle $\mathcal{O}$ and the sampled oracle $\mathcal{O}'$ on elements that are not in Q.

We revise the attack and its analysis to deal with the above obstacles. As in Rudich's attack, the algorithm $\mathcal{A}$ considers the collection of all oracles that are consistent with Q. However, for dealing with the third challenge, it then chooses one of these oracles *uniformly at random* and does not pick just an arbitrarily one as in Rudich's attack. We then show that with all but an exponentially-small probability, there is no overlap between the range of the sampled oracle $\mathcal{O}'$ and the true oracle $\mathcal{O}$, and therefore the hybrid oracle $\widetilde{\mathcal{O}}$ can almost always be constructed in an injective manner. Then, dealing with the first challenge, we show that Eval does not give P a significant capability as one may imagine. Intuitively, this is due to the fact that $\mathcal{O}$ is length increasing, and therefore its range is very sparse. As a result, it is hard to sample a valid image of $\mathcal{O}$ without first querying it, and almost any Eval query can be simulated by the construction P itself. Finally, due to the dependencies between the oracles, for dealing with the second challenge, the algorithm $\mathcal{A}$ will have to sample additional, carefully-chosen, queries that do not necessarily appear in the evaluations $P^{\Gamma}(x^*) = y^*$ or $P^{\Gamma'}(x') = y^*$, but are related to the set of queries that appears in these evaluations. This results in a rather involved proof, where we carefully define this set of queries, and extend the analysis accordingly.

As expected, our proof does not extend to constructions that are not domain invariant. For example, in such constructions for two distinct (injective) functions Γ and Γ', the domain of the permutations P^{Γ} and $P^{\Gamma'}$ may be completely distinct, and this forces additional restrictions on the number of oracles Γ' that are "valid" (i.e., can be used to construct the hybrid oracle $\widetilde{\Gamma}$ as above). As a result, while in the original proof of Rudich all of the oracles Γ' that the adversary may pick are valid, and while in our case all but some exponentially-small amount of oracles Γ' are valid, here the number of valid oracles may be significantly smaller and therefore the attack may succeed with only a negligibly small probability.

Attacking Non-Domain-Invariant Permutation Families in the Random-Oracle Model. At a first sight, it seems that a natural approach towards ruling out non-domain-invariant families relative to a random oracle, is to reduce them to the case of a single permutation. That is, the adversary receives some index α of some permutation in the family, together with the challenge element $y^* \in \mathcal{D}_\alpha^f$ which it needs to invert (note that now the respective domain $\mathcal{D}_\alpha^f$ may depend on both f and α). A natural approach is to apply Rudich's attack to the single permutation $P^f(\alpha, \cdot)$.

However, this approach seems somewhat insufficient due to the following reasons. First, since the construction is not domain invariant, the set of valid indices depends on the underlying primitive, and the set of valid indices for the true oracle f may be completely different than the set of valid indices for the oracle f' that will be sampled by $\mathcal{A}$ in each iteration (e.g., α might even not be a valid index with respect to the sampled f').

Second, when $\mathcal{A}$ inverts y^* relative to f', it may be that the pre-image x' that it finds is not even in the domain $\mathcal{D}_\alpha^f$ of the permutation $P^f(\alpha, \cdot)$ that it needs to invert. That is, it may be that even when the index α is valid relatively to both f and f', the domain of the permutation indexed by α relative to f is completely different than the domain relative to f'. One can try restricting $\mathcal{A}$ to sampling x' from the domain $\mathcal{D}_\alpha^f$, but conditioning on $P^{f'}(\alpha, x') = y^*$ it is not clear that such an x' even exists (and, even if it exists, $\mathcal{A}$ would typically need an exponential number of queries to f for finding it – since $\mathcal{A}$ has no "simple" representation of the sets $\mathcal{D}_\alpha^f$ and $\mathcal{D}_\alpha^{f'}$).

Finally, even when x' is the pre-image of y^* relative to f' and x^* is the pre-image of y^* relative to f, we have no guarantee that neither x' or x^* are even in the domain of the permutation indexed by α when considering the hybrid oracle $\tilde{f}$. Therefore, the fact that $P^f(\alpha, x^*) = P^{f'}(\alpha, x')$ and $x^* \neq x'$ may not indicate any contradiction.

In Sect. 4 we show how to overcome these obstacles. Intuitively, when sampling some function f' and the element x', the algorithm $\mathcal{A}$ samples in addition two "certificates" that ensure that α is a valid index relative to f', and that x' is in the respective domain. These certificates include the randomness used by the index sampling and input sampling procedures of the permutation family, as well as all oracle queries and answers that are involved in the execution of these two procedures. We later use these certificates when defining the hybrid function $\tilde{f}$, and thus ensure that α is a valid index relative to $\tilde{f}$ and that x' is in the respective domain. Similarly, relative to the true oracle f, there exist some other certificates (which are unknown to $\mathcal{A}$), that ensure that α and x^* are valid, and are considered as well when defining the hybrid $\tilde{f}$. Only then we can conclude the existence of a hybrid oracle $\tilde{f}$ relative to which there exist an index α and two distinct inputs x^* and x' in the domain of α such that $P^{\tilde{f}}(\alpha, x^*) = P^{\tilde{f}}(\alpha, x')$.

1.4 Paper Organization

The remainder of this paper is organized as follows. In Sect. 2 we introduce the cryptographic primitives under consideration in this paper, oracle-aided one-way permutation families and indistinguishability obfuscation for oracle-aided circuits, as well as some standard notation. In Sect. 3 we present our negative result for constructing domain-invariant one-way permutation families from indistinguishability obfuscation and one-way functions. Then, in Sect. 4 we present our negative result for constructing one-way permutation families from one-way functions.

2 Preliminaries

In this section we present the notation and basic definitions that are used in this work. For a distribution X we denote by $x \leftarrow X$ the process of sampling a value x from the distribution X. Similarly, for a set $\mathcal{X}$ we denote by $x \leftarrow \mathcal{X}$ the process of sampling a value x from the uniform distribution over $\mathcal{X}$. For an integer $n \in \mathbb{N}$ we denote by $[n]$ the set $\{1, \ldots, n\}$. A function $\mathsf{negl} : \mathbb{N} \to \mathbb{R}^+$ is *negligible* if for every constant $c > 0$ there exists an integer N_c such that $\mathsf{negl}(n) < n^{-c}$ for all $n > N_c$. Throughout the paper, we denote by n the security parameter.

2.1   Oracle-Aided One-Way Permutation Families

We consider the standard notion of a one-way permutation family (see, for example, [37]) when naturally generalized to the setting of oracle-aided algorithms (as required within the context of black-box reductions [43,54]). We start by formalizing the notion of an oracle-aided permutation family, and then introduce the standard one-wayness requirement.

Definition 2.1. *Let* (Gen, Samp, P) *be a triplet of oracle-aided polynomial-time algorithms. We say that* (Gen, Samp, P) *is* an oracle-aided permutation family relative to an oracle* Γ *if the following properties are satisfied:*

- **Index Sampling:** $\mathsf{Gen}^{\Gamma}(\cdot)$ *is a probabilistic algorithm that takes as input the security parameter* 1^n *and produces a distribution over indices* α. *For every* $n \in \mathbb{N}$ *we denote by* $\mathcal{I}_n^{\Gamma}$ *the support of the distribution* $\mathsf{Gen}^{\Gamma}(1^n)$, *and we let* $\mathcal{I}^{\Gamma} \stackrel{\text{def}}{=} \bigcup_{n \in \mathbb{N}} \mathcal{I}_n^{\Gamma}$.
- **Input Sampling:** $\mathsf{Samp}^{\Gamma}(\cdot)$ *is a probabilistic algorithm that takes as input an index* $\alpha \in \mathcal{I}^{\Gamma}$, *and produces a uniform distribution over a set denoted* $\mathcal{D}_\alpha^{\Gamma}$.
- **Permutation Evaluation:** *For any index* $\alpha \in \mathcal{I}^{\Gamma}$, $\mathsf{P}^{\Gamma}(\alpha, \cdot)$ *is a deterministic algorithm that computes a permutation over the set* $\mathcal{D}_\alpha^{\Gamma}$.

Definition 2.2. *An oracle-aided permutation family* (Gen, Samp, P) *is* one way relative to an oracle* Γ *if for any probabilistic polynomial-time algorithm* $\mathcal{A}$ *there exists a negligible function* $\mathsf{negl}(\cdot)$ *such that*

$$\Pr\left[\mathcal{A}^{\Gamma}(\alpha, \mathsf{P}^{\Gamma}(\alpha, x)) = x\right] \leq \mathsf{negl}(n)$$

for all sufficiently large $n \in \mathbb{N}$, *where the probability is taken over the choice of* $\alpha \leftarrow \mathsf{Gen}^{\Gamma}(1^n)$, $x \leftarrow \mathsf{Samp}^{\Gamma}(\alpha)$, *and over the internal randomness of* $\mathcal{A}$.

2.2   Indistinguishability Obfuscation for Oracle-Aided Circuits

We consider the standard notion of indistinguishability obfuscation [6,31] when naturally generalized to oracle-aided circuits (i.e., circuits that may contain oracle gates in addition to standard gates). We first define the notion of functional equivalence relative to a specific function (provided as an oracle), and then we define the notion of an indistinguishability obfuscation for a class of oracle-aided circuits. In what follows, when considering a class $\mathcal{C} = \{\mathcal{C}_n\}_{n \in \mathbb{N}}$ of oracle-aided circuits, we assume that each $\mathcal{C}_n$ consists of circuits of size at most n.

Definition 2.3. *Let* C_0 *and* C_1 *be two oracle-aided circuits, and let* f *be a function. We say that* C_0 *and* C_1 *are* functionally equivalent relative to f, *denoted* $C_0^f \equiv C_1^f$, *if for any input* x *it holds that* $C_0^f(x) = C_1^f(x)$.

Definition 2.4. *A probabilistic polynomial-time algorithm* $i\mathcal{O}$ *is an* indistinguishability obfuscator *relative to an oracle* Γ *for a class* $\mathcal{C} = \{\mathcal{C}_n\}_{n \in \mathbb{N}}$ *of oracle-aided circuits if the following conditions are satisfied:*

– **Functionality.** *For all $n \in \mathbb{N}$ and for all $C \in \mathcal{C}_n$ it holds that*

$$\Pr\left[C^\Gamma \equiv \widehat{C}^\Gamma \; : \; \widehat{C} \leftarrow i\mathcal{O}^\Gamma(1^n, C)\right] = 1.$$

– **Indistinguishability.** *For any probabilistic polynomial-time distinguisher $D = (D_1, D_2)$ there exists a negligible function $\mathsf{negl}(\cdot)$ such that*

$$\mathsf{Adv}^{i\mathcal{O}}_{\Gamma, i\mathcal{O}, D, \mathcal{C}}(n) \stackrel{\text{def}}{=} \left| \Pr\left[\mathsf{Exp}^{i\mathcal{O}}_{\Gamma, i\mathcal{O}, D, \mathcal{C}}(n) = 1\right] - \frac{1}{2}\right| \leq \mathsf{negl}(n)$$

for all sufficiently large $n \in \mathbb{N}$, where the random variable $\mathsf{Exp}^{i\mathcal{O}}_{\Gamma, i\mathcal{O}, D, \mathcal{C}}(n)$ is defined via the following experiment:

1. $b \leftarrow \{0, 1\}$.
2. $(C_0, C_1, ,) \leftarrow D_1^\Gamma(1^n)$ *where* $C_0, C_1 \in \mathcal{C}_n$ *and* $C_0^\Gamma \equiv C_1^\Gamma$.
3. $\widehat{C} \leftarrow i\mathcal{O}^\Gamma(1^n, C_b)$.
4. $b' \leftarrow D_2^\Gamma(, , \widehat{C})$.
5. *If $b' = b$ then output 1, and otherwise output 0.*

3 Impossibility for Constructions Based on $i\mathcal{O}$ and One-Way Functions

In this section we present our negative result for domain-invariant constructions of a one-way permutation family from from a one-way function and an indistinguishability obfuscator. In Sect. 3.1 we formally define the class of constructions to which our negative result applies. Then, in Sect. 3.2 we present the structure of our proof, which is provided in Sects. 3.3–3.4.

3.1 The Class of Constructions

We consider fully black-box constructions of a one-way permutation family from a one-way function f and an indistinguishability obfuscator for all oracle-aided circuits C^f. Following [3], we model these primitives as two independent building blocks due to the following reasons. First, although indistinguishability obfuscation is known to imply one-way functions under reasonable assumptions [45], this enables us to prove an unconditional result. Second, and more importantly, this enables us to capture the common techniques that have been used so far in constructions based on indistinguishability obfuscation. As discussed in Sect. 1.3, these include, in particular, *non-black-box* techniques such as the punctured programming approach of Sahai and Waters [58] and its variants.

We now formally define the class of constructions considered in this section, tailoring our definitions to the specific primitives under consideration. We remind the reader that two oracle-aided circuits, C_0 and C_1, are functionally equivalent relative to a function f, denoted $C_0^f \equiv C_1^f$, if for any input x it holds that $C_0^f(x) = C_1^f(x)$ (see Definition 2.3). The following definition is based on those of [3] (which, in turn, are motivated by [36,46,54]).

Definition 3.1. *A fully black-box construction of a one-way permutation family from a one-way function and an indistinguishability obfuscator for the class* $\mathcal{C} = \{C_n\}_{n\in\mathbb{N}}$ *of all polynomial-size oracle-aided circuits, consists of a triplet of oracle-aided probabilistic polynomial-time algorithms* (Gen, Samp, P)*, an oracle-aided algorithm* M *that runs in time* $T_M(\cdot)$*, and functions* $\epsilon_{M,1}(\cdot)$ *and* $\epsilon_{M,2}(\cdot)$*, such that the following conditions hold:*

- **Correctness:** *For any functions* f $i\mathcal{O}$ *such that* $i\mathcal{O}(C;r)^f \equiv C^f$ *for all* $C \in \mathcal{C}$ *and* $r \in \{0,1\}^*$*, the triplet* (Gen, Samp, P) *is a permutation family relative to the oracle* $(f, i\mathcal{O})$ *(as in Definition 2.1).*
- **Black-Box Proof of Security:** *For any function* f*, for any function* $i\mathcal{O}$ *such that* $i\mathcal{O}(C;r)^f \equiv C^f$ *for all* $C \in \mathcal{C}$ *and* $r \in \{0,1\}^*$*, for any oracle-aided algorithm* $\mathcal{A}$ *that runs in time* $T_\mathcal{A} = T_\mathcal{A}(n)$*, and for any function* $\epsilon_\mathcal{A} = \epsilon_\mathcal{A}(n)$*, if*

$$\Pr\left[\mathcal{A}^{f,i\mathcal{O}}(\alpha, \mathsf{P}^{f,i\mathcal{O}}(\alpha, x)) = x\right] \geq \epsilon_\mathcal{A}(n)$$

for infinitely many values of $n \in \mathbb{N}$*, where the probability is taken over the choice of* $\alpha \leftarrow \mathsf{Gen}^{f,i\mathcal{O}}(1^n)$*,* $x \leftarrow \mathsf{Samp}^{f,i\mathcal{O}}(\alpha)$*, and over the internal randomness of* $\mathcal{A}$*, then either*

$$\Pr\left[M^{\mathcal{A},f,i\mathcal{O}}(f(x)) \in f^{-1}(f(x))\right] \geq \epsilon_{M,1}\left(T_\mathcal{A}(n) \cdot \epsilon_\mathcal{A}^{-1}(n)\right) \cdot \epsilon_{M,2}(n)$$

for infinitely many values of $n \in \mathbb{N}$*, where the probability is taken over the choice of* $x \leftarrow \{0,1\}^n$ *and over the internal randomness of* M*, or*

$$\left|\Pr\left[\mathsf{Exp}^{i\mathcal{O}}_{(f,i\mathcal{O}),i\mathcal{O},M^\mathcal{A},\mathcal{C}}(n) = 1\right] - \frac{1}{2}\right| \geq \epsilon_{M,1}\left(T_\mathcal{A}(n) \cdot \epsilon_\mathcal{A}^{-1}(n)\right) \cdot \epsilon_{M,2}(n)$$

for infinitely many values of $n \in \mathbb{N}$ *(see Definition 2.4 for the description of the experiment* $\mathsf{Exp}^{i\mathcal{O}}_{(f,i\mathcal{O}),i\mathcal{O},M^\mathcal{A},\mathcal{C}}(n)$*).*

The "Security Loss" Functions. Black-box constructions are typically formulated with a reduction algorithm M that runs in *polynomial* time and offers a *polynomial* security loss. In our setting, as we are interested in capturing constructions that may be based on super-polynomial security assumptions, we allow the algorithm M to run in arbitrary time $T_M(n)$ and to have an arbitrary security loss.

In general, the security loss of a reduction is a function of the adversary's running time $T_\mathcal{A}(n)$, of its success probability $\epsilon_\mathcal{A}(n)$, and of the security parameter $n \in \mathbb{N}$. Following Luby [46] and Goldreich [36], we simplify the presentation by considering Levin's unified security measure $T_\mathcal{A}(n) \cdot \epsilon_\mathcal{A}^{-1}(n)$. Specifically, our definition captures the security loss of a reduction by considering an "adversary-dependent" security loss $\epsilon_{M,1}(T_\mathcal{A}(n) \cdot \epsilon_\mathcal{A}^{-1}(n))$, and an "adversary-independent" security loss $\epsilon_{M,2}(n)$. By considering arbitrary security loss functions, we are indeed able to capture constructions that rely on super-polynomial security assumptions. For example, in the recent construction of Bitansky et al. [13] (and in various other recent constructions based on indistinguishability obfuscation), the adversary-dependent loss is polynomial whereas the adversary-independent loss is sub-exponential[6].

Domain-Invariant Constructions. We now define the notion of *domain invariance* which allows us to refine the above class of constructions. Recall that for an oracle-aided permutation family (Gen, Samp, P) and for any oracle Γ, we denote by $\mathcal{I}_n^\Gamma$ the support of the distribution $\mathsf{Gen}^\Gamma(1^n)$ for every $n \in \mathbb{N}$, and we let $\mathcal{I}^\Gamma \overset{\text{def}}{=} \bigcup_{n \in \mathbb{N}} \mathcal{I}_n^\Gamma$ (i.e., $\mathcal{I}^\Gamma$ is the set of all permutation indices). In addition, for any permutation index $\alpha \in \mathcal{I}^\Gamma$ we denote by $\mathcal{D}_\alpha^\Gamma$ the domain of the permutation $\mathsf{P}^\Gamma(\alpha, \cdot)$.

Definition 3.2. *An oracle-aided one-way permutation family* (Gen, Samp, P) *is* domain invariant *relative to a set $\mathfrak{S}$ of oracles if there exist sequences $\{\mathcal{I}_n\}_{n \in \mathbb{N}}$ and $\{\mathcal{D}_\alpha\}_{\alpha \in \mathcal{I}}$ such that for every oracle $\Gamma \in \mathfrak{S}$ the following conditions hold:*

1. *$\mathcal{I}_n^\Gamma = \mathcal{I}_n$ for every $n \in \mathbb{N}$ (i.e., a permutation index α is either valid with respect to all oracles in $\mathfrak{S}$ or invalid with respect to all oracles in $\mathfrak{S}$).*
2. *$\mathcal{D}_\alpha^\Gamma = \mathcal{D}_\alpha$ for every $\alpha \in \bigcup_{n \in \mathbb{N}} \mathcal{I}_n$ (i.e., the domain of $\mathsf{P}^\Gamma(\alpha, \cdot)$ is the same for all $\Gamma \in \mathfrak{S}$).*

3.2 Proof Overview and the Oracle Γ

Our result in this section is obtained by presenting a distribution over oracles Γ relative to which the following two properties hold: (1) there is no domain-invariant one-way permutation family (Gen, Samp, P), and (2) there exist an *exponentially-secure* one-way function f and an *exponentially-secure* indistinguishability obfuscator $i\mathcal{O}$ for the class of all polynomial-size oracle-aided circuits C^f. Equipped with the notation and terminology introduced in Sect. 3.1, we prove the following theorem:

Theorem 3.3. *Let* (Gen, Samp, P, $M, T_M, \epsilon_{M,1}, \epsilon_{M,2}$) *be a fully black-box domain-invariant construction of a one-way permutation family from a one-way function f and an indistinguishability obfuscator for the class of all polynomial-size oracle-aided circuits C^f. Then, at least one of the following propertied holds:*

1. *$T_M(n) \geq 2^{\zeta n}$ for some constant $\zeta > 0$ (i.e., the reduction runs in exponential time).*
2. *$\epsilon_{M,1}(n^c) \cdot \epsilon_{M,2}(n) \leq 2^{-n/4}$ for some constant $c > 1$ (i.e., the security loss is exponential).*

In particular, the theorem implies that if the running time $T_M(\cdot)$ of the reduction is sub-exponential and the adversary-dependent security loss $\epsilon_{M,1}(\cdot)$ is polynomial as in the vast majority of constructions (and, in particular, as in

---

[6] This is also the situation, for example, when using "complexity leveraging" for arguing that any selectively-secure identity-based encryption scheme is in fact adaptively secure.

the construction of Bitansky et al. [13]), then the adversary-independent security loss $\epsilon_{M,2}(\cdot)$ must be exponential (thus ruling out even constructions that rely on sub-exponential security assumptions – as discussed in Sect. 3.1).

In what follows we describe the oracle Γ (more accurately, the distribution over such oracles), and then explain the structure of our proof.

The Oracle Γ. The oracle Γ is a triplet $\left(f, \mathcal{O}, \mathsf{Eval}^{f,\mathcal{O}}\right)$ that is sampled from a distribution $\mathfrak{S}$ defined as follows:

- **The Function $f = \{f_n\}_{n \in \mathbb{N}}$.** For every $n \in \mathbb{N}$, the function f_n is a uniformly chosen function $f_n : \{0,1\}^n \to \{0,1\}^n$.
  Looking ahead, we will prove that f is a one-way function relative to Γ.
- **The Functions $\mathcal{O} = \{\mathcal{O}_n\}_{n \in \mathbb{N}}$ and $\mathsf{Eval}^{f,\mathcal{O}} = \{\mathsf{Eval}_n^{f,\mathcal{O}}\}_{n \in \mathbb{N}}$.** For every $n \in \mathbb{N}$ the function $\mathcal{O}_n$ is an injective function $\mathcal{O}_n : \{0,1\}^{2n} \to \{0,1\}^{10n}$ chosen uniformly at random. The function $\mathsf{Eval}_n^{f,\mathcal{O}}$ on input $(\widehat{C}, x) \in \{0,1\}^{10n} \times \{0,1\}^n$ finds the unique pair $(C, r) \in \{0,1\}^n \times \{0,1\}^n$ such that $\mathcal{O}_n(C, r) = \widehat{C}$, where C is an oracle-aided circuit and r is a string (uniqueness is guaranteed since $\mathcal{O}_n$ is injective). If such a pair exists, it evaluates and outputs $C^f(x)$, and otherwise it outputs $\bot$.
  Looking ahead, we will use $\mathcal{O}$ and Eval for realizing an indistinguishability obfuscator $i\mathcal{O}$ relative to Γ for the class of all polynomial-size oracle-aided circuits C^f.

The Structure of Our Proof. Our proof consists of two parts: (1) showing that relative to Γ there is no domain-invariant one-way permutation family, and (2) showing that relative to Γ the function f is an *exponentially-secure* one-way function and that the pair $(\mathcal{O}, \mathsf{Eval})$ can be used for implementing an *exponentially-secure* indistinguishability obfuscator for oracle-aided circuits C^f.

The vast majority of our effort in this proof is in showing that relative to Γ there is no domain-invariant one-way permutation family. Specifically, as for the second part, our oracle Γ is somewhat similar to the oracle introduced by [3], relative to which they proved the existence of an exponentially-secure one-way function and an exponentially-secure indistinguishability obfuscator. The main difference between the oracles is that the function $\mathcal{O}$ in their case is a *permutation*, whereas in our case it is an *injective length-increasing* function. Since our aim here is to rule out constructions of one-way permutations, then clearly we cannot allow $\mathcal{O}$ to be a permutation. This requires us to revisit the proof of [3] and generalize it to the case where $\mathcal{O}$ is injective and length increasing.

In what follows, we say that an algorithm $\mathcal{A}$ that has oracle access to Γ is a q-query algorithm if it makes at most q queries to Γ, and each of its queries to Eval consists of a circuit of size at most q.

Part 1: Inverting any Domain-Invariant Construction. Building upon and generalizing the work of Rudich [57], we show that relative to the oracle Γ there are no domain-invariant one-way permutations families. As discussed in

Sect. 1.3, Rudich presented an attacker that inverts any *single* domain-invariant permutation that has oracle access to a random function. Here we need to deal with constructions that have oracle access to a significantly more structured functionality[7], and that are permutation *families*. Nevertheless, inspired by the main ideas underlying Rudich's attacker we prove the following theorem in Sect. 3.3:

Theorem 3.4 (Simplified). *Let* $(\mathsf{Gen}, \mathsf{Samp}, \mathsf{P})$ *be an oracle-aided domain-invariant permutation family. Then, there exist a polynomial* $q(\cdot)$ *and a* q-*query algorithm* $\mathcal{A}$ *such that*

$$\Pr\left[\mathcal{A}^\Gamma(\alpha, \mathsf{P}^\Gamma(\alpha, x)) = x\right] \geq 1 - 2^{-10}$$

for any $n \in \mathbb{N}$, *where the probability is taken over the choice of* $\Gamma \leftarrow \mathfrak{S}$, $\alpha \leftarrow \mathsf{Gen}^\Gamma(1^n)$, $x \leftarrow \mathsf{Samp}^\Gamma(\alpha)$, *and over the internal randomness of* $\mathcal{A}$. *Moreover, the algorithm* $\mathcal{A}$ *can be implemented in polynomial time given access to a* PSPACE-*complete oracle.*

Part 2: The Existence of a One-Way Function and an Indistinguishability Obfuscator. As discussed above, by refining the proof of [3] we prove that f is an exponentially-secure one-way function relative to Γ, and we construct an exponentially-secure indistinguishability obfuscator $i\mathcal{O}$. Our obfuscator is defined as follows: For obfuscating an oracle-aided circuit $C \in \{0,1\}^n$ (i.e., we denote by $n = n(C)$ the bit length of C's representation), the obfuscator $i\mathcal{O}$ samples $r \leftarrow \{0,1\}^n$ uniformly at random, computes $\widehat{C} = \mathcal{O}_n(C, r)$, and outputs the circuit $\mathsf{Eval}(\widehat{C}, \cdot)$. That is, the obfuscated circuit consists of a single Eval gate with hardwired input $\widehat{C}$. We prove the following theorem in the full version of this paper [4]:

Theorem 3.5 (Simplified). *For any oracle-aided* $2^{n/4}$-*query algorithm* $\mathcal{A}$ *it hold that*

$$\Pr\left[\mathcal{A}^\Gamma(f(x)) \in f^{-1}(f(x))\right] \leq 2^{-n/2}$$

and

$$\left|\Pr\left[\mathsf{Exp}_{\Gamma,i\mathcal{O},\mathcal{A},\mathcal{C}}^{i\mathcal{O}}(n) = 1\right] = 1 - \frac{1}{2}\right| \leq 2^{-n/4}$$

for all sufficiently large $n \in \mathbb{N}$, *where the probability is taken over the choice of* $\Gamma \leftarrow \mathfrak{S}$ *and internal randomness of* $\mathcal{A}$ *for both cases, in addition to the choice of* $x \leftarrow \{0,1\}^n$ *in the former case and to the internal randomness of the challenger in the latter case.*

3.3 Attacking Domain-Invariant Permutation Families Relative to Γ

We show that relative to the oracle Γ there are no domain-invariant one-way permutations families. As discussed in Sect. 1.3, Rudich presented an attacker

---

[7] For example, there are dependencies between $\mathcal{O}$, Eval and f which allow Eval to query $\mathcal{O}$ for a exponential number of times.

that inverts any *single* domain-invariant permutation that has oracle access to a random function. Here we need to deal with constructions that have oracle access to a significantly more structured functionality. We prove the following theorem:

Theorem 3.6. *Let* (Gen, Samp, P) *be an oracle-aided permutation family that is domain invariant relative to the support of the distribution* $\mathfrak{S}$. *Then, there exist a polynomial* $q(\cdot)$ *and a* q-*query algorithm* $\mathcal{A}$ *such that*

$$\Pr\left[\mathcal{A}^{\Gamma}(\alpha, \mathsf{P}^{\Gamma}(\alpha, x^*)) = x^*\right] \geq 1 - 2^{-10}$$

for any $n \in \mathbb{N}$, *where the probability is taken over the choice of* $\Gamma \leftarrow \mathfrak{S}$, $\alpha \leftarrow \mathsf{Gen}^{\Gamma}(1^n)$, $x^* \leftarrow \mathsf{Samp}^{\Gamma}(\alpha)$, *and over the internal randomness of* $\mathcal{A}$. *Moreover, the algorithm* $\mathcal{A}$ *can be implemented in polynomial time given access to a* PSPACE-*complete oracle.*

We first provide additional notation definitions that we require for the proof of the above theorem, and then we provide its formal proof.

The Event spoof. The event spoof will help up show that the oracle Eval does not provide the construction with any significant capabilities. We formally define this event and then state an important claim that will help up to prove our theorem.

Definition 3.7. *For any oracle-aided algorithm* M, *consider the following event* spoof$_n$ *that may occur during an execution of* $M^{\Gamma}(1^n)$: *The algorithm makes a query* $\mathsf{Eval}_n(\widehat{C}, a)$ *with* $|\widehat{C}| = 10n$ *whose output is not* $\perp$, *yet* $\widehat{C}$ *was not an output of a previous* $\mathcal{O}_n$-*query.*

In the full version of this paper [4] we prove the following claim:

Claim 3.8. *For any* $n \in \mathbb{N}$, *for any* f *and* $\mathcal{O}_{-n} = \{\mathcal{O}_m\}_{m \in \mathbb{N}, m \neq n} m$ *and for any* q-*query algorithm* M, *the probability that* spoof$_n$ *occurs in an execution of* $M^{\Gamma}(1^n)$ *satisfies*

$$\Pr_{\mathcal{O}_n}\left[\text{ spoof}_n\right] \leq q \cdot 2^{-8n} .$$

Notation. Denote by $\mathcal{T}$ the support of the distribution $\mathfrak{S}$ from which our oracle $\Gamma = (f, \mathcal{O}, \mathsf{Eval}^{f, \mathcal{O}})$ is sampled. Note that the oracle Eval is fully determined given f and $\mathcal{O}$, and therefore it is enough to consider the choice of the latter only. For every $n \in \mathbb{N}$ we let $\mathcal{I}_n$ denote the support of $\mathsf{Gen}^{\Gamma}(1^n)$, which is the same for every $\Gamma \in \mathcal{T}$ due to the domain invariant assumption, and we let $\mathcal{I} = \bigcup_{n \in \mathbb{N}} \mathcal{I}_n$. In addition, we let $\mathcal{D} = \{D_\alpha\}_{\alpha \in \mathcal{I}}$ be the set of domains (which is again the same for any $\Gamma \in \mathcal{T}$).

We let $\mathsf{Partial}(\Gamma')$ denote the set of oracle queries that our adversary $\mathcal{A}$ will sample in each iteration. We let Q denote the set of actual queries that made by $\mathcal{A}$ to the true oracle Γ. We write, e.g., $[\mathcal{O}_n(C, r) = \widehat{C}] \in Q$ to denote that Q

contains an $\mathcal{O}_n$-query with input (C, r) and output $\widehat{C}$. Likewise, $[f_n(x) = y] \in$ Partial(Γ') denotes that there is some f_n query in Partial(Γ') with input x and output y. We also use the symbol $\star$ to indicate an arbitrary value, for instance $[\mathsf{Eval}(\widehat{C}, a) = \star] \in Q$ denotes that $\mathcal{A}$ made an Eval call to Γ on the pair $(\widehat{C}, a)$, but we are not interested in the value that was returned by the oracle.

The Set of Queries/Answers that the Adversary Samples. Our adversary $\mathcal{A}$ will sample in each iteration some oracle queries/answers Partial$(\Gamma') = (f', \mathcal{O}', \mathsf{Eval}')$ that are consistent with the actual queries Q it made so far. However, since the oracles $(f, \mathcal{O}, \mathsf{Eval})$ have some dependencies, we want that these dependencies will appear explicitly in the set of queries/answers that the adversary samples (looking ahead, by doing so, we will be able to construct a hybrid oracle $\widetilde{\Gamma}$). Formally, we define:

Definition 3.9 (Consistent Oracle Queries/Answers). *Let* Partial$(\Gamma') = (f', \mathcal{O}', \mathsf{Eval}')$ *be a set of queries/answers. We say it is* consistent *if for every* $m \in \mathbb{N}$ *it holds that:*

1. *For every query* $\left[\mathsf{Eval}_m(\widehat{C}, \star) = \star\right] \in \mathsf{Eval}'$, *there exists a query* $\left[\mathcal{O}_m(\star) = \widehat{C}\right] \in \mathcal{O}'$.
2. *For every query* $\left[\mathsf{Eval}_m(\widehat{C}, a) = \beta\right] \in \mathsf{Eval}'$ *with* $|\widehat{C}| = 10m$ *and* $|a| = m$, *let* $\left[\mathcal{O}_m(C, r) = \widehat{C}\right] \in \mathcal{O}'$ *that is guaranteed to exist by the previous requirement. Then, the oracle* f' *contains also queries/answers sufficient for the evaluation of* $C^{f'}(a)$, *and the value of this evaluation is indeed* β.

Augmented Oracle Queries. For the analysis, we consider the queries that are associated with the execution of $\mathsf{P}^{\Gamma}(\alpha, x^*) = y^*$, for some $\alpha \in \mathcal{I}$. In fact, the set that we consider may contain some additional queries that do not necessarily appear in the execution of $\mathsf{P}^{\Gamma}(\alpha, x^*)$, but are still associated with this execution. Let $\mathsf{RealQ}(\Pi, \Gamma, \alpha, x^*)$ denote the set of actual queries to Γ in the evaluation of $\mathsf{P}^{\Gamma}(\alpha, x^*)$. We define:

Definition 3.10 (Augmented Oracle Queries). *The set of extended queries, denoted* $\mathsf{AugQ}(\Pi, \Gamma, x^*)$, *consists of the following queries:*

1. *All the queries in* $\mathsf{RealQ}(\Pi, \Gamma, \alpha, x^*)$.
2. *For every query* $[\mathsf{Eval}_m(\widehat{C}, a) = \beta] \in \mathsf{RealQ}(\Pi, \Gamma, \alpha, x^*)$ *with* $|\widehat{C}| = 10m$, $|a| = m$ *and* $b \neq \bot$, *let* $C, r \in \{0, 1\}^m$ *be the unique pair such that* $\mathcal{O}_m(C, r) = \widehat{C}$. *Then, the set* $\mathsf{AugQ}(\Pi, \Gamma, x^*)$ *contains all the f-queries/answers sufficient to for the evaluation of* $C^f(a)$.

Note that these additional queries correspond to the consistent queries/answers that the adversary samples in the attack, as in Definition 3.9. We do not explicitly require the first requirement of Definition 3.9 here. This is because our analysis focuses on the case where there is no Eval query on an obfuscated circuit $\widehat{C}$ that is not an output of a previous $\mathcal{O}$-query.

Looking ahead, all the circuits that will be evaluated by the oracle Eval are of some polynomial size in the security parameter, and therefore each evaluation adds some polynomial number of oracle queries to f. Therefore, the overall size of $\mathsf{AugQ}(\Pi, \Gamma, x^*)$ is some polynomial. Let $\ell = \ell(n) > n$ be an upper bound of $|\mathsf{AugQ}(P, \widetilde{\Gamma}, x)|$ for all possible $\widetilde{\Gamma} \in \mathcal{T}$ and all $x \in D_\alpha$.

Equipped with the above notation and definitions, we are now ready to prove Theorem 3.6.

Proof of Theorem 3.6. Let $\Pi = (\mathsf{Gen}, \mathsf{Samp}, \mathsf{P})$ be an oracle-aided permutation family that is domain invariant relative to the support of the distribution $\mathfrak{S}$. Consider the following oracle-aided algorithm $\mathcal{A}$:

The Algorithm $\mathcal{A}$.

- **Input:** *An index $\alpha \in \mathcal{I}$ and a value $y^* \in D_\alpha$.*
- **Oracle access:** *The oracle Γ.*
- **The algorithm:**
  1. *Initialize an empty list Q of oracle queries/answers to Γ (looking ahead, the list Q will always be consistent with the true oracle Γ).*
  2. **Avoiding spoof$_m$ for small** m. *Let $t = \log(16\ell)$. The adversary $\mathcal{A}$ queries the oracle f_m on all inputs $|x| = m$ for all $m \leq t$. It queries $\mathcal{O}_m(C, r)$ for all $|C| = |r| = m \leq t$; and queries $\mathsf{Eval}_m(\widehat{C}, a)$ on all $m \leq t$ with $|\widehat{C}| = m/10$ and $|a| = m$. Denote this set of queries by Q^*.*
  3. *Run the following for $\ell + 1$ iterations:*
     (a) **Simulation phase:** *$\mathcal{A}$ finds a value $x' \in D_\alpha$ and a set $\mathsf{Partial}(\Gamma')$ of consistent oracle queries/answers that is consistent with the list of queries/answers Q, such that $\mathsf{P}^{\mathsf{Partial}(\Gamma')}(\alpha, x') = y^*$ as follows:* [8]
        i. *$\mathcal{A}$ samples an oracle $\Gamma' = (f', \mathcal{O}', \mathsf{Eval}')$ uniformly at random from the set of all oracles that are consistent with Q. That is, f' and $\mathcal{O}'$ are sampled uniformly at random conditioned on Q, and then Eval' is defined accordingly.*
        ii. *$\mathcal{A}$ inverts y^* relative to Γ'. Specifically, $\mathcal{A}$ enumerates over $\mathcal{D}_\alpha$ and find the unique input $x' \in D_\alpha$ for which $\mathsf{P}^{\Gamma'}(\alpha, x') = y^*$.*
        iii. *$\mathcal{A}$ sets $\mathsf{Partial}(\Gamma')$ to be all the queries in Q, and all the queries included in the evaluation of $\mathsf{P}^{\Gamma'}(\alpha, x')$.*
     (b) **Evaluation phase:** *The adversary evaluates $\mathsf{P}^{\Gamma}(\alpha, x')$. If the output of the evaluation is y^*, it halts and outputs x'.*
     (c) **Update phase:** *Otherwise, $\mathcal{A}$ makes all the queries in $\mathsf{Partial}(\Gamma') \setminus Q$ to the true oracle Γ, and continues to the next iteration.*
  4. *In case the adversary has not halted yet, it outputs $\perp$.*

---

[8] Note that the set of queries/answers $\mathsf{Partial}(\Gamma')$ may be inconsistent with the true oracle Γ on all queries $\mathsf{Partial}(\Gamma') \setminus Q$.

Analysis. We show that in each iteration the adversary either finds x^* or learns some query associated with the evaluation $\mathsf{P}^\Gamma(\alpha, x^*)$. We now define these two "bad" events and show that they occur with small probability. We then proceed to the analysis conditioned that these two bad events do not occur.

The Event spoof. For any $m \in \mathbb{N}$, define spoof_m to be the event where

$$\left[\mathsf{Eval}_m(\widehat{C}, a) \neq \bot\right] \in \mathsf{AugQ}(\Pi, \Gamma, x^*)$$

but

$$\left[\mathcal{O}_m(\star, \star) = \widehat{C}\right] \notin \mathsf{AugQ}(\Pi, \Gamma, x^*) \cup Q^* \ .$$

Let $\mathsf{spoof}_\Gamma = \bigvee_m \mathsf{spoof}_m$. By construction, Q^* contains all possible $\mathcal{O}_m$-queries for every $m \leq t$, and therefore spoof_m cannot occur for $m \leq t$. Moreover, by Claim 3.8, we have that

$$\Pr[\ \mathsf{spoof}_\Gamma\ ] \leq \Pr\left[\ \bigvee_m \mathsf{spoof}_m\ \right] \leq \sum_{m=t}^\infty \Pr[\ \mathsf{spoof}_m\ ]$$

$$\leq \sum_{m=\log 16\ell}^\infty \ell \cdot 2^{-8m} \leq 2 \cdot \ell \cdot 2^{-8\log 16\ell} \leq 2^{-31}$$

Let spoof'_m be the event where the adversary $\mathcal{A}$ queries the real oracle Γ some query $[\mathsf{Eval}_m(\widehat{C}, \star)]$, receives a value differ than $\bot$, but $\widehat{C}$ was not an output of Γ on some previous query of $\mathcal{A}$ to $\mathcal{O}_m$. Let $\mathsf{spoof}_{\mathcal{A}} = \bigvee_m \mathsf{spoof}'_m$. Similarly to the above, the probability of $\mathsf{spoof}_{\mathcal{A}}$ is bounded by 2^{-31}. Finally, we let $\mathsf{spoof} = \mathsf{spoof}_\Gamma \vee \mathsf{spoof}_{\mathcal{A}}$, and this probability is bounded by 2^{-30}.

The Event fail. The second bad event that we consider is the event fail. This event occurs whenever $\mathcal{A}$ samples an oracle Γ' that has some contradiction with the oracle Γ, and therefore the hybrid oracle $\widetilde{\Gamma}$ cannot be constructed.

Let $\mathcal{T}(Q)$ be the set of all oracles Γ' that are consistent with Q (namely, each query in Q is answered the same for all $\Gamma' \in \mathcal{T}(Q)$, with the same answer as Γ). In each iteration, the adversary $\mathcal{A}$ samples the oracle Γ' which is consistent with the true oracle queries Q. Let $\mathcal{T}$-admissible denote the set of "valid" oracles that $\mathcal{A}$ may sample; the set $\mathcal{T}$-admissible contains all oracles $\Gamma' = (f', \mathcal{O}', \mathsf{Eval}')$ such that:

- Γ' *is consistent with* Q.
- Γ' *avoids the outputs of* $\mathcal{O}$. For every $m \in \mathbb{N}$, the true oracle $\mathcal{O}_m$ and the sampled oracle $\mathcal{O}'_m$ should have disjoint outputs (except for the queries in Q). Formally, let $Q_m^{\mathcal{O}} = \{x \in \{0,1\}^{2m} \mid [\mathcal{O}_m(x) = \star] \in Q\}$. Then, we require that for every $x, y \notin Q_m^{\mathcal{O}}$ it holds that $\mathcal{O}_m(x) \neq \mathcal{O}'_m(y)$.
- Γ' *avoids invalid* Eval-*queries.* That is, for every $[\mathsf{Eval}_m(\widehat{C}, a) = \bot] \in \mathsf{AugQ}(\Pi, \Gamma, x^*)$, with $|\widehat{C}| = 10m$, for every $C, r \in \{0,1\}^m$ it holds that $\mathcal{O}'_m(C, r) \neq \widehat{C}$.

Notice that the first two conditions relate to the set of queries Q, whereas the third condition relates to the set $\mathsf{AugQ}(\Pi, \Gamma, x^*)$. Moreover, note that the second condition defines $2^{2m} - |Q|$ outputs of $\mathcal{O}'_m$ that are invalid, and the third condition defines at most q invalid outputs. Therefore, there are overall at most 2^{2m} outputs of $\mathcal{O}'_m$ that are invalid.

Note that between iterations, the set Q varies. We define by $\mathsf{Invalid}\text{-}\mathsf{Im}_m^{(i)}$ the set of all invalid outputs for $\mathcal{O}'_m$, in the ith iteration. In all iterations, the set $\mathsf{Invalid}\text{-}\mathsf{Im}_m^{(i)}$ is bounded by 2^{2m}.

Let $\mathsf{fail}_m^{(i)}$ denote the event where $\mathcal{A}$ samples an invalid oracle $\mathcal{O}'_m$ in some iteration i. Let $\mathsf{fail}^{(i)} = \bigvee_m \mathsf{fail}_m^{(i)}$, and let $\mathsf{fail} = \bigvee_i \mathsf{fail}^{(i)}$. For every m, we have that:

$$\Pr_{\mathcal{O}'_m}\left[\, \mathsf{fail}_m^{(i)} \,\right] = \Pr_{\mathcal{O}'_m}\left[\, \exists x \in \{0,1\}^{2m} \text{ s.t. } \mathcal{O}'_m(x) \in \mathsf{Invalid}\text{-}\mathsf{Im}_m^{(i)} \,\right]$$

$$\leq 2^{2m} \cdot \frac{\left|\mathsf{Invalid}\text{-}\mathsf{Im}_m^{(i)}\right|}{2^{10m} - 2^{2m}} \leq 2^{-5m} \,.$$

As a result, we get that the probability that sampling $\mathcal{O}$ fails for some length $m > t$ is bounded by

$$\Pr_{\mathcal{O}'}\left[\, \mathsf{fail}^{(i)} \,\right] \leq \sum_{m=t}^{\infty} 2^{-5m} \leq 2 \cdot 2^{-5t} \,.$$

We therefore conclude that the probability that in some of the $\ell + 1$ iterations, the adversary $\mathcal{A}$ samples some oracle $\Gamma' \notin \mathcal{T}$-admissible is bounded by

$$\Pr[\, \mathsf{fail} \,] \leq \sum_{i=1}^{\ell+1} \Pr\left[\, \mathsf{fail}^{(i)} \,\right] \leq (\ell+1) \cdot 2 \cdot 2^{-5t} = 2(\ell+1) \cdot \left(2^{-4} \cdot \ell^{-1}\right)^5 \leq 2^{-19} \,,$$

where recall that $t = \log(16\ell)$. We are now ready for the main claim of the analysis.

Claim 3.11. *Assume that* fail *and* spoof *do not occur. Then, in every iteration at least one of the following occurs:*

1. $\mathcal{A}$ *finds the pre-image* x^* *such that* $\mathsf{P}^\Gamma(\alpha, x^*) = y^*$.
2. *During the update phase* $\mathcal{A}$ *queries* Γ *with at least one of the queries in* $\mathsf{AugQ}(\Pi, \Gamma, x^*)$.

Proof. Assume that neither one of the above conditions hold. Then, we show that there exists an oracle $\widetilde{\Gamma} \in \mathcal{T}$ that behaves like the true oracle Γ on $\mathsf{P}^{\widetilde{\Gamma}}(\alpha, x^*) = \mathsf{P}^\Gamma(\alpha, x^*) = y^*$, and on the other hand, it behaves like Γ' in the evaluation of $\mathsf{P}^{\widetilde{\Gamma}}(\alpha, x') = \mathsf{P}^{\mathsf{Partial}(\Gamma')}(\alpha, x') = y^*$. According to this oracle $\widetilde{\Gamma}$, the following hold:

1. Since Π is a domain-invariant construction, and since $\widetilde{\Gamma} \in \mathcal{T}$, there exists some randomness $r \in \{0,1\}^*$ such that $\mathsf{Gen}^{\widetilde{\Gamma}}(1^n; r) = \alpha$.

2. Since Π is a domain-invariant construction, it holds that $\mathrm{Im}(\mathsf{Samp}^{\widetilde{\Gamma}}(\alpha)) = \mathrm{Im}(\mathsf{Samp}^{\Gamma}(\alpha)) = \mathrm{Im}(\mathsf{Samp}^{\mathsf{Partial}(\Gamma')}(\alpha)) = D_\alpha$. As a result, there exists some randomness $r' \in \{0,1\}^*$ such that $\mathsf{Samp}^{\widetilde{\Gamma}}(\alpha; r') = x'$ and $\mathsf{Samp}^{\widetilde{\Gamma}}(\alpha; r^*) = x^*$.

3. As mentioned above, $\mathsf{P}^{\widetilde{\Gamma}}(\alpha, x') = y^*$ and $\mathsf{P}^{\widetilde{\Gamma}}(\alpha, x^*) = y^*$.

Since the first condition in the statement does not hold, we conclude that $x' \neq x^*$ but still $\mathsf{P}^{\widetilde{\Gamma}}(\alpha, x') = \mathsf{P}^{\widetilde{\Gamma}}(\alpha, x^*)$, in contradiction to the assumption that $\mathsf{P}^{\widetilde{\Gamma}}(\alpha, \cdot)$ defines a permutation.

We now show that the oracle $\widetilde{\Gamma} = (\widetilde{f}, \widetilde{\mathcal{O}}, \widetilde{\mathsf{Eval}})$ as above can be constructed. Recall that we assume that the both conditions of the statement of the claim do not hold, and therefore in particular it holds that $\mathsf{AugQ}(\Pi, \Gamma, x^*) \cap \mathsf{Partial}(\Gamma') \subseteq Q$.

The Oracle $\widetilde{f}$. Note that for every $m \leq t$, the set of queries Q^* contains all the functions $\{f_m\}_{m \leq t}$ and thus agrees completely with f (i.e., also with f'). We therefore set $\widetilde{f}_m = f_m$.

For every $m > t$, we define the function $\widetilde{f}_m$ as follows. For every x such that $[f_m(x) = y'] \in \mathsf{AugQ}(\Pi, \Gamma, x^*)$, we set $\widetilde{f}_m(x) = y'$. For every $[f_m(x) = y] \in \mathsf{Partial}(\Gamma')$, we set $\widetilde{f}_m(x) = y$. Since $\mathsf{AugQ}(\Pi, \Gamma, x^*) \cap \mathsf{Partial}(\Gamma') \subseteq Q$, we have that there is no contradiction, i.e., there are no input x and outputs y, y' such that $y \neq y'$ and $[f_m(x) = y'] \in f'$ and $[f_m(x) = y] \in \mathsf{AugQ}(\Pi, \Gamma, x^*)$. For any other value $x \notin \mathsf{Partial}(\Gamma') \cap \mathsf{AugQ}(\Pi, \Gamma, x^*)$, we set $\widetilde{f}_m(x) = 0^m$.

Before we continue to define the oracle $\widetilde{\mathcal{O}}$, we first define some set of output values that $\widetilde{\mathcal{O}}$ will have to avoid. For every $m > t$, we define the set $\mathsf{avoid}\text{-}\mathcal{O}_m$ as

$$\mathsf{avoid}\text{-}\mathcal{O}_m \stackrel{\mathrm{def}}{=} \left\{ \widehat{C} \in \{0,1\}^{10m} \mid \exists\, [\mathsf{Eval}_m(\widehat{C}, \star) = \star] \in \mathsf{AugQ}(\Pi, \Gamma, x^*) \cup \mathsf{Partial}(\Gamma') \right\} .$$

The Oracle $\widetilde{\mathcal{O}}$. The oracle is already defined for every $m \leq t$. For every $m > t$, we define the function $\widetilde{\mathcal{O}}_m$ as follows. For every $[\mathcal{O}_m(x) = y] \in \mathsf{AugQ}(\Pi, \Gamma, x^*)$, we set $\widetilde{\mathcal{O}}_m(x) = y$. Likewise, for every $[\mathcal{O}_m(x) = y] \in \mathsf{Partial}(\Gamma')$, we set $\widetilde{\mathcal{O}}_m(x) = y$. Since $\mathsf{AugQ}(\Pi, \Gamma, x^*) \cap \mathsf{Partial}(\Gamma') \subseteq Q$, we have that there is no contradiction, that is, there is no pre-image that has two possible outputs. Moreover, since fail does not occur, it holds that $\Gamma' \in \mathcal{T}$-admissible, the two functions $\mathcal{O}_m$ and (the partially defined function) $\mathcal{O}'_m$ do not evaluate to the same output, and so the partially defined function $\widetilde{\mathcal{O}}_m$ is injective. We continue to define $\widetilde{\mathcal{O}}_m$ on the additional values, such that $\widetilde{\mathcal{O}}_m$ is injective and avoids the set $\mathsf{avoid}\text{-}\mathcal{O}_m$.

The Oracle $\widetilde{\mathsf{Eval}}$. We define the oracle $\widetilde{\mathsf{Eval}}$ using the oracles $\widetilde{f}$ and $\widetilde{\mathcal{O}}$ exactly as the true oracle Eval is defined using the true oracles f and $\mathcal{O}$. We now show that $\widetilde{\mathsf{Eval}}$ is consistent with $\mathsf{AugQ}(\Pi, \Gamma, x^*)$ and $\mathsf{Partial}(\Gamma')$. That is, that every query $[\mathsf{Eval}_m(\star, \star)] \in \mathsf{AugQ}(\Pi, \Gamma, x^*) \cup \mathsf{Partial}(\Gamma')$ has the same answer with $\widetilde{\mathsf{Eval}}$, and therefore $\mathsf{P}^{\Gamma}(\alpha, x^*) = \mathsf{P}^{\widetilde{\Gamma}}(\alpha, x^*)$ and $\mathsf{P}^{\Gamma'}(x\alpha, x') = \mathsf{P}^{\widetilde{\Gamma}}(\alpha, x')$. We have:

1. Assume that there exists $[\mathsf{Eval}(\widehat{C}, a) = \beta] \in \mathsf{Eval}'$ for some $\beta \neq \bot$. Since the oracle $\mathsf{Partial}(\Gamma') = (f', \mathcal{O}', \mathsf{Eval}')$ is consistent (recall Definition 3.9), then there exists a query $[\mathcal{O}_m(C, r) = \widehat{C}] \in \mathsf{Partial}(\Gamma')$ and f' contains all the necessary queries/answers for the evaluation of $C^{f'}(a)$, and it also holds that $C^{f'}(a) = \beta$. However, since any $(f', \mathcal{O}')$-queries in $\mathsf{Partial}(\Gamma')$ has the exact same answer with $(\widetilde{f}, \widetilde{\mathcal{O}})$, it holds that $C^{\widetilde{f}}(a) = \beta$ and $\widetilde{\mathcal{O}}(C, r) = \widehat{C}$, and so, from the definition of $\widetilde{\mathsf{Eval}}$ it holds that $\widetilde{\mathsf{Eval}}(\widehat{C}, a) = \beta$ as well.

2. Assume that there exists $[\mathsf{Eval}(\widehat{C}, a) = \beta] \in \mathsf{AugQ}(\Pi, \Gamma, x^*)$ for some $\beta \neq \bot$. Since the event spoof does not occur, there exists a query $[\mathcal{O}(C, r) = \widehat{C}] \in \mathsf{AugQ}(\Pi, \Gamma, x^*)$ as well, and $\mathsf{AugQ}(\Pi, \Gamma, x^*)$ contains also all the f-queries necessary for the evaluation $C^f(a)$. Since these queries appear in $\mathsf{AugQ}(\Pi, \Gamma, x^*)$, it holds that $\widetilde{f}$ and $\widetilde{\mathcal{O}}$ agree on the same queries, and therefore $\widetilde{\mathsf{Eval}}(\widehat{C}, a) = \beta$, as well.

3. For every query $[\mathsf{Eval}(\widehat{C}, a) = \bot] \in \mathsf{Partial}(\Gamma') \cup \mathsf{AugQ}(\Pi, \Gamma, x^*)$ we show that $\widetilde{\mathsf{Eval}}(\widehat{C}, a) = \bot$ as well. Specifically, it suffices to show that there do not exist C and r for which $\widetilde{\mathcal{O}}(C, r) = \widehat{C}$. Assume towards a contradiction that there exist such C and r, then there is inconsistency only if $\widetilde{\mathcal{O}}(C, r) = \widehat{C}$ but $[\mathsf{Eval}(\widehat{C}, a) = \bot] \in \mathsf{Partial}(\Gamma') \cup \mathsf{AugQ}(\Pi, \Gamma, x^*)$. However, this cannot occur since the oracles $\mathcal{O}$ and $\mathcal{O}'$ do not contradict, and $\widetilde{\mathcal{O}}$ avoids all Eval-queries in both $\mathsf{Partial}(\Gamma')$ and $\mathsf{AugQ}(\Pi, \Gamma, x^*)$, since it avoids the set avoid-$\mathcal{O}$.

This completes the proof of claim 3.11.

From the previous claim we conclude that:

$$\Pr_{\substack{\Gamma \leftarrow \mathfrak{S} \\ \alpha \leftarrow \mathsf{Gen}^\Gamma(1^n) \\ x^* \leftarrow \mathsf{Samp}^\Gamma(\alpha)}} \left[\mathcal{A}^\Gamma(\alpha, \mathsf{P}^\Gamma(\alpha, x^*)) = x^* \mid \overline{\mathsf{fail}} \wedge \overline{\mathsf{spoof}} \right] = 1 .$$

Since $\Pr[\,\mathsf{fail}\,] + \Pr[\,\mathsf{spoof}\,] \leq 2^{-10}$, it holds that:

$$\Pr_{\substack{\Gamma \leftarrow \mathfrak{S} \\ \alpha \leftarrow \mathsf{Gen}^\Gamma(1^n) \\ x^* \leftarrow \mathsf{Samp}^\Gamma(\alpha)}} \left[\mathcal{A}^\Gamma(\alpha, \mathsf{P}^\Gamma(\alpha, x^*)) = x^* \right] \geq 1 - 2^{-10}.$$

Finally, we observe that $\mathcal{A}$ makes at most a polynomial number of oracle queries to Γ, and all other computations that are done by $\mathcal{A}$ can be done using a polynomial number of queries to a PSPACE-complete oracle (as in the work of Impagliazzo and Rudich [43]): In each iteration, sampling x' and $\mathsf{Partial}(\Gamma')$ can be done in polynomial space, requires access only to Q which is of polynomial size, and does not require access to Γ. ∎

3.4 Proof of Theorem 3.3

Equipped with the proofs of Theorems 3.4 and 3.5, we are now ready to prove Theorem 3.3.

Proof of Theorem 3.3. Let $(\mathsf{Gen}, \mathsf{Samp}, \mathsf{P}, M, T_M, \epsilon_{M,1}, \epsilon_{M,2})$ be a fully black-box construction of a domain-invariant one-way permutation family from a one-way function f and an indistinguishability obfuscator $i\mathcal{O}$ for the class $\mathcal{C}$ of all oracle-aided polynomial-size circuits C^f (recall Definition 3.2). Theorem 3.4 guarantees the existence of an oracle-aided algorithm $\mathcal{A}$ that runs in polynomial time $T_{\mathcal{A}}(n)$ such that

$$\Pr\left[\mathcal{A}^{\mathsf{PSPACE},\Gamma}(\alpha, \mathsf{P}^{\Gamma}(\alpha, x)) = x\right] \geq \epsilon_{\mathcal{A}}(n)$$

for any $n \in \mathbb{N}$, where $\epsilon_{\mathcal{A}}(n) = 1 - 2^{-10}$, and the probability is taken over the choice of $\Gamma \leftarrow \mathfrak{S}$, $\alpha \leftarrow \mathsf{Gen}^{\Gamma}(1^n)$, $x \leftarrow \mathsf{Samp}^{\Gamma}(\alpha)$, and over the internal randomness of $\mathcal{A}$. Definition 3.1 then states that there are two possible cases to consider: $\mathcal{A}$ can be used either for inverting the one-way permutation f or for breaking the indistinguishability obfuscator $i\mathcal{O}$.

In the first case we obtain from Definition 3.1 that

$$\Pr\left[M^{\mathcal{A}^{\mathsf{PSPACE}},\Gamma}(f(x)) \in f^{-1}(f(x))\right] \geq \epsilon_{M,1}\left(T_{\mathcal{A}}(n) \cdot \epsilon_{\mathcal{A}}^{-1}(n)\right) \cdot \epsilon_{M,2}(n)$$

for infinitely many values of $n \in \mathbb{N}$, where the probability is taken over the choice of $x \leftarrow \{0,1\}^n$ and over the internal randomness of M. The algorithm M may invoke $\mathcal{A}$ on various security parameters (i.e., in general M is not restricted to invoking $\mathcal{A}$ only on security parameter n), and we denote by $\ell(n)$ the maximal security parameter on which M invokes $\mathcal{A}$ (when M itself is invoked on security parameter n). Thus, viewing $M^{\mathcal{A}}$ as a single oracle-aided algorithm that has access to a PSPACE-complete oracle and to Γ, its running time $T_{M^{\mathcal{A}}}(n)$ satisfies $T_{M^{\mathcal{A}}}(n) \leq T_M(n) \cdot T_{\mathcal{A}}(\ell(n))$ (this follows since M may invoke $\mathcal{A}$ at most $T_M(n)$ times, and the running time of $\mathcal{A}$ on each such invocation is at most $T_{\mathcal{A}}(\ell(n))$). In particular, viewing $M' \overset{\text{def}}{=} M^{\mathcal{A}^{\mathsf{PSPACE}}}$ as a single oracle-aided algorithm that has oracle access to Γ, implies that M' is a q-query algorithm where $q(n) = T_{M^{\mathcal{A}}}(n)$.[9] Theorem 3.5 then implies that either $2^{n/4} \leq q(n)$ or $\epsilon_{M,1}\left(T_{\mathcal{A}}(n) \cdot \epsilon_{\mathcal{A}}^{-1}(n)\right) \cdot \epsilon_{M,2}(n) \leq 2^{-n/2}$. In the first sub-case, noting that $\ell(n) \leq T_M(n)$, we obtain that

$$2^{n/4} \leq q(n) = T_{M^{\mathcal{A}}}(n) \leq T_M(n) \cdot T_{\mathcal{A}}(\ell(n)) \leq T_M(n) \cdot T_{\mathcal{A}}(T_M(n)).$$

The running time $T_{\mathcal{A}}(n)$ of the adversary $\mathcal{A}$ (when given access to a PSPACE-complete oracle) is some fixed polynomial in n, and therefore $T_M(n) \geq 2^{\zeta n}$ for some constant $\zeta > 0$. In the second sub-case, we have that $\epsilon_{M,1}(T_{\mathcal{A}}(n)) \cdot \epsilon_{M,2}(n) \leq 2^{-n/2}$, and since $T_{\mathcal{A}}(n)$ is some fixed polynomial in n (and $\epsilon_{\mathcal{A}}(n)$ is a constant) we obtain that $\epsilon_{M,1}(n^c) \cdot \epsilon_{M,2}(n) \leq 2^{-n/2}$ for some constant $c > 1$.

In the second case we obtain from Definition 3.1 that

$$\left|\Pr\left[\mathsf{Exp}^{i\mathcal{O}}_{\Gamma, i\mathcal{O}, M^{\mathcal{A}^{\mathsf{PSPACE}}}, \mathcal{C}}(n) = 1\right] - \frac{1}{2}\right| \geq \epsilon_{M,1}\left(T_{\mathcal{A}}(n) \cdot \epsilon_{\mathcal{A}}^{-1}(n)\right) \cdot \epsilon_{M,2}(n)$$

---

[9] Recall that an algorithm that has oracle access to Γ is a q-query algorithm if it makes at most q queries to Γ, and each of its queries to Eval consists of a circuit of size at most q.

for infinitely many values of $n \in \mathbb{N}$, where $\Gamma \leftarrow \mathfrak{S}$. As in the first case, viewing $M' \overset{\text{def}}{=} M^{\mathcal{A}^{\text{PSPACE}}}$ as a single oracle-aided algorithm that has oracle access to Γ, implies that M' is a q-query algorithm where $q(n) = T_{M^{\mathcal{A}}}(n)$. Theorem 3.5 then implies that either $2^{n/4} \leq q(n)$ or $\epsilon_{M,1}\left(T_{\mathcal{A}}(n) \cdot \epsilon_{\mathcal{A}}^{-1}(n)\right) \cdot \epsilon_{M,2}(n) \leq 2^{-n/4}$. As in the first case, this implies that either $T_M(n) \geq 2^{\zeta n}$ for some constant $\zeta > 0$, or $\epsilon_{M,1}(n^c) \cdot \epsilon_{M,2}(n) \leq 2^{-n/4}$ for some constant $c > 1$. ∎

4   Impossibility for Constructions Based on One-Way Functions

As discussed in Sect. 1.3, the known impossibility results for constructing one-way permutations based on one-way functions [44,50,57] fall short in two aspects. First, these results rule out constructions of a *single* one-way permutation, and do not rule out constructions of a one-way permutation *family*. Second, these results rule out constructions that are *domain invariant* (recall Definition 3.2), and do not rule out constructions that are *not* domain invariant (such as the construction of Bitansky et al. [13]).

In this section we resolve this surprising gap by ruling out *all* fully black-box constructions of one-way permutation *families* from one-way functions – even constructions that are *not* domain invariant. In what follows we first formally define this class of reductions, and then state and prove our result.

Definition 4.1. *A fully black-box construction of a one-way permutation family from a one-way function consists of a triplet of oracle-aided probabilistic polynomial-time algorithms* (Gen, Samp, P), *an oracle-aided algorithm M that runs in time $T_M(\cdot)$, and functions $\epsilon_{M,1}(\cdot)$ and $\epsilon_{M,2}(\cdot)$, such that the following conditions hold:*

- **Correctness:** *For any function f the triplet* (Gen, Samp, P) *is a permutation family relative to f (as in Definition 2.1).*
- **Black-Box Proof of Security:** *For any function f, for any oracle-aided algorithm $\mathcal{A}$ that runs in time $T_{\mathcal{A}} = T_{\mathcal{A}}(n)$, and for any function $\epsilon_{\mathcal{A}} = \epsilon_{\mathcal{A}}(n)$, if*

$$\Pr\left[\mathcal{A}^f(\alpha, \mathsf{P}^f(\alpha, x)) = x\right] \geq \epsilon_{\mathcal{A}}(n)$$

for infinitely many values of $n \in \mathbb{N}$, where the probability is taken over the choice of $\alpha \leftarrow \mathsf{Gen}^f(1^n)$, $x \leftarrow \mathsf{Samp}^f(\alpha)$, and over the internal randomness of $\mathcal{A}$, then

$$\Pr\left[M^{f,\mathcal{A}}\left(f(x)\right) \in f^{-1}(f(x))\right] \geq \epsilon_{M,1}\left(T_{\mathcal{A}}(n) \cdot \epsilon_{\mathcal{A}}^{-1}(n)\right) \cdot \epsilon_{M,2}(n)$$

for infinitely many values of $n \in \mathbb{N}$, where the probability is taken over the choice of $x \leftarrow \{0,1\}^n$ and over the internal randomness of M.

The above definition clearly captures constructions that are not domain invariant. First, it allows the support of the distribution $\mathsf{Gen}^f(1^n)$ to depend on f. Second, for each permutation index α that is produced by $\mathsf{Gen}^f(1^n)$, it allows the domain of the permutation $\mathsf{P}^f(\alpha, \cdot)$ to depend on f. For this general class of reductions we prove the following theorem:

Theorem 4.2. *Let* $(\mathsf{Gen}, \mathsf{Samp}, \mathsf{P}, M, T_M, \epsilon_{M,1}, \epsilon_{M,2})$ *be a fully black-box construction of a one-way permutation family from a one-way function. Then, at least one of the following properties holds:*

1. $T_M(n) \geq 2^{\zeta n}$ *for some constant* $\zeta > 0$ *(i.e., the reduction runs in exponential time).*
2. $\epsilon_{M,1}(n^c) \cdot \epsilon_{M,2}(n) \leq 2^{-\beta n}$ *for some constants* $c > 1$ *and* $\beta > 0$ *(i.e., the security loss is exponential).*[10]

Towards proving Theorem 4.2 we generalize the attack presented in Sect. 1.3 from inverting any *single* oracle-aided *domain-invariant* permutation to inverting any oracle-aided one-way permutation *family* – even such families that are not domain invariant. In the full version of this paper [4], we prove the following theorem:

Theorem 4.3. *Let* $(\mathsf{Gen}, \mathsf{Samp}, \mathsf{P})$ *be a triplet of oracle-aided probabilistic polynomial-time algorithms that is a permutation family relative to any oracle* f. *Then, there exists an oracle-aided algorithm* $\mathcal{A}$ *that makes a polynomial number of oracle queries such that for any function* f *it holds that*

$$\Pr\left[\mathcal{A}^f(\alpha, \mathsf{P}^f(\alpha, x)) = x\right] = 1$$

for any $n \in \mathbb{N}$, *where the probability is taken over the choice of* $\alpha \leftarrow \mathsf{Gen}^f(1^n)$ *and* $x \leftarrow \mathsf{Samp}^f(\alpha)$, *and over the internal randomness of* $\mathcal{A}$. *Moreover, the algorithm* $\mathcal{A}$ *can be implemented in polynomial time given access to a* PSPACE-*complete oracle.*

References

1. Ananth, P., Brakerski, Z., Segev, G., Vaikuntanathan, V.: From selective to adaptive security in functional encryption. In: Gennaro, R., Robshaw, M. (eds.) CRYPTO 2015. LNCS, vol. 9216, pp. 657–677. Springer, Heidelberg (2015)
2. Ananth, P., Jain, A.: Indistinguishability obfuscation from compact functional encryption. In: Gennaro, R., Robshaw, M. (eds.) CRYPTO 2015. LNCS, vol. 9215, pp. 308–326. Springer, Berlin (2015)
3. Asharov, G., Segev, G.: Limits on the power of indistinguishability obfuscation and functional encryption. In: Proceedings of the 56th Annual IEEE Symposium on Foundations of Computer Science (2015, To appear). https://eprint.iacr.org/2015/341.pdf

---

[10] In particular, if the adversary-dependent security loss $\epsilon_{M,1}(\cdot)$ is polynomial, then the adversary-independent security loss $\epsilon_{M,2}(\cdot)$ is exponential.

4. Asharov, G., Segev, G.: On constructing one-way permutations from indistinguishability obfuscation. Cryptology ePrint Archive, Report 2015/752 (2015). http://eprint.iacr.org/2015/752.pdf

5. Baecher, P., Brzuska, C., Fischlin, M.: Notions of black-box reductions, revisited. In: Sako, K., Sarkar, P. (eds.) ASIACRYPT 2013, Part I. LNCS, vol. 8269, pp. 296–315. Springer, Heidelberg (2013)

6. Barak, B., Goldreich, O., Impagliazzo, R., Rudich, S., Sahai, A., Vadhan, S.P., Yang, K.: On the (im)possibility of obfuscating programs. J. ACM **59**(2), 6 (2012)

7. Barak, B., Mahmoody-Ghidary, M.: Merkle puzzles are optimal — an $O(n^2)$ query attack on any key exchange from a random oracle. In: Halevi, S. (ed.) CRYPTO 2009. LNCS, vol. 5677, pp. 374–390. Springer, Heidelberg (2009)

8. Bellare, M., Stepanovs, I., Tessaro, S.: Poly-many hardcore bits for any one-way function and a framework for differing-inputs obfuscation. In: Sarkar, P., Iwata, T. (eds.) ASIACRYPT 2014, Part II. LNCS, vol. 8874, pp. 102–121. Springer, Heidelberg (2014)

9. Bitansky, N., Canetti, R., Cohn, H., Goldwasser, S., Kalai, Y.T., Paneth, O., Rosen, A.: The impossibility of obfuscation with auxiliary input or a universal simulator. In: Garay, J.A., Gennaro, R. (eds.) CRYPTO 2014, Part II. LNCS, vol. 8617, pp. 71–89. Springer, Heidelberg (2014)

10. Bitansky, N., Canetti, R., Kalai, Y.T., Paneth, O.: On virtual grey box obfuscation for general circuits. In: Garay, J.A., Gennaro, R. (eds.) CRYPTO 2014, Part II. LNCS, vol. 8617, pp. 108–125. Springer, Heidelberg (2014)

11. Bitansky, N., Paneth, O.: ZAPs and non-interactive witness indistinguishability from indistinguishability obfuscation. In: Dodis, Y., Nielsen, J.B. (eds.) TCC 2015, Part II. LNCS, vol. 9015, pp. 401–427. Springer, Heidelberg (2015)

12. Bitansky, N., Paneth, O., Rosen, A.: On the cryptographic hardness of finding a nash equilibrium. In: Proceedings of the 56th Annual IEEE Symposium on Foundations of Computer Science (2015, To appear). https://eprint.iacr.org/2014/1029.pdf

13. Bitansky, N., Paneth, O., Wichs, D.: Perfect structure on the edge of chaos. In: Kushilevitz, E., Malkin, T., (eds.) TCC 2016-A, Part I. LNCS, vol. 9563, pp. 474–502. Springer, Heidelberg (2016)

14. Bitansky, N., Vaikuntanathan, V.: Indistinguishability obfuscation from functional encryption. In: Proceedings of the 56th Annual IEEE Symposium on Foundations of Computer Science (2015, To appear). https://eprint.iacr.org/2014/163.pdf

15. Blum, M., Micali, S.: How to generate cryptographically strong sequences of pseudo-random bits. SIAM J. Comput. **13**(4), 850–864 (1984)

16. Bogdanov, A., Brzuska, C.: On basing size-verifiable one-way functions on NP-hardness. In: Dodis, Y., Nielsen, J.B. (eds.) TCC 2015, Part I. LNCS, vol. 9014, pp. 1–6. Springer, Heidelberg (2015)

17. Boneh, D., Zhandry, M.: Multiparty key exchange, efficient traitor tracing, and more from indistinguishability obfuscation. In: Garay, J.A., Gennaro, R. (eds.) CRYPTO 2014, Part I. LNCS, vol. 8616, pp. 480–499. Springer, Heidelberg (2014)

18. Brakerski, Z., Katz, J., Segev, G., Yerukhimovich, A.: Limits on the power of zero-knowledge proofs in cryptographic constructions. In: Ishai, Y. (ed.) TCC 2011. LNCS, vol. 6597, pp. 559–578. Springer, Heidelberg (2011)

19. Brakerski, Z., Rothblum, G.N.: Virtual black-box obfuscation for all circuits via generic graded encoding. In: Lindell, Y. (ed.) TCC 2014. LNCS, vol. 8349, pp. 1–25. Springer, Heidelberg (2014)

20. Brzuska, C., Farshim, P., Mittelbach, A.: Random-oracle uninstantiability from indistinguishability obfuscation. In: Dodis, Y., Nielsen, J.B. (eds.) TCC 2015, Part II. LNCS, vol. 9015, pp. 428–455. Springer, Heidelberg (2015)
21. Canetti, R., Goldwasser, S., Poburinnaya, O.: Adaptively secure two-party computation from indistinguishability obfuscation. In: Dodis, Y., Nielsen, J.B. (eds.) TCC 2015, Part II. LNCS, vol. 9015, pp. 557–585. Springer, Heidelberg (2015)
22. Canetti, R., Lin, H., Tessaro, S., Vaikuntanathan, V.: Obfuscation of probabilistic circuits and applications. In: Dodis, Y., Nielsen, J.B. (eds.) TCC 2015, Part II. LNCS, vol. 9015, pp. 468–497. Springer, Heidelberg (2015)
23. Canetti, R., Kalai, Y.T., Paneth, O.: On obfuscation with random oracles. In: Dodis, Y., Nielsen, J.B. (eds.) TCC 2015, Part II. LNCS, vol. 9015, pp. 456–467. Springer, Heidelberg (2015)
24. Chang, Y., Hsiao, C., Lu, C.: The impossibility of basing one-way permutations on central cryptographic primitives. J. Cryptology 19(1), 97–114 (2006)
25. Chung, K., Lin, H., Mahmoody, M., Pass, R.: On the power of nonuniformity in proofs of security. In: Proceedings of the 4th Innovations in Theoretical Computer Science Conference, pp. 389–400 (2013)
26. Chung, K., Lin, H., Pass, R.: Constant-round concurrent zero-knowledge from indistinguishability obfuscation. Cryptology ePrint Archive, Report 2014/991 (2014)
27. Dachman-Soled, D., Katz, J., Rao, V.: Adaptively secure, universally composable, multiparty computation in constant rounds. In: Dodis, Y., Nielsen, J.B. (eds.) TCC 2015, Part II. LNCS, vol. 9015, pp. 586–613. Springer, Heidelberg (2015)
28. Dachman-Soled, D., Lindell, Y., Mahmoody, M., Malkin, T.: On the black-box complexity of optimally-fair coin tossing. In: Ishai, Y. (ed.) TCC 2011. LNCS, vol. 6597, pp. 450–467. Springer, Heidelberg (2011)
29. Dachman-Soled, D., Mahmoody, M., Malkin, T.: Can optimally-fair coin tossing be based on one-way functions? In: Lindell, Y. (ed.) TCC 2014. LNCS, vol. 8349, pp. 217–239. Springer, Heidelberg (2014)
30. Garg, S., Gentry, C., Halevi, S., Raykova, M.: Two-round secure MPC from indistinguishability obfuscation. In: Lindell, Y. (ed.) TCC 2014. LNCS, vol. 8349, pp. 74–94. Springer, Heidelberg (2014)
31. Garg, S., Gentry, C., Halevi, S., Raykova, M., Sahai, A., Waters, B.: Candidate indistinguishability obfuscation and functional encryption for all circuits. In: Proceedings of the 54th Annual IEEE Symposium on Foundations of Computer Science, pp. 40–49 (2013)
32. Garg, S., Polychroniadou, A.: Two-round adaptively secure MPC from indistinguishability obfuscation. In: Dodis, Y., Nielsen, J.B. (eds.) TCC 2015, Part II. LNCS, vol. 9015, pp. 614–637. Springer, Heidelberg (2015)
33. Gennaro, R., Gertner, Y., Katz, J., Trevisan, L.: Bounds on the efficiency of generic cryptographic constructions. SIAM J. Comput. 35(1), 217–246 (2005)
34. Gertner, Y., Malkin, T., Myers, S.: Towards a separation of semantic and CCA security for public key encryption. In: Vadhan, S.P. (ed.) TCC 2007. LNCS, vol. 4392, pp. 434–455. Springer, Heidelberg (2007)
35. Gertner, Y., Malkin, T., Reingold, O.: On the impossibility of basing trapdoor functions on trapdoor predicates. In: Proceedings of the 42nd Annual IEEE Symposium on Foundations of Computer Science, pp. 126–135 (2001)
36. Goldreich, O.: On security preserving reductions - revised terminology. Cryptology ePrint Archive, Report 2000/001 (2000)
37. Goldreich, O.: Foundations of Cryptography - Volume 1: Basic Techniques. Cambridge University Press, Cambridge (2001)

38. Goldwasser, S., Gordon, S.D., Goyal, V., Jain, A., Katz, J., Liu, F.-H., Sahai, A., Shi, E., Zhou, H.-S.: Multi-input functional encryption. In: Nguyen, P.Q., Oswald, E. (eds.) EUROCRYPT 2014. LNCS, vol. 8441, pp. 578–602. Springer, Heidelberg (2014)

39. Haitner, I., Hoch, J.J., Reingold, O., Segev, G.: Finding collisions in interactive protocols - tight lower bounds on the round and communication complexities of statistically hiding commitments. SIAM J. Comput. 44(1), 193–242 (2015)

40. Håstad, J., Impagliazzo, R., Levin, L.A., Luby, M.: A pseudorandom generator from any one-way function. SIAM J. Comput. 28(4), 1364–1396 (1999)

41. Hohenberger, S., Sahai, A., Waters, B.: Replacing a random oracle: full domain hash from indistinguishability obfuscation. In: Nguyen, P.Q., Oswald, E. (eds.) EUROCRYPT 2014. LNCS, vol. 8441, pp. 201–220. Springer, Heidelberg (2014)

42. Hsiao, C.-Y., Reyzin, L.: Finding collisions on a public road, or do secure hash functions need secret coins? In: Franklin, M. (ed.) CRYPTO 2004. LNCS, vol. 3152, pp. 92–105. Springer, Heidelberg (2004)

43. Impagliazzo, R., Rudich, S.: Limits on the provable consequences of one-way permutations. In: Proceedings of the 21st Annual ACM Symposium on Theory of Computing, pp. 44–61 (1989)

44. Kahn, J., Saks, M., Smyth, C.D.: The dual BKR inequality and Rudich's conjecture. Comb. Probab. Comput. 20(2), 257–266 (2011)

45. Komargodski, I., Moran, T., Naor, M., Pass, R., Rosen, A., Yogev, E.: One-way functions and (im)perfect obfuscation. In: Proceedings of the 55th Annual IEEE Symposium on Foundations of Computer Science, pp. 374–383 (2014)

46. Luby, M.: Pseudorandomness and Cryptographic Applications. Princeton University Press, Princeton (1996)

47. Mahmoody, M., Maji, H.K., Prabhakaran, M.: On the power of public-key encryption in secure computation. In: Lindell, Y. (ed.) TCC 2014. LNCS, vol. 8349, pp. 240–264. Springer, Heidelberg (2014)

48. Mahmoody, M., Pass, R.: The curious case of non-interactive commitments – on the power of black-box vs. non-black-box use of primitives. In: Safavi-Naini, R., Canetti, R. (eds.) CRYPTO 2012. LNCS, vol. 7417, pp. 701–718. Springer, Heidelberg (2012)

49. Matsuda, T.: On the impossibility of basing public-coin one-way permutations on trapdoor permutations. In: Lindell, Y. (ed.) TCC 2014. LNCS, vol. 8349, pp. 265–290. Springer, Heidelberg (2014)

50. Matsuda, T., Matsuura, K.: On black-box separations among injective one-way functions. In: Ishai, Y. (ed.) TCC 2011. LNCS, vol. 6597, pp. 597–614. Springer, Heidelberg (2011)

51. Naor, M., Yung, M.: Universal one-way hash functions and their cryptographic applications. In: Proceedings of the 21st Annual ACM Symposium on Theory of Computing, pp. 33–43 (1989)

52. Pass, R., Tseng, W.-L.D., Venkitasubramaniam, M.: Towards non-black-box lower bounds in cryptography. In: Ishai, Y. (ed.) TCC 2011. LNCS, vol. 6597, pp. 579–596. Springer, Heidelberg (2011)

53. Rabin, M.O.: Digitalized signatures and public-key functions as intractable as factorization. Technical report 212, Laboratory for Computer Science, Massachusetts Institute of Technology (1979)

54. Reingold, O., Trevisan, L., Vadhan, S.P.: Notions of reducibility between cryptographic primitives. In: Naor, M. (ed.) TCC 2004. LNCS, vol. 2951, pp. 1–20. Springer, Heidelberg (2004)

55. Rivest, R.L., Shamir, A., Adleman, L.M.: A method for obtaining digital signatures and public-key cryptosystems. Commun. ACM **21**(2), 120–126 (1978)
56. Rompel, J.: One-way functions are necessary and sufficient for secure signatures. In: Proceedings of the 22nd Annual ACM Symposium on Theory of Computing, pp. 387–394 (1990)
57. Rudich, S.: Limits on the provable consequences of one-way functions. Ph.D. thesis, EECS Department, University of California, Berkeley (1988)
58. Sahai, A., Waters, B.: How to use indistinguishability obfuscation: deniable encryption, and more. In: Proceedings of the 46th Annual ACM Symposium on Theory of Computing, pp. 475–484 (2014)
59. Simon, D.R.: Findings collisions on a one-way street: can secure hash functions be based on general assumptions? In: Nyberg, K. (ed.) EUROCRYPT 1998. LNCS, vol. 1403, pp. 334–345. Springer, Heidelberg (1998)
60. Waters, B.: A punctured programming approach to adaptively secure functional encryption. In: Gennaro, R., Robshaw, M. (eds.) CRYPTO 2015. LNCS, vol. 9216, pp. 678–697. Springer, Heidelberg (2015)
61. Wee, H.M.: One-way permutations, interactive hashing and statistically hiding commitments. In: Vadhan, S.P. (ed.) TCC 2007. LNCS, vol. 4392, pp. 419–433. Springer, Heidelberg (2007)

Contention in Cryptoland: Obfuscation, Leakage and UCE

Mihir Bellare[1]([✉]), Igors Stepanovs[1], and Stefano Tessaro[2]

[1] Department of Computer Science and Engineering,
University of California San Diego, San Diego, USA
mihir@eng.ucsd.edu,
http://cseweb.ucsd.edu/~mihir/,
https://sites.google.com/site/igorsstepanovs/
[2] Department of Computer Science,
University of California Santa Barbara, Santa Barbara, USA
http://www.cs.ucsb.edu/~tessaro/

Abstract. This paper addresses the fundamental question of whether or not different, exciting primitives now being considered actually exist. We show that we, unfortunately, cannot have them all. We provide results of the form $\neg \mathbf{A} \vee \neg \mathbf{B}$, meaning one of the primitives $\mathbf{A}, \mathbf{B}$ cannot exist. (But we don't know which.) Specifically, we show that: (1) **VGBO** (Virtual Grey Box Obfuscation) for all circuits, which has been conjectured to be achieved by candidate constructions, cannot co-exist with Canaletto's 1997 **AI-DHI** (auxiliary input DH inversion) assumption, which has been used to achieve many goals including point-function obfuscation (2) **iO** (indistinguishability obfuscation) for all circuits cannot co-exist with **KM-LR-SE** (key-message leakage-resilient symmetric encryption) (3) **iO** cannot co-exist with hash functions that are **UCE** secure for computationally unpredictable split sources.

1 Introduction

Cryptographic theory is being increasingly bold with regard to assumptions and conjectures. This is particularly true in the area of obfuscation, where candidate constructions have been provided whose claim to achieve a certain form of obfuscation is either itself an assumption [31] or is justified under other, new and strong assumptions [12,34,41]. This is attractive and exciting because we gain new capabilities and applications. But it behoves us also to be cautious and try to ascertain, not just whether the assumptions are true, but whether the goals are even achievable.

But how are we to determine this? The direct route is cryptanalysis, and we have indeed seen some success [27,28,33,39]. But cryptanalysis can be difficult and runs into major open complexity-theoretic questions. There is another rewarding route, that we pursue here. This is to seek and establish relations that we call *contentions*. These take the form $\neg A \vee \neg B$ where A, B are different primitives or assumptions. This shows that A, B are not *both* achievable, meaning they cannot co-exist. We may not know which of the two fails, but at least

© International Association for Cryptologic Research 2016
E. Kushilevitz and T. Malkin (Eds.): TCC 2016-A, Part II, LNCS 9563, pp. 542–564, 2016.
DOI: 10.1007/978-3-662-49099-0_20

one of the two must, which is valuable and sometimes surprising information. Indeed, many intriguing contentions of this form have been provided in recent work [5,11,14,20–22,32,36]. For example, we know that the following cannot co-exist: "Special-purpose obfuscation" and diO [32]; Multi-bit point function obfuscation and iO [22]; extractable one-way functions and iO [14].

In this paper we begin by addressing the question of whether VGBO (Virtual Grey Box Obfuscation) for all circuits is possible, as conjectured by BCKP [13, Section 1.1]. We show that this is in contention with the AI-DHI assumption of [15,25]. We go on to show that iO is in contention with certain forms of leakage resilient encryption and UCE.

1.1   VGBO and AI-DHI

We show that ¬VGBO ∨ ¬AI-DHI. That is, Virtual Grey Box Obfuscation (VGBO) of all circuits is in contention with Canaletto's 1997 AI-DHI (Auxiliary-Input Diffie-Hellman Inversion) assumption [15,25]. One of the two (or both) must fail. Let us now back up to provide more information on the objects involved and the proof.

The study of obfuscation began with VBBO (Virtual Black Box Obfuscation) [4,37], which asks that for any PT adversary $\mathcal{A}$ given the obfuscated circuit, there is a PT simulator $\mathcal{S}$ given an oracle for the original circuit, such that the two have about the same probability of returning 1. The impossibility of VBBO [4,16,35] has lead to efforts to define and achieve weaker forms of obfuscation. VGBO [10] is a natural relaxation of VBBO allowing the simulator $\mathcal{S}$ to be computationally unbounded but restricted to polynomially-many oracle queries. This bypasses known VBBO impossibility results while still allowing interesting applications. Furthermore BCKP [12,13] show that VGBO for NC^1 is achievable (under a novel assumption). They then say "existing candidate indistinguishability obfuscators for all circuits [3,19,31] may also be considered as candidates for VGB obfuscation, for all circuits" [13, Section 1.1]. This would mean, in particular, that VGBO for all circuits is achievable. In this paper we ask if this "VGB conjecture" is true.

The AI-DHI assumption [15,25] says that there is an ensemble $\mathcal{G} = \{\mathbb{G}_\lambda : \lambda \in \mathbb{N}\}$ of prime-order groups such that, for r, s chosen at random from $\mathbb{G}_\lambda$, no polynomial-time adversary can distinguish between (r, r^x) and (r, s), even when given auxiliary information a about x, as long as this information a is "x-prediction-precluding," meaning does not allow one to just compute x in polynomial time. The assumption has been used for oracle hashing [25], AIPO (auxiliary-input point-function obfuscation) [15] and zero-knowledge proofs [15].

Our result is that ¬VGBO ∨ ¬AI-DHI. That is, either VGBO for all circuits is impossible or the AI-DHI assumption is false. To prove this, we take any ensemble $\mathcal{G} = \{\mathbb{G}_\lambda : \lambda \in \mathbb{N}\}$ of prime-order groups. For random x, we define a way of picking the auxiliary information a such that (1) a is x-prediction-precluding, but (2) there is a polynomial-time adversary that, given a, can distinguish between (r, r^x) and (r, s) for random r, s. Consider the circuit C_x that on input u, v returns 1 if $v = u^x$ and 0 otherwise. The auxiliary information a will be a VGB obfuscation $\overline{C}$ of C_x. Now (2) is easy to see: the adversary, given challenge (u, v), can win by returning $\overline{C}(u, v)$. But why is (1) true? We use the assumed VGB security of the obfuscator to reduce (1) to showing that *no, even unbounded,* simulator, given an oracle for C_x, can extract x in a polynomial number of queries. This is shown through an information-theoretic argument that exploits the group structure.

The natural question about which one would be curious is, which of VGBO and AI-DHI is it that fails? This is an intriguing question and we do not know the answer at this point.

1.2   Key-Message Leakage Resilience

DKL [30] and CKVW [26] provide key leakage resilient symmetric encryption (K-LR-SE) schemes. This means they retain security even when the adversary has auxiliary information about the key, as long as this information is key-prediction-precluding, meaning does not allow one to compute the key. We consider a generalization that we call key-message leakage resilient symmetric encryption (KM-LR-SE). Here the auxiliary information is allowed to depend not just on the key but also on the message, the requirement however still being that it is key-prediction-precluding, meaning one cannot compute the key from the auxiliary information. The enhancement would appear to be innocuous, because the strong semantic-security style formalizations of encryption that we employ in any case allow the adversary to have a priori information about the message. However, we show that this goal is impossible to achieve if iO for all circuits is possible. That is, we show in Theorem 3 that $\neg$iO $\vee$ $\neg$KM-LR-SE. Since iO seems to be growing to be more broadly accepted, this indicates that KM-LR-SE is not likely to exist. We think this may be of direct interest from the perspective of leakage resilience, but its main importance for us is as a tool to establish new negative results for UCE as discussed in Sect. 1.3 below. The proof of Theorem 3 is a minor adaptation of the proof of BM [22] ruling out MB-AIPO under iO.

1.3   UCE for Split Sources

UCE is a class of assumptions for function families introduced in BHK [6] with the goal of instantiating random oracles. For a class **S** of algorithms called sources, BHK define UCE[**S**] security of a family of functions. The parameterization is necessary because security is not achievable for the class of all sources. BHK and subsequent work [5,6,20,23,29,40] have considered several restricted classes of sources and, based on the assumption of UCE security for these, been able to instantiate random oracles to obtain secure, efficient instantiations for primitives including deterministic public-key encryption, message-locked encryption, encryption secure for key-dependent messages, encryption secure under related-key attacks, adaptive garbling, hardcore functions and IND-CCA public-key encryption.

However UCE here has functioned as an assumption. We know little about its achievability. The basic foundational question in this area is, for which source classes $\mathbf{S}$ is UCE[$\mathbf{S}$] security achievable? The first step towards answering this was taken by BFM [20], who showed that $\neg$iO $\vee$ $\neg$UCE[$\mathbf{S}^{\mathrm{cup}}$]. That is, iO for all circuits is in contention with UCE security relative to the class $\mathbf{S}^{\mathrm{cup}}$ of all computationally unpredictable sources. (These are sources whose leakage is computationally unpredictable when their queries are answered by a random oracle.) This lead BHK [6] to propose restricting attention to "split" sources. Such sources can leak information about an oracle query and its answer separately, but not together. This circumvents the BFM attack. Indeed, UCE[$\mathbf{S}^{\mathrm{cup}} \cap \mathbf{S}^{\mathrm{splt}}$] appeared plausible even in the presence of iO. However in this paper we show $\neg$iO $\vee$ $\neg$UCE[$\mathbf{S}^{\mathrm{cup}} \cap \mathbf{S}^{\mathrm{splt}}$], meaning iO and UCE[$\mathbf{S}^{\mathrm{cup}} \cap \mathbf{S}^{\mathrm{splt}}$] security cannot co-exist. The interpretation is that UCE[$\mathbf{S}^{\mathrm{cup}} \cap \mathbf{S}^{\mathrm{splt}}$]-secure function families are unlikely to exist. We obtain our $\neg$iO $\vee$ $\neg$UCE[$\mathbf{S}^{\mathrm{cup}} \cap \mathbf{S}^{\mathrm{splt}}$] result by showing that UCE[$\mathbf{S}^{\mathrm{cup}} \cap \mathbf{S}^{\mathrm{splt}}$] $\Rightarrow$ KM-LR-SE, meaning we can build a key-message leakage resilient symmetric encryption scheme given any UCE[$\mathbf{S}^{\mathrm{cup}} \cap \mathbf{S}^{\mathrm{splt}}$]-secure function family. But we saw above that $\neg$iO $\vee$ $\neg$KM-LR-SE and can thus conclude that $\neg$iO $\vee$ $\neg$UCE[$\mathbf{S}^{\mathrm{cup}} \cap \mathbf{S}^{\mathrm{splt}}$].

BM2 [23] show that UCE[$\mathbf{S}^{\mathrm{cup}} \cap \mathbf{S}^{\mathrm{splt}} \cap \mathbf{S}^{1}$] security —$\mathbf{S}^{\mathrm{cup}} \cap \mathbf{S}^{\mathrm{splt}} \cap \mathbf{S}^{q}$ is the class of computationally unpredictable split sources making q oracle queries— is achievable. (They assume iO and AIPO.) Our $\neg$iO $\vee$ $\neg$UCE[$\mathbf{S}^{\mathrm{cup}} \cap \mathbf{S}^{\mathrm{splt}}$] result does not contradict this since our source makes a polynomial number of oracle queries. Indeed our result complements the BM2 one to show that a bound on the number of source oracle queries is necessary for a positive result. Together these results give a close to complete picture of the achievability of UCE for split sources, the remaining open question being achievability of UCE[$\mathbf{S}^{\mathrm{cup}} \cap \mathbf{S}^{\mathrm{splt}} \cap \mathbf{S}^{q}$] for constant $q > 1$.

We note that we are not aware of any applications assuming UCE[$\mathbf{S}^{\mathrm{cup}} \cap \mathbf{S}^{\mathrm{splt}}$]. Prior applications have used either UCE[$\mathbf{S}^{\mathrm{cup}} \cap \mathbf{S}^{\mathrm{splt}} \cap \mathbf{S}^{1}$] or quite different classes like UCE[$\mathbf{S}^{\mathrm{sup}}$] —$\mathbf{S}^{\mathrm{sup}}$ is the class of statistically unpredictable sources [6,20]— and neither of these is at risk from our results. However our $\neg$iO $\vee$ $\neg$UCE[$\mathbf{S}^{\mathrm{cup}} \cap \mathbf{S}^{\mathrm{splt}}$] result is of interest towards understanding the achievability of UCE assumptions and the effectiveness of different kinds of restrictions (in this case, splitting) on sources. The achievability of UCE[$\mathbf{S}^{\mathrm{cup}} \cap \mathbf{S}^{\mathrm{splt}}$] security was an open problem from prior work.

1.4   Discussion and Related Work

The idea of using an obfuscated circuit as an auxiliary input to obtain contention results has appeared in many prior works [5,11,14,20–22,32,36]. Some of the contentions so established are between "Special-purpose obfuscation" and diO [32], between MB-AIPO and iO [22] and between extractable one-way functions and iO [14]. Our work follows in these footsteps.

KM-LR-SE can be viewed as a symmetric encryption re-formulation of MB-AIPO following the connection of the latter to symmetric encryption established by CKVW [26]. The main change is in the correctness condition. We formulate a weak correctness condition, which is important for our application to UCE. In its absence, our negative result for split-source UCE would only be for injective functions, which is much weaker. With this connection in mind, the proof of Theorem 3, as we have indicated above, is a minor adaptation of the proof of BM [22] ruling out MB-AIPO under iO. Our result about KM-LR-SE is thus not of technical novelty or interest but we think this symmetric encryption re-formulation of BM [22] is of interest from the leakage resilience perspective and as a tool to obtain more negative results, as exemplified by our application to UCE.

In independent and concurrent work, BM3 [24] show $\neg \text{iO} \vee \neg \text{UCE}[\mathbf{S}^{\text{s-cup}}]$, where $\mathbf{S}^{\text{s-cup}}$ is the class of *strongly* computationally unpredictable sources as defined in [22]. But the latter show that $\mathbf{S}^{\text{cup}} \cap \mathbf{S}^{\text{splt}}$ is a *strict* subset of $\mathbf{S}^{\text{s-cup}}$. This means that our $\neg \text{iO} \vee \neg \text{UCE}[\mathbf{S}^{\text{cup}} \cap \mathbf{S}^{\text{splt}}]$ result is strictly stronger than the $\neg \text{iO} \vee \neg \text{UCE}[\mathbf{S}^{\text{s-cup}}]$ result of [24]. (Under iO, our result rules out UCE security for a smaller, more restricted class of sources.)

Our results on UCE, as with the prior ones of BFM [20], are for the basic setting, where there is a single key or single user [6]. BHK [6] also introduce a multi-key (multi-user) setting. Some negative results about this are provided in [8].

2 Preliminaries

Notation. We denote by $\lambda \in \mathbb{N}$ the security parameter and by 1^λ its unary representation. If $x \in \{0,1\}^*$ is a string then $|x|$ denotes its length, $x[i]$ denotes its i-th bit, and $x[i..j] = x[i] \ldots x[j]$ for $1 \leq i \leq j \leq |x|$. We let ε denote the empty string. If s is an integer then $\text{Pad}_s(\text{C})$ denotes circuit C padded to have size s. We say that circuits C_0, C_1 are equivalent, written $\text{C}_0 \equiv \text{C}_1$, if they agree on all inputs. If $\mathbf{x}$ is a vector then $|\mathbf{x}|$ denotes the number of its coordinates and $\mathbf{x}[i]$ denotes its i-th coordinate. If X is a finite set, we let $x \leftarrow_\$ X$ denote picking an element of X uniformly at random and assigning it to x. Algorithms may be randomized unless otherwise indicated. Running time is worst case. "PT" stands for "polynomial-time," whether for randomized algorithms or deterministic ones. If A is an algorithm, we let $y \leftarrow A(x_1, \ldots; r)$ denote running A with random coins r on inputs $x_1, \ldots$ and assigning the output to y. We let $y \leftarrow_\$ A(x_1, \ldots)$ be the result of picking r at random and letting $y \leftarrow A(x_1, \ldots; r)$. We let $[A(x_1, \ldots)]$ denote the set of all possible outputs of A when invoked with inputs $x_1, \ldots$. We say that $f \colon \mathbb{N} \to \mathbb{R}$ is negligible if for every positive polynomial p, there exists $\lambda_p \in \mathbb{N}$ such that $f(\lambda) < 1/p(\lambda)$ for all $\lambda > \lambda_p$. We use the code based game playing framework of [7]. (See Fig. 1 for an example.) By $\text{G}^{\mathcal{A}}(\lambda)$ we denote the event that the execution of game G with adversary $\mathcal{A}$ and security parameter λ results in the game returning true.

Auxiliary Information Generators. Many of the notions we consider involve the computational unpredictability of some quantity even given "auxiliary information" about it. We abstract this out via our definition of an *auxiliary information generator* X. The latter specifies a PT algorithm X.Ev that takes 1^λ to return a *target* $k \in \{0,1\}^{\text{X.tl}(\lambda)}$, a *payload* $m \in \{0,1\}^{\text{X.pl}(\lambda)}$ and an *auxiliary information* a, where X.tl, X.pl$\colon \mathbb{N} \to \mathbb{N}$ are the target and payload length functions associated to X, respectively. Consider game PRED of Fig. 1 associated to X and a predictor adversary $\mathcal{Q}$. For $\lambda \in \mathbb{N}$ let $\mathsf{Adv}^{\mathrm{pred}}_{\mathsf{X},\mathcal{Q}}(\lambda) = \Pr[\mathrm{PRED}^{\mathcal{Q}}_{\mathsf{X}}(\lambda)]$. We say that X is *unpredictable* if $\mathsf{Adv}^{\mathrm{pred}}_{\mathsf{X},\mathcal{Q}}(\cdot)$ is negligible for every PT adversary $\mathcal{Q}$. We say that X is *uniform* if $\mathsf{X}.\mathsf{Ev}(1^\lambda)$ picks the target $k \in \{0,1\}^{\mathsf{X}.\mathsf{tl}(\lambda)}$ and the payload $m \in \{0,1\}^{\mathsf{X}.\mathsf{pl}(\lambda)}$ uniformly and independently. Note that the auxiliary information a may depend on both the target k and the payload m, but unpredictability refers to recovery of the target k alone.

Game $\mathrm{PRED}^{\mathcal{Q}}_{\mathsf{X}}(\lambda)$	Game $\mathrm{PRG}^{\mathcal{R}}_{\mathsf{R}}(\lambda)$	Game $\mathrm{IO}^{\mathcal{O}}_{\mathsf{Obf},\mathsf{S}}(\lambda)$
$(k,m,a) \leftarrow_\$ \mathsf{X}.\mathsf{Ev}(1^\lambda)$	$b \leftarrow_\$ \{0,1\}$	$b \leftarrow_\$ \{0,1\}$
$k' \leftarrow_\$ \mathcal{Q}(1^\lambda, a)$	$m \leftarrow_\$ \{0,1\}^{\mathsf{R}.\mathsf{sl}(\lambda)}$	$(\mathsf{C}_0, \mathsf{C}_1, aux) \leftarrow_\$ \mathsf{S}(1^\lambda)$
Return $(k = k')$	$y_1 \leftarrow \mathsf{R}.\mathsf{Ev}(1^\lambda, m)$	$\overline{\mathsf{C}} \leftarrow_\$ \mathsf{Obf}(1^\lambda, \mathsf{C}_b)$
	$y_0 \leftarrow_\$ \{0,1\}^{2 \cdot \mathsf{R}.\mathsf{sl}(\lambda)}$	$b' \leftarrow_\$ \mathcal{O}(1^\lambda, \overline{\mathsf{C}}, aux)$
	$b' \leftarrow_\$ \mathcal{R}(1^\lambda, y_b)$	Return $(b = b')$
	Return $(b = b')$	

Fig. 1. Games defining unpredictabilty of auxiliary information generator X, PR-security of pseudorandom generator R and iO-security of obfuscator Obf relative to circuit sampler S.

PRGs. A pseudorandom generator R [17,44] specifies a deterministic PT algorithm $\mathsf{R}.\mathsf{Ev}$ where $\mathsf{R}.\mathsf{sl}: \mathbb{N} \to \mathbb{N}$ is the seed length function of R such that $\mathsf{R}.\mathsf{Ev}(1^\lambda, \cdot): \{0,1\}^{\mathsf{R}.\mathsf{sl}(\lambda)} \to \{0,1\}^{2 \cdot \mathsf{R}.\mathsf{sl}(\lambda)}$ for all $\lambda \in \mathbb{N}$. We say that R is PR-secure if the function $\mathsf{Adv}^{\mathrm{pr}}_{\mathsf{R},\mathcal{R}}(\cdot)$ is negligible for every PT adversary $\mathcal{R}$, where for $\lambda \in \mathbb{N}$ we let $\mathsf{Adv}^{\mathrm{pr}}_{\mathsf{R},\mathcal{R}}(\lambda) = 2\Pr[\mathrm{PRG}^{\mathcal{R}}_{\mathsf{R}}(\lambda)] - 1$ and game PRG is specified in Fig. 1.

Obfuscators. An *obfuscator* is a PT algorithm Obf that on input 1^λ and a circuit C returns a circuit $\overline{\mathsf{C}}$ such that $\overline{\mathsf{C}} \equiv \mathsf{C}$. (That is, $\overline{\mathsf{C}}(x) = \mathsf{C}(x)$ for all x.) We refer to the latter as the *correctness condition*. We will consider various notions of security for obfuscators, including VGBO and iO.

Indistinguishability Obfuscation. We use the BST [9] definitional framework which parameterizes security via classes of circuit samplers. Let Obf be an obfuscator. A *sampler* in this context is a PT algorithm S that on input 1^λ returns a triple $(\mathsf{C}_0, \mathsf{C}_1, aux)$ where $\mathsf{C}_0, \mathsf{C}_1$ are circuits of the same size, number of inputs and number of outputs, and aux is a string. If $\mathcal{O}$ is an adversary and $\lambda \in \mathbb{N}$ we let $\mathsf{Adv}^{\mathrm{io}}_{\mathsf{Obf},\mathsf{S},\mathcal{O}}(\lambda) = 2\Pr[\mathrm{IO}^{\mathcal{O}}_{\mathsf{Obf},\mathsf{S}}(\lambda)] - 1$ where game $\mathrm{IO}^{\mathcal{O}}_{\mathsf{Obf},\mathsf{S}}(\lambda)$ is defined in Fig. 1. Now let $\boldsymbol{S}$ be a class (set) of circuit samplers. We say that Obf is $\boldsymbol{S}$-secure if $\mathsf{Adv}^{\mathrm{io}}_{\mathsf{Obf},\mathsf{S},\mathcal{O}}(\cdot)$ is negligible for every PT adversary $\mathcal{O}$ and every circuit sampler $\mathsf{S} \in \boldsymbol{S}$. We say that circuit sampler S produces equivalent circuits if there exists a negligible function ν such that $\Pr\left[\, \mathsf{C}_0 \equiv \mathsf{C}_1 \;:\; (\mathsf{C}_0, \mathsf{C}_1, aux) \leftarrow_\$ \mathsf{S}(1^\lambda)\,\right] \geq 1 - \nu(\lambda)$ for all $\lambda \in \mathbb{N}$. Let $\boldsymbol{S}_{\mathrm{eq}}$ be the class of all circuit samplers that produce equivalent circuits. We say that Obf is an indistinguishability obfuscator if it is $\boldsymbol{S}_{\mathrm{eq}}$-secure [4,31,42].

Function Families. A family of functions F specifies the following. PT key generation algorithm F.Kg takes 1^λ to return a key $fk \in \{0,1\}^{\text{F.kl}(\lambda)}$, where F.kl: $\mathbb{N} \to \mathbb{N}$ is the key length function associated to F. Deterministic, PT evaluation algorithm F.Ev takes 1^λ, key $fk \in [\text{F.Kg}(1^\lambda)]$ and an input $x \in \{0,1\}^{\text{F.il}(\lambda)}$ to return an output F.Ev$(1^\lambda, fk, x) \in \{0,1\}^{\text{F.ol}(\lambda)}$, where F.il, F.ol: $\mathbb{N} \to \mathbb{N}$ are the input and output length functions associated to F, respectively. We say that F is *injective* if the function F.Ev$(1^\lambda, fk, \cdot)$: $\{0,1\}^{\text{F.il}(\lambda)} \to \{0,1\}^{\text{F.ol}(\lambda)}$ is injective for every $\lambda \in \mathbb{N}$ and every $fk \in [\text{F.Kg}(1^\lambda)]$.

UCE Security. Let us recall the Universal Computational Extractor (UCE) framework of BHK [6]. Let H be a family of functions. Let $\mathcal{S}$ be an adversary called the *source* and $\mathcal{D}$ an adversary called the *distinguisher*. We associate to them and H the game UCE$_H^{\mathcal{S},\mathcal{D}}(\lambda)$ in the left panel of Fig. 2. The source has access to an oracle HASH and we require that any query x made to this oracle have length H.il(λ). When the challenge bit b is 1 (the "real" case) the oracle responds via H.Ev under a key hk that is chosen by the game and *not* given to the source. When $b = 0$ (the "random" case) it responds as a random oracle. The source then leaks a string L to its accomplice distinguisher. The latter *does* get the key hk as input and must now return its guess $b' \in \{0,1\}$ for b. The game returns true iff $b' = b$, and the uce-advantage of $(\mathcal{S}, \mathcal{D})$ is defined for $\lambda \in \mathbb{N}$ via Adv$_{H,\mathcal{S},\mathcal{D}}^{\text{uce}}(\lambda) = 2\Pr[\text{UCE}_H^{\mathcal{S},\mathcal{D}}(\lambda)] - 1$. If **S** is a class (set) of sources, we say that H is UCE[**S**]-secure if Adv$_{H,\mathcal{S},\mathcal{D}}^{\text{uce}}(\cdot)$ is negligible for all sources $\mathcal{S} \in \mathbf{S}$ and all PT distinguishers $\mathcal{D}$.

Game UCE$_H^{\mathcal{S},\mathcal{D}}(\lambda)$	Game PRED$_{\mathcal{S}}^{\mathcal{P}}(\lambda)$	Source $\mathcal{S}^{\text{HASH}}(1^\lambda)$
$b \leftarrow_\$ \{0,1\}$; $hk \leftarrow_\$ \text{H.Kg}(1^\lambda)$	$X \leftarrow \emptyset$	$(L_0, \mathbf{x}) \leftarrow_\$ \mathcal{S}_0(1^\lambda)$
$L \leftarrow_\$ \mathcal{S}^{\text{HASH}}(1^\lambda)$	$L \leftarrow_\$ \mathcal{S}^{\text{HASH}}(1^\lambda)$	For $i = 1, \ldots, |\mathbf{x}|$ do
$b' \leftarrow_\$ \mathcal{D}(1^\lambda, hk, L)$	$x' \leftarrow_\$ \mathcal{P}(1^\lambda, L)$	$\quad \mathbf{y}[i] \leftarrow_\$ \text{HASH}(\mathbf{x}[i])$
Return $(b' = b)$	Return $(x' \in X)$	$L_1 \leftarrow_\$ \mathcal{S}_1(1^\lambda, \mathbf{y})$
		$L \leftarrow (L_0, L_1)$
HASH(x)	HASH(x)	Return L
If $T[x] = \bot$ then	If $T[x] = \bot$ then	
$\quad$ If $b = 0$ then $T[x] \leftarrow_\$ \{0,1\}^{\text{H.ol}(\lambda)}$	$\quad T[x] \leftarrow_\$ \{0,1\}^{\text{H.ol}(\lambda)}$	
$\quad$ Else $T[x] \leftarrow \text{H.Ev}(1^\lambda, hk, x)$	$X \leftarrow X \cup \{x\}$	
Return $T[x]$	Return $T[x]$	

Fig. 2. Games defining UCE security of function family H, unpredictability of source $\mathcal{S}$, and the split source $\mathcal{S} = \text{Splt}[\mathcal{S}_0, \mathcal{S}_1]$ associated to $\mathcal{S}_0, \mathcal{S}_1$.

It is easy to see that UCE[**S**]-security is not achievable if **S** is the class of all PT sources [6]. To obtain meaningful notions of security, BHK [6] impose restrictions on the source. A central restriction is unpredictability. A source is unpredictable if it is hard to guess the source's HASH queries even given the leakage, in the *random case* of the UCE game. Formally, let $\mathcal{S}$ be a source and $\mathcal{P}$ an adversary called a predictor and consider game PRED$_{\mathcal{S}}^{\mathcal{P}}(\lambda)$ in the middle panel of Fig. 2. For $\lambda \in \mathbb{N}$ we let $\mathsf{Adv}^{\mathsf{pred}}_{\mathcal{S},\mathcal{P}}(\lambda) = \Pr[\mathrm{PRED}^{\mathcal{P}}_{\mathcal{S}}(\lambda)]$. We say that $\mathcal{S}$ is computationally unpredictable if $\mathsf{Adv}^{\mathsf{pred}}_{\mathcal{S},\mathcal{P}}(\cdot)$ is negligible for all PT predictors $\mathcal{P}$, and let $\mathbf{S}^{\mathsf{cup}}$ be the class of all PT computationally unpredictable sources. We say that $\mathcal{S}$ is statistically unpredictable if $\mathsf{Adv}^{\mathsf{pred}}_{\mathcal{S},\mathcal{P}}(\cdot)$ is negligible for all (not necessarily PT) predictors $\mathcal{P}$, and let $\mathbf{S}^{\mathsf{sup}} \subseteq \mathbf{S}^{\mathsf{cup}}$ be the class of all PT statistically unpredictable sources.

BFM [20] show that UCE[$\mathbf{S}^{\mathsf{cup}}$]-security is not achievable assuming that indistinguishability obfuscation is possible. This has lead applications to either be based on UCE[$\mathbf{S}^{\mathsf{sup}}$] or on subsets of UCE[$\mathbf{S}^{\mathsf{cup}}$], meaning to impose further restrictions on the source. UCE[$\mathbf{S}^{\mathsf{sup}}$], introduced in [6,20], seems at this point to be a viable assumption. In order to restrict the computational case, one can consider split sources as defined in BHK [6]. Let $\mathcal{S}_0, \mathcal{S}_1$ be algorithms, neither of which have access to any oracles. The *split source* $\mathcal{S} = \mathsf{Splt}[\mathcal{S}_0, \mathcal{S}_1]$ associated to $\mathcal{S}_0, \mathcal{S}_1$ is defined in the right panel of Fig. 2. Algorithm $\mathcal{S}_0$ returns a pair $(L_0, \mathbf{x})$. Here $\mathbf{x}$ is a vector over $\{0,1\}^{\mathsf{H.il}(\lambda)}$ all of whose entries are required to be distinct. (If the entries are not required to be distinct, collisions can be used to communicate information between the two components of the source, and the BFM [20] attack continues to apply, as pointed out in [23].) The first adversary creates the oracle queries for the source $\mathcal{S}$, the latter making these queries and passing the replies to the second adversary to get the leakage. In this way, neither $\mathcal{S}_0$ nor $\mathcal{S}_1$ have an input-output pair from the oracle, limiting their ability to create leakage useful to the distinguisher. A source $\mathcal{S}$ is said to belong to the class $\mathbf{S}^{\mathsf{splt}}$ if there exist PT $\mathcal{S}_0, \mathcal{S}_1$ such that $\mathcal{S} = \mathsf{Splt}[\mathcal{S}_0, \mathcal{S}_1]$, meaning is defined as above. The class of interest is now UCE[$\mathbf{S}^{\mathsf{cup}} \cap \mathbf{S}^{\mathsf{splt}}$], meaning UCE-security for computationally unpredictable, split sources.

Another way to restrict a UCE source is by limiting the number of queries it can make. Let $\mathbf{S}^q$ be the class of sources making $q(\cdot)$ oracle queries. This allows to consider $\mathbf{S}^{\mathsf{cup}} \cap \mathbf{S}^{\mathsf{splt}} \cap \mathbf{S}^1$, a class of computationally unpredictable split sources that make a single query. BM2 [23] show that UCE[$\mathbf{S}^{\mathsf{cup}} \cap \mathbf{S}^{\mathsf{splt}} \cap \mathbf{S}^1$]-security is achievable assuming iO and AIPO.

3   VGBO and the AI-DHI Assumption

BCKP [12,13] conjecture that existing candidate constructions of iO also achieve VGBO and thus in particular that VGB obfuscation for all circuits is possible. Here we explore the plausibility of this "VGB conjecture." We show that it implies the failure of Canaletto's AI-DHI assumption. Either this assumption is false or VGBO for all circuits is not possible. (In fact, our result refers to an even weaker VGBO assumption.) That is, the long-standing AI-DHI assumption and VGBO are in contention; at most one of these can exist. We start by defining VGBO and recalling the AI-DHI assumption, and then give our result and its proof. We then suggest a weakening of AI-DHI that we call AI-DHI2 that is parameterized by a group generator. We show that our attack on AI-DHI extends to rule out AI-DHI2 for group generators satisfying a property we call verifiability. However there may be group generators that do not appear to be verifiable, making AI-DHI2 a potential alternative to AI-DHI.

Game $\mathrm{VGB1}^{\mathcal{A}}_{\mathsf{Obf,Smp}}(\lambda)$	Game $\mathrm{VGB0}^{\mathcal{S}}_{\mathsf{Smp},q}(\lambda)$
$\mathrm{C} \leftarrow_{\$} \mathsf{Smp}(1^\lambda)$	$\mathrm{C} \leftarrow_{\$} \mathsf{Smp}(1^\lambda) \,;\, i \leftarrow 0$
$\overline{\mathrm{C}} \leftarrow_{\$} \mathsf{Obf}(1^\lambda, \mathrm{C})$	$b' \leftarrow_{\$} \mathcal{S}^{\mathrm{CIRC}}(1^\lambda)$
$b' \leftarrow_{\$} \mathcal{A}(1^\lambda, \overline{\mathrm{C}})$	Return $(b' = 1)$
Return $(b' = 1)$	
	$\underline{\mathrm{CIRC}(x)}$
	$i \leftarrow i + 1$
	If $i > q(\lambda)$ then return $\perp$
	$y \leftarrow \mathrm{C}(x)\,;\,$ Return y

Fig. 3. Games defining VGB security of obfuscator Obf.

VGBO. Let Obf be an obfuscator as defined in Sect. 2. We define what it means for it to be a VGB obfuscator. We will use a weak variant of the notion used in some of the literature [10,12], which strengthens our results since they are negative relations with starting point VGBO.

A *sampler* Smp in this context is an algorithm that takes 1^λ to return a circuit C. Let q be a polynomial, $\mathcal{A}$ an adversary and $\mathcal{S}$ a (not necessarily PT) algorithm called a simulator. For $\lambda \in \mathbb{N}$ let

$$\mathsf{Adv}^{\mathsf{vgb}}_{\mathsf{Obf,Smp},q,\mathcal{A},\mathcal{S}}(\lambda) = \left| \Pr\left[\, \mathrm{VGB1}^{\mathcal{A}}_{\mathsf{Obf,Smp}}(\lambda) \,\right] - \Pr\left[\, \mathrm{VGB0}^{\mathcal{S}}_{\mathsf{Smp},q}(\lambda) \,\right] \right|$$

where the games are in Fig. 3. Let SAMP be a set of samplers. We say that Obf is a VGB obfuscator for SAMP if for every PT adversary $\mathcal{A}$ there exists a (not necessarily PT) simulator $\mathcal{S}$ and a polynomial q such that $\mathsf{Adv}^{\mathsf{vgb}}_{\mathsf{Obf,Smp},q,\mathcal{A},\mathcal{S}}(\cdot)$ is negligible for all $\mathsf{Smp} \in \mathsf{SAMP}$.

We note that [12] use a VGB variant stronger than the above where the advantage measures the difference in probabilities of $\mathcal{A}$ and $\mathcal{S}$ guessing a predicate $\pi(\mathrm{C})$, rather than just the probabilities of outputting one, which is all we need here. Also note that our VGB definition is vacuously achievable whenever $|\mathsf{SAMP}| = 1$, since $\mathcal{S}$ can simulate game $\mathrm{VGB1}^{\mathcal{A}}_{\mathsf{Obf,Smp}}(\lambda)$ for any fixed choice of $\mathcal{A}$ and Smp. Our applications however use a SAMP of size 2.

The AI-DHI Assumption. Let $\mathcal{G} = \{\mathbb{G}_\lambda : \lambda \in \mathbb{N}\}$ be an ensemble of groups where for every $\lambda \in \mathbb{N}$ the order $p(\lambda)$ of group $\mathbb{G}_\lambda$ is a prime in the range $2^{\lambda-1} < p(\lambda) < 2^\lambda$. We assume that relevant operations are computable in time polynomial in λ, including computing $p(\cdot)$, testing membership in $\mathbb{G}_\lambda$ and performing operations in $\mathbb{G}_\lambda$. By $\mathbb{G}_\lambda^*$ we denote the non-identity members of the group, which is the set of generators since the group has prime order. An auxiliary information generator X for $\mathcal{G}$ is an auxiliary information generator as per Sect. 2 with the additional property that the target k returned by $\mathsf{X.Ev}(1^\lambda)$ is in $\mathbb{Z}_{p(\lambda)}$ (i.e. is an exponent) and the payload m is ε (i.e. is effectively absent).

Now consider game AIDHI of Fig. 4 associated to $\mathcal{G}$, X and an adversary $\mathcal{A}$. For $\lambda \in \mathbb{N}$ let $\mathsf{Adv}^{\mathsf{aidhi}}_{\mathcal{G},\mathsf{X},\mathcal{A}}(\lambda) = 2\Pr[\mathrm{AIDHI}^{\mathcal{A}}_{\mathcal{G},\mathsf{X}}(\lambda)] - 1$. We say that $\mathcal{G}$ is AI-DHI-secure if $\mathsf{Adv}^{\mathsf{aidhi}}_{\mathcal{G},\mathsf{X},\mathcal{A}}(\cdot)$ is negligible for every unpredictable X for $\mathcal{G}$ and every PT adversary $\mathcal{A}$

Game $\text{AIDHI}_{\mathcal{G},\mathsf{X}}^{\mathcal{A}}(\lambda)$	Game $\text{AIDHI2}_{\mathsf{GG},\mathsf{X}}^{\mathcal{A}}(\lambda)$
$b \leftarrow\!\!{}_\$ \{0,1\}$; $(k,\varepsilon,a) \leftarrow\!\!{}_\$ \mathsf{X}.\mathsf{Ev}(1^\lambda)$	$b \leftarrow\!\!{}_\$ \{0,1\}$; $(k,\varepsilon,a) \leftarrow\!\!{}_\$ \mathsf{X}.\mathsf{Ev}(1^\lambda)$
$g \leftarrow\!\!{}_\$ \mathbb{G}_\lambda^*$	$\langle \mathbb{G} \rangle \leftarrow\!\!{}_\$ \mathsf{GG}(1^\lambda)$; $g \leftarrow\!\!{}_\$ \mathsf{Gen}(\mathbb{G})$
$K_1 \leftarrow g^k$; $K_0 \leftarrow\!\!{}_\$ \mathbb{G}_\lambda$	$K_1 \leftarrow g^k$; $K_0 \leftarrow\!\!{}_\$ \mathbb{G}$
$b' \leftarrow\!\!{}_\$ \mathcal{A}(1^\lambda, g, K_b, a)$	$b' \leftarrow\!\!{}_\$ \mathcal{A}(1^\lambda, \langle \mathbb{G} \rangle, g, K_b, a)$
Return $(b = b')$	Return $(b = b')$

Fig. 4. Games defining the AI-DHI assumption and the AI-DHI2 assumption.

$\mathcal{A}$. The AI-DHI assumption [15,25] is that there exists a family of groups $\mathcal{G}$ that is AI-DHI secure.

¬ **VGBO** ∨¬ **AI-DHI.** The following says if VGB obfuscation is possible then the AI-DHI assumption is false: there exists *no* family of groups $\mathcal{G}$ that is AI-DHI secure. Our theorem only assumes a very weak form of VGB obfuscation for a class with two samplers (given in the proof).

Theorem 1. *Let $\mathcal{G}$ be a family of groups. Then there is a pair $\mathsf{Smp}, \mathsf{Smp}_0$ of PT samplers (defined in the proof) such that if there exists a VGB-secure obfuscator for the class $\mathsf{SAMP} = \{\mathsf{Smp}, \mathsf{Smp}_0\}$, then $\mathcal{G}$ is not AI-DHI-secure.*

Proof (Theorem 1). Let Obf be the assumed obfuscator. Let X be the auxiliary information generator for $\mathcal{G}$ defined as follows:

Algorithm $\mathsf{X}.\mathsf{Ev}(1^\lambda)$	Circuit $\mathsf{C}_{1^\lambda,k}(g, K)$
$k \leftarrow\!\!{}_\$ \mathbb{Z}_{p(\lambda)}$	If $(g \notin \mathbb{G}_\lambda^*$ or $K \notin \mathbb{G}_\lambda)$ then return 0
$\overline{\mathsf{C}} \leftarrow\!\!{}_\$ \mathsf{Obf}(1^\lambda, \mathsf{C}_{1^\lambda,k})$	If $(g^k = K)$ then return 1
Return $(k, \varepsilon, \overline{\mathsf{C}})$	Else return 0

The auxiliary information $a = \overline{\mathsf{C}}$ produced by X is an obfuscation of the circuit $\mathsf{C}_{1^\lambda,k}$ shown on the right above. The circuit has 1^λ and the target value k embedded inside. The circuit takes inputs g, K and checks that the first is a group element different from the identity —and thus a generator— and the second is a group element. It then returns 1 if g^k equals K, and 0 otherwise.

We first construct a PT adversary $\mathcal{A}^*$ such that $\mathsf{Adv}_{\mathcal{G},\mathsf{X},\mathcal{A}^*}^{\text{aidhi}}(\cdot)$ is non-negligible. On input $1^\lambda, g, K_b, \overline{\mathsf{C}}$, it simply returns $\overline{\mathsf{C}}(g, K_b)$. That is, it runs the obfuscated circuit $\overline{\mathsf{C}}$ on g and K_b and returns its outcome. If the challenge bit b in game $\text{AIDHI}_{\mathcal{G},\mathsf{X}}^{\mathcal{A}^*}(\lambda)$ is 1 then the adversary always outputs $b' = 1$. Otherwise, the adversary outputs $b' = 1$ with probability $1/p(\lambda)$. We have $\mathsf{Adv}_{\mathcal{G},\mathsf{X},\mathcal{A}^*}^{\text{aidhi}}(\lambda) = 1 - 1/p(\lambda) \geq 1 - 2^{1-\lambda}$, which is not negligible.

We now show that the constructed auxiliary information generator X is unpredictable. In particular, for any PT adversary $\mathcal{Q}$ we construct a PT adversary $\mathcal{A}$ and samplers $\mathsf{Smp}, \mathsf{Smp}_0$ such that for all simulators $\mathcal{S}$ and all polynomials q,

$$\mathsf{Adv}_{\mathsf{X},\mathcal{Q}}^{\text{pred}}(\lambda) \leq \mathsf{Adv}_{\mathsf{Obf},\mathsf{Smp},q,\mathcal{A},\mathcal{S}}^{\text{vgb}}(\lambda) + \mathsf{Adv}_{\mathsf{Obf},\mathsf{Smp}_0,q,\mathcal{A},\mathcal{S}}^{\text{vgb}}(\lambda) + \frac{q(\lambda)}{2^{\lambda-1}} . \qquad (1)$$

Concretely, the adversary $\mathcal{A}$ and the samplers $\mathsf{Smp}, \mathsf{Smp}_0$ operate as follows:

Adversary $\mathcal{A}(1^\lambda, \overline{\mathsf{C}})$	Algorithm $\mathsf{Smp}(1^\lambda)$	Algorithm $\mathsf{Smp}_0(1^\lambda)$
$k' \leftarrow_s \mathcal{Q}(1^\lambda, \overline{\mathsf{C}})$	$k \leftarrow_s \mathbb{Z}_{p(\lambda)}$	Return C_0
$\bar{g} \leftarrow_s \mathbb{G}_\lambda^*$	Return $\mathsf{C}_{1^\lambda, k}$	
Return $\overline{\mathsf{C}}(\bar{g}, \bar{g}^{k'})$		

In Smp_0, the circuit C_0 takes as input a pair of group elements g, g' from $\mathbb{G}_\lambda$ and always returns 0.

To show Eq. (1), we first note that by construction

$$\mathsf{Adv}_{\mathsf{X},\mathcal{Q}}^{\mathsf{pred}}(\lambda) = \Pr\left[\, \mathrm{VGB1}_{\mathsf{Obf},\mathsf{Smp}}^{\mathcal{A}}(\lambda) \,\right], \tag{2}$$

because an execution of $\mathrm{PRED}_{\mathsf{X}}^{\mathcal{Q}}(\lambda)$ results in the same output distribution as in $\mathrm{VGB1}_{\mathsf{Obf},\mathsf{Smp}}^{\mathcal{A}}(\lambda)$. The only difference is that in the latter, the check of whether the guess is correct is done via the obfuscated circuit $\overline{\mathsf{C}}$. Now, for all simulators $\mathcal{S}$ and polynomials q, we can rewrite Eq. (2) as

$$\begin{aligned} \mathsf{Adv}_{\mathsf{X},\mathcal{Q}}^{\mathsf{pred}}(\lambda) = {}&\Pr\left[\, \mathrm{VGB1}_{\mathsf{Obf},\mathsf{Smp}}^{\mathcal{A}}(\lambda) \,\right] - \Pr\left[\, \mathrm{VGB0}_{\mathsf{Smp},q}^{\mathcal{S}}(\lambda) \,\right] \\ &+ \Pr\left[\, \mathrm{VGB0}_{\mathsf{Smp},q}^{\mathcal{S}}(\lambda) \,\right] - \Pr\left[\, \mathrm{VGB0}_{\mathsf{Smp}_0,q}^{\mathcal{S}}(\lambda) \,\right] \\ &+ \Pr\left[\, \mathrm{VGB0}_{\mathsf{Smp}_0,q}^{\mathcal{S}}(\lambda) \,\right] - \Pr\left[\, \mathrm{VGB1}_{\mathsf{Obf},\mathsf{Smp}_0}^{\mathcal{A}}(\lambda) \,\right] \\ &+ \Pr\left[\, \mathrm{VGB1}_{\mathsf{Obf},\mathsf{Smp}_0}^{\mathcal{A}}(\lambda) \,\right]. \end{aligned}$$

To upper bound $\mathsf{Adv}_{\mathsf{X},\mathcal{Q}}^{\mathsf{pred}}(\lambda)$, we first note that

$$\Pr\left[\, \mathrm{VGB1}_{\mathsf{Obf},\mathsf{Smp}}^{\mathcal{A}}(\lambda) \,\right] - \Pr\left[\, \mathrm{VGB0}_{\mathsf{Smp},q}^{\mathcal{S}}(\lambda) \,\right] \le \mathsf{Adv}_{\mathsf{Obf},\mathsf{Smp},q,\mathcal{A},\mathcal{S}}^{\mathsf{vgb}}(\lambda)$$

and

$$\Pr\left[\, \mathrm{VGB0}_{\mathsf{Smp}_0,q}^{\mathcal{S}}(\lambda) \,\right] - \Pr\left[\, \mathrm{VGB1}_{\mathsf{Obf},\mathsf{Smp}_0}^{\mathcal{A}}(\lambda) \,\right] \le \mathsf{Adv}_{\mathsf{Obf},\mathsf{Smp}_0,q,\mathcal{A},\mathcal{S}}^{\mathsf{vgb}}(\lambda).$$

Moreover, we have $\Pr\left[\, \mathrm{VGB1}_{\mathsf{Obf},\mathsf{Smp}_0}^{\mathcal{A}}(\lambda) \,\right] = 0$ by constructon. Namely, adversary $\mathcal{A}$ never outputs 1 in game $\mathrm{VGB1}_{\mathsf{Obf},\mathsf{Smp}_0}^{\mathcal{A}}(\lambda)$, since it is given an obfuscation of the constant zero circuit C_0.

We are left with upper bounding the difference between $\Pr\left[\, \mathrm{VGB0}_{\mathsf{Smp},q}^{\mathcal{S}}(\lambda) \,\right]$ and $\Pr\left[\, \mathrm{VGB0}_{\mathsf{Smp}_0,q}^{\mathcal{S}}(\lambda) \,\right]$. Note that $\mathcal{S}$ is allowed to issue at most $q(\lambda)$ queries to the given circuit, which is either $\mathsf{C}_{1^\lambda, k}$ for a random $k \leftarrow_s \mathbb{Z}_{p(\lambda)}$ or C_0. Denote by Hit the event that $\mathcal{S}$ makes a query (g, K) in $\mathrm{VGB0}_{\mathsf{Smp},q}^{\mathcal{S}}(\lambda)$ such that $g^k = K$. Then, by a standard argument,

$$\Pr\left[\, \mathrm{VGB0}_{\mathsf{Smp},q}^{\mathcal{S}}(\lambda) \,\right] - \Pr\left[\, \mathrm{VGB0}_{\mathsf{Smp}_0,q}^{\mathcal{S}}(\lambda) \,\right] \le \Pr\left[\, \mathsf{Hit} \,\right].$$

To compute $\Pr\left[\, \mathsf{Hit} \,\right]$, we move from $\mathrm{VGB0}_{\mathsf{Smp},q}^{\mathcal{S}}(\lambda)$ to the simpler $\mathrm{VGB0}_{\mathsf{Smp}_0,q}^{\mathcal{S}}(\lambda)$, where all of $\mathcal{S}$'s queries are answered with 0. We extend the latter game to sample a random key $k \leftarrow\!\!{}_{\$} \mathbb{Z}_{p(\lambda)}$, and we define Hit$'$ as the event in this game that for one of $\mathcal{S}$'s queries (g, K) we have $g^k = K$. It is not hard to see that $\Pr[\text{Hit}']$ and $\Pr[\text{Hit}]$ are equal, as both games are identical as long as none of such queries occur. Since there are at most $q(\lambda)$ queries, and exactly one k can produce the answer 1 for these queries, the union bound yields

$$\Pr[\text{Hit}] = \Pr[\text{Hit}'] \leq \frac{q(\lambda)}{p(\lambda)} \leq \frac{q(\lambda)}{2^{\lambda-1}} \, ,$$

which concludes the proof. □

The AI-DHI2 Assumption. We now suggest a relaxation AI-DHI2 of the AI-DHI assumption given above. The idea is that for each value of λ there is not one, but many possible groups. Formally, a *group generator* is a PT algorithm GG that on input 1^λ returns a description $\langle \mathbb{G} \rangle$ of a cyclic group $\mathbb{G}$ whose order $|\mathbb{G}|$ is in the range $2^{\lambda-1} < |\mathbb{G}| < 2^\lambda$. We assume that given 1^λ, $\langle \mathbb{G} \rangle$, relevant operations are computable in time polynomial in λ, including performing group operations in $\mathbb{G}$ and picking at random from $\mathbb{G}$ and from the set Gen$(\mathbb{G})$ of generators of $\mathbb{G}$. An auxiliary information generator X for GG is an auxiliary information generator as per Sect. 2 with the additional property that the target k returned by $\text{X.Ev}(1^\lambda)$ is in $\mathbb{Z}_{2^{\lambda-1}}$ —this makes it a valid exponent for *any* group $\mathbb{G}$ such that $\langle \mathbb{G} \rangle \in [\text{GG}(1^\lambda)]$— and the payload m is ε (i.e. is effectively absent).

Now consider game AIDHI2 of Fig. 4 associated to GG, X and an adversary $\mathcal{A}$. For $\lambda \in \mathbb{N}$ let $\text{Adv}^{\text{aidhi2}}_{\text{GG,X},\mathcal{A}}(\lambda) = 2 \Pr[\text{AIDHI2}^{\mathcal{A}}_{\text{GG,X}}(\lambda)] - 1$. We say that GG is AI-DHI2-secure if $\text{Adv}^{\text{aidhi2}}_{\text{GG,X},\mathcal{A}}(\cdot)$ is negligible for every unpredictable X for GG and every PT adversary $\mathcal{A}$. The (new) AI-DHI2 assumption is that there exists a group generator GG which is AI-DHI2 secure.

BP [15] give a simple construction of AIPO from AI-DHI. It is easy to extend this to use AI-DHI2.

A *verifier* for group generator GG is a deterministic, PT algorithm GG.Vf that can check whether a given string d is a valid description of a group generated by the generator GG. Formally, GG.Vf on input $1^\lambda, d$ returns true if $d \in [\text{GG}(1^\lambda)]$ and false otherwise, for all $d \in \{0, 1\}^*$. We say that GG is *verifiable* if it has a verifier and additionally, in time polynomial in 1^λ, $\langle \mathbb{G} \rangle$, where $\langle \mathbb{G} \rangle \in [\text{GG}(1^\lambda)]$, one can test membership in $\mathbb{G}$ and in the set Gen$(\mathbb{G})$ of generators of $\mathbb{G}$. The following extends Theorem 1 to say that if VGBO is possible then no verifiable group generator is AI-DHI2 secure.

Theorem 2. *Let* GG *be a verifiable group generator. Then there is a pair* Smp, Smp$_0$ *of PT samplers such that if there exists a VGB-secure obfuscator for the class* SAMP $= \{\text{Smp}, \text{Smp}_0\}$, *then* GG *is not AI-DHI2-secure.*

We omit a full proof, as it is very similar to the one of Theorem 1. We only note that to adapt the proof, we require $\text{X.Ev}(1^\lambda)$ to output a random k in $\mathbb{Z}_{2^{\lambda-1}}$ together with the obfuscation of the following circuit $C_{1^\lambda, k}$. The circuit $C_{1^\lambda, k}$ takes as input a string d expected to be a group description, together with two strings g and K. It first runs GG.Vf on input $1^\lambda, d$ to check whether $d \in [\text{GG}(1^\lambda)]$, returning 0 if the check fails. If the check succeeds, so that we can write $d = \langle \mathbb{G} \rangle$, it further checks that $g \in \text{Gen}(\mathbb{G})$ and $K \in \mathbb{G}$, returning 0 if this fails. Finally the circuit returns 1 if and only if $g^k = K$ in the group $\mathbb{G}$. The crucial point is that for every valid input (d, g, K), there is at most one $k \in \mathbb{Z}_{2^{\lambda-1}}$ which satisfies $g^k = K$ in the group described by d. This uses the assumption that $\mathbb{G}$ is cyclic.

Many group generators are cyclic and verifiable. For example, consider a generator GG that on input 1^λ returns a description of $\mathbb{G} = \mathbb{Z}_p^*$ for a safe prime $p = 2q - 1$. (That is, q is also a prime.) The verifier can extract p, q from $\langle \mathbb{G} \rangle$ and check their primality in PT. For such generators, we may prefer not to assume AI-DHI2-security, due to Theorem 2. However there are group generators that do not appear to be verifiable and where Theorem 2 thus does not apply. One must be careful to note that this does *not* mean that VGBO would not rule out AI-DHI2 security for these group generators. It just means that our current proof method may not work. Still at this point, the AI-DHI2 assumption, which only says there is *some* group generator that is AI-DHI2-secure, seems plausible.

Discussion. As we indicated, one of the main applications of AI-DHI was AIPO [15], and furthermore this connection is very direct. If VGB is in contention with AI-DHI, it is thus natural to ask whether it is also in contention with AIPO. We do not know whether or not this is true. One can also ask whether VGB is in contention with other, particular AIPO constructions, in particular the one of BP [15] based on the construction of Wee [43]. Again, we do not know the answer. We note that alternative constructions of AIPO and other forms of point-function obfuscation are provided in [8].

4   KM-Leakage Resilient Encryption

We refer to a symmetric encryption scheme as K-leakage-resilient if it retains security in the presence of any leakage about the key that leaves the key computationally unpredictable [30]. Such schemes have been designed in [26,30]. Here, we extend the model by allowing the leakage to depend not just on the key but also on the message, still leaving the key computationally unpredictable. The extension seems innocuous, since the indistinguishability style formalizations used here already capture the adversary having some information about the message. But Theorem 3 shows that KM-leakage-resilience is in contention with iO. The interpretation is KM-leakage-resilience is not achievable.

Theorem 3 is of direct interest with regard to understanding what is and is not achievable in leakage-resilient cryptography. But for us its main importance will be as a tool to rule out UCE for computationally unpredictable split sources assuming iO in Sect. 5.

We use standard definitions of indistinguishability obfuscation [2,4,18,31,42] and pseudorandom generators [17,44], as recalled in Sect. 2. We now start by formalizing KM-leakage resilience.

KM-Leakage Resilient Encryption. Let a symmetric encryption scheme SE specify the following. PT encryption algorithm SE.Enc takes 1^λ, a key

Game $\text{IND}_{\text{SE},X}^{\mathcal{A}}(\lambda)$	Game $\text{DEC}_{\text{SE}}(\lambda)$
$b \leftarrow\!\!\text{\$}\ \{0,1\}$	$k \leftarrow\!\!\text{\$}\ \{0,1\}^{\text{SE.kl}(\lambda)}$
$(k, m_1, a) \leftarrow\!\!\text{\$}\ \text{X.Ev}(1^\lambda)$	$m \leftarrow\!\!\text{\$}\ \{0,1\}^{\text{SE.ml}(\lambda)}$
$m_0 \leftarrow\!\!\text{\$}\ \{0,1\}^{\text{SE.ml}(\lambda)}$	$c \leftarrow\!\!\text{\$}\ \text{SE.Enc}(1^\lambda, k, m)$
$c \leftarrow\!\!\text{\$}\ \text{SE.Enc}(1^\lambda, k, m_b)$	$m' \leftarrow \text{SE.Dec}(1^\lambda, k, c)$
$b' \leftarrow\!\!\text{\$}\ \mathcal{A}(1^\lambda, a, c)$	Return $(m = m')$
Return $(b = b')$	

Fig. 5. Games defining X-KM-leakage resilience of symmetric encryption scheme SE and decryption correctness of symmetric encryption scheme SE.

$k \in \{0,1\}^{\text{SE.kl}(\lambda)}$ and a message $m \in \{0,1\}^{\text{SE.ml}(\lambda)}$ to return a ciphertext c, where $\text{SE.kl}, \text{SE.ml} \colon \mathbb{N} \to \mathbb{N}$ are the key length and message length functions of SE, respectively. Deterministic PT decryption algorithm SE.Dec takes $1^\lambda, k, c$ to return a plaintext $m \in \{0,1\}^{\text{SE.ml}(\lambda)}$. Note that there is a key length but no prescribed key-generation algorithm.

For security, let X be an auxiliary information generator with $\text{X.tl} = \text{SE.kl}$ and $\text{X.pl} = \text{SE.ml}$. Consider game $\text{IND}_{\text{SE},X}^{\mathcal{A}}(\lambda)$ of Fig. 5 associated to SE, X and adversary $\mathcal{A}$. The message m_0 is picked uniformly at random. The adversary $\mathcal{A}$ must determine which message has been encrypted, given not just the ciphertext but auxiliary information a on the key and message m_1. For $\lambda \in \mathbb{N}$ we let $\text{Adv}_{\text{SE},X,\mathcal{A}}^{\text{ind}}(\lambda) = 2\Pr[\text{IND}_{\text{SE},X}^{\mathcal{A}}(\lambda)] - 1$. We say that SE is X-KM-leakage resilient if the function $\text{Adv}_{\text{SE},X,\mathcal{A}}^{\text{ind}}(\cdot)$ is negligible for all PT adversaries $\mathcal{A}$. This is of course not achievable if a allowed the adversary to compute k, so we restrict attention to unpredictable X. Furthermore, weakening the definition, we restrict attention to uniform X, meaning k and m_1 are uniformly and independently distributed. Thus we say that SE is KM-leakage-resilient if it is X-KM-leakage resilient for all unpredictable, uniform X.

The above requirement is strong in that security is required in the presence of (unpredictable) leakage on the key and first message. But beyond that, in other ways, it has been made weak, because this strengthens our negative results. Namely, we are only requiring security on random messages, not chosen ones, with the key being uniformly distributed, and the key and the two messages all being independently distributed. Furthermore, in contrast to a typical indistinguishability definition, the adversary does not get the messages as input.

The standard correctness condition would ask that $\text{SE.Dec}(1^\lambda, k, \text{SE.Enc}(1^\lambda, k, m)) = m$ for all $k \in \{0,1\}^{\text{SE.kl}(\lambda)}$, all $m \in \{0,1\}^{\text{SE.ml}(\lambda)}$ and all $\lambda \in \mathbb{N}$. We call this perfect correctness. We formulate and use a weaker correctness condition because we can show un-achievability even under this and the weakening is crucial to our applications building KM-leakage-resilient encryption schemes to obtain further impossibility results. Specifically, we require correctness only for random messages and random keys with non-negligible probability. Formally, consider game $\mathrm{DEC}_{\mathsf{SE}}(\lambda)$ of Fig. 5 associated to SE, and for $\lambda \in \mathbb{N}$ let $\mathsf{Adv}^{\mathsf{dec}}_{\mathsf{SE}}(\lambda) = \Pr[\mathrm{DEC}_{\mathsf{SE}}(\lambda)]$ be the decryption correctness function of SE. We require that $\mathsf{Adv}^{\mathsf{dec}}_{\mathsf{SE}}(\cdot)$ be non-negligible.

¬iO ∨ ¬KM-LR-SE. The following says that KM-leakage-resilient symmetric encryption is not achievable if iO and PRGs (which can be obtained from one-way functions [38]) exist:

Theorem 3. *Let* SE *be a symmetric encryption scheme. Let* Obf *be an indistinguishability obfuscator. Let* R *be a PR-secure PRG with* R.sl = SE.ml. *Assume that* $2^{-\mathsf{SE.kl}(\lambda)}$ *and* $2^{-\mathsf{R.sl}(\lambda)}$ *are negligible. Then there exists a uniform auxiliary information generator* X *such that the following holds: (1)* X *is unpredictable, but (2)* SE *is not* X-*KM-leakage resilient.*

The proof is a minor adaptation of the proof of BM [22] ruling out MB-AIPO under iO. Following BM [22], the idea is that the auxiliary information generator X picks a key k and message m uniformly and independently at random and lets C be the circuit that embeds k and the result y of the PRG on m. On input a ciphertext c, circuit C decrypts it under k and then checks that the PRG applied to the result equals y. The auxiliary information is an obfuscation $\overline{C}$ of C. The attack showing claim (2) of Theorem 3 is straightforward but its analysis is more work and exploits the security of the PRG. Next one shows that iO-security of the obfuscator coupled with security of the PRG implies claim (1), namely the unpredictability of X. For completeness we provide a self-contained proof in Appendix A. A consequence of Theorem 3 is the following.

Corollary 4. *Let* SE *be a symmetric encryption scheme such that* SE.ml$(\cdot) \in \Omega((\cdot)^\epsilon)$ *for some constant* $\epsilon > 0$. *Assume the existence of an indistinguishability obfuscator and a one-way function. Then* SE *is not KM-leakage resilient.*

Proof (Corollary 4). The assumption on SE.ml implies that there exists a PR-secure PRG R with R.sl = SE.ml [38]. To conclude we apply Theorem 3.   □

Related Work. CKVW [26] show that symmetric encryption with weak keys satisfying a wrong key detection property is equivalent to MB-AIPO. Wrong key detection, a form of robustness [1], asks that, if you decrypt, under a certain key, a ciphertext created under a different key, then the result is ⊥. This is not a requirement for KM-LR-SE. However, implicit in the proof of Theorem 3 is a connection between KM-LR-SE and a form of MB-AIPO with a relaxed correctness condition.

5   UCE for Split Sources

BFM [20] showed that UCE[$\mathsf{S}^{\mathrm{cup}}$]-security is not possible if iO exists. We improve this to show that UCE[$\mathsf{S}^{\mathrm{cup}} \cap \mathsf{S}^{\mathrm{splt}}$]-security is not possible if iO exists. We obtain this by giving a construction of a KM-leakage-resilient symmetric encryption scheme from UCE[$\mathsf{S}^{\mathrm{cup}} \cap \mathsf{S}^{\mathrm{splt}}$] and then invoking our above-mentioned result. Definitions of UCE-secure function families [6] are recalled in Sect. 2.

UCE[$S^{cup} \cap S^{splt}$] $\Rightarrow$ KM-LR-SE. We give a construction of a KM-leakage resilient symmetric encryption scheme from a UCE[$S^{cup} \cap S^{splt}$] family H, which will allow us to rule out such families under iO. Assume for simplicity that H.il is odd, and let $\ell = (\text{H.il} - 1)/2$. We call the symmetric encryption scheme SE = H&C[H] that we associate to H the Hash-and-Check scheme. It is defined as follows. Let SE.kl(λ) = SE.ml(λ) = $\ell(\lambda)$ for all $\lambda \in \mathbb{N}$. Let the encryption and decryption algorithms be as follows:

Algorithm SE.Enc($1^\lambda, k, m$)	Algorithm SE.Dec($1^\lambda, k, (hk, \mathbf{y})$)
$hk \leftarrow_\$ \text{H.Kg}(1^\lambda)$	For $i = 1, \ldots, |\mathbf{y}|$ do
For $i = 1, \ldots, |m|$ do	If ($\text{H.Ev}(1^\lambda, hk, k\|1\|\langle i\rangle_{\ell(\lambda)}) = \mathbf{y}[i]$)
$\mathbf{y}[i] \leftarrow \text{H.Ev}(1^\lambda, hk, k\|m[i]\|\langle i\rangle_{\ell(\lambda)})$	Then $m[i] \leftarrow 1$ else $m[i] \leftarrow 0$
Return ($hk, \mathbf{y}$)	Return m

Here $\langle i\rangle_{\ell(\lambda)} = 1^i 0^{\ell(\lambda)-i}$ denotes a particular, convenient encoding of integer $i \in \{1, \ldots, \ell(\lambda)\}$ as a string of $\ell(\lambda)$ bits, and $m[i]$ denotes the i-th bit of m. The ciphertext ($hk, \mathbf{y}$) consists of a key hk for H chosen randomly and anew at each encryption, together with the vector $\mathbf{y}$ whose i-th entry is the hash of the i-th message bit along with the key and index i. This scheme will have perfect correctness if H is injective, but we do not want to assume this. The following theorem says that the scheme is KM-leakage resilient and also has (somewhat better than) weak correctness under UCE-security of H.

Theorem 5. *Let* H *be a family of functions that is* UCE[$S^{cup} \cap S^{splt}$]*-secure. Assume* H.il($\cdot$) $\in \Omega((\cdot)^\epsilon)$ *for some constant* $\epsilon > 0$ *and* $2^{-\text{H.ol}(\cdot)}$ *is negligible. Let* SE = H&C[H]. *Then (1) symmetric encryption scheme* SE *is KM-leakage resilient, and (2)* $1 - \text{Adv}_{SE}^{dec}(\cdot)$ *is negligible.*

Proof (Theorem 5). Assuming for simplicity as in the construction that H.il is odd, let $\ell(\cdot) = (\text{H.il}(\cdot) - 1)/2$. We now prove part (1). Let X be an unpredictable, uniform auxiliary information generator. Let $\mathcal{A}$ be a PT adversary. We build a PT source $\mathcal{S} \in S^{cup} \cap S^{splt}$ and a PT distinguisher $\mathcal{D}$ such that

$$\text{Adv}_{SE,X,\mathcal{A}}^{ind}(\lambda) \leq 2 \cdot \text{Adv}_{H,\mathcal{S},\mathcal{D}}^{uce}(\lambda) \qquad (3)$$

for all $\lambda \in \mathbb{N}$. The assumption that H is UCE[$S^{cup} \cap S^{splt}$]-secure now implies part (1) of the theorem.

We proceed to build $\mathcal{S}, \mathcal{D}$. We let $\mathcal{S}$ be the split source $\mathcal{S} = \text{Splt}[\mathcal{S}_0, \mathcal{S}_1]$, where algorithms $\mathcal{S}_0, \mathcal{S}_1$ are shown below, along with distinguisher $\mathcal{D}$:

Algorithm $\mathcal{S}_0(1^\lambda)$	Algorithm $\mathcal{S}_1(1^\lambda, \mathbf{y})$	Distinguisher $\mathcal{D}(1^\lambda, hk, L)$
$(k, m_1, a) \leftarrow_\$ \text{X.Ev}(1^\lambda)$	Return $\mathbf{y}$	$((d, a), \mathbf{y}) \leftarrow L$; $c \leftarrow (hk, \mathbf{y})$
$m_0 \leftarrow_\$ \{0,1\}^{\ell(\lambda)}$; $d \leftarrow_\$ \{0,1\}$		$d' \leftarrow_\$ \mathcal{A}(1^\lambda, a, c)$
For $i = 1, \ldots, \ell(\lambda)$ do		If ($d = d'$) then $b' \leftarrow 1$
$\mathbf{x}[i] \leftarrow k\|m_d[i]\|\langle i\rangle_{\ell(\lambda)}$		Else $b' \leftarrow 0$
Return $((d, a), \mathbf{x})$		Return b'

Here $\mathcal{S}_0$ calls the auxiliary information generator X to produce a key, a plaintext message and the corresponding auxiliary input. It then picks another plaintext message and the challenge bit d at random, and lets $\mathbf{x}$ consist of the inputs on which the hash function would be applied to create the challenge ciphertext. It leaks the challenge bit and auxiliary information. Algorithm $\mathcal{S}_1$ takes as input the result $\mathbf{y}$ of oracle HASH on $\mathbf{x}$, and leaks the entire vector $\mathbf{y}$. The distinguisher gets the leakage from both stages, together with the key hk. Using the latter, it can create the ciphertext c, which it passes to $\mathcal{A}$ to get back a decision. Its output reflects whether $\mathcal{A}$ wins its game.

Letting b denote the challenge bit in game $\mathrm{UCE}_{\mathsf{H}}^{\mathcal{S},\mathcal{D}}(\lambda)$, we claim that

$$\Pr[\,b' = 1 \mid b = 1\,] = \frac{1}{2} + \frac{1}{2}\mathsf{Adv}_{\mathsf{SE},\mathsf{X},\mathcal{A}}^{\mathsf{ind}}(\lambda) \quad \text{and} \quad \Pr[\,b' = 1 \mid b = 0\,] = \frac{1}{2}\,,$$

from which Eq. (3) follows. The first equation above should be clear from the construction. For the second, when $b = 0$, we know that HASH is a random oracle. But the entries of $\mathbf{x}$ are all distinct, due to the $\langle i \rangle_{\ell(\lambda)}$ components. So the entries of $\mathbf{y}$ are uniform and independent, and in particular independent of the challenge bit d.

This however does not end the proof: We still need to show that $\mathcal{S} \in \mathbf{S}^{\mathrm{cup}} \cap \mathbf{S}^{\mathrm{splt}}$. We have ensured that $\mathcal{S} \in \mathbf{S}^{\mathrm{splt}}$ by construction. The crucial remainig step is to show that $\mathcal{S} \in \mathbf{S}^{\mathrm{cup}}$. This will exploit the assumed unpredictability of X. Let $\mathcal{P}$ be a PT predictor. We build PT adversary $\mathcal{Q}$ such that

$$\mathsf{Adv}_{\mathcal{S},\mathcal{P}}^{\mathsf{pred}}(\lambda) \leq \mathsf{Adv}_{\mathsf{X},\mathcal{Q}}^{\mathsf{pred}}(\lambda) \tag{4}$$

for all $\lambda \in \mathbb{N}$. The assumption that X is unpredictable now implies that $\mathcal{S} \in \mathbf{S}^{\mathrm{cup}}$. The construction of $\mathcal{Q}$ is as follows:

Adversary $\mathcal{Q}(1^\lambda, a)$

For $i = 1, \ldots, \ell(\lambda)$ do $\mathbf{y}[i] \leftarrow\!\!{}_\$ \{0,1\}^{\mathsf{H.ol}(\lambda)}$

$d \leftarrow\!\!{}_\$ \{0,1\}$; $x' \leftarrow\!\!{}_\$ \mathcal{P}(1^\lambda, ((d,a), \mathbf{y}))$; $k \leftarrow x'[1..\ell(\lambda)]$; Return k

---

Adversary $\mathcal{Q}$ computes leakage $((d,a), \mathbf{y})$ distributed exactly as it would be in game $\mathrm{PRED}_{\mathcal{S}}^{\mathcal{P}}(\lambda)$, where HASH is a random oracle. It then runs $\mathcal{P}$ to get a prediction x' of some oracle query of $\mathcal{S}$. If game $\mathrm{PRED}_{\mathcal{S}}^{\mathcal{P}}(\lambda)$ returns true, then x' must have the form $k \| m_d[i] \| \langle i \rangle_{\ell(\lambda)}$ for some $i \in \{1, \ldots, \ell(\lambda)\}$, where k, d are the key and challenge bit, respectively, chosen by $\mathcal{S}$. Adversary $\mathcal{Q}$ can then win its $\mathrm{PRED}_{\mathsf{X}}^{\mathcal{Q}}(\lambda)$ game by simply returning k, which establishes Eq. (4).

This completes the proof of part (1) of the theorem. We prove part (2) by building a PT source $\mathcal{S} \in \mathbf{S}^{\mathrm{sup}} \cap \mathbf{S}^{\mathrm{splt}}$ and a PT distinguisher $\mathcal{D}$ such that

$$1 - \mathsf{Adv}_{\mathsf{SE}}^{\mathsf{dec}}(\lambda) \leq \mathsf{Adv}_{\mathsf{H},\mathcal{S},\mathcal{D}}^{\mathsf{uce}}(\lambda) + \frac{\ell(\lambda)}{2^{\mathsf{H.ol}(\lambda)}} \tag{5}$$

for all $\lambda \in \mathbb{N}$. But we have assumed that H is $\mathrm{UCE}[\mathbf{S}^{\mathrm{cup}} \cap \mathbf{S}^{\mathrm{splt}}]$-secure, so it is also $\mathrm{UCE}[\mathbf{S}^{\mathrm{sup}} \cap \mathbf{S}^{\mathrm{splt}}]$-secure. We have also assumed $2^{-\mathsf{H.ol}(\cdot)}$ is negligible. Part (2) of the theorem follows.

We proceed to build $\mathcal{S}, \mathcal{D}$. We let $\mathcal{S}$ be the split source $\mathcal{S} = \mathsf{Splt}[\mathcal{S}_0, \mathcal{S}_1]$, where algorithms $\mathcal{S}_0, \mathcal{S}_1$ are shown below, along with distinguisher $\mathcal{D}$:

Algorithm $\mathcal{S}_0(1^\lambda)$	Algorithm $\mathcal{S}_1(1^\lambda, \mathbf{y})$	Distinguisher $\mathcal{D}(1^\lambda, hk, (\varepsilon, \mathbf{y}))$
$k \leftarrow\!\!{\$}\ \{0,1\}^{\ell(\lambda)}$	Return $\mathbf{y}$	$b' \leftarrow 0$
For $i = 1, \ldots, \ell(\lambda)$ do		For $i = 1, \ldots, \ell(\lambda)$ do
$\quad \mathbf{x}[2i-1] \leftarrow k\|1\|\langle i\rangle_{\ell(\lambda)}$		$\quad$ If $(\mathbf{y}[2i-1] = \mathbf{y}[2i])$
$\quad \mathbf{x}[2i] \leftarrow k\|0\|\langle i\rangle_{\ell(\lambda)}$		$\quad$ Then $b' \leftarrow 1$
Return $(\varepsilon, \mathbf{x})$		Return b'

Letting b denote the challenge bit in game $\mathrm{UCE}_{\mathsf{H}}^{\mathcal{S},\mathcal{D}}(\lambda)$, we claim that

$$\Pr[\, b' = 1 \mid b = 1\,] \geq 1 - \mathsf{Adv}_{\mathsf{SE}}^{\mathrm{dec}}(\lambda) \quad \text{and} \quad \Pr[\, b' = 1 \mid b = 0\,] \leq \frac{\ell(\lambda)}{2^{\mathsf{H.ol}(\lambda)}} \,,$$

from which Eq. (5) follows. The first equation above is true because decryption errors only happen when hash outputs collide for different values of the message bit. For the second, when $b = 0$, we know that HASH is a random oracle. But the entries of $\mathbf{x}$ are all distinct. So the entries of $\mathbf{y}$ are uniform and independent. The chance of a collision of two entries is thus $2^{-\mathsf{H.ol}(\lambda)}$, and the equation then follows from the union bound.

$\mathcal{S}$ is a split source by construction. To conclude the proof we need to show that $\mathcal{S} \in \mathbf{S}^{\mathrm{sup}}$. In the case HASH is a random oracle, the distinctness of the oracle queries of $\mathcal{S}$ means that the entries of $\mathbf{y}$ are uniformly and independently distributed. Since there is no leakage beyond $\mathbf{y}$, the leakage gives the predictor $\mathcal{P}$ no extra information about the entries of $\mathbf{x}$. The uniform choice of k by $\mathcal{S}$ means that $\mathsf{Adv}_{\mathcal{S},\mathcal{P}}^{\mathrm{pred}}(\cdot) \leq 2^{-\ell(\cdot)}$, even if $\mathcal{P}$ is not restricted to PT. But our assumption on $\mathsf{H.il}(\cdot)$ in the theorem statement implies that $2^{-\ell(\cdot)}$ is negligible. $\quad\square$

$\neg \mathbf{iO} \vee \neg \mathbf{UCE}[\mathbf{S}^{\mathrm{cup}} \cap \mathbf{S}^{\mathrm{splt}}]$. In the BFM [20] iO-based attack on $\mathrm{UCE}[\mathbf{S}^{\mathrm{cup}}]$, the source builds a circuit which embeds an oracle query x and its answer y, and outputs an obfuscation of this circuit in the leakage. Splitting is a restriction on sources introduced in BHK [6] with the aim of preventing such attacks. A split source cannot build the BFM circuit because the split structure denies it the ability to leak information that depends both on a query and its answer. Thus, the BFM attack does not work for $\mathrm{UCE}[\mathbf{S}^{\mathrm{cup}} \cap \mathbf{S}^{\mathrm{splt}}]$. However, we show that in fact $\mathrm{UCE}[\mathbf{S}^{\mathrm{cup}} \cap \mathbf{S}^{\mathrm{splt}}]$-security is still not achievable assuming iO. This is now a simple corollary of Theorems 3 and 5 that in particular was the motivation for the latter:

Theorem 6. *Let* H *be a family of functions such that* $\mathsf{H.il}(\cdot) \in \Omega((\cdot)^\epsilon)$ *for some constant* $\epsilon > 0$ *and* $2^{-\mathsf{H.ol}(\cdot)}$ *is negligible. Assume the existence of an indistinguishability obfuscator and a one-way function. Then* H *is not* $\mathrm{UCE}[\mathbf{S}^{\mathrm{cup}} \cap \mathbf{S}^{\mathrm{splt}}]$-*secure.*

BM2 [23] show that $\mathrm{UCE}[\mathbf{S}^{\mathrm{cup}} \cap \mathbf{S}^{\mathrm{splt}} \cap \mathbf{S}^1]$-security is achievable assuming iO and AIPO. Our negative result of Theorem 6 does not contradict this, and in fact complements it to give a full picture of the achievability of UCE security for split sources.

Acknowledgments. Bellare and Stepanovs were supported in part by NSF grants CNS-1116800, CNS-1228890 and CNS-1526801. Tessaro was supported in part by NSF grant CNS-1423566. This work was done in part while Bellare and Tessaro were visiting the Simons Institute for the Theory of Computing, supported by the Simons Foundation and by the DIMACS/Simons Collaboration in Cryptography through NSF grant CNS-1523467.

We thank Huijia Lin for discussions and insights. We thank the TCC 2016-A reviewers for extensive and insightful comments.

A Proof of Theorem 3

The construction and proof follow [22]. We specify uniform auxiliary information generator X as follows:

Algorithm $X.\mathsf{Ev}(1^\lambda)$	Circuit $C_{1^\lambda,k,y}(c)$
$k \leftarrow_{\$} \{0,1\}^{\mathsf{SE.kl}(\lambda)}$	$m \leftarrow \mathsf{SE.Dec}(1^\lambda, k, c)$
$m \leftarrow_{\$} \{0,1\}^{\mathsf{SE.ml}(\lambda)}$; $y \leftarrow \mathsf{R.Ev}(1^\lambda, m)$	$y' \leftarrow \mathsf{R.Ev}(1^\lambda, m)$
$C \leftarrow \mathsf{Pad}_{s(\lambda)}(C_{1^\lambda,k,y})$; $\overline{C} \leftarrow_{\$} \mathsf{Obf}(1^\lambda, C)$	If $(y = y')$ then return 1
Return $(k, m, \overline{C})$	Else return 0

The circuit $C_{1^\lambda,k,y}$ takes as input a ciphertext c, decrypts it under the embedded key k to get back a $\mathsf{SE.ml}(\lambda)$-bit message m, applies the PRG to m to get a string y', and returns 1 iff y' equals the embedded string y. The auxiliary information generator creates this circuit as shown and outputs its obfuscation.

We define polynomial s so that $s(\lambda)$ is an upper bound on $\max(|C^1_{1^\lambda,k,y}|, |C^2|)$ where the circuits are defined in Fig. 6 and the maximum is over all $k \in \{0,1\}^{\mathsf{SE.kl}(\lambda)}$ and $y \in \{0,1\}^{2 \cdot \mathsf{R.sl}(\lambda)}$. Let us first present an attack proving part (2) of the theorem. Below we define an adversary $\mathcal{A}$ against the X-KM-leakage resilience of SE and an adversary $\mathcal{R}$ against the PR-security of R:

Adversary $\mathcal{A}(1^\lambda, \overline{C}, c)$	Adversary $\mathcal{R}(1^\lambda, y)$
$b' \leftarrow \overline{C}(c)$	$k \leftarrow_{\$} \{0,1\}^{\mathsf{SE.kl}(\lambda)}$; $m_0 \leftarrow_{\$} \{0,1\}^{\mathsf{SE.ml}(\lambda)}$
Return b'	$c \leftarrow_{\$} \mathsf{SE.Enc}(1^\lambda, k, m_0)$; $m \leftarrow \mathsf{SE.Dec}(1^\lambda, k, c)$
	$y' \leftarrow \mathsf{R.Ev}(1^\lambda, m)$
	If $(y' = y)$ then $g' \leftarrow 1$ else $g' \leftarrow 0$; Return g'

Adversary $\mathcal{A}$ has input 1^λ, the auxiliary information (leakage) which here is the obfuscated circuit $\overline{C}$, and a ciphertext c. It simply computes and returns the bit $\overline{C}(c) = C_{1^\lambda,k,y}(c)$. For the analysis, consider game $\mathrm{IND}^{\mathcal{A}}_{\mathsf{SE},X}(\lambda)$ of Fig. 5. If the challenge bit b is 1 and the decryption performed by $\overline{C}$ is correct then $y' = y$, so

$$\Pr[\, b' = 1 \mid b = 1 \,] \geq \mathsf{Adv}^{\mathsf{dec}}_{\mathsf{SE}}(\lambda) \,. \tag{6}$$

In the case $b = 0$, the corresponding analysis in [22] for the insecurity of MB-AIPO relied on the fact that PRGs have low collision probability on random seeds. This will not suffice for us because of our weak correctness condition. The latter means that when $b = 0$, we do not know that $\mathsf{SE.Dec}(1^\lambda, k, c)$ equals m_0

and indeed have no guarantees on the distribution of decrypted plaintext message. Instead, we directly exploit the assumed PR-security of the PRG. Thus, consider game $\mathrm{PRG}_R^{\mathcal{R}}(\lambda)$ with adversary $\mathcal{R}$ as above. Letting g denote the challenge bit in the game, we have

$$\mathsf{Adv}_{R,\mathcal{R}}^{\mathsf{pr}}(\lambda) = \Pr[\, g' = 1 \mid g = 1\,] - \Pr[\, g' = 1 \mid g = 0\,]$$

$$\geq \Pr[\, b' = 1 \mid b = 0\,] - 2^{-2 \cdot \mathsf{R.sl}(\lambda)} \,. \tag{7}$$

From Eqs. (7) and (6), we have

$$\mathsf{Adv}_{\mathsf{SE},\mathsf{X},\mathcal{A}}^{\mathsf{ind}}(\lambda) = \Pr[\, b' = 1 \mid b = 1\,] - \Pr[\, b' = 1 \mid b = 0\,]$$

$$\geq \mathsf{Adv}_{\mathsf{SE}}^{\mathsf{dec}}(\lambda) - \mathsf{Adv}_{R,\mathcal{R}}^{\mathsf{pr}}(\lambda) - 2^{-2 \cdot \mathsf{R.sl}(\lambda)} \,. \tag{8}$$

Our weak correctness condition implies that the first term of Eq. (8) is non-negligible. On the other hand, the second and third terms are negligible. This means $\mathsf{Adv}_{\mathsf{SE},\mathsf{X},\mathcal{A}}^{\mathsf{ind}}(\cdot)$ is not negligible, proving claim (2) of Theorem 3.

Games G_0–G_3

$k \leftarrow_{\$} \{0,1\}^{\mathsf{SE.kl}(\lambda)}$; $m \leftarrow_{\$} \{0,1\}^{\mathsf{SE.ml}(\lambda)}$

  $y \leftarrow \mathsf{R.Ev}(1^\lambda, m)$; $\overline{C} \leftarrow_{\$} \mathsf{Obf}(1^\lambda, \mathsf{Pad}_{s(\lambda)}(C^1_{1^\lambda,k,y}))$ $/\!\!/ \ G_0$

  $y \leftarrow_{\$} \{0,1\}^{2 \cdot \mathsf{R.sl}(\lambda)}$; $\overline{C} \leftarrow_{\$} \mathsf{Obf}(1^\lambda, \mathsf{Pad}_{s(\lambda)}(C^1_{1^\lambda,k,y}))$ $/\!\!/ \ G_1$

  $y \leftarrow_{\$} \{0,1\}^{2 \cdot \mathsf{R.sl}(\lambda)}$; $\overline{C} \leftarrow_{\$} \mathsf{Obf}(1^\lambda, \mathsf{Pad}_{s(\lambda)}(C^2))$ $/\!\!/ \ G_2$

$k' \leftarrow_{\$} \mathcal{Q}(1^\lambda, \overline{C})$; Return $(k = k')$

Circuit $C^1_{1^\lambda,k,y}(c)$	Circuit $C^2(c)$
$m \leftarrow \mathsf{SE.Dec}(1^\lambda, k, c)$; $y' \leftarrow \mathsf{R.Ev}(1^\lambda, m)$	Return 0
If $(y = y')$ then return 1 else return 0	

Fig. 6. Games for proof of part (1) of Theorem 3.

We proceed to prove part (1) of the theorem statement. Let $\mathcal{Q}$ be a PT adversary. Consider the games and associated circuits of Fig. 6. Lines not annotated with comments are common to all three games. Game G_0 is equivalent to $\mathrm{PRED}_{\mathsf{X}}^{\mathcal{Q}}(\lambda)$, so

$$\mathsf{Adv}_{\mathsf{X},\mathcal{Q}}^{\mathsf{pred}}(\lambda) = \Pr[G_2] + (\Pr[G_0] - \Pr[G_1]) + (\Pr[G_1] - \Pr[G_2]) \,. \tag{9}$$

We have $\Pr[G_2] = 2^{-\mathsf{SE.kl}(\lambda)}$, where the latter is assumed to be negligible, because k is uniformly random and the circuit $\overline{C}$ that is passed to adversary $\mathcal{Q}$ does not depend on k. We now show that $\Pr[G_i] - \Pr[G_{i+1}]$ is negligible for $i \in \{0,1\}$, which by Eq. (9) implies that $\mathsf{Adv}_{\mathsf{X},\mathcal{Q}}^{\mathsf{pred}}(\cdot)$ is negligible and hence proves the claim.

First, we construct a PT adversary $\mathcal{R}$ against PRG R, as follows:

Adversary $\mathcal{R}(1^\lambda, y)$

$k \leftarrow_{\$} \{0,1\}^{\mathsf{SE.kl}(\lambda)}$; $\overline{C} \leftarrow_{\$} \mathsf{Obf}(1^\lambda, \mathsf{Pad}_{s(\lambda)}(C^1_{1^\lambda,k,y}))$; $k' \leftarrow_{\$} \mathcal{Q}(1^\lambda, \overline{C})$

If $(k = k')$ then return 1 else return 0

We have $\Pr[G_0] - \Pr[G_1] = \mathsf{Adv}^{\mathsf{pr}}_{\mathsf{R},\mathcal{R}}(\lambda)$, where the advantage is negligible by the assumed PR-security of R.

Next, we construct a circuit sampler S and an iO-adversary $\mathcal{O}$, as follows:

Circuit Sampler $\mathsf{S}(1^\lambda)$	Adversary $\mathcal{O}(1^\lambda, \overline{C}, aux)$
$k \twoheadleftarrow \{0,1\}^{\mathsf{SE.kl}(\lambda)}$; $y \twoheadleftarrow \{0,1\}^{2 \cdot \mathsf{R.sl}(\lambda)}$	$k \leftarrow aux$; $k' \twoheadleftarrow \mathcal{Q}(1^\lambda, \overline{C})$
$C_1 \leftarrow \mathsf{Pad}_{s(\lambda)}(C^1_{1^\lambda,k,y})$; $C_0 \leftarrow \mathsf{Pad}_{s(\lambda)}(C^2)$	If $(k = k')$ then return 1
$aux \leftarrow k$; return (C_0, C_1, aux)	Else return 0

It follows that $\Pr[G_1] - \Pr[G_2] = \mathsf{Adv}^{\mathsf{io}}_{\mathsf{Obf},\mathsf{S},\mathcal{O}}(\lambda)$. We now show that $\mathsf{S} \in \boldsymbol{S}_{\mathsf{eq}}$, and hence $\mathsf{Adv}^{\mathsf{io}}_{\mathsf{Obf},\mathsf{S},\mathcal{O}}(\lambda)$ is negligible by the assumed iO-security of Obf. Specifically, note that $C^1_{1^\lambda,k,y}$ and C^2 are not equivalent only if y belongs to the range of R, which contains at most $2^{\mathsf{R.sl}(\lambda)}$ values. However, y is sampled uniformly at random from a set of size $2^{2 \cdot \mathsf{R.sl}(\lambda)}$. It follows that

$$\Pr\left[\, C_0 \equiv C_1 \;\; : \;\; (C_0, C_1, aux) \twoheadleftarrow \mathsf{S}(1^\lambda) \,\right] \geq 1 - 2^{-\mathsf{R.sl}(\lambda)},$$

where $2^{-\mathsf{R.sl}(\lambda)}$ is assumed to be negligible, and hence $\mathsf{S} \in \boldsymbol{S}_{\mathsf{eq}}$.

References

1. Abdalla, M., Bellare, M., Neven, G.: Robust encryption. In: Micciancio, D. (ed.) TCC 2010. LNCS, vol. 5978, pp. 480–497. Springer, Heidelberg (2010)
2. Ananth, P., Boneh, D., Garg, S., Sahai, A., Zhandry, M.: Differing-inputs obfuscation and applications. Cryptology ePrint Archive, Report 2013/689 (2013). http://eprint.iacr.org/2013/689
3. Barak, B., Garg, S., Kalai, Y.T., Paneth, O., Sahai, A.: Protecting obfuscation against algebraic attacks. In: Nguyen, P.Q., Oswald, E. (eds.) EUROCRYPT 2014. LNCS, vol. 8441, pp. 221–238. Springer, Heidelberg (2014)
4. Barak, B., Goldreich, O., Impagliazzo, R., Rudich, S., Sahai, A., Vadhan, S.P., Yang, K.: On the (im)possibility of obfuscating programs. In: Kilian, J. (ed.) CRYPTO 2001. LNCS, vol. 2139, pp. 1–18. Springer, Heidelberg (2001)
5. Bellare, M., Hoang, V.T.: Resisting randomness subversion: fast deterministic and hedged public-key encryption in the standard model. In: Oswald, E., Fischlin, M. (eds.) EUROCRYPT 2015. LNCS, vol. 9057, pp. 627–656. Springer, Heidelberg (2015)
6. Bellare, M., Hoang, V.T., Keelveedhi, S.: Instantiating Random Oracles via UCEs. In: Canetti, R., Garay, J.A. (eds.) CRYPTO 2013, Part II. LNCS, vol. 8043, pp. 398–415. Springer, Heidelberg (2013)
7. Bellare, M., Rogaway, P.: The security of triple encryption and a framework for code-based game-playing proofs. In: Vaudenay, S. (ed.) EUROCRYPT 2006. LNCS, vol. 4004, pp. 409–426. Springer, Heidelberg (2006)
8. Bellare, M., Stepanovs, I.: Point-function obfuscation: a framework and generic constructions. In: Kushilevitz, E., Malkin, T., (eds.) TCC 2016-A, Part II. LNCS, vol. 9563, pp. 565–594. Springer, Heidelberg (2016)
9. Bellare, M., Stepanovs, I., Tessaro, S.: Poly-many hardcore bits for any one-way function and a framework for differing-inputs obfuscation. In: Sarkar, P., Iwata, T. (eds.) ASIACRYPT 2014, Part II. LNCS, vol. 8874, pp. 102–121. Springer, Heidelberg (2014)

10. Bitansky, N., Canetti, R.: On Strong Simulation and Composable Point Obfuscation. In: Rabin, T. (ed.) CRYPTO 2010. LNCS, vol. 6223, pp. 520–537. Springer, Heidelberg (2010)
11. Bitansky, N., Canetti, R., Cohn, H., Goldwasser, S., Kalai, Y.T., Paneth, O., Rosen, A.: The impossibility of obfuscation with auxiliary input or a universal simulator. In: Garay, J.A., Gennaro, R. (eds.) CRYPTO 2014, Part II. LNCS, vol. 8617, pp. 71–89. Springer, Heidelberg (2014)
12. Bitansky, N., Canetti, R., Kalai, Y.T., Paneth, O.: On virtual grey box obfuscation for general circuits. In: Garay, J.A., Gennaro, R. (eds.) CRYPTO 2014, Part II. LNCS, vol. 8617, pp. 108–125. Springer, Heidelberg (2014)
13. Bitansky, N., Canetti, R., Kalai, Y.T., Paneth, O.: On virtual grey box obfuscation for general circuits. Cryptology ePrint Archive, Report 2014/554 (2014). http://eprint.iacr.org/2014/554
14. Bitansky, N., Canetti, R., Paneth, O., Rosen, A.: On the existence of extractable one-way functions. In: Shmoys, D.B. (ed.) 46th ACM STOC, pp. 505–514. ACM Press, May / June (2014)
15. Bitansky, N., Paneth, O.: Point obfuscation and 3-round zero-knowledge. In: Cramer, R. (ed.) TCC 2012. LNCS, vol. 7194, pp. 190–208. Springer, Heidelberg (2012)
16. Bitansky, N., Paneth, O., On the impossibility of approximate obfuscation and applications to resettable cryptography. In: Boneh, D., Roughgarden, T., Feigenbaum, J. (eds.) 45th ACM STOC, pp. 241–250. ACM Press, June 2013
17. Blum, M., Micali, S.: How to generate cryptographically strong sequences of pseudorandom bits. SIAM J. Comput. 13(4), 850–864 (1984)
18. Boyle, E., Chung, K.-M., Pass, R.: On extractability obfuscation. In: Lindell, Y. (ed.) TCC 2014. LNCS, vol. 8349, pp. 52–73. Springer, Heidelberg (2014)
19. Brakerski, Z., Rothblum, G.N.: Virtual black-box obfuscation for all circuits via generic graded encoding. In: Lindell, Y. (ed.) TCC 2014. LNCS, vol. 8349, pp. 1–25. Springer, Heidelberg (2014)
20. Brzuska, C., Farshim, P., Mittelbach, A.: Indistinguishability Obfuscation and UCEs: The Case of Computationally Unpredictable Sources. In: Garay, J.A., Gennaro, R. (eds.) CRYPTO 2014, Part I. LNCS, vol. 8616, pp. 188–205. Springer, Heidelberg (2014)
21. Brzuska, C., Farshim, P., Mittelbach, A.: Random-oracle uninstantiability from indistinguishability obfuscation. In: Dodis, Y., Nielsen, J.B. (eds.) TCC 2015, Part II. LNCS, vol. 9015, pp. 428–455. Springer, Heidelberg (2015)
22. Brzuska, C., Mittelbach, A.: Indistinguishability Obfuscation versus Multi-bit Point Obfuscation with Auxiliary Input. In: Sarkar, P., Iwata, T. (eds.) ASIACRYPT 2014, Part II. LNCS, vol. 8874, pp. 142–161. Springer, Heidelberg (2014)
23. Brzuska, C., Mittelbach, A.: Using indistinguishability obfuscation via UCEs. In: Sarkar, P., Iwata, T. (eds.) ASIACRYPT 2014, Part II. LNCS, vol. 8874, pp. 122–141. Springer, Heidelberg (2014)
24. Brzuska, C., Mittelbach, A.: Universal computational extractors and the superfluous padding assumption for indistinguishability obfuscation. Cryptology ePrint Archive, Report 2015/581 (2015). http://eprint.iacr.org/2015/581
25. Canetti, R.: Towards realizing random oracles: hash functions that hide all partial information. In: Kaliski Jr, B.S. (ed.) CRYPTO 1997. LNCS, vol. 1294, pp. 455–469. Springer, Heidelberg (1997)
26. Canetti, R., Tauman Kalai, Y., Varia, M., Wichs, D.: On Symmetric Encryption and Point Obfuscation. In: Micciancio, D. (ed.) TCC 2010. LNCS, vol. 5978, pp. 52–71. Springer, Heidelberg (2010)

27. Cheon, J.H., Han, K., Lee, C., Ryu, H., Stehlé, D.: Cryptanalysis of the multilinear map over the integers. In: Oswald, E., Fischlin, M. (eds.) EUROCRYPT 2015. LNCS, vol. 9056, pp. 3–12. Springer, Heidelberg (2015)

28. Coron, J.-S., Lepoint, T., Tibouchi, M.: Cryptanalysis of two candidate fixes of multilinear maps over the integers. Cryptology ePrint Archive, Report 2014/975 (2014). http://eprint.iacr.org/2014/975

29. Dodis, Y., Ganesh, C., Golovnev, A., Juels, A., Ristenpart, T.: A formal treatment of backdoored pseudorandom generators. In: Oswald, E., Fischlin, M. (eds.) EUROCRYPT 2015. LNCS, vol. 9056, pp. 101–126. Springer, Heidelberg (2015)

30. Dodis, Y., Kalai, Y.T., Lovett, S.: On cryptography with auxiliary input. In: Mitzenmacher, M. (ed.) 41st ACM STOC, pp. 621–630. ACM Press, May / June (2009)

31. S. Garg, C. Gentry, S. Halevi, M. Raykova, A. Sahai, Waters, B.: Candidate indistinguishability obfuscation and functional encryption for all circuits. In: 54th FOCS, pp. 40–49. IEEE Computer Society Press October 2013

32. Garg, S., Gentry, C., Halevi, S., Wichs, D.: On the implausibility of differing-inputs obfuscation and extractable witness encryption with auxiliary input. In: Garay, J.A., Gennaro, R. (eds.) CRYPTO 2014, Part I. LNCS, vol. 8616, pp. 518–535. Springer, Heidelberg (2014)

33. Gentry, C., Halevi, S., Maji, H.K., Sahai, A.: Zeroizing without zeroes: Cryptanalyzing multilinear maps without encodings of zero. Cryptology ePrint Archive, Report 2014/929 (2014). http://eprint.iacr.org/2014/929

34. Gentry, C., Lewko, C.A., Sahai, A., Waters, B.: Indistinguishability obfuscation from the multilinear subgroup elimination assumption. Cryptology ePrint Archive, Report 2014/309 (2014). http://eprint.iacr.org/2014/309

35. Goldwasser, S., Kalai, Y.T.: On the impossibility of obfuscation with auxiliary input. In: 46th FOCS, pp. 553–562. IEEE Computer Society Press, October 2005

36. Green, M.D., Katz, J., Malozemoff, A.J., Zhou, H.-S.: A unified approach to idealized model separations via indistinguishability obfuscation. Cryptology ePrint Archive, Report 2014/863 (2014). http://eprint.iacr.org/2014/863

37. Hada, S.: Zero-knowledge and code obfuscation. In: Okamoto, T. (ed.) ASIACRYPT 2000. LNCS, vol. 1976, pp. 443–457. Springer, Heidelberg (2000)

38. Håstad, J., Impagliazzo, R., Levin, L.A., Luby, M.: A pseudorandom generator from any one-way function. SIAM J. Comput. 28(4), 1364–1396 (1999)

39. Lee, H.T., Seo, J.H.: Security analysis of multilinear maps over the integers. In: Garay, J.A., Gennaro, R. (eds.) CRYPTO 2014, Part I. LNCS, vol. 8616, pp. 224–240. Springer, Heidelberg (2014)

40. Matsuda, T., Hanaoka, G.: Chosen ciphertext security via UCE. In: Krawczyk, H. (ed.) PKC 2014. LNCS, vol. 8383, pp. 56–76. Springer, Heidelberg (2014)

41. Pass, R., Seth, K., Telang, S.: Indistinguishability obfuscation from semantically-secure multilinear encodings. In: Garay, J.A., Gennaro, R. (eds.) CRYPTO 2014, Part I. LNCS, vol. 8616, pp. 500–517. Springer, Heidelberg (2014)

42. Sahai, A., Waters, B.: How to use indistinguishability obfuscation: deniable encryption, and more. In: Shmoys, D.B. (ed.) 46th ACM STOC, pp. 475–484. ACM Press, May / June 2014

43. Wee, H.: On obfuscating point functions. In: Gabow, H.N., Fagin, R. (eds.) 37th ACM STOC, pp. 523–532. ACM Press, May 2005

44. Yao, A.C.-C.: Theory and applications of trapdoor functions (extended abstract). In: 23rd FOCS, pp. 80–91. IEEE Computer Society Press, November (1982)

Point-Function Obfuscation: A Framework and Generic Constructions

Mihir Bellare$^{(\boxtimes)}$ and Igors Stepanovs

Department of Computer Science and Engineering,
University of California, San Diego, USA
mihir@eng.ucsd.edu
http://cseweb.ucsd.edu/mihir/,
https://sites.google.com/site/igorsstepanovs/

Abstract. We give a definitional framework for point-function obfuscation in which security is parameterized by a class of algorithms we call target generators. Existing and new notions are captured and explained as corresponding to different choices of this class. This leads to an elegant question: Is it possible to provide a generic construction, meaning one that takes an arbitrary class of target generators and returns a point-function obfuscator secure for it? We answer this in the affirmative with three generic constructions, the first based on indistinguishability obfuscation, the second on deterministic public-key encryption and the third on universal computational extractors. By exploiting known constructions of the primitives assumed, we obtain new point-function obfuscators, including many under standard assumptions. We end with a broader look that relates different known and possible notions of point function obfuscation to each other and to ours.

1 Introduction

In the theory of point-function obfuscation (PO), there are many different goals and definitions. It is (at least to us) hard territory to navigate. Meanwhile, there are few constructions; indeed, there are fewer constructions than there are definitions. And the ones that exist use strong assumptions. We try to bring some structure and unity to this area via a parameterized definitional framework, generic constructions and relations between definitions.

1.1 The State of Point-Function Obfuscation

A point function with target $k \in \{0,1\}^*$ is the circuit $\mathbf{I}_k$ that on input $k' \in \{0,1\}^{|k|}$ returns 1 if $k' = k$ and 0 otherwise. A point-function obfuscator Obf takes input $\mathbf{I}_k$ and returns another circuit $\overline{\mathrm{P}}$ that is functionally equivalent to $\mathbf{I}_k$, meaning on input $k' \in \{0,1\}^{|k|}$ it also returns 1 if $k' = k$ and 0 otherwise. Security requires that $\overline{\mathrm{P}}$ hides k. We now discuss the state of the area with regard to both definitions and constructions.

© International Association for Cryptologic Research 2016
E. Kushilevitz and T. Malkin (Eds.): TCC 2016-A, Part II, LNCS 9563, pp. 565–594, 2016.
DOI: 10.1007/978-3-662-49099-0_21

Definitions. The theory of PO contains a large number of different goals and definitions. Sometimes there is auxiliary information [14,21,35], other times not [23,27,39,47]. Sometimes security pertains to a single target, other times to many [25]. Sometimes the formalization is a VBB-style simulation based one, other times indistinguishability based. Within each category, there are variants, for example, for indistinguishability, whether the necessary unpredictability condition on targets should be for polynomial-time or unbounded adversaries, and with negligible or sub-exponential advantage. And this list is not complete.

While from one perspective there are too many definitions, from other perspectives there are too few. Think of different elements that have been considered (for example whether or not auxiliary information is present, one target or many, polynomial-time or unbounded predictability adversaries, ..., in the context of an indistinguishability-based definition) as dimensions or axes in a multi-dimensional space. Then definitions in the literature can be seen as capturing some points in this space. But there is no systematic attempt to look in some unified way at all the points in this space. There is a connection that does not seem to have been explicitly made and pursued, namely that definitionally, there is little to no difference between PO and deterministic public-key encryption DPKE [3,4,17] or other forms of entropic security [30,36]. Existing systematic and in-depth consideration of DPKE definitions and relations between them [4,17] can be exploited to obtain semantic-security formalizations of PO that address issues with current definitions, and also to obtain definitional relations.

Constructions. Existing constructions use strong assumptions and achieve only some of the goals. A primary construction is from the AI-DHI (Auxiliary-Input Diffie-Hellman Inversion) assumption [14,23]. Calling it a construction is a bit of a stretch; the security just amounts to the assumption. The latter cannot co-exist with VGBO (Virtual Grey Box Obfuscation) [10]. That doesn't mean it is wrong (perhaps VGBO does not exist) but it would be preferable to base PO on assumptions not in contention with VGBO. Wee [47] provides a construction based on a fixed permutation about which a novel, strong uninvertibility assumption is made. He only proves security in the absence of auxiliary information, and GK [35] show that the construction does not in fact provide security in the presence of auxiliary information. However BP [14] specify an extension of Wee's construction with a family of permutations rather than a fixed one, and show, under a novel assumption called Assumption 2.1 in their paper, that it achieves security with targets that are hard to predict given the auxiliary information. BP [14] explain that Assumption 2.1 asks for (a weak form of) extractability, making it a strong assumption in light of the impossibility of related extractable primitives [13]. DKL [29] use a novel assumption they call LSN to give a construction for targets that are exponentially hard to predict given the auxiliary information. BHK [6] give a construction for statistically hard to predict targets and no auxiliary information based on a multi-key version of their UCE assumption. There are simple constructions in the ROM [39].

In summary, there are few (standard-model) constructions and those that exist all use strong and sometimes novel assumptions. Also, each construction achieves a different variant of the goal and it is hard to visualize, or say in a concise way, what has been done. The framework that we now discuss provides language to do this.

1.2 Contributions in Brief

We pick one, simple indistinguishability-based definitional template. Using this, we provide a framework parameterized by a class $\mathbf{X}$ of objects we call *target generators*, giving a definition of what it means for a point-function obfuscator to be IND[$\mathbf{X}$] secure. This allows us to recover and explain different notions in the literature as each corresponding to a choice of $\mathbf{X}$, and also obtain many natural new ones, points in the above-mentioned multi-dimensional space that had not been explicitly considered.

This taxonomy leads to a compelling and general new question: Is it possible to find a *generic construction*, meaning a compiler that given an arbitrary $\mathbf{X}$ returns a point-function obfuscator secure relative to it? We answer this in the affirmative by providing three such generic constructions. As a consequence we obtain new constructions for both old and new forms of PO.

We then step back to consider other definitions of PO. These include existing simulation and indistinguishability style notions, as well as new, semantic security style ones emanating from the above-mentioned connection to DPKE. We formulate these also in a parameterized framework and then provide relations (implications and separations) between these notions and our IND notion.

We now look at these three contributions in more detail.

1.3 Definitional Framework

Recall that a point-function obfuscator Obf takes input $\mathbf{I}_k$ and returns another circuit $\overline{\mathrm{P}}$ that is functionally equivalent to $\mathbf{I}_k$. Security requires that $\overline{\mathrm{P}}$ hides k. We define a *target generator* X as a polynomial-time algorithm that on input the security parameter returns a vector $\mathbf{k}$ of target points together with auxiliary information a. We measure security of a candidate point-function obfuscator Obf relative to X. To do this, we associate to an adversary $\mathcal{A}$ its advantage $\mathsf{Adv}^{\mathrm{ind}}_{\mathrm{Obf},\mathsf{X},\mathcal{A}}(\cdot)$ in guessing the challenge bit b in the following game. We run X to get $(\mathbf{k}, a)$. We let $\overline{\mathbf{P}}$ be the vector obtained by independently obfuscating $\mathbf{I}_k$ for each of the targets k from $\mathbf{k}$ ($b = 1$) or by obfuscating the same number of random, independent targets ($b = 0$). The input to $\mathcal{A}$ is $\overline{\mathbf{P}}$ and a. Now we let $\mathbf{X}$ be a class (set) of target generators X and say that obfuscator Obf is IND[$\mathbf{X}$]-secure if $\mathsf{Adv}^{\mathrm{ind}}_{\mathrm{Obf},\mathsf{X},\mathcal{A}}(\cdot)$ is negligible for all polynomial time $\mathcal{A}$ and all $\mathsf{X} \in \mathbf{X}$. See Sect. 3 for a formal definition.

What we have here is a notion of point-function obfuscation parameterized by a class of target generators. We view the latter as knobs. By turning these knobs (defining specific classes) we can capture specific restrictions, and by intersecting classes we can combine them, allowing us to speak precisely yet concisely about different variant notions that are unified in this way.

IND[$\mathbf{X}$]-security is not achievable for all $\mathbf{X}$. For example, X could pick $\mathbf{k}[1]$ to be the string of all zeroes, and the adversary could test whether or not $\overline{P}$ returns 1 on input that string. The minimal requirement for security is that the target points produced by X are unpredictable given a. In Sect. 3 we formalize a prediction game and advantage so that we can define the classes $\mathbf{X}^{\mathrm{cup}}$, $\mathbf{X}^{\mathrm{seup}}$ and $\mathbf{X}^{\mathrm{sup}}$ of computationally, sub-exponentially and statistically unpredictable target generators. We let $\mathbf{X}^{q(\cdot)}$ denote the class of target generators outputting $q(\cdot)$ target points and $\mathbf{X}^{\varepsilon}$ the class of target generators that produce no auxiliary information. (Formally it is the empty string.)

Already we can characterize prior work in a precise way. IND[$\mathbf{X}^{\mathrm{cup}} \cap \mathbf{X}^{\varepsilon} \cap \mathbf{X}^1$] is plain point-function obfuscation [23,27,39,47], where there is just one target point, no auxiliary information, and unpredictability is computational. IND[$\mathbf{X}^{\mathrm{cup}} \cap \mathbf{X}^1$] is AIPO [14,20], where there is again one target point, but auxiliary information is now present, while unpredictability continues to be computational. IND[$\mathbf{X}^{\mathrm{cup}}$] is composable AIPO [25], where there are many arbitrarily correlated target points, auxiliary information is present, and unpredictability is computational. DKL [29] achieve IND[$\mathbf{X}^{\mathrm{sup}} \cap \mathbf{X}^1$], where there is a single target that is statistically hard to predict given the auxiliary information. BHK [6] achieve IND[$\mathbf{X}^{\mathrm{sup}} \cap \mathbf{X}^{\varepsilon}$], where there are multiple targets, unpredictability is statistical, and there is no auxiliary information. Other prior notions can be captured in similar ways, and many natural new notions emerge as well.

1.4   Generic Constructions

As we saw above, constructions so far have been ad hoc, targeting different security goals and using strong, novel assumptions to achieve them. The above framework allows us to frame a compelling question, namely whether there are generic constructions. By this we mean that we are handed an arbitrary class $\mathbf{X}$ of target generators and asked to craft an obfuscator that is IND[$\mathbf{X}$]-secure. If we can do this, we can, in one unified swoop, obtain constructions for a wide variety of forms of PO, not only ones considered in the past, but also new ones.

In this paper we provide three such generic constructions. The first is based on indistinguishability obfuscation, the second on deterministic public-key encryption and the third on (multi-key) UCE.

One natural objection at this point is that we know that IND[$\mathbf{X}$] is not achievable for some choices of $\mathbf{X}$. For example, assuming iO, this is true for $\mathbf{X} = \mathbf{X}^{\mathrm{cup}}$, meaning composable PO. (This follows by combining [20,24].) So how can our constructions achieve IND[$\mathbf{X}$] for any given $\mathbf{X}$? In fact, they do, and this, interestingly, yields new negative results, ruling out the primitives we start from for those particular values of $\mathbf{X}$. We will explain further below.

PO from iO. The emergence of candidate constructions for iO (indistinguishability obfuscation) [12,33,34,43] raised a natural hope, namely that one could obtain PO from iO. But this has not happened. Despite the many powerful applications of iO, constructing point-function obfuscation from it has surprisingly evaded effort.

We show that iO plus a OWF yields PO. More precisely, we show iO + OWF[$\mathbf{X}$] $\Rightarrow$ IND[$\mathbf{X}$]: Given iO and a family of functions that is one-way relative to $\mathbf{X}$ as defined in Sect. 5.1 we can construct an obfuscator that is IND[$\mathbf{X}$]-secure. The construction, result and proof are in Sect. 5.1. The idea is that to obfuscate $\mathbf{I}_k$ we pick at random a key fk for the OWF F (formally, the latter is a family of functions) and let $y = \mathsf{F}(fk, k)$. We consider the circuit C that hardwires fk, y and on input k' returns 1 if $\mathsf{F}(fk, k') = y$ and 0 otherwise. We then apply an indistinguishability obfuscator to C to produce the obfuscated point function. The security proof is a sequence of hybrids. Although we assume only iO, we exploit diO [1,2,16] in the proof in a manner similar to [9]. We will need it for circuits that differ only on one input, and in this case the result of BCP [16] says that an iO-secure obfuscator is also diO-secure, so the assumption remains iO. As part of the proof we state and prove a lemma reducing (d)iO on polynomially-many, related circuits to the usual single-circuit case. We note that to guarantee the usual (perfect) correctness condition of a PO, we require the OWF to be injective.

We highlight the simplest case of this result as still being novel and of interest. Namely, given iO and an ordinary injective OWF, we achieve plain point-function obfuscation, IND[$\mathbf{X}^{\mathrm{cup}} \cap \mathbf{X}^{\varepsilon} \cap \mathbf{X}^1$] in our notation. Previous constructions have been under assumptions that at this point seem less accepted than iO, and Wee [47] gives various arguments as to why this goal is hard under standard assumptions. Also on the negative side, combining our result with [20,24] allows us, under iO, to rule out OWF[$\mathbf{X}^{\mathrm{cup}}$] (one-way functions secure for polynomially-many, computationally unpredictable correlated inputs), at least in the injective case.

PO from DPKE. Deterministic public key encryption (DPKE) [3] was motivated by applications to efficient searchable encryption [3]. It cannot provide IND-CPA security. Instead, BBO [3] provide a definition of a goal called PRIV which captures the best-possible security that encryption can provide subject to being deterministic. At this point many constructions of DPKE are known for various variant goals [3–5,15,17,32,38,42,45,48,50].

We show how to leverage these for point-function obfuscation via our second generic construction. We show that PRIV1[$\mathbf{X}$] $\Rightarrow$ IND[$\mathbf{X}$]. That is, given a deterministic public-key encryption scheme that is PRIV1 secure relative to $\mathbf{X}$ we can build a point-function obfuscator secure relative to the same class in a simple and natural way. Namely to obfuscate $\mathbf{I}_k$ we pick at random a public key pk and the associated secret key sk for the DPKE scheme and let c be the encryption of k under pk. The point-function obfuscation is the circuit C that hardwires pk, c and on input k', returns 1 if the encryption of k' under pk equals c, and 0 otherwise. The fact that the encryption is deterministic is used crucially to define the circuit. (The latter must be deterministic.) The secret key sk is discarded and not used in the construction. We note that we only require security of the

DPKE scheme for a single message (PRIV1) so the negative result of Wichs [49] does not apply. The construction, result and proof are in Sect. 5.2.

From the LTDF-based DPKE scheme of BFO [15] and LTDFs from [31,37,44, 48,51] we now get $\text{IND}[\mathbf{X}^{\text{sup}} \cap \mathbf{X}^{\varepsilon} \cap \mathbf{X}^{1}]$-secure obfuscators under a large number of standard assumptions. We also get $\text{IND}[\mathbf{X}^{\text{seup}} \cap \mathbf{X}^{1}]$-secure obfuscators under the DLIN, Subgroup Indistinguishability and LWE assumptions via [17,48,50]. On the negative side we can rule out $\text{PRIV1}[\mathbf{X}^{\text{cup}}]$-secure DPKE under iO via [20,24].

PO from UCE. UCE [6] is a class of assumptions on function families crafted to allow instantiation of random oracles in certain settings. UCE security is parameterized so that we have UCE[**S**] security of a family of functions for different choices of classes **S** of algorithms called sources. The parameterization is necessary because security is not achievable for the class of all sources. Different applications rely on UCE relative to different classes of sources [5,6,18,21,28,41].

In this work we use the multi-key version of UCE, abbreviated mUCE [6]. We show how to associate to any given class **X** of target generators a class $\mathbf{S}^{\mathbf{X}}$ of sources such that $\text{mUCE}[\mathbf{S}^{\mathbf{X}}] \Rightarrow \text{IND}[\mathbf{X}]$, meaning we can build a point-function obfuscator secure for **X** given a family of functions that is $\text{mUCE}[\mathbf{S}^{\mathbf{X}}]$-secure. The definition of $\mathbf{S}^{\mathbf{X}}$ is given in Sect. 5.3. But what is most relevant here is that the strength of UCE-framework assumptions is very sensitive to the choice of class of sources that parameterizes the particular assumption, and $\mathbf{S}^{\mathbf{X}}$ has good properties in this regard. The sources are what are called "split" in [6], and they inherit the unpredictability attributes of the target generators. $\text{mUCE}[\mathbf{S}^{\mathbf{X}}]$-security is not achievable for all choices of **X** but the assumption is valid as far as we know for many choices of **X**, yielding new constructions.

1.5   Alternative Notions and Relations Between Notions

Above, we fixed one, basic definitional template, which we called IND, and then parameterized it by classes **X** of target generators to get notions $\text{IND}[\mathbf{X}]$. However, there are other possible choices for the basic template, some emanating from the literature, and others from the definitional similarity of PO with DPKE. We consider parameterized versions of some of these and relate them to each other and to IND. Specifically we define and consider the following (see Sect. 6 for formal definitions):

- SIM[**X**]: (Simulation) The first definitions for PO simply restricted VBB security [2] to the class of point functions [25,35,39,47]. With SSS[**X**] we give an **X**-parameterized version of this.
- SIND[**X**]: (Strong Indistinguishability) Recall that in $\text{IND}[\mathbf{X}]$, the adversary decision bit is produced as a function of the vector $\overline{\mathbf{P}}$ of obfuscated point functions and the auxiliary information a. In SIND[**X**], this bit is not the final decision, but is passed to another adversary who produces the final decision based on it and the target vector itself. This is a parameterized version of the definition of [23].
- CSS[**X**]: (Comparison-based semantic security) This is an analogue of comparision based semantic security for boolean functions for DPKE [4] in which the

adversary needs to compute some predicate on the target vector and auxiliary information.

- SSS[**X**]: (Simulation-based semantic security) This is an analogue of simulation based semantic security for boolean functions for DPKE [4] in which a simulator with an oracle for the point functions must compute a predicate on the target vectors and auxiliary information.

Figure 8 shows the relations between five parameterized notions of PO, namely the four above and our original IND[**X**].

1.6 Discussion and Further Related Work

In concurrent and independent work, BM3 [22] take first steps towards a parameterized definition for point-function obfuscation, with separate definitions for the basic and composable cases. They also show that injective mUCE-secure function families for strongly unpredictable sources making one oracle query per key implies composable AIPO (both for computational and statistical unpredictability), which is a special case of our mUCE result.

Multi-bit auxiliary-input point-function obfuscation (MB-AIPO) [11,25,40] allows one to obfuscate the circuit $\mathbf{I}_{k,m}$ that on input k' returns m if $k = k'$ and $\perp$ otherwise, where k, m are strings. CD [25] show that composable AIPO implies MB-AIPO. MB-AIPO was subsequently used in BP [14] and MH [40]. BM1 [20] show that if iO is possible then MB-AIPO is not. MB-AIPO seems to be quite a bit stronger than AIPO itself and in particular this result does not rule out AIPO.

In Sect. 5.1 we define OWF[**X**], one-wayness of a function family relative to a class of target generators, the targets here being the inputs to the OWF. We note that OWF[$\mathbf{X}^{\mathrm{sup}} \cap \mathbf{X}^\varepsilon$] (inputs are statistically unpredictable and there is no auxiliary information) is the notion of a one-way correlation intractable hash (CIH) function family as per GOR [36].

Our parameterized PRIV1[**X**] notions of security for DPKE schemes apply equally to function families and thus recover, via particular choices of **X**, some of the security notions for CIH function families from GOR [36]. In these cases, since our DPKE-based constructions of PO do not require that decryption in the DPKE scheme is polynomial-time, CIH function families meeting the corresponding notions suffice as well.

Seeing that prior work can be characterized in terms of intersections of certain basic classes in our framework makes apparent that so far the literature has considered only a few points from the larger space of all possible intersections. A systematic consideration of the full space (which is lacking) would surface other notions of interest and give a coherent picture of the area.

2 Notation and Standard Definitions

Notation. We denote by $\lambda \in \mathbb{N}$ the security parameter and by 1^λ its unary representation. We let ε denote the empty string. If s is an integer then $\mathsf{Pad}_s(C)$

denotes circuit C padded to have size s. We say that circuits C_0, C_1 are equivalent, written $C_0 \equiv C_1$, if they agree on all inputs. If $\mathbf{x}$ is a vector then $|\mathbf{x}|$ denotes the number of its coordinates and $\mathbf{x}[i]$ denotes its i-th coordinate. We write $x \in \mathbf{x}$ as shorthand for $x \in \{\mathbf{x}[1], \ldots, \mathbf{x}[|\mathbf{x}|]\}$. If X is a finite set, we let $x \leftarrow_{\$} X$ denote picking an element of X uniformly at random and assigning it to x. Algorithms may be randomized unless otherwise indicated. Running time is worst case. "PT" stands for "polynomial-time," whether for randomized algorithms or deterministic ones. If A is an algorithm, we let $y \leftarrow A(x_1, \ldots ; r)$ denote running A with random coins r on inputs $x_1, \ldots$ and assigning the output to y. We let $y \leftarrow_{\$} A(x_1, \ldots)$ be the result of picking r at random and letting $y \leftarrow A(x_1, \ldots ; r)$. We let $[A(x_1, \ldots)]$ denote the set of all possible outputs of A when invoked with inputs $x_1, \ldots$. We say that $f \colon \mathbb{N} \to \mathbb{R}$ is negligible if for every positive polynomial p, there exists $\lambda_p \in \mathbb{N}$ such that $f(\lambda) < 1/p(\lambda)$ for all $\lambda > \lambda_p$. We use the code based game playing framework of [7]. (See Fig. 3 for an example.) By $G^{\mathcal{A}}(\lambda)$ we denote the event that the execution of game G with adversary $\mathcal{A}$ and security parameter λ results in the game returning true.

Obfuscators. An *obfuscator* is a PT algorithm Obf that on input 1^λ and a circuit C returns a circuit $\overline{C}$. If $\mathbf{C}$ is an n-vector of circuits then $\mathsf{Obf}(1^\lambda, \mathbf{C})$ denotes the vector $(\mathsf{Obf}(1^\lambda, \mathbf{C}[1]), \ldots, \mathsf{Obf}(1^\lambda, \mathbf{C}[n]))$ formed by applying Obf independently to each coordinate of $\mathbf{C}$. The *correctness condition* of obfuscator Obf requires that for every circuit C, every $\lambda \in \mathbb{N}$ and every $\overline{C} \in [\mathsf{Obf}(1^\lambda, C)]$ we have $\overline{C} \equiv C$ (meaning $\overline{C}(x) = C(x)$ for all x). We also call the latter a *perfect* correctness condition and we require that it holds for all obfuscators. We consider various notions of security for obfuscators, namely indistinguishability obfuscation and variants of point-function obfuscation, including AIPO.

Indistinguishability Obfuscation. Although our results need only iO, we use diO [1,2,16] in the proof, applying BCP [16] to then reduce the assumption to iO. To give the definitions compactly, we use the definitional framework of BST [9] which allows us to capture iO variants (including diO) via classes of circuit samplers. Let Obf be an obfuscator. A *sampler* in this context is a PT algorithm S that on input 1^λ returns a triple (C_0, C_1, aux) where C_0, C_1 are circuits of the same size, number of inputs and number of outputs, and aux is a string. If $\mathcal{O}$ is an adversary and $\lambda \in \mathbb{N}$ we let $\mathsf{Adv}^{\mathsf{io}}_{\mathsf{Obf},\mathsf{S},\mathcal{O}}(\lambda) = 2\Pr[\mathrm{IO}^{\mathcal{O}}_{\mathsf{Obf},\mathsf{S}}(\lambda)] - 1$ where game $\mathrm{IO}^{\mathcal{O}}_{\mathsf{Obf},\mathsf{S}}(\lambda)$ is defined in Fig. 1. Now let $\boldsymbol{S}$ be a class (set) of circuit samplers. We say that Obf is $\boldsymbol{S}$-secure if $\mathsf{Adv}^{\mathsf{io}}_{\mathsf{Obf},\mathsf{S},\mathcal{O}}(\cdot)$ is negligible for every PT adversary $\mathcal{O}$ and every circuit sampler $\mathsf{S} \in \boldsymbol{S}$. We say that circuit sampler S produces equivalent circuits if there exists a negligible function ν such that $\Pr\big[\, C_0 \equiv C_1 \; : \; (C_0, C_1, aux) \leftarrow_{\$} \mathsf{S}(1^\lambda)\,\big] \geq 1 - \nu(\lambda)$ for all $\lambda \in \mathbb{N}$. Let $\boldsymbol{S}_{\mathrm{eq}}$ be the class of all circuit samplers that produce equivalent circuits. We say that Obf is an indistinguishability obfuscator if it is $\boldsymbol{S}_{\mathrm{eq}}$-secure [2,33,46].

We say that a circuit sampler S is difference secure if $\mathsf{Adv}^{\mathsf{diff}}_{\mathsf{S},\mathcal{D}}(\cdot)$ is negligible for every PT adversary $\mathcal{D}$, where $\mathsf{Adv}^{\mathsf{diff}}_{\mathsf{S},\mathcal{D}}(\lambda) = \Pr[\mathrm{DIFF}^{\mathcal{D}}_{\mathsf{S}}(\lambda)]$ and game $\mathrm{DIFF}^{\mathcal{D}}_{\mathsf{S}}(\lambda)$ is defined in Fig. 1. Difference security of S means that given C_0, C_1, aux it is hard to find an input on which the circuits differ [1,2,16]. Let $\boldsymbol{S}_{\mathrm{diff}}$ be the class of

Game $\text{DIFF}_{\mathsf{S}}^{\mathcal{D}}(\lambda)$	Game $\text{IO}_{\text{Obf},\mathsf{S}}^{\mathcal{O}}(\lambda)$
$(C_0, C_1, aux) \leftarrow_{\$} \mathsf{S}(1^\lambda)$	$b \leftarrow_{\$} \{0, 1\}$; $(C_0, C_1, aux) \leftarrow_{\$} \mathsf{S}(1^\lambda)$
$x \leftarrow_{\$} \mathcal{D}(C_0, C_1, aux)$	$\overline{C} \leftarrow_{\$} \text{Obf}(1^\lambda, C_b)$; $b' \leftarrow_{\$} \mathcal{O}(1^\lambda, \overline{C}, aux)$
Return $(C_0(x) \neq C_1(x))$	Return $(b = b')$

Fig. 1. Games defining difference-security of circuit sampler S and iO-security of obfuscator Obf relative to circuit sampler S.

all difference-secure circuit samplers. We say that circuit sampler S produces d-differing circuits, where $d\colon \mathbb{N} \to \mathbb{N}$, if for all $\lambda \in \mathbb{N}$ circuits C_0 and C_1 differ on at most $d(\lambda)$ inputs with an overwhelming probability over $(C_0, C_1, aux) \leftarrow_{\$} \mathsf{S}(1^\lambda)$. Let $\boldsymbol{S}_{\text{diff}}(d)$ be the class of all difference-secure circuit samplers that produce d-differing circuits, so that $\boldsymbol{S}_{\text{eq}} \subseteq \boldsymbol{S}_{\text{diff}}(d) \subseteq \boldsymbol{S}_{\text{diff}}$. The interest of this definition is the following result of BCP [16] that we use:

Proposition 1. *If d is a polynomial then any $\boldsymbol{S}_{\text{eq}}$-secure circuit obfuscator is also an $\boldsymbol{S}_{\text{diff}}(d)$-secure circuit obfuscator.*

Function Families. A family of functions F specifies the following. PT key generation algorithm F.Kg takes 1^λ to return a key $fk \in \{0,1\}^{\mathsf{F.kl}(\lambda)}$, where F.kl: $\mathbb{N} \to \mathbb{N}$ is the key length function associated to F. Deterministic, PT evaluation algorithm F.Ev takes 1^λ, key $fk \in [\mathsf{F.Kg}(1^\lambda)]$ and an input $x \in \{0,1\}^{\mathsf{F.il}(\lambda)}$ to return an output $\mathsf{F.Ev}(1^\lambda, fk, x) \in \{0,1\}^{\mathsf{F.ol}(\lambda)}$, where F.il, F.ol: $\mathbb{N} \to \mathbb{N}$ are the input and output length functions associated to F, respectively. We say that F is *injective* if the function $\mathsf{F.Ev}(1^\lambda, fk, \cdot)\colon \{0,1\}^{\mathsf{F.il}(\lambda)} \to \{0,1\}^{\mathsf{F.ol}(\lambda)}$ is injective for every $\lambda \in \mathbb{N}$ and every $fk \in [\mathsf{F.Kg}(1^\lambda)]$. Notions of security for function families that we use are mUCE and OWF, the latter defined in Sect. 5.1.

UCE Framework. We recall the Universal Computational Extractor (UCE) framework of BHK [6]. We will use what BHK call the multi-key version of UCE (mUCE). It is an extension of the more commonly used UCE notion for a single key, meaning that it implies the latter. Meanwhile, no implications in the other direction (from single-key to multi-key) are known.

Let H be a family of functions. Let $\mathcal{S}$ be an adversary called the *source* and $\mathcal{D}$ an adversary called the *distinguisher*. Consider game $\text{mUCE}_{\mathsf{H}}^{\mathcal{S},\mathcal{D}}(\lambda)$ in the left panel of Fig. 2. Associated to $\mathcal{S}$ is a polynomial $\mathcal{S}.\text{nk}$ that indicates how many keys $\mathcal{S}$ uses. The source has access to an oracle HASH. A query to HASH consists of an index i of a key and the actual input x, which is a string required to have length $\mathsf{H.il}(\lambda)$. When the challenge bit b is 1 (the "real" case) the oracle responds via H.Ev under a key $\mathbf{hk}[i]$ that is chosen by the game and *not* given to the source. When $b = 0$ (the "random" case) it responds as a random oracle. The source then leaks a string L to its accomplice distinguisher. The latter *does* get the key vector $\mathbf{hk}$ as input and must now return its guess $b' \in \{0, 1\}$ for b. The game returns true iff $b' = b$. The advantage of $(\mathcal{S}, \mathcal{D})$ against the mUCE security of H is defined for $\lambda \in \mathbb{N}$ via $\text{Adv}_{\mathsf{H},\mathcal{S},\mathcal{D}}^{\text{m-uce}}(\lambda) = 2\Pr[\text{mUCE}_{\mathsf{H}}^{\mathcal{S},\mathcal{D}}(\lambda)] - 1$.

If **S** is a class (set) of sources, we say that H is mUCE[**S**]-secure if $\mathsf{Adv}^{\text{m-uce}}_{\mathsf{H},\mathcal{S},\mathcal{D}}(\cdot)$ is negligible for all sources $\mathcal{S} \in \mathbf{S}$ and all PT distinguishers $\mathcal{D}$.

$\text{mUCE}^{\mathcal{S},\mathcal{D}}_{\mathsf{H}}(\lambda)$	$\text{mSPRED}^{\mathcal{P}}_{\mathcal{S}}(\lambda)$
For $i = 1,\ldots,\mathcal{S}.\mathsf{nk}(\lambda)$ do $\mathbf{hk}[i] \leftarrow\!\!{\scriptstyle\$}\, \mathsf{H.Kg}(1^\lambda)$	$X \leftarrow \emptyset$; $L \leftarrow\!\!{\scriptstyle\$}\, \mathcal{S}^{\text{HASH}}(1^\lambda)$
$b \leftarrow\!\!{\scriptstyle\$}\, \{0,1\}$; $L \leftarrow\!\!{\scriptstyle\$}\, \mathcal{S}^{\text{HASH}}(1^\lambda)$	$x \leftarrow\!\!{\scriptstyle\$}\, \mathcal{P}(1^\lambda, L)$
$b' \leftarrow\!\!{\scriptstyle\$}\, \mathcal{D}(1^\lambda, \mathbf{hk}, L)$	Return $(x \in X)$
Return $(b = b')$	
	$\underline{\text{HASH}(i,x)}$
$\underline{\text{HASH}(i,x)}$	If not $(1 \leq i \leq \mathcal{S}.\mathsf{nk}(\lambda))$ then
If not $(1 \leq i \leq \mathcal{S}.\mathsf{nk}(\lambda))$ then return $\bot$	Return $\bot$
If $T[i,x] = \bot$ then	If $T[i,x] = \bot$ then
If $b = 0$ then $T[i,x] \leftarrow\!\!{\scriptstyle\$}\, \{0,1\}^{\mathsf{H.ol}(\lambda)}$	$T[i,x] \leftarrow\!\!{\scriptstyle\$}\, \{0,1\}^{\mathsf{H.ol}(\lambda)}$
Else $T[i,x] \leftarrow \mathsf{H.Ev}(1^\lambda, \mathbf{hk}[i], x)$	$X \leftarrow X \cup \{x\}$
Return $T[i,x]$	Return $T[i,x]$

Fig. 2. Games defining mUCE security of function family H and unpredictability of source $\mathcal{S}$.

It is easy to see that mUCE[**S**]-security is not achievable if **S** is the class of all PT sources [6]. To obtain meaningful notions of security, BHK [6] impose restrictions on the source. A central restriction is unpredictability. A source is unpredictable if it is hard to guess the source's HASH queries even given the leakage, in the *random case* of the mUCE game. Formally, let $\mathcal{S}$ be a source and $\mathcal{P}$ an adversary called a predictor and consider game $\text{mSPRED}^{\mathcal{P}}_{\mathcal{S}}(\lambda)$ in Fig. 2. For $\lambda \in \mathbb{N}$ we let $\mathsf{Adv}^{\text{m-spred}}_{\mathcal{S},\mathcal{P}}(\lambda) = \Pr[\text{mSPRED}^{\mathcal{P}}_{\mathcal{S}}(\lambda)]$. We say that $\mathcal{S}$ is computationally unpredictable if $\mathsf{Adv}^{\text{m-spred}}_{\mathcal{S},\mathcal{P}}(\cdot)$ is negligible for all PT predictors $\mathcal{P}$, and let $\mathbf{S}^{\text{cup}}$ be the class of all PT computationally unpredictable sources. We say that $\mathcal{S}$ is statistically unpredictable if $\mathsf{Adv}^{\text{m-spred}}_{\mathcal{S},\mathcal{P}}(\cdot)$ is negligible for all (not necessarily PT) predictors $\mathcal{P}$, and let $\mathbf{S}^{\text{sup}} \subseteq \mathbf{S}^{\text{cup}}$ be the class of all PT statistically unpredictable sources. We say that $\mathcal{S}$ is sub-exponentially unpredictable if there is an $\epsilon > 0$ such that for any PT predictor $\mathcal{P}$ there is a $\lambda_{\mathcal{P}}$ such that $\mathsf{Adv}^{\text{m-spred}}_{\mathcal{S},\mathcal{P}}(\lambda) \leq 2^{-\lambda^\epsilon}$ for all $\lambda \geq \lambda_{\mathcal{P}}$, and let $\mathbf{S}^{\text{seup}} \subseteq \mathbf{S}^{\text{cup}}$ be the class of all PT sub-exponentially unpredictable sources.

BFM [18] show that UCE-framework security notions (both single-key and multi-key) are not achievable for $\mathbf{S}^{\text{cup}}$ assuming that indistinguishability obfuscation exists. This has lead applications to impose further restrictions on the source by using either $\mathbf{S}^{\text{sup}}$ or subsets of $\mathbf{S}^{\text{cup}}$. Assumptions based on $\mathbf{S}^{\text{sup}}$, introduced in [6,18], at this point seem to be a viable. In order to restrict the computational case, one can consider split sources as defined in BHK [6]. Such sources can leak information about oracle queries and answers separately, but not together. We let $\mathbf{S}^{\text{splt}}$ denote the class of split sources. Another way to restrict a source is by limiting the number of queries it can make. Let $\mathbf{S}^{n,q}$ be the class of sources

$\mathcal{S}$ such that $\mathcal{S}.\mathsf{nk}(\cdot) \leq n(\cdot)$ and $\mathcal{S}$ makes at most $q(\cdot)$ queries to each key. In particular $\mathbf{S}^{1,1}$ is the class of sources that use only one key and make only one query to it.

3   Point-Function Obfuscation Framework

The literature considers many different variants of point function obfuscation. Here we provide a definitional framework that unifies these concepts and allows us to obtain not just known but also new variants of point function obfuscation as special cases. The framework parameterizes the security of a point-obfuscator by a class of algorithms we call target generators. Different notions of point obfuscation then correspond to different choices of this class. We start by defining target generators.

Target Generators. A *target generator* X specifies a PT algorithm $\mathsf{X}.\mathsf{Ev}$ that takes 1^λ to return a *target vector* $\mathbf{k}$ and *auxiliary information* $a \in \{0,1\}^*$. The entries of $\mathbf{k}$ are the targets, each of length $\mathsf{X}.\mathsf{tl}(\lambda)$, and the vector itself has length $\mathsf{X}.\mathsf{vl}(\lambda)$, where $\mathsf{X}.\mathsf{tl}, \mathsf{X}.\mathsf{vl}: \mathbb{N} \to \mathbb{N}$ are the target length and target-vector length functions associated to X, respectively.

Game $\mathrm{IND}_{\mathsf{Obf},\mathsf{X}}^{\mathcal{A}}(\lambda)$	Game $\mathrm{PRED}_{\mathsf{X}}^{\mathcal{Q}}(\lambda)$	Game $\mathrm{TRIV}_{\mathsf{X}}^{\mathcal{A}}(\lambda)$
$b \leftarrow_\$ \{0,1\}$	$(\mathbf{k}, a) \leftarrow_\$ \mathsf{X}.\mathsf{Ev}(1^\lambda)$	$b \leftarrow_\$ \{0,1\}$
$(\mathbf{k}_1, a_1) \leftarrow_\$ \mathsf{X}.\mathsf{Ev}(1^\lambda)$	$k \leftarrow_\$ \mathcal{Q}(1^\lambda, a)$	$(\mathbf{k}_1, a_1) \leftarrow_\$ \mathsf{X}.\mathsf{Ev}(1^\lambda)$
For $i = 1, \ldots, \mathsf{X}.\mathsf{vl}(\lambda)$ do	Return $(\exists i : \mathbf{k}[i] = k)$	For $i = 1, \ldots, \mathsf{X}.\mathsf{vl}(\lambda)$ do
$\quad \mathbf{k}_0[i] \leftarrow_\$ \{0,1\}^{\mathsf{X}.\mathsf{tl}(\lambda)}$		$\quad \mathbf{k}_0[i] \leftarrow_\$ \{0,1\}^{\mathsf{X}.\mathsf{tl}(\lambda)}$
$\overline{\mathbf{P}} \leftarrow_\$ \mathsf{Obf}(1^\lambda, \mathbf{I}_{\mathbf{k}_b})$		$b' \leftarrow_\$ \mathcal{A}(1^\lambda, \mathbf{k}_b, a_1)$
$b' \leftarrow_\$ \mathcal{A}(1^\lambda, \overline{\mathbf{P}}, a_1)$		Return $(b = b')$
Return $(b = b')$		

Fig. 3. Games defining IND security of point-function obfuscator Obf relative to target generator X, unpredictabilty of target generator X and triviality of target generator X.

Point-Function Obfuscation. If k is a bit-string then $\mathbf{I}_k: \{0,1\}^{|k|} \to \{0,1\}$ denotes a canonical representation of the circuit that on input $k' \in \{0,1\}^{|k|}$ returns 1 if $k = k'$ and 0 otherwise. It is assumed that given $\mathbf{I}_k$, one can compute k in time linear in $|k|$. A circuit C is called a *point circuit* if there is a k, called the circuit target, such that $\mathrm{C} \equiv \mathbf{I}_k$. If $\mathbf{k}$ is an n-vector of strings then we let $\mathbf{I}_{\mathbf{k}} = (\mathbf{I}_{\mathbf{k}[1]}, \ldots, \mathbf{I}_{\mathbf{k}[n]})$.

Let Obf be an obfuscator, as defined in Sect. 2. Its correctness condition guarantees that on input $1^\lambda, \mathbf{I}_k$, it returns a point circuit with target k, which is the condition for calling it a *point-function obfuscator*. We say that Obf has target length $\mathsf{Obf}.\mathsf{tl}: \mathbb{N} \to \mathbb{N}$ if its correctness condition is only required on inputs $\mathbf{I}_k$ with $k \in \{0,1\}^{\mathsf{Obf}.\mathsf{tl}(\lambda)}$.

Security of Point-Function Obfuscation. We now define security of point-function obfuscator relative to a class of target generators. We will then consider various choices of these classes.

Consider game IND of Fig. 3 associated to a point-function obfuscator Obf, a target generator X and an adversary $\mathcal{A}$, such that Obf.tl = X.tl. For $\lambda \in \mathbb{N}$ let $\mathsf{Adv}^{\mathrm{ind}}_{\mathsf{Obf},\mathsf{X},\mathcal{A}}(\lambda) = 2\Pr[\mathrm{IND}^{\mathcal{A}}_{\mathsf{Obf},\mathsf{X}}(\lambda)] - 1$. The game generates a target vector $\mathbf{k}_1$ and corresponding auxiliary information a_1 via X. It also samples a target vector $\mathbf{k}_0$ uniformly at random, containing $\mathsf{X.vl}(\lambda)$ elements each of length $\mathsf{X.tl}(\lambda)$. It then obfuscates the targets in the challenge vector $\mathbf{k}_b$ via Obf to produce $\overline{\mathbf{P}}$ which, as per our notation, will be the vector $(\mathsf{Obf}(1^\lambda, \mathbf{I}_{\mathbf{k}_b[1]}), \ldots, \mathsf{Obf}(1^\lambda, \mathbf{I}_{\mathbf{k}_b[\mathsf{X.vl}(\lambda)]}))$ formed by independently obfuscating the targets in the target vector. Given $\overline{\mathbf{P}}$ and a_1, adversary $\mathcal{A}$ outputs a bit b', and wins the game if this equals b, meaning it guesses whether the target vector that was obfuscated was the one corresponding to auxiliary information a_1 or one independent of it.

Let $\mathbf{X}$ be a class (set) of target generators. We say that Obf is IND[$\mathbf{X}$]-secure if $\mathsf{Adv}^{\mathrm{ind}}_{\mathsf{Obf},\mathsf{X},\mathcal{A}}(\cdot)$ is negligible for every PT $\mathcal{A}$ and every $\mathsf{X} \in \mathbf{X}$. We now capture different notions in the literature, as well as new ones, by considering particular classes $\mathbf{X}$. At the end of this section we will present what we call the triviality theorem, showing how the definition is vacuous for some classes, and discuss its implications. We will further discuss alternative security definitions for point-function obfuscation in Sect. 6.

Classes of Target Generators. One important (and necessary) condition on a target generator is unpredictability. To define this, consider game PRED of Fig. 3 associated to X and a predictor adversary $\mathcal{Q}$. For $\lambda \in \mathbb{N}$ let $\mathsf{Adv}^{\mathrm{pred}}_{\mathsf{X},\mathcal{Q}}(\lambda) = \Pr[\mathrm{PRED}^{\mathcal{Q}}_{\mathsf{X}}(\lambda)]$. The game generates a target vector $\mathbf{k}$ and associated auxiliary information a. The adversary $\mathcal{Q}$ gets a and wins if it can predict any entry of the vector $\mathbf{k}$.

The first dimension along which point-function obfuscators are classified is the type of unpredictability, encompassing two sub-dimensions: the success probability of predictors (may be required to be negligible or sub-exponential) and their computational power (PT and computationally unbounded are the popular choices, but one could also consider sub-exponential time). Some relevant classes are the following:

- $\mathbf{X}^{\mathrm{cup}}$ — Class of computationally unpredictable target generators — $\mathsf{X} \in \mathbf{X}^{\mathrm{cup}}$ if $\mathsf{Adv}^{\mathrm{pred}}_{\mathsf{X},\mathcal{Q}}(\cdot)$ is negligible for all PT predictor adversaries $\mathcal{Q}$.
- $\mathbf{X}^{\mathrm{seup}}$ — Class of sub-exponentially unpredictable target generators — $\mathsf{X} \in \mathbf{X}^{\mathrm{seup}}$ if there exists $0 < \epsilon < 1$ such that for every PT predictor adversary $\mathcal{Q}$ there is a $\lambda_{\mathcal{Q}}$ such that $\mathsf{Adv}^{\mathrm{pred}}_{\mathsf{X},\mathcal{Q}}(\lambda) \leq 2^{-\lambda^\epsilon}$ for all $\lambda \geq \lambda_{\mathcal{Q}}$.
- $\mathbf{X}^{\mathrm{sup}}$ — Class of statistically unpredictable target generators — $\mathsf{X} \in \mathbf{X}^{\mathrm{sup}}$ if $\mathsf{Adv}^{\mathrm{pred}}_{\mathsf{X},\mathcal{Q}}(\cdot)$ is negligible for all (even computationally unbounded) predictor adversaries $\mathcal{Q}$.

Another dimension is the number of target points in the target vector, to capture which, for any polynomial $q \colon \mathbb{N} \to \mathbb{N}$, we let

- $\mathbf{X}^{q(\cdot)}$ — Class of generators producing $q(\cdot)$ target points — $X \in \mathbf{X}^{q(\cdot)}$ if X.vl $= q$. An important special case is $q(\cdot) = 1$.

Another important dimension is auxiliary information, which may be present or absent (the latter, formally means it is the empty string), to capture which we let

- $\mathbf{X}^{\varepsilon}$ — Class of generators with no auxiliary information — $X \in \mathbf{X}^{\varepsilon}$ if $a = \varepsilon$ for all $(\mathbf{k}, a) \in [\text{X.Ev}(1^{\lambda})]$ and all $\lambda \in \mathbb{N}$.

We can recover notions from the literature as follows:

- $\text{IND}[\mathbf{X}^{\text{cup}} \cap \mathbf{X}^{\varepsilon} \cap \mathbf{X}^{1}]$ — This is basic point-function obfuscation, secure for a single computationally unpredictable target point, and no auxiliary information is allowed. It is achieved in [23,27,39,47].
- $\text{IND}[\mathbf{X}^{\text{cup}} \cap \mathbf{X}^{1}]$ — This is AIPO [14,35], secure for a single computationally unpredictable target point in the presence of auxiliary information. It is achieved under the AI-DHI assumption by Canetti [23], and using the extended construction of Wee [47] by BP [14].
- $\text{IND}[\mathbf{X}^{\text{cup}}]$ — This is composable AIPO [25], meaning that it is secure for arbitrarily many correlated target points that are computationally unpredictable in the presence of auxiliary information. BM1 [20] showed that this notion cannot co-exit with iO in the presence of OWFs.
- $\text{IND}[\mathbf{X}^{\text{sup}} \cap \mathbf{X}^{\varepsilon}]$ — This is composable point-function obfuscation, secure for arbitrarily many correlated target points that are statistically unpredictable, and no auxiliary information is allowed. It is achieved from $\text{mUCE}[\mathbf{S}^{\text{sup}}]$ in BHK [6].

Furthermore, DKL [29] achieve $\text{IND}[\mathbf{X}^{\text{sup}} \cap \mathbf{X}^{1}]$ from the LSN (i.e. auxiliary-input LPN) assumption and BM3 [22] build $\text{IND}[\mathbf{X}^{\text{sup}}]$ from $\text{mUCE}[\mathbf{S}^{\text{s-sup}} \cap \mathbf{X}^{1}]$. Here $\mathbf{S}^{\text{s-sup}}$ denotes a subclass of $\mathbf{S}^{\text{sup}} \cap \mathbf{S}^{\text{splt}}$ that is used to denote sources with "strong statistical unpredictability", as defined in BM2 [21]. We note that some of the above results achieve notions that are stronger than IND. Such notions are discussed and defined in Sect. 6.

Triviality Theorem. The $\text{IND}[\mathbf{X}]$ definition has the peculiar property of trivializing for some choices of $\mathbf{X}$. For example, let X be a target generator that returns a vector of random, independent targets and auxiliary information $a = \varepsilon$ the empty string. Then *any* point-function obfuscator Obf is $\text{IND}[\{X\}]$-secure. This is true because game IND in this case samples $\mathbf{k}_0, \mathbf{k}_1$ from the same distribution and the information provided to the adversary $\mathcal{A}$ is thus independent of the challenge bit. Before discussing and assessing what this means for the definition, we provide a general *triviality theorem* that characterizes for what choices of $\mathbf{X}$ this phenomenon happens.

Consider game TRIV of Fig. 3 associated to a target generator X and an adversary $\mathcal{A}$. For $\lambda \in \mathbb{N}$ let $\text{Adv}_{X,\mathcal{A}}^{\text{triv}}(\lambda) = 2 \Pr[\text{TRIV}_X^{\mathcal{A}}(\lambda)] - 1$. We say that X is trivial if $\text{Adv}_{X,\mathcal{A}}^{\text{triv}}(\cdot)$ is negligible for every PT $\mathcal{A}$. An example of trivial X is the one given above. Let $\mathbf{X}^{\text{triv}}$ be the class of all trivial target generators, and say that a class $\mathbf{X}$ is trivial if $\mathbf{X} \subseteq \mathbf{X}^{\mathrm{triv}}$. The proof of the following triviality theorem follows directly from the definitions of games IND and TRIV and is omitted.

Theorem 2. *Let* $\mathbf{X} \subseteq \mathbf{X}^{\mathrm{triv}}$ *be a class of target generators. Let* Obf *be any point-function obfuscator. Then* Obf *is* IND[$\mathbf{X}$]*-secure.*

This can be viewed as a defect of the IND definition, but whether or not this is true is debatable. The IND definition has been successfully employed in applications [14,21]. In these cases, $\mathbf{X} = \mathbf{X}^{\mathrm{cup}} \cap \mathbf{X}^1$, a class to which Theorem 2 does not apply. This indicates that the classes of target generators arising in applications are naturally not trivial. And the constructions we give in Sect. 5 cover such non-trivial classes. Thus we are on the whole unsure whether or not Theorem 2 should be viewed as a definitional weakness. In Sect. 6 we will provide alternative security definitions for PO that avoid this type of triviality theorem and are meaningful for all choices of target generators. But if an application can be obtained via IND, then it seems preferable, since this definition is simpler and easier to use and, from Sect. 5, we have more constructions for it.

4   (d)iO for Multi-circuit Samplers

We state and prove a lemma we will use that may be of independent interest. We extend the standard definition of circuit samplers from Sect. 2 to get *multi-circuit samplers*, which are samplers that may produce a vector of circuit pairs (but still only a single auxiliary information string). We also extend the security definition of differing-inputs obfuscation to work with respect to multi-circuit samplers. We then use a hybrid argument to show that the security of the latter is implied by the standard definition of differing-inputs obfuscation for circuit samplers that produce only a single pair of circuits. This result will be used for our iO-based construction of a point-function obfuscator, BCP [16] being applied to move from diO to iO. (We stress that diO is used as a tool but not as an assumption in our results.)

iO for Multi-circuit Samplers. A multi-circuit sampler is a PT algorithm S with an associated circuit-vector length function S.vl: $\mathbb{N} \to \mathbb{N}$. Algorithm S on input 1^λ returns a triple $(\mathbf{C}_0, \mathbf{C}_1, aux)$ where aux is a string and $\mathbf{C}_0, \mathbf{C}_1$ are

Game $\mathrm{MDIFF}_{\mathsf{S}}^{\mathcal{D}}(\lambda)$	Game $\mathrm{MIO}_{\mathsf{Obf},\mathsf{S}}^{\mathcal{O}}(\lambda)$
$(\mathbf{C}_0, \mathbf{C}_1, aux) \leftarrow\!\!{\scriptstyle\$}\, \mathsf{S}(1^\lambda)$	$b \leftarrow\!\!{\scriptstyle\$}\, \{0,1\}$; $(\mathbf{C}_0, \mathbf{C}_1, aux) \leftarrow\!\!{\scriptstyle\$}\, \mathsf{S}(1^\lambda)$
$x \leftarrow\!\!{\scriptstyle\$}\, \mathcal{D}(\mathbf{C}_0, \mathbf{C}_1, aux)$	$\overline{\mathbf{C}} \leftarrow\!\!{\scriptstyle\$}\, \mathsf{Obf}(1^\lambda, \mathbf{C}_b)$; $b' \leftarrow\!\!{\scriptstyle\$}\, \mathcal{O}(1^\lambda, \overline{\mathbf{C}}, aux)$
Return $(\exists i \ : \ \mathbf{C}_0[i](x) \neq \mathbf{C}_1[i](x))$	Return $(b = b')$

Fig. 4. Games defining difference-security of multi-circuit sampler S and iO-security of obfuscator Obf relative to multi-circuit sampler S.

circuit vectors of length $S.\mathsf{vl}(\lambda)$, such that circuits $\mathbf{C}_0[i]$ and $\mathbf{C}_1[i]$ are of the same size, number of inputs and number of outputs for every $i \in \{1, \ldots, S.\mathsf{vl}(\lambda)\}$.

Consider game MIO of Fig. 4 associated to an obfuscator Obf, a multi-circuit sampler S and an adversary $\mathcal{O}$. For $\lambda \in \mathbb{N}$ let $\mathsf{Adv}^{\text{m-io}}_{\mathsf{Obf},\mathsf{S},\mathcal{O}}(\lambda) = 2 \Pr[\mathrm{MIO}^{\mathcal{O}}_{\mathsf{Obf},\mathsf{S}}(\lambda)] - 1$. Let $\boldsymbol{S}$ be a class of multi-circuit samplers. We say that Obf is $\boldsymbol{S}$-secure if $\mathsf{Adv}^{\text{m-io}}_{\mathsf{Obf},\mathsf{S},\mathcal{O}}(\cdot)$ is negligible for every multi-circuit sampler $\mathsf{S} \in \boldsymbol{S}$ and every PT adversary $\mathcal{O}$.

Consider game MDIFF of Fig. 4 associated to a multi-circuit sampler S and an adversary $\mathcal{D}$. For $\lambda \in \mathbb{N}$ let $\mathsf{Adv}^{\text{m-diff}}_{\mathsf{S},\mathcal{D}}(\lambda) = \Pr[\mathrm{MDIFF}^{\mathcal{D}}_{\mathsf{S}}(\lambda)]$. We say that a multi-circuit sampler S is difference secure if $\mathsf{Adv}^{\text{m-diff}}_{\mathsf{S},\mathcal{D}}(\cdot)$ is negligible for every PT adversary $\mathcal{D}$. Let $\boldsymbol{S}_{\text{m-diff}}$ be the class of all difference-secure multi-circuit samplers and let $d\colon \mathbb{N} \to \mathbb{N}$. We say that multi-circuit sampler S produces d-differing circuits if circuits $\mathbf{C}_0[i]$ and $\mathbf{C}_1[i]$ differ on at most $d(\lambda)$ inputs with an overwhelming probability over $(\mathbf{C}_0, \mathbf{C}_1, aux) \in [\mathsf{S}(1^\lambda)]$, for all $\lambda \in \mathbb{N}$ and all $i \in \{1, \ldots, S.\mathsf{vl}(\lambda)\}$. Let $\boldsymbol{S}_{\text{m-diff}}(d)$ be the class of all difference-secure multi-circuit samplers that produce d-differing circuits. The proof of the following lemma is provided in [8].

Lemma 3. *Let $d\colon \mathbb{N} \to \mathbb{N}$. Let* Obf *be an $\boldsymbol{S}_{\text{diff}}(d)$-secure obfuscator. Then* Obf *is also an $\boldsymbol{S}_{\text{m-diff}}(d)$-secure obfuscator.*

5 Generic Constructions of PO

Prior constructions have targeted $\mathrm{IND}[\mathbf{X}]$ for specific choices of $\mathbf{X}$ in ad hoc ways and used non-standard assumptions. In this section we provide constructions that are generic. This means they take an arbitrary, given class $\mathbf{X}$ of target generators and return a point-function obfuscator that is $\mathrm{IND}[\mathbf{X}]$-secure.

5.1 PO from iO

OWFs. Consider game OWF of Fig. 5 associated to a function family F, a target generator X with $X.\mathsf{tl} = F.\mathsf{il}$, and an adversary $\mathcal{F}$. For $\lambda \in \mathbb{N}$ let $\mathsf{Adv}^{\text{owf}}_{\mathsf{F},\mathsf{X},\mathcal{F}}(\lambda) = \Pr[\mathrm{OWF}^{\mathcal{F}}_{\mathsf{F},\mathsf{X}}(\lambda)]$. Let $\mathbf{X}$ be a class of target generators with target length $F.\mathsf{il}$. Let $X^{1\text{ur}}$ be the target generator with $X^{1\text{ur}}.\mathsf{vl}(\cdot) = 1$ and $X^{1\text{ur}}.\mathsf{tl} = F.\mathsf{il}$, where the target is sampled from a uniform distribution and the auxiliary information is always empty, meaning $a = \varepsilon$. We say that F is $\mathrm{OWF}[\mathbf{X}]$-secure if $\mathsf{Adv}^{\text{owf}}_{\mathsf{F},\mathsf{X},\mathcal{F}}(\cdot)$ is negligible for all PT adversaries $\mathcal{F}$ and all $X \in \mathbf{X} \cup \{X^{1\text{ur}}\}$. Relevant classes $\mathbf{X}$ are the same as for PO. The standard notion of a OWF is recovered as $\mathbf{X} = \emptyset$, meaning that F is secure only with respect to $X^{1\text{ur}}$.

The definition of CD [24] is the special case of ours with vectors of length one. That of FOR [32], like ours, considers evaluations of the function on multiple inputs, but in their case the key for the evaluations is the same and there is no auxiliary input, while in our case the key is independently chosen for each evaluation and auxiliary inputs may be present. We stress that we require only

Game $\mathrm{OWF}_{\mathsf{F},\mathsf{X}}^{\mathcal{F}}(\lambda)$	Game $\mathrm{PRIV1}_{\mathsf{DPKE},\mathsf{X}}^{\mathcal{A}}(\lambda)$
$(\mathbf{k},a) \leftarrow_{\$} \mathsf{X}.\mathsf{Ev}(1^\lambda)$	$b \leftarrow_{\$} \{0,1\}$; $(\mathbf{k}_1,a) \leftarrow_{\$} \mathsf{X}.\mathsf{Ev}(1^\lambda)$
For $i = 1,\ldots,\mathsf{X}.\mathsf{vl}(\lambda)$ do	For $i = 1,\ldots,\mathsf{X}.\mathsf{vl}(\lambda)$ do
$\quad \mathbf{fk}[i] \leftarrow_{\$} \mathsf{F}.\mathsf{Kg}(1^\lambda)$	$\quad \mathbf{k}_0[i] \leftarrow_{\$} \{0,1\}^{\mathsf{DPKE}.\mathsf{ml}(\lambda)}$
$\quad \mathbf{y}[i] \leftarrow \mathsf{F}.\mathsf{Ev}(1^\lambda,\mathbf{fk}[i],\mathbf{k}[i])$	$\quad (\mathbf{pk}[i],\mathbf{sk}[i]) \leftarrow_{\$} \mathsf{DPKE}.\mathsf{Kg}(1^\lambda)$
$k \leftarrow_{\$} \mathcal{F}(1^\lambda,\mathbf{fk},\mathbf{y},a)$	$\quad \mathbf{c}[i] \leftarrow \mathsf{DPKE}.\mathsf{Enc}(1^\lambda,\mathbf{pk}[i],\mathbf{k}_b[i])$
Return $(\exists i\ :\ \mathsf{F}.\mathsf{Ev}(1^\lambda,\mathbf{fk}[i],k) = \mathbf{y}[i])$	$b' \leftarrow_{\$} \mathcal{A}(1^\lambda,\mathbf{pk},\mathbf{c},a)$; Return $(b = b')$

Fig. 5. Games defining one-wayness of function family F relative to target generator X and PRIV1-security of deterministic public-key encryption scheme DPKE relative to target generator X.

one-wayness; we do *not* require extractability. The latter is a much stronger assumption [13].

We now show that indistinguishability obfuscation can be used to build a IND[**X**]-secure point-function obfuscator for an arbitrary target generator class **X** from any OWF[**X**]-secure function family.

Construction. Let F be a family of functions. Let $\mathsf{Obf}_{\mathsf{io}}$ be an obfuscator. We construct a point-function obfuscator Obf with $\mathsf{Obf}.\mathsf{tl} = \mathsf{F}.\mathsf{il}$ as follows:

Algorithm $\mathsf{Obf}(1^\lambda,\mathbf{I}_k)$	Circuit $\mathsf{C}_{1^\lambda,fk,y}(k')$
$fk \leftarrow_{\$} \mathsf{F}.\mathsf{Kg}(1^\lambda)$; $y \leftarrow \mathsf{F}.\mathsf{Ev}(1^\lambda,fk,k)$	If $(y = \mathsf{F}.\mathsf{Ev}(1^\lambda,fk,k'))$ then return 1
$\overline{\mathsf{P}} \leftarrow_{\$} \mathsf{Obf}_{\mathsf{io}}(\mathsf{C}_{1^\lambda,fk,y})$; Return $\overline{\mathsf{P}}$	Else return 0

Theorem 4. *Let* F *be an injective family of functions. Let* **X** *be a class of target generators with target length* $\mathsf{F}.\mathsf{il}$. *Assume that* F *is* OWF[**X**]-*secure. Let* $\mathsf{Obf}_{\mathsf{io}}$ *be an indistinguishability obfuscator. Then* Obf *constructed above from* F *and* $\mathsf{Obf}_{\mathsf{io}}$ *is a* IND[**X**]-*secure point-function obfuscator.*

Proof (Theorem 4). The injectivity of F implies that Obf satisfies the correctness condition of a point-function obfuscator. We now prove security.

Let $\mathsf{X} \in \mathbf{X}$ be a target generator. Let $\mathcal{A}$ be a PT adversary. Consider the games and the associated circuits of Fig. 6, where s is defined as follows. For any λ let $s(\lambda)$ be a polynomial upper bound on $\max(|\mathsf{C}_{1^\lambda,fk,y}^1|)$, where the maximum is over all $fk \in [\mathsf{F}.\mathsf{Kg}(1^\lambda)]$ and $y \in \{0,1\}^{\mathsf{F}.\mathsf{ol}(\lambda)}$. Lines not annotated with comments are common to all games.

Game G_0 is equivalent to $\mathrm{IND}_{\mathsf{Obf},\mathsf{X}}^{\mathcal{A}}(\lambda)$. The inputs to adversary $\mathcal{A}$ in game G_1 do not depend on the challenge bit b, so we have $\Pr[\mathsf{G}_1] = 1/2$. It follows that

$$\mathsf{Adv}_{\mathsf{Obf},\mathsf{X},\mathcal{A}}^{\mathsf{ind}}(\lambda) = 2 \cdot \Pr[\mathsf{G}_0] - 1 = 2 \cdot (\Pr[\mathsf{G}_0] - \Pr[\mathsf{G}_1]).$$

The first equality holds by the definition of IND, and the second equality holds because of $\Pr[\mathsf{G}_1] = 1/2$. We now show that $\Pr[\mathsf{G}_0] - \Pr[\mathsf{G}_1]$ is negligible, meaning that $\mathsf{Adv}_{\mathsf{Obf},\mathsf{X},\mathcal{A}}^{\mathsf{ind}}(\cdot)$ is also negligible. This proves the the theorem.

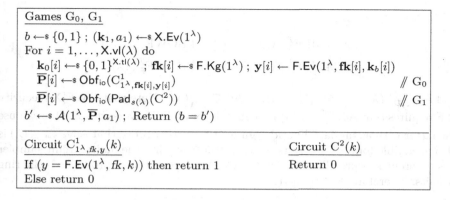

Fig. 6. Games for proof of Theorem 4.

We construct a multi-circuit sampler S and a PT iO-adversary $\mathcal{O}$ as follows:

Multi-circuit Sampler $S(1^\lambda)$	Adversary $\mathcal{O}(1^\lambda, \overline{\mathbf{C}}, aux)$
$d \leftarrow\!\!{}_{\$} \{0,1\}$; $(\mathbf{k}_1, a_1) \leftarrow\!\!{}_{\$} X.\mathsf{Ev}(1^\lambda)$	$(d, a_1) \leftarrow aux$
For $i = 1, \ldots, X.\mathsf{vl}(\lambda)$ do	$d' \leftarrow\!\!{}_{\$} \mathcal{A}(1^\lambda, \overline{\mathbf{C}}, a_1)$
$\quad \mathbf{k}_0[i] \leftarrow\!\!{}_{\$} \{0,1\}^{X.\mathsf{tl}(\lambda)}$	If $(d = d')$ then return 1
$\quad \mathbf{fk}[i] \leftarrow\!\!{}_{\$} F.\mathsf{Kg}(1^\lambda)$; $\mathbf{y}[i] \leftarrow F.\mathsf{Ev}(1^\lambda, \mathbf{fk}[i], \mathbf{k}_d[i])$	Else return 0
$\quad \mathbf{C}_1[i] \leftarrow C^1_{1^\lambda, \mathbf{fk}[i], \mathbf{y}[i]}$; $\mathbf{C}_0[i] \leftarrow \mathsf{Pad}_{s(\lambda)}(C^2)$	
$aux \leftarrow (d, a_1)$; Return $(\mathbf{C}_0, \mathbf{C}_1, aux)$	

We have $\Pr[G_0] - \Pr[G_1] = \mathsf{Adv}^{\mathsf{m\text{-}io}}_{\mathsf{Obf}_{io}, S, \mathcal{O}}(\lambda)$ by construction. Next, we show that $S \in \mathbf{S}_{\mathsf{m\text{-}diff}}(1)$. According to Proposition 1 (the result of BCP [16]), any indistinguishability obfuscator is also an $\mathbf{S}_{\mathsf{diff}}(1)$-secure obfuscator. And according to Lemma 3, any $\mathbf{S}_{\mathsf{diff}}(1)$-secure obfuscator is an $\mathbf{S}_{\mathsf{m\text{-}diff}}(1)$-secure obfuscator. It follows that $\mathsf{Adv}^{\mathsf{m\text{-}io}}_{\mathsf{Obf}_{io}, S, \mathcal{O}}(\cdot)$ is negligible by the iO-security of Obf_{io}.

Let X^{ur} be the target generator with $X^{ur}.\mathsf{vl} = X.\mathsf{vl}$ and $X^{ur}.\mathsf{tl} = F.\mathsf{il}$, where the targets are sampled independently, from a uniform distribution and auxiliary information is always $a = \varepsilon$. Given any PT difference adversary $\mathcal{D}$ against multi-circuit sampler S, we build PT adversaries $\mathcal{F}_0$ and $\mathcal{F}_1$ against the OWF-security of F relative to target generators X^{ur} and X, respectively. The constructions are as follows:

Adversary $\mathcal{F}_0(1^\lambda, \mathbf{fk}, \mathbf{y}, a)$	Adversary $\mathcal{F}_1(1^\lambda, \mathbf{fk}, \mathbf{y}, a)$
$d \leftarrow 0$; $(\mathbf{k}_1, a_1) \leftarrow\!\!{}_{\$} X.\mathsf{Ev}(1^\lambda)$	$d \leftarrow 1$
For $i = 1, \ldots, |\mathbf{y}|$ do	For $i = 1, \ldots, |\mathbf{y}|$ do
$\quad \mathbf{C}_1[i] \leftarrow C^1_{1^\lambda, \mathbf{fk}[i], \mathbf{y}[i]}$	$\quad \mathbf{C}_1[i] \leftarrow C^1_{1^\lambda, \mathbf{fk}[i], \mathbf{y}[i]}$
$\quad \mathbf{C}_0[i] \leftarrow \mathsf{Pad}_{s(\lambda)}(C^2)$	$\quad \mathbf{C}_0[i] \leftarrow \mathsf{Pad}_{s(\lambda)}(C^2)$
$aux \leftarrow (d, a_1)$; $x \leftarrow\!\!{}_{\$} \mathcal{D}(\mathbf{C}_1, \mathbf{C}_0, aux)$	$aux \leftarrow (d, a)$; $x \leftarrow\!\!{}_{\$} \mathcal{D}(\mathbf{C}_1, \mathbf{C}_0, aux)$
Return x	Return x

Let d denote the value sampled by multi-circuit sampler S in game $\mathsf{MDIFF}^{\mathcal{D}}_S(\lambda)$. Then we have

$$\Pr[\text{MDIFF}_\mathsf{S}^\mathcal{D}(\lambda) \mid d = 0] = \Pr[\text{OWF}_{\mathsf{F,X^{ur}}}^{\mathcal{F}_0}(\lambda)],$$
$$\Pr[\text{MDIFF}_\mathsf{S}^\mathcal{D}(\lambda) \mid d = 1] = \Pr[\text{OWF}_{\mathsf{F,X}}^{\mathcal{F}_1}(\lambda)].$$

and $\text{Adv}_{\mathsf{S},\mathcal{D}}^{\text{m-diff}}(\lambda) = \frac{1}{2}(\text{Adv}_{\mathsf{F,X^{ur}},\mathcal{F}_0}^{\text{owf}}(\lambda) + \text{Adv}_{\mathsf{F,X},\mathcal{F}_1}^{\text{owf}}(\lambda))$. Note that $\text{OWF}[\mathbf{X}]$-security of F requires that $\text{Adv}_{\mathsf{F,X^{ur}},\mathcal{F}}^{\text{owf}}(\lambda)$ is negligible for all PT adversaries $\mathcal{F}$. One can use the latter with a standard hybrid argument to further prove that $\text{Adv}_{\mathsf{F,X^{ur}},\mathcal{F}_0}^{\text{owf}}(\lambda)$ is also negligible for all PT adversaries $\mathcal{F}_0$. It follows that the multi-circuit sampler S is difference-secure. The injectivity of F also implies that S produces 1-differing circuits. Therefore, $\mathsf{S} \in \boldsymbol{S}_{\text{m-diff}}(1)$.

5.2   PO from DPKE

Our next generic construction is based on deterministic public-key encryption [3]. As before we aim to provide point-function obfuscation secure for any given class of target generators. We are able to do this assuming the existence of a deterministic public-key encryption scheme that is secure relative to the same class viewed as a class of message generators. We can then exploit known constructions of deterministic public-key encryption to get a slew of point-function obfuscators based on standard assumptions. We begin with a parameterized definition of security for deterministic public-key encryption.

DPKE. A deterministic public-key encryption scheme DPKE [3] specifies the following. PT key generation algorithm DPKE.Kg takes 1^λ to return a public encryption key pk and a secret decryption key sk. Deterministic PT encryption algorithm DPKE.Enc takes 1^λ, pk and a plaintext message $k \in \{0,1\}^{\text{DPKE.ml}(\lambda)}$ to return a ciphertext c, where $\text{DPKE.ml}: \mathbb{N} \to \mathbb{N}$ is the message length function associated to DPKE. Deterministic decryption algorithm DPKE.Dec takes 1^λ, sk, c to return plaintext message k. We do not require the decryption algorithm to be PT but we do require decryption correctness, namely that for all $\lambda \in \mathbb{N}$, all $(pk, sk) \in [\text{DPKE.Kg}(1^\lambda)]$ and all $k \in \{0,1\}^{\text{DPKE.ml}(\lambda)}$ we have $\text{DPKE.Dec}(1^\lambda, sk, \text{DPKE.Enc}(1^\lambda, pk, k)) = k$.

Now consider game PRIV1 of Fig. 5 associated to a deterministic public-key encryption scheme DPKE, a target generator X satisfying X.tl = DPKE.ml, and an adversary $\mathcal{A}$. For $\lambda \in \mathbb{N}$ let $\text{Adv}_{\mathsf{DPKE,X},\mathcal{A}}^{\text{priv1}}(\lambda) = 2\Pr[\text{PRIV1}_{\mathsf{DPKE,X}}^{\mathcal{A}}(\lambda)] - 1$. If $\mathbf{X}$ is a class of target generators then we say that DPKE is $\text{PRIV1}[\mathbf{X}]$-secure if $\text{Adv}_{\mathsf{DPKE,X},\mathcal{A}}^{\text{priv1}}(\cdot)$ is negligible for all PT adversaries $\mathcal{A}$ and all $\mathsf{X} \in \mathbf{X}$.

This definition reflects what BBO [3] call the multi-user setting where there are many, independent public keys. However, in our case, only a single message is encrypted under each key. The single-key version of this is called PRIV1 in the literature, so we retained the name in moving to the multi-user setting. The definition is in the indistinguishability style of [4,15] rather than the semantic security style of [3]. These definitions however did not allow auxiliary inputs. We are allowing those following BS [17]. Finally, while prior definitions require unpredictability of the message distribution, ours is simply parameterized by the latter. Prior definitions are captured as special cases, meaning they can be recovered as PRIV1[$\mathbf{X}$] for some choice of $\mathbf{X}$.

Construction. Let DPKE be a deterministic public-key encryption scheme. We construct an obfuscator Obf with Obf.tl = DPKE.ml as follows:

Algorithm $\mathsf{Obf}(1^\lambda, \mathbf{I}_k)$	Circuit $\mathsf{C}_{1^\lambda, pk, c}(k)$
$(pk, sk) \leftarrow_\$ \mathsf{DPKE.Kg}(1^\lambda)$	If $(\mathsf{DPKE.Enc}(1^\lambda, pk, k) = c)$
$c \leftarrow \mathsf{DPKE.Enc}(1^\lambda, pk, k)$; Return $\mathsf{C}_{1^\lambda, pk, c}$	Then return 1 else return 0

The construction is simple. To obfuscate $\mathbf{I}_k$ we pick a new key pair for the deterministic public-key encryption scheme and return a circuit that embeds the public key pk as well as the encryption c of the target point k. The circuit, given a candidate target point k', re-encrypts it under the embedded public key pk and checks that the ciphertext so obtained matches the embedded ciphertext c. Note that the determinism of DPKE.Enc is used crucially to ensure that the circuit is deterministic. For randomized encryption, one cannot check that a message corresponds to a ciphertext by re-encryption. The secret key sk is discarded and not used in the construction, but its existence will guarantee correctness of the point-function obfuscator.

Result. We show that this is a generic construction. Namely, a point-function obfuscator for a given class $\mathbf{X}$ of target generators can be obtained if we have a deterministic public-key encryption scheme secure for the same class.

Theorem 5. *Let* DPKE *be a deterministic public-key encryption scheme and* $\mathbf{X}$ *a class of target generators such that* X.tl = DPKE.ml *for all* X $\in \mathbf{X}$. *Assume* DPKE *is* PRIV1[$\mathbf{X}$]-*secure. Let* Obf *be as defined above. Then* Obf *is a* IND[$\mathbf{X}$]-*secure point-function obfuscator.*

Proof (Theorem 5). The correctness of Obf follows from the decryption correctness of DPKE, and it does not require the decryption algorithm DPKE.Dec to be PT. We now prove that Obf is IND[$\mathbf{X}$]-secure.

Let X $\in \mathbf{X}$ be a target generator with X.tl = DPKE.ml. Let $\mathcal{A}$ be PT adversary against the IND security of Obf relative to X. We construct a PT adversary $\mathcal{B}$ against the PRIV1 security of DPKE relative to X as follows:

Adversary $\mathcal{B}(1^\lambda, \mathbf{pk}, \mathbf{c}, a)$	Circuit $\mathsf{C}_{1^\lambda, pk, c}(k)$
For $i = 1, \ldots, |\mathbf{c}|$ do $\overline{\mathbf{P}}[i] \leftarrow \mathsf{C}_{1^\lambda, \mathbf{pk}[i], \mathbf{c}[i]}$	If $(\mathsf{DPKE.Enc}(1^\lambda, pk, k) = c)$
$b' \leftarrow_\$ \mathcal{A}(1^\lambda, \overline{\mathbf{P}}, a)$; Return b'	Then return 1 else return 0

We have $\mathsf{Adv}^{\mathrm{priv1}}_{\mathsf{DPKE}, \mathsf{X}, \mathcal{B}}(\lambda) = \mathsf{Adv}^{\mathrm{ind}}_{\mathsf{Obf}, \mathsf{X}, \mathcal{A}}(\lambda)$ by construction. Hence, for any X $\in \mathbf{X}$ the IND-security of Obf relative to X follows from the assumed PRIV1-security of DPKE relative to X.

In applying Theorem 5 to get point function obfuscators, the first case of interest is $\mathbf{X} = \mathbf{X}^{\mathrm{sup}} \cap \mathbf{X}^\varepsilon \cap \mathbf{X}^1$. In this case, PRIV1[$\mathbf{X}$]-secure deterministic public-key encryption is a standard form of the latter for which many constructions are known. The central construction, due to BFO [15], is from lossy trapdoor functions (LTDFs). But the latter can be built from a wide variety of standard assumptions [31,37,44,48,51]. Thus we get $\text{IND}[\mathbf{X}^{\text{sup}} \cap \mathbf{X}^\varepsilon \cap \mathbf{X}^1]$-secure point-function obfuscators under the same assumptions. The second case of interest is $\mathbf{X} = \mathbf{X}^{\text{seup}} \cap \mathbf{X}^1$. Unlike in the first case, there is now auxiliary information, but it leaves the targets sub-exponentially unpredictable. Constructions of $\text{PRIV1}[\mathbf{X}]$-secure deterministic public-key encryption are known under standard assumptions including DLIN, Subgroup Indistinguishability and LWE [17,48,50]. Accordingly we get $\text{IND}[\mathbf{X}^{\text{seup}} \cap \mathbf{X}^1]$-secure point-function obfuscators under the same assumptions. BH [5] obtain PRIV-secure DPKE from $\text{UCE}[\mathbf{S}^{\text{sup}}]$, which via Theorem 5 yields $\text{IND}[\mathbf{X}^{\text{sup}} \cap \mathbf{X}^\varepsilon \cap \mathbf{X}^1]$ under $\text{UCE}[\mathbf{S}^{\text{sup}}]$.

Theorem 5 also yields negative results. Assume iO exists. Then we know that there do not exist point function obfuscators that are $\text{IND}[\mathbf{X}^{\text{cup}}]$-secure [20]. Theorem 5 then implies that there also do not exist deterministic public-key encryption schemes that are $\text{PRIV1}[\mathbf{X}^{\text{cup}}]$-secure.

CIH function families as per GOR [36] do not seem to have a unique associated security notion. Rather the authors discuss a few choices. Our parameterized PRIV definitions above apply to function families as well and can be viewed as providing more security notions for CIH function families. These function families can also be used in our PO construction above as long as they are injective.

5.3   PO from UCE

Our next generic construction is based on UCE, a class of assumptions on function families from [6]. We use the multi-key version of the UCE assumption, denoted mUCE. As before we aim to provide point-function obfuscation secure for any given class of target generators. We are able to do this with mUCE by associating to the class of target generators a class of sources. The existence of an mUCE-secure function family relative to the latter suffices to construct a point-function obfuscator secure relative to the former.

Construction. Let H be a family of functions. Associate to it a point-function obfuscator Obf defined as follows. Let $\text{Obf.tl} = \text{H.il}$, and

Algorithm $\text{Obf}(1^\lambda, \mathbf{I}_k)$	Circuit $C_{1^\lambda, hk, y}(k')$
$hk \leftarrow\!\!{\$}\ \text{H.Kg}(1^\lambda); \ y \leftarrow \text{H.Ev}(1^\lambda, hk, k)$	$y' \leftarrow \text{H.Ev}(1^\lambda, hk, k')$
Return $C_{1^\lambda, hk, y}$	If $(y = y')$ then return 1 else return 0

The construction is simple and natural. The point-function obfuscation of $\mathbf{I}_k$ is a circuit that embeds the hash y of target k under a freshly-chosen key hk also embedded in the circuit, and, given a candidate target k', checks whether its hash under hk equals the embedded hash value.

Source Classes. To state the result, we need a few definitions. Associate to a target generator X a source $\mathcal{S}^{\mathsf{X}}$ defined as follows:

Source $\mathcal{S}^{\mathsf{X}}(1^\lambda)$

$d \leftarrow\!\!{}_\$\ \{0,1\}; \ (\mathbf{k}_1, a_1) \leftarrow\!\!{}_\$\ \mathsf{X}.\mathsf{Ev}(1^\lambda)$
For $i = 1, \ldots, \mathsf{X}.\mathsf{vl}(\lambda)$ do $\mathbf{k}_0[i] \leftarrow\!\!{}_\$\ \{0,1\}^{\mathsf{X}.\mathsf{tl}(\lambda)}$
For $i = 1, \ldots, \mathsf{X}.\mathsf{vl}(\lambda)$ do $\mathbf{y}[i] \leftarrow\!\!{}_\$\ \mathrm{HASH}(i, \mathbf{k}_d[i])$
$L \leftarrow ((d, a_1), \mathbf{y})$; Return L

The number of keys for this source is $\mathcal{S}^{\mathsf{X}}.\mathsf{nk} = \mathsf{X}.\mathsf{vl}$, the number of points in the target vector. Now let $\mathbf{X}$ be a class of target generators and let $\mathbf{S}^{\mathbf{X}} = \{\ \mathcal{S}^{\mathsf{X}} : \mathsf{X} \in \mathbf{X}\ \}$ be the corresponding class of sources. We will show that the construction above is IND[$\mathbf{X}$]-secure assuming H is mUCE[$\mathbf{S}^{\mathbf{X}}$]-secure. To appreciate what this provides we now discuss the assumption further.

Assumptions in the UCE framework are very sensitive to the class of sources for which security is assumed. Accordingly one tries to restrict sources in different ways. In this regard $\mathbf{S}^{\mathbf{X}} = \{\ \mathcal{S}^{\mathsf{X}} : \mathsf{X} \in \mathbf{X}\ \}$ has some good attributes as we now discuss, referring to definitions of classes of mUCE sources recalled in Sect. 2.

The first attribute is that the sources in $\mathbf{S}^{\mathbf{X}}$ are what BHK [6] call "split," so that $\mathbf{S}^{\mathbf{X}} \subseteq \mathbf{S}^{\mathsf{splt}}$. "Split" means that the leakage is a function of the oracle queries and answers separately, but not both together. (Above, (d, a_1) depends only on the oracle queries, and $\mathbf{y}$ depends only on the answers.) The second attribute is that the sources make only one query per key. (In particular when there is only one target point, the source makes only one query overall.) That is, $\mathbf{S}^{\mathbf{X}} \subseteq \mathbf{S}^{n,1}$ if $\mathcal{S}.\mathsf{nk}(\cdot) \leq n(\cdot)$ for all $\mathcal{S} \in \mathbf{S}^{\mathbf{X}}$. The third attribute is that the source class inherits the unpredictability properties of the target generator class. Thus if $\mathbf{X} \subseteq \mathbf{X}^{\mathsf{cup}}$ then $\mathbf{S}^{\mathbf{X}} \subseteq \mathbf{S}^{\mathsf{cup}}$ consists of computationally unpredictable sources; if $\mathbf{X} \subseteq \mathbf{X}^{\mathsf{sup}}$ then $\mathbf{S}^{\mathbf{X}} \subseteq \mathbf{S}^{\mathsf{sup}}$ consists of statistically unpredictable sources; and if $\mathbf{X} \subseteq \mathbf{X}^{\mathsf{seup}}$ then $\mathbf{S}^{\mathbf{X}} \subseteq \mathbf{S}^{\mathsf{seup}}$ consists of sources that are sub-exponentially unpredictable.

We warn that mUCE[$\mathbf{S}^{\mathbf{X}}$]-security is not achievable for all choices of $\mathbf{X}$. The value of our result is that it is entirely general, reducing IND security for a given $\mathbf{X}$ to a question of mUCE security for a related class of sources, and we can then investigate the latter separately. In this way we get many new constructions.

Result. The following theorem shows that our construction above provides secure point-function obfuscation in a very general and modular way, namely the point-function obfuscator is secure relative to a class of target generators if H is mUCE-secure relative to the corresponding class of sources. After stating and proving this general result we will look at some special cases of interest.

Theorem 6. *Let* H *be an injective family of functions. Let* $\mathbf{X}$ *be a class of target generators such that* $\mathsf{X}.\mathsf{tl} = \mathsf{H}.\mathsf{il}$ *for all* $\mathsf{X} \in \mathbf{X}$. *Assume* H *is* mUCE[$\mathbf{S}^{\mathbf{X}}$]-*secure. Let* Obf *be as defined above. Then* Obf *is a* IND[$\mathbf{X}$]-*secure point-function obfuscator.*

Function family H is assumed to be injective in order to meet the perfect correctness condition of a point-function obfuscator, and it is not important for security. In [8] we show that *non-injective* mUCE is sufficient to construct a point-function obfuscator that satisfies a relaxed correctness condition and achieves the same security as above.

Proof (Theorem 6). Correctness of the obfuscator follows from the assumed injectivity of H, meaning that the output of $\mathsf{Obf}(1^\lambda, \mathbf{I}_k)$ is always a point circuit with target k. We now prove that Obf is IND[$\mathbf{X}$]-secure.

Let $\mathsf{X} \in \mathbf{X}$ be any target generator with $\mathsf{X.tl} = \mathsf{H.il}$. Let $\mathcal{S}^\mathsf{X}$ be the corresponding source as defined above. Let $\mathcal{A}$ be a PT adversary against the IND-security of Obf relative to X. We define a PT distinguisher $\mathcal{D}$ as follows:

Distinguisher $\mathcal{D}(1^\lambda, \mathbf{hk}, L)$	Circuit $\mathrm{C}_{1^\lambda, hk, y}(k')$
$((d, a_1), \mathbf{y}) \leftarrow L$	$y' \leftarrow \mathsf{H.Ev}(1^\lambda, hk, k')$
For $i = 1, \ldots, |\mathbf{y}|$ do $\overline{\mathbf{P}}[i] \leftarrow \mathrm{C}_{1^\lambda, \mathbf{hk}[i], \mathbf{y}[i]}$	If $(y = y')$ then return 1
$d' \leftarrow_\$ \mathcal{A}(1^\lambda, \overline{\mathbf{P}}, a_1)$	Else return 0
If $(d = d')$ then return 1 else return 0	

Let b denote the challenge bit in game $\mathrm{mUCE}_\mathsf{H}^{\mathcal{S}^\mathsf{X}, \mathcal{D}}(\lambda)$, and let b' denote the bit returned by $\mathcal{D}$ in the same game. We claim that

$$\Pr[\, b' = 1 \mid b = 1 \,] = \Pr\left[\, \mathrm{IND}_{\mathsf{Obf}, \mathsf{X}}^\mathcal{A}(\lambda) \,\right] \quad \text{and} \quad \Pr[\, b' = 1 \mid b = 0 \,] = \frac{1}{2} \, .$$

The first equation holds by construction. The second equation is true because $\mathcal{D}$ runs $\mathcal{A}$ with inputs that are independent of the challenge bit d. Namely, for $b = 0$ the entries in $\mathbf{y}$ are uniform and independent, since the source $\mathcal{S}$ makes only one query per key index. We have $\mathsf{Adv}_{\mathsf{H}, \mathcal{S}^\mathsf{X}, \mathcal{D}}^{\mathrm{m\text{-}uce}}(\lambda) = \mathsf{Adv}_{\mathsf{Obf}, \mathsf{X}, \mathcal{A}}^{\mathrm{ind}}(\lambda)/2$. Therefore, for any $\mathsf{X} \in \mathbf{X}$ the IND security of Obf relative to X follows from the assumed $\mathrm{mUCE}[\{\mathcal{S}^\mathsf{X}\}]$-security of H.

Negative Results for Multi-key UCE. Let $n \colon \mathbb{N} \to \mathbb{N}$ be a polynomial such that $n(\cdot) \in \Omega((\cdot)^\epsilon)$. Theorem 6 allows us to conclude that $\mathrm{mUCE}[\mathbf{S}^{\mathrm{cup}} \cap \mathbf{S}^{\mathrm{splt}} \cap \mathbf{S}^{n,1}]$-secure injective function families do not exist under certain assumptions. This is a simple corollary of the prior results which show that MB-AIPO can not co-exist with iO [20,25]. We now explain our claim in more details.

Theorem 6 shows that the existence of $\mathrm{mUCE}[\mathbf{S}^{\mathrm{cup}} \cap \mathbf{S}^{\mathrm{splt}} \cap \mathbf{S}^{n,1}]$-secure injective function families implies $\mathrm{IND}[\mathbf{X}^{\mathrm{cup}} \cap \mathbf{X}^n]$-secure point-function obfuscation. Note that the latter is a composable AIPO as per CD [25]. CD [25] show that composable AIPO can be used to construct MB-AIPO, which is an obfuscation that is secure for functions that map a target point to a multi-bit output (as opposed to an output in $\{0, 1\}$). Finally, BM1 [20] show that MB-AIPO cannot co-exist with iO, assuming one-way functions. These results imply the following:

Corollary 7. *Let* H *be an injective function family. Let* $n \colon \mathbb{N} \to \mathbb{N}$ *be a polynomial such that* $n(\cdot) \in \Omega((\cdot)^\epsilon)$ *for some constant* $\epsilon > 0$. *Assume the existence of one-way functions and indistinguishability obfuscation. Then* H *is not* $\mathrm{mUCE}[\mathbf{S}^{\mathrm{cup}} \cap \mathbf{S}^{\mathrm{splt}} \cap \mathbf{S}^{n,1}]$-*secure.*

In a concurrent and independent work, BM3 [22] discuss a similar impossibility result for $\mathrm{mUCE}[\mathbf{S}^{\mathrm{s\text{-}cup}} \cap \mathbf{S}^{n,1}]$-security. Here $\mathbf{S}^{\mathrm{s\text{-}cup}}$ is a class of UCE sources introduced in (BM2) [21] who also show that $\mathbf{S}^{\mathrm{cup}} \cap \mathbf{S}^{\mathrm{splt}} \subsetneq \mathbf{S}^{\mathrm{s\text{-}cup}}$. We note that impossibility of mUCE[$\mathbf{S}^{\mathrm{cup}} \cap \mathbf{S}^{\mathrm{splt}} \cap \mathbf{S}^{n,1}$]-secure function families is a stronger result because it concerns a smaller class of sources.

No other impossibility results are known for mUCE exclusively, but any negative results for (single-key) UCE also apply to mUCE. Specifically, BFM [18] give an iO-based attack on UCE[$\mathbf{S}^{\mathrm{cup}}$]. And BST [10] show that UCE[$\mathbf{S}^{\mathrm{cup}} \cap \mathbf{S}^{\mathrm{splt}}$]-secure function families do not exist assuming the existence of OWFs and iO, which is a strictly stronger impossibility result than the latter. The result by BST [10] implies that mUCE[$\mathbf{S}^{\mathrm{cup}} \cap \mathbf{S}^{\mathrm{splt}} \cap \mathbf{S}^{1,p}$]-secure function families do not exist for a polynomial $p(\cdot) \in \Omega((\cdot)^\epsilon)$, but we currently do not know whether this notion is comparable to mUCE[$\mathbf{S}^{\mathrm{cup}} \cap \mathbf{S}^{\mathrm{splt}} \cap \mathbf{S}^{n,1}$].

Related Work. One special case of Theorem 6 is when $\mathbf{X} = \mathbf{X}^{\mathrm{cup}} \cap \mathbf{X}^1$, so that IND[$\mathbf{X}$] is AIPO. The theorem and the remarks preceding it imply that we get this assuming mUCE[$\mathbf{S}^{\mathrm{cup}} \cap \mathbf{S}^{\mathrm{splt}} \cap \mathbf{S}^{1,1}$]-security. This special case of our result was independently and concurrently obtained in [22]. Note that BM2 [21] showed that mUCE[$\mathbf{S}^{\mathrm{cup}} \cap \mathbf{S}^{\mathrm{splt}} \cap \mathbf{S}^{1,1}$]-security is achievable assuming iO and AIPO. It follows from our result that mUCE[$\mathbf{S}^{\mathrm{cup}} \cap \mathbf{S}^{\mathrm{splt}} \cap \mathbf{S}^{1,1}$] and AIPO are equivalent, assuming iO.

6 Alternative Security Notions for PO

In Sect. 3 we defined IND security of point-function obfuscation. It extends security notions that were used for variants of AIPO in BP [14], MH [40], MB1 [20] and MB3 [22]. The main difference is that IND is parameterized with a class of target generators, allowing us to unify the treatment of AIPO from the literature.

In this section we provide several alternative security notions for point-function obfuscation, and show relations between them and IND. Specifically, we extend the security notion introduced by Canetti [23] as well as the notions of average-case [26,29] and worst-case [2,25,35,39,47] simulation-based security for point-function obfuscation. Similar to IND, our extended notions are parameterized with classes of target generators. We also define a novel security notion, called computational semantic security, by adapting the corresponding definition that was used for DPKE in [4] to the setting of point-function obfuscation and parameterizing it in the same way as above. Finally, we discuss the security achieved by our PO constructions from Sect. 5 with respect to the new notions.

Strong Indistinguishability. Consider game SIND of Fig. 7 associated to a point-function obfuscator Obf, a target generator X, an adversary $\mathcal{A}$ and a distinguisher $\mathcal{D}$, such that $\mathcal{A}$ returns an output in $\{0,1\}$ and Obf.tl = X.tl. For $\lambda \in \mathbb{N}$ let $\mathrm{Adv}^{\mathrm{sind}}_{\mathrm{Obf},\mathsf{X},\mathcal{A},\mathcal{D}}(\lambda) = 2\Pr[\mathrm{SIND}^{\mathcal{A},\mathcal{D}}_{\mathrm{Obf},\mathsf{X}}(\lambda)] - 1$. Let $\mathbf{X}$ be a class of target generators. We say that Obf is SIND[$\mathbf{X}$]-secure if $\mathrm{Adv}^{\mathrm{sind}}_{\mathrm{Obf},\mathsf{X},\mathcal{A},\mathcal{D}}(\cdot)$ is negligible for every $\mathsf{X} \in \mathbf{X}$, every PT $\mathcal{A}$ and every PT $\mathcal{D}$. The difference between our definitions of IND and SIND is that the latter also runs a distinguisher in the last stage of the game, which makes this definition meaningful even for trivial target generators (as defined in Sect. 3). Our definition of SIND extends the security notion used for oracle hashing by Canetti [23], parameterizing it with classes of

Game SIND$_{\mathsf{Obf},X}^{\mathcal{A},\mathcal{D}}(\lambda)$	Game CSS$_{\mathsf{Obf},X}^{\mathcal{A}}(\lambda)$	Game SSS$_{\mathsf{Obf},X}^{\mathcal{A},\mathcal{S},\mathcal{P}}(\lambda)$
$b \leftarrow\!\!{\scriptstyle\$}\ \{0,1\}$	$b \leftarrow\!\!{\scriptstyle\$}\ \{0,1\}$	$b \leftarrow\!\!{\scriptstyle\$}\ \{0,1\}$
$(\mathbf{k}_1, a_1) \leftarrow\!\!{\scriptstyle\$}\ \mathsf{X.Ev}(1^\lambda)$	$(\mathbf{k}_1, a_1) \leftarrow\!\!{\scriptstyle\$}\ \mathsf{X.Ev}(1^\lambda)$	$(\mathbf{k}, a) \leftarrow\!\!{\scriptstyle\$}\ \mathsf{X.Ev}(1^\lambda)$
For $i = 1, \ldots, \mathsf{X.vl}(\lambda)$ do	For $i = 1, \ldots, \mathsf{X.vl}(\lambda)$ do	$\overline{\mathbf{P}} \leftarrow\!\!{\scriptstyle\$}\ \mathsf{Obf}(1^\lambda, \mathbf{I_k})$
$\quad \mathbf{k}_0[i] \leftarrow\!\!{\scriptstyle\$}\ \{0,1\}^{\mathsf{X.tl}(\lambda)}$	$\quad \mathbf{k}_0[i] \leftarrow\!\!{\scriptstyle\$}\ \{0,1\}^{\mathsf{X.tl}(\lambda)}$	$p \leftarrow\!\!{\scriptstyle\$}\ \mathcal{P}(1^\lambda, \mathbf{k}, a)$
$\overline{\mathbf{P}} \leftarrow\!\!{\scriptstyle\$}\ \mathsf{Obf}(1^\lambda, \mathbf{I_{k_b}})$	$\overline{\mathbf{P}} \leftarrow\!\!{\scriptstyle\$}\ \mathsf{Obf}(1^\lambda, \mathbf{I_{k_b}})$	If $(b = 1)$ then
$d \leftarrow\!\!{\scriptstyle\$}\ \mathcal{A}(1^\lambda, \overline{\mathbf{P}}, a_1)$	$t \leftarrow\!\!{\scriptstyle\$}\ \mathcal{A}_1(1^\lambda, \mathbf{k}_1, a_1)$	$\quad p' \leftarrow\!\!{\scriptstyle\$}\ \mathcal{A}(1^\lambda, \overline{\mathbf{P}}, a)$
$b' \leftarrow\!\!{\scriptstyle\$}\ \mathcal{D}(1^\lambda, \mathbf{k}_1, a_1, d)$	$t' \leftarrow\!\!{\scriptstyle\$}\ \mathcal{A}_2(1^\lambda, \overline{\mathbf{P}}, a_1)$	Else $p' \leftarrow\!\!{\scriptstyle\$}\ \mathcal{S}^{\mathbf{I_k}}(1^\lambda, a)$
Return $(b = b')$	If $(t = t')$ then $b' \leftarrow 1$	If $(p = p')$ then $b' \leftarrow 1$
	Else $b' \leftarrow 0$	Else $b' \leftarrow 0$
	Return $(b = b')$	Return $(b = b')$

Fig. 7. Games defining SIND security, CSS security and SSS security of point-function obfuscator Obf relative to target generator X.

target generators. Another difference is that SIND samples target vectors $\mathbf{k}_0, \mathbf{k}_1$ from distributions that are potentially different, whereas [23] used the same distribution for both. Note that adversary $\mathcal{A}$ cannot be allowed to return an output of an arbitrary length because then it would be able to return $\overline{\mathbf{P}}$, hence making the security trivially unachievable.

Computational Semantic Security. Consider game CSS of Fig. 7 associated to a point-function obfuscator Obf, a target generator X and an adversary $\mathcal{A} = (\mathcal{A}_1, \mathcal{A}_2)$ such that algorithms $\mathcal{A}_1, \mathcal{A}_2$ return outputs in $\{0,1\}$ and Obf.tl = X.tl. For $\lambda \in \mathbb{N}$ let $\mathsf{Adv}_{\mathsf{Obf},X,\mathcal{A}}^{\mathsf{css}}(\lambda) = 2\Pr[\mathsf{CSS}_{\mathsf{Obf},X}^{\mathcal{A}}(\lambda)] - 1$. Let $\mathbf{X}$ be a class of target generators. We say that Obf is CSS[$\mathbf{X}$]-secure if $\mathsf{Adv}_{\mathsf{Obf},X,\mathcal{A}}^{\mathsf{css}}(\cdot)$ is negligible for every $X \in \mathbf{X}$ and every PT $\mathcal{A}$. This is an adaptation of the definition of *computational semantic security* for DPKE from [4], which we further parameterize with classes of target generators. It asks that adversary $\mathcal{A}$ can not use an obfuscation $\overline{\mathbf{P}}$ of $\mathbf{k}_1$ to compute any partial information about the latter, even in the presence of auxiliary information a_1. This provides us with a better intuition about the desired security of point-function obfuscation, as opposed to the less intuitive definition of SIND.

Simulation-Based Semantic Security. We consider two different definitions of simulation-based semantic security. Informally, both definitions require that for every PT adversary $\mathcal{A}$ that receives as input an obfuscation of some point-function $\mathbf{I}_k$, there exists a PT simulator with only an oracle access to $\mathbf{I}_k$, such that the output distribution of the former is indistinguishable from that of the latter. The two definitions differ in the way how $\mathbf{I}_k$ is chosen. One option is to quantify over all possible point-functions that can be produced by a particular target generator. For this purpose, we extend the definitions of *worst-case* security [2,25,35,39,47] for point-function obfuscation. We use SIM to denote our new security notion. An alternative approach is to use target generator X in order to sample point-functions. This follows the definitions of *average-case* security

[26,29] for point function obfuscation, and we use SSS to denote our extended security notion.

Consider game SSS of Fig. 7 associated to a point-function obfuscator Obf, a target generator X, an adversary $\mathcal{A}$, a simulator $\mathcal{S}$ and a predicate algorithm $\mathcal{P}$, such that algorithms $\mathcal{A}, \mathcal{S}, \mathcal{P}$ return outputs in $\{0, 1\}$ and Obf.tl = X.tl. For $\lambda \in \mathbb{N}$ let $\mathrm{Adv}^{\mathrm{sss}}_{\mathrm{Obf}, \mathrm{X}, \mathcal{A}, \mathcal{S}, \mathcal{P}}(\lambda) = 2 \Pr[\mathrm{SSS}^{\mathcal{A}, \mathcal{S}, \mathcal{P}}_{\mathrm{Obf}, \mathrm{X}}(\lambda)] - 1$. Let $\mathbf{X}$ be a class of target generators. We say that Obf is SSS[$\mathbf{X}$]-secure if for every target generator $\mathrm{X} \in \mathbf{X}$ and every PT $\mathcal{A}$ there exists PT $\mathcal{S}$ such that $\mathrm{Adv}^{\mathrm{sss}}_{\mathrm{Obf}, \mathrm{X}, \mathcal{A}, \mathcal{S}, \mathcal{P}}(\cdot)$ is negligible for every PT $\mathcal{P}$. Informally, this security notion requires that for every adversary $\mathcal{A}$ there exists a simulator $\mathcal{S}$ such that if $\mathcal{A}$ can use obfuscations $\mathbf{I}_{\mathbf{k}}$ to compute any property (function) $\mathcal{P}$ of $\mathbf{k}$, then $\mathcal{S}$ can do the same using only an oracle access to $\mathbf{I}_{\mathbf{k}}$ (meaning that $\mathcal{S}$ has oracle access to each of $\mathbf{I}_{\mathbf{k}[1]}, \ldots, \mathbf{I}_{\mathbf{k}[n]}$ for $n = |\mathbf{k}|$). This is required to hold even when $\mathcal{A}, \mathcal{S}, \mathcal{P}$ receive as input some auxiliary information a about $\mathbf{k}$.

SIM Security. Next, we define the SIM-security of PO. Let $\mathbf{X}$ be a class of target generators. Let Obf be a point-function obfuscator. We say that Obf is SIM[$\mathbf{X}$]-secure if for every target generator $\mathrm{X} \in \mathbf{X}$ and every PT adversary $\mathcal{A}$ there exists a PT simulator $\mathcal{S}$ and a negligible function $\mu \colon \mathbb{N} \to \mathbb{N}$ such that

$$\left| \Pr[\mathcal{A}(1^\lambda, \mathrm{Obf}(1^\lambda, \mathbf{I}_{\mathbf{k}}), a) = \mathcal{P}(\mathbf{k}, a)] - \Pr[\mathcal{S}^{\mathbf{I}_{\mathbf{k}}}(1^\lambda, a) = \mathcal{P}(\mathbf{k}, a)] \right| \leq \mu(\lambda)$$

for every $\lambda \in \mathbb{N}$, every $(\mathbf{k}, a) \in [\mathrm{X.Ev}(1^\lambda)]$ and every PT predicate algorithm $\mathcal{P}$ that returns an output in $\{0, 1\}$.

In the above definition of SIM-security, predicate $\mathcal{P}$ can be substituted with a constant function, resulting in an equivalent definition (as noted in [2,35,47]). In contrast, this is not true for the definition of SSS-security. Replacing $\mathcal{P}$ with a constant function will allow $\mathcal{S}$ to run X in order to generate fresh $(\mathbf{k}, a)$, obfuscate $\mathbf{I}_{\mathbf{k}}$ to get $\overline{\mathbf{P}}$, and simulate $\mathcal{A}$ on $\overline{\mathbf{P}}, a$. As a result, every obfuscator would be vaciously SSS-secure for any class of target generators $\mathbf{X}$.

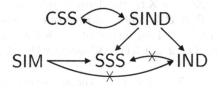

Fig. 8. Relations between security notions for point-function obfuscation.

Relations Between Security Notions. Figure 8 shows relations between the security notions for point-function obfuscation that are discussed in this paper. Consider any two security notions A and B. An arrow from A to B means that any A[$\mathbf{X}$]-secure point-function obfuscator is also B[$\mathbf{X}$]-secure, for every class of target generators $\mathbf{X}$. A crossed arrow going from A to B means that there exists an obfuscator Obf and a class of target generators $\mathbf{X}$ such that Obf is A[$\mathbf{X}$]-secure but not B[$\mathbf{X}$]-secure.

Implications SIM → SSS and SIND → IND trivially follow from our definitions of the corresponding security notions. The proofs for all other implications and separations shown in Fig. 8 are provided in [8]. The only relations that are missing in the figure (and can not be deduced using transitivity) are those between SIM and both of SSS, CSS. We leave it as an open question to show the remaining relations between these security notions.

Security of Our PO Constructions. Let $\mathbf{X}$ be a class of target generators. In Sect. 5 we showed how to build a point-function obfuscator that is IND[$\mathbf{X}$]-secure, based on any of the following: a OWF[$\mathbf{X}$]-secure function family and an iO, or a PRIV1[$\mathbf{X}$]-secure DPKE, or an mUCE[$\mathbf{S^X}$]-secure function family for $\mathbf{S^X}$ as defined in Sect. 5.3. We do not know how to adapt our constructions to achieve SIM[$\mathbf{X}$]-security. But each of our construction achieves CSS[$\mathbf{X}$]-security, requiring only minimal changes in the used assumptions.

We now provide some intuition about our claim. Recall that game CSS computes $t \in \{0, 1\}$ by running $\mathcal{A}_1(1^\lambda, \mathbf{k}_1, a_1)$, and subsequently compares it to the output of $\mathcal{A}_2(1^\lambda, \overline{\mathbf{P}}, a_1)$. This is different from game IND where the adversary consists only of an algorithm $\mathcal{A}(1^\lambda, \overline{\mathbf{P}}, a_1)$. The difficulty of adapting proofs of IND[$\mathbf{X}$]-security to achieve CSS[$\mathbf{X}$]-security is that in the latter $\mathbf{k}_1$ (required to run $\mathcal{A}_1$) and $\overline{\mathbf{P}}$ (required to run $\mathcal{A}_2$) are usually available in different stages of the security proof, meaning that one has to find a way to pass around the value of t (which depends on $\mathbf{k}_1$) across the stages. We resolve this by pushing t into the auxiliary information of target generators that parametrize our security notions.

Let $\mathbf{X}$ be a class of target generators. Let $\mathbf{P}$ be the set of all PT predicate algorithms $\mathcal{P}$ such that $\mathcal{P}(1^\lambda, \cdot, \cdot): \{0, 1\}^* \times \{0, 1\}^* \to \{0, 1\}$ for all $\lambda \in \mathbb{N}$. For any $\lambda \in \mathbb{N}$, $\mathsf{X} \in \mathbf{X}$ and $\mathcal{P} \in \mathbf{P}$ let $\mathsf{X}^\mathcal{P}$ be defined as follows:

$$\text{Source } \mathsf{X}^\mathcal{P}(1^\lambda)$$
$$\overline{(\mathbf{k}, a) \leftarrow_\$ \mathsf{X}(1^\lambda); \ \beta \leftarrow_\$ \mathcal{P}(1^\lambda, \mathbf{k}, a); \ \text{Return } (\mathbf{k}, (a, \beta))}$$

where $\mathsf{X}^\mathcal{P}.\mathsf{vl} = \mathsf{X}.\mathsf{vl}$ and $\mathsf{X}^\mathcal{P}.\mathsf{tl} = \mathsf{X}.\mathsf{tl}$. We define a new class of target generators $\mathbf{X}' = \{ \mathsf{X}^\mathcal{P} : \mathsf{X} \in \mathbf{X}, \mathcal{P} \in \mathbf{P} \}$. Then each of our constructions from Sect. 5 achieves CSS[$\mathbf{X}$]-security, based on either of the following: a OWF[$\mathbf{X}'$]-secure function family and an iO, or a PRIV1[$\mathbf{X}'$]-secure DPKE, or an mUCE[$\mathbf{S^{X'}}$]-secure function family.

Note that for any $\mathsf{X} \in \mathbf{X}$ and $\mathcal{P} \in \mathbf{P}$, the construction of $\mathsf{X}^\mathcal{P}$ expands the auxiliary information of X only by a single bit. This means that $\mathsf{X}^\mathcal{P}$ inherits the unpredictability properties of X. Namely, for any $\lambda \in \mathbb{N}$, $\mathsf{X} \in \mathbf{X}$, $\mathcal{P} \in \mathbf{P}$ and any PT adversary $\mathcal{R}$ we can construct a PT adversary $\mathcal{Q}$ such that $\Pr[\mathrm{PRED}_\mathsf{X}^\mathcal{Q}(\lambda)] \geq \frac{1}{2} \Pr[\mathrm{PRED}_{\mathsf{X}^\mathcal{P}}^\mathcal{R}(\lambda)]$ for all $\lambda \in \mathbb{N}$. Adversary $\mathcal{Q}$ would attempt to guess the extra bit of information and then simulate $\mathcal{R}$. The same approach can be used to show that any OWF[$\mathbf{X}$]-secure function family is also OWF[$\mathbf{X}'$]-secure, recovering the construction of CSS[$\mathbf{X}$]-secure PO directly from a OWF[$\mathbf{X}$]-secure function family and an iO.

Definitional Choices. All of our security notions for point-function obfuscation require that adversaries return single-bit outputs. This is consistent with the prior work. Specifically, simulation-based definitions in the prior literature always compare the outputs of adversary and simulator to either a predicate [27,35] or a constant [2,25,26,29,39,47]. However, it would be more intuitive to not restrict the size of outputs returned by adversaries in games CSS, SSS and SIM. The goal of these adversaries can be thought as to compute some "property" of the target vector, and there is no reason to limit it to a single bit.

The initial work on obfuscation [2] discusses various definitional choices and chooses to use the weakest of them to achieve stronger impossibility results. Subsequent work continues to use definitions of the same style even for positive results. We are not aware of any follow-up discussion on alternative definitions.

Some of our implications from Fig. 8 might change if adversaries in games CSS, SSS and SIM are allowed to return multiple-bit outputs. In particular, note that our definitions of CSS and SSS are similar to those that were used for DPKE schemes in BFOR [4], who showed them to be equivalent for multiple-bit outputs in their setting. We leave it as an open problem to extend our definitions to allow outputs of an arbitrary size.

Acknowledgments. Bellare and Stepanovs were supported in part by NSF grants CNS-1116800, CNS-1228890 and CNS-1526801. This work was done in part while Bellare was visiting the Simons Institute for the Theory of Computing, supported by the Simons Foundation and by the DIMACS/Simons Collaboration in Cryptography through NSF grant CNS-1523467. We thank Stefano Tessaro and Arno Mittelbach for discussions and insights. Extensive and insightful comments by the TCC 2016-A reviewers lead to considerable changes and additions to the paper including Theorem 2, Corollary 7 and Sect. 6.

References

1. Ananth, P., Boneh, D., Garg, S., Sahai, A., Zhandry, M.: Differing-inputs obfuscation and applications. Cryptology ePrint Archive, Report 2013/689 (2013). http://eprint.iacr.org/2013/689
2. Barak, B., Goldreich, O., Impagliazzo, R., Rudich, S., Sahai, A., Vadhan, S.P., Yang, K.: On the (im)possibility of obfuscating programs. In: Kilian, J. (ed.) CRYPTO 2001. LNCS, vol. 2139, pp. 1–18. Springer, Heidelberg (2001)
3. Bellare, M., Boldyreva, A., O'Neill, A.: Deterministic and efficiently searchable encryption. In: Menezes, A. (ed.) CRYPTO 2007. LNCS, vol. 4622, pp. 535–552. Springer, Heidelberg (2007)
4. Bellare, M., Fischlin, M., O'Neill, A., Ristenpart, T.: Deterministic encryption: definitional equivalences and constructions without random oracles. In: Wagner, D. (ed.) CRYPTO 2008. LNCS, vol. 5157, pp. 360–378. Springer, Heidelberg (2008)
5. Bellare, M., Hoang, V.T.: Resisting randomness subversion: fast deterministic and hedged public-key encryption in the standard model. In: Oswald, E., Fischlin, M. (eds.) EUROCRYPT 2015. LNCS, vol. 9057, pp. 627–656. Springer, Heidelberg (2015)
6. Bellare, M., Hoang, V.T., Keelveedhi, S.: Instantiating random oracles via UCEs. In: Canetti, R., Garay, J.A. (eds.) CRYPTO 2013, Part II. LNCS, vol. 8043, pp. 398–415. Springer, Heidelberg (2013)

7. Bellare, M., Rogaway, P.: The security of triple encryption and a framework for code-based game-playing proofs. In: Vaudenay, S. (ed.) EUROCRYPT 2006. LNCS, vol. 4004, pp. 409–426. Springer, Heidelberg (2006)

8. Bellare, M., Stepanovs, I.: Point-function obfuscation: a framework and generic constructions. Cryptology ePrint Archive, Report 2015/703 (2015). http://eprint.iacr.org/2015/703

9. Bellare, M., Stepanovs, I., Tessaro, S.: Poly-many hardcore bits for any one-way function and a framework for differing-inputs obfuscation. In: Sarkar, P., Iwata, T. (eds.) ASIACRYPT 2014, Part II. LNCS, vol. 8874, pp. 102–121. Springer, Heidelberg (2014)

10. Bellare, M., Stepanovs, I., Tessaro, S.: Contention in cryptoland: obfuscation, leakage and uce. In: Kushilevitz, E., Malkin, T. (eds.) TCC 2016-A, Part II. LNCS, vol. 9563, pp. 542–564. Springer, Heidelberg (2016)

11. Bitansky, N., Canetti, R.: On strong simulation and composable point obfuscation. In: Rabin, T. (ed.) CRYPTO 2010. LNCS, vol. 6223, pp. 520–537. Springer, Heidelberg (2010)

12. Bitansky, N., Canetti, R., Kalai, Y.T., Paneth, O.: On virtual grey box obfuscation for general circuits. In: Garay, J.A., Gennaro, R. (eds.) CRYPTO 2014, Part II. LNCS, vol. 8617, pp. 108–125. Springer, Heidelberg (2014)

13. Bitansky, N., Canetti, R., Paneth, O., Rosen, A.: On the existence of extractable one-way functions. In: Shmoys, D.B. (ed.) 46th ACM STOC, pp. 505–514. ACM Press (May/June 2014)

14. Bitansky, N., Paneth, O.: Point obfuscation and 3-round zero-knowledge. In: Cramer, R. (ed.) TCC 2012. LNCS, vol. 7194, pp. 190–208. Springer, Heidelberg (2012)

15. Boldyreva, A., Fehr, S., O'Neill, A.: On notions of security for deterministic encryption, and efficient constructions without random oracles. In: Wagner, D. (ed.) CRYPTO 2008. LNCS, vol. 5157, pp. 335–359. Springer, Heidelberg (2008)

16. Boyle, E., Chung, K.-M., Pass, R.: On extractability obfuscation. In: Lindell, Y. (ed.) TCC 2014. LNCS, vol. 8349, pp. 52–73. Springer, Heidelberg (2014)

17. Brakerski, Z., Segev, G.: Better security for deterministic public-key encryption: the auxiliary-input setting. In: Rogaway, P. (ed.) CRYPTO 2011. LNCS, vol. 6841, pp. 543–560. Springer, Heidelberg (2011)

18. Brzuska, C., Farshim, P., Mittelbach, A.: Indistinguishability obfuscation and UCEs: the case of computationally unpredictable sources. In: Garay, J.A., Gennaro, R. (eds.) CRYPTO 2014, Part I. LNCS, vol. 8616, pp. 188–205. Springer, Heidelberg (2014)

19. Brzuska, C., Farshim, P., Mittelbach, A.: Random-oracle uninstantiability from indistinguishability obfuscation. In: Dodis, Y., Nielsen, J.B. (eds.) TCC 2015, Part II. LNCS, vol. 9015, pp. 428–455. Springer, Heidelberg (2015)

20. Brzuska, C., Mittelbach, A.: Indistinguishability obfuscation versus multi-bit point obfuscation with auxiliary input. In: Sarkar, P., Iwata, T. (eds.) ASIACRYPT 2014, Part II. LNCS, vol. 8874, pp. 142–161. Springer, Heidelberg (2014)

21. Brzuska, C., Mittelbach, A.: Using indistinguishability obfuscation via UCEs. In: Sarkar, P., Iwata, T. (eds.) ASIACRYPT 2014, Part II. LNCS, vol. 8874, pp. 122–141. Springer, Heidelberg (2014)

22. Brzuska, C., Mittelbach, A.: Universal computational extractors and the superfluous padding assumption for indistinguishability obfuscation. Cryptology ePrint Archive, Report 2015/581 (2015). http://eprint.iacr.org/2015/581

23. Canetti, R.: Towards realizing random oracles: hash functions that hide all partial information. In: Kaliski Jr, B.S. (ed.) CRYPTO 1997. LNCS, vol. 1294, pp. 455–469. Springer, Heidelberg (1997)
24. Canetti, R., Dakdouk, R.R.: Extractable perfectly one-way functions. In: Aceto, L., Damgård, I., Goldberg, L.A., Halldórsson, M.M., Ingólfsdóttir, A., Walukiewicz, I. (eds.) ICALP 2008, Part II. LNCS, vol. 5126, pp. 449–460. Springer, Heidelberg (2008)
25. Canetti, R., Dakdouk, R.R.: Obfuscating point functions with multibit output. In: Smart, N.P. (ed.) EUROCRYPT 2008. LNCS, vol. 4965, pp. 489–508. Springer, Heidelberg (2008)
26. Canetti, R., Tauman Kalai, Y., Varia, M., Wichs, D.: On symmetric encryption and point obfuscation. In: Micciancio, D. (ed.) TCC 2010. LNCS, vol. 5978, pp. 52–71. Springer, Heidelberg (2010)
27. Canetti, R., Micciancio, D., Reingold, O.: Perfectly one-way probabilistic hash functions (preliminary version). In: 30th ACM STOC, pp. 131–140. ACM Press (May 1998)
28. Dodis, Y., Ganesh, C., Golovnev, A., Juels, A., Ristenpart, T.: A formal treatment of backdoored pseudorandom generators. In: Oswald, E., Fischlin, M. (eds.) EUROCRYPT 2015. LNCS, vol. 9056, pp. 101–126. Springer, Heidelberg (2015)
29. Dodis, Y., Kalai, Y.T., Lovett, S.: On cryptography with auxiliary input. In: Mitzenmacher, M. (ed.) 41st ACM STOC, pp. 621–630. ACM Press (May/June 2009)
30. Dodis, Y., Smith, A.: Entropic security and the encryption of high entropy messages. In: Kilian, J. (ed.) TCC 2005. LNCS, vol. 3378, pp. 556–577. Springer, Heidelberg (2005)
31. Freeman, D.M., Goldreich, O., Kiltz, E., Rosen, A., Segev, G.: More constructions of lossy and correlation-secure trapdoor functions. In: Nguyen, P.Q., Pointcheval, D. (eds.) PKC 2010. LNCS, vol. 6056, pp. 279–295. Springer, Heidelberg (2010)
32. Fuller, B., O'Neill, A., Reyzin, L.: A unified approach to deterministic encryption: new constructions and a connection to computational entropy. J. Cryptology 28(3), 671–717 (2015)
33. Garg, S., Gentry, C., Halevi, S., Raykova, M., Sahai, A., Waters, B.: Candidate indistinguishability obfuscation and functional encryption for all circuits. In: 54th FOCS, pp. 40–49. IEEE Computer Society Press (October 2013)
34. Gentry, C., Lewko, A., Sahai, A., Waters, B.: Indistinguishability obfuscation from the multilinear subgroup elimination assumption. Cryptology ePrint Archive, Report 2014/309 (2014). http://eprint.iacr.org/2014/309
35. Goldwasser, S., Kalai, Y.T.: On the impossibility of obfuscation with auxiliary input. In: 46th FOCS, pp. 553–562. IEEE Computer Society Press (October 2005)
36. Goyal, V., O'Neill, A., Rao, V.: Correlated-input secure hash functions. In: Ishai, Y. (ed.) TCC 2011. LNCS, vol. 6597, pp. 182–200. Springer, Heidelberg (2011)
37. Hemenway, B., Ostrovsky, R.: Building lossy trapdoor functions from lossy encryption. In: Sako, K., Sarkar, P. (eds.) ASIACRYPT 2013, Part II. LNCS, vol. 8270, pp. 241–260. Springer, Heidelberg (2013)
38. Koppula, V., Pandey, O., Rouselakis, Y., Waters, B.: Deterministic public-key encryption under continual leakage. Cryptology ePrint Archive, Report 2014/780 (2014). http://eprint.iacr.org/2014/780
39. Lynn, B.Y.S., Prabhakaran, M., Sahai, A.: Positive results and techniques for obfuscation. In: Cachin, C., Camenisch, J.L. (eds.) EUROCRYPT 2004. LNCS, vol. 3027, pp. 20–39. Springer, Heidelberg (2004)

594   M. Bellare and I. Stepanovs

40. Matsuda, T., Hanaoka, G.: Chosen ciphertext security via point obfuscation. In: Lindell, Y. (ed.) TCC 2014. LNCS, vol. 8349, pp. 95–120. Springer, Heidelberg (2014)
41. Matsuda, T., Hanaoka, G.: Chosen ciphertext security via UCE. In: Krawczyk, H. (ed.) PKC 2014. LNCS, vol. 8383, pp. 56–76. Springer, Heidelberg (2014)
42. Mironov, I., Pandey, O., Reingold, O., Segev, G.: Incremental deterministic public-key encryption. In: Pointcheval, D., Johansson, T. (eds.) EUROCRYPT 2012. LNCS, vol. 7237, pp. 628–644. Springer, Heidelberg (2012)
43. Pass, R., Seth, K., Telang, S.: Indistinguishability obfuscation from semantically-secure multilinear encodings. In: Garay, J.A., Gennaro, R. (eds.) CRYPTO 2014, Part I. LNCS, vol. 8616, pp. 500–517. Springer, Heidelberg (2014)
44. Peikert, C., Waters, B.: Lossy trapdoor functions and their applications. In: Ladner, R.E., Dwork, C. (eds.) 40th ACM STOC, pp. 187–196. ACM Press (May 2008)
45. Raghunathan, A., Segev, G., Vadhan, S.: Deterministic public-key encryption for adaptively chosen plaintext distributions. In: Johansson, T., Nguyen, P.Q. (eds.) EUROCRYPT 2013. LNCS, vol. 7881, pp. 93–110. Springer, Heidelberg (2013)
46. Sahai, A., Waters, B.: How to use indistinguishability obfuscation: deniable encryption, and more. In: Shmoys, D.B. (ed.) 46th ACM STOC, pp. 475–484. ACM Press (May/June 2014)
47. Wee, H.: On obfuscating point functions. In: Gabow, H.N., Fagin, R. (eds.) 37th ACM STOC, pp. 523–532. ACM Press (May 2005)
48. Wee, H.: Dual projective hashing and its applications — lossy trapdoor functions and more. In: Pointcheval, D., Johansson, T. (eds.) EUROCRYPT 2012. LNCS, vol. 7237, pp. 246–262. Springer, Heidelberg (2012)
49. Wichs, D.: Barriers in cryptography with weak, correlated and leaky sources. In: Kleinberg, R.D. (ed.) ITCS 2013, pp. 111–126. ACM (January 2013)
50. Xie, X., Xue, R., Zhang, R.: Deterministic public key encryption and identity-based encryption from lattices in the auxiliary-input setting. In: Visconti, I., De Prisco, R. (eds.) SCN 2012. LNCS, vol. 7485, pp. 1–18. Springer, Heidelberg (2012)
51. Xue, H., Li, B., Lu, X., Jia, D., Liu, Y.: Efficient lossy trapdoor functions based on subgroup membership assumptions. In: Abdalla, M., Nita-Rotaru, C., Dahab, R. (eds.) CANS 2013. LNCS, vol. 8257, pp. 235–250. Springer, Heidelberg (2013)

Author Index